ISBN 978-1-334-47725-6
PIBN 10759219

A PLEA FOR GREATER SIMPLICITY IN THE LANGUAGE OF SCIENCE.

BY

T. A. RICKARD.

[*Reprinted from* SCIENCE, *N. S., Vol. XV., No. 369, Pages 132–139, January 24, 1902.*]

[*Reprinted from* SCIENCE, *N. S., Vol. XV., No 369, Pages 132–139, January 24, 1902.*]

*A PLEA FOR GREATER SIMPLICITY IN THE LANGUAGE OF SCIENCE.**

SCIENTIFIC ideas are with difficulty soluble in human speech. Man, in his contemplation of the flux of phenomena at work all about him, is embarrassed by the want of a vehicle of thought adequate for expression, as a child whose stammering accents do not permit him to tell his mother the new ideas which suddenly crowd upon him when he meets with something alien to his experience.

Our knowledge of the mechanism of nature has been undergoing a process of growth, much of which has been sudden. It is not surprising, therefore, that the incompletely formed ideas of science should become translated into clumsy language and that inexact thinking should be evidenced by vagueness of expression. This inexactness is often veiled by the liberal use of sonorous Greek-Latin words.

The growth of knowledge has required an increase in the medium of intellectual exchange. New conceptions have called for new terms. Sir Courtenay Boyle has pointed out that the purity of a nation's coinage is properly safeguarded, while the verbal coinage of its national language is subject to no control. Specially qualified persons prepare the standards of gold and silver. This insures the absolute purity of the measures of commercial exchange and gives the English sovereign and the American gold piece, for example, an assured circulation along all the avenues of commerce. It is not so with the standards of speech. The nation debases

*A paper read before Section E of the American Association for the Advancement of Science, August 28, 1901.

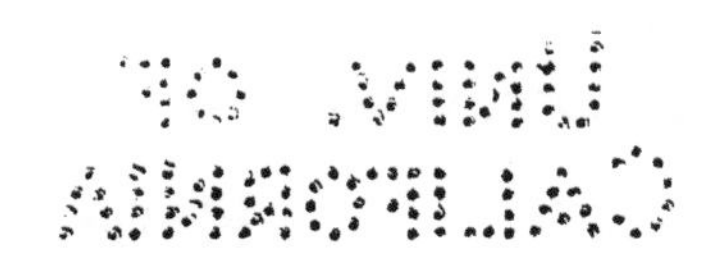

its language with slang, with hybrid and foreign words, the impure alloys and the cheap imports of its verbal coinage, mere tokens which should not be legal tender on the intellectual exchanges. France has an academy which in these matters has much of the authority given to the Mint, whose assayers test our metal coins; but in our country the mintage of words is wholly unrestricted, and, as a consequence, the English language, circulating as it does to all the four corners of the globe, has received an admixture of fragments of speech taken from various languages, just as the currency with which one is paid at the frontier, where empires meet, includes the coinage of several governments, each of which passes with an equally liberal carelessness.

Science ignores geographical lines and bemoans the babel of tongues which hinders the free interchange of ideas between all the peoples of the earth. Nevertheless, the international character of technical literature is suggested by the fact that three languages, French, German and English, are practically recognized as the standard mediums of intellectual exchange. One of these affords the most lucid solvent of thought, another is the speech of the most philosophical of European people and the third goes with world-wide dominion, so that each has a claim to become the recognized language of science. The brotherhood of thinking men will have been fully recognized when all agree to employ the same tongue in their intercourse, but such a 'far-off divine event' is not within the probabilities of the present, consequently there remains only for us to make the best of our own particular language and to safeguard its purity, so that when it goes abroad the people of other countries may at least be assured that they are not dealing with the debased currency of speech.

Barrie has remarked that in this age the man of science appears to be the only one who has anything to say—and the only one who does not know how to say it. It is far otherwise in politics, an occupation which numbers among its followers a great many persons who have the ability for speaking. far beyond anything worth the saying that they have to say. Nor is it so in the arts, the high priests of which, according to Huxley, have 'the power of expression so cultivated that their sensual caterwauling may be almost mistaken for the music of the spheres.' In science there is a language as of coded telegrams, by the use of which a limited amount of information is conveyed through the medium of six-syllabled words. Even when not thus overburdened with technical terms it is too often the case that scientific conceptions are conveyed in a raw and unpalatable form, mere indigestible chunks of knowledge, as it were, which are apt to provoke mental dyspepsia. Why, I ask, should the standard English prose of the day be a chastened art and the writing of science, in a great scientific era, merely an unkempt dressing of splendid ideas? The luminous expositions of Huxley, the occasional irradiating imagery of Tyndall, the manly speech of Le Conte, and of a very few others, all serve simply to emphasize the fact that the literature of scientific research as a whole is characterized by a flat and ungainly style, which renders it distasteful to all but those who have a great hunger for learning.

To criticism of this sort the professional scientist can reply that he addresses himself not to the public at large, but to those who are themselves engaged in similar research, and he may be prompted to add to this the further statement that he cannot pitch the tone of his teaching so as to reach the unsensitive intelligence of persons who lack a technical education. Furthermore, he will claim that he cannot do without the use of the terms to which objection is made. However, in condemning the needless employment of bombastic words of classical origin, in place of plain English, I do not wish to be understood as attacking all technical terms. They are a necessary evil. Some of them are instruments of precision invented to cover particular scientific ideas. Old words have associations which sometimes unfit them to express new conceptions and therefore fresh words are coined. The complaint lodged against the pompous, ungainly wordiness of a large part of the scientific writing of the day is that it is an obstacle to the spread of knowledge.

Let us consider the subject as it is thus presented. In the first place, does the excessive use of technical terms impede the advancement of science? I think it does. It kills the grace and purity of the literature by means of which the discoveries of science are made known. Ruskin, himself a most accurate observer of nature, and also a geologist, said that he was stopped from pursuing his studies 'by the quite frightful inaccuracy of the scientific people's terms, which is the consequence of their always trying to write mixed Latin and English, so losing the grace of the one and the sense of the other.' But grace of diction is not needed, it may well be said; that is true, and it is also true that a clear, forceful, unadorned mode of expression is more difficult of attainment and more desirable in the teaching of science than either grace or fluency of diction. One must not, as Huxley himself remarks, 'varnish the fair face of Truth with that pestilent cosmetic, rhetoric,' and Huxley most assuredly solved the problem of how to avoid rhetorical cosmetics and yet convey deep reasoning on the most complex of subjects in addresses which are not only as clear as a trout stream, but are also brightened by warm touches of humanity, keen wit and the glow of his own courageous manhood. Nevertheless, though clearness of expression be the first desired, yet grace is not to be scorned. When you have a teaching to convey, it is well to employ all the aids which will enable you to get a sympathetic hearing. Man lives not by bread alone, much less by stones. He likes his mental food garnished with a sauce. Let the cooking be good, of course, but a *chef* knows the value of a well-seasoned adjunct to the best dish.

Our language is capable of a grace and a finish greater than we give it credit. That it is possible to write on geology, for instance, in the most exquisite simple English has been proved by Ruskin, whose 'Deucalion' and 'Modern Painters' contain many pages describing accurately the details of the structure of rocks and mountains, and dealing with their geological features in language which is marked by the most sparing use of words which have not an Anglo-Saxon origin.

The next aspect of the enquiry is whether the language of science, apart from the view of mere grace of style in literature, is not likely, in its present everyday form, to delay the advance of knowledge by its very obscurity. Leaving the reader's feelings out of the argument, for the present, it seems obvious that the whole purpose of science, namely, the search after truth, which is best advanced by accuracy of observation and exactness of statement, is hindered by a phraseology which sometimes means very much but oftener means very little, and, on the whole, is most serviceable when required as a cloak for ignorance. To distinguish between what we know and what we think we know, to comprehend accurately the little that we do know, surely these are the foundations of scientific progress. If a man knows what a thing really is, he can say so, describing it, for example, as being black or white; if he does not know, he masks his ignorance by stating in a few Greek or Latin terms that it partakes of the general quality of grayness. Writers get into the habit of using words that they do not clearly understand themselves and which, as a consequence, must fail in conveying an exact meaning to their readers. Many persons who possess only the smattering of a subject are apt to splash all over it with words of learned sound which are more quickly acquired, of course, than the reality of knowledge. Huxley said that if a man does really know his subject "he will be able to speak of it in an easy language and with the completeness of conviction, with which he talks of an ordinary everyday matter. If he does not, he will be afraid to wander beyond the limits of the technical phraseology which he has got up." If any scientific writer should complain that simplicity of speech is impracticable in dealing with essentially technical subjects, I refer him to the course of lectures delivered by Huxley to workingmen, lectures which conveyed original investigations of the greatest importance in language which was as easily understood by his audience as it was accurate when regarded from a purely professional standpoint.

Science has been well defined as 'organized common sense'; let us then express its findings in something better than a mere jargon of speech and avoid that stupidity which Samuel Johnson, himself an archsinner in this respect, has fitly described as 'the immense pomposity of sesquipedalian verbiage.' George Meredith, a great mint-master of words, has recorded his objection to 'conversing in tokens not standard coin.' Indeed the clumsy latinity of much of our scientific talk is an inheritance from the schoolmen of the past; it is the degraded currency of a period when the vagaries of astrology and alchemy found favor among intelligent men.

Vagueness of language produces looseness of knowledge in the teacher as well as the pupil. Huxley, in referring to the use of such comprehensive terms as 'development' and 'evolution,' remarked that words like these were mere 'noise and smoke,' the important thing being to have a clear conception of the idea signified by the name. Examples of this form of error are easy to find. The word 'dynamic' has a distinct meaning in physics, but it is ordinarily employed in the loosest possible manner in geological literature. Thus, the origin of a perplexing ore deposit was recently imputed to the effects produced by the 'dynamic power' which had shattered a certain mountain. 'Dynamic' is of Greek derivation and means powerful, therefore a 'powerful power' had done this thing; but in physics the word is used in the sense of active, as opposed to 'static' or stationary, and it implies motion resulting from the application of force. In the case quoted, and in many similar instances, the word 'agency' or 'activity' would serve to interpret the hazy idea of the writer, and there is every reason to infer, from the context, that he substituted the term 'dynamic power' merely as a frippery of speech. It is much easier to talk grandiloquently about a 'dynamic power' which perpetrates unutterable things and reconstructs creation in the twinkling of an eye than it is to make a careful study of a region, trace its structural lines and decipher the relations of a complicated series of faults. When this has been done and a writer uses comprehensive words to summarize activities which he has expressly defined and described, then indeed he has given a meaning to such words which warrants him in the use of them.

In this connection it is amusing to remember how Ruskin attacked Tyndall for a similar indiscretion. The latter had referred to a certain theory which was in debate, and had said that it, and the like of it, was 'a dynamic power which operates against intellectual stagnation.' Ruskin commented thus: "How a dynamic power differs from an undynamic one, and, presumably, also, a potestatic dynamis from an unpotestatic one—and how much more scientific it is to say, instead of—that our spoon stirs our porridge—that it 'operates against the stagnation of our porridge,' Professor Tyndall trusts the reader to recognize with admiration."

Among geological names there is that comfortable word 'metasomatosis' and its offspring of 'metasomatic interchange' 'metasomatic action,' 'metasomatic origin,' etc., etc. To a few who employ the term to express a particular manner in which rocks undergo change, it is a convenient word for a definite idea, but for the greater number of writers on geological subjects it is a wordy cloud, a nebular phrase, which politely covers the haziness of their knowledge concerning a certain phenomenon. When you don't know what a thing is, call it a 'phenomenon'! Instances of mere vulgarity of scientific language are too numerous to mention. 'Auriferous' and 'argentiferous' are ugly words. They are unnecessary ones also. The other day a metallurgical specialist spoke of 'auriferous amalgamation' as though any process in which mercury is used could be gold-bearing unless it was part of the program that somebody should add particles of gold to the ore under treatment. A mining engineer, of the kind known to the press as an expert, described a famous lode as traversing 'on the one hand a feldspathic tufaceous rock' and 'on the other hand a metamorphic matrix of a somewhat argillo-arenaceous composition.' This is scientific nonsense, the mere travesty of speech. To those who care to dissect the terms used it is easily seen that the writer of them could make nothing out of the rocks he had examined, save the fact that they were decomposed and that the rock which he described last might have been almost anything, for all he said of it; for his description, when translated, means literally a changed matter of a somewhat clayey-sandy composition, which, in Anglo-Saxon, is m-u-d! The 'somewhat' is the one useful word in the sentence. Such language may be described in the terms of mineralogy as metamorphosed English pseudomorphic after blatherskite. Some years ago, when I was at a small mine near Georgetown, in Colorado, a professor visited the underground workings and was taken through them. He immediately began to make a display of verbal fireworks which bewildered the foreman and the other miners whom he met in the mine, all save one, a little Cornishman, who, bringing him a bit of the clay which accompanied one of the walls of the lode, said to him, 'What do 'ee call un, you?' The professor replied, 'It is the argillaceous remnant of an antediluvian world.' Quick as a flash came the comment, 'That's just what I told me pardner.' He was not deceived by the vapor of words.

Next consider the position of the reader. It is scarcely necessary at this date to plead for the cause of technical education and the generous bestowal of the very best that there is of scientific knowledge. The great philosophers of that New Reformation which marked the era of the publication of 'The Origin of Species' have given most freely to all men of their wealth of learning and research. When these have given so much we might well be less niggardly with our small change and cease the practice of distributing, not good wholesome intellectual bread, but the mere stones of knowledge, the hard fossils of what were once stimulating thoughts. In the ancient world the Eleusinian mysteries were withheld from the crowd and knowledge was the possession of a few. Do the latter day priests of science desire to imitate the attendants of the old Greek temples and confine their secrets to a few of the elect by the use of a formalism which is the mere abracadabra of speech? Among certain scientific men there is a feeling that scientists should address themselves only to fellow scientists, and that to become an expositor to the unlearned is to lose caste among the learned. It is the survival of the narrow spirit of the dark ages, before modern science was born. There are not many, however, who dare confess to such a creed, although their actions may occasionally endorse it. On the whole, modern science is nothing if not catholic in its generosity. 'To promote the increase of natural knowledge and to forward the application of scientific methods of investigation to all the problems of life' was the avowed purpose of the greatest of the philosophers of the Victorian era.

There are those who are full of a similar good will, but they fail in giving effect to it because they are unable to use language which can be widely understood. In its very infancy geology was nearly choked with big words, for Lyell, the father of modern geology, said, seventy years ago, that the study of it was 'very easy, when put into plainer language than scientific writers choose often unnecessarily to employ.' At this day even the publications of the Geological Surveys of the United States and the Australian colonies, for example, are occasionally restricted in usefulness by erring in this respect, and as I yield to none in my appreciation of the splendid service done to geology and to mining by these surveys, I trust my criticism will be accepted in the thoroughly friendly spirit with which it is offered. It seems to me that one might almost say that certain of these extremely valuable publications are 'badly' prepared because they seem to overlook the fact that they are, of course, intended to aid the mining community in the first place and the public, whether lay or scientific, only secondarily. From a wide experience among those engaged in mining I can testify that a large part of the literature thus prepared is useless to them and that no one regrets it more deeply than they, because there is a marked tendency among this class of workers to appreciate the assistance which science can give. Take, for example, a sentence like the following, extracted from one of the recent reports of the U. S. Geological Survey. "The ore forms a series of imbricating lenses, or a stringer lead, in the slates, the quartz conforming as a rule to the carunculated schistose structures, though occasionally breaking across laminæ, and sometimes the slate is so broken as to form a reticulated deposit." This was written by one of our foremost geologists and, when translated, the sentence is found to convey a useful fact, but is it likely to be clear to anyone but a traveling dictionary? A thoroughly literary man might know the exact meaning of the two or three very unusual words which are employed in this statement, but the question is, will it be of any use whatever even to a fairly educated miner, or be understood by those who pay for the preparation of such literature, namely, the taxpayers? An example of another kind is afforded by a Tasmanian geologist who recently described certain ores as due to 'the effects of a reduction in temperature of the hitherto liquefied hydroplutonic solutions, and their consequent regular precipitation.' These solutions, it is further stated, presumably for the guidance of those who wield the pick, 'ascended in the form of metallic superheated vapors which combined eventually with ebullient steam to form other aqueous solutions, causing geyser-like discharges at the surface, aided by subterranean and irrepressible pressure.' At the same time certain 'dynamical forces' were very busy indeed and 'eventuated in the opening of fissures'—of which one can only regret that they did not swallow up the author as Nathan and Abiram were once engulfed in the sight of all Israel.

It will be well to contrast these two examples of exuberant verbosity because the first befogs the statement of a scientific observation of value, made by an able man, while the second cloaks the ignorance of a charlatan, who masquerades his nonsense in the trappings of wisdom. Here you have an illustration of the harmfulness of this kind of language, which obscures truth and falseness alike, to the degradation of science and the total confusion of those of the unlearned who are searching after information.

Let the writer on scientific matters learn the derivation of the words he uses and then translate them literally into English before he uses them, and thereby avoid the unconscious talking of nonsense. If he knows not the exact meaning of the terms which offer themselves to his pen, let him avoid them and trust to the honest aid of his own language. 'Great part of the supposed scientific knowledge of the day is simply bad English, and vanishes the moment you translate it,' says Ruskin.

The examples already given illustrate this. 'Every Englishman has, in his native tongue, an almost perfect instrument of literary expression,' so says Huxley, and he illustrated his own saying. Huxley and Ruskin were wide apart in many things and yet they agreed in this. Ruskin proved abundantly that the language of Shakespeare and the Bible can be used as a weapon of expression keen as a Damascus saber when it is freed from the rust of classic importations, which make it clumsy as a crowbar.

There is yet another reason against the excessive use of Greek-English words, in particular. Greece is not a remnant of extinct geography, but an existing land with a very active people and a living language. The terms which paleontology has borrowed from the Greek may be returned by the Greeks to us. And, as Ruskin points out, "What you, in compliment to Greece call a 'Dinotherium,' Greece, in compliment to you, must call a 'Nasty-beastium,' and you know the interchange of compliments can't last long."

In all seriousness, however, is it too much to ask that such technical terms as are considered essential shall not be used carelessly, and that in publications intended for an untechnical public, as are most government reports, an effort be made to avoid them and, where unavoidable, those which are least likely to be understood shall be translated in footnotes. Even as regards the transactions of scientific societies, I believe that those of us who are active members have little to lose and much to gain by confining the use of our clumsy terminology to cover ideas which we cannot otherwise express. By doing so we shall contribute, I earnestly believe, to that advancement of science which we all have at heart.

The words which, at first, are the exclusive privilege of the specialist, gradually extend into wider use, following in the wake of that diffusion of scientific knowledge which is one of the objects of this Association. We believe that to get alongside facts, to apply the best knowledge available, to seek truth for its own sake, is as essential to the well-being of the individual life as it is to the success of a machine shop, and as beneficial to the community as it is to a smelting works.

In furtherance of this principle we must remember that language in relation to ideas is a solvent, the purity and clearness of which affect that which it bears in solution. Whewell, in 'The Philosophy of the Inductive Sciences,' has expressed this view of the matter with noble eloquence. 'Language,' he said, 'is often called an instrument of thought, but it is also the nutriment of thought; or rather, it is the atmosphere in which thought lives; a medium essential to the activity of our speculative powers, although invisible and imperceptible in its operation, and an element modifying, by its qualities and changes, the growth and complexion of the faculties which it feeds.'

In considering the subject from this standpoint, there is borne in upon the mind a suggestion which carries our thought far beyond the confines of the matter under discussion. Such power of speech as man possesses is a faculty which appears to divide him from all other living things, while at the same time the imperfection of it weighs him down continually with the sense of an essential frailty. To be able to express oneself perfectly would be divine, to be unable to make oneself understood *is* human. In 'Man's Place in Nature,' Huxley points out that the endowment of intelligible speech separates man from the brutes which are most like him, namely, the anthropoid apes, whom he otherwise resembles closely in substance and in structure. This endowment enables him to transmit the experience which in other animals is lost with each individual life; it has enabled him to organize his knowledge and to hand it down to his descendants, first by word of mouth and then by written words. If the experience thus recorded were properly utilized, instead of being largely disregarded, then man's advancement in knowledge and conduct would enable him to emphasize, much more than it is permitted him at present, his superiority over the dumb brutes. Considered from this standpoint language is a factor in the evolution of the race and an instrument which works for ethical progress. It is a gift most truly divine which should be cherished as the ladder which has permitted of an ascent from the most humble beginnings and leads to the heights of a loftier destiny, when man, ceasing to stammer forth in accents which are but the halting expression of swift thought, shall photograph his mind in the fulness of speech, and, neither withholding what he wants to say nor saying what he wants to withhold, shall be linked to his fellow by the completeness of a perfect communion of ideas.

DENVER. T. A. RICKARD.

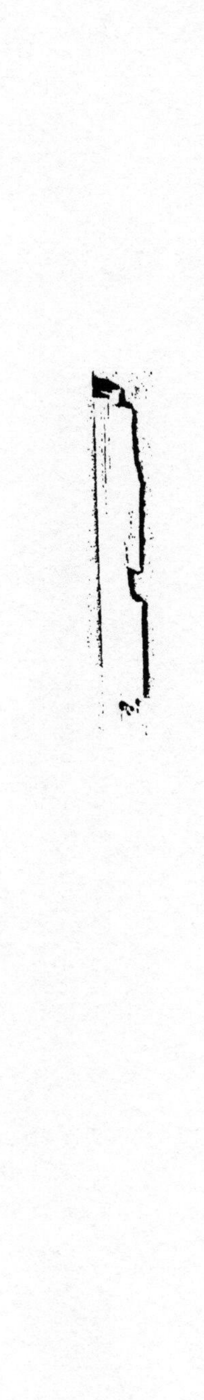

2

"The Minerals which accompany Gold, and their bearing upon the Richness of Ore Deposits."

By T. A. Rickard, M.Inst. M.M.

In the general advance of knowledge anatomy has distanced medicine. The surgeon works with confident skill while the diagnosis of the physician is yet a faltering guess. It is thus in our own profession. The structure of the rocky envelope of the earth has been in many localities so carefully deciphered that the mining engineer can unravel the geology of an ore deposit with a success to which the mines of Leadville, Bendigo and Gympie bear ungrudging testimony. When, however, he endeavours to ascertain the causes which have determined the presence of gold at one spot and its absence at another he hesitates, and, saturated with the brutal experience of widely separated regions, he confesses that it is a phenomenon yet to be explained.

The working of gold mines in different countries has yielded an accumulation of scattered evidence which needs only scant examination to emphasise how incomplete and contradictory it is. The absence of accurate knowledge on this subject has encouraged the growth of attractive fallacies, in the combating of which the whole matter often comes up for informal discussion wherever mining men congregate. During recent journeyings over the goldfields of West Australia, I found that many an old fallacy had sprung again into vigorous life amid the congenial atmosphere of a community over which the windy breath of a boom had but lately passed.

It may seem a thankless task to oppose those plausible theories, which become rampant only when facts are scarce, yet I am convinced that in the search after truth the first step must be to sift the little that we really do know from the much that we think we know; the first is science, the second is popular knowledge. The old industry of mining was formerly guided by the venerable rule of thumb, and would forever have remained but a blind sort of groping in the dark had it not obtained the willing aid of the younger science of geology.

The immediate problem to which my most recent experience has called attention may be summarised thus:—In examining a lode only incompletely developed and in a new country, what is the evidence on which a correct estimate of the prospective value of the mine may best be based? The two most common answers would be a plain denial of any man's ability to see further into the ground than the point of his pick, and against this obvious surrender would come the reply that the best indication would be found in the presence of particular minerals in association with the gold.

It is this question of indicative minerals which I purpose discussing. The experience of certain mining districts has gone to prove that gold is notably accompanied by particular minerals, and this to such an extent that they are considered to assure the richness of an ore in which the gold itself cannot visibly be discerned. These "indicative minerals," as they may be termed, are not the same in every locality. A few examples, such as have come within my own experience, may be quoted. Every mining engineer can add to the list.

In Boulder county, Colorado, roscoelite (a vanadium mica) is closely associated with the tellurides of gold (calaverite and sylvanite). This fact is rendered of additional interest because the same rare mineral has been recognised by me in the telluride ores of Kalgoorlie, in West Australia.* At San Andreas, in California, uranium ochre (the yellow oxide of uranium) is found to distinguish the pockets of specimen gold ore to such an extent as to serve as a guide in prospecting. In several parts of Arizona, in Yuma, Yavapai and Pinal counties, especially the last, vanadinite and descloizite (both vanadates of lead) characterise ores rich in gold and silver. Wulfenite (the molybdate of lead) often accompanies the vanadium minerals, and has been noticed in the ores of two celebrated lodes, the Comstock in Nevada and the Vulture in Arizona. The association of gold with these lead ores is notable, because the particular minerals mentioned are all of great beauty and delicacy. The same may be said of crocoite

* Roscoelite also occurs generously in the veins of the district (El Dorado) where gold was first discovered in California.

(the chromate of lead), which characterises several gold veins in the North Coolgardie goldfield, especially at Menzies.* The more common sulphide, galena, is also an accessory mineral in the richest mines at Menzies, and accompanies coarse native gold. At Niagara, Pinyalling, and Wagiemoola, three widely separated localities in West Australia, I found gold closely associated with tourmaline in the form of acicular crystals in contact with coarse gold, and also in a condition of minute diffusion forming dark blotches in the white quartz. The gold ores of the Mysore in India carry tourmaline.

Instances might be multiplied, but it would be to no purpose. Those already cited will serve as a sufficient text for the discussion of the subject.

Such occurrences as these would seem at first sight to afford a much needed aid to the explorer. It is unfortunately easy to prove that as evidence they, and any number more of them, are of very uncertain value in a new mining territory, because they are contradicted by similar testimony of a negative kind, which compels us to regard them as mere coincidences. Take the case of zinc blende. One locality was quoted where this mineral is a sure sign of a generous amount of gold and silver in the ore. The Morgan Mine, in Wales, affords, I understand, corroborative testimony. At the East Murchison Mine, in West Australia, this mineral has been found to be an almost unerring guide in separating rich from poor ore. But in Arizona it is a common experience that the impoverishment of lodes in depth, when it does occur, is concomitant with an increasing percentage of zinc. Other regions echo this unpleasant fact. Broken Hill knows it. Similarly, there is the beautiful mineral fluorite or fluorspar. The association of fluorite with the tellurides of gold was early recognised in both the Boulder and Cripple Creek districts of Colorado, and the purple tint imparted by this mineral was speedily hailed as a distinction peculiar to rich veins. Later experience, notably in Park County, has proved that poor ores are favoured with fluorite no less than the bonanzas. I have mentioned a locality where rhodochrosite is esteemed a favourable mineral. But while it is thus characteristic of rich ore at Rico, in Colorado it is a negligible factor in certain lodes at Butte City, Montana. Again, consider calcite. When it is seen amid the gold bearing quartz of California it is recognised with regret, because it so often means a falling off in values, while at Kalgoorlie the same mineral charac-

* Dana mentions that crocoite is associated with gold in the quartz of Niznhi Tagilsk in the Urals.

terises ores rich in calaverite (the telluride of gold), and at Rhuda in Transylvania, a very valuable gold vein has been worked whose matrix was essentially calcite.

We are all familiar with mines in which iron pyrites is so intimately associated with the gold that a lessening of the one means a diminution in the other, but there are also cases where an excessive percentage of pyrites coincides with impoverishment. Moreover, there are lodes in which the coarse cubes of pyrites are less favourable to the presence of gold than the finely crystalline variety, but there are also those in which the reverse is true. Much in the same way, there used to be an idea that coarse cubical galena was less silver-bearing than the fine grained kind, but this as a generalisation has long since been exploded. Thus, therefore, it requires but little sifting of this sort of evidence to emphasise its contradictory character.

The rich lodes of the same district frequently differ widely in their mineralisation.* Poor veins often carry the ores considered characteristic of the rich ones. The neglect of the former causes this fact to be overlooked. When the field of comparison is enlarged from mines to whole districts the divergence of evidence becomes tenfold emphasised.

In matters like these the experience of each mining engineer is a personal equation by which every theory must be eventually tested. Out of the whole sad wreck of glittering generalisations on this particular subject, only one or two have survived my own particular trial of them, and even they, I fear, await the destructive testimony which may at any moment be found in the development of new mining regions. When gold occurs in pyrrhotite ores it has been as yet invariably proved to be in immediate association with a small, often overlooked, percentage of copper pyrites. The testimony of Montana, Colorado, and British Columbia is at one in this deduction. Again, while many veins carrying coarse gold encased in white quartz persist to great depths, and in this respect Bendigo does especially set at naught the dictum of American experience, nevertheless I have not known gold quartz to be persistent when wholly barren of the sulphides of the baser metals, while, on the other hand, I do remember innumerable examples of ore quite destitute of pyrites, galena, and blende which proved particularly short lived. I might venture

* This word is not of vulgar coinage. It comes through the French, who use it as we do, to express the fact that a rock carries minerals of economic value, that is, ore. "Minéralisé" is the equivalent of "mineralised," and is related rather to the word "minerai" (ore) than to "mineral."

one other. There is no better indirect evidence of the size of a body of gold ore than uniformity in the distribution of the gold. A patchy occurrence is less likely than homogeneity to indicate continuity or size; samples which vary between narrow limits are more encouraging than those which swing between wide extremes of richness and poverty. It is with a desire to avoid a merely destructive attitude that I have dared to offer these three observations.

While therefore the evidence in support of the value of the supposed indicative minerals is, as we have seen, of a very contradictory nature, it also has another feature which must not be overlooked. At its best, the aid of these minerals would be delusive, because, even if it proved the invariable association of gold with particular minerals, it could not predicate the actual amount of that gold. That a lode should carry gold is quite insufficient to the mining engineer, whose operations for its profitable extraction require that it should be there in paying quantity. This is the difference between science and business. The union of the two creates an industry. One might discuss in a learned and entertaining manner, as Stelzner has done, the suggestiveness of the association of gold with such a compound as the silicate of boron and aluminium (tourmaline), because it indicates a deep-seated origin. The presence in notable quantity of the fluoride of calcium (fluorspar) might prompt speculations of a vapourous and corroding kind. But the proof of gold having either a profound or a gaseous origin, were it attained by the geologist or chemist, would not permit the mining engineer to infer that the gold persists in *paying* quantity to any particular depth. An ore which carries 2 dwt. per ton in a region where the conditions are such as to require the equivalent of 15 dwt. to be expended in the extraction and reduction of a ton, is to all practical intent as valueless as one which is wholly barren.

The question of indicative minerals bears many points of resemblance to that of the plants which have been observed to distinguish the soil enriched by particular ores. The *Viola lutea* was supposed to be peculiar to the soil covering the zinc deposits of Westphalia, and was subsequently recognised growing on the outcrop of the zinc ores of the Horn Silver Mine in Utah. It became known as the "zinc plant." Similarly there is a so-called "lead plant," the *Amorpha canescens*, which characterises the lead deposits of Michigan, Wisconsin, and Illinois. These plants are local varieties rather than a distinct species, their colour being affected by their absorption of the particular metallic ingredient in the soil.* Their occurrence has long ceased to have anything more than academic interest.

If then we cannot accept the belief that certain minerals are indicative of the plentiful occurrence of gold, what shall be said for the idea that they give an assurance of persistence in depth? Every one has read serious statements to the effect that this or that mine was of undoubted value because its ores contained particular minerals, the presence of which promised that the vein would increase in richness in depth. In young mining regions such ideas find a fertile soil. In West Australia the finding of tellurides in an ore is now generally considered to permit of the inference that the lode will go down to a great depth. The early discoveries of extraordinary pockets of native gold in veins of white quartz, such as made the Londonderry famous, proved sporadic and bunchy to a distressing degree. When, therefore, the telluride ores of Kalgoorlie became recognised, and were found to occur in bodies of magnificent size and very satisfactory persistence, the conclusion was jumped at that if tellurides were only present in an ore, continuity in depth became thereby guaranteed. *Post hoc, ergò propter hoc.*

In these matters a very small portion of ascertained truth is swamped amid a mass of supposition quite unworthy of the name of theory. A theory embodies an underlying principle of universal application. The idea which to-day dominates the mining of Westralia is a vain imagining, delusive as a promoter's dream.

Those who are acquainted with the history of the mining of tellurides need not be told how ill founded is the statement that these particular minerals characterise lodes of peculiar permanence. The combinations of tellurium with gold and silver have proved of notable commercial importance in three mining regions, namely, Transylvania, Colorado, and West Australia. There are other districts where the mining of them is an incident in the working of ordinary gold ores. Such is the case in certain localities in California, South Dakota, and the North Island of New Zealand. In none of these, however, have they been indicative of any special persistence in the richness of the veins, and their occurrence has been merely an added obstacle to the successful extraction of the gold.

It was in 1802 that Klaproth discovered the presence of a new element, tellurium, in the ores of Zalathna, and so led to the recognition of a large number of the compounds of that metal

* See "Indicative Plants," by R. W. Raymond, *Trans. Amer. Inst. of M.E.*, vol. xv, p. 644.

with gold and silver. Zalathna, Nagyag, and Verospatak were gold mining centres when Transylvania was the Roman province of Dacia and was ruled by Trajan. The very complex ores of these districts have been a puzzle to the metallurgists of many generations. The veins penetrate young volcanic rocks (andesites) and Tertiary limestones, sandstones, and conglomerates. This very ancient mining country exhibits to-day a very forceful example of impoverishment in depth, while the refractory character of the ores has tried the resources of the smelting establishments of Schemnitz, Zalathna, and Freiberg.

In Colorado the tellurides were recognised as early as 1874 in Boulder county, more especially at the mines of Magnolia, Salina, and Sunshine. An experience of nearly twenty-five years has proved to the miners of that county that these ores occur there in comparatively small bodies of remarkable richness but of very irregular and uncertain behaviour. In two other districts of the same State, namely, in Hinsdale county and amid the La Plata mountains, valuable mines, carrying ores rich in the tellurides of gold and silver, have been worked during the past ten years, but their record corroborates that of Boulder. There remains the more important goldfield of Cripple Creek, where veins penetrating a remarkable complex of volcanic rocks have proved so persistently rich to a depth now approaching one thousand feet, that they have obliterated the reputation which this class of ore won in the three older parts of the same State.

In West Australia tellurides are being mined in three localities —Redhill, Bardoc, Kalgoorlie—and the wonderful development of the last has given a new impulse to the whole industry of the colony.

Thus, one hears much of Cripple Creek, and, lately, of Kalgoorlie. In both districts very rich lodes, characterised by telluride ores, have been opened up, with results so eloquent as to silence the story of the more numerous localities where these particular minerals have been only a metallurgical obstacle associated with ore bodies of no satisfactory continuity. As a consequence, investors are prepared to swallow the rhetorical confectionery of an irresponsible press, and to believe that a new era has dawned for any neglected region in which these tellurides are now for the first time recognised. Believe me, the compounds of tellurium are far more widely dispersed in gold ores than is generally supposed. I have detected them in several mines when a low extraction in the stamp mill suggested an unusual difficulty in the ore, and it is certain that during the next few years the greater familiarity with these minerals will lead to their recognition in so many unsuspected localities, that they will cease to be a mineralogical curiosity.*

The idea, now very prevalent, that at Kalgoorlie especially the presence of tellurides guarantees persistent richness is contradicted by several facts, the most notorious of which should be the fact that the very mine in which they were first recognised (by Mr. J. C. Moulden, in May, 1896), and subsequently acclaimed by the local press as heralding sure prosperity, has proved unprofitable; indeed, the particular ore-body in which the first telluride was seen has been demonstrated by later development to be an isolated patch leading to nothing of any moment. Moreover, the lodes at Kalgoorlie vary in the amount of tellurides which they carry without any proportionate difference in their richness. The gold occurs not only in chemical combination with tellurium, but also in its ordinary native condition. Further, tellurides occur which do not contain gold. The so-called "black tellurium" of certain mines is the rare telluride of mercury, called "coloradoite," from the locality of its first discovery. Native tellurium also exists. Again, some of the veins even below the zone of oxidation are so free from tellurium as not to differ from ordinary gold-bearing reefs. Yet there is no reason to suppose that these are less rich or less persistent than those which are characterised by a notable percentage of tellurides.

It is not too much to say, therefore, that the observations gathered from the working of telluride ores in various parts of the world refute the idea that their occurrence has any particular bearing on the question of persistence in depth; while, on the other hand, experience has frequently demonstrated that their presence is an important and objectionable factor from a metallurgical standpoint, because it increases the cost of gold extraction.

Thus we are brought to face the general question of the enrichment of ores in depth. It is, however, outside the scope of this contribution. As a far-reaching fallacy, persisting in spite of the accumulated experience of many mining regions, it possesses a pestilent vitality, which must at times astonish those who from the actual direction of mining operations have seen so much evidence to the contrary. I am referring now to gold mines only,

* The sulphide ore of Mount Morgan, Queensland, carries tellurides. The fact was recently determined by Mr. E. S. Simpson, Government Assayer for West Australia. How much of the early trouble in the treatment of the oxidised ores is explained by this discovery? It also recalls to me the resemblance between the dull brown gold of some of the Mount Morgan ore and that of the first discoveries at Cripple Creek in 1892.

because the facts relating to baser metals, which by oxidation become soluble, differ in detail. Gold and tin are in this respect unlike copper and silver. While, therefore, I do not wish on this occasion to reopen the whole subject, I cannot forbear referring to the question as it has come up for discussion in West Australia, where I have lately been.

In that country the assertion is made with tiresome iteration that veins become richer below the water level, because in the oxidised zone the ore has been impoverished by the leaching out of the gold. It is also held by many who pose as having authority and not as the scribes, that if the outcrop give evidence of notable mineralisation it is reasonable to expect almost barren quartz to become valuable when the sulphides are reached in the ordinary course of deeper mining. If these pleasant doctrines are questioned, you are bidden to go to Kalgoorlie, where, so they say, you will be forced by the evidence there obtainable to admit the fact that lodes which now attract the attention of the financial world, were too poor to meet expenses until the mine workings penetrated into the unoxidized ore. Indeed, at Kalgoorlie especially this idea has a strong foothold, especially among stock-jobbers.

The matter seemed to me to be one of great interest. If the facts really did indicate a general enrichment of the veins as they approached the drainage level of the district, which is also the water level of the mines, then Kalgoorlie offered a striking exception to the experience of other goldfields. If, moreover, a satisfactory theory were forthcoming to account for these unusual facts, then the uncertain chemistry of ore occurrence would receive invaluable aid. But the prettily coloured bubble was dissipated as soon as an earnest investigation was commenced. The history of the development of the lodes, and a few quiet conversations with the able men who have come from elsewhere to direct the big mines, were sufficient to stamp this as another resurrection of a fallacy that was old before the Phœnicians came to Cornwall.

To say that mines get richer in depth is, in such a region as West Australia, a cruel cynicism. If anyone is inclined to believe it, let him wander over the desert, and count the idle stamp mills which lie rusting in the sweltering sun, and the long succession of abandoned shafts which now serve only to water the passing camel train. By mere repetition of an untruth you may effect persuasion, but you do not alter the falsity of it.

If you omit the manifest failures and turn to the production of the more successful mines in the Menzies, Murchison, Coolgardie, and Norseman goldfields, you will still find that it is a work of supererogation to attempt a serious discussion of the statement that the veins have improved in depth. Incomplete as the Government statistics are, and vitiated by a total lack of system in the determination of the actual tonnage treated at the mills, yet they, too, tender an emphatic denial.*

The alleged facts are further explained by a general hypothesis that the gold in the oxidised ore has been leached out so that it is beneath its normal richness, which will be found unimpaired below the water level. This is stated to be evidenced by the removal of the iron pyrites, the casts of the crystals of which now appear as cavities in the quartz. In some of these gold is found, but in others it is absent, proving, so it is said, that the gold has been removed from the cavities which are now empty. The occurrence of films of fine gold, called "paint gold," on the faces of fractures is instanced as an illustration of secondary reactions. Finally, the alkaline composition of the waters in the mines is quoted in proof of their solvent power.

Although the chemistry of the oxidised zone is far from being thoroughly understood, yet in its incompleteness it is sufficient to disprove these arguments. When iron pyrites is decomposed the sulphide becomes a sulphate, and this in turn, by further oxidation in the presence of water, is resolved into sulphuric acid and the hydrated oxide of iron. The native sulphur frequently seen in the cavities left by the removal of the cubes of pyrites is not a direct product of decomposition, but is traceable to later secondary reactions in which the organic matter of the surface has served as a reducing agent. The gold intimately mixed with the pyrites, and probably originally deposited with it, remains, because the noble metal is insoluble to the waters which dissolved the iron.

The occurrence of gold in pyrites is now less of a mystery than it used to be. Microscopic examination has disclosed the existence of the gold in the planes of the pyrites, and the leaching of the gold by cyanide solutions without any apparent deformation of the pyrites is no less suggestive. It is indeed true that in some of the casts, left by the removal of the sulphides gold does not occur, but this is frequently due to the fact that it is very difficult to break a piece of cellular quartz without shaking the gold out of the cavities in which it lies loosely. Hence their emptiness

* Of course the grand totals are misleading, since they include the production of new mines, and do not indicate the shutting down of unprofitable ones. The yields of individual mines must be consulted.

may mean nothing. At other times this may indeed be due to the real absence of the gold, and against this observation we then balance another, namely, that the pyrites as we find it unaltered below the water level also varies in its gold contents, so that in the same vein it is sometimes barren and sometimes rich, accounting in this way for the uncertain presence of free gold in the oxidised zone. The "paint gold," frequently noticed in the gossan, appears to be the result of secondary and comparatively recent reactions. It is, however, only a proof of precipitation, and therefore presupposes a previous leaching, but not necessarily in the vadose region. It evidences local enrichment rather than impoverishment.

Of the solvent power of surface waters upon the gold there is only supposition, and this supposition must first overcome the fact of the occurrence of such organic matter in the ground approaching the surface as would reduce any known salts of gold should they be in solution. In the Sugarloaf Mine, near Kunanalling, I saw the roots of trees at 74 ft., and in the Great Boulder Main Reef Mine, at Kalgoorlie, I saw some which had penetrated the rock to 85 ft. below the surface. In the former case the oxidation of the enclosing rock ceases at 130 ft. and in the latter at 175 ft. The region is an unusually dry one, which has undergone erosion with extreme slowness; and this is doubtless the reason why roots in search of moisture should penetrate so deeply.

The waters of the mines are all brackish; they are many times more salt than the sea. Analyses made at the Great Boulder Proprietary Mine show an average in solids of 8·9 per cent., and of this 6·2 was chloride of sodium, 0·45 chloride of magnesia, and 0·73 sulphate of lime. The water from the Lane shaft gave the maximum of 11·9 per cent. solids, containing 8·8 NaCl, 0·51 $MgCl_2$, and 1·1 $CaSO_4$.*

Certain results obtained in the treatment of the zinc precipitate in the cyanide works of the Associated Mines have led the mill manager, Mr. Grayson, to deduce the solubility of gold in sulphuric acid when tellurium oxide is also present, and thus to offer a chemical theory for the alleged leaching of gold in the surface ores. This explanation, as it stands, is questionable, because the solubility of gold under the stated conditions is not known. When, however, free chlorine is also present such solubility would occur.† The telluric oxide, in the presence of salt and sulphuric

* For these figures I am indebted to the courtesy of Mr. Richard Hamilton, the mine manager.

† As pointed out by Mr. Richard Pearce, in a recent letter to the writer.

acid, would take the part given in the laboratory to the black oxide of manganese. The occurrence, therefore, of waters rich in chlorides, of sulphuric acid derived from decomposing pyrites and of tellurium oxide obtained from the oxidation of tellurides, are all recognisable amid the conditions at Kalgoorlie, and thus afford a theory which, while it has no very strong evidence to commend it, is at least tenable from a chemical standpoint.

In many cases the supposed enrichment in depth can be easily explained. For example, a shaft passes, in course of sinking, from poor into rich ore. The lode, it would appear, has become better in depth. The fact is, that the shaft was started off the oreshoot, which has a pitch such as to bring it across the line of the shaft at a certain level. The accompanying diagram will illustrate this point.

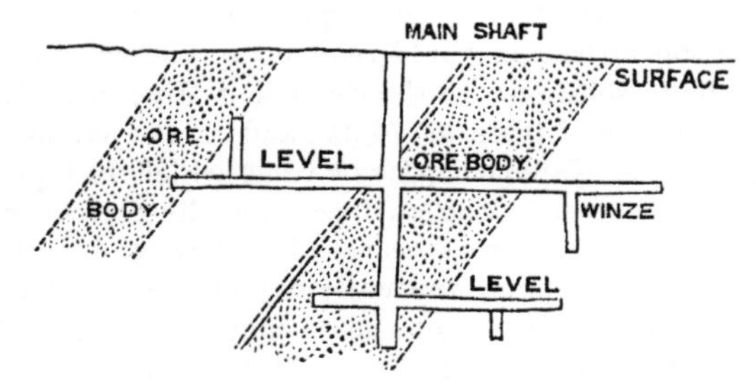

It would be just as reasonable to argue that the ore occurred in vertical bands because the levels pass in and out of the shoots. In another suggestive instance a vertical shaft was sunk in a very wide lode which is not quite perpendicular. The shaft was started on the hanging wall side, where the ore is now known to be always poor, and in depth it approached the footwall, where the lode is richest. It was assumed, for a time, that an enrichment in depth characterised the lode. The sketch (p. 13) will explain this occurrence.

The idea of general enrichment in depth at Kalgoorlie arises from an ignorance of the history of the early development of the region. The discovery of the big ore-bodies was not made in the infancy of the district, but came in the wake of that preliminary digging which usually precedes systematic exploration. Patrick Hannan pegged out the first claim on April 12th, 1892. At that time nothing was known save the existence of a few superficial deposits on the flat overlooked by Cassidy Hill and Mount Charlotte. These deposits were of the kind termed "dry blowers' patches." They may be described as the alluvium of a waterless country. Their substance varies. They may consist merely of a few inches of sand and clay lying on the decomposed rock surface, but elsewhere they may have a thickness of many feet,

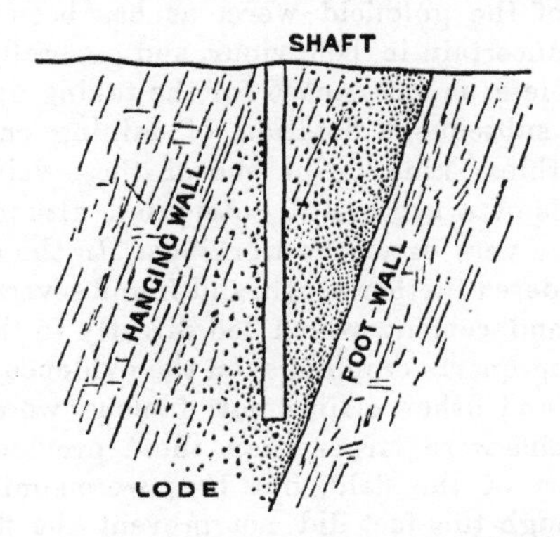

and include a lower portion which is so compact as to be called "cement." In the absence of the transporting power of running water, the wind has been an active agent in sifting the *débris* caused by that disintegration of the surface which is mainly traceable to the variation between the heat of day and the cold of night. The finer rock particles are thus separated from the larger fragments of hard quartz. And although the wind is but a feeble agent as compared with the mountain stream, yet, owing to the prevalence of the constant and violent draughts of a high plateau, such as the interior of West Australia, the sum of its activities during long periods of time is capable of producing noteworthy results. The miner imitates the ways of nature, and in default of sluicing he winnows the dirt by a process known as "dry blowing." The "dry blower" of West Australia is brother to the "gulch miner" of California and the "alluvial digger" of Victoria.

From the deposits which are thus accumulated gold has been extensively won, and occasional patches of extreme richness have been encountered. The heaviest gold is found resting close to the underlying rock surface, and therefore, as in ordinary alluvial mining, the working of these deposits often leads to the discovery of the veins from which the gold was shed. Thus many "dry blowers' patches" have been found adjacent to very valuable lodes, and in the case of the alluvium of the Kalgoorlie flats there were found small veins which led to the first mining. This early mining was confined to that portion of the district which is close to the present town of Kalgoorlie, and which is distinct from the scene of the great mining developments of the past two years, 2 miles further to the southward. The veins found in this northern part of the goldfield were, as has been stated, small; they were also uncertain in behaviour, and generally poor. They served nevertheless as an excuse for the taking up of numerous leases and the subsequent flotation of mining enterprises of a thoroughly worthless kind. Not one of these veins has as yet become the basis of a successful enterprise, although they have been followed by very extensive workings. In the meanwhile the prospectors wandered further south, and found several rich patches of surface dirt and cement, which led directly to the discovery of one or two strong quartz croppings, on the evidence of which the Great Boulder and other claims now famous were pegged out. These quartz veins were larger than those previously worked in the northern part of the field, but they were similarly poor and uncertain, although this fact did not prevent the flotation of the Great Boulder Proprietary Company. The story of this splendid mine is typical of that of several of its neighbours. The first explorations were confined to a vein which outcropped in the eastern part of the lease, and the shaft sunk upon it is the one known as No. 3, or Gamble North's. The vein was soon proved to be of no importance, becoming thin and poor at a depth of 75 ft. A trench was then started to cut a big ironstone outcrop which forms a hillock behind the present office. This ironstone reef was found to contain only 12 dwt. of gold per ton, an amount too small for profitable operations at that time. When, however, the trench referred to was extended farther west it penetrated another vein which had no outcrop. This proved to be the eastern portion of the magnificent lode which subsequently gave such great value to the mine. It was rich from the start. A wide open-cut now bears confirmatory evidence, and the least observant cannot help noticing that the stopes have been extended from the underground workings up to the daylight. The section afforded by the opencut exhibits the fact that 2½ ft. of cement form a cap over a lode nearly 20 ft. wide. The Great Boulder Perseverance had a similar beginning. The first work was done on what is known now as the old No. 1 Shaft, carrying a comparatively poor vein which was abandoned when the rich lode in the neighbouring Lake View Consols Mine was traced into the Perseverance ground. This lode, which gave the value to this company's property, also failed to outcrop. A big opencut now explains the reason to be similar to that noted in the case of the Great Boulder. The surface workings on the Ivanhoe, Kalgurli, and other adjacent mines exhibit similar conditions, big lodes capped by cement and long lines of stopes breaking through into daylight.

In the meantime the extensive development of the goldfield has afforded an explanation for the chequered nature of the early explorations. It has been found that the quartzose lodes are uniformly poor, and that the rich ones have a chloritic and magnesian matrix which renders them susceptible to easy degradation and erosion. Thus it is easy to understand the current misunderstanding. The first veins worked were naturally those which had outcrops. They outcropped because they carried much quartz and were harder than the encasing country. They also happened to be poor; therefore the first attempts at mining were unprofitable. On the other hand the rich lodes had a composition into which the carbonates of lime and magnesia entered largely, rendering them softer than the rock enclosing them, and therefore they suffered erosion at least as rapidly as the surrounding country. The detrital deposits of the surface capped the tops of the rich ore-bodies, and it was not until an accidental exploration had pierced the cap of this "cement" that the first of the soft and rich lodes was uncovered. The discovery of the others followed in due course.

The idea of an enrichment in depth was based on the fact that the mines which were at first unprofitable became subsequently marvellous ore producers. Careless observation and that wishing which is father to much loose thinking served to bolster up an erroneous idea, and to spread the statement that the ore became richer at the water level. It is a matter which does not affect the reputation of the really good mines so much as that of the forlorn hopes, the poverty of which is excused by a lack of depth, thus leading directly to the spending of much money on the foundation of a fallacy.

From careful investigation and the sifting of much evidence I am forced to the conclusion that Kalgoorlie is no exception to a common experience, and that where rare enrichments occurred in the sinking of a shaft, the fact can be traced to the structural relations of the ore deposits. I regret the conclusion. It would have encouraged those in other regions had there at last been found a district where nature had placed the best ore where man could with most difficulty reach it, and, apart from its economic aspects, the occurrence would have been one of great scientific interest.

Thus the tellurides, like rhodochrosite, tourmaline, zinc blende, pyrites and a host of other supposed indicative minerals, must be discarded as helps no better than the will-o'-the-wisp which leads the wanderer into a morass worse than the darkness itself. Better no guide at all than a false one.

Shareholders and investors may be tempted to inquire whether I would go so far as to deny the possibility of veterans in the profession having such ability and. experience as would permit them to come to safe conclusions as to the prospective value of a mine from a mere examination into the character of the ore. To this I would answer that I believe the most experienced, the very Ulysses among mining engineers, would be the first to emphatically disclaim such short cuts to the valuation of mines. A moil and a 4-lb. hammer are of more use than a book full of sounding theories; a careful sampling of the workings is of more immediate utility than a treatise on mineralogy. Successful mining must be based on facts; all the rhetoric and fond imagining in the world cannot alter them. It is the province of the mining engineer to determine the facts, to get alongside them, as Huxley would say, and when a theory comes floating by leave it, as Joseph left his coat in the hands of the harlot, and flee. Experience has proved that indicative minerals are delusive; so let them go. But the mineral contents of an ore, though they have no bearing on the mining, decide its metallurgical treatment, and therefore need careful examination. The importance of the character of the ore from this point of view is too often overlooked by those enthusiastic mineralogists who permit themselves to make the most sweeping deductions on matters of much greater uncertainty.

Careful sampling is worth a bushel of suppositions, and the painstaking determination of the working costs is better than any amount of geological generalisation. Mining is not a scientific pursuit, although at times it may to the observant have seemed to be either that or one big insanity. But mining is an industry. The good sense which financial men have of late years contributed to its operations has done much to bring it from a windy mistiness to the solid footing of sound business The main purpose is not to develop the waste places of the earth, nor to spoil the scenery of the mountainous ones, but simply to win a profit by extracting ores out of the ground. It is a plain matter of profit and loss. On the one side is the value of the gold in the ore, and on the other is the cost of the processes needed to obtain it. To arrive at the former there is only one way, namely, to sample the workings systematically. The result will be reliable in proportion to the care taken. Any shirking of difficult places in the mine, any avoidance of hard portions of the vein, any assistance from untrustworthy hands, will vitiate the result. Against this must be placed the costs of operation. Here it is that experience is needed. The sampling is largely mechanical, like ordinary assaying, and requires patience and care more than anything else. In the estimate of the costs there must be included many items of expenditure, such as the breaking of the ore, the development work, the equipment, the milling, the management; and to arrive at these the previous actual charge of mines is the only proper preparation. Then comes the question of the quantity of ore available or likely to become available by further exploration. This is the *pons asinorum* of mining. That which some describe as ore in sight is often really ore out of sight. The over sanguine estimation of ore reserves has ruined more enterprises than all the bad management and over-capitalisation of which complaints are daily made. It is ever a difficult matter and requires a cool judgment, wide experience, and a careful investigation into the circumstances and structure of each particular mine.

When the value and tonnage of the ore available have been arrived at, and when working costs have been determined, then the engineer has the greater part of the evidence needed to submit to the client whom he is advising. The other data which will influence an opinion are more variable in their character. The geological conditions may affect the distribution of the ore bodies and, consequently, the cost of mining, and the mineralogical composition of the ore may determine the expense of milling. As such they must not be overlooked, but the padding of a report with a large amount of geological disquisition, where it is not necessary to a comprehension of the facts of the case, is very nearly an impertinence, seeing that it is not expected that it will be understood by the person or persons for whose guidance the report is written.

In concluding this contribution I would express the hope that this paper may lead to a useful discussion. If there are any technical men who seriously entertain the idea of an enrichment in depth due to the leaching of gold in the oxidised ores, then it would be of much service to the industry if they would frame a defence and an explanation of views which daily experience must otherwise condemn as nonsense. Nor in denouncing one generalisation would I make the equally grave mistake of advocating its opposite, namely, that all mines must necessarily become poorer in depth. There may be causes, founded on geological structure, why a change in the value of the ore in a vein may take place in any direction, upward or downward, in dip or in strike, and it is an undoubted fact that there have been instances where the deepening of the workings has led to the discovery of new ore bodies. Bendigo is a telling illustration.* The gold-bearing quartz occurs along the anticlinal axes of sedimentary rocks. In sinking, a succession of saddle formations is penetrated. No single one of these has any notable vertical extent, yet the series as a whole is wonderfully persistent. This explains why the last resort of a perplexed mine manager is to advise the sinking of the main shaft. At Bendigo the advice is well founded, but when the managers from this district go to West Australia and recommend deeper exploration every time they encounter poor ore, they do so without regard to the total unlikeness of the conditions. Again, it cannot be denied that in certain regions alternations of comparative richness and poverty appear to coincide with the penetration of the workings through successive zones of rock. The Gympie district is a case in point. There the veins cut through a series of shales, limestones, conglomerates, and sandstones, amid which there are several beds of black slate. The gold occurs in paying quantity only when the veins are traversing the slates. Every district deserves a study unbiassed by the record of its neighbours; each mine must be taken on its merits and inspected without prejudice. It is, however, one thing to examine a mine with the assumption that gold veins in general become enriched in depth, and it is quite another thing to recognise that while it may occur in a particular case, it is an expectation which experience does not justify. When, however, such enrichment or impoverishment does occur, experience suggests to us that its cause is to be sought for rather in the geological structure of the encasing rocks than in the merely coincident presence of certain minerals found associated with the gold.

* See "The Bendigo Goldfield," by the writer, in vol. xx, *Trans. Amer. Inst. of Mining Engineers.*

Harrison and Sons, Printers in Ordinary to Her Majesty, St. Martin's Lane.

of Proceedings.—Vol. VIII.

"The Cripple Creek Goldfield."

A PAPER READ BEFORE THE

UTION OF MINING AND METALLURGY,

ON WEDNESDAY, 15TH NOVEMBER, 1899,

BY

CKARD, M.Inst.M.M., State Geologist of Colorado.

"The Cripple Creek Goldfield."

By T. A. Rickard., M.Inst.M.M., State Geologist of Colorado.

Introductory.

The romantic history of mining records many names which had an odd sound until fame familiarised them. Ballarat, the Cœur d'Alene, Bendigo, the Yuba, Leadville, Broken Hill, Klondyke, and Kalgoorlie, for example, have been words to conjure with, although in the beginnings of the particular mining districts designated by them they were spelt with difficulty and pronounced with uncertainty. In 1892, when Cripple Creek began to be talked about among the clubs and banks of Denver, as a new locality where important discoveries of gold had been made, the name seemed only provocative of derision. It is said to have originated from the fact that at a certain point on the course of the little stream there was a morass in which straying cattle wandered and, in their efforts to extricate themselves, were occasionally lamed. So says one of the survivors of the band of men who once tended the herds that grazed on the hills now pierced with many shafts. However, the name needs no apology to-day, the magic baptism of golden discovery has made it sound as alliterative and impressive as the most exacting historian could demand.

The locality of the great goldfield is full of romantic suggestion, because its very situation, overshadowed by the granite battlements of Pike's Peak, recalls the fact that it has fulfilled the expectations of an older generation. The wave of immigration which, after the financial panic of 1857, swept westward until it broke against the ramparts of the Rocky Mountains, was the vanguard of a new civilisation destined to dispossess the Indian and the buffalo. In their progress across the prairies, the pioneers of that advance ever sought with shaded eyes for the first glimpse of the beacon mountain whose white crest on the far horizon gave promise of the land of gold. That beckoning guide was the granite peak which Lieut

Zebulon Pike had reconnoitred in the first years of the century. "Pike's Peak or bust," the motto of the adventurers of 1857 and 1858, sounds but mock heroic in our ears, but it expressed something of the mingled humour and daring of the men who first pierced the unknown wilderness which was then the borderland of the territory of Kansas.

The immigration which marked the beginning of Colorado's history, was thus known as "the Pike's Peak excitement." The rallying call of the pioneers expressed a delusion. No gold discoveries of any moment were made at that time in the canyons or on the hills surrounding the peak. It was in the mountainous region 70 miles northward, now known as Gilpin County, that the first beginning was made. On the 6th of May, 1859, John Hamilton Gregory found the outcrop of the lode which still bears his name and the working of which celebrated the commencement of the mining industry of Colorado. This discovery led to the development of Gilpin county, which was contemporaneous with the opening up of Clear Creek and Boulder. In 1860 the placers of California gulch and Breckenridge were discovered; in 1875, Leadville; in 1879, Aspen; in 1882, Red Mountain; in 1889, Creede; and so on, until the mining districts of Colorado were distributed all over its wonderful complex of mountain land.

Amid all these rapidly succeeding discoveries, and the extraordinary activity which followed each of them, the silence of the quiet hills surrounding Pike's Peak remained unbroken. Suddenly, in the spring of 1884, rumours were circulated of a great discovery south of the peak. During the darkness of an April night a horde of prospectors stole swiftly away in obedience to vague hints which had been scattered among the saloons of Leadville and surrounding camps. Each party aimed to be the first on the ground. The dawn of the next day found an excited crowd of four thousand men gathered at the base of Mt. Pisgah. The incident has since become known in local history as the Mt. Pisgah fiasco.

Among the hills which like a flock of sheep cluster around the southern base of Pike's Peak, there is a dark cone standing in solitude among its smaller brethren. This is Mt. Pisgah. In 1884 the miners who rushed thither could find no gold in workable quantity save in the prospect holes made by the original locators. Salting was suspected, the man who had instigated the rush was conspicuous by absence, an accomplice was caught with a bottle of yellow stuff in his pocket. It was not whisky, but its modern antidote, the chloride of gold. Man had endeavoured to remedy nature's seeming niggardliness and the rock had been artificially enriched. Angry feelings found vent in threats of lynching, but in the failure to lay hands upon the real perpetrators of the fraud, the affair broke up in a big picnic and a general drunk. A little digging had been done, one or two veins were uncovered, but the comparative poverty of the ore only added bitterness to the general disappointment. The crowd disappeared as quickly as it had come. The hillsides resumed the quiet aspect of the cattle range for which they seemed best fitted. The incident was over.

The vicissitudes of mining are proverbial. No district illustrates it more forcibly than Cripple Creek. It provokes one to cynicism to think how near the deluded prospectors of 1884 were to the eve of big discoveries. The dark front of Mt. Pisgah now overshadows the very streets of the town of Cripple Creek, with its 20,000 inhabitants, and on the ridges opposite the lines of smoking chimneys bespeak a long succession of productive mines.

Not until 1893 did Cripple Creek come to the front; two years before, the name had begun to be mentioned in mining circles, but the Mt. Pisgah excitement had discredited the district and the contemporaneous discoveries of large silver lodes at Creede, at the headwaters of the Rio Grande, diverted attention for a time. The closing of the Indian mints, in June, 1893, prostrated the silver mining industry, which at that period was far more important to the state than its gold production.* As soon, however, as the silver market became disorganized, all the activity of a most energetic mining community was concentrated upon its gold resources. The experienced men of Leadville and Aspen, where silver-lead mining predominated, turned with the energy of despair to the new gold-field which previously they had pooh-poohed. A new impetus was given to exploratory work, Cripple Creek underwent rapid development and the result soon became apparent in the opening up of several rich mines and in a production of gold which sprang from $583,000 in 1892 to $2,100,000 in the year following.

A few notes regarding the events which led up to the present era of prosperity and productiveness will fittingly close this introductory history. For several years before 1891 prospectors had wandered over the hills from Colorado Springs and Florissant, treading in the footsteps of the men of 1858. Indeed, there is a story current in the camp to this day that on the top of Bull Cliffs,† above the Victor

* In 1892, Colorado produced 26,350,000 oz. of silver, valued at $23,082,600, and 256,410 oz. of gold, worth $5,300,000. In 1898 the State produced 1,138,584 oz. of gold, worth $23,534,531.

† The names of several hills and gulches are reminders of the fact that the district was once given up to the quiet herds of cattle.

Mine, there was found a shallow shaft, which had been dug by the pioneers of the Pike's Peak excitement. Among the earliest of the gold-seekers was Robert Womack, who once owned a small ranch in the district. He sold it to Bennett and Myers, the proprietors at that time of the cattle range, which covered a large part of the area now forming the environs of the town of Cripple Creek. For many years, between 1880 and 1890, Bob Womack lived in the district, doing occasional work for Bennett and Myers, and spending his spare time in prospecting. He had previously had some experience in Gilpin County, and knew gold ore when he saw it. In the course of desultory digging, he found several veins, and when he would turn up at intervals, at Colorado Springs, he exhibited pieces of float (surface ore) as evidence of his discoveries; but having a reputation for honesty rather than shrewdness, his statements made little impression. For many years he worked on a hole in Poverty Gulch without staking a claim in proper form. There seemed no need to do so; no one came to disturb him; the whole hill country was at that time fenced in so as to serve as a summer range for cattle. The cowboys and herdsmen looked good-naturedly at Bob's digging, but did not consider it of any moment. In December, 1890, E. M. de la Vergne and F. F. Frisbee came up from Colorado Springs to prospect. George Carr, who was in charge of the ranch belonging to Bennett and Myers, showed them around the district. The hills were under snow, and only a few bare spots permitted of any prospecting. On Guyot Hill, in Eclipse and in Poverty Gulches, they found evidences of gold veins, and samples were taken away. These averaged about 2 oz. of gold per ton. Encouraged by their first visit, De La Vergne and Frisbee returned early in February, 1891. They found Bob Womack at work in Poverty Gulch. He had sunk a shaft to a depth of 48 ft., and encountered good ore. The claim he had pegged out was called the "Chance," and a number of stakes indicated that he had relocated it six years in succession without recording the fact or complying with the conditions of the mining law in regard to the amount of assessment work required annually. When he found the newcomers were making inquiries, he relocated the claim as the "El Paso," and De La Vergne, finding another lode, heavy in iron pyrites, to the west of Womack's vein, located a claim, which he called the "El Dorado." It was recorded a few days later, and in the certificate the district was called for the first time by the name which it still bears, Cripple Creek. Although these locations had been made, little actual mining was done upon them for some time afterwards. Womack absented himself. Frisbee saw that there was a good deal of surface ore which could easily be removed, so while Womack was away, he sent 1,100 lb. by wagon to the Pueblo Smelting and Refining Company, who gave returns at the rate of $200 per ton. This was in August, 1891. Frisbee induced Womack to give him a bond and option on the El Paso for $5,000. Shortly afterwards it was transferred to Messrs. Lennox and Giddings, who still own it, as a part of a very successful mine, the Gold King.

In May, Frisbee and De La Vergne happened to be at Colorado Springs and met W. S. Stratton, to whom they showed certain assays of ores brought down by them from Cripple Creek. Stratton was a house-builder and carpenter by trade, but in the intervals of his regular occupation he had been prospecting for fully 20 years previous to this date; he had learnt the use of the blow-pipe, and was familiar with the outlines of mineralogy and geology, in fact an energetic, well-informed man, thoroughly equipped for prospecting work of any kind. At that time he had been searching for cryolite, a mineral from which the metal aluminium is obtained, and had a camp on the Little Beaver, on the Cripple Creek side of Pike's Peak. After the meeting with De La Vergne and Frisbee, he went to Cripple Creek and camped there. Stratton met Bob Womack and went around seeing the little work done by the latter and his associates. Among those who were prospecting in the vicinity was Dick Houghton, an old mountaineer, prospector, and specimen hunter, whose labours have enriched many museums. One day Houghton brought down a piece of rock from the Lone Star claim on Gold Hill, and, meeting Stratton in Poverty Gulch, he told him he had found some galena (the sulphide of lead). Stratton examined it with his magnifying glass and expressed doubts as to its being galena, and in looking at the ore he saw little cubes of rusty gold, one of which had been scratched by being carried in Houghton's pocket so as to expose a bright surface. They went down to Stratton's tent, and he pulled out his blow-pipe and made a test which proved that it was gold. Neither of these men knew at that time that the bright silvery mineral, which Houghton thought to be galena, was sylvanite, the telluride of gold and silver. It is not known who first recognised the tellurides of Cripple Creek, but there is a story that a miner made a camp fireplace with some pieces of rock which carried this silvery mineral, and that the heat of his cooking operations roasted the ore so as to bring out the fact that it carried gold. *Se non è vero è ben trovato.* Stratton went up and located the claim adjacent to Houghton's; it was named the Gold King, and is now a part of the Gold and Globe property. On the 5th of June, Stratton, accompanied by Fred

Troutman, went to the ridge above Battle Mountain, and seeing the willows at the head of Wilson Creek (where now the town of Goldfield is situated), they inferred the presence of water. They descended the hill and got a drink; then climbing the hill behind the spring, they found loose pieces of rock, one of which was broken open and found to be smothered in gold. The owner of the Independence says that this was the only time he got really excited. Camp was moved from Cripple Creek next day and pitched close to the spring. A search was begun for the lode which had shed so goodly a float. Trenches were dug; but Stratton had an idea at this time that veins with a north and south direction were the ones which carried rich ores, and so his trenches were dug at right angles to this course, with the result that they paralleled the veins actually existing there, and since developed into the Legal Tender, Lillie, and Vindicator Mines. They found nothing. An old ranchman, Billy Fernay, came along about this time and brought some float which he had found on the hill below, now called Battle Mountain. Stratton liked the look of it, so Fernay located it for Stratton, Troutman, and himself, calling it the Black Diamond. It is now one of the claims included within the territory of the Portland Mine. Next day Stratton went down to see the vein and tried to make the course of it accord with the line of the ridge. This led him down the hill to a big outcrop of granite. It was the Independence vein, which had already been seen by many, including nearly all of those whose names have been mentioned. The path from one ranch to another went close by, and all the cattle men who had any idea of prospecting had looked at it. Every one had condemned it as worthless granite. Fernay pointed it out to Stratton, but he also did as the others had done. On examining the outcrop he remarked the absence of any metallic mineral and of vein quartz such as he had been accustomed to in the San Juan region, and therefore concluded that it was an unlikely looking rock. And so it really was, for it was granite without the ordinary gold-bearing minerals visible in it, differing indeed from the granite of the dome of Pike's Peak in being less fresh in appearance, brown instead of pink, and marked by dark spots where the mica had been decomposed. Some of it yet remains in place, inviting the observation of those who may wonder why the great lode was so long disregarded. Stratton overlooked it, but not irretrievably. Two days later, John R. McKinnie, who was one of the first prospectors in the district, came to their camp, and so did Charlie Love, a ranchman from Beaver Park, who had pointed out the big outcrop to many of the prospectors. The latter asked McKinnie if he had seen it, but the matter was allowed to drop. Stratton remembered the incident when, on the morning of the Fourth of July, he was at Colorado Springs, whither he had gone with five samples for assay. The assays gave only three or four dollars per ton at the best, notwithstanding that he had obtained good results by panning. It suddenly occurred to him that the granite outcrop must be *the* lode. He had found gold in the loose fragments of porphyry lying upon the south face of Battle Mountain near the granite outcrop, but he had been unable to trace its source to any vein in the porphyry formation. Acting on the impulse, he took a horse immediately, and on arrival found Troutman ready to leave in order to celebrate the Fourth at Colorado Springs. Stratton made two locations: the Washington and the Independence. I doubt if any man ever celebrated the Fourth of July to better advantage. Some pieces of the outcrop of granite ore were broken, and Troutman took them to be assayed at the Springs, while Stratton awaited the result. Troutman returned on horseback next day with the assay certificate, proving the ore to be worth $380 per ton! The rest of the story is simple. It records the steady development of one of the richest mines ever uncovered by the miner's pick.

I might go on to tell the story of the Portland and that of other mines, now celebrated, which made beginnings no less romantic; but the above few notes will give an idea of the manner in which the Cripple Creek goldfield was first opened up. Let me emphasize the fact that it was not the result of accident, nor the work of ignorant men, but it was the accomplishment of experienced miners who knew what they were doing, and had the energy and ability to do it. It is now a part of the true and romantic story of beneficent endeavour which turned the barren prairies into the granary of a continent, and made the snowclad mountains an empire whose tribute of gold and silver transcends that of the Cæsars'.

Physical Features of the District.

The known gold-bearing portion of the district covers an area four miles long by three miles wide, occupying a group of hills which rise from 300 ft. to 1,000 ft. above the general surface, and have an average altitude of 10,500 to 11,000 ft. above the sea. The drainage of the region flows into the Arkansas river, whose gateway into the plains is at Cañon City. The general slope is southward, and the sunny aspect incident to this configuration of the surface has caused the hill-sides to be clad with sufficient grass, and rendered them, at one time, despite the high altitude, a good pasturage for cattle.

Few mining camps have so picturesque a situation, and Cripple Creek is further notable because the picturesque is not obtained at any sacrifice of accessibility. The beauty of the panoramic view to be obtained from most of the mines is not due to mere ruggedness or to the ordinary grandeur of a mountainous country; it is traceable to a position upon the slopes flanking Pike's Peak, which permits of an uninterrupted view of snow-clad ranges a hundred miles away. It is a panorama rather than a picture. In front are hills like giants tumbled in troubled sleep, whose feet touch the plateau of the South Park. To the left are the Arkansas hills that confine the river of the same name to its tumultuous gorge; further south is the Wet Mountain valley, and beyond that the long magnificent serrated range of the Sangre de Cristo, telling of the shattered dream of Spanish conquest, of which no trace now survives, save in the occasional name of a peak or a stream. These remind a practical age of the priestly warriors and the warlike priests who once sought to win the golden treasures of a land whose aboriginal people have almost passed away. The wind blows the snow of the Sangre de Cristo into streaming banners, and the clouds like vanquished legions sweep across the far horizon. Turning northward, the valley of the Arkansas can be seen dividing the mountains which overlook Leadville. Further to the right are the beautiful Kenosha hills, at the headwaters of the Platte, and beyond them are snow-clad peaks rising above 14,000 ft. The details of the view are lost in the vastness of it, which impresses the observer no less because he is surrounded by a noisy murmur of trains, steam whistles, wagons, and machinery, which tell of the activity going on about him. Still, there is a nobility of human endeavour and successful achievement no less impressive than the beauty of snow-clad peak and silver summits.

The physical condition of the surface had much to do with the chequered history of the district. Owing to the southern exposure and the comparative absence of a protecting growth of trees, the rocks, which mostly possess a fissile structure, have been shattered by frost so as to overspread the solid formation with a thickness of débris to which the tufted grass has given a further covering. Water, owing to its expansion between 4° C. and the freezing point, is a ceaselessly destructive agent. When it penetrates the cracks and crannies of the rocks it serves as a wedge shattering their stony substance with a resistless power. The heat of day and the cold of night, the warmth of summer and the snows of winter, alike aid this disintegrating process. A high altitude and a south slope afford the conditions most favourable to such action. Thus it came about that the district of Cripple Creek is largely covered with the shattered rock which the miners call "wash," incorrectly, however, because it is not composed of rounded waterworn material, but of angular fragments which, if not in place, are not far from their original position, having slid down the hill slope in obedience to the laws of gravity. This shattering of the rock-surface has caused one very important and, in Cripple Creek's case, far-reaching result. There are no outcrops. Ordinarily, the veins of gold-ore stand above the surface with that boldness which caused the Australian miner to term them "reefs," and the Californian to call them "ledges." The ore, as will be seen when discussing the geology of the gold-field, is essentially mineralised and enriched rock, comparatively devoid of the quartz composing the typical lodes of other districts in America or Australia, and consequently it shares with the rock the tendency to undergo easy shattering. Solid veinstone, therefore, rarely survives amid the general disintegration, the outcrop of the Independence being a very notable exception.

The first discoveries in mining are usually due to the finding of outcrops ; in the absence of them, deep explorations are rarely undertaken. Deep ravines often afford good natural sections of the rock formation. The Cripple Creek district was as deficient in the one feature as the other. The absence of steep declivities and consequent rock faces was characteristic of the pastoral landscape, and the angular debris covering the rounded hildsides made digging difficult. For these reasons, although the district was traversed by many thousands of prospectors at successive epochs, the existence of rich lodes was not surmised until a very recent date, and many experienced miners failed of success at first because they encountered conditions unfamiliar to them. Among the early arrivals, in 1891 and 1892, were miners from Gilpin, Leadville, and Aspen, men of knowledge in their own habitat, but unable to understand the peculiar vein structure which they saw at Cripple Creek. It was the adverse opinion of these men, rather than the views of geologists or scientific observers, which injured the reputation of the goldfield in the beginning of its development.

The Geology of the Goldfield.

The intimate relationship between geological structure and ore occurrence has nowhere been more forcibly emphasised than in Colorado. This is largely due to the labours of the United States

Geological Survey, whose monographs on Leadville and Cripple Creek (particularly the former) have been of inestimable assistance to the mining industry.

The dissemination of accurate knowledge regarding the geology of mining districts has been aided by the increasing simplicity and clearness of the language employed to convey it. There was a time when geological data were expressed through the distorted medium of six-syllabled words, and scientific men appeared to follow the example of the hierophants of the ancient temples, who spoke to the populace only in language incomprehensible to them. The application of science to commerce, of geology to the industry of mining, has led to the recognition of the fact that to talk to business men in a hybrid Greek-Latin jargon, is only an impertinence. If mining be not a business, it is a vain delusion. The language of technical science when it bears upon business must be made intelligible to business men, otherwise it will remain a mere abrakadabra of speech.*

The mines of Cripple Creek are situated in a complex of volcanic rocks, occurring amid the mass of granite whose culminating point is Pike's Peak. These volcanic rocks found a passage through the underlying granite in the comparatively recent period known to science as the Miocene, an early part of the last of the three great subdivisions of geological time. The granite was formed in the very dawn of time, out of the substance of it the mighty foundations of Pike's Peak were upbuilt and the crest of the mountain was chiselled. It is the basal rock of the region and at one time probably formed the floor of the ancient seas which received the sediments now composing the sandstones and limestones flanking the Front range. The granite is of a particular type, known, because of its prevalence in this locality, as the Pike's Peak granite. It is coarsely crystalline and its three ingredients, the minerals quartz, mica, and felspar, are easily distinguishable by the unaided eye. A beautiful red tint, mainly due to the colour of the felspar, characterises it and renders it recognizable by the least observant.

* I have in the present contribution laid myself open to the criticism of technical men by avoiding the use of technical terms and by a seemingly unnecessary translation of those the use of which was unavoidable. It may seem to be a work of supererogation, yet because an institution of mining and metallurgy does its most useful work when it transmits the recorded observations of its members to those who are engaged in the industries the benefit of which it promotes, therefore it seemed better to me to give explanations needless to scientific men than to be unintelligible to those to whom these observations may be of interest, however slight, and of use, however insignificant.

Long subsequent to the formation of the granite and the sedimentary rocks which were laid down upon it, there began an elevatory movement supposed to be traceable to the re-adjustment of the earth's exterior to its cooling and shrinking interior. Accompanying this movement there occurred a general fracturing of the rocks thus affected, so as to permit volcanic matter to force a way upward, after the manner of water rising through cracks in the overlying ice. The volcanic rock thus brought to the surface of the granite slowly filled the hollows of its uneven surface, and spread over a large area since then diminished by the patient forces of atmospheric erosion, which during the long period of time separating the Miocene from the present day, have slowly sculptured the hills and valleys of the district.

A glance at the coloured geological map of the goldfield exhibits a great variety of volcanic rocks. The principal of these is andesite breccia.* The very nature of the breccia suggests the violence of the volcanic action which brought it to the surface of the granite. The miners call the breccia "porphyry" from its apparent resemblance to the rocks of that class with which they were previously familiar in the Leadville and in the Gilpin County mines. The porphyry of Leadville is quartz-felsite, that of Gilpin is quartz-andesite. Porphyry† is an adjective-noun and refers to the structure rather than to the composition of a rock, so that there is "granite-porphyry," "diorite-porphyry," "andesite-porphyry," &c., the term being applied to rock of igneous origin in which particular minerals are distinguishable amid the ground mass of the rock so as to give it a speckled appearance. The Cripple Creek breccia has this appearance, but it is due to the fact that it is made up of a heterogeneous mass of rock particles of every size, from the most minute powder to fragments as large as a man's head. These con-

* "Andesite" is derived from Andes, the mountain range where this rock is especially prevalent. "Breccia" is a word of Italian origin, and means "broken." It is a term applied to rocks which are made up of fragmentary material.

† "Porphyry" comes to us through the Greek word *porphyra*, signifying purple. It was first used to designate a beautiful rock of this type which the Romans obtained from the quarries of Gebel Dokhan, on the shores of the Red Sea. This original "porphyry" called by the Italians "*porfido rosso antico*," had, according to Zirkel, a beautiful blood-red ground mass speckled with small snow-white and rose-red crystals of felspar. But the first meaning of the term which depended on the colour has long been lost in another meaning which refers to the structure. A rock is a "porphyry," or, more correctly, is "porphyritic," when some particular constituent mineral, very often felspar, stands out well defined from the general ground-mass, as in the western miners' familiar "bird's-eye porphyry."

sist mainly of andesite, but the other rocks are included, especially near the edges of the volcanic vent. Some of this material is mere volcanic dust, called tuff,* which, when consolidated under pressure and cemented by silicious waters, becomes compacted into a dense hard substance difficult to distinguish from a true crystalline rock; so that it is not to be wondered that the miners often label it with an incorrect name.

The breccia lies in the uneven hollows of the ancient surface of the granite, and probably fills a large part of the vent through which it was ejected. The thickness of the breccia has not been proved, nor has the exact position of the vent been discovered, although there is evidence, in the composition of the rocks, indicating the approximate position of it to be just west of the town of Goldfield and near the locality covered by the Hull City placer. The mine workings have shown the thickness of the breccia to be over a thousand feet in several places; but as these are for the most part near the edge of the mass there is every probability that the maximum depth of this formation is several times one thousand feet in the vicinity of the point of extrusion.

The accompanying sketch of the Cripple Creek volcano is largely diagrammatic, but it will serve to convey a general idea of the geological structure of the district.

The breccia is penetrated and traversed by later volcanic rocks, of which phonolite† is the most important in its relation to the occurrence of ore. Until recent years phonolite was not known as a rock species save as forming the Wolf rock in Cornwall, and, therefore, its association with great mineral wealth at Cripple Creek has been one of the most interesting features of the development of that district. The phonolite occurs for the most part in dykes, that is to say, in approximately vertical sheets which traverse the older formations, the granite and the breccia, in various directions, and are probably united, at depths far beyond the reach of human exploration, to larger masses of rock having a similar composition, just as the cracks in ice are filled with a liquid similar to that beneath.

These dykes follow such lines of weakness in the older rocks as

* "Tuff" comes from the Italian *tufa*. Vesuvius is responsible for the Italian nomenclature of many volcanic products.

† "Phonolite" is derived from two Greek words, *phonē*, signifying sound, and *lithos*, meaning stone. It owes this name to the fact that it rings when struck by a hammer. This is due to its hardness and close texture. It is also called "clinkstone." The essential constituents of phonolite are nepheline and the glassy variety of felspar termed sanidine.

were developed into fractures during those periods when the rocks underwent strains, the latter being considered to be the result of the slow wrinkling of the earth's crust due to its readjustment over

FIG. 1.

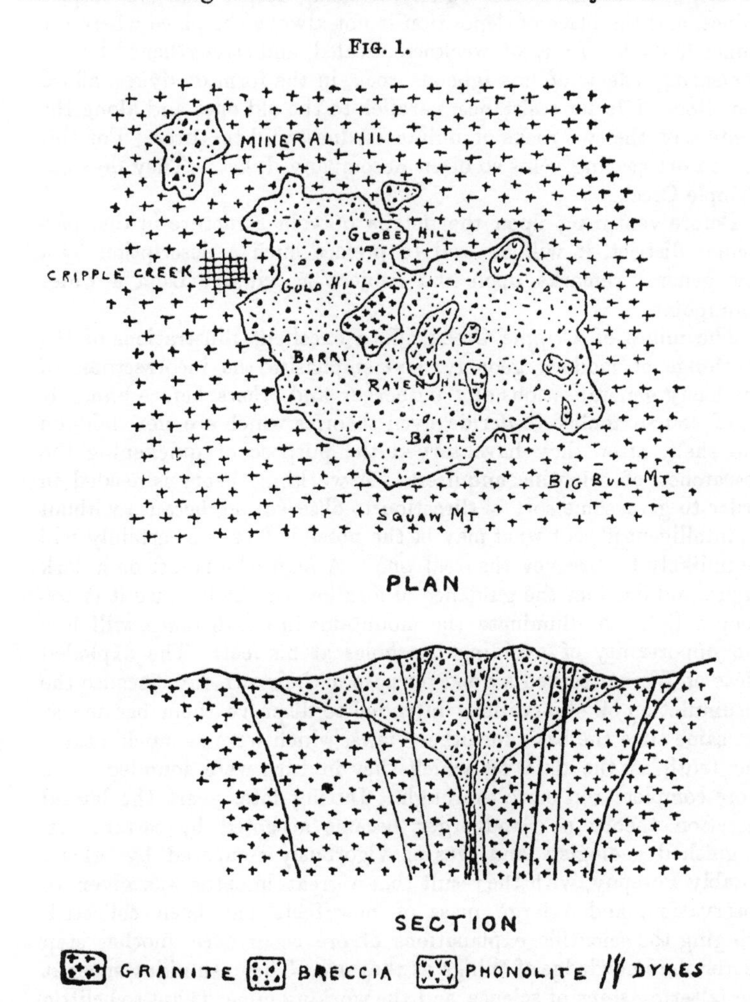

a cooling and shrinking interior. The phonolite rose in a mobile, if not molten, condition through the fractures thus formed after the manner of water rising through the cracks in the overlying ice. The structural conditions thus created gave a direction to the subsequent circulation of underground waters. The deposition of ore is the result of such circulation, the mineral-bearing solutions being the vehicle by which the metals are leached out of the rocks and laid down elsewhere in such a concentrated form and within such a distance of the surface as to render them valuable to man. The place of origin is surmised, but vaguely, as being deeper than our deepest mines, and the place of deposition is not always the place where the miner finds it. Lines of weakness, healed and strengthened by the cementing effects of hot igneous rock, in the form of dykes, afford new lines of lesser resistance, parallel to the old ones and along the contact of the two rocks of unlike hardness and texture. For this reason ore-bearing veins so often accompany dykes. They do so at Cripple Creek.

Before venturing upon the details of vein structure in this particular district, it will be well to preface such a description by a few general remarks upon the subject as viewed from a wider standpoint.

The mines of Cripple Creek afford excellent illustrations of the teachings of modern geology, and emphasise the incorrectness of the hasty generalisations of Sandberger and others before him. It is, of course, foolish to deride even theories which are now laid on the shelf, after they have served the purpose of quickening the researches of scientific enquirers. A working theory is needed in order to give some sort of direction to observation, because without an intelligent idea of what may be the possible fact one certainly will be unlikely to discover the real one. A man who is lost on a dark night, and declines the guidance of a tallow candle because it is too poor a light to illuminate the mountains in the distance, will lose the opportunity of avoiding the holes at his feet. The exploded ideas of Werner served a good purpose in their own day, because the accumulation of observations intended to disprove them became at the same time the basis for other views, which were as much nearer the truth as the data on which they in turn were founded, were more complete and more thorough. During later years the lateral secretion theory of Sandberger became accepted by certain distinguished geologists, and was as vigorously combated by others, notably Posepny, with the result that a great impetus was given to observation, and a large mass of new facts has been collected, bringing the scientific explanations of ore occurrence another step nearer the knowledge of things as they are. The unthinking sneer at the faltering steps of science, and the working miner is apt to belittle the aid which geology gives him. The amount of accurate knowledge of any kind so far attained by the human race is small when compared to the bulky mass of that which they think they know. Each intelligent observation is one step nearer the attainment of truth, and those who direct the working of mines will get more aid from geology when they contribute the necessary data without which the occurrence of ore will remain merely a maze of tangled phenomena.

It is well to begin a discussion of a subject by defining the terms employed. A "lode" is something that leads a miner, the words "lode" and "lead" having an identical Saxon origin. Australian miners designate a small continuous vein connecting larger ore-bodies as a "leader." "Lode" is therefore a comprehensive term covering many diverse forms of ore occurrence. The word "vein" has a more restricted usage, and describes those lodes in which the ore is supposed to occur in a tabular form, occupying continuous planes which are approximately vertical, and traverse the rocks like interminable sheets of paper set on edge, that is, they are supposed to fill simple fractures made in a perfectly homogeneous material. The term was originally borrowed from the human anatomy, and the oldest writers have used the simile of the rock veined with the precious metal. Nature does not recognise the definitions of the technical dictionary and in mining practice it has been found that regularity of structure is the exception rather than the rule. The geologist of 50 years ago, when the science was more the product of the library and the laboratory than of actual observation underground, conceived the ore as having filled gaping fissures in the rock, comparable to the crevasses of a glacier, and when he had noted the dissimilarity between the ore and the encasing rock, he imagined the former to have been due to an upwelling of molten metallic matter. The ideas of the present day are still slightly tainted by the imaginations of the past and the terms of an obsolete philosophy continue to cling to our nomenclature.*

Modern investigations, based on accurate chemical knowledge, as well as geological observation, have all gone to prove that gold ores are not the product of direct volcanic action, but that they have been conveyed to the place where the miner finds them through the medium of water, the metals having been dissolved, in various chemical combinations, by underground solutions, and precipitated along those fractures in the rocks which have been first lines of least resistance, and then lines of maximum circulation. The mineral solutions cannot have come from indefinite depths, because the

* Thus the terms "fissure-vein," "vein-filling," "vein-walls," &c., carry with them suggestions which are misleading to untechnical persons, and are the heritage of ideas now, I trust, recognised as untenable in the light of later evidence.

increase of heat (1° F. for every 48 ft. of descent) observable in the sinking of shafts and boreholes indicates that at a horizon of about 20,000 ft. below the present surface water would become dissociated* into its constituent gases. It is considered probable, from the evidence yielded by certain classes of lodes, particularly those of nickel ores,† that volcanic action serves to bring the metals from these great depths to that zone of the earth's exterior wherein solvent waters can circulate. The experience of gold mining corroborates this view, the association of volcanic rocks with bodies of valuable ore having become almost proverbial.

It is not surprising, therefore, that this very fact has tended to cause a confusion of ideas between volcanic action and lode formation.

In a railway cutting between the towns of Cripple Creek and Anacoda, there is a bit of nature's testimony which will be of service in getting a clear idea of the essential characteristics of gold-bearing veins as compared with dykes of volcanic rock. The accompanying drawings will help the description. In Fig. 2 there is afforded an excellent illustration of simple dyke structure. The dyke in this case is composed of basalt; it is from 9 to 15 in. in width, and can be easily traced as an irregular dark band traversing the coarse-grained pink granite. The dyke is very well defined, exhibiting clean-cut lines of demarcation from the enclosing granite, and it is evident from the contour of the walls that it occupies a fault fissure. The outline of the east wall corresponds exactly to that of the western one, the movement of the latter having been upward, causing a displacement equal to about 14 in. It is a clean-cut fissure in the granite, filled with foreign material, a basic volcanic rock, which probably welled upward in a mobile condition, filling the fissure as it was formed, so as at no time to permit of a vacuity. Compare this with Fig. 3, which is a sketch of a gold-bearing vein, situated at a distance of a few yards from the dyke illustrated in Fig. 2. The country is the same, viz., granite, but in this instance the vein filling is not foreign matter, but essentially rock in place; it is granite, altered, indeed, but easily recognisable, in spite of the kaolinisation of the felspar, and the partial removal of the mica. There are no clearly defined boundaries between the decomposed vein matter and the enclosing country, nor is there any evidence of faulting. The lines of fracture shown in the granite are the joints of that rock, and those which are observable in the vein itself are

* I refer to the critical point, which is 773° F. At this temperature water cannot, however great the pressure, retain its liquid form.

† As indicated by the researches of J. H. L. Vogt.

not continuous, but rather a closely knit series of little breaks, which have afforded a passage for a liquid more subtle than the

Fig. 2.

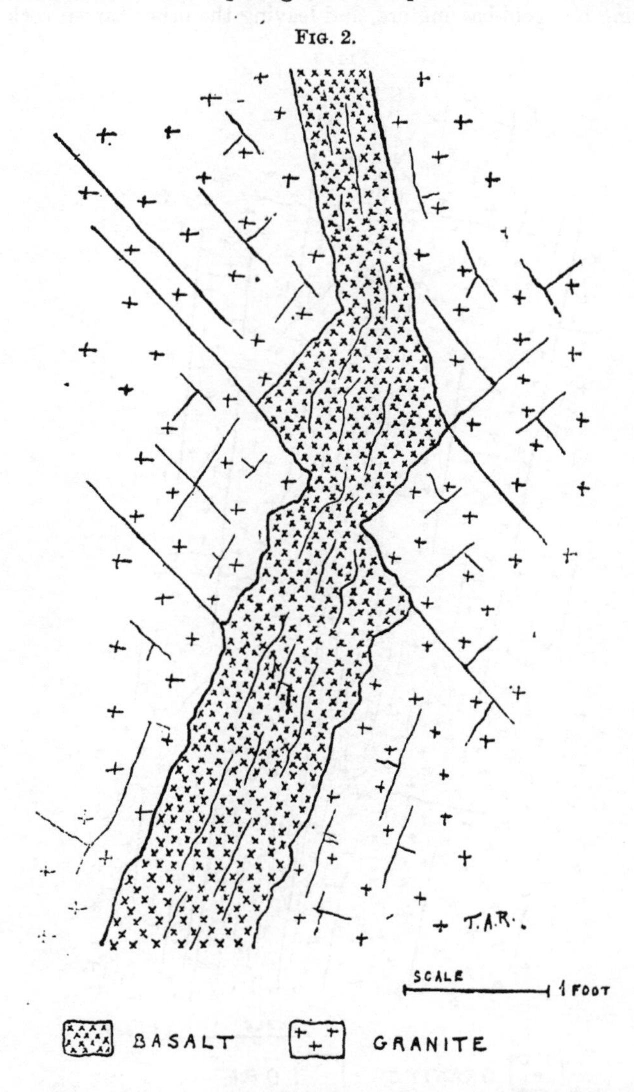

basalt. The vein occupies a line of maximum porosity along which water, more searching than any molten lava, has found a way, decomposing the soluble ingredients of the rock, and depositing a minute quantity of gold, insufficient to make the decomposed granite of the vein differ essentially from the outer country, but rendering one gold-bearing ore, and leaving the other barren rock.

Fig. 3.

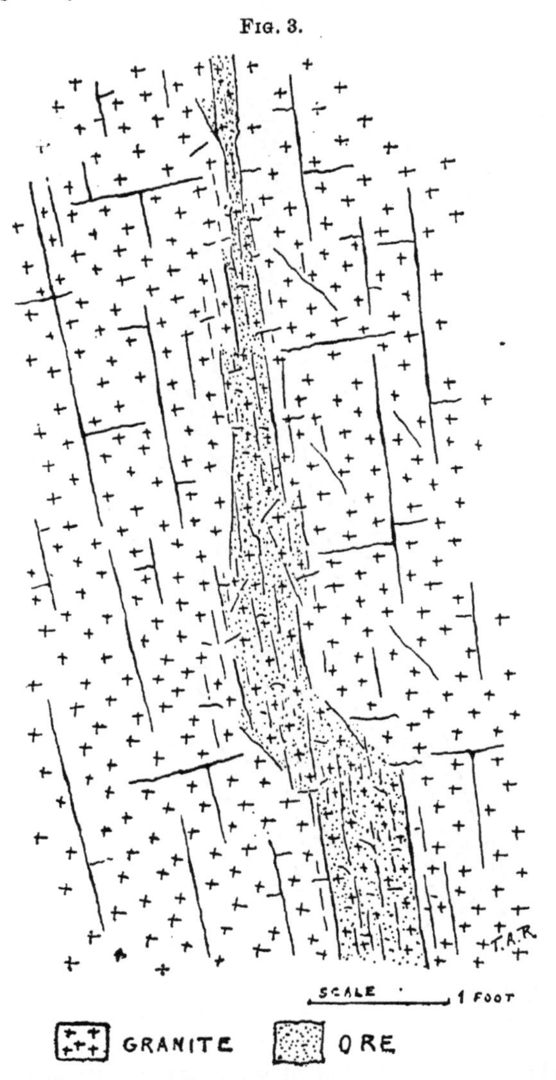

Here we have a dyke compared with a vein and volcanic agencies brought into strong contrast with aqueous action. The faulting along the fissure followed by the dyke is easily seen, but no evidences of such movement can be discerned along the seam of altered granite, which forms the gold vein. Nevertheless there must have been some movement, however slight, because a crack or break is not made evident, can be considered only as latent, until the two faces of it are caused, by that very shifting, so to disagree as to produce the irregularities which, when linked together, form the visible line of fracture. Even the joints in the solid granite require such [an explanation, and however insignificant the shifting may be, it marks the adjustment of the rock to the effects of stresses, traceable in this case to the volcanic energies which extruded the large masses of breccia forming the characteristic feature of the geology of Cripple Creek. Permit me to repeat, however insignificant this shifting may have been, it made the rocks pervious to underground mineral-bearing solutions, and where it occurred it developed a series of united passages, which afforded a line of maximum porosity, permitting of the circulation of gold-bearing waters, and the subsequent precipitation of the metal dissolved in them.

The Cripple Creek district exhibits a great variety of vein structure, and in order to afford a general idea of the conditions under which the ore is found it will be necessary to select one or two typical examples. Many of the veins are essentially mineralised dykes, that is, a part, or even occasionally the whole width of the dyke, is sufficiently rich to be regarded as pay ore, and the boundaries of the dyke then become the walls of the lode. The Moose vein will exemplify this type. The accompanying sketch, Fig. 4, was made at the 350-ft. level. From D to F is the width of the dyke, which in this case is nepheline basalt. It traverses the andesite breccia, which is indicated at AA. The pay ore extends from E to F, a width of 10 in., and it is distinguished from the remainder, EG, comparatively barren portion of the dyke, a dark bluish grey rock, by being iron stained and seamed with reddish brown threads in which gold and tellurides occur. The multiple fracturing, parallel to the walls of the dyke, is a characteristic feature of such lodes, and experience has shown that there is reason to expect the lode to consist of rich ore when it becomes threaded with minute seams following these lines of fracture. This feature can be described as a sheeting of the rock; it is a very important factor in ore deposition.

Another type is presented by those veins which accompany the phonolite dykes. This is very characteristic of the Cripple Creek district. The direction followed by a majority of the veins, especially on Battle Mountain and Bull Hill, conforms to that of a system of phonolite dykes. While a particular vein may not adhere continuously to the line of the dyke with which it is associated, nor the ore itself be found in the substance of the dyke, when they

Fig. 4.

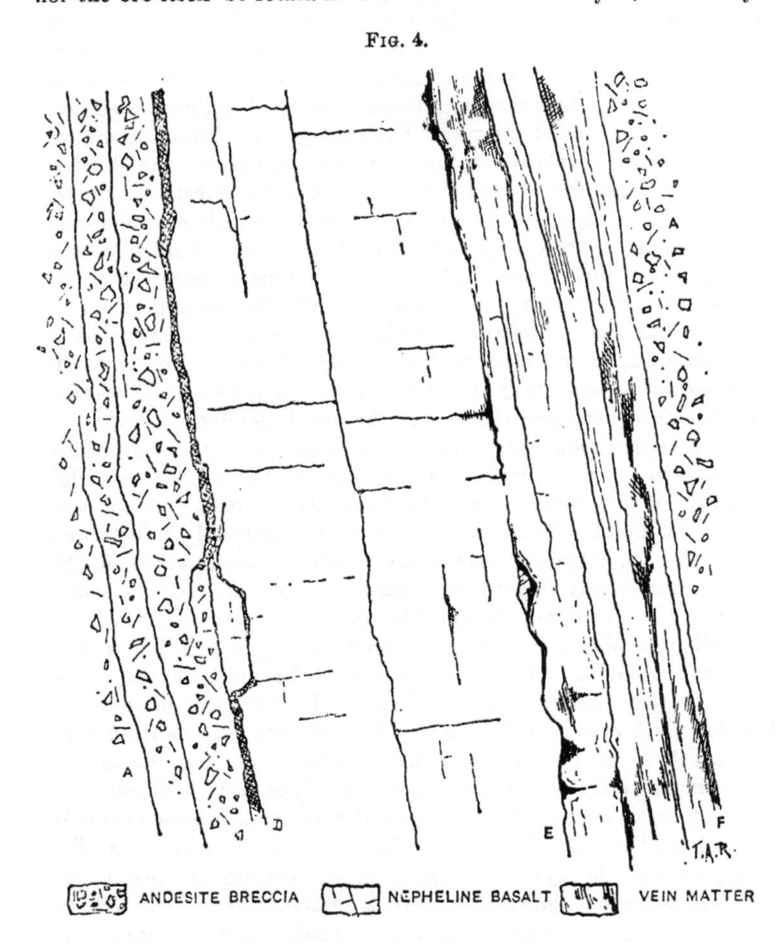

are together, yet the behaviour of the vein is intimately connected with that of the dyke, and the ore occurrence is modified by both. The accompanying sketch, Fig. 5, recently made in the Independence Mine, will illustrate this. The lode AB is shown to be in breccia, and follows a phonolite dyke, BC, which is 2 ft. wide. The ore is confined to the breccia, and if the phonolite is enriched it is so to a slight extent only, and the enrichment is confined to the planes of fracture near the lode and does not extend into the body of the dyke. The ore is essentially andesite breccia, rendered gold-bearing by the

FIG. 5.

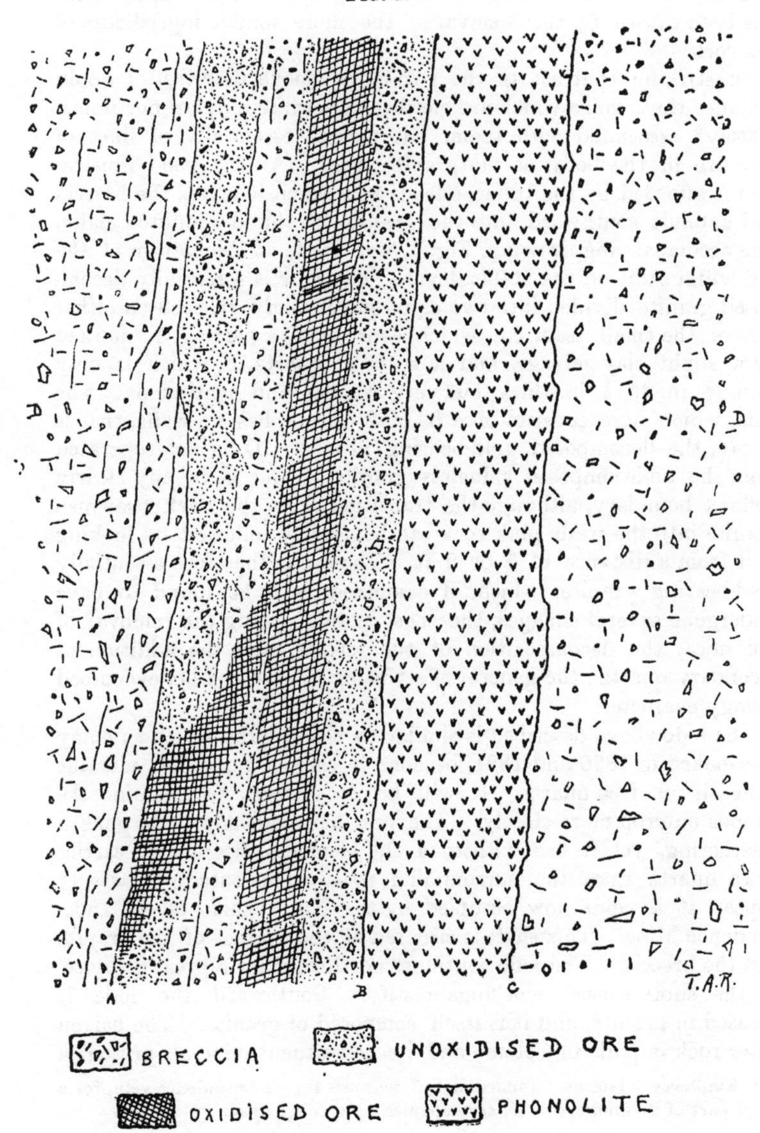

penetration of solutions which have circulated along the line of contact between the phonolite and the breccia. The gold occurs not only in the seams which follow the fracture in the breccia, but it is also found in the heart of it, where a spongy vesicular appearance has been caused by the removal of the more soluble ingredients of the rock.

In striking contrast to the two types above described I would instance those veins which are neither mineralised dykes nor immediately* associated with them, but consist essentially of lines of fracture in the body of the granite itself. A very good illustration is afforded by the Independence vein when it leaves the breccia and extends southward into the granite of the Washington claim. The accompanying drawing, Fig. 6, was made at the bottom of the old whim shaft on the 100-ft. level. The lode is essentially decomposed granite divided into two equal portions of 2 ft. each on either side of the small seam EF, which consists of a parting, E, marked by a slight clay selvage, and a seam of quartz, F, which is only from $\frac{1}{4}$ in. to $\frac{1}{2}$ in. thick, but very regular and continuous. The lode, which here consists of 4 ft. of 2 oz. ore, has no walls, that is to say, the decomposed gold-bearing granite, CD, is not separated from the undecomposed valueless granite, A and B, by any clearly defined boundary, although the transition from the dark brownish granite into the fresh pink rock can be clearly followed by looking at it from a distance of 5 or 6 ft. The ore is therefore essentially gold-bearing granite, which, if examined, will be found to have undergone several changes, the most notable being the removal of the mica, the decomposition of the felspar, and the addition of secondary quartz; the general effect being to give it a honeycombed spongy character.

The lode above described is similar to the outcrop which so many overlooked in 1890 and 1891, because it carried no visible metallic minerals or free quartz. It will be of interest to examine the famous outcrop more closely. The reef has evidently undergone a weathering, yet it has withstood the elements, because it carries more quartz than the surrounding rock. The outcrop ceases to appear at a point now occupied by the No. 1 shaft of the Independence Mine. This shaft is situated at the contact of the granite and the breccia. The lode continues northward for a great distance, as the subterranean workings testify. Southward the lode is encased in granite, and it is itself composed of granite. The barren outer rock is pink in colour, and its constituent minerals present a

* Emphasis is laid on "immediately" because the Independence vein, for a large part of its known course, accompanies a phonolite dyke.

fresh unaltered appearance. The gold-bearing portion, comprised within the lode, has a dull brown tinge; it is noticeable that the mica is absent, having evidently been leached out, leaving patches of chlorite and iron-stained spots in which free gold can occasionally

Fig. 6.

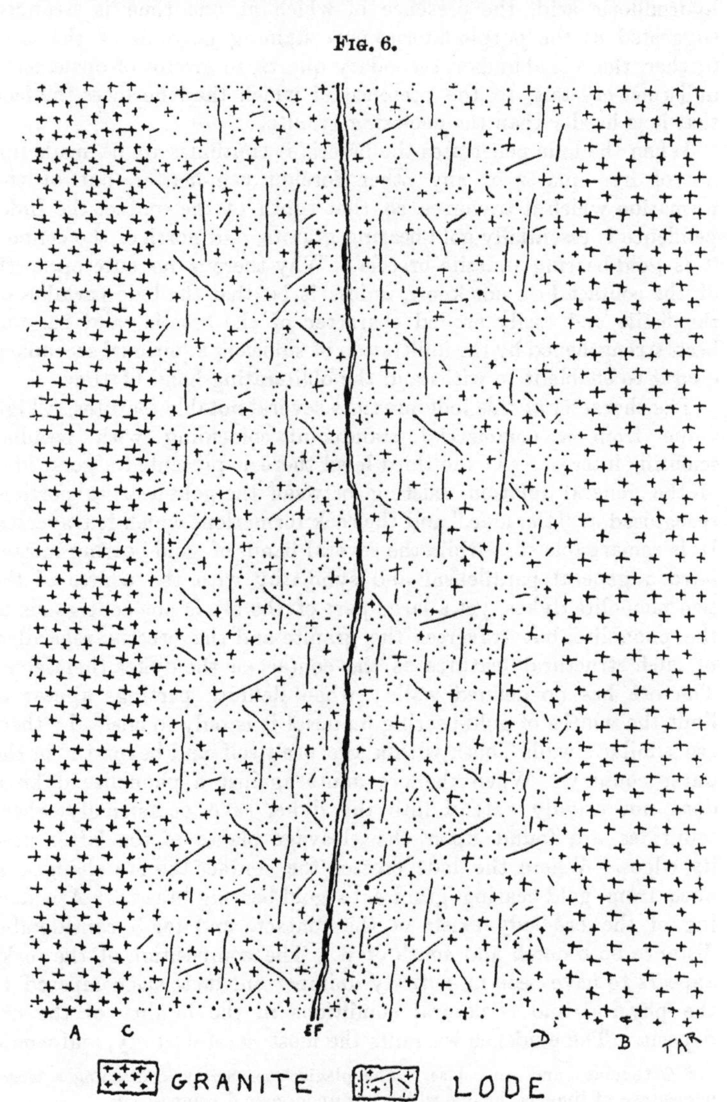

be distinguished; of the two felspars* of the original rock the more soluble has become kaolinised, while the other remains so undecomposed as to preserve much of the usual appearance of the granite; the original quartz has been removed, probably by the action of the hydrofluoric acid, the presence of which at one time is strongly suggested by the purple fluorite now staining portions of the ore; further, there is abundant secondary quartz, in groups of opalescent, indistinct crystals, to the presence of which the lode owes the fact that it is harder than the enclosing granite.

When the lode penetrates the breccia it continues to be made up, not of free quartz or any other foreign vein-matter, but of the formation which it traverses, so that south of the contact the Independence is essentially gold-bearing granite, and north of the contact it is gold-bearing andesite breccia. Why there is no outcrop north of the contact I do not know, unless it be that the lode partakes of the fissile and easily eroded character of the breccia, and has not been strengthened by the infiltration of sufficient additional secondary quartz to enable it to withstand the obliterating hand of time.

The character of this lode presents several notable features, a high value from a commercial standpoint coinciding with peculiar scientific interest. At the third level there is presented a good idea of the general relation existing between the gold-bearing portion, recognised as "the lode," and the rock formations which it traverses. It is remarkable that while the several bands of gold-bearing matter have a general parallelism and sympathy with the course of the two phonolite dykes, yet a large part of the ore is not contiguous to the phonolite, but traverses the granite and the breccia, regardless of such structural features as the contact or the dykes themselves. The ore has no defined walls; when definite partings appear to limit the widths of gold-bearing material it is only to mislead; there are similar "walls" both within the ore itself and beyond it in the outer country. When the ore abuts against a phonolite dyke it does not usually extend into the dyke, save occasionally where tellurides are found upon the cleavage planes of the latter near its edges. Where the lode crosses the contact the ore changes at once from gold-bearing granite to gold-bearing breccia. A widening of the ore-body exists at the contact, and for a considerable distance both north and south of it. The composition of the rocks appears to have been an entirely unimportant factor as compared to the physical and structural conditions in the locality of the ore deposit. The evidence warrants the most careful study, and would

* Orthoclase and oligoclase. The plagioclase variety, containing a larger percentage of lime, is the one which has undergone decomposition.

furnish the text for a long discussion of the essential features of ore deposition. Few examples of lode structure so well emphasise the general truth of those modern ideas of ore deposition which are advocated by the teachings of Posepny. It is interesting to accentuate this fact, because there was an erroneous impression conveyed, in the early days of Cripple Creek, by the writers of the daily press, that the gold-field set at naught the accepted teachings of geological science. The idea was due to the careless utterances of local men, and never emanated from any recognised authorities. The converse is true, and should be emphasised, namely, no modern mining region so clearly establishes that explanation of ore deposition, which is based upon the recognition of the laws governing the circulation of underground waters, under conditions traceable to the geological structure of the rocks.

The Ores and Minerals.

In discussing the geological features of the district, frequent emphasis has been laid upon the fact, that the ore is usually only rock, whether granite, phonolite, or breccia, which has become impregnated with gold-bearing minerals to a slight extent as regards percentage, but to a notable degree as regards commercial value. In 1893, when W. S. Stratton, the owner of the Independence, sent several carloads of rich ore from his mine to the Denver smelters, the officials at the works thought a blunder had been made and that loads of ballast had been inadvertently consigned to them. The ore was obviously granite and it required a trained eye, such as that of Dr. Richard Pearce, manager of the Boston and Colorado Smelting Company, to detect the fact that the mica of the granite had been largely removed, leaving small, iron-stained spots amid which there were disseminated dull yellow specks of gold. A glance into the ore-bins of the chlorination establishments will exhibit a mixture of broken rock, which looks more like the spoil of a barren cross-cut than the yield of a rich stope. The petrographer, in looking over this material, could easily label the rock, and if not initiated, he would wonder whether it could be gold-bearing. On being assured that it was valuable as such, he would take a few pieces, and break them open so as to examine a fresh surface, and it would not be long before he would see on the planes of fracture, evidences of richness.

The gold occurs either in the native condition or as a telluride, and is found distributed among the interstices of the rock, lining the fractures or penetrating the substance of it in threads of varying minuteness. In lodes traversing the granite, the gold, or the tellurides containing it, will be scattered amid the porous cavities due to the removal of certain more soluble portions of the rock; in phonolite, the values will be found more frequently along fractures than in the heart of it. This renders the last mentioned class of ore very difficult of estimation. In the andesite breccia, the component fragments of which are so heterogeneous, the physical character of the rock varies considerably, and the gold values will partake of an irregular sporadic distribution.

The distinguishing feature of the ores of the district, is the occurrence of tellurides. The discoveries of these uncommon minerals, produced nearly as many misconceptions a few years ago as the finding of similar minerals at Kalgoorlie in West Australia Tellurides seemed to create as much confusion among the miners of both countries, and to be as puzzling as the finding of the *Ornithorhyncus paradoxus* (or duckbilled *Platypus*) to the naturalists at the beginning of this century. Colorado, however, was better prepared for the telluride discoveries of Cripple Creek, than Australia was to exploit those of Kalgoorlie. The latter were unprecedented, the former only followed the line of previous, partly forgotten, experiences in the Boulder and La Plata districts of the same State.

A few general notes concerning the composition of these complex ores will be of interest. And first a few definitions are necessary. The "tellurides" are compounds of tellurium with certain metals, and they are so termed just as the compounds of sulphur are called "sulphides"; thus, for example, calaverite is the telluride of gold, while iron pyrites is the sulphide of iron. The name "tellurium" is derived from the Latin *tellus*, meaning the earth, and was chosen in opposition to that of the element selenium, which comes from the Greek word, *selēnē*, the moon. Tellurium is a non-metallic element with a metallic lustre, it acts as an acid base, just as selenium and sulphur: the three forming a recognized chemical group having common affinities. If tellurium were a metal, as is often supposed, its combination with gold or silver would only be possible as an alloy.

It is usually assumed that the precious metals enter into the composition of tellurides, and a high commercial value is taken for granted in speaking of this species of minerals. Nevertheless, there are many varieties which are valueless to the gold miner, such as the telluride of mercury, called Coloradoite, which occurs at Kalgoorlie, as well as at Magnolia in Boulder County, Colorado; there are also the telluride of bismuth, called Tetradymite, and the telluride of lead, called Altaite, and the telluride of nickel, Melonite, all of them found occasionally in the mines of Boulder County, as well as in the ancient mining region of Transylvania, where most of them were first recognised. These tellurides of the baser metals occur, for the most part, in quantities so restricted as to have no commercial value and are rather to be considered as curiosities much desired by the mineral collector.

In the mining camps it is a frequent custom to speak of tellurium when it is intended to refer not to that element, but to its compounds with gold or silver. Tellurium itself occurs native, that is, in a free state, just as gold, silver or copper. It is tin-white in colour, it is soft but brittle, and is usually found in a granular massive condition. A commercial value of $3·50 per ounce is ordinarily quoted for tellurium, but the demand for this rare earth (as the chemist terms it) is very slight, and a few shipments of it would quickly demoralize the market. No native tellurium has yet been found in the Cripple Creek district, but in Boulder County, 90 miles further north, it occurs among gold ores in a free state, and a mass weighing 25 lb. was found in the John Jay Mine about twenty years ago.

Certain telluride materials very much resemble the common ores of silver, the sulphide of silver (called argentite), for example, is difficult to distinguish from hessite, the telluride of the same metal, silver. The instance is quoted in order to refer to a simple test applicable to any doubtful cases. Remove a small bit of the suspected mineral with the point of an old knife, and put it in a porcelain dish or a white saucer. Add three or four drops of strong sulphuric acid, and heat over a lamp. Should tellurium enter into the composition of the suspected mineral, a beautiful purple will suffuse the colourless acid. The miner's time-honoured test is to put the ore in the fire of a blacksmith's forge and roast it. Tellurium fuses at a comparatively low temperature, and becomes volatilised, passing off in white fumes of telluric oxide. If the telluride mineral contains gold, the latter will remain in the form of globules. Even the precious telluride hidden in the seams of the piece of ore will be exuded as a yellow perspiration. The miner calls this process "sweating," and the reason for it becomes obvious when the results are observed.

Experienced miners, such as those of the Boulder district of Colorado, who have been working amid these particular ores for twenty-five years, know very well, from the disappointing returns of many a shipment to the smelter, that there are several minerals closely resembling tellurides which are not rich in gold, and therefore that a rough test, by roasting, is occasionally necessary. The mineral may go off in fumes when heated, leaving no gold or silver behind, as native tellurium does. Consequently, the Boulder miners are apt to call everything native tellurium which in the process of roasting disappears entirely or leaves a residue in which the precious metals are not recognisable. But among the latter would be included the tellurides of lead, bismuth, and nickel; and to the former belongs not native tellurium alone, but also the telluride of mercury, because mercury also readily volatilises under the conditions of the roasting test. This mistake was made at Kalgoôrlie, where I found the telluride of mercury being labelled "black tellurium," for the reason just referred to. The mineral combination of tellurium and mercury is rare; it was first found at the Mountain Lion mine at Magnolia, in Boulder County, and a simple test for it is to roast a particle of it in a glass tube. The mercury is volatilised at the hot lower end of the tube, and is condensed at the cool upper end in the form of minute metallic globules, which are readily recognisable as quicksilver.

The principal telluride minerals found in the Cripple Creek ores are sylvanite, calaverite, and petzite. Sylvanite, named after the place of its discovery, the historic goldfield of Transylvania, is the most characteristic of Cripple Creek ores, very beautiful specimens of it having been obtained at, among others, the Independence, Portland, Moon, Anchor, and Anchoria-Leland Mines. It is a double telluride, containing gold and silver, an average composition being 28 per cent. gold, 16 per cent. silver, and 56 per cent. tellurium. Sylvanite is a brilliant silvery-white mineral, having a characteristic crystalline habit, to which it owes its other name, "graphic tellurium," the arrangement of the crystals resembling Arabic lettering.

On account of the large percentage of gold in the composition of the richest ores of Cripple Creek, it has been concluded that the prevalent mineral is not a silver-bearing telluride, but that the gold occurs combined with tellurium alone in the form of the mineral calaverite. Calaverite is the simple telluride of gold, pure specimens contain 44·5 per cent. gold and 55·5 per cent. tellurium. It is named after the county of Calaveras, in California, where it was first found at the Stanislaus Mine. It usually carries from 2 to 3 per cent. of silver, which must then be considered as an impurity. The purest varieties have a bronze-yellow colour. It is the characteristic mineral of the rich ores of Kalgoorlie, and beautiful specimens of it are common in the Great Boulder, Associated, and Lake View Mines of that celebrated district. Calaverite is rather difficult to distinguish from iron pyrites; the difference in colour is slight, but the former is easily cut by a knife, while the latter will not permit of it. The cubic crystalline habit of pyrites will usually help to make it known, because calaverite rarely occurs in any other than a massive form. The best specimens obtained in the Cripple Creek district came from the Work Mine.

Reference has been made to the uncertainty as to the identity of the particular telluride mineral which carries the values in many of the Cripple Creek ores. It is very probable that future investigations will discover the existence of a new variety, peculiar to the district, in which many of the physical characteristics of sylvanite will be united to a composition so rich in gold as to resemble calaverite.

Petzite, named after the German chemist Petz, is, like sylvanite, a double telluride of gold and silver, its average composition being 25 per cent. gold, 42 per cent. silver, and 33 per cent. tellurium. It is much darker than sylvanite, being steel-grey to iron-black ; it is also slightly harder and more brittle. The best specimens of petzite found in the district came from the Geneva claim on Gold Hill, about four years ago. Recently the writer found petzite in the ore of the Porter Gold King Mine

All these tellurides are distinguished from the baser minerals, with which they may be occasionally confounded, by their rich lustre.

The lodes of Cripple Creek are further characterised by the presence of fluorite or fluorspar (the fluoride of calcium), a beautiful purple mineral, which is so notably associated with the ores of the district as to have led to the idea that it could be accepted as an indication of the richness of the veins in which it was found. But, like most similar attempts at short cuts to knowledge of this kind, the generalisation has proved fallacious. There are several large lodes of very low grade ore in the district which are purple with fluorite, and there are some very rich ores almost devoid of it. The association of the gold and the fluorspar points to a similarity, and possibly a contemporaneity, of origin ; but this fact does not, and could not, predicate whether the quantity of gold present will give the ore an average value of $2 or of $200 per ton. As a matter of science, both kinds of ore may be considered gold-bearing, as a matter of business, one spells losses and the other dividends.

In the upper levels of the mines, within the reach of surface waters, the tellurides are decomposed and the gold has been liberated. In the first years of development, certain mines, such as the Pike's Peak and Garfield Grouse, on Bull Hill, afforded specimens of native gold, the dull brown lustreless appearance of which was a puzzling variation. At Kalgoorlie, likewise, a brittle spongy variety became known as "mustard gold." In each instance, when such a specimen is scratched with a knife or burnished, the bright gleam of pure gold is made apparent. These varieties of native gold are directly traceable to their origin from tellurides, which have been decomposed by oxidation, the gold often retaining the form of the original mineral, and being a perfect skeleton of a former mode of occurrence. Beautiful pseudomorphs, as they are called, of gold after sylvanite,* are common in the ores of Cripple Creek.

Below the depth, which ranges from 100 to 400 ft., reached by surface waters, the unaltered tellurides appear in all their untarnished beauty. At a further depth, from 500 to 700 ft., the ores become more complex, because of the increasing percentage of baser minerals, chiefly iron pyrites, but including also galena (the sulphide of lead) and stibnite (the sulphide of antimony). This change is important chiefly from a metallurgical standpoint, because the increase of sulphur renders the roasting of the ores more expensive. The general question of the probable changes to be encountered as the mines become deeper will be discussed under another heading, at the close of this description of the district.

The Treatment of the Ores.

Most of the gold of commerce has been won from simple ores, those in which the precious metal occurs in a native condition encased in quartz. Stamp-milling, accompanied by amalgamation, is the process ordinarily employed to reduce them. Another type, of greater importance every year, is represented by the ores in which the gold occurs intimately associated with iron pyrites, but so minutely disseminated as to be rarely visible even under the microscope, and therefore inviting the supposition that it is chemically combined in a condition as yet not understood. A modified form of stamp-milling is employed for some of these ores, but for the most part they afford a field for great diversity of practice, including chemical and smelting processes of many kinds. The third type of gold mining, and in many respects it is the most modern, is represented by ores in which the gold is known to be chemically combined in definite proportions with the element tellurium. Several mining regions produce these telluride ores. Transylvania in Europe, Colorado in the United States, and West

* At the Gold King Mine, in Poverty Gulch, I recently secured specimens carrying crystals of gold pseudomorphic after krennerite, a mineral resembling calaverite, but occurring in prismatic crystals.

Australia in the Antipodes are the principal localities. Transylvania has an interest which is chiefly historic, but Cripple Creek in Colorado, and Kalgoorlie in West Australia, are writing their names in glittering figures on the records of the present. The new phases of metallurgical practice inaugurated by these recent discoveries of telluride ores have given a great impetus to this technical science, the aid of which is now thoroughly appreciated by the miner and the capitalist alike.

The changes which have occurred since 1891 in the treatment of the ores produced by the mines of Cripple Creek are a fitting illustration of the fact that the development of metallurgical practice, like all true progress, is a slow evolution from simple beginnings to a full fruition. It is a common fallacy to suppose that processes of ore treatment are unexpectedly discovered by ruminating chemists or revealed from on high to the millman, and frequent paragraphs go the round of an ill-informed press to the effect that this or that spectacled professor has lit on a new combination in physics or chemistry which is to revolutionise the existing methods of ore reduction by extracting 100 per cent. at a negligible cost. The progress of metallurgical practice resembles organic evolution in that it does not advance *per saltum*. The chlorination process was first applied, in 1857, by Deetkin, in California. It took a quarter of a century of patient endeavour to place it on a safe commercial basis. MacArthur and Forrest's application of potassium cyanide was used, at the Crown Mine in New Zealand, as early as 1889, but the chemical reactions which occur in the cyanide process are still incompletely understood to-day, and the development of this method of ore treatment cannot be said to have ceased.

The first ore broken at Cripple Creek was carried on the backs of prospectors to Colorado Springs or Florissant, and forwarded to the smelters of Pueblo and Denver. Such small shipments of selected ore usually mark the birth of our western mining districts. Subsequently larger lots in wagons were similarly consigned, but the costs were high and the proportion of ore capable of yielding a profit under such conditions was small. The miner felt the necessity of extracting the gold by milling it nearer the place whence it came. He fell back on his previous experience in other localities and put up a stamp-mill.

The stamp-mills were built on the model of those of Gilpin County, the oldest gold mining district in Colorado, and therefore had light stamps, 400 to 500 lb.; a slow drop, 30 to 35 per minute; a long drop, 17 to 20 in.; and a deep discharge, 12 to 15 in. The ore was broken by hammers and fed by hand into the batteries, except in the case of the Rosebud mill, which was provided with rock-breakers and automatic feeders. Amalgamation, on copper plates, both inside and outside of the mortars, followed the stamping. During the three years succeeding the first discoveries the following plants were at work :—

Name of mill.	Locality.	Date of erection.	No. of stamps.
Lawrence	Lawrence	1892	20
Summit	Gillett	1892	30
Gold and Globe	Cripple Creek	1892	40
Beaver Park	Beaver Park	1893	20
Colorado Springs	Beaver Park	1893	25
Denver	Beaver Park	1893	20
Hartzell	Anaconda	1893	20
Gold Geiser	Cripple Creek	1892	15
Cranmer	Arequa	1893	20
Rosebud	Mound City	1893	60

The total is no less than 270 stamps. In April, 1897, only 50 out of this number were dropping, and to-day they are all idle, having been replaced by large leaching plants shortly to be described.

The first ores came of course from the surface, or near it, and were therefore oxidised. The gold occurred in a free state, having been liberated from its combination as a telluride. This last fact was not known and not appreciated even when first it became known. It affected the milling most seriously, because gold having this origin is coated with a film of the tellurite of iron, which is a serious obstacle to amalgamation. The extraction in the stamp-mills was soon found to be low, and attempts were made to remedy the incompleteness of the treatment by employing bumping tables and Frue vanners to arrest the gold escaping amid the tailings. When this failed, blankets were added, and these in time became a recognised addition to the mills. Even then the best extraction barely reached 50 per cent. of the gold in the ore, as shown by assays, and a great deal of good ore was wasted in an ineffectual effort to win a profit. The usual rate for treatment was $3 per ton, delivered at the mill.

Early in the development of the district there came men who recognized the unsuitability of stamp milling for the treatment of telluride ores, and in 1893, W. S. Morse, of the Russell Lixiviation Works, at Aspen, made the experiments which formed a basis for the erection of a chlorination plant built by Edward Holden, at Lawrence, 2 miles from the town of Cripple Creek, at the close of the same year, 1893. Holden purchased an old stamp-mill at Lawrence and altered it to a chlorination plant which, although it was crude and incomplete, successfully demonstrated the suitability of the process employed. This was barrel chlorination on the model of the practice of South Carolina and Dakota. The erection of the first well-designed plant, using this process, was begun at Gillett, in August, 1894, and completed in the following January. This had a capacity of 50 tons per day.

The cyanide process had been already tried, with results not wholly satisfactory, because of the variable composition of the ores and the inexperience of the men who organised the first milling enterprise. The local company, controlling the MacArthur-Forrest patents, built a cyanide mill of 40 tons capacity, at Brodie, in 1892. It was remodelled in 1894. In 1895 the Metallic Extraction Company's mill was built near Florence. It has been enlarged from time to time and now has a capacity of 8,000 tons per month. The cyanide process made less progress than barrel chlorination, so that while two mills now employ this method, six use the other. The Colorado-Philadelphia chlorination mill was built at Colorado city in 1896. It has a capacity of 6,000 tons per month. The El Paso mill, with a capacity of 3,000 per month was built at Florence in 1897. The Gillett mill doubled its capacity in 1898. Others were erected as shown on the accompanying list.

The Mills of the Cripple Creek District.

Name.	Locality.	Date of erection.	Capacity per day.
Chlorination—			tons.
Lawrence	Cripple Creek	1893	30
Gillett	Gillett	1894	80
Colorado-Philadelphia	Colorado City	1896	200
El Paso	Florence	1897	100
Colorado	Arequa	1897	75
Kilton	Florence	1897	50
Cyanidation—			
Brodie	Cripple Creek	1892	60
Metallic Extraction	Florence	1895	270
Page	Florence	1896	20

The Lawrence mill was burnt down in 1896, and the owners of it built the El Paso, at Florence, immediately afterwards. The

Page mill was erected to use a secret modification of the cyanide process, but after a brief activity it became idle.

The trend of events indicates that the future growth of the milling practice will favour an increase in the chlorination mills rather than in cyanide establishments. At first, cyanidation was conducted upon raw ores, but this gave good results only when they were oxidised. Roasting is now considered a necessary preliminary and it has removed the advantage which in the earlier years of the district cyanidation possessed over chlorination, a process always preceded by the roasting of the material subjected to it.

The chlorination practice of the district has undergone no radical changes during recent years. It is typical barrel chlorination. In the matter of the recovery of the gold from solution there is a difference, the Gillett mill, for instance, using charcoal as a precipitant, while the Colorado-Philadelphia plant uses sulphuretted hydrogen. The mills at Florence enjoy an important advantage in the use of oil residuum, instead of coal, in their roasting furnaces. The residuum comes from the distillation of petroleum and costs about 1 dollar per barrel now, although a few years ago the price was much less. There is an increased demand for it because it affords a fuel peculiarly adapted for roasting, permitting of a nice and quick adjustment of the temperature of the furnace.

During recent years the metallurgical practice in the chlorination plants has remained fairly constant, the improvements being in the direction of large capacity and better mechanical arrangements rather than in any changes in the chemical department.

The cost of treatment at one of the larger mills, having a capacity of 3,000 tons per month, is, per ton of ore treated:—

Labour and salaries	$1·20
Chemicals and supplies	0·78
Fuel	0·65
Wear and tear	0·55
Incidentals	0·28
	$3·46

Add to this interest on investment, general expenses and depreciation of plant, and the total costs will approximate $4·00 per ton. The above figures are based on chloride of lime costing $2·40, and sulphuric acid $1·25 per 100 lb. Roasting alone costs from 45 to 60 cents per ton. This item has grown as the mines have become deeper, the sulphur contents having increased from an average of 1 per cent. in 1895 to 2¼ per cent. at the present time. The improvements in other departments of the milling have more than balanced this change.

The ores are altered phonolite, andesite breccia or granite, and therefore have a composition similar to these rocks save in an increased prcentage of quartz. A representative analysis may be quoted as

Alumina	29·94
Silica	63·13
Lime	0·70
Iron oxide	3·66
Iron sulphide	2·64
Magnesia	trace
Manganese oxide	0·40
Sulphur	0·96

Two interesting features of Colorado practice have been brought out by the treatment of the Cripple Creek ores, the first being the greater use of mechanical roasters, such as the Pearce, Ropp, and other furnaces, and, secondly, a modification of the methods employed to sample the ores. The material produced by the mines is, as compared with the typical ores of other districts, high grade and very variable in its gold contents. This variability is due to its mode of occurrence as a telluride in minute seams irregularly scattered, and, of course, extremely rich, rendering it difficult to apply the law of averages and obtain a satisfactory sample of a large lot of ore. In its passage from the hands of the miner to those of the smelter, the ore usually goes through the sampling works, the owner of which may be considered a broker whose business it is to see that both parties in the transaction get those values in the ore which one sells and the other buys. It was soon found that Cripple Creek ores were most unsatisfactory to sample, the results being unreliable and erratic. This became remedied in process of time by crushing a larger part of any particular lot of ore and taking pains to pulverize the final pulp sent to the assayer to 100 mesh,* instead of 60- or 80-mesh. It has also been found that, on account of the variability of the ore, it is good business for the mine owner to let all his product pass through the sampler on its way to the smelter, and this has led to the erection of half a dozen sampling works in the Cripple Creek district. The usual cost of sampling is 75 cents per ton.

* That is, 10,000 holes per square inch.

The Mines.

Cripple Creek is interesting to the geologist; it is fascinating to the financier. When viewed from the latter standpoint it presents many aspects which render it unique among modern gold mining districts. The mines of Cripple Creek have paid their own way from grass roots, that is to say, they have not grown up by the expenditure of working capital in development. Working muscle has made a 10-ft. hole into a productive mine. There is, therefore, an instant contrast between the financial features of the industry when compared with the goldfields of the Transvaal or West Australia.

It may be objected that all mines are first made by the pick of the miner and are then enlarged into important enterprises by the investment of capital in machinery and development. Nevertheless the contrast above suggested is not a strained one, because it is a fact that the mines of Kalgoorlie and Johannesburg were merely tracts of land when they became the basis for large company flotations. The shafts were sunk afterwards by the companies organised in London, or elsewhere, and the ore was uncovered by workings which had no existence at the time of the first organisation of the enterprise. Companies were brought out with a large subscribed capital, a portion of which was paid out in the purchase of property and another portion was set aside as a working capital, that is, a fund out of which to pay the expenses of equipment and that amount of development necessary to enable the mine to maintain a steady production. It is rarely the case that a mine floated on the London market is able to pay the increased dividends necessitated by an increased capitalisation without some preliminary vigorous development such as will permit of an enlarged output without exhausting the ore reserves. In the case of the deep-level mines of the Rand, the working capital is a princely sum, sufficient to sink a shaft a couple of thousand feet, or even more, and erect a mill of two or three hundred stamps. The finances of South African mining are a study apart and the result of conditions previously unknown in gold mining, although paralleled in the coal industry. It represents a wonderful development of modern practice, such as renders the digging for gold a steady business very much unlike the feverish uncertainties of the days of 1851 in California and Australia.

Even in districts which have not the extensive low.grade deposits of the Rand, it is necessary to provide for a rainy day, and English capitalists are accustomed to the idea of furnishing a certain amount of money as a reserve fund to tide over the lean places in a mine's development. The practice is a good one, and to a great extent offsets the large capitalisation of most British incorporations. The American buys his mine and proceeds to work it for all it is worth, the dividends are pocketed, and when the rich ore is exhausted there comes a collapse. The shares of American companies are usually non-assessable; the shareholders, who receive the dividends, will often refuse to contribute fresh capital, and the mine may be closed down until, after an interval, some one with a good deal of pluck re-opens it. But this is rare. As a consequence many properties which made a magnificent record are now idle, and this idleness means, not inactivity alone, but the rusting of the machinery and the falling in of the workings, rendering it a very costly matter to re-open the mine. The English practice of heavy capitalisation has been censured often enough, but there is this to be said for the methods of London investors, that by subscribing an adequate working capital they often permit the mine manager to enlarge the capabilities of a property so as to enable it to hold up cheerfully under its increased load and continue exploratory work during those intervals when the production does not meet the operating expenses.

Our ways of flotation in Colorado are much more simple, and this very simplicity would ruin any but the best of mines. A mining claim is in the possession of three or four working miners, who secured it originally by location. It does not pay its way, although there is evidence that it is worthy of development. They go to a couple of merchants in the nearest town, and each of these pays the wages of one working miner to represent him. The prospect hole becomes a shaft, and a little good ore is discovered. A company is organised by a lawyer or a stockbroker. The capitalisation will be a million shares of a nominal value of a dollar, but a rating of 10 or 20 cents per share. The owners give 10 per cent. of the stock to the incorporator. Work is resumed with the aid of the money obtained by selling a small block of shares. If the mine does not pay, it is leased or some stock is sold. At the best it is a hand-to-mouth policy, which succeeds in spite of its weak points, because of the richness of the mine.

In other cases the original working miner is bought out for a few hundred dollars, and the new owners go ahead subscribing the money required to meet expenses, month by month, until the patience of one or more is exhausted, and either the impatient minority sells to the plucky majority or the entire party gives up the fight for riches, and awaits a buyer who may happen along at a time when the camp is excited by new discoveries on neighbouring claims.

The history of Cripple Creek, however, records comparatively few abandonments, because the veins have proved so generally productive that their yield has sufficed to keep the work of development going ahead, even if it did not meet all the expenses. As soon as a small mine becomes organised into a company, it is listed on the stock exchanges of Denver and Colorado Springs as a "prospect," and when it becomes a regular producer it is promoted to the society of "mines," the two classes being kept distinct on the call-lists of the exchanges.

A mine which does not pay goes through eras of greater or less activity, according as the owners of it can scrape together the money needed to continue the work. Idleness is possible because the mining regulations do not entail forfeiture on that account. The claim can be "patented," that is to say, the owner obtains a title in fee simple from the U.S Government when he has done a certain amount of work, nominally rated equal to the expenditure of $500, and the Government surveyor has prepared the map needed for record in the bureau of the surveyor-general. Labour conditions such as obtain in the English colonies are unknown in Colorado, and the continuous operation of a mine cannot therefore be enforced under the mining laws.

This is itself an important feature of mining in this State, and it will appeal at once to those who have been shareholders of Australian mining companies. I remember, when at Coolgardie in 1897, what a serious matter the labour conditions were to the West Australian companies which had acquired an extensive acreage during boom times, and were compelled to work a large force of men during a period of financial stringency. The regulations required the working of claims on which no ore had been found on pain of forfeiture, and, therefore, enforced the expenditure of money regardless of the nature of the work to be done. This clause in the mining law, which compels the working of a certain number of men on each claim, known in Australia as the "labour covenants," has its good features, the chief being that it prevents good mining claims from remaining undeveloped by those who, from choice or necessity, would like to pursue a dog-in-the-manger policy. In a lively mining district, like Cripple Creek, it is not needed, and would only impose irritating obligations, but in the older camps of the United States it has led to the severance from the public domain of large areas, which remain unexplored on account of the poverty or want of energy of absentee owners, who will not exploit the mines, and refuse to sell them save at prohibitive prices. It must be remembered that an idle mine is only a hole in the ground, and there is nothing more worthless. The problem is a far reaching one, and is only mentioned incidentally because it is still agitating the mining departments of the colonial governments.

The leasing system is still an important factor in Cripple Creek mining. Three or four years ago it was more prevalent. The owners of claims who were too poor to develop them, or too timid, cut them up into several sub-divisions and leased them. The lessees might be working miners, who engaged the service of others under them, or the mine might be leased to outside parties, who hired workmen in the ordinary way. The period of the leases varied from six months to two years, and the royalties were 15 to 20 per cent. of the net returns, that is, the receipts from the mill or smelter, after the costs of transport and treatment are deducted. Leasing is the last resort of the perplexed mine-owner, and is a confession of inability to work one's property, yet in the early days, because the mines were so often owned by men wholly unfamiliar with mining, the system served a good purpose, and many rich mines owe their first beginnings to the enterprise of a lessee.

The history of the mines is a pleasant one, encouraging to those who have been led by the experience of other districts to consider mining only as a reckless gamble in which blind luck outweighs intelligence. The infrequent transfer of mines, which has characterised the story of Cripple Creek up to the present, is due to the fact that the conditions for successful mining are present to an unusual degree. In the first place no deep sinking was needed to reach the ore. Although outcrops were rare, and discovery was thus delayed, it soon became evident that the tops of the ore-bearing veins could be reached by trenching. The hills became spotted with yellow heaps of rock, and it looked in 1894 and 1895 as if they had been invaded by gophers of a larger growth. Ore was encountered at depths of from 10 to 30 ft. It was rich. Cripple Creek has always been a high grade camp; the average yield for several years exceeded 3 oz. per ton, and last year, notwithstanding an enormous increase in the low grade milling ore, it was fully 2 oz. Therefore, even in the earliest period of development, the ore could meet the costs of transport to the smelters. But competing lines of railroad were soon pushed into the district, and facilities for cheap mining were obtained long before the goldfield had reached a quarter of its present importance. This fact permitted poor men to work the mines, because it gave them a market for their ores; muscle and energy were sufficient, capital was not needed to the extent known in less favoured localities. Even to this day no single mine has a reduction plant of its own; it is not needed. As the camp grew, the chlorination and cyanide mills erected in or near the district began to compete with the smelters, and prevented the latter from levying excessive rates of treatment; and, lastly, the demoralisation of the silver market already referred to earlier in this description as having occurred at the time of the closing of the Indian mints in 1893, which put a severe check on what was then the chief industry of Colorado, did at the same time aid Cripple Creek immensely by diverting the energies of a very resourceful population to the development of the mines of its newly discovered goldfield.

Statistics.

From the first beginnings in 1859 to the close of 1898 Colorado has produced—

Gold...............	9,512,242 oz.	$196,618,054
Silver	362,526,541 „	326,482,532
Lead..............	1,064,762 tons	87,131,457
Copper	43,089 „	10,742,167

The production for 1898 is estimated to have been—

			Proportion of U.S. output.
Gold.........	1,138,584 oz.	$23,534,531	34 per cent.
Silver	23,502,601 „	13,690,265	38 „
Lead.........	56,708 tons	4,117,043	26
Copper......	5,435	1,304,504	2 „

At the present time the mineral production of the State, inclusive of coal and iron, has a value of $55,000,000 per annum.

There are 30,000 men employed in the mines of Colorado out of a total population of 520,000.

The smelting industry of Colorado has been an all-important factor in the growth of its mining districts. There are ten large smelting works in the State (at Denver, Pueblo, Leadville, and Durango), and several smaller plants whose activity is intermittent. The value of the entire product of the smelters amounted in 1898 to $61,000,000. This included the product from ores which came from neighbouring States and from British Columbia.

At the present time 3,500 tons are smelted daily in Colorado. The annual tonnage consists of 900,000 tons of ore, 300,000 tons of iron flux, and 185,000 tons of limestone, the balance of the lime needed being obtained in the form of ores, which carry more of it than is required for their own treatment. Aspen provides the bulk of the latter material.

The charges for treatment depend upon the composition of the ore, particularly the ratio of iron to that of silica. On a neutral basis the charge is $4·25 for oxidised ores and $7 per ton for sulphide ores. On highly silicious ores, containing 60 to 70 per cent. silica, the charge is $8 per ton, flat. $19 per oz. is paid for the gold in ores carrying up to 2 oz. per ton, and $19·50 per oz. for those which are richer.

Until 1892 silver mining was the most important industry; since that year gold mining has been in the ascendant. The production of gold has grown steadily and rapidly, as the following figures testify:—

1892	256,410 oz.	$5,539,000
1893	364,151 „	7,527,000
1894	462,009 „	9,549,730
1895	656,021 „	13,559,954
1896	738,618 „	15,267,234
1897	947,249 „	19,579,637
1898	1,138,584 „	23,534,531

This increase is to be credited chiefly to the development of Cripple Creek, the gain from that district having been the largest part of the improvement made by the State in its entirety. The output of the Cripple Creek district has been:—

1891	100 oz.	$2,060
1892	2,821 „	583,010
1893	104,000 „	2,150,000
1894	140,710 „	2,908,702
1895	332,800 „	6,879,137
1896	363,400 „	7,512,911
1897	490,500 „	10,139,708
1898	653,410 „	13,507,349

The analysis of the figures of the last three years exhibits the growth in the tonnage of ores treated by the chlorination and cyanide mills as compared with the smelters:—

	Smelting ore.		Milling ore.	
	Tons.	Value.	Tons.	Value.
1896...........	84,659	$6,045,319	77,388	$1,467,592
1897...........	94,287	5,697,788	181,885	4,441,920
1898...........	110,036	7,137,366	251,862	6,369,983

The smelters charge a minimum rate of $6·50* on Cripple Creek

* Smelter rates have gone up since this was written, pending the readjustment of the ore market, recently deranged by a labour strike which compelled several of the works to be inactive for three months.

ores, the mine owner paying the cost of transport by rail, which item varies from $3 to $5 according to the richness of the ore.

The chlorination and cyanide plants have the following scale of charges :—

$7	per ton on ores carrying up to		½ oz. per ton.	
7·75	,,	,,	¾	,,
8·50			1	
9			1¼	
10	,,		1½	
10·50	,,	,,	2	,,

Ores carrying more than 2 oz. per ton are subject to smelter rates. The mills pay $20 per ounce for the gold. In the above charges is included the cost of transport, varying from 90 cents to $1·50 per ton, which is met by the mill owners. Thus on a 2-oz. ore the smelter charge would be $6·50 plus the freight. $4, making $10·50, this being equal to the milling rate, plus transport, on the same ore. The consequence is that ores carrying less than 2 oz. go to the mills and the richer stuff to the smelters.

The dividends declared by the mines of Cripple Creek during 1898 amounted to $2,596,144, but to this figure must be added the profit made by mines not owned by public companies and also that of the numerous leasing parties, bringing the total dividends to fully $3,000,000.

Out of the 121 American mines quoted as dividend payers on the list of the New York Stock Exchange, 38 are in Colorado, and 23 are situated in the Cripple Creek district. A few of the principal ones among the latter are quoted below, together with the details of their production.

	Depth area.		Production in 1898.			Dividends in 1898.	Total dividends to Oct. 1, 1899.
	Feet.	Acres	Tons.	Value.	Per ton.		
				$	$	$	$
Independence ...	920	112	8,378	459,576	54·85	220,949	**3,062,164**
Portland	903	183	27,798	1,879,681	67·62	570,000	**2,377,080**
Isabella	735	100	13,548	565.279	41·72	None.	**472,500**
Gold Coin	680	(150)	31,512	1,051,149	33·33	130,000	**240,000**
Moon-Anchor ...	643	14	8,252	352,329	42·67	135,000	**261,000**
Elkton	518	30	17,183	536,265	31·21	220,000	**686,960**
Victor	1000	11	34,775	952,134	30·28	350,000	**1,555,000**
Golden Cycle	820	40	22,342	415,863	18·61	45,000	**228,500**
Vindicator	650	103	19,329	576,518	29·83	126,875	**253,750**
Lillie	620	7½	9,957	453,987	45·59	142,140	**256,610**

Some additional notes concerning a few of the big mines will not be out of place. The Independence is one of the pioneer mines of the district. It was pegged out by W. S. Stratton on the 4th of July, 1891, and remained under his sole ownership until May 1st, 1899, when he transferred it to an English company, incorporated under the name of Stratton's Independence, Limited. This great mine has never yet been pushed to its full capacity for production. In 1895 eight men, while engaged in purely development work, broke ore which netted an average of $155,000 per month for seven months in succession. Of the total area covered by the property, not one quarter has as yet been explored. Up to May 1st, 1899, the mine had produced 44,224 tons, having a gross value of $4,071,860 and yielding a profit of $2,574,164. During the first quarter of the new company there were produced 9,222 tons, having an average gold content of 4·02 oz., and a total gross value of $708,106. A dividend of £100,000 was distributed on account of operations from May 1st to July 31st. The ore reserves already opened up ensure a continuance of this rate of production for many years.

The Portland adjoins the Independence on the north. Of its entire acreage only a little over 8 per cent. has been as yet explored. During 1898 the total receipts were $1,890,641, while the expenditure was $881,833, and the resulting profit $1,008,808. Dividends to the amount of $570,000 were distributed, and additional claims were acquired. The surplus carried forward was $668,000. At the present time dividends at the rate of 24 per cent. are being paid on the share capital of $3,000,000. The total production of this mine to the end of 1898 has been 109,591 tons, having a value of $6,427,523. The average per ton is $58·65. There have been expended on permanent equipment $228,213, and on purchase of adjoining claims $619,953. Nevertheless the total dividends have already been $2,377,080, and there is reason to believe that the mine is yet young, and destined to do bigger things.

The Isabella became famous early in 1899 on account of extraordinarily rich discoveries which permitted of a resumption of quarterly dividends. Three separate shipments of over 50 tons each averaged 45 oz. of gold per ton. This ore was obtained 20 ft. above the ninth level, and the news of it caused a wild speculation in the shares of the Company.

Next door to the Isabella is the Victor Mine, which is said to have been bought in 1893 for $52,000 by Messrs. Moffat and Smith, who still control its operations. The dividends paid since then have an aggregate value 22 times the purchase price of six years ago. The profit earned in 1898 was $323,724. The total output to the end of that year has been 12,242 tons of smelting ore and 82,249 tons of milling ore, having a total value of $2,161,186.

Regarding other mines above mentioned, it remains but to add that most of them have made much larger profits than the dividends indicate, the want of reserve capital causing the expenditure of part of the profits for the acquisition of adjoining territory or the carrying forward of a surplus to be used as necessity arises.

Future Prospects.

In the eighth year of its existence Cripple Creek produced 13½ millions, of which 22 per cent. was distributed in the form of dividends. Is this to be the height of achievement? It is less difficult to foretell the career of a young man than the future of a mere child. Cripple Creek has grown to full manhood, having passed safely through the ills of its adolescence, and has developed a distinct character of such stability that one is justified in prophesying a career of increasing success.

The present boundaries of the productive portion of the district are approximately identical with the area occupied by the andesite breccia, and the other rocks of the volcanic complex. This covers about ten square miles. It is undoubtedly the proper territory for further exploration, and no part of it offers greater promise than the line of contact separating the breccia from the granite. Nevertheless the geological conditions outside this circumscribed area are such as to forbid hasty conclusions discouraging to prospecting in the granite, which surrounds the district on all sides.

The recent discoveries of pay-ore in Grassy Gulch, and on Copper Mountain are suggestive of the probable extension of the boundaries of the goldfield. It is known that the phonolite dykes which are associated with the productive lodes of the central area extend into the outer granite. Indeed several rich mines are wholly in the granite, at distance varying from a few hundred feet to over 2,000 feet from the breccia. The Strong and Gold Coin are notable examples, while among those most distant from the main mass of the breccia may be mentioned the Orizaba and Prince Albert on Beacon Mountain, and the Sweet and Caledonia, west of Anaconda. It must not be thought that these mines exhibit an ore occurrence wholly distinct from the main volcanic complex; such an idea would be unwarranted, because, while they are situated far outside the mass of andesite breccia, the gold-bearing lodes, which they develop, are associated with phonolite, in dykes and masses. The geology of Beacon Hill is especially characterised by a very large intrusion of that rock.

It is not extraordinary that the vicinity of the contact of the granite with the breccia should be a favourable environment for large bodies of ore. The contact must have been, at all times, first a line of weakness, next a line of movement and consequent fracturing, and finally a line of water circulation. Thus there existed the conditions which experience and observation indicate as being most favourable to mineralisation, especially when to these is added the penetration, across the contact, of bodies of volcanic rock, such as the dykes of andesite, basalt, and phonolite. Although comparatively little ore has been found lying immediately upon the contact, there is plenty of proof, in the Independence, Portland, and Granite Mines, for example, of the fact that the lodes crossing it have been beneficiated. Similarly the phonolite dykes, when they pass out into the granite, have afforded a line of weakness along which fracturing has subsequently occurred, forming shattered planes and lines of maximum porosity permitting of the circulation of the underground solutions which precipitated the precious metal. Every added page of evidence only further confirms the view, held by the writer early in the infancy of the camp, that Cripple Creek corroborates to a striking degree the most modern explanations of ore deposition chiefly associated with the name of Posepny.

What of the deep? Will increasing depth be accompanied by impoverishment? This is not asked with the timidity of a few years ago when the lodes had only been followed two or three hundred feet in vertical descent, and it was foreseen that they would eventually cut into the granite under the breccia. At that time the future of the district was uncertain, and many cautious men held back in fear of unfavourable developments. It is obvious that the mines near the edge of the depression occupied by the breccia, will penetrate into granite by sinking their shafts or by extending their levels, as illustrated in the accompanying sketch, Fig. 7. This has occurred notably in the case of the Independence and Portland Mines, which reach the granite on the southern and western sides of the territory owned by them. It is very satisfactory to be able to record the fact, that magnificient ore-bodies have been found in these two properties upon veins which have been followed downward into the underlying granite. Nor should this be surprising in view of the discoveries made during recent years in the granite south and west of the contact with the breccia. If good ore is found in the granite at a horizontal distance of 2,000 ft. from the mass of breccia, why should it not be also found at a similar vertical distance below the same formation? All the evidence to hand goes to show that wherever the dykes have broken through the rocks of the volcanic complex, they have permitted of ore deposition, and that therefore the possibilities of

FIG. 7.

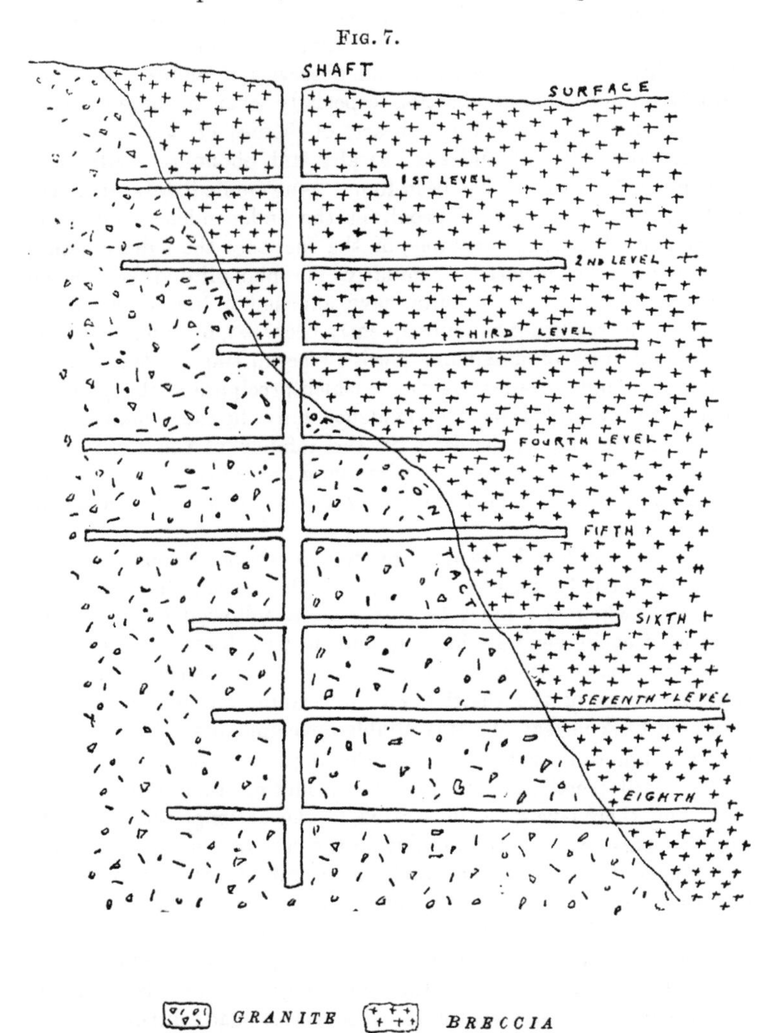

future discovery are not limited to the present restricted known productive area.

If the mines of the district were old, deep, worked out, one would be timid of foretelling a continued and advancing yield of gold, because the brutal facts of experience do not countenance the popular idea of increasing richness with depth; but the mines of Cripple Creek are young, comparatively shallow, and only fractionally explored; each month's development, whether in a lateral or a vertical direction, uncovers ore reserves previously unsuspected, save by those to whom painstaking observation has given a clue.

There is every probability that the near future will see the introduction of capital from the outside, because the production of the district has attracted the attention of the larger financial centres. Nor is this anything but desirable. If the sale of the mines meant the retirement of the present owners, their withdrawal from mining operations and the transfer of the properties to absentee capitalists, the benefits of the change might well be questioned; but it is a fact that those who control the best mines also own several smaller undeveloped claims, and the result of a sale of any of the big properties would mean the liberating of large sums of money to be used for the opening up of promising young mines, which now are either idle or incompletely explored. The mining men of Colorado have been bred in the atmosphere of gold-seeking, and the acquirement of riches usually only leads to larger operations. It is unnecessary to cite examples. They are a part of local history.

If the growing reputation of the district should thus lead to the investment of large sums of money, the immediate result would be greatly to augment activity in exploratory work, and in the wake of this greater development there will come, inevitably, discoveries of an importance eclipsing those, the yield of which is the basis of the present productiveness.

The steady betterment in economic conditions tends continually to decrease the working costs and to make a commensurate addition to the tonnage of ore available for exploitation. This is the story of all modern mining regions. The Rand illustrates it; Kalgoorlie accentuates it. There is no immediate probability of any radical change in methods of ore reduction or transport, but there is a tendency in several directions to diminish the expense of handling low grade ores. Two railroads tap the district, and the competition prevents rates of transport being unduly high. The building of a third line is probable, should the production of the camp increase as it promises. Electric haulage is likely to become a factor in reducing this item of expenditure. The increased capacity of the mills, due to their steady enlargement and the better arrangement of the machinery, is permitting them to buy ores which at one time were considered to have no commercial value. Thus five years ago the minimum rate for freight and treatment was $12 per ton, while to-day it is $7 per ton. During the same period the smelter rates of treatment have dropped from $15 to $6·50 per ton. What a reduction such as this means, and how great a tonnage of low grade material it transfers from the category of rock to ore, will be appreciated by those who have watched the growth of other districts.

The writer has seen the development of more than one of the great goldfields of the globe, and looking back upon the brief history of that development, he recognises that those which have proved permanent, share certain characteristics in common. In the general persistence of the ore bodies, in the size and continuity of the lode channels, and in the economic conditions favouring an easy realisation of the values contained in the ores, Cripple Creek exhibits the features of a great goldfield, and affords the promise of a future which will eclipse the achievement of the first eight years of its existence.

Harrison and Sons, Printers in Ordinary to Her Majesty, St. Martin's Lane.

SUBJECT TO REVISION.

[TRANSACTIONS OF THE AMERICAN INSTITUTE OF MINING ENGINEERS.]

The Cripple Creek Volcano.

BY T. A. RICKARD, STATE GEOLOGIST OF COLORADO, DENVER, COLO.

(Washington Meeting, February, 1900.)

THE Cripple Creek district occupies a cluster of foot-hills on the south side of Pike's Peak and is a portion of an extensive, though uneven, plateau which unites the eastern range of the Rocky mountains with the Sangre de Cristo. It is essentially a small volcanic area, of about 20 square miles, amid the granite of the Front range. But though, when regarded as a rich mining district, it may be considered as an isolated area,* yet, geologically, it is, as Whitman Cross has pointed out, only an outlying portion of a much larger volcanic region, which† stretches to the south and west, around Silver Cliff and the Rosita hills, forming the picturesque country cut by the deep cañons of the Arkansas river and its tributaries.

The mines are situated amid a volcanic complex, consisting of tuffs and breccias which have been penetrated by an extensive system of dikes and other intrusive masses. The prevailing formation is an andesite breccia, which lies upon the worn surface of the granite and fills the deep basin around a volcanic vent. The breccia, since its deposition, has been broken into by several eruptions of phonolite and, later still, by a series of thin dikes of basalt and other allied rocks of a highly basic composition.

The successive sedimentary formations which, elsewhere in Colorado, lie upon the basal granite, are not represented in the district; whatever sediments were laid down before the volcanic period must have been removed by erosion, and there is very little evidence which affords a datum-line whereby the geological age of the volcanic eruptions can be determined. Whit-

* The main mining belt of Colorado is 30 or 40 miles to the west, and extends through Boulder, Gilpin and Summit counties, into Leadville, and then southwestward, through Aspen, into the San Juan region.

† "Geology of the Rosita Hills, Custer County, Colo.," by Whitman Cross. *Proceedings of the Colorado Scientific Society*, vol. iii., 1890.

man Cross has referred the breccia* of Cripple Creek to the close of the Eocene period or to the early Miocene.† This is done by correlating the small deposit of grit which occurs on Straub mountain, and is the only sedimentary formation in the district, with the lake-beds at Florissant, 15 miles to the north. These celebrated fossil-beds belong to the late Eocene; they are largely made up of volcanic dust and are covered by breccia similar to that of Cripple Creek. Moreover, they are overlain by rhyolite identical in character with that which forms the floor of the gravel-deposit on Straub mountain.

The granite which forms the basal rock of the region is usually described as Archean. It is probably Algonkian. It is, elsewhere, overlain by Upper Cambrian strata; and it has been found to include fragments of quartzite which are believed to be of pre-Cambrian or Algonkian age. Therefore the granite is not necessarily Archean, but, to quote Dr. Cross, "older than the only Cambrian rocks as yet identified in Colorado."‡

Within the Cripple Creek area the granite differs in appearance from the rock which generally prevails in the Pike's Peak region. It is a well defined reddish biotite (black mica) granite, and, instead of the microcline (feldspar) which ordinarily characterizes the Pike's Peak formation, it carries orthoclase in prominent tabular crystals.§ The quartz is usually

* The miners call the breccia "porphyry." This term is derived from the Greek word "porphyra," meaning purple. It was first applied to the beautiful dark-red rock which the Romans obtained from the quarries of Ghebel Dokhan, near the shores of the Red Sea. This original "porphyry" would now be classed as a porphyrite. However, the original name, which depended upon color, has long since lost its force in another meaning, which refers to the structure. The original "porphyry," according to Zirkel, was speckled with snow-white and rose-red crystals of feldspar in a blood-red ground-mass. Hence the term became applied to rocks in which some particular mineral, frequently feldspar or quartz, stands out well defined from the general matrix, giving it a spotted look, as for example in the familiar "bird's-eye porphyry" of the western miner. The term is often employed as though it covered a particular species of rock, while in fact it is merely a descriptive adjective-noun, covering any kind of crystalline rock having a mottled appearance due to the predominant development of one of its constituent minerals in individual crystals.

† "Geology and Mining Industries of the Cripple Creek District, Colorado," by Whitman Cross and R. A. F. Penrose, Jr., 16th *Ann. Rep. U. S. Geol. Sur.*, Part II, p. 18, 1895. The writer owes a great deal of his descriptive geology to this valuable monograph.

‡ *Op. cit.*, p. 17.

§ I obtained some fine twin crystals (Carlsbad type), 2½ inches long, from the top of Bull Hill.

iron-stained, and the mica shows the commencement of a change into green chlorite. The oligoclase, which occurs as a subordinate feldspar, shows a ready tendency to decomposition, especially near the ore-bodies in the mines.

Turning to the examination of the breccia,* we find that it consists of a consolidated mass of fragmentary material having a coarseness comparable to gravel. Occasionally the pieces are as large as a man's hand; but these are rare. More frequently the breccia is very fine, and then comes under the designation of "tuff." It is mainly composed of augite-andesite, in which smoky-brown prismatic crystals of apatite occur. The feldspar is kaolinized, and the dark silicates, such as augite or biotite, have undergone destruction by leaching. While the mass of this formation is made up of andesite, it exhibits, locally, a good deal of fragmental phonolite and granite, the latter more particularly in the vicinity of the contact. Fine pyrite is distinguishable in most specimens. In the upper workings of the mines the kaolinization of the feldspar has given the rock a bleached appearance; and at the surface the iron oxides, derived from the alterations of the pyrite, have stained it yellow or red. Fluorite is a frequent constituent, wherever the breccia is penetrated by the gold-bearing veins, and colors it a dark purple. The finer tuffs are often so silicified as to be undistinguishable, except under the microscope, from massive rocks. The decomposed breccia also exhibits the healing effects due to the infiltration of secondary quartz; and, when it is included within the boundaries of any of the ore-deposits, the soluble ingredients have been removed to such an extent as to leave often only a pumice-like remnant of interlacing quartz.

Penetrating the mass of the breccia and extending into the surrounding granite, there is an intricate series of dikes, chiefly of phonolite.† This is usually a light-colored rock of even

---

* "Breccia" and "tuff" are both words of Italian origin. We owe many terms describing volcanic materials to the study of Vesuvius. "Breccia" means "broken." It is applied to rocks made up of a consolidation of angular fragments. "Tuff" is employed both for a rock built up of fragmental material of smaller size than that composing a "breccia," and also for the rock resulting from the mud caused by the action of water upon volcanic dust. In the latter sense, the original Italian word "tufa" is made use of by many authorities.

† "Phonolite" is derived from the Greek words "phone," sound, and "lithos," stone. The close texture and even grain of the phonolite causes it to give a ring-

texture and sherd-like fracture; but the conditions of its occurrence, in thin and in thick dikes, in almost horizontal sheets, and in shapeless intrusive cores, are so varied as to have induced a great many modifications in its physical characteristics. The essential constituents are sanidine (the glassy variety of orthoclase) and nepheline. The crystals of the latter are occasionally sufficiently developed to give the rock a porphyritic appearance. Besides the normal phonolite there are allied rocks, such as trachytic* phonolite and nepheline-syenite,† occurring under various structural conditions. There are also dikes of andesite,‡ similar in character to the earlier rock which, in its fragmental form, composes the bulk of the breccia. Finally, crossing these rocks, and therefore last in the sequence of eruption, are the dikes of nepheline-basalt with which important ore-bodies are associated on Raven hill and Battle mountain.

The Cripple Creek district represents the ground-floor of a volcano,§ the superstructure of which has been removed by erosion. Let us consider what this means. Among the gains of modern science there is none more striking than the elucidation of the causes which bring about the terrifying phenomena of volcanic action. A hundred years ago, an active volcano excited superstitious fear, and was regarded only as a catastrophic interruption to the order of nature. Since then, the patient researches of such men as Spallanzani, Scrope and Judd have enabled us to recognize in these activities the orderly operations of forces subject to definite laws.

---

ing sound when struck with a hammer. In England it is often called "clinkstone."

* "Trachyte" is from the Greek word meaning "rough." The rock usually has an uneven fracture, due to the angular sanidine and the porosity of the groundmass.

† "Syenite" comes from the Greek Syene, the town in Egypt now known as Assouan. It is a curious fact that it has, comparatively recently, been found that the rock at Assouan is not a typical syenite, which is a variety of granite containing very little quartz, with hornblende replacing the mica. It is really a red granite, very much resembling that of Pike's Peak. The Egyptians quarried it for their obelisks, and out of it they built the Temple of the Sphinx at Ghiseh.

‡ "Andesite" is derived from Andes, the mountain range in South America where it is particularly prevalent.

§ The word "volcano" is Italian. It was the name given to one of the Lipari Islands in the Mediterranean, where quiet eruptive action has been going on since the time of the ancients, who considered the little mountain-island as the forge of the Roman god Vulcan.

The scope of scientific investigation has included not only the observation of existing* volcanic action, but also the examination of the remains of extinct volcanoes. The structure of the latter has thrown light on the behavior of the former. As the story of the development of forms of life now extinct, but preserved in fossil-beds and recorded for us by the palæontologist, advanced our insight into the structure of living things, while biology repaid the aid thus received from palæontology by contributing the clues through which the incomplete evidence of the rocks was so correlated as to demonstrate the sequence of strata, so the study of the volcanoes of to-day led geologists to recognize the results of similar action in masses of rock, the eruptive origin of which was previously unsuspected, and, in turn, the deciphering of the skeletons of extinct volcanoes advanced the understanding of those which have survived. Natural sections gave the requisite testimony. Atmospheric erosion, acting through vast periods of time, has cut into the mass of many of the ancient volcanoes of the earth so as to uncover their anatomy. The dissection, by Professor Judd, of the old volcano of Mull,† in the Western Isles of Scotland, is an excellent example of this method of research. Occasionally mine-workings, or excavations made for other purposes, afford valuable evidence as to the internal structure of volcanic mountains. The Kammerbuhl, in Bohemia, is a curious instance.‡ It is a small hill, apparently of no particular interest, but, nevertheless, it was once the subject of a hot scientific discussion. The poet Goëthe took part in the dispute, and persuaded a friend, Count Sternberg, to drive a tunnel into the hill with a view to settling the question of its origin. The result justified Goëthe's claim that it was "a pocket edition of a volcano." It was found that the hill consisted of a mass of volcanic scoria,§ through the center of which passed a plug of basalt. The plug obviously occupied the choked-up vent of the volcano, from which proceeded a lava-stream which had flowed over the flank of the hill. Fig 1 illustrates this statement.

---

* There are about 350 active volcanic vents on the surface of the earth at the present time.

† *Quarterly Journal of the Geological Society*, vol. xxx., p. 220, etc.

‡ See Judd's "Volcanoes," pp. 112–114.

§ "Scoria" is a Latin word, unchanged. It is used especially for coarsely vesicular lava, but often for fragmental lava in general.

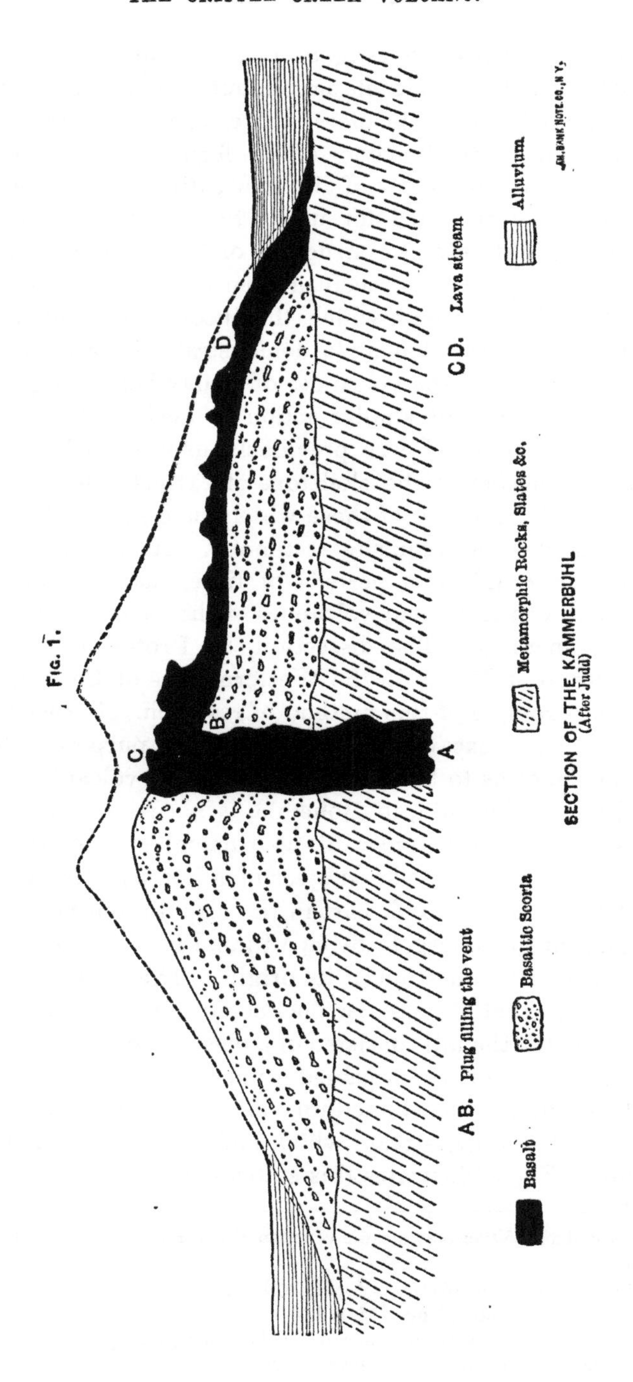

SECTION OF THE KAMMERBUHL
(After Judd)

At Cripple Creek, the mine-workings afford a good deal of information concerning the underground structure of the region. It is hoped that an inquiry into the history of the volcano which determined the interesting character of the district will contribute towards a clearer comprehension of the geology of the mines.

The operations of nature in the past are inferred from the observation of those which take place to-day. The intensity may vary; the forces are the same. This is the corner-stone of modern geology as laid down by Lyell. The volcanic complex at Cripple Creek is to be understood in the light of the evidence gathered for us by the patient investigators who have stood by the side of the craters of Stromboli, Vesuvius and Kilauea.

The conclusions of those who have made a specialty of this branch of geology* may be summarized thus: The explosive violence of volcanic eruptions is due to the access of water to the fused rock within the conduit of the volcano; but, as it appears that this water is not contained within the substance† of the lava‡ emitted during the tranquil emissions succeeding the first paroxysmal outburst, it is inferred that the water is not the primary cause of volcanic action, which originates at a depth greater than that to which it is believed that water can penetrate. The evidence collected is not complete; but it warrants a reasonable conjecture that volcanoes owe their origin to the contraction, caused by the cooling,§ of the earth's crust upon a yielding substratum, separating the solid outermost shell from an equally solid nucleus.|| While, therefore, the force which

---

* Among the best literature on the subject may be mentioned *Volcanos* by Poulett-Scrope, *Volcanoes* by Judd, and *Characteristics of Volcanoes* by Dana.

† That is, "occluded."

‡ Lava is Italian for "stream." It is from the same Latin root as lave, lavatory, etc. Although the term is usually applied to the fused material emitted by a volcano, it is often employed in referring to the same rock after it has become consolidated, especially when the rock has not been specifically classified.

§ "Secular refrigeration."

|| The question of the condition of the earth's interior is too large for extended reference. Besides the standard text-books on geology, the reader will find much suggestive matter in Osmond Fisher's "Physics of the Earth's Crust," and in Prestwich's "Controverted Questions of Geology." There is also a summary of the evidence regarding this subject in the address of Sir William Thompson (Lord Kelvin) before the meeting of the British Association in 1876. The general conclusions of science have been lately expressed, in a popular way, by Pro-

pushes large quantities of fused rock to the exterior of the earth has, probably, a deep-seated origin, nevertheless, the immediate cause of the uncertainty, the violence and the magnificent energy of volcanic action is traceable to the effects produced by water coming into contact with the lava as it approaches the surface.

The destructive energy of a volcano may be likened to a boiler-explosion; volcanoes may be considered the safety-valves of creation. The mass of incandescent rock which is slowly being squeezed upward meets a large volume of water which flashes into steam with a sudden expansion causing the most astounding results. One cubic foot of water yields 1700 cubic feet of steam. It is accepted by specialists that, whatever the ultimate origin of volcanic action may be, the surface effects are due to the explosive escape of accumulations of steam suddenly released from pressure. This explanation is based upon accurate observation of the quiet workings of the miniature volcanoes of the Lipari Islands, in the Mediterranean, and upon the evidence obtained during the more violent, apparently paroxysmal, outbursts of Vesuvius, Etna, Tarawera, Kilauea and Krakatoa.

The volcanic rocks of the Cripple Creek district have come up through the granite. It underlies them all; they rest upon it, and can be seen penetrating it in the form of dikes. Previous to the first eruption, the granite must have presented a weather-worn surface, such as characterizes the high hills. Ever since its first emergence from the ocean this region has been undergoing an intermittent elevatory movement, which culminated in making the Front range. Erosion had been continuous, but the uplift more than counterbalanced such wearing away; and the granite hills had been slowly raised far above the Cretaceous seas which washed their edges in the era preceding that to which the eruption is assigned. The forces which had done this work were of the most patient kind; their

---

fessor Milne, thus: "The earth became solid under two influences; it began to solidify at the surface by cooling, the crust growing thicker and thicker; and it began to solidify at the center by pressure, the core growing larger and larger. This double phenomenon of solidifying continued until a solid outer shell and a solid inner core came close together in what may be called the critical region of the earth, the region which feeds lava to volcanoes."

manifestation had about it nothing of a violent or paroxysmal character; time was an essential element of the process.

At the close of the Eocene period this apparent equilibrium was disturbed. The foundations of the granite hills trembled. Slight tremors were followed by earthquakes, and these were the precursors of greater violence.

Earthquakes usually precede an eruption. In certain volcanic regions, such as the north island of New Zealand and Japan, the minor shocks, designated as "tremors," are of daily occurrence. They represent the vibrations set up by the sudden generation of steam from water coming into contact with the upwelling lava. It is water-vapor, and not smoke, which is emitted by volcanoes.* This water is derived from the surface, having sunk into the soil, permeating the more porous sedimentary rocks, lodging in the crevices of unstratified formations, and becoming stored underground, as mine-explorations testify. Where the country surrounding the volcanic vent has become covered with the products of previous eruptions, the loose character of the soil, resulting from the disintegration of scoriaceous lava, facilitates the descent of the rains, and tends to the accumulation of large quantities of water. Moreover, the vibrations set up by the superheated steam cause fissures which allow distant bodies of water, from subterranean reservoirs, fresh-water lakes,† or, if the volcano be situated near the coast, the ocean itself, to be let down‡ suddenly into the volcanic vent and into explosive contact with the incandescent lava. Humboldt§ found small fishes in the water

---

* Volcanoes are not necessarily mountains. The mountain is the result of the volcano, and not *vice versa.* It is the accumulation of the material ejected from the vent which slowly builds up the cone. Many emissions of lava occur at the base of mountain ranges and have quietly overspread the surface from fissures, much as water rises through cracks in the ice and overspreads it, when a heavy wagon presses it down. The lava-plains of the Snake river, traversing Oregon and Idaho, afford an example; so does the Deccan (India), where successive, nearly horizontal, flows, covering an area of 200,000 square miles, have reached a thickness of 6000 feet.

† Lake Rotomahana was drained at the time of the eruption of Tarawera, in New Zealand, in 1886.

‡ Mosely (*Notes by a Naturalist on the Challenger*, p. 503) mentions that, in 1877, when on board the "Challenger," he saw the sea-water actually pouring down into a fissure formed in the bed of the sea off the Hawaiian coast. The fissure was traced to the shore and three miles inland. This occurrence was connected with the volcanic activity of Mauna Loa, on the neighboring island of Hawaii.

§ *Controverted Questions of Geology*, Prestwich, p. 116.

emitted from fissures caused by earthquakes in the Andes. Diatoms, the microscopic forms of life characterizing the deep sea, have been found in the volcanic ejections of the Pacific islands. Where volcanoes are in proximity to the ocean it has been found that among the emanations from the lava there exist, not only chlorides, but also sea-salt itself. A sudden diminution of the water-supply in wells and springs near Naples has been repeatedly observed to presage the eruption of Vesuvius.

There is therefore ample evidence that water does penetrate into the conduit of the volcano, and that it is originally derived from the surface. As against the contrary belief, namely, that the water-vapor accompanying eruptions is an essential constituent of the lava, and therefore shares with it a deep-seated origin, there is the following evidence. It has been found, as the result of a large number of accurate observations in wells, shafts and bore-holes, that the temperature underground increases 1° F. for every 48 feet of descent.* At 7776 feet, the boiling-point, and at 34,700 feet, the critical point, 773° F., of water, would be reached. The expansive force of steam increases rapidly with the temperature, so that at 773° F. it would be equal to the pressure of 350 atmospheres.† This is termed the "critical point," because, at this temperature, water, however great the pressure to which it is subjected, can no longer exist as a liquid, but becomes dissociated into its constituent gases. Although the exact conditions which obtain at these great depths cannot be known with certainty, nevertheless, all the evidence goes to show that there is a limit set to the descent of surface-water by the rapid increase in the expansive force of its vapor, due to the rising temperature. Prestwich‡ put the maxim limit at 6 to 7 miles, and Delesse§ estimated it

* "On Underground Temperatures," Sir Joseph Prestwich. *Proceedings of the Royal Society*, February, 1885. Of course this increment of 1 degree per 48 feet can only apply to the outermost portion of the earth. Beyond a few miles of depth there must exist conditions of which very little can be inferred. There, the enormous pressure probably counteracts the expansive effects of heightened temperature, and upsets many of the conclusions of physics which hold good near the surface.

† Which, at 15 lbs. per square inch for each atmosphere, amounts to about 2½ tons per square inch.

‡ *Controverted Questions of Geology*, p. 93.

§ *Bulletin Société Géol. de France*, vol. xix., p. 64.

at 60,000 feet, or about 11 miles. Moreover, experience goes to show that the water encountered in mines is the drainage from the surface. Deep mines are usually dry ones. I may instance the deepest metal-mines, the Calumet-Hecla and Tamarack, in the Lake Superior region, and the "180," "New Chum-Victoria," and neighboring shafts, at Bendigo, in Australia.

The evidence obtainable concerning the first eruption of the Cripple Creek volcano is necessarily very meager. The first vent must have been formed at some point along one of the fractures caused by the earthquake-shocks; the lava, in forcing for itself a way to the surface, being aided by the force of the expanding steam. The pressure required to break a passage through the overlying rocks is stupendous; and, as a consequence, when the steam accompanying the lava is finally, and very suddenly, released from that pressure, on its immediate arrival at the surface, it escapes with explosive energy, and with projectile discharges which may reach to an astonishing height. Thus, when the outburst of Krakatoa, an island near Java, occurred in 1883, the finer fragments ascended skyward 10 miles, and were recognizable* in the atmosphere of London. The winds carried the dust of Krakatoa round the world, and thus gave rise to the extraordinary sunsets observed in the autumn following.

The material ejected during the first outburst of a volcano consists of fragments of rock torn from the sides of the vent. The extinct volcano of the Kammerbuhl, already mentioned in this paper,† exhibits pieces of burnt slate within the mass of the scoria forming its cone. The underlying formation consists of slates and other metamorphic rocks.

The Cripple Creek volcano first ejected fragments of granite. These were probably small in size, and became further reduced by colliding with each other as they were discharged, so that they fell to earth in showers of particles like gravel. Of this first eruption there is little trace now, unless the grits of Straub and Grouse mountains be the remnants, as is probable,‡

* The writer, then a student at the Royal School of Mines, saw this volcanic ash under the microscope after it had been collected from the London atmosphere, which hardly needed solid contributions from such a distant source.

† See Fig. 1 and the corresponding text.

‡ This is the opinion of Whitman Cross. See page 71, *op. cit.*

of the *débris* accumulated at that time. Material resembling this must certainly have covered the surface around the vent, until the larger portion of it was washed away. The steam which, in enormous volumes, accompanies the first outbursts of volcanic action, becomes condensed as soon as it issues into the cold air and forms rain-clouds, the downpouring of which frequently removes the accumulations formed at the inital stage of the volcano. The floods which succeed eruptions are due to the super-saturation of the atmosphere with the water-vapor emitted by the volcano. Such floods are more feared by the dwellers around Vesuvius, for example, than the lava-streams, the destructive effects of which are comparatively restricted. It was the formation of a liquid mud, by the action of heavy rains on the fine material, called "tufa," which buried the city of Herculaneum.

It is unlikely that sufficient data will ever be forthcoming to give an exact presentation of the chief vent of the Cripple Creek volcano, unless one of the millionaires, enriched by the gold he has won from the mines, shall prove as public-spirited as Goëthe's friend, and shall undertake the requisite explorations. Yet some very interesting evidence on this point is available. A miniature vent* exists near the town of Victor, and the railroad has cut through it, so as to furnish the section of it shown in Fig. 2. As a hand-specimen may exemplify the structure of a mountain range, so this small vent typifies many of the characteristics of the orifice probably existing in the earlier stages of the Cripple Creek volcano.

This vent occurs in the massive granite of Squaw mountain, about 1700 feet south of the main breccia-formation of Battle mountain. In the railroad-cut, where it is to be seen, it has a width of 35 feet. It is filled with fragments of granite and the gravel derived from the brecciation of granite. The edges are not particularly well-defined, because the face of the enclosing rock is shattered. The most peculiar feature of the section is presented by pellets, nodules and rounded fragments of dark-red scoriaceous lava, which occur all through the material filling the vent. At the edges, rounded inclusions† of this lava can be

* Whitman Cross describes this vent on page 77 of his Cripple Creek report.

† The largest are 1 to 1½ inches in size. On microscopic examination Professor Kemp found it difficult to determine the exact petrographic character of this lava. "It shows only alteration products in some parallel arrangement, but not in sig-

seen in the mass of fragmentary granite; and in the center the lava, by reason of oxidation, forms a red granular matrix, in which large pieces of granite are separately discernible. The

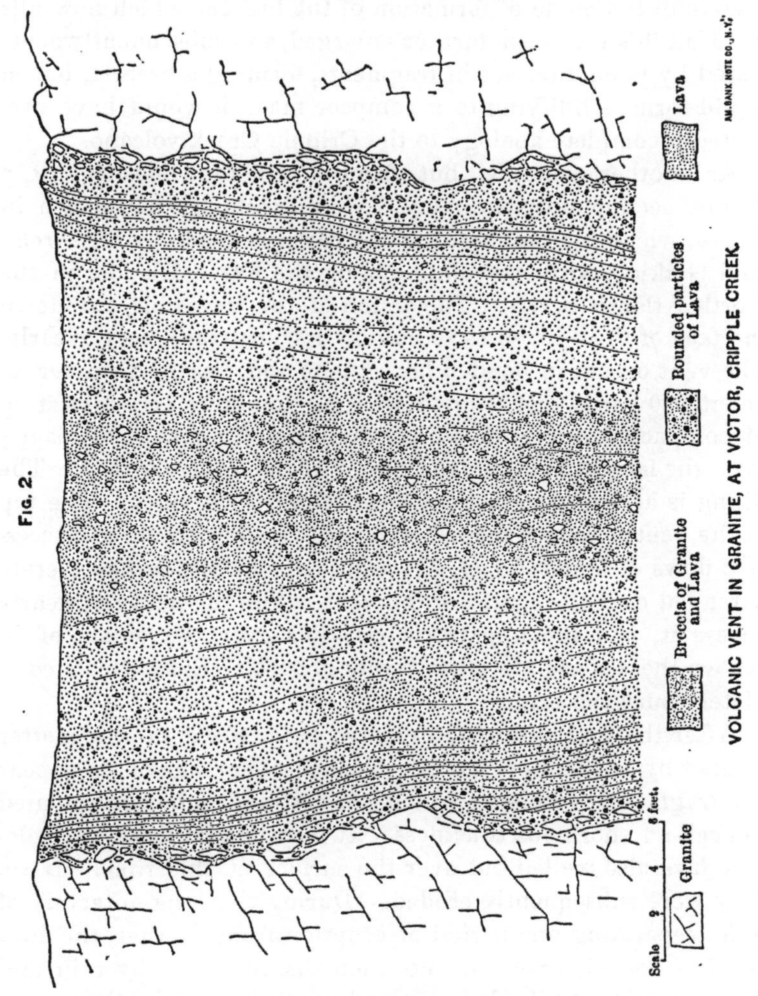

VOLCANIC VENT IN GRANITE, AT VICTOR, CRIPPLE CREEK.

material, especially near the edges, has a laminated structure, parallel to the sides of the vent. These laminations vary in thickness according to the coarseness of the material.

---

nificant condition." In both of the specimens I sent to him he detected large scales of biotite. Having in view this fact, and the character of the material, it seems most probable that the lava closely resembles the rock of the basaltic dike in which the neighboring Anna Lee ore-chimney was found to occur.

This illustration is of great interest. The vent is in granite, as was the first vent of the volcano. It is now filled with breccia, as, at one time, that was. The shattering of the sides is suggestive of the mode of formation of the breccia, which now fills it. Had this vent been further enlarged, and subsequently penetrated by phonolite, not in fragments, forming a breccia, but in liquid form, solidifying to a compact mass, it would have presented a complete analogy to the Cripple Creek volcano.

As another example, but from a different locality, of a natural section of a small vent, I would instance that shown in Fig. 3, which represents a drawing recently made by Sir Archibald Geikie,* while traveling among the Faroe Islands, in the North Atlantic ocean. The action of the waves has cut down the face of the cliff, so as to exhibit its structure very clearly. The vent occurs in banded lava (A A) and has a diameter of about 100 yards. It is filled with agglomerate (B) consisting of compacted *débris*, in which lie large fragments of slaggy lava, the largest being in the center of the former orifice. The filling is arranged in distinct layers toward the sides. The top of the vent is saucer-shaped, and is covered with three successive flows of basalt (D, G, E); of these, the lowest has merely extended over the center of the vent, while the next (D) nearly covers it, and the uppermost (G) lies over the whole of it. Above these there are other layers of basalt (E, F) which completely bury the orifice.

After the first outburst, a change took place in the matter ejected by the Cripple Creek volcano; there began to appear the fragmentary andesite which was destined to be accumulated to such an enormous thickness. It may be that flows of andesitic lava also welled out over the surface at this time. If so, they were subsequently eroded. During the long intervals of quiet separating one period of eruption from another, the lava became cooled, cracked, and then disintegrated by rain and frost, so as to be broken up and carried away by the mountain streams to form a part of the alluvium of the valleys. Thus the superficial flows were removed; but the corresponding bodies of lava which consolidated underground, when the extrusion at the surface had ceased, are now, thanks to that very

* "The Tertiary Basalt Plateaux of North Western Europe." *Quarterly Journal, Geological Society,* vol. lii., page 344.

erosion, to be seen as bodies of andesite rock in several parts of the district, notably on the eastern side of Battle mountain and near Legal Tender Hill, above Goldfield.*

The fragmentary eruption of andesite continued. At this time the volcano must have been a splendid sight, especially by

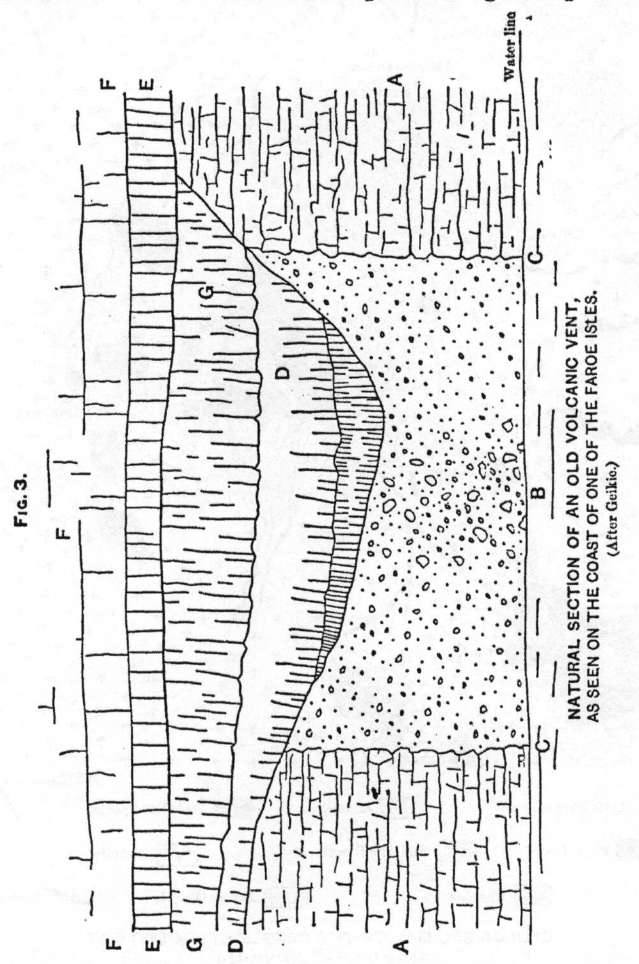

Fig. 3.

NATURAL SECTION OF AN OLD VOLCANIC VENT, AS SEEN ON THE COAST OF ONE OF THE FAROE ISLES.
(After Geikie.)

night. It was so late in geological time that Pike's Peak was already a giant among its fellows, and towered in lonely grandeur above the lesser hills where the eruption was taking

* As the accompanying geological map (Fig. 4) of the district illustrates. This map is a reduced copy of the colored map published by the United States Geological Survey. The ideal section which I have drawn (Fig. 5) is taken in an east and west line across the southern part of the area.

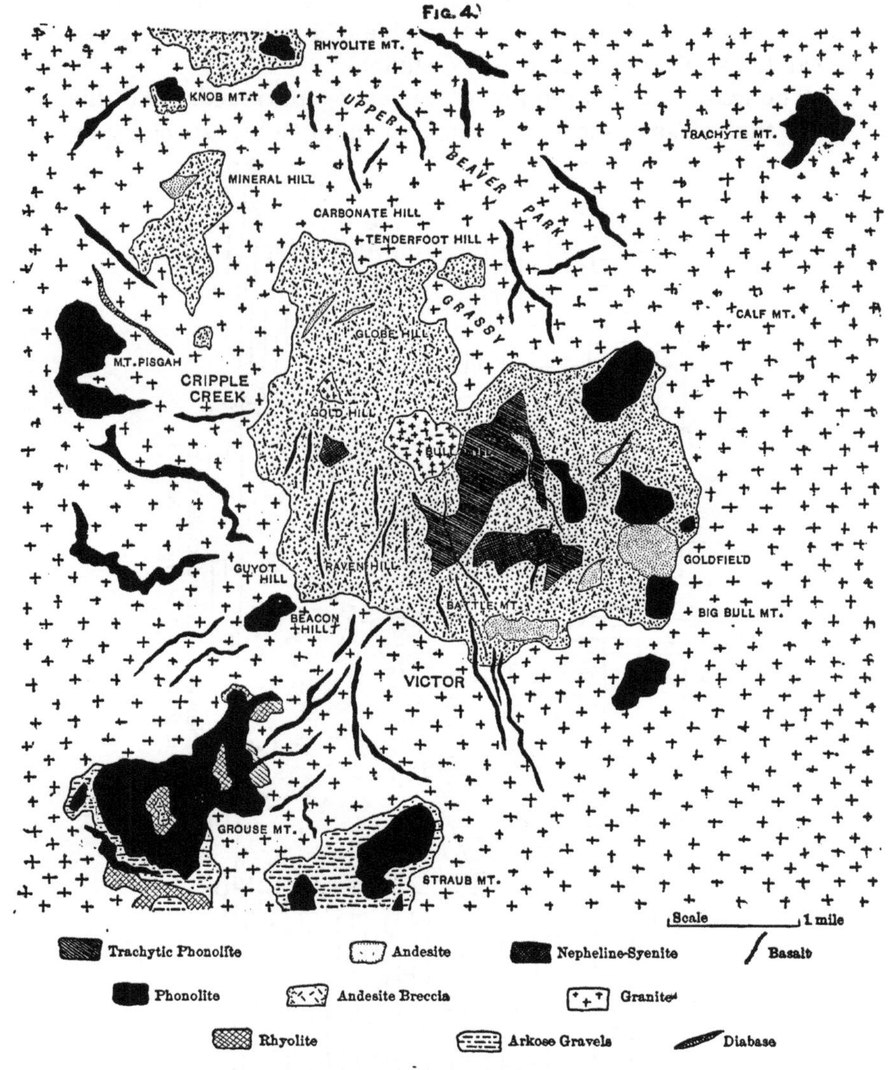

GEOLOGICAL MAP OF THE CRIPPLE CREEK DISTRICT.
(After U.S. Geological Survey.)

place. The shifting lights of the volcano were reflected by the snow-fields of the peak. Those lights were due to the glow of the incandescent lava in the crater thrown upon the clouds* of

* Professor Judd very aptly likens this effect to that caused when, at night, the engineer of a locomotive pulls open his furnace door and permits the light of the fire to be thrown upon the stream of vapor issuing from the funnel.

watery vapor which hovered overhead. To this appearance* were added lightning-flashes. The steam issuing through the

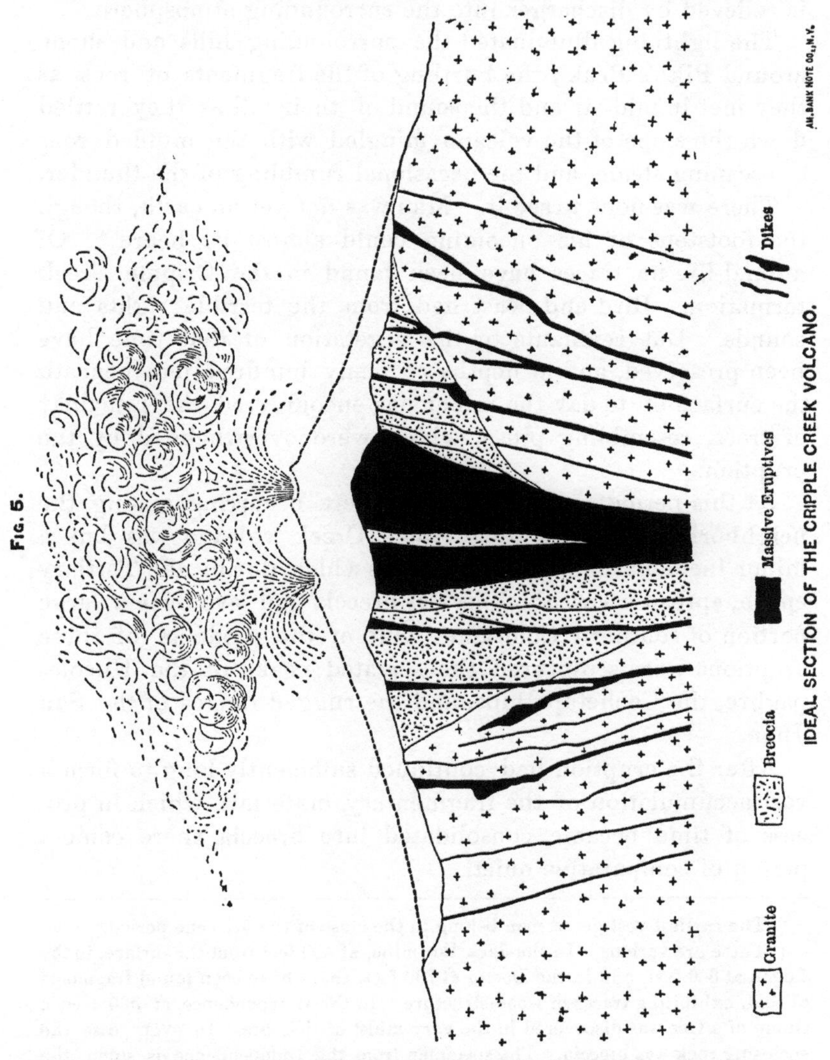

FIG. 5.

IDEAL SECTION OF THE CRIPPLE CREEK VOLCANO.

orifice of a volcano is highly charged with electricity, generated by its upward rush, and the friction of the colliding particles

* The old idea of a volcano was a mountain which spouted fire, ashes and smoke. The "fire" is the reflection referred to above; the "smoke" is vapor; the "ashes" are lava rendered vesicular or pumice-like by the bubbles of steam penetrating fused rock.

of solid matter ejected with the steam contributes further in producing a condition of intense electrical excitement. This is relieved by discharges into the surrounding atmosphere.

The lightning illuminated the surrounding hills and shone around Pike's Peak; the hurtling of the fragments of rock as they met in mid-air and the sound of their fall as they rattled down the slope of the volcano mingled with the muffled roar of escaping steam and the occasional rumbling of the thunder.

There was none to see it. Man was not yet on earth, though the footsteps of his oncoming could almost be heard.* Of animal life no traces have been found in the Cripple Creek formation. Bird and beast fled from the terrible sights and sounds. But remnants of the vegetation of that time have been preserved, and at depths of many hundred feet beneath the surface of to-day the miner has encountered the remains† of trees, resembling pines, which were overwhelmed by the eruption.

At this period similar outbursts were occurring among the neighboring hills, for the Cripple Creek volcano was but a minor incident among the eruptions which, during the Tertiary epoch, spread a vast thickness of breccia and lava over a large portion of southern Colorado. Out of the products of these eruptions were sculptured the serrated peaks of the Uncompaghre, the Cochetopa hills, and the rugged ranges of the San Juan.

After the eruption had continued sufficiently long to form a vast accumulation of the fragmentary materials, which in process of time became consolidated into breccia, there came a period of comparative quiet.

---

* The earliest vestiges of man belong to the close of the Miocene period.

† These are various. In the Jack Pot mine, at 400 feet from the surface, in the Logan at 600 feet, and in the Doctor at 700 feet, there have been found fragments of coal, exhibiting traces of wood-structure In the Independence, at 500 feet, a stump of a tree was discovered in the very midst of rich ore In every case the enclosing rock was breccia. The specimen from the Independence is stone, the others are coal. In the former case, the tree-portion must have become buried under conditions free from access of air, and must have been subjected subsequently to the action of siliceous waters, which gradually replaced the fiber of the wood with a mineral precipitate. In the other cases, the tree must have become enclosed within the breccia and subjected to a slow oxidizing action which carbonized the wood, without permitting it to burn freely. Otherwise, it would have been destroyed, leaving only ashes As it was, it became coal, carrying 60 per cent. carbon, and having the other characteristics of a typical lignite.

The Cripple Creek volcano must have formed a conspicuous mountain. This is inferred from the nature of the material ejected. The size and shape of the cones formed by the emissions of a volcano depend upon the condition in which they are emitted. Limpid lavas, like those of the Hawaiian volcanoes, form extremely flat cones. Mauna Loa, for example, has a height of 13,675 feet above the sea, with a base of over 70 miles, the slopes having, according to Dana, an angle of 4° to 6° only. The great volcanic cones of the Andes are made up of a much less liquid lava, and, according to Whymper, have slopes which range from 27° to 37°. Mount Shasta, in California, which is built up of similar rocks, stands, according to Whitney, at an angle ranging from 28° to 32°. A cone such as that formed around the vent of the Cripple Creek volcano, which emitted vast quantities of fragmental material alternating with occasional lava-flows of the more viscid type, would partake of the character of the well-known pûys or peaks of Auvergne, which dot the surface of that part of south-central France, in shapes resembling a candle-extinguisher. Breccia and lava together make steeper cones than lava or breccia separately; therefore the Cripple Creek volcano, when at its maximum height, must have appeared as a steep mountain.

Projectile discharges were succeeded by tranquil emissions of lava. The bodies of massive andesite in the southeastern part of the district may represent such extrusions. They were marked by an absence of the violence which accompanied the earlier outbursts, due, perhaps, to a diminution in the quantity of escaping steam and a lessening of the pressure upon that which remained. The earlier ejectamenta of a volcano are scoriaceous and vesicular; that is, they have been penetrated and torn by the explosive escape of superheated water-vapor, while the lava characterizing the later stages of activity is compact and homogeneous. The creation of a vent serves as a safety-valve in releasing the tremendous pressure of the steam, due to its sudden expansion when coming into contact with incandescent fused rock. Attendant upon the relief given to that pressure, are all the terrifying phenomena of the first outburst. Subsequently the force of the eruption diminishes. The lava ceases to be violently projected by escaping high-pressure steam. The underground waters near the conduit have become used up.

The rise of the lava underground, followed by its protrusion at the surface, becomes a quiet process, which must be referred to a more deep-seated cause, namely, the local readjustment of the earth's crust, causing the fused rock to ooze out slowly. Many lava streams have a glacier-like movement. They seldom progress more than 3 miles per day, and often require a year to advance a few miles.* Observers have described the flows of lava which follow the first eruption as welling out "with the tranquility of a water-spring,"† as "proceeding in silence,"‡ as "being effected quietly and without noise."§ All this is in vivid contrast to the paroxysmal outburst which marks the first stage of volcanic activity. The difference is to be referred to the relative quantity of steam taking part in the process of eruption.

The period of quiet may have been, and probably was, succeeded by a complete, though temporary, cessation of activity. This interval may have persisted for several hundred years. Geology is lavish of time. The inaction was due to the diminution of pressure consequent on the withdrawal of the lava in the conduit of the volcano. Such a result would be brought about by the shifting of the center of eruption to another place along the line of fissure. The island of Vulcano, in the Lipari group, affords an excellent example of such a change of vent. Among the extinct craters of Auvergne|| in south-central France, similar instances are numerous. (See Fig. 6.) The first conduit of the Cripple Creek volcano became plugged up by material which had failed of ejection. Other minor vents may have been formed on the flanks of the mountain which had been slowly formed by the long continuance of discharges. When, after an interval, a vigorously active condition was resumed, the second eruption, in all probability, took place through a new vent, produced, as the original one had been, by a fissuring of the rock immediately over congested masses of steam due to the water which had accumulated during the interval of inaction.

---

* Dana. † Scrope. ‡ Fouqué. § *Ibid.*

|| The writer cannot claim to have any special knowledge of volcanoes, but he is familiar with the volcanic region of Auvergne, in south-central France, has seen Vesuvius, and has traveled in the volcanic parts of New Zealand, and also in Oahu, one of the Hawaiian Islands.

There is evidence indicating that the Cripple Creek volcano had several vents. One existed near the present site of the Hull City placer; another must have been situated near Anaconda. The original position of the orifice of an extinct volcano can be inferred from the composition of the rocks. The lava which cools rapidly in the open air assumes the character of a glassy substance,* containing only a few embryonic

FIG. 6.

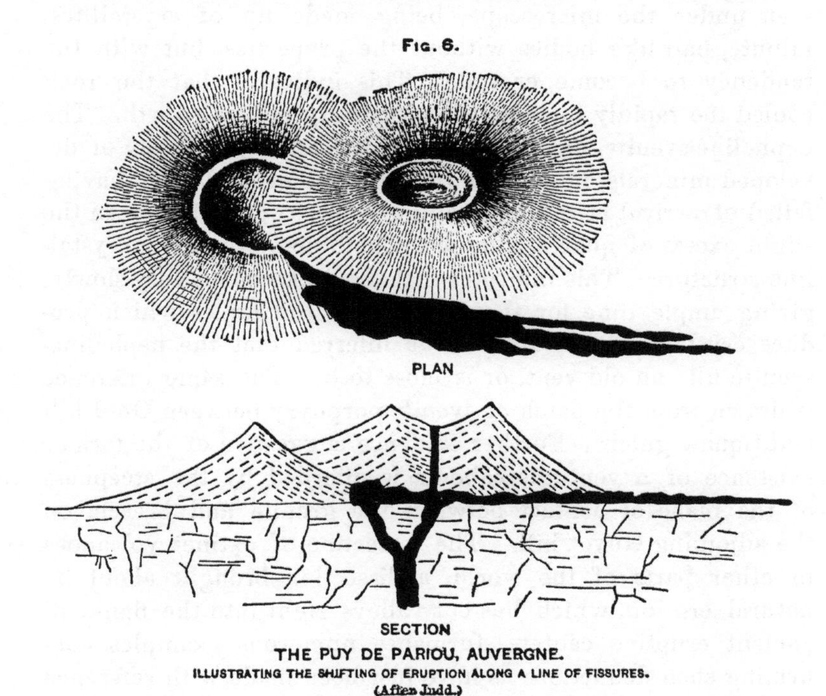

THE PUY DE PARIOU, AUVERGNE.
ILLUSTRATING THE SHIFTING OF ERUPTION ALONG A LINE OF FISSURES.
(After Judd.)

crystals, but that which cools slowly underground, and while still subjected to great pressure, is developed into completely crystalline rock. Experiments with smelter-slags, and a microscopic examination of the resulting material, have confirmed this proposition. In this way the lava streams which have issued from the vent are distinguishable from the material

---

* On June 3, 1840, a stream of lava from Kilauea reached the sea, after having flowed over the island of Hawaii for a distance of 11 miles. "The burning lava, on meeting the waters, was shivered like melted glass into millions of particles, which were thrown up in clouds that darkened the sky and fell like a storm of hail over the surrounding country."—DANA, *Characteristics of Volcanoes*, page 63.

which has solidified in the throat of the volcano. The nepheline-syenite near the Lillie and Vindicator mines is the granular equivalent of the phonolite which occurs so plentifully all over the district. The phonolite and the syenite have a similar chemical composition, but their texture is very different. This is due to the fact that in the former a crystalline structure has not been fully developed, but the ground-mass or matrix, as seen under the microscope, being made up of crystallites, minute, hair-like bodies without the properties, but with the tendency to become, crystals. This indicates that the rock cooled too rapidly to permit of proper crystalline growth. The nepheline-syenite, on the contrary, is made up entirely of developed minerals, no part of the original ground-mass having failed of arrival at true crystalline maturity; so that even the slight excess of quartz, though uncombined, presents a crystalline structure. This indicates that the rock cooled very slowly, giving ample time for the full play of the forces which produce crystallization. It is to be inferred that the nepheline-syenite fills an old vent, or is close to it. The same inference is drawn from the patch of syenite-porphyry between Gold hill and Squaw gulch. Further evidence suggestive of the former existence of a vent thereabouts is afforded by the steepness of the plane of contact between the granite and breccia on the adjoining Guyot hill. The dissection of extinct volcanoes in other parts of the world, a dissection brought about by natural erosion, which has cut valleys right into the flanks of ancient eruptive centers, furnishes numerous examples confirming such deductions as have just been made with reference to the vents of the Cripple Creek volcano. Even in Great Britain, which has not known volcanic disturbances during the time covered by the brief record of human life, there are abundant proofs concerning the shifting of vents and the resulting relations between perfected and undeveloped rock-types.

The occurrence of several vents would not be unusual. Volcanoes are not mere bores through which eruptive discharges take place. Where one single vent survives, it may be considered to represent the centralization of energy due to the choking-up of many other openings along the line of fissure formed at the time of the first manifestation of activity. This is well illustrated in the accompanying sketch of Mount Etna, as it appeared in 1865. (See Fig. 7.) The expansive force of the steam, to which the violence of the initial stages of volcanic action is due, tends to radiate from the central point of energy so as to form cracks, the character and extent of which will vary according to the structure of the rocks through which the shocks are propagated. At certain points along these cracks, or at the crossing of two of them, openings are formed, permitting eruptive discharge. Those openings which are immediately above the points of greatest pressure, will survive longest; the others become plugged up with the material they are unable to eject. One vent usually remains as the center of

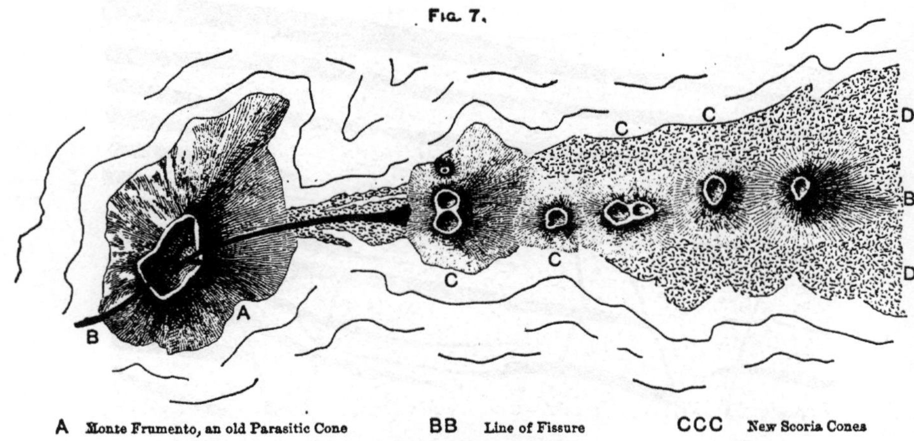

A Monte Frumento, an old Parasitic Cone BB Line of Fissure CCC New Scoria Cones
DD Lava, from the small Scoria Cones
FISSURE FORMED ON THE FLANKS OF MT. ETNA IN 1865.
(After Judd.)

energy. The others become extinct, until an increase of eruptive activity finds a single conduit insufficient, and thus necessitates the obtaining of relief at other points. Lava-flows do not necessarily take place at the central vent. Many of the largest flows known to have occurred among the Hawaiian volcanoes, for example, have emanated, not from the crater at the top of the mountain, but far down upon its flank. In certain instances, as at Kilauea,* the larger number of discharges have been subterranean.

The subterranean discharges of lava are of peculiar interest

* Dana. That of July, 1840, started at a point 16 miles distant from the crater.

to the miner, because they are among the factors which he has found by experience to influence the distribution of the ores which he seeks. They are to be seen both in natural sections, afforded by ravines, and in those other sections of the rocks which are presented underground in the mines. The accompanying drawing,* after Fouqué (see Fig. 8), of a natural section seen on the slope of the old volcano of Santorin, will be suggestive. These intrusions take a variety of forms. Such as seek out the lines of weakness presented by the bedding-planes of sedimentary rocks, or the lines of successive deposi-

FIG. 8.

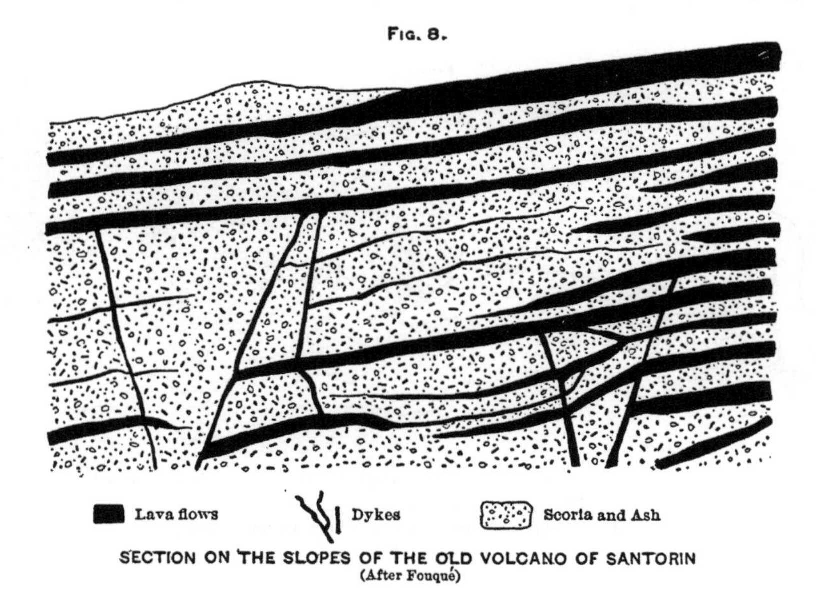

SECTION ON THE SLOPES OF THE OLD VOLCANO OF SANTORIN
(After Fouqué)

tion of fragmentary volcanics, form sheets. In England, such an intrusive sheet is termed a "sill." An instance is illustrated in the accompanying section, obtained in the western islands of Scotland by Sir Archibald Geikie.† (See Fig. 9.) The intrusive masses of porphyry (quartz-felsite) which, at Leadville, penetrate the sedimentaries, afford an example which is of peculiar interest on account of the remarkable ore-deposits found at the contact of the porphyry with the Carboniferous limestone.

* From *Santorin et ses Éruptions.*

† "The Tertiary Basalt-Plateaux of Northwestern Europe."—*Quarterly Journal Geol. Soc.*, vol. lii., p. 377.

Those subterranean flows of lava which do not find a ready passage, either in a lateral or a vertical direction, tend, when thus impeded, to congest locally, so as to form huge under-

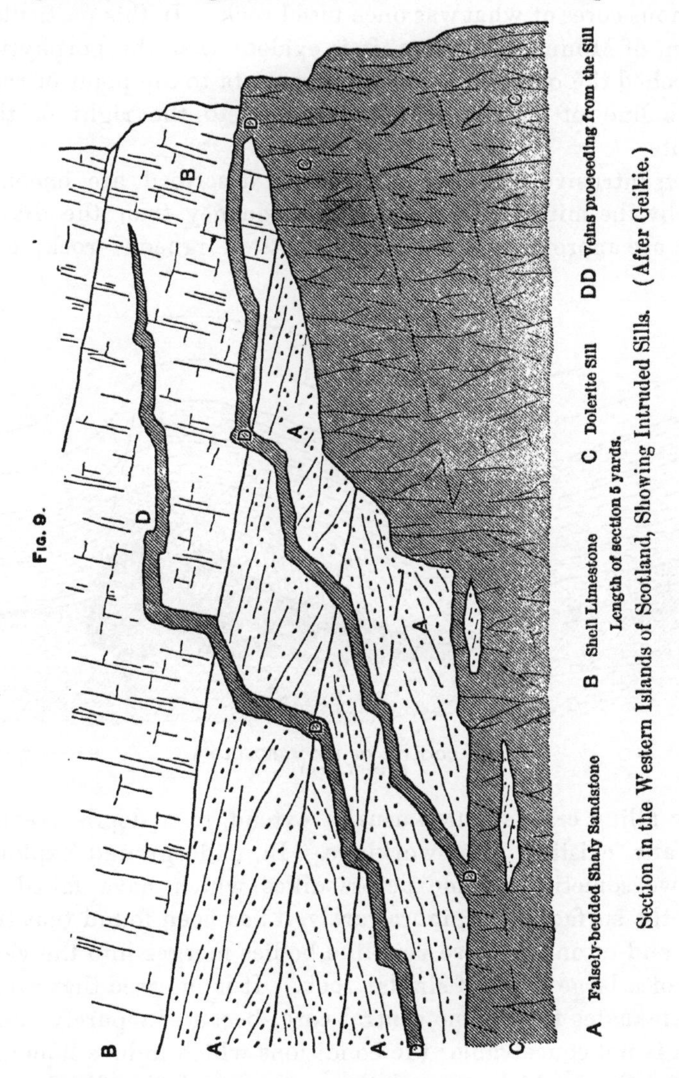

Section in the Western Islands of Scotland, Showing Intruded Sills. (After Geikie.)

ground blisters which are sometimes large enough to arch the overlying strata into dome-shaped hills. Such "laccolites,"* as they are termed, were first recognized as a type by G. K.

* From the Greek *lakkos*, cistern, and *lithos*, stone. Laccolith would be a better form.

Gilbert.* Since then, Whitman Cross† has described similar occurrences in southwestern Colorado. The accompanying drawing (Fig. 10) represents his ideal section of one of these enormous cores of what was once fused rock. In this particular section, of Mount Marcellina, it is evident that the porphyrite has arched the overlying coal-bearing strata to the point of rupture, a line of fracture being indicated to the right of the laccolite.

Such intrusive masses as have been described are encountered by the miner with much less frequency than the dikes, which are approximately vertical sheets of igneous rock, evi-

FIG. 10.

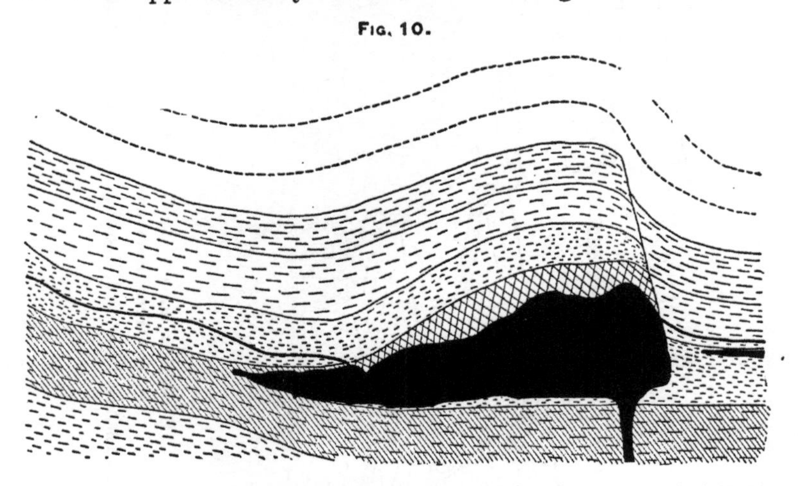

THE LACCOLITE OF MT. MARCELLINA
(After W. Cross)

AM. BANK NOTE CO., N. Y.

dently filling cracks which usually extend to a depth greater than any existing mine-workings. In underground explorations we sometimes come across dikes which have failed to reach the surface; and, more rarely, it has been found that the lower end of one of these vein-like bodies merges into the very heart of a large mass of similar rock. It is inferred that every dike emanates from some central core, because a purely local origin is not conceivable; the conditions which induce liquefaction and those which compel the upthrust of the fused rock are alike referable to factors created by the forces at work

* "Report on the Geology of the Henry Mountains," 1877.

† "The Laccolitic Mountain Groups of Colorado, Utah and Arizona." 14*th* *Annual Report, U. S. Geol. Survey.*

within large masses of rock. The water coming up through a crack in the ice is referred to the body of it beneath that ice. The fused rock rises in the fractures caused by earth-movements much as the water fills the cracks in the ice; that is, no gaping crevasse is necessarily formed, but the lava rises and occupies the fracture as it is formed; it follows it, it does not make it, although the movement which makes the crack and the pressure which squeezes the lava into it may both be traceable, far back, to a common cause.

The behavior of dikes often affords striking evidence of their adaptability to the structure of the rocks they traverse. They seek out lines of least resistance, and thus frequently make evident structural features which otherwise would have been merely latent. The accompanying drawing (Fig. 11) illustrates this. It represents a flat surface of granite (in West Australia) traversed by a dike of dolerite, which has evidently utilized for its passage the lines of fracture produced along a sheer-zone in the granite. The very low conductivity of lavas (as of smelter-slag) may explain their ability to pass through rock-fractures for great distances. The edges of a dike would cool instantly, but in so doing would afford a protection to the central portion, the liquidity of which would thus tend to be maintained.

The bulk of the material thrown up by the Cripple Creek volcano was fragmentary, and became the great mass of breccia now constituting the leading geological feature of the district. The earliest lavas extruded were of medium fusibility: namely, andesite, and then phonolite. On reaching the surface, they formed streams, the exterior of which became promptly chilled to a black-looking slag, to which escaping steam gave a cindery structure. The lava rolled down the slope of the volcano with the utmost slowness, making clinking sounds such as are heard when the workmen empty the slag-pots over the dump of a smelter. Such lava-streams weather very easily; their exterior, by the contraction of the surface due to cooling, becomes porous, and water penetrates into the mass of them, disintegrating them so that they are readily carried away by the rains of spring.

The last extrusions of the Cripple Creek volcano were of basalt. These were more limpid, and must have formed

FIG. 11.

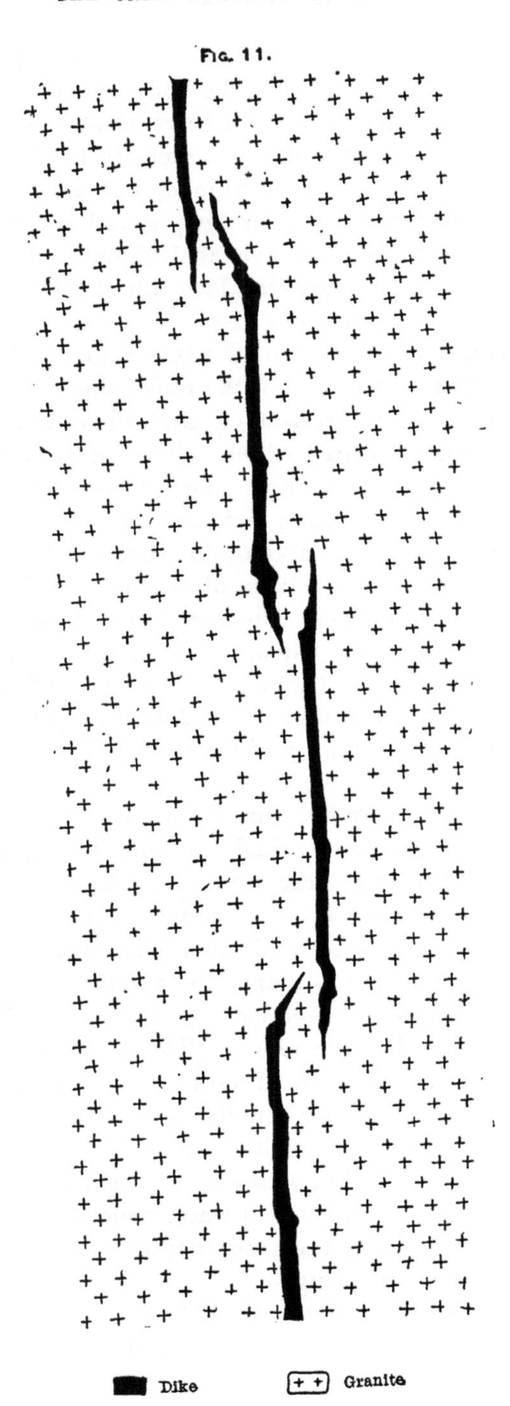

streams which traveled much farther than the less fusible phonolite and andesite. The basalt, judging from the behavior of similiar lava-flows actually observed in the Hawaiian Islands and elsewhere, would progress rapidly down the slopes of the mountain and overwhelm the forests which, probably, clothed the lower portions of the Cripple Creek volcano, setting them on fire and adding greatly to the aspects of destruction presented by the scene. Upon cooling, these basaltic flows would be cleft asunder by symmetrical series of cracks forming prismatic columns grouped like the pillars of a Gothic cathedral. But where they were not protected by a later covering of rock, the ruthless hand of decay attacked them also, the frost of many thousand years shivered the straight columns; and the freshets of spring swept the remnants into the torrents which fed the Arkansas river.

The successive periods of activity in the life of the Cripple Creek volcano are marked by the sequence of lavas extruded. This sequence is indicated by structural relations, the older extrusions being penetrated by the younger. But this is not all. The crystalline structure and the chemical composition of the rocks resulting from the cooling and consolidating of the successive lavas exhibit differences which have been found to be closely analogous to those presented by similar successions of rock at other volcanic centers, both in the United States and in Europe. The earliest lava extruded by the Cripple Creek volcano was andesite. Then came the phonolite, and, lastly, the basalt. These three rocks represent types which vary in their chemical composition and in their consequent fusibility. Basalt fuses at about 2250° F.; certain varieties fuse at about 2000° F. What is usually termed a "white heat" is equivalent to a temperature of 2100° F*. The least fusible rocks are of the granite and trachyte class; they fuse with difficulty at about 2700° F. To the intermediate type belong the andesites, which fuse at about 2520° F.† The relative fusibility of these rocks is dependent upon the fusibility of their chief constituent, feld-

---

* According to the latest determinations by Henry M. Howe. According to Pouillet's experiments, gold melts at 2192° and silver at 1832° F. See *Eng. and Min. Journal*, Jan. 20, 1900, p. 75.

† These are the temperatures derived from the experiments of Carl Barus. See Dana's *Manual of Geology*, p. 273.

spar, the variety in the basalts being labradorite, the most fusible of the feldspars. Moreover, in basalt there is present a good deal of augite, a still more fusible mineral, and a large percentage of iron which, as in smelter-slags, contributes directly to fusibility. The trachytes are largely made up of orthoclase, the least fusible of the feldspars. The andesites are intermediate in composition and of medium fusibility, their characteristic feldspar being oligoclase.

This fusibility used to be expressed in terms of "acid" and "basic character," the rocks high in silica and low in iron being at one extreme, and those low in silica and high in iron at the other. But Dana has pointed out* that this does not express it correctly, fusibility being dependent not so much on the percentage of silica as upon the amount of alkali, namely, potash and soda.† Thus the rocks rich in alkaline feldspars are the most fusible. Free quartz exists in most rocks; and the percentage of it, which is far from uniform among the members of any particular type, increases the acid character of the rock, so that it becomes a secondary factor in determining fusibility. Similarly iron occurs as an oxide (magnetite) in all rocks, to an insignificant degree in the granites, but in the basalts and gabbros freely, so as to form an important ingredient, giving them their dark coloring. This large percentage of iron contributes to easy fusibility; indeed, certain basalts are known to become so limpid that they can be taken up in a spoon attached to the end of a cane.‡

This question of fusibility would be of slight importance were it not for one interesting fact, namely: it has been observed in several volcanic regions that lava of intermediate composition, such as andesite, is succeeded by those of the extreme types, namely, the very alkaline or comparatively non-alkaline rocks, such as basalt and rhyolite, respectively. This was the case at the Cripple Creek volcano. It has been inferred from these facts that in the earlier stages of volcanic activity the lavas are mingled together underground, and that during the period of eruption the heavier portion separates

---

* *Manual of Mineralogy and Petrography*, p. 437.

† Dana, *Characteristics of Volcanoes*, p. 146.

‡ This was actually done in the case of the basaltic lava of Kilauea, in Hawaii. Coan. *American Journal of Science.*

from the lighter, causing two diverse products to be separately emitted.

Eventually (it may have been several thousand years after the first manifestation of activity) the volcanic energies became wearied, and lava ceased to appear at the surface. The readjustment of the earth's crust, at this particular locality, had been accomplished, and a condition of equilibrium supervened. The lava sank beneath the level of the crater, and, on cooling, plugged up the conduit, as was the case, for instance, at the Kammerbuhl.* The sinking of the lava may have gone further, so that the withdrawal from the upper part of the mountain, formed by the ejections of the volcano, may have caused extensive subsidence and created deep fissures. Such was the case at Kilauea in 1832 and 1840.† Those who are engaged in mining at Cripple Creek are aware of the existence of numerous large cavities underground, particularly in the southern part of Bull hill and the northwestern portion of Battle mountain. In the Logan mine the orifice of a very large cavity was recently encountered while sinking the shaft.‡ The sudden flows of water which have embarrassed some of the mines are due to the unexpected drainage of such openings. It is worthy of note that these especially characterize the trachyte-phonolite and those rock-masses which represent the lavas extruded last.

After the volcanic energies had declined, there followed a long period of smothered activity, evidenced by geysers and hot springs. Steam continued to escape, but gently. There was none of the violence of the earlier period. Heated water accompanied the steam, instead of fused rock. The hot lava still existing at greater depth served to give expansive force to the surface waters which found their way, by seepage, through the overlying deposits of volcanic material. The steam and hot water now emitted, at some spots quietly as a thermal spring and elsewhere intermittently as a geyser,§ probably carried a

* Judd's *Volcanoes*, p. 114. See also p. 5 of this paper.

† Dana's *Characteristics of Volcanoes*, p. 124

‡ The miners heard the inrush of air caused when they tapped the cavity and promptly left their work, to go to the surface. It is probable that the pumping operations of the neighboring Portland mine had drained the water, which at one time had filled the cavity, leaving it void.

§ Geyser is an Icelandic word, and means "gusher." A thermal spring which spouts or gushes out above the surface is a geyser.

good deal of mineral matter in solution. A wonderful work is accomplished in this quiet way, because such activities extend over enormous periods of time. Professor Judd has shown that the hot spring at Bath (England), although an apparently unimportant geological agent, brings daily to the surface 180,000 gallons of water at a constant temperature of 120° F. This spring was doing its dùty at the time of the Roman invasion of England, and it is estimated that since that time it has brought up, in solution, enough material to form a good sized volcanic cone.*

The Cripple Creek district exhibits abundant evidence of hydrothermal action. This is particularly the case in the northwestern part of the mining area. The breccia of the upper parts of Globe and Ironclad hills, penetrated by the workings of the Deerhorn, Summit, South Park, Plymouth Rock and other mines, is much decomposed, and has a loose, crumbly character. It is seamed to an unusual degree with irregular fractures, lined with secondary minerals, among which crystalline gypsum and amorphous kaolin are the most common. In the Deerhorn shaft there is evidence of a more definite kind. At a depth of 240 feet the shaft cuts into a mass of gypsum, and from that point to the bottom, 575 feet below the surface, it has been sunk in the midst of what appears to be a series of extinct thermal springs. The accompanying drawing, Fig. 12, will illustrate the occurrence.

The workings are very extensive in a direction at right-angles to the plane of the section followed by the illustration, and connect with the adjoining mines. Breccia and tuff compose the prevailing rock.† No distinct dikes are visible near the Deerhorn shaft; it is probable, judging from the composition of the breccia at several points, that several intrusions do exist, but that they have been so shattered in place as to be scarcely distinguishable from the original breccia which surrounds them. At the third level, and at the succeeding levels, there are three distinct narrow-pointed cones of com-

---

* Page 219 of *Volcanoes*.

† The breccia was found by Professor Kemp to contain undoubted fragments of kaolinized granite, decomposed orthoclase being easily recognizable. The breccia of Cripple Creek, although fragmentary andesite predominates, everywhere exhibits a scattering of granite particles, which in places become so numerous as to give it a truly granitic character.

FIG 12.

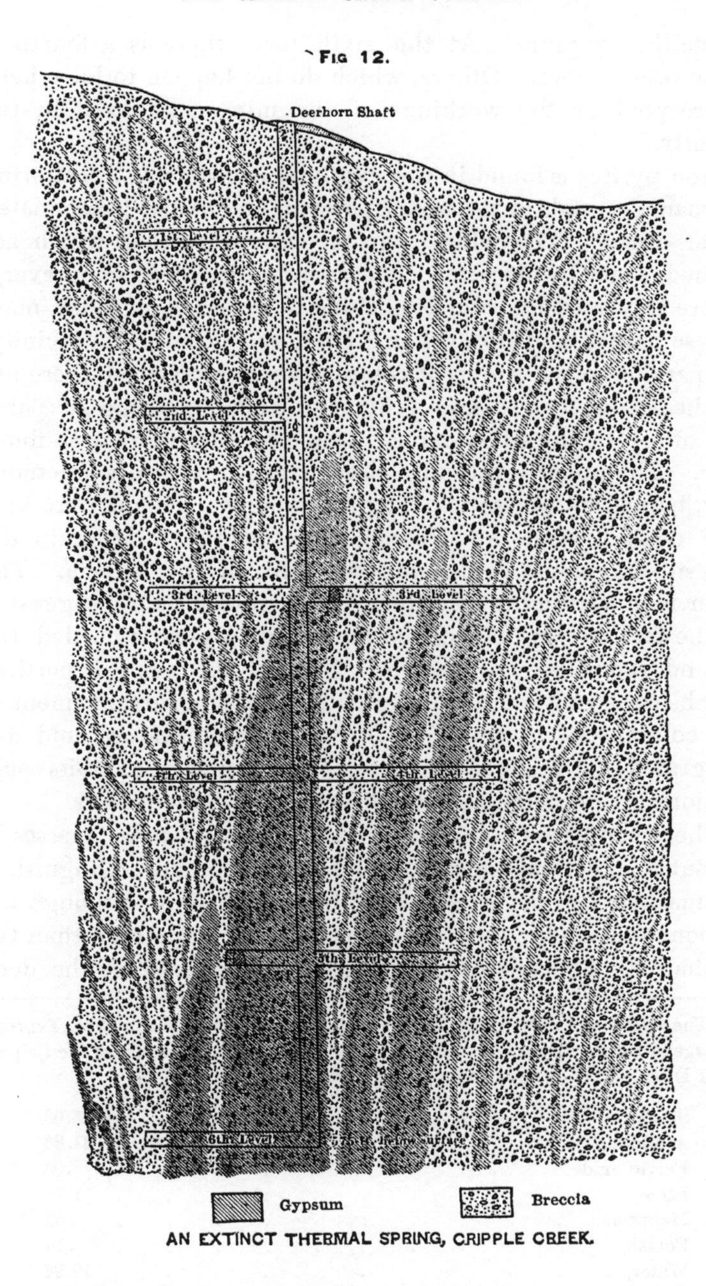

AN EXTINCT THERMAL SPRING, CRIPPLE CREEK.

pact white gypsum which, at 15 to 25 feet from their apex, graduate into chimney-like masses of breccia cemented by crystalline gypsum. At the sixth level there is a fourth of these occurrences. Others, which do not happen to have been intercepted by the workings of the mine, may exist in the vicinity.

Iron pyrites is found in the gypsum; it occurs as a scattering of coarse crystals in the upper portion, and finely disseminated lower down. The white gypsum carries patches which are stained pink by fluorite. The surrounding breccia is everywhere traversed by color-bands due to layers of gypsum, manganese oxides, and iron ocher. Scattered through the vicinity, but parallel to the group of columns above described, there are patches, as well as seams, of fluorite sand, consisting of particles of crystalline silica stained purple by admixture with fluorspar. The upper levels also show bands of a white unctuous clay, named "Chinese talc" by the miners. This is pure kaolin,* derived from the decomposition of the feldspar in the andesite fragments composing the bulk of the breccia. The latter is in a crumbly condition, its character being suggested by the fact that in driving the levels only a pick is needed, the ground requiring no blasting. Beyond the central portion, which has structural lines sympathetic to the arrangement of the columns of gypsum, the breccia is still stained and disintegrated for a great distance, and in places exhibits suggestions of the neighborhood of other thermal conduits.

There can be no doubt as to the nature of these masses of gypsum. Thermal springs which have become extinguished are marked by just such accumulations of lime, although the carbonate is, under such circumstances, more common than the sulphate.† The flows of hot water encountered in the deep

* The following is Dr. Hillebrand's analysis, made by him for Prof. Penrose. See page 128 of the Report on "The Geology and Mining Industries of the Cripple Creek District," *U. S. Geological Survey*, 16*th Ann. Rep.* Part II.

Silica,	45.08
Alumina,	31.83
Ferric oxide,	.95
Lime,	1.76
Magnesia,	.59
Potash,	.14
Water,	19.96
	100.31

† Siliceous deposits characterize geysers.

workings of the Comstock carried a notable percentage of gypsum. Last April, while examining certain copper-mines near Hawthorne, in Nevada, the writer came across a group of similar vents, marking the site of former thermal springs. The conduits, in this case, occurred in lime-shales, and were still open to a considerable depth, as was proved by dropping stones into them. They were surrounded by a compact chimney of carbonate of lime, which had also overspread the enclosing rock.

Recurring to the conditions observed in the Deerhorn shaft, it would seem that the rising hot waters, in their approach to the surface, were unable to maintain a defined channel through the breccia higher than the level marked by the tops of the cones of gypsum. This might be caused, first, by the fact that the vapors ascending above the subterranean springs disintegrated the breccia so as to destroy its cohesion, and changed it from a compact rock to loose material. The most potent factor, however, was probably the diffusion of the ascending waters into the drainage of the surface, the effect of which would be encountered at this horizon. The condition of the breccia and the wide area which has undergone disintegration favor this view.

It is in accord with facts observed in other regions that the vents which permitted the emission of lava-flows should be in one part of the volcanic area (in this case the southern portion) while the escape of hot waters which marked the time when the volcanic energies were waning should have occurred in another part, in this case the northern and northeastern. The lava had healed lines of weakness; it had cemented the fractures produced by the earlier paroxysmal efforts of the volcano; and therefore the thermal waters found a better chance of exit elsewhere. With the hot waters which found their way to the surface during the closing period of the volcanic cycle there were emanations of gas. Sulphuretted hydrogen was probably emitted, sulphurous acid gas, and, in all likelihood, carbonic acid gas also, although not all of these were to be found at one place or at one time. The volcano had now reached the "solfatara" stage.* These acid gases played an important part in

* "Solfatara" is from the Italian "solfo," meaning sulphur. It is a name given to one of the small volcanoes, near Naples, which is in a condition such as marks the dying out of volcanic activity.

altering the volcanic rocks, and were, possibly, a factor in the process of ore-deposition which was beginning. The vapor of hydrofluoric acid was also among the agencies at work. This is inferred from the large amount of fluorite, the fluoride of calcium, which occurs all over the district, and more especially in the gold-bearing lodes. Fluorite is not common in volcanic regions, although it is found in the lava of Vesuvius. The action of hydrofluoric acid on feldspars containing lime would form fluorite. It would also convert gypsum in a similar way. Fouqué has shown that the action of hydrofluoric acid in the liquid state is to decompose, first, uncrystalline silicates or glasses, then feldspar and other acid silicates, then quartz, and lastly, basic silicates. Whether the vapor of hydrofluoric acid would act in the same way is uncertain, although it is possible that in this case quartz might be attacked in preference to the feldspar. This is a matter of interest, because in examining specimens of granite which have been converted into ore (by the addition of gold-bearing tellurides) it is observable that the original quartz of the granite has been attacked while the orthoclase remains comparatively fresh.

This last stage of the Cripple Creek volcano is of great importance to the mining geology of the region. It extended over an enormous period, coinciding, roughly speaking, with that which has elapsed since the time to which is ascribed the first evidence of the existence of the human genus, and it afforded, to an unusual degree, those particular conditions which are considered to favor the deposition of precious ores. During this time, also, the breccia, with its finer portions, the tuff, became solidified. The pressure of the overlying masses of lava which at one time covered it, and the chemical solutions, which deposited fresh crystalline substances in the interspaces, converted the scoriaceous material into a compact mass, which eventually became solid rock as we now see it. The cooling of the intrusive bodies of lava caused them to contract, and thus developed lines of weakness along which the energies of the volcanic center developed fractures permitting the subsequent prolonged circulation of underground waters. The readjustment of this particular portion of the earth's exterior, which followed the cessation of volcanic eruptions, and the partial settling of the entire mass forming the Cripple Creek volcano, must have formed an extensive system of ruptures, which afforded lines of maximum porosity along which the gold-bearing solutions found passage-ways. Thus the hot waters which are supposed to dissolve out the metals from the deep-seated rocks were permitted to ascend toward the surface, where the release from pressure and the lowering of temperature forced them to precipitate their contents.

The activity of the geysers ceased; the warmth of the water bubbling from the springs gradually diminished; and at length the last vestige of the volcanic fires passed away. The mountain became as cold as the snow which mantled it each winter, and as still as the darkness enshrouding it nightly.

SUBJECT TO REVISION.

[TRANSACTIONS OF THE AMERICAN INSTITUTE OF MINING ENGINEERS.]

The Lodes of Cripple Creek.

BY T. A. RICKARD, DENVER, COLORADO.

(New Haven Meeting, October, 1902.)

A. Introductory.

In a former paper*. the writer has described the essential features of the general geology of the Cripple Creek region. In the present account it is intended to examine into the occurrence of the ores, the value of which has made this district the most important among existing American gold-fields. The production of Cripple Creek from its discovery, in 1891, to the close of 1901, has reached a valuation of fully $125,000,000. During the past year (1901) the output amounted to $17,285,470. In 1900 it was $18,174,681.

The first discoveries, which led to the development of the district, were made in the spring of 1891, but it was not until 1893 that vigorous work was commenced. A great impetus was then given to the exploration for gold on account of the sudden drop in the market-price of silver, caused by the closing of the Indian mints in the summer of that year. This induced an energetic population from the older silver-mining camps of Colorado to go to the new gold-field, which was then beginning to attract attention. Prospecting, at first, was hindered by the comparative absence of outcrops, due to the fact that the surface of the hills is covered with a considerable thickness of shattered rock, resulting from the action of frost at a high altitude; but so much indiscriminate digging was done that a number of rich veins were uncovered, and this stimulated the search for others. The advanced condition of the mining industry of Colorado offered unusual facilities for exploration and reduction; progress was therefore rapid, with the result that the district soon achieved great prominence.

* "The Cripple Creek Volcano," *Trans.*, xxx., 367–403. It is proper that reference should also be made to the more authoritative monograph of the U. S. Geological Survey, namely, "Geology and Mining Industries of the Cripple Creek District, Colorado," by Whitman Cross and R. A. F. Penrose, Jr. 1895.

B. Geological Character of the District.

The geological environment of the gold-bearing veins can be outlined briefly. The district occupies the ground-floor of a volcano, the superstructure of which has been removed by erosion. This basal wreck of material erupted during the Tertiary period now survives as a complex of volcanic rocks, filling the hollows and occupying the plug of a basin which is surrounded by the granite of Pike's Peak. The volcanic area of Cripple Creek occupies about nine square miles, and consists, for the most part, of breccia, in which andesite predominates. Penetrating the breccia in every direction are numerous dikes, composed of various rocks, those of basalt and phonolite being the most notable, on account of their close association with the occurrence of ore.

The mine-workings have reached a maximum depth of 1400 ft. Added depth appears to have affected the persistence of the ore to the same extent as experience elsewhere would lead one to expect. The veins situated near the edge of the breccia have in several cases been followed downward in their penetration of the underlying granite, and it has been demonstrated that some of the ore-bodies have continued from the upper into the lower geological horizon. The distribution of these ore-bodies offers the same perplexing problems as in other gold-fields. Extensive developments, due to very successful mining, have, however, afforded a great deal of interesting evidence, the consideration of which may contribute toward the better understanding of the economic geology of the district.

The mines exhibit examples of a great diversity of lode-structure. This diversity is mainly traceable to the complexity of the enclosing rocks. The variations in ore-occurrence due to this fact explain the vicissitudes which marked the early history of the district, and the recognition of them should promote the success of future exploratory work.

During the past five years, while examining a dozen of the principal mines, the writer has gathered many examples of vein-structure which are herewith submitted as testimony bearing upon the ever-fascinating problem of ore-occurrence. As a poor witness may sometimes furnish a good lawyer with an illuminating bit of evidence, so the writer hopes that this testimony may be of service to the geological philosophers who are engaged in the study of ore-deposits.

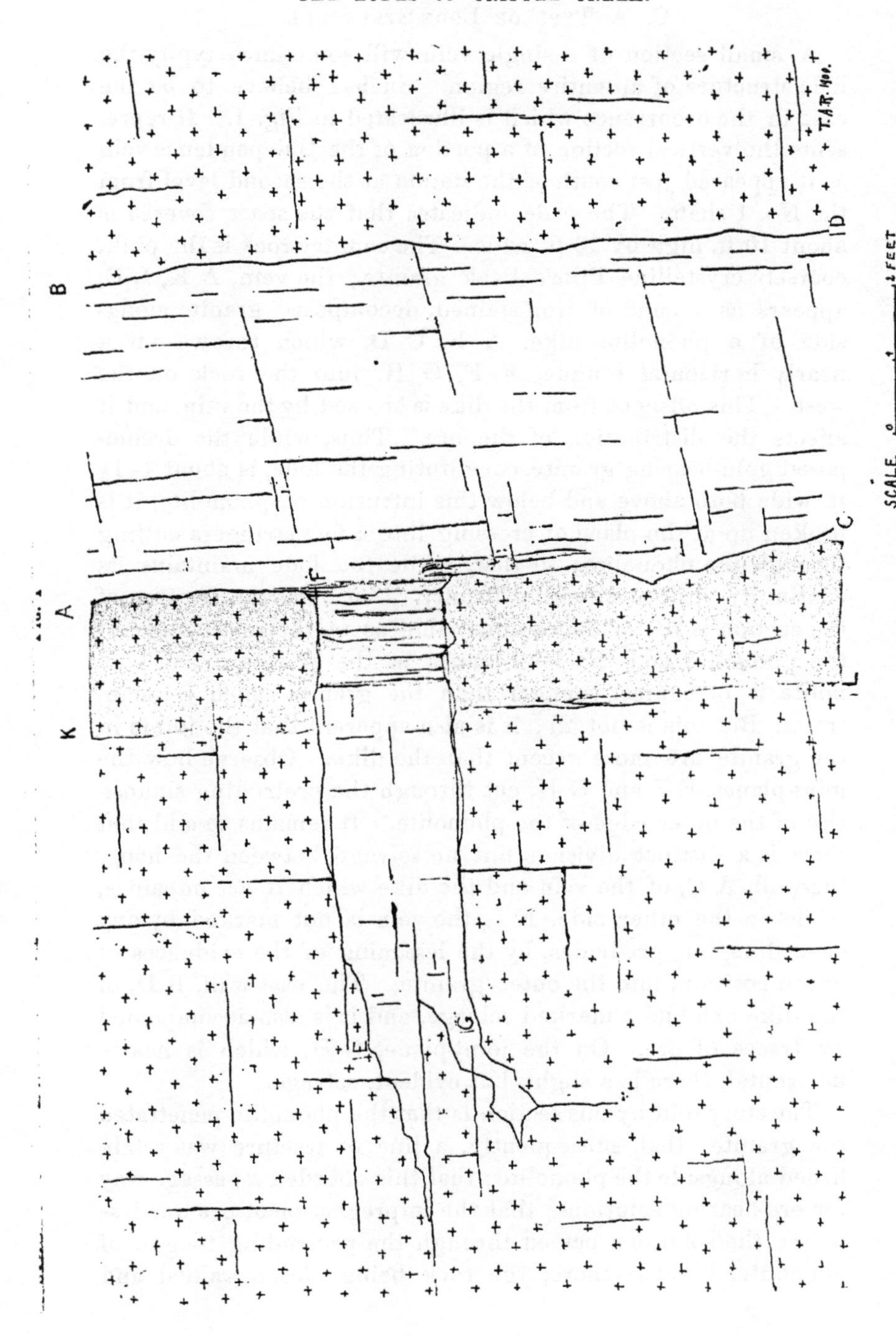
B
A
K
D
C
L
F
G
SCALE 0 1 2 FEET

C. A Type of Lode-Structure.

A small section of a single vein will sometimes typify the lode-structure of an entire region. Such I believe to be the case in the occurrence which is illustrated in Fig. 1. It represents the vertical section of a portion of the Independence vein as it appeared just south of the station at the second level from the No. 1 shaft. The scale indicates that the space covered is about 10 ft. high by 16 ft. wide. The country-rock is the pink, coarsely crystalline Pike's Peak granite; the vein, A K, L C, appears as a band of iron-stained, decomposed granite alongside of a phonolite dike, A B, C D, which throws out a nearly horizontal tongue, E F, G H, into the rock on the west. This offshoot from the dike is crossed by the vein, and it affects the distribution of the ore. Thus, while the decomposed gold-bearing granite, constituting the lode, is about 1–1½ ft. wide both above and below this intrusion of phonolite, it is broken up at the place of crossing into a few stringers cutting through the phonolite; so that, while the lode maintains its continuity, it does so with difficulty. An important feature of the section is the evidence obtainable as to the relative age of the phonolite and the joint-planes in the granite. The phonolite is, of course, younger than the granite which it penetrates. But this is not all; it is also apparent that the joints in the granite are more recent than the dike. Observe how the joint-planes, E F and G H, cut through the protruding sinuosities of the outer edge of the phonolite. It remains to add that there is a distinct division, but no selvage, between the hanging-wall, A C, of the vein and the dike which it accompanies, while on the other side, K L, the vein is not marked by any clear line, but graduates, by the lessening of the evidences of decomposition, into the outer granite. The east wall, B D, of the dike exhibits a marked selvage, and it is also accompanied by traces of ore. On the joint-plane, G H, which is nearly horizontal, there is a slight, but evident, selvage.

The story told by this section is that the phonolite penetrated the granite; that, subsequently, a line of fracture was established alongside the phonolite; that this afforded a passage-way for ore-bearing solutions; that the impregnation of ore was less where the solutions passed through the protruding tongue of phonolite, because there, the rock being closer-grained and more fissile than the granite, it was broken by a very few decided cracks, rather than by an irregular multiple fracturing such as determined the diffused mineralization constituting the lode in the granite; further, it is evident that the jointing of

FIG. 2

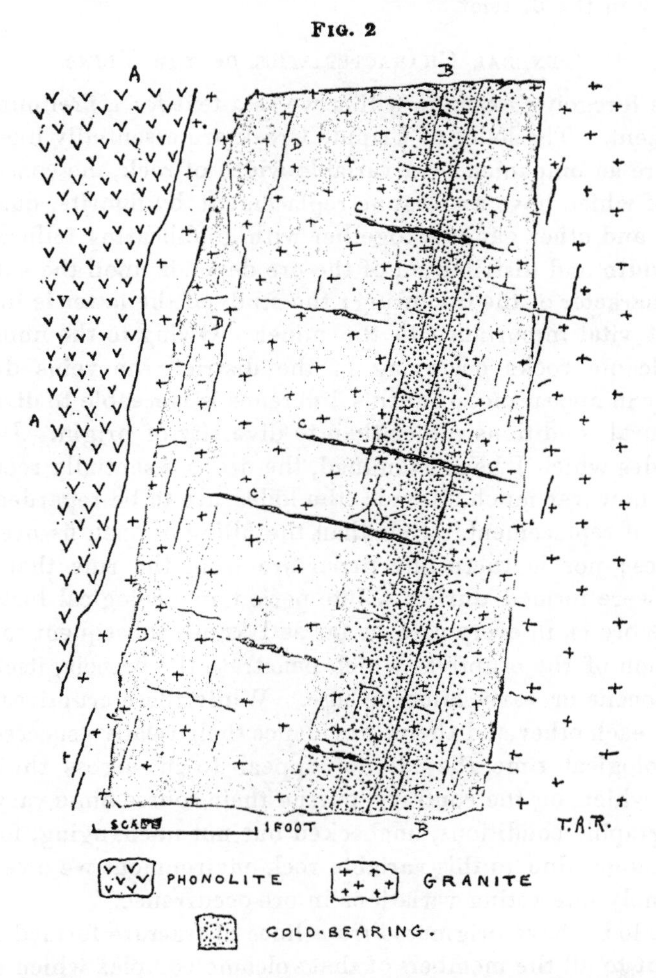

the granite, due to a condition of strain, must have occurred subsequent to the intrusion of the phonolite, because the joints cut through the edges of the phonolite;* finally, it is rendered

* Of course it is possible that the cracks through the phonolite are merely the prolongation of the joints, and that this extension of the latter may have been of later origin; but this is not my interpretation of the evidence as I have seen it.

very probable that the jointing of the granite and the fracturing now identified with the lode were contemporaneous, both being the result of mechanical stresses connected with the earth-movements which followed the last stages of volcanic activity in the district.

D. General Characteristics of the Veins.

The foregoing example will serve as a text for a preliminary statement. The lodes of Cripple Creek are essentially lines of fracture accompanied by a variable width of rock, the constituents of which have undergone replacement by fluorite, quartz, pyrite and other gangue, together with gold-bearing tellurides. The width and distribution of the ore depends upon the extent and character of the fracturing; the study of the latter is therefore of vital importance to the miner. Owing to the number of volcanic rocks occurring in the district, the veins differ greatly in appearance; but this difference is traceable to diverse structural conditions rather than to diversity of origin. In all the lodes which I have examined, the ore is essentially rock in place, however much altered; the lodes are to be regarded as bands of replacement, rather than the filling of open fissures or crevices; nor is there any departure from the rule that the lodes were formed during a late period in geological history, for the ore is, in every case, as far as I know, subsequent to the intrusion of the eruptives which penetrate the breccia, itself of late Eocene or early Miocene age. While these eruptives cut across each other and thereby evidence their relative succession in geological time, they do not appear to cut across the ore-veins, which, on the contrary, pursue their course amid varying petrographic conditions, unchecked but not unchanging, for to the changes due to this variable rock-environment we owe the extremely interesting variations in ore-occurrence.

The lodes have originated from lines of fracture formed subsequent to all the members of that volcanic complex which constitutes the gold-field. These fractures are the outward manifestation of lines of weakness, or of such comparative weakness as is the equivalent of least resistance; therefore, it is not at all surprising that the veins frequently follow the planes of contact between rocks which are unlike in hardness, and especially the very close-grained eruptives when these traverse the coarse-textured breccia.

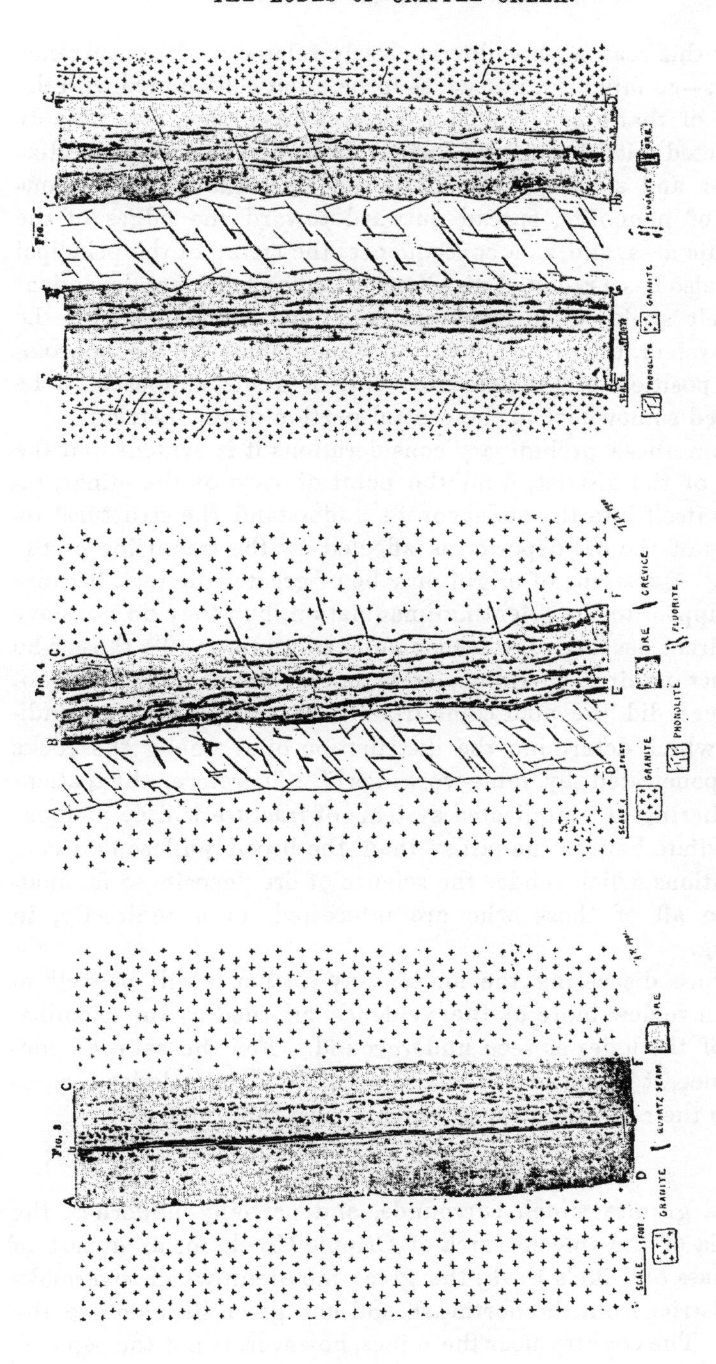

THREE STUDIES OF THE INDEPENDENCE VEIN

For this reason phonolite is closely related to the occurrence of ore,—so much so, indeed, as to be often confounded with the origin of that ore. The later dikes of basalt are also notably connected with important lodes. It is possible to generalize further and state that the dikes, especially the very numerous dikes of phonolite, spread outward toward the edges of the volcanic area, and, as a consequence, the strike of the principal veins also has a radiated distribution. The latter are dependent in their strike upon structural relations coincident with the dike system, the arrangement of which points to the approximate position of the volcanic vent, or vents, supposed to be situated somewhere in the central portion of the district.

From these preliminary considerations it is evident that the study of the district, from the point of view of the miner, resolves itself into the endeavor to understand the structural relations of the ore-deposits as affected by the fracturing of the rocks. Questions of origin may be of greater interest, because they appeal to the scientific imagination, but they do not have any direct bearing on the economics of mining. To those who conduct mining operations it is not so much a question of "Where did the gold come from?" as "What are the conditions which determine the distribution of it among the rocks now penetrated by mine-workings?" Therefore, the patient deciphering of complicated systems of fracture will be of more immediate help to the miner than the broad philosophic considerations which render the science of ore-deposits so fascinating to all of those who are interested, even indirectly, in mining.

Before discussing the matter any further it will be well to pass in review more of the evidence afforded by the examination of the lodes as seen underground. For the sake of convenience, it is considered advisable to divide the lodes according to their encasing rock, which is also their matrix.

E. Lodes in Granite.

The granite which surrounds, and partially underlies, the breccia of the volcanic area of Cripple Creek forms a part of the mass of Pike's Peak, the great mountain which overlooks the district from the northeast, and has given its name to the rock. The country near the mines, however, is not the same as

FIG. 6.

Anaconda Lode as seen in an Open Cut in 1897.

the Pike's Peak granite, orthoclase replacing, to a large extent, the microcline of the typical rock. The Cripple Creek granite may be described as a pink, rather coarsely crystalline rock, the most prominent constituents of which are biotite (black mica) and orthoclase (feldspar), the latter occurring in large, tabular crystals. The color of the rock is due to that of the orthoclase which has been stained red by iron oxide. Fluorite occurs sparingly, but it is interesting on account of its association with the ores of the district. In different portions of the district the granite presents variations, the most important of which is a very fine-grained rock that, in the form of dikes, penetrates the coarser variety, and is evidently of more recent date than the basal rock of the region.*

Fig. 2 illustrates a gold-bearing lode in granite. It represents the heading of the 160-ft. level in the Hallett and Hamburg claims, near Victor, as seen in April, 1897. The ore appears as an ill-defined band, B B, about 1 ft. wide, which is wholly in granite, but at the same time is only 2 ft. distant from a large phonolite dike, A A. The alteration of the granite, which marks the course of the ore, follows a series of short, overlapping seams, parallel to the line of the dike, and also extends into the surrounding rock along the cross-joints, C C. The width of 2 ft. of granite which separates the vein from the dike exhibits partial alteration along the seams, D D, and carries a feeble scattering of ore. The granite under the vein is fresh and unaltered. In the ore-bearing rock the mica is notably absent, the granite is honeycombed by decomposition, and of the two constituent feldspars, oligoclase and orthoclase, the former is kaolinized. Iron pyrite bespatters the gold-bearing portion, and, by its partial oxidation, stains it dark red.

Fig. 3 exhibits the Independence lode in the Washington claim, which is a portion of the Stratton's Independence property. The lode, which farther north traverses the breccia and is closely associated with a phonolite dike, is seen here as a band of decomposed granite subdivided equally by a central thread of quartz. The ore is essentially granite. The portion A C, D F, is 4 ft. wide, and carries a little over 3 ounces

---

* This is discussed by Whitman Cross in "Geology and Mining Industries of the Cripple Creek District, Colorado," *U. S. Geol. Survey*, page 23.

of gold per ton. The width of 4 ft. which carries gold, and is therefore ore, has no parting or wall separating it from the outer rock, which is granite, and is regarded as waste, but is distinguished from the latter in many ways. The outer granite is fresh and unaltered, exhibiting with great clearness its constituent minerals, reddish quartz, black biotite-mica and pink orthoclase-feldspar. The inner gold-bearing rock is much altered by decomposition and replacement; the orthoclase alone appears to have survived the general destruction; the mica has been removed, and, in its stead, chlorite can be seen in green patches; the original crystalline quartz is largely gone, and the presence of purple fluorite suggests that hydrofluoric acid may have been a primary agent in that removal; secondary hydrous quartz fills many of the interstices between the crystalline constituents of the rock;* in iron-stained cavities free gold can be seen by the aid of a pocket-lens, and the gold is observed to have the dark, lusterless appearance which characterizes it when derived from the oxidation of tellurides. The entire width of this gold-bearing decomposed granite is heavily iron-stained by the oxides resulting from the disintegration of the small crystals of iron pyrite, which can still be seen, in an unaltered condition, scattered throughout the same lode, at lower levels. In the center of the band of ore there is a distinct parting, B E, which is separated from a persistent thread of white quartz, only about a quarter of an inch in width, by a slight selvage of red clay. At a distance, the lode appears as a distinct broad band of iron-stained granite, and it is only by closer examination that the boundaries of it are seen to consist, not of "walls" or of any such evident demarkation, but merely of a transition from decomposed into undecomposed granite.

The Independence vein is illustrated again in Fig. 4, which was obtained on the same level as Fig. 3, but 300 ft. farther to the north, where the vein lies against a phonolite dike, A B—

* This description is founded on the examination of hand-specimens by the aid of a pocket-lens. To those who desire to go into the matter further, there is the detailed description of Mr. Lindgren, together with microscopic sections of this very ore (made from specimens which I gave to Mr. Emmons in 1900), to be found in that most important contribution entitled "Metasomatic Processes in Fissure-Veins," by Waldemar Lindgren, *Trans.*, xxx., 578–692, especially page 655.

D E. The thread of quartz, shown at B E in Fig. 3, is to be seen again as a larger, but less regular seam, B E, in Fig. 4. In this case it is characterized by cavities, or "vughs," as the miners call them, which gave evidence,* at the time they were first encountered by the workings, that they had served as water-holes along a line of underground circulation. This may be considered as marking the line of the original fracture which determined the course of the lode. In Fig. 3 it was in the center of the ore; in Fig. 4 it marks the western limit, and separates the ore-bearing granite from the phonolite. The latter is 20 inches wide, and regular. The lode consists, as in the preceding instance, of highly-altered granite, which is richest along the contact with the dike, and shades off eastward (from B E toward C F) along an irregular wavy line, C F, which in no case has any of the characteristics of a "wall" or defined separation between what is gold-bearing ore and what is barren rock.

These characteristics are repeated in other sections which I have sketched underground. The lodes in the granite are frequently remarkable for absence of definition, as was instanced in Fig. 4, this being due to the evenly granular texture of the rock. Within the zone of oxidation the boundaries of a granite lode are made manifest by the red-brown stain, due to the decomposition of iron pyrite; but below the zone of surface-waters it becomes difficult to distinguish country-rock from ore without frequent assays. When the granite has undergone impregnation it is usually porous, by reason of the removal of the microcline and some of the quartz of the original rock. Whatever biotite, hornblende and epidote it contained are absent from the ore, and their decomposition-product, chlorite, is in evidence. Microscopic sections† indicate that secondary valencianite (a form of orthoclase feldspar) and sericite (hydrous mica) have been formed, and that iron pyrite, fluorite and the tellurides have been deposited within the cavities produced by the removal of parts of the granite, and also along the cracks which traverse the rock. Occasional specimens, obtained from the stopes, exhibit these changes on a scale visible

---

* By containing water and by being lined with slime.

† See Lindgren, *Trans.*, xxx., 656, and *Genesis of Ore-Deposits*, p. 576.

without any lens. Thus I possess a piece of ore characterized by large crystals of pinkish feldspar (or orthoclase) and a little silvery mica (muscovite) held together apparently in a cavernous mass of quartz, through which bright specks of calaverite are scattered. The quartz contains spots of green, earthy chlorite and a very few minute cubes of fluorite. White quartz, as a distinct veinstone, does not characterize these lodes to the extent usually observable in gold-lodes elsewhere, although secondary quartz is everywhere found penetrating the altered veinstone. Pyrite is also a constant companion of the gold-bearing tellurides, and fluorite is readily to be seen save where, near the surface, it has been decomposed. The occurrence of fluorite has suggested many theories, but the fact that it forms an original constituent of the granite of the Pike's Peak region and the prevalence, in the same neighborhood, of such fluorine-bearing minerals as cryolite and topaz,* renders it dangerous to draw inferences connecting it with the ore-forming agencies. There is the appearance of probability about the idea that the secondary fluorite of the lodes was derived by the circulating waters from the granite of a lower horizon, and it may be mentioned that Fouqué showed that hydrofluoric acid in a liquid state has a notable effect on silica and silicates by first decomposing the uncrystallized silicates, or glasses, and then acting similarly on feldspars and other acid silicates, then on quartz, and lastly upon the basic silicates. Whether hydrofluoric vapor acts similarly is an open question. Under such circumstances, quartz might be attacked before feldspar.

F. Veins in Andesite and Andesite Breccia.

The andesite of Cripple Creek is usually an augite-mica-andesite. It is distinguished by having apatite as one of its constituent minerals. Although this andesite forms the principal element of the breccia, the latter is notably irregular in its composition. Phonolite is sometimes locally predominant, and near the edge of the volcanic area the breccia contains a large proportion of fragments of granite. The breccia, since

* Florissant and the Pike's Peak region generally are celebrated for specimens of topaz. W. S. Stratton, the discoverer and former owner of the Independence mine, was prospecting for cryolite, as a source of aluminum, just previous to his first trip to Cripple Creek.

it was laid down as a product of violently explosive volcanic eruption, has become decomposed and cemented. According to Whitman Cross, the decomposition has led to the "total destruction of the dark silicates," such as the augite, hornblende and biotite present in the original fragments, and in the removal, by leaching, of the compounds resulting from this decomposition.* The result has been to change the breccia from dark crumbling material into a bleached compact mass which, in process of time, by reason of pressure and waters containing kaolin, silica and other cementing substances, has become consolidated into a hard, massive rock. In the vicinity of the lodes the effects of siliceous solutions are rendered apparent by the impregnation of quartz to such a degree as to obscure the original fragmentary nature of the rock and make the finer-grained breccia, or tuff, resemble phonolite in texture and appearance. The variations in the andesite and andesite-breccia are responsible for corresponding changes of lode-structure, as will be presently illustrated.

Fig. 6 is a photograph of the Anaconda lode, as seen in an open-cut, in July, 1897. The lode at this place forms a part of an andesite dike traversing the breccia. The dike exhibits a multiplication of fractures parallel to its walls, and along these lines of cleavage there occur seams of quartz and fluorite carrying tellurides.† In the surface-workings, the gold liberated from the tellurides occurred pseudomorphic after sylvanite, distributed in yellow patches amid purple fluorite, affording specimens of great beauty.‡

When, as rarely happens at Cripple Creek, the lode consists of massive ore notably separated from its encasing rock, it will be found that such definition of structure is due to the presence of fluorite and secondary quartz which have so filled up the interstices of the decomposed breccia as quite to obscure its original character. Fig. 7 represents the main lode of the Gold King mine, in Poverty Gulch. At the time the drawing

---

* "Geology and Mining Industries of the Cripple Creek District, Colorado," *U. S. Geol. Survey*, page 52.

† The general characteristics of the ores of this district have been separately described. "The Telluride Ores of Cripple Creek and Kalgoorlie," by T. A. Rickard, *Trans.*, xxx., 708–718.

‡ See "Further Notes on Cripple Creek Ores," by Richard Pearce, *Proc. Colo. Scientific Society*, vol. v., pp. 15–18.

FIG. 7

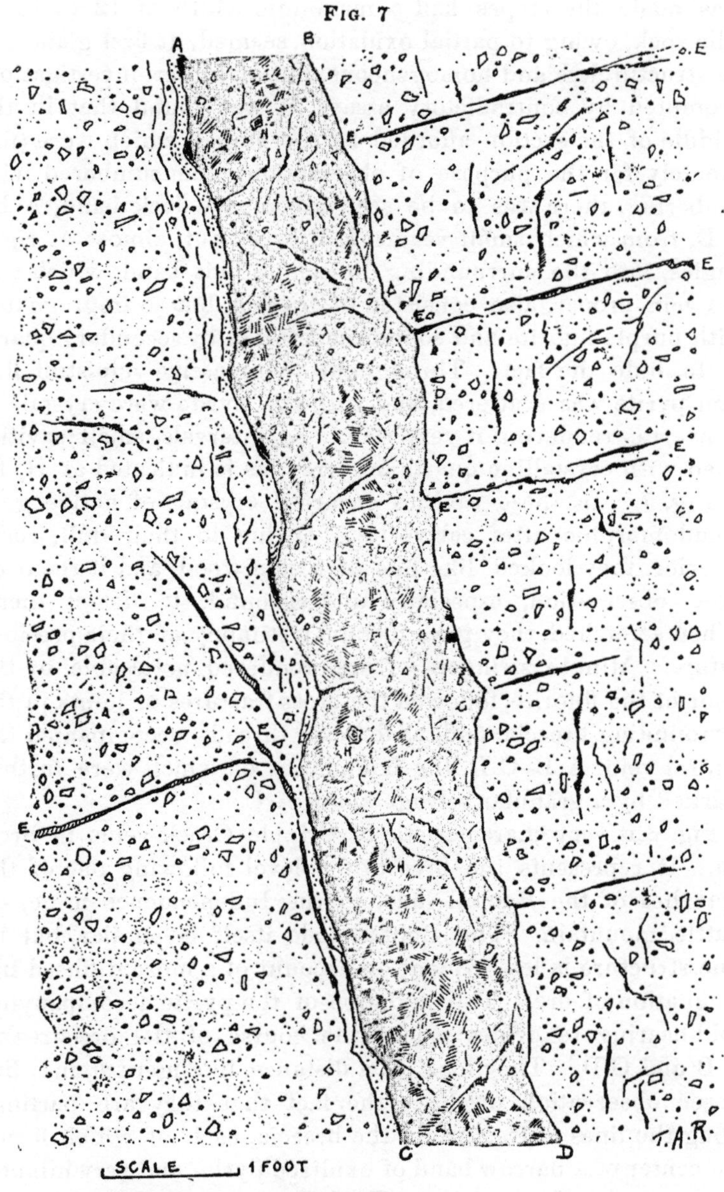
A
B
C
D
E
E
E
E
E
E
H
H
H
T.A.R.
SCALE 1 FOOT
ORE
ANDESITE BRECCIA
FLUORITE
SELVAGE

was made the stopes had a maximum width of 12 to 18 ft. The rock, owing to partial oxidation, seemed, at first glance, to be structureless and homogeneous, but on closer investigation, prompted by contradictory assays, it was found that in the middle of the section afforded by the stopes, which were (fortunately for the purpose of observation) unencumbered with timbering, there was to be seen a compact dark band, A B–C D, from which small seams, E E, went out, almost at right angles, into the surrounding breccia. This band, which was the vein proper, also consisted of breccia, but so impregnated with purple fluorite and so interpenetrated by secondary quartz as to hide the fact. Throughout the massive vein-stuff fine iron pyrite was scattered, and with the pyrite were crystals of calaverite, rendering it very rich. At intervals, small cavities lined with crystalline quartz occurred, as is indicated at H, H, in Fig. 7. In these cavities occurred crystals of native gold pseudomorphic after calaverite. Owing to their dull, rusty exterior, they looked like bits of rotten wood, and it required close observation, especially underground, to detect them. When scratched, they gave instant testimony of their precious nature. Minute stringers of ore, rendered noticeable by the color of the fluorite, followed the cross-fractures or joints in the surrounding breccia, and, as delicate threads, accompanied the central vein, A B–C D, the boundaries of which were further marked by a granular selvage along A C.

Fig. 8 is very characteristic of Cripple Creek veins in breccia. It represents the breast of a level following one of the branches of the Bobtail lode, in the Independence mine, on Battle mountain. The Bobtail lode itself is similar, but its true structure is less evident on account of a more diffused impregnation of ore. The oxidation of the pyrite accompanying gold-bearing tellurides marks the course of the ore-streaks, A B and C D. They look, at a distance, like mere stains; but closer observation discloses the fact that they are partings along the lines of fracture in the breccia, each of which forms the center of a narrow band of oxidized pyrite and very minute, bright specks of calaverite. The latter is seen, under the magnifying glass, to be in process of decomposition, the oxidation of the tellurium of the telluride in the presence of decomposing pyrite having resulted in the formation of the tellurite of

FIG. 8.

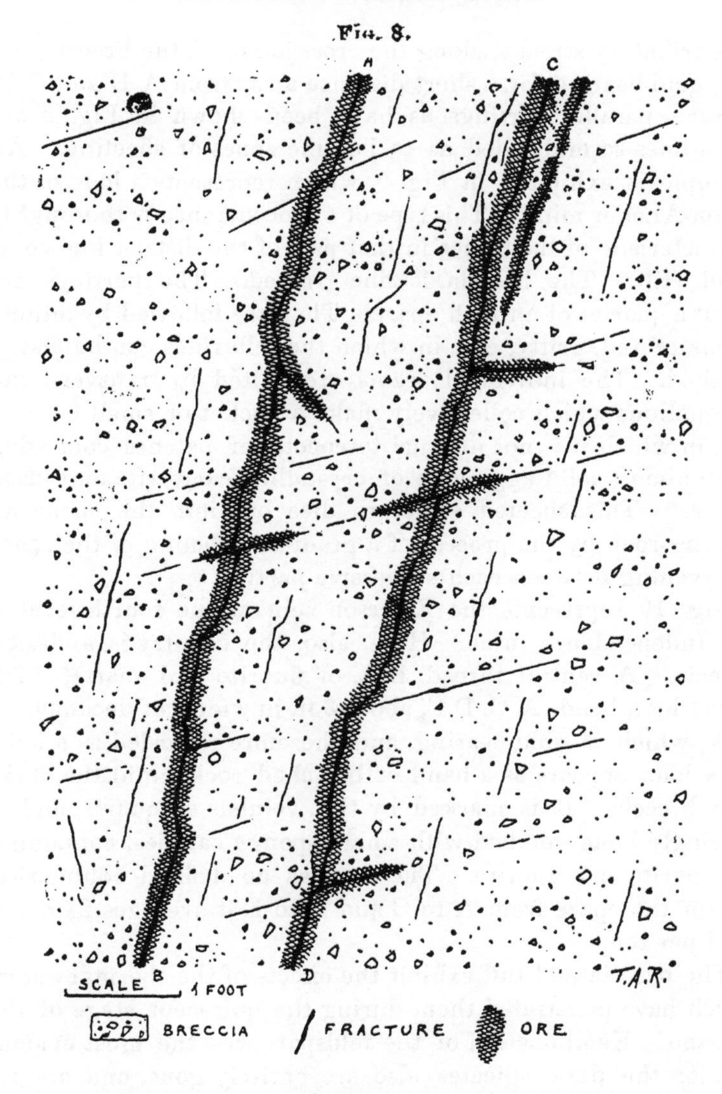

IMPREGNATION ALONG PARALLEL FRACTURES

iron* and the liberation of the gold in a brown amorphous condition, resembling yellow paint which has become tarnished.

---

* This alteration product has a definite chemical composition, as has been determined by F. C. Knight. See "A Suspected New Mineral from Cripple Creek," *Proceedings of the Colorado Scientific Society*, vol. v., pp. 66–71, October 1, 1894. Mr. Knight's analysis gave a percentage of Fe_2O_3, 32.72; TeO_2, 65.45; and H_2O, 1.83. The physical characteristics ascertained were, a light-brown color, a dull luster, a brilliant and uneven fracture, a hardness between 3 and 4, and a bright yellow streak.

The tributary streaks, along the cross-joints of the breccia, are also gold-bearing for a short distance away from A B and C D.

Such parallel partings as have been shown in Fig. 8 are sometimes so multiplied as to become zones of sheeting. An example is exhibited in Fig. 9, which represents a lode in the Moon-Anchor mine. This type of ore-occurrence is thoroughly characteristic of the mines in that part of the district known as Gold Hill. The breccia is fine-grained. The partings are about a quarter of an inch apart. They are followed by minute seams of red, gritty, clay in which the tellurides can be distinguished. The individual seams are united by transverse impregnations which collectively make a pocket or small body of ore, in which it is not unusual to encounter patches consisting of an almost solid aggregate of crystalline calaverite and krennerite.* This sheeted structure dies out into the enclosing country-rock by the process of a gradual widening of the space intervening between each successive parting.

Fig. 10 represents the Emerson vein at the fourth level of the Independence mine. Here, also, the country is andesite-breccia. A central thread, B E, of fluorite and quartz, is followed by a band, A C, D F, about 3 ft. in width, of decomposed rock, which is gold-bearing, and therefore regarded as a lode. This lode appears as a band of bleached rock amid the dark-gray breccia. It is marked by thin veinlets of quartz, and is sparingly honeycombed with small, spongy cavities, containing iron pyrite and fluorite. The ore has no defined boundaries, but in the space from A to C and D to F it averages 2½ oz. of gold per ton.

The breccia and tuff exhibit the effects of the thermal waters which have penetrated them during the quiescent stage of the volcano. Kaolinization of the feldspars was the most evident result; the dark silicates also are entirely gone, and are replaced by white mica.† During the subsequent period, when the ore-deposits were in process of formation, these decomposed fragmentary rocks, already partially cemented by their kaolinization, became further consolidated by siliceous solutions, so as

---

* Krennerite is a telluride of gold, approximating calaverite as regards composition, but differing from the latter in possessing a perfect cleavage. It was named after Professor J. A. Krenner, of Buda-Pesth.

† As observed both by Lindgren and by Cross.

to be changed into a compact hard rock. Underground, near the veins, by reason of the bleaching, due to decomposition, the

Fig. 9.

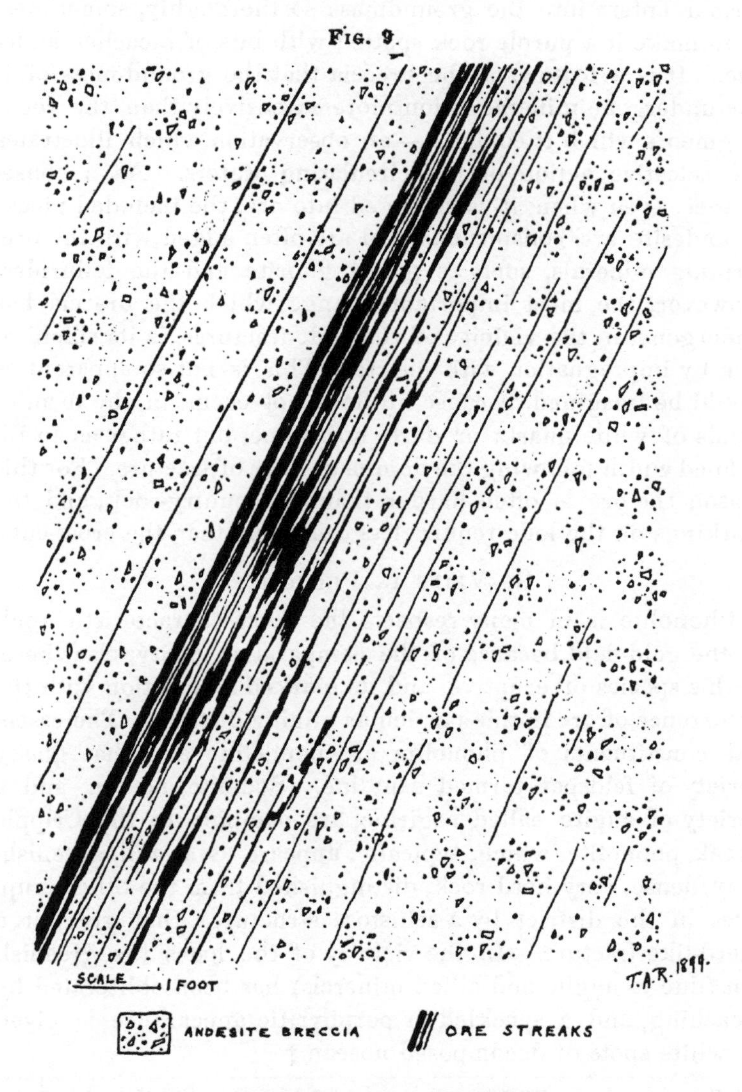

ORE ALONG SHEETED ZONE.

breccia has the mottled look which the miners recognize by the term, "porphyry." As in the case of the granite, secondary minerals are readily found wherever the breccia has been changed into ore and the groundmass of the rock is seen to have undergone substitution by fluorite and pyrite. The former enters into the groundmass so thoroughly, sometimes, as to make it a purple rock spotted with bits of bleached andesite. It is a feature of the breccia that the ground-mass of it has undergone mineralization more extensively than the rock-fragments which it contains,—an observation which illustrates the selective action of the circulating waters. As a consequence, even when it is changed into ore, the included pieces of andesite are conspicuous, and are often edged with the ore-forming minerals, such as pyrite, fluorite and the tellurides. However, the most important change which the breccia has undergone, in the vicinity of the lode-fractures, is its silicification by impregnation with quartz. This is not so apparent as would be imagined, because it does not occur in the form of bands of white quartz or dark hornstone, but rather as an ill-defined width following the dominant lines of fracture. For this reason the ore is often harder than the country-rock, and the workings on the lode require less timbering than the cross-cuts.

G. Veins in Phonolite.

Phonolite is in many respects the most characteristic rock of the gold-field because of the comparative rarity, elsewhere, of this species of eruptive, and its marked association with the occurrence of ore in this particular mining district. The essential constituents of phonolite are nepheline and that glassy variety of feldspar termed sanidine. Sodalite, nosean, and a variety of augite called aegirine, are common to the Cripple Creek phonolite, which, typically, appears as a dull greenish-gray, dense, very hard rock, distinguished from the other eruptives in the district by a schistose structure* that gives it a sherd-like fracture. In the vicinity of the lodes the greenish tint (due to augite and allied minerals) has been obliterated by bleaching, and a speckled or porphyritic appearance is given by white spots of decomposed nosean.†

* Due, according to Whitman Cross, to the fluidal arrangement of the tabular feldspars. *Op. cit.*, p. 33.

† This is especially a characteristic of the so-called Independence dike, as seen at the third and fourth levels of that mine. Professor Judd, F.R.S., from specimens which I sent to him, labelled this rock distinctively a "nosean phonolite." For further discussion of the varieties of phonolite occurring in the region, the reader must refer to Dr. Whitman Cross's interesting descriptions in the "Geology and Mining Industries of the Cripple Creek District, Colorado," pp. 34 to 41.

Phonolite is normally a fine-grained close-textured rock, and for this reason the alteration which it has undergone, wherever it has been in the passage-way of the ore-depositing agencies,

FIG. 10

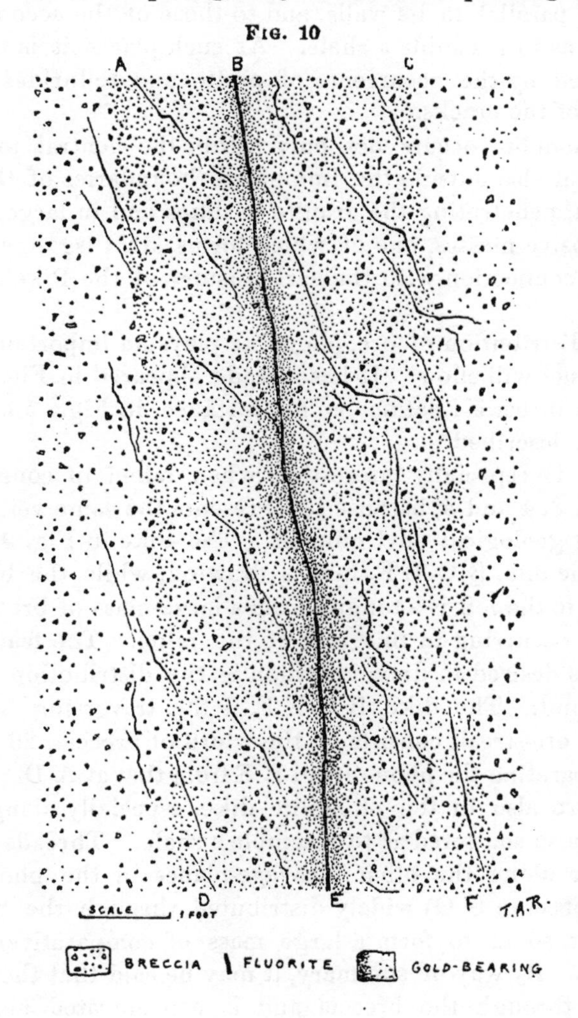

IMPREGNATION FOLLOWING A SINGLE FRACTURE

is rendered more striking. The Independence dike, which accompanies a very rich vein, has been so corroded by the solutions as to be a spongy-looking porphyritic rock, especially wherever it happens to have lain in the path of the ore-bearing solutions which found a way along the vein-fracture. Where not thus altered by impregnation, the phonolite is often so cleaved by cracks parallel to its walls, and to those of the accompanying lode, as to resemble a shale. At such places it is usually mineralized by the occurrence of pyrite and tellurides along the faces of the cracks.

The phonolite occurs not only within the central mass of breccia, but also outside the immediate boundaries of the volcanic area, penetrating the granite in dikes and in large, irregular, intrusive masses, one of which forms Mt. Pisgah, so celebrated in connection with the early history of the Pike's Peak region.*

In the distribution of ore, phonolite plays an important part, as the sequel will show. It has already appeared in Fig. 4, but not in so direct a relation as, for instance, in Figs. 5 and 11, now to be described.

Fig. 11 is especially interesting when taken in connection with Figs. 3, 4 and 5, because it represents the same vein amid a different geological environment. The dike in Fig. 4 is the same as the one, B C–E F, in this drawing, while the body of phonolite to the left is a tongue from a large mass of brecciated phonolite occurring in this part of the mine. The feature to which it is desired to draw attention is the distribution of ore at this point. The dike, B C–E F, is the traversing breccia. The main ore-streak consists of the width of breccia, 26 inches across, separating the phonolite at B E from that at A D. There is some ore also in the phonolite dike, especially along C F, where it is so shattered as to resemble a shale. Threads of ore also occur along the other cleavage-planes in the phonolite, and are (observe G G) widely distributed through the breccia to the east, so as to form a large mass of comparatively low-grade ore. By way of summary, it may be said that the ore is scattered through the breccia and is concentrated near the edge of the phonolite, occurring in the body of the latter only where it happens to be shattered.

In Fig. 5 the Independence vein is again illustrated, as it

* The Mt. Pisgah story is told in my earlier paper, "The Cripple Creek Gold-field." *Proceedings of the Institution of Mining and Metallurgy*, London. Vol. viii., pp. 50–51.

appeared in the raise between the 800-ft. and the 700-ft. levels, in June, 1899. The same dike of phonolite, E F–G H, having

FIG. 11

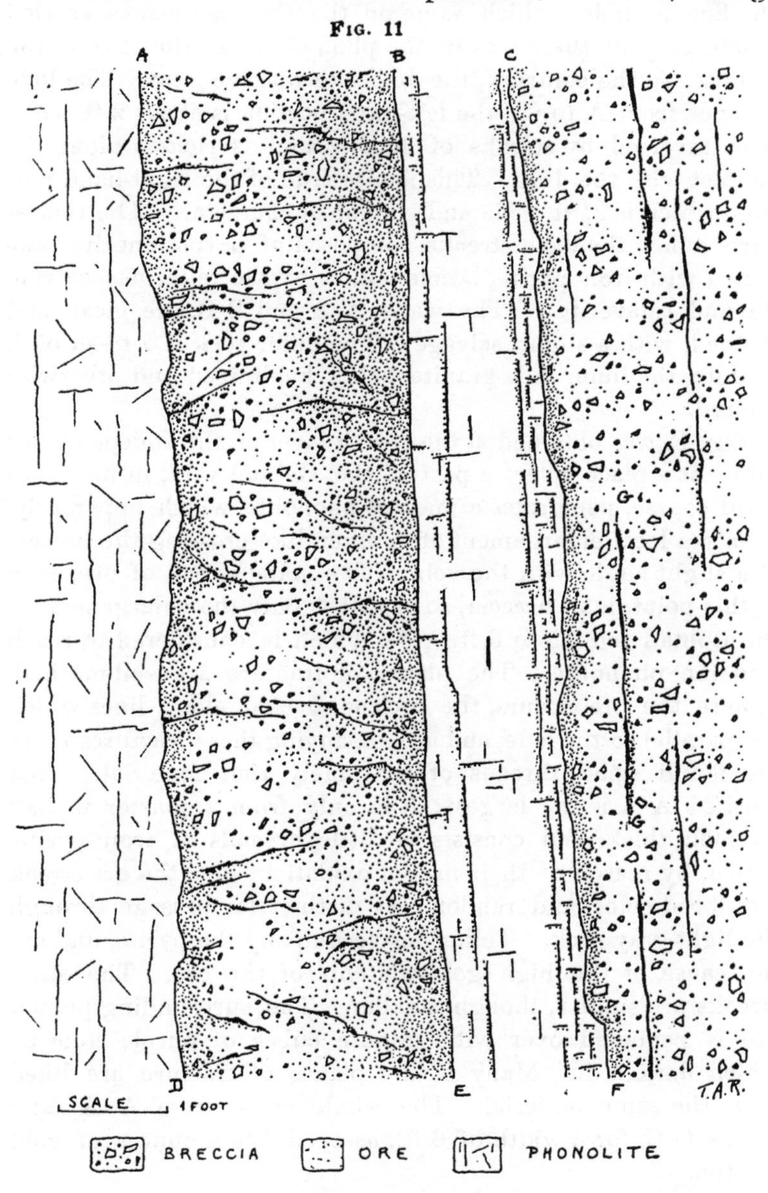

here a width of from 18 to 20 inches, occupies the center of the band of gold-bearing granite, A C–B D, which is the lode.

When sampling the ore, previous to shipment, it was found that the dike only yielded fines or "screenings," indicating that the fine particles, which came off the cleavage-planes, carried whatever gold there was in the phonolite. In the case of the granite, on the contrary, the bulk-ore was the best. The lode extends from A to C; the left-hand portion is about 2 ft. wide, it is traversed by streaks of fluorite, and is richest along the contact with the dike. This is also true of the right-hand portion, which is 2½ ft. wide and similar in character. The richest parts follow the dark streaks composed of purple fluorite associated with iron pyrite, accompanied by the tellurides, sylvanite and calaverite.* The walls, A A and B B, are clean and defined, with a slight selvage. The outer rock is a fresh pink granite, the inner lode-granite being kaolinized and otherwise altered.

Fig. 12 was obtained at the fourth level of the Independence mine, at a place where a part of the Bobtail vein, in its northward course, penetrates a mass of phonolite, which, apparently, is only a local enlargement of another dike crossing the breccia at a slight angle with the vein. The main streak of ore, even at this point, is in breccia, so that it is only the enlargement of the Bobtail lode, 5 to 6 ft. wide, which is considered to reach into the phonolite. The planes of fracture are well-marked. Where the ore occurs, the rock is sheeted along lines which are parallel to the lode, and accompanying these fractures there are found small threads of extremely rich material. That marked A A is the largest; it is only from a quarter to half an inch thick, and consists of little crystals of iron pyrite, which, by reason of their partial oxidation, give the ore-streak a dull-red color, and render very distinct its passage through the light-gray rock. Tellurides accompany the pyrite, and are the cause of the high gold-contents of the ore. The other streaks are similar, though smaller. The surrounding phonolite is peppered over with minute cubes of purple fluorite, which darken it. Many of the planes of fracture are lined with the same material. The whole mass, from A A, and across C C, for a width of 6 ft., assayed 3 to 4 ounces of gold per ton.

---

* And especially a massive granular telluride, found, by Mr. W. E. Ford, to be a variety which, as regards composition, is intermediate between calaverite and sylvanite.

H. Veins in Basalt, Trachytic Phonolite, etc.

Numerous dikes of nepheline basalt, the last product of the Cripple Creek volcano, occur in the district, especially in its southwestern portion. The Raven, Elkton, Anna Lee, Black

Fig. 12.

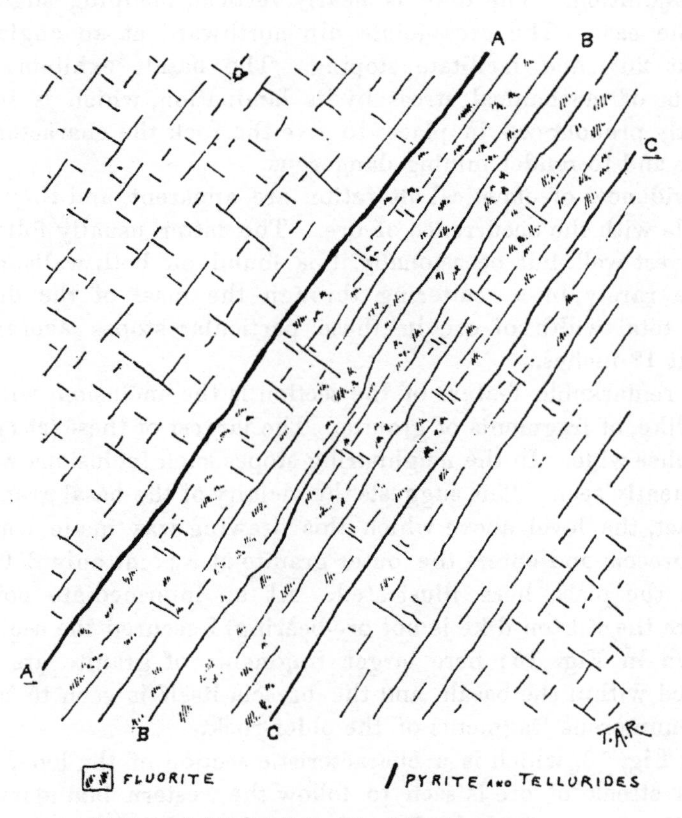

RICH STREAKS IN PHONOLITE

Diamond, Moose, Bertha, Trail and other mines contain lodes which are an integral portion of such dikes. They are usually much decomposed, by reason of their basic constitution, and do not make any showing at the surface. Of the several lodes associated with these basalt dikes, the Elkton is the most interesting. A characteristic section is given in Fig. 13, which illustrates this lode as seen in the stopes above the fifth level, in May, 1899. The dike is nearly 4 ft. wide, from A C to B D, and traverses the andesitic breccia of Raven Hill. It shows a distinct lamination parallel to the walls. This is very strong along the outer edges of the basalt, where it is also bleached and decomposed. The central portion of the dike appears as a hard, dark-gray mottled rock, marked by evident cross-jointing. The dike is nearly vertical, inclining slightly to the east. The cross-joints dip northward at an angle of about 20°, and facilitate stoping. The basalt exhibits the effects of mechanical stress by its lamination, which is sufficiently pronounced in places to give the rock the character of shale and to render mining dangerous.

Evidences of chemical alteration are apparent, and they coincide with the occurrence of ore. The latter usually follows the west wall, but occasionally it is found on both walls, and more rarely in a scattering through the mass of the dike. The total width of ore in these particular stopes averaged about 18 inches.

A remarkable feature of the section is the inclusion, within the dike, of fragments of granite. The largest of these (at Q) is 4 inches wide. In the neighboring stopes such inclusions were frequently seen. This suggests the vicinity of the basal granite. In fact, the level above which this drawing was made leaves the breccia and enters the outer granite at a point only 850 ft. from the place here illustrated. At an intermediate point, where the Elkton dike is not ore-bearing, I secured the section shown in Fig. 15; here larger fragments of granite are included within the basalt, and the breccia itself is seen to contain numerous fragments of the older rock.

In Fig. 13, which is a characteristic section of the lode, the main streak of ore is seen to follow the western boundary of the dike, and to include the rock on either side of that line, so as to obscure it. There is no parting or selvage to mark the line of division between the basalt and the breccia; that is, it is, as a miner would express it, "a frozen contact." The ore spreads across into both. Both alike exhibit the destruction of their original soluble constituents and the replacement by veinstone, especially fluorite. The breccia is bleached by the kaolinization of the feldspar, and is honeycombed with cavities which contain water-quartz in various forms, especially hyalite. The tellurides, sylvanite and calaverite, are scattered through this decomposed breccia and extend into the adjoin-

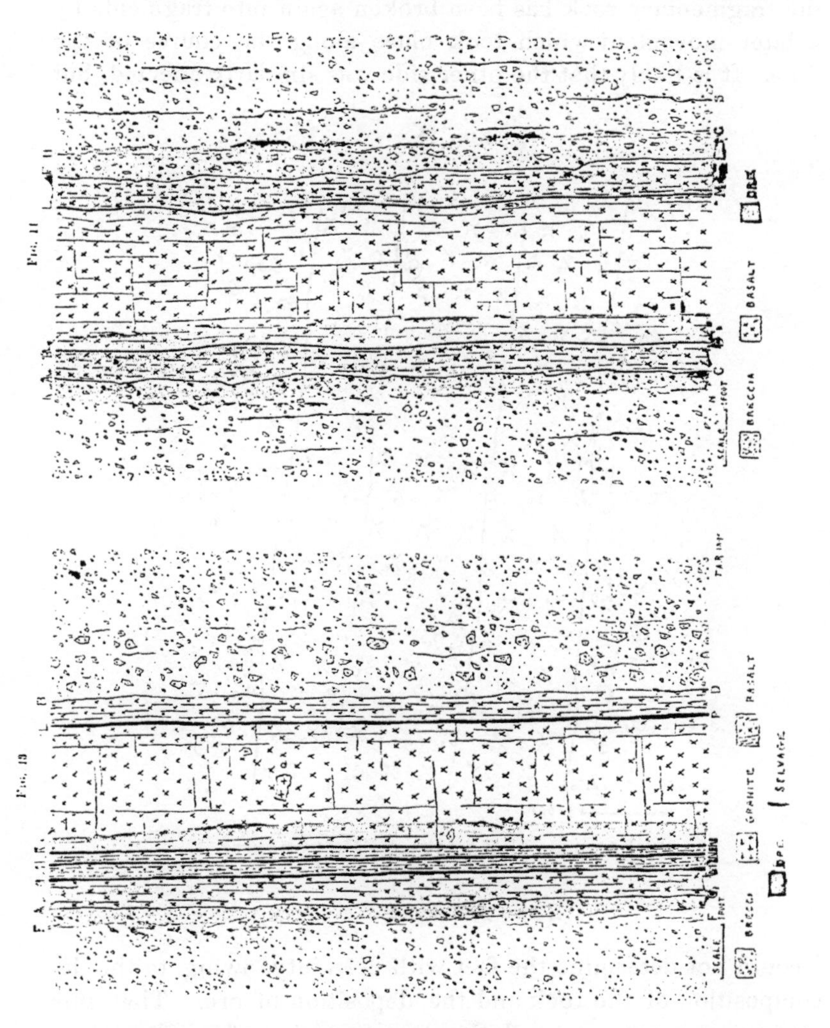

ing basalt. This latter is also bleached to a dull gray, and is seamed with ore along the faces of the cleavages. Although neither the walls of the dike nor the ore-streaks themselves are indicated by selvages, it is noteworthy that several very distinct partings, followed by clay-seams, traverse the basalt in lines parallel to its strike. They are shown at G M, H N, K O and L. P.

This section presents another interesting feature. It will be noticed that on the east side the breccia is brecciated; that is, the fragmentary rock has been broken again into fragments by a later movement which took place along the course of the dike. It is likely that the other side was similarly affected, but

FIG. 15

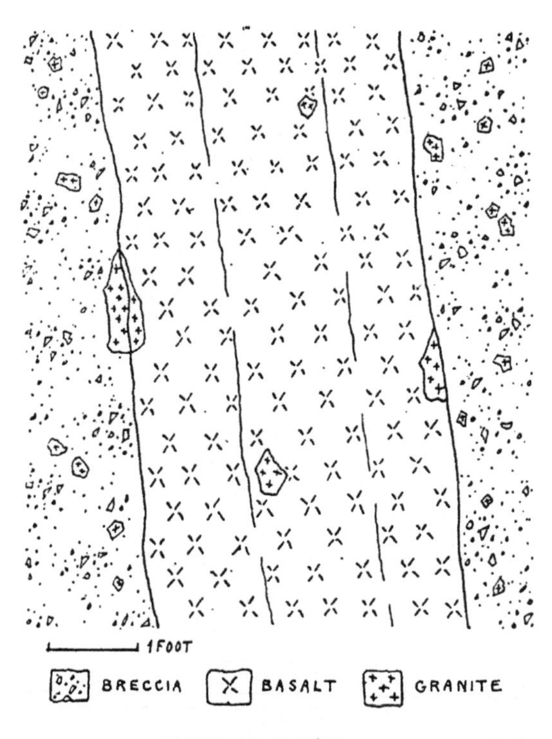

ELKTON DIKE

I could not determine the fact with certainty, owing to the decomposition of the rock and the deposition of ore. That this movement bears some relation to the period of ore-formation is most likely; that it occurred subsequent to the complete consolidation of the dike is rendered certain by another section, shown in Fig. 16, obtained in a neighboring level of the same mine. Here a lateral offshoot, A B, of basalt is seen to be clearly broken by vertical movement. The central portion of the dike is dark green, with secondary chlorite, and is speckled by feldspar phenocrysts. The dike, C D–E F, only 11 inches wide at this point, also exhibits a banded structure along the sides, suggesting a differentiation between the core and its

FIG. 16

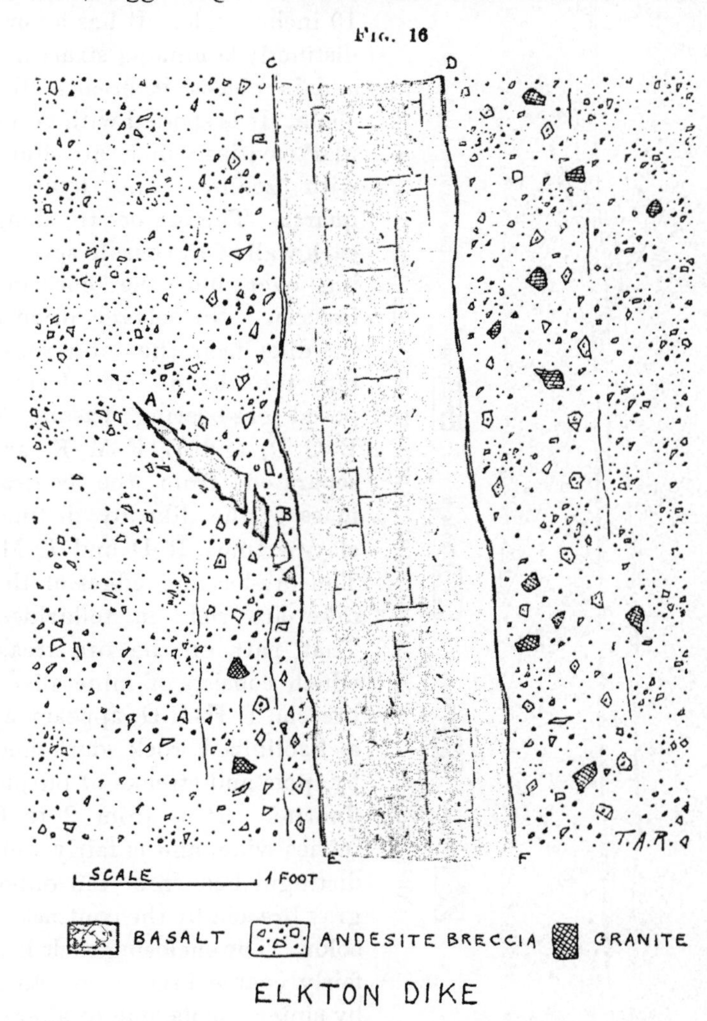

ELKTON DIKE

edges, due to a less complete crystalline development, consequent upon rapid cooling at the time of intrusion into the breccia. The latter is seen to contain numerous fragments of granite, for this section, also, was secured at a distance of 500 feet only from the granite rim.

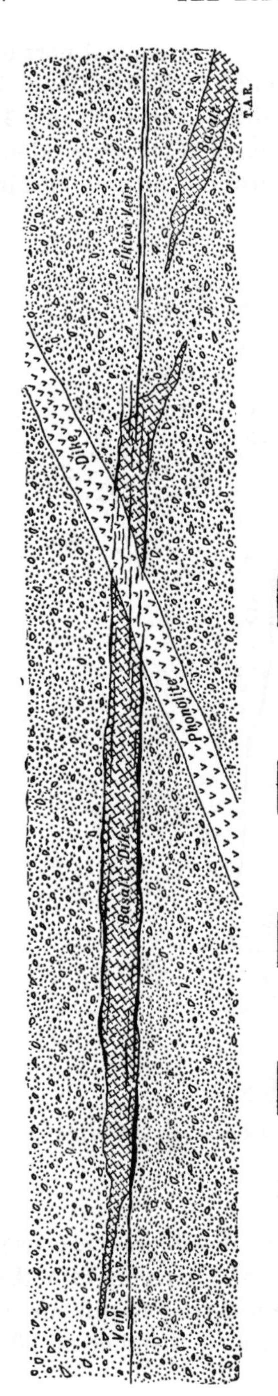

Breccia Basalt Phonolite Vein Dike and Vein nearly Vertical

FIG. 17. COMPOSITE SKETCH SHOWING CHARACTERISTICS OF THE ELKTON VEIN. PLAN

Another section is illustrated in Fig. 14, obtained in the same stopes as Fig. 13, but about 50 ft. farther north. The dike, A C–F E, is 3 ft. 10 inches wide. It has a very distinctly laminated structure, and in places it breaks like shale.. It is spotted with vesicular cavities which are lined with zeolites and hydrous quartz. The ore occurs along each wall of the basalt, spreading over into the encasing breccia. No selvage divides the dike from the outer rock, but the bands of decomposed ore-bearing basalt, A B–C D, and L F–M E, are separated from the central mass of the dike by distinct clay-partings, B D and L M. The ore-bearing edges of the dike are rich in tellurides. That part of the ore-streak which consists of mineralized breccia, F H–E G, appears as a kaolinized rock darkened by spots and streaks of purple fluorite. It is from 2 to 6 inches wide, and is fairly well distinguished from the outer gray breccia by the contrast of color. The enclosing rock is a fairly coarse breccia, marked by sintery spots, due to alteration. These are frequently ore-bearing, by reason of tellurides. The slips or parallel fractures, S S, also carry a little ore upon their faces, and permit the rock to be mined at a profit, because it yields "screenings" or fines, which are rich enough to be sent to the smelter.

In Fig. 17 there is given a characterization of the chief features of the Elkton lode-structure. The vein is remarkably straight; in the breccia it appears as one or more small fractures carrying tellurides, accompanied by chlorite; when the vein encounters the basalt dike, it follows the latter as long as the basalt maintains a direction similar to the strike of the vein; when the basalt is crossed by a later dike of phonolite, the vein-fractures persist across the phonolite, and spread so as to make a large width of gold-bearing rock.

The trachytic phonolite of Cripple Creek occurs in large intrusive masses, which are penetrated by later dikes of phonolite and basalt. It has a decided porphyritic habit by reason of the occurrence of large orthoclase crystals in a dense groundmass. The workings of the Legal Tender (or Golden Cycle) property are in this rock, and afford examples of lode-structure. In Figs. 18 and 19 the Harrison vein is illustrated as it is seen in this mine.

The Harrison vein consists of a band of shattered country in the trachytic phonolite; the center of it is marked by a leader, usually very rich, bordered by fractured rock having a very variable width. This leader is shown at B E in Fig. 18 and A B in Fig. 19.

In the first example, secured just above the 6th level, B E appeared underground as a streak of crushed rock, 3 to 4 inches wide, and dark-red in color by reason of the oxidation of pyrite. The band of brecciated and bleached rock on the hanging-wall side, from A to B and from E to D, contained numerous little threads and spots of pyrite, accompanied by just sufficient gold tellurides to make it low-grade ore. The corresponding band on the foot-wall, from B to C and from E to F, was less decomposed and exhibited more clearly a defined system of fractures, the latter being lined with fine-grained iron pyrite. In Fig. 19, A B appeared underground as a seam, 1 to $1\frac{1}{2}$ inches wide, of white, gritty mud, very rich in gold. The hanging-wall, from A to C and from B to D, is brecciated and ore-bearing, as is the corresponding band, A E–B F, on the foot-wall; but the latter, being less fractured, is also less rich than the other side of the lode.

FIG. 18

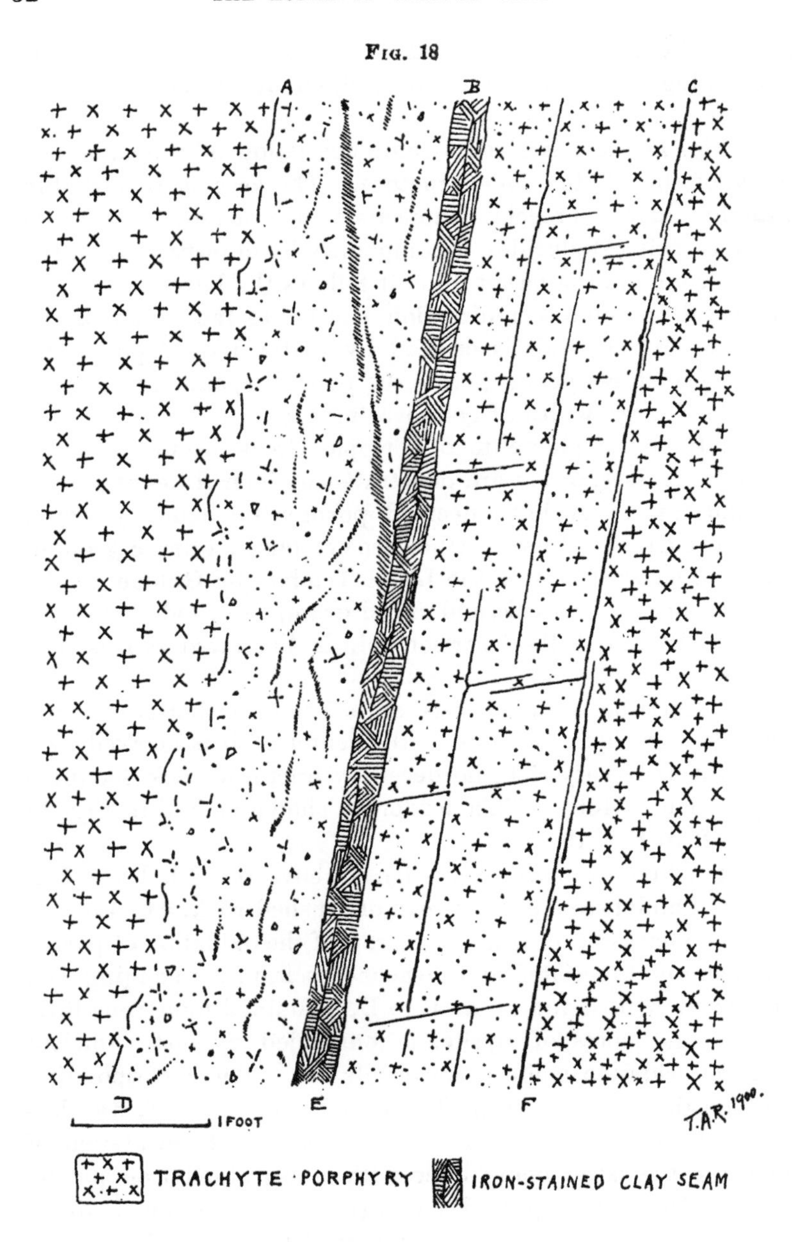

HARRISON VEIN

The widest ore found on the Harrison vein is found at places where spurs or subordinate veins joined the main lode.

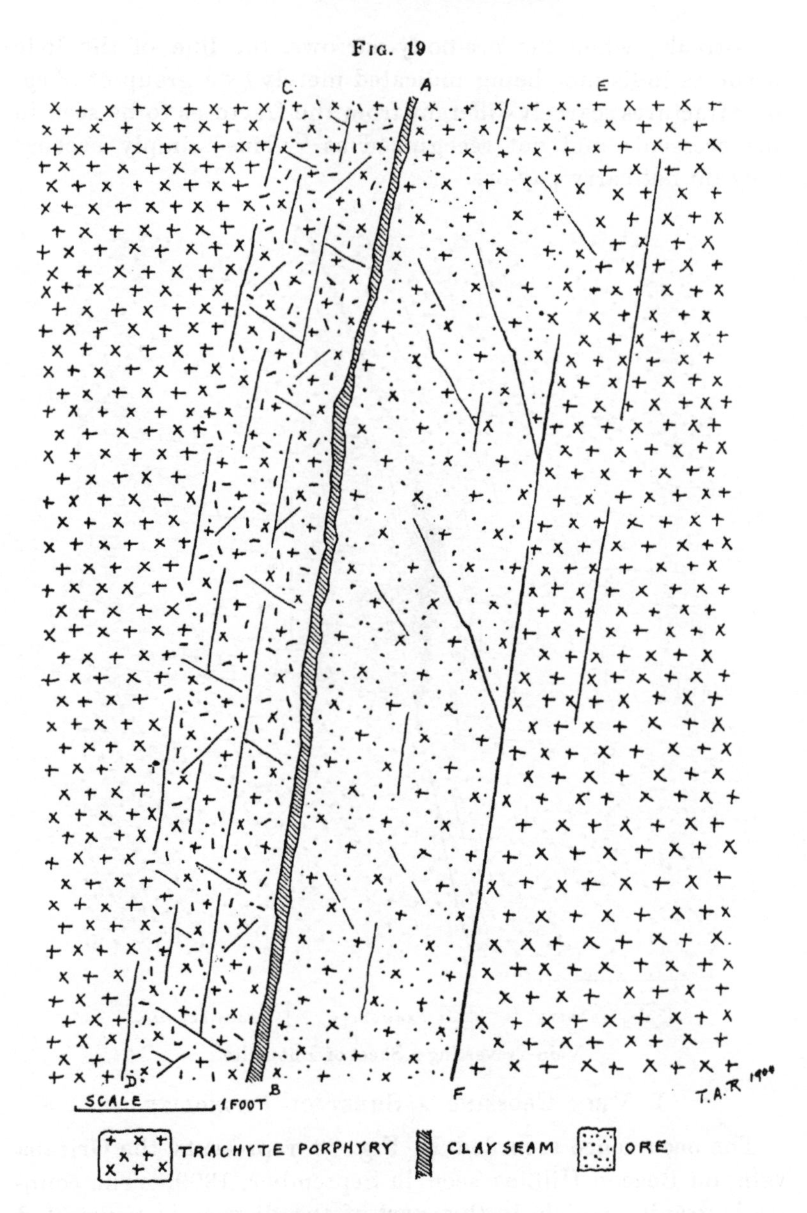

Fig. 19

This gives the ore-bodies a disconnected character, a series of linked enlargements, rather than the appearance of a persistent ore-streak; when the ore-body narrows, the line of the lode becomes indistinct, being indicated merely by a group of irregular fractures scarcely different from the fractures to be seen in the crosscuts, and not recognized as "veins" simply because they do not carry pay-ore.

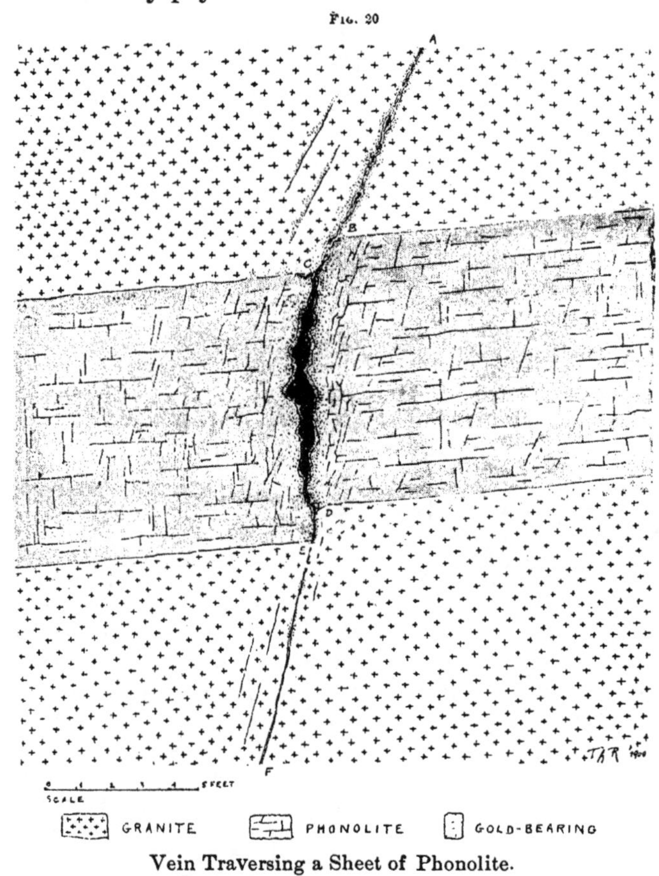

Vein Traversing a Sheet of Phonolite.

I. Vein Crossing a Sheet of Phonolite.

The occurrence recorded in Fig. 20 represents the Orizaba vein, on Beacon Hill, as seen in September, 1899. The country is granite, which, in this part of the district, is penetrated by several sheets and intrusive cores of phonolite. The Orizaba vein cuts through at least two of these flat intrusions, and in doing so presents several interesting features. Fig. 21 illustrates the relation between the geological structure and the mine-workings at 220 ft. below the surface. The phonolite is from 9 to 11 ft. thick, dipping flatly both northward and westward. At the south winze the phonolite is cut at a point 4½ ft. down, and is found to extend thence to a depth of 13½ ft. The remainder of the winze, to the bottom, at 23¾ ft., is in granite. The north winze, which is 77 ft. distant from the other, penetrates phonolite at 13½ ft., and strikes granite again at the bottom, 23 ft. down. The phonolite is faulted about 14 in., B to C and D to E, Fig. 20, by the fracture which marks the line, A B D F, of the vein. In the granite the vein appears as a narrow seam, but in its passage through the phonolite it opens out and forms a series of cavities which are lined with long prismatic crystals of sylvanite, encrusted with quartz, affording specimens of great beauty, and, of course, of extraordinary richness. The tellurides also impregnate the rock encasing the fracture.

FIG. 21

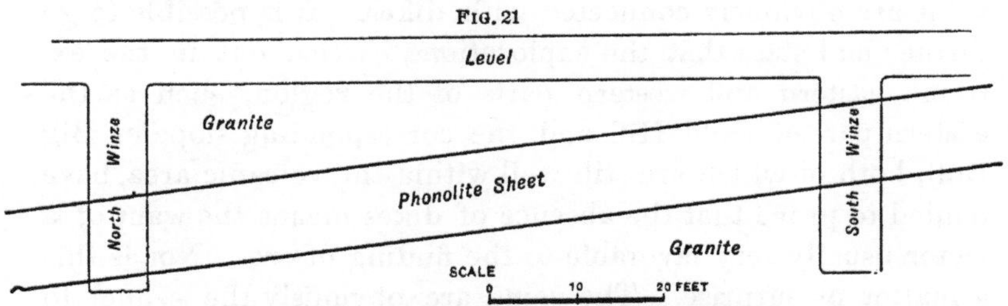

This shows the Relation of the Phonolite Sheet (Fig. 20) to the Workings.

In crossing the phonolite, the lode-fracture straightens up; in leaving it, and passing into the granite again, it flattens. While traversing the 9 to 10-ft. sheet of phonolite, and for a farther distance of 15 to 20 ft. into the granite above the phonolite, the vein carries very rich ore, forming a flat body, the pitch of which conforms to the dip of the phonolite. Farther up, the vein becomes impoverished, until, at a point 85 ft. above the 220-ft. level, it encounters another flat sheet of phonolite, characterized by a repetition of the conditions just described. In the stopes above the 220-ft. level the granite is mined not only for the 4 or 5 inches of vein proper, but also for as much as 5 or 6 ft. into the hanging-wall, which is traversed by telluride threads, parallel to the line of the vein. In the phonolite the vein is characterized by "vughs" or cavities, 2 to 8 inches wide, yielding an average of about 5 inches of rich ore. This rich ore does not continue downward into the granite under the phonolite; the vein thins out, becoming a mere thread amid a series of parallel seams, which give the granite a schistose character for a width of one foot.

J. General Observations.

The occurrence of ore in the Cripple Creek district is intimately related to the distribution of fractures. As a rule, the veins, in their strike and dip, exhibit an evident sympathy with the dikes, more particularly those of phonolite, which are also the most numerous. It is true that the locality of Bull Hill is crowded with a very large number of veins which appear to be independent of dikes, but it is a fact, proved by the experience of mining in the district, that these numerous veins are less persistent and carry ore-bodies which are more uncertain than those, for example, of Battle Mt. and Raven Hill, where the veins are obviously connected with dikes. It is possible to go further and state that the explorations carried out in the extreme eastern and western parts of the region, such as the eastern part of Gold Hill and the corresponding slope of Big Bull, both of which are still well within the volcanic area, have tended to prove that the absence of dikes means the want of a factor usually very favorable to the finding of ore. Nor is this a matter of surprise. The veins are obviously the sequel to the volcanic activity which occurred in this region, and it is a reasonable deduction that the agency of ore-precipitation was linked to that of the thermal waters which marked the last stage of the dying volcano.

The principal veins, such as those which have made rich mines out of the territory controlled by the Independence, Portland, Strong, Gold Coin, Granite, Anaconda, Elkton, Gold King and other companies, either follow dikes or have a course lying closely parallel to them. It is noticeable that the later fracture constituting the vein is apt to be straighter than the older line of fracture occupied by the dike; so that the vein may be compared to a road, alongside a river, which avoids the excessive bends of the latter and keeps a course as straight as is consistent with a given general direction, namely, that of the river. The Independence main lode illustrates this observation, as the accompanying plan will indicate. (See Fig. 22.)

This exhibits a portion of the first level where the vein is mostly in the granite. It will be seen that when the vein-fracture encounters the phonolite dike it follows the latter; or, looking at the occurrence from the opposite standpoint, when

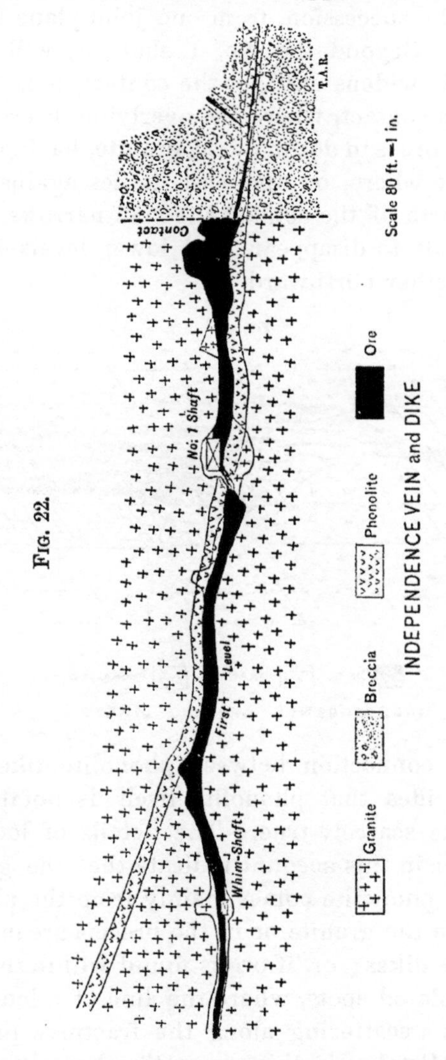

FIG. 22.

the phonolite makes a sharp turn to the west, the vein maintains its general strike. Near No. 1 shaft the vein crosses the dike. This is shown in greater detail in Fig. 23, at E–D, where it is evident that the vein-fracture persists in its course despite the fact that it has to cross the sharp bend made by the dike. At the place of crossing, E D, there are one or two small stringers of ore which serve to connect the lode on either side of the phonolite. Between C and E the ore, here 4 ft. wide, narrows down, by steplike succession, from one joint-plane in the granite to the next. Beyond the No. 1 shaft, as will be seen in Fig. 22, the lode widens until, at the contact, it is 10 to 15 ft. across. At this contact, where the overlying breccia rests on the granite, the ore is in decomposed granite, having no marked boundaries save where, on the east, it lies against the dike. Immediately north of the contact the vein narrows and crosses the phonolite only to disappear. At lower levels it has been traced much farther northward.

Fig. 23

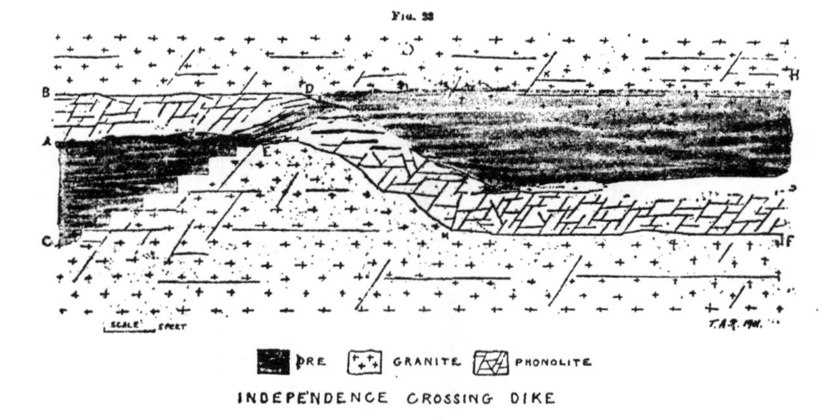

INDEPENDENCE CROSSING DIKE

The obvious connection between phonolite dikes and veins has led to the idea that phonolite itself is notably an ore-carrier. This is scarcely true. The details of lode-structure described earlier in this account indicate that the gold-ore obtained from the phonolite comes mainly from the places where the ore-bodies in the granite or in the breccia are in immediate contact with the dikes; or, if ore is found within the phonolite itself, it is at isolated spots, where the line of a lode crosses a dike and makes a scattering along the fractures produced by the crossing. Mistakes in this connection have been made by confounding phonolite with certain finer-grained breccias and tuffs which, by impregnation with quartz, have put on the appearance of a close-textured crystalline rock. This happens frequently in the mines on Gold Hill. Where valuable ore is really taken from phonolite, it will be found that its true character is sufficiently indicated by the fact that it is secured in the form of "screenings." The fine, powdery material obtained by passing the ore over a wire-screen (which separates all the larger fragments, and concentrates the small proportion of rich ore), occurs as an encrustation, of fluorite and tellurides, which lines the numerous cracks of a phonolite dike. That shattering of the dense, fine-grained siliceous rock appears to be an essential factor in the deposition of the ore. In this connection it is a notable fact that the phonolite dikes are found occasionally to widen into tongues or cores of large dimensions, which, if they lie on the strike of a series of veins, are so much shattered as to permit of a generous dissemination of ore. Such an ore-body, of noteworthy size and richness, occurs in the Independence mine at the second, third and fourth levels, pitching northward more flatly than the contact with which it was originally supposed to coincide, causing a confusion of ideas very detrimental to the development of the mine.

The chief characteristic of the ore-bodies of Cripple Creek is that they are essentially impregnations spreading outward from lines of fracture; therefore, it is not surprising that the distribution of ore is affected by the changes in country. Prof. Penrose has pointed out that "the character of the fissures of the district is much affected by the nature of the rocks they intersect,"* and it is but a further step to connect this observation with the distribution of ore. In discussing this aspect of the inquiry it is preferable to avoid the use of terms such as "fissure," with its old associations of open crevasses and gaping cavities, and to risk the weariness of iteration by employing the word "fracture," which goes no further than to suggest dislocation or breaking without necessarily bringing in the idea of an open space, because the experience of mining in the Cripple Creek district is all against the theory that open spaces are the necessary adjuncts to ore-deposition; and Mr. Becker's well-known dictum, made in connection with the Comstock, that "the first condition for the formation of a quartz body is

* "Geology and Mining Industries of the Cripple Creek District," *U. S. Geol. Survey*, p. 143.

an opening to receive it,"* is daily contradicted by the underground workings of this district, as it has been by many others with which I am familiar. The open spaces which are frequently encountered, especially in the mines situated on the south slope of Bull Hill and on Ironclad Hill, are, as a rule,† notably unfavorable to the finding of ore; and while certain veins, among which the Bobtail, in the Independence mine, may be instanced, are indeed marked by frequent "vughs" or cavities, these cavities do not contain ore, nor is their presence a factor in connection with the distribution of good ore; quite otherwise, all such "pot-holes" are disliked by the mine-foremen, because they coincide with lean places; and by this is understood not only poverty in respect of gold-bearing minerals, but also an absence of gangue, such as quartz, fluorite, and the other constituents of the ores of the region. This all goes to show that the term "fissure" and the associations which go with it are to be avoided in an attempt to convey the real nature of the lode-structure, because it cannot be emphasized too much that the ores of the district occur as a dispersion into the rock where it is traversed by lines of fracture, not only in the fracture itself, which is often only a mere parting, but also along the minor cracks and porosities of the enclosing rock.

The physical and chemical characters of the rock—the first more than the second—are, consequently, a primary factor in determining the shape and extent of the impregnation which constitutes the lode. Close-grained rocks are apt to be more fissile and less porous; coarse-textured rocks are likely to break in a larger, more irregular manner, but they are often more penetrable by solutions. The phonolite and the breccia exhibit extreme divergence in this respect. Granite resembles the breccia because it is coarsely granular, but its jointings are more regular; basalt resembles the phonolite in texture, but,

---

* Quoted from the Comstock monograph.

† An interesting exception was recently encountered on the seventh level of the Elkton mine, where a cavity, having maximum dimensions of 20 x 20 x 35 ft., was found to contain ore and water. The latter was struck in such quantities as seriously to impede operations; the ore occurred in a mass of brecciated rock, lining the fractures and penetrating the shattered country, especially near the periphery of the mass. It looked like a thermal spring which had become choked before reaching the surface.

on account of its basic composition, the weak places in it are more readily searched out by corroding waters.

This "structural dependency" (as it may be termed) of the lodes in relation to the rock is manifested not only by the character of the vein-fractures but also in the manner of their impregnation by ore. Thus, while the bands of ore in granite are often wanting in walls, yet on the whole they are less lacking in definition than the breccia, because structural lines in the granite, such as joint-planes, have served as barriers to indefinite impregnation. On the other hand, the breccia, not being composed of a crystalline granular material, but being built up by fragments which are confusedly mingled together, has no defined system of joints, and the width of the impregnation which constitutes the ore will be determined by purely local conditions of fracture; the lodes may have no walls, and, when they are limited by such boundaries, these are apt to be cracks sympathetic with the main vein-fractures and parallel to them. In the phonolite the limits are set up by the sheeting of that rock; the ore, which is infrequently found in phonolite, occurs then as a lining or powdery encrustation upon the faces of the laminæ, and not, as a rule, within the matrix of the rock itself. As the lamination is parallel to the walls of these dikes, the width of ore also has a shape conforming to them.

Faulting on a large scale is conspicuously absent in the mine-workings; that is to say, there is no evidence of extensive movement since the ore was deposited. Penrose has recorded his opinion that "the fissures represent fault-planes of slight displacement."* The amount of this displacement is not measurable, because, as a matter of fact, the multiplicity of the fracturing and the subsequent precipitation of ore has obscured it most effectually, and it is this character of multiple fracture which pervades the entire structure of the volcanic area. To what extent these fractures are merely shrinkage-cracks and to what extent they represent radical movements due to the readjustment around a volcanic orifice, I cannot say. In so far as the question concerns ore-deposition, one can emphasize the fact that all breaks, from a crack to a *crevasse*, are the outward and visible signs of displacement; for without displacement

---

* "Geology and Mining Industries of the Cripple Creek District," *U. S. Geol. Survey*, p. 153.

the fracture is only latent, and a latent fracture has no possibilities for ore-deposition and does not concern the mining geologist. Since the ore was formed conditions of comparative rest must have supervened, for this is a direct inference from the comparative absence of faults. This fact, taken with the known age of the volcano, which broke out so late in geological time as the end of the Eocene or early in the Miocene period, points to the recency of the agencies which made the ore-deposits.

Note by the Secretary.—Comments or criticisms upon all papers, whether private corrections of typographical or other errors or communications for publication as "Discussions," or independent papers on the same or a related subject, are earnestly invited.

SUBJECT TO REVISION.

[TRANSACTIONS OF THE AMERICAN INSTITUTE OF MINING ENGINEERS.]

Vein-Walls.

BY T. A. RICKARD, DENVER, COLORADO.

(Pittsburgh Meeting, February, 1896.)

FROM time immemorial the fissure-vein has been held the simplest type of ore-deposit. The prominence given to it by Cotta and his disciples, from their study of the mines of the *Erzgebirge*, is impressed upon technical literature; and, in in consequence, the ores which carry the valuable metals have been supposed to occur mainly in fissures, cleaving the rocks in diverse directions, and the noblest type of vein has been deemed that which cut across the country independent of its structure, whether evidenced as bedding, foliation or cleavage, and which was identified with rents produced in the rocky crust of the earth.

As so conceived, the vein was a fissure filled with ore, extending through the country for a varying distance, and continued downward to a depth more or less proportionate to its longitudinal extent. The vein-material was bounded by an encasement of rock, and those immediate surfaces which limited it on either side were called "walls."

These primary conceptions have become modified by the experience of modern mining in widely separated regions. The study of lode-formations has led to the recognition of notable departures from the supposed normal structure of the veins of Saxony and Cornwall, the two classic homes of early economic geology.

Typically the walls of a vein are conceived as parallel rock-planes enclosing the ore; the upper one being called the "hanging," and the lower the "foot-wall."*

Walls are rarely alike. Even where a vein traverses a homogeneous formation, such as a massive crystalline rock, it is usually found that the surface which bounds it underneath

* The French equivalents are *le toit*, "the roof," and *le mur*, literally, "the wall." In German, *das Hangende* and *das Liegende*.

differs from that which limits it overhead. This is to be ascribed to the effect of the agencies which brought about the deposition of the ore. The action of underground waters tends at first to affect both equally; but in many cases probably the solutions, as they slowly ascend along the line of fissuring, are prevented from penetrating into the encasing rock by the occurrence of an impermeable covering of clay, due to abrasion, which may line either wall, but, because of gravity, generally accompanies the under one. Similarly we are justified in supposing that the deposition of a mineral deposit may form a coating which would serve to protect the foot-wall from the corroding effects of chemical action. The activity of the mineral-bearing current thus becomes diverted in its greatest intensity toward the upper wall, where the decomposition of the rock-surface may be followed by its disintegration so as to cause the exposure of fresh faces for further dissolution.

Illustrations of these conditions may be seen in Figs. 1 and 2. The first is reproduced from a sketch made June 25, 1895, in the lower level of the Union and Companion mine at Cornucopia, Union county, Oregon. It represents the breast of the north drift on the west vein. The country, a fine-grained granite, is not visibly altered under the foot-wall; but along the hanging it exhibits an alteration of its more soluble ingredients. There is a slight selvage, D, separating the granite from the pay-ore, C, which is about 10 inches thick, and consists of ribbons of quartz, impregnated with pyrite and alternating with strips of altered country. A distinct parting, unaccompanied by any apparent selvage, divides this streak of ore from one, B, below it, which is twice as thick, but much less gold-bearing. This part, B, of the vein consists of white quartz, carrying occasional patches of pyrite, and marked by large inclusions of slightly altered country, arranged along the foot-wall, where a thin selvage separates them from the outer granite. The evidence of vein-structure embodied in this figure permits diverse interpretations. The upper pay-streak, C, appears to me to be country-rock, in place, decomposed, fractured, and silicified, with accompanying precipitation of gold. The central wall may have been the original hanging-wall. The present foot-wall is sufficiently distinct; but the occurrence of the pieces of enclosed country leads me to believe that at an earlier stage

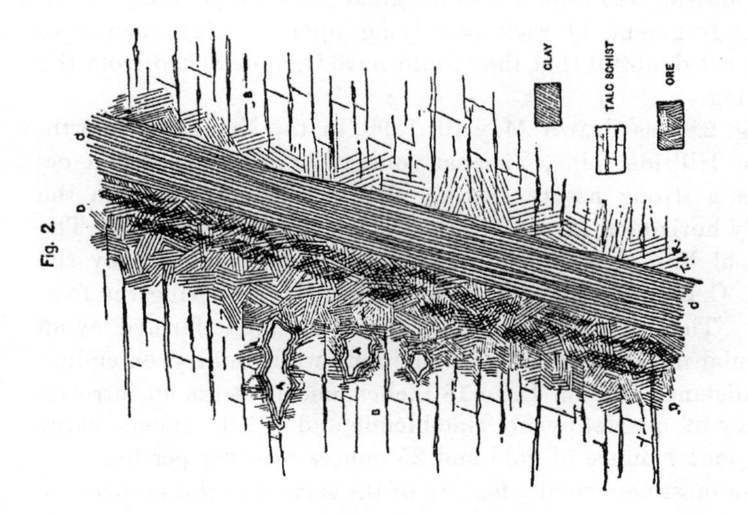

HILLSIDE MINE, ARIZONA.

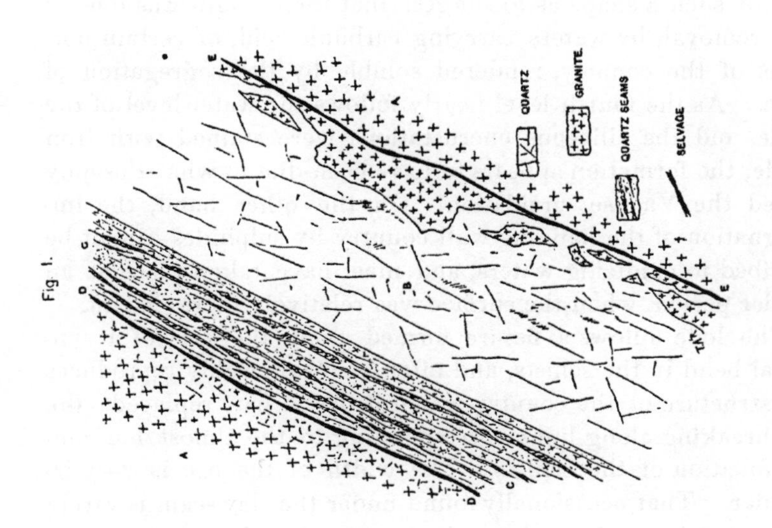

UNION AND COMPANION MINE, OREGON.

the foot-wall was broken and irregular; the shape and position of the fragments of rock now lying upon it being such as to render it doubtful that they could have been detached from the hanging.

Fig. 2* was drawn May 10, 1893, in the No. 4 level, north, of the Hillside mine, Yavapai county, Arizona. The lode occupies a strong fissure, cutting almost vertically through the nearly horizontal layers of a quartzose talc schist, B B. The original line of fracturing is probably now occupied by the seam, C, 6 inches thick, of white talcose clay, covering the foot-wall. The ore-bearing portion, D, of the lode is formed by an irregular mineralization of the hanging-wall country, extending to a distance of from 15 to 18 inches, and presents an intricate medley of quartz, pyrite, zinc-blende and a little galena, carrying about 1 ounce of gold and 25 ounces of silver per ton.

The most noteworthy feature of the section is the occurrence in the hanging, on the outer confines of the main ore-streak, of several irregular cavities, A A, whose inner surface is covered by a series of siliceous coatings, evidently deposited by mineralizing waters that have circulated through them. Along the outcrop of the lode, at Wikiup Point, there occur hollows in the schists, of a character similar to those above described, and of such a shape as to suggest that their origin was due to the removal, by waters carrying carbonic acid, of certain portions of the country, rendered soluble by the segregation of lime. As the fourth level nearly follows the water-level of the mine, and the siliceous encrustations were stained with iron oxide, the formation appears to have been due to what Posepny called the Vadose circulation. On the other hand, the impregnation of the hanging-wall country by sulphides cannot be ascribed to oxidizing waters, and must have taken place at an earlier period, when the surface was relatively more distant.

The lode follows a fissure formed along the axis of a synclinal bend in the schists, and often very noticeably reproduces the structure of the country which it has in part replaced; the ore breaking along lines corresponding to the almost horizontal foliation of the schists. The width of the ore is very irregular. That occasionally found under the clay seam is rarely rich enough to mine; the main pay-streak being that portion

* See also *Trans.*, xxiv., 945.

of the vein bounded underneath by the clay and extending into the hanging until the mineralization becomes so meager that "ore" becomes "country-rock."

Fig. 3.

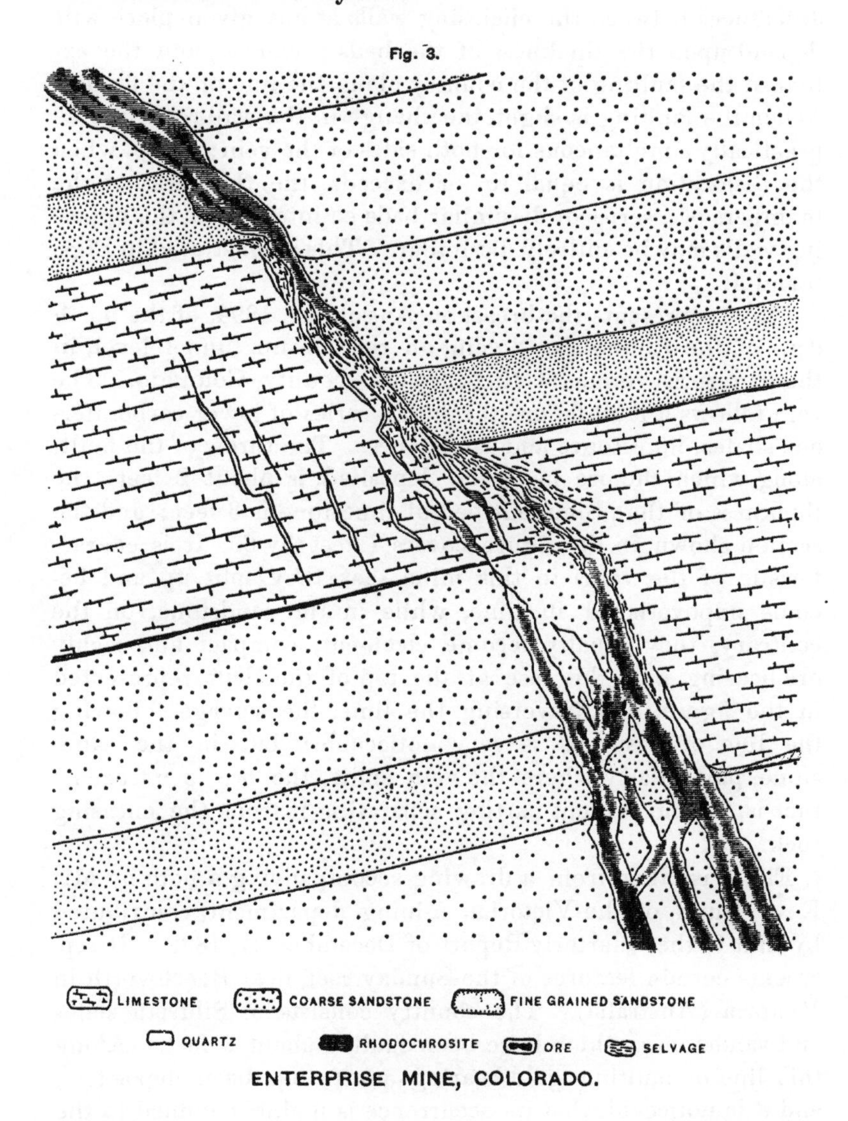

ENTERPRISE MINE, COLORADO.

When a vein occurs in a formation composed of several kinds of rock it may cut across the lines of parting and be labelled a "true fissure;" or it may conform to them, and become a "bedded vein," if the two beds happen to be similar, or a "contact-vein," if they are dissimilar. It is evident that, when a vein crosses the bedding of a series of sedimentary rocks, the differences between the enclosing walls at any given place will depend upon the thickness of the beds traversed, and the extent of the faulting of the country along the line of the fissure. When the faulting is slight, the change in the wall-rock will be practically simultaneous for both sides of the vein; while, when the dislocation is equal to, or exceeds, the thickness of the members of a series of dissimilar beds so intersected, the opposing walls may be entirely dissimilar. This is illustrated in Figs. 3 and 4.

Fig. 3 represents the breast on August 14, 1894, of the north drift of the Jumbo No. 2 vein, on the Group tunnel level, in the Enterprise mine, at Rico, Dolores county, Colorado. The vein follows a fault-fissure through a series of lower carboniferous shales, limestones and sandstones. The throw of the fault, along which the ore has been deposited, is about 2 feet; the thickness of the prominent bed of limestone is 3 feet; and the section shown in the figure covers 7 feet by 6. It is characteristic of the veins in this mine that they split up and become impoverished in lime, while in the sandstone, on the contrary, they usually become clean-cut, compact and richly ore-bearing, as is the case at the top of the drift represented in the figure. In traversing the lime, the selvage following the line of fissuring is very noticeable; but in the sandstone, particularly where the vein splits, the ore is "frozen," that is, has no evident parting separating it from the encasing rock.

Fig. 4 is taken from a drawing accompanying a note by Mr. E. J. Dunn, of the Victorian mining department, contributed by him to the Quarterly Report of December 31, 1888. It represents certain features of the Sunday reef, near Beechworth in Victoria (Australia). The country consists of Silurian slates and sandstones, which have been faulted about 2 feet. Along this line of faulting gold-bearing quartz has been deposited; and it is noticeable that its occurrence is mainly confined to the under side of the sandstone, while under the slate it disappears and gives place to fluccan or clay. I would suggest that the lenticular shape of the quartz-bodies indicates that the spaces

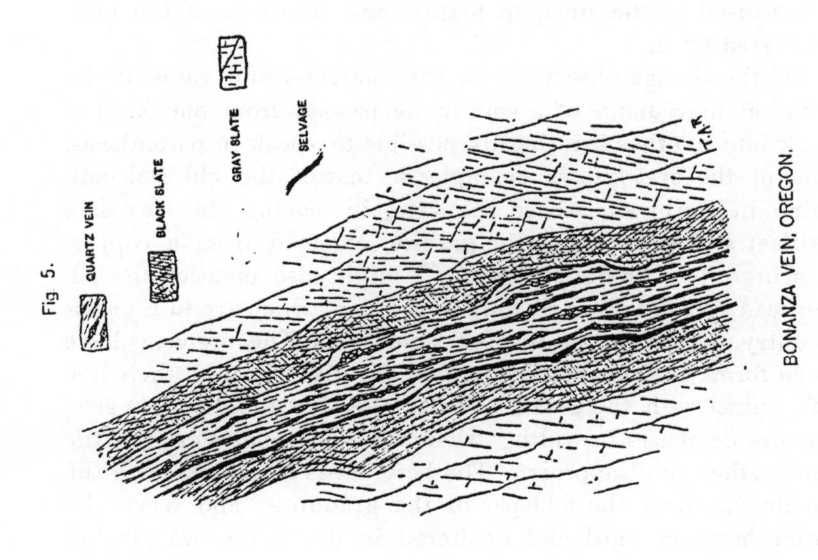

Fig. 5.

BONANZA VEIN, OREGON.

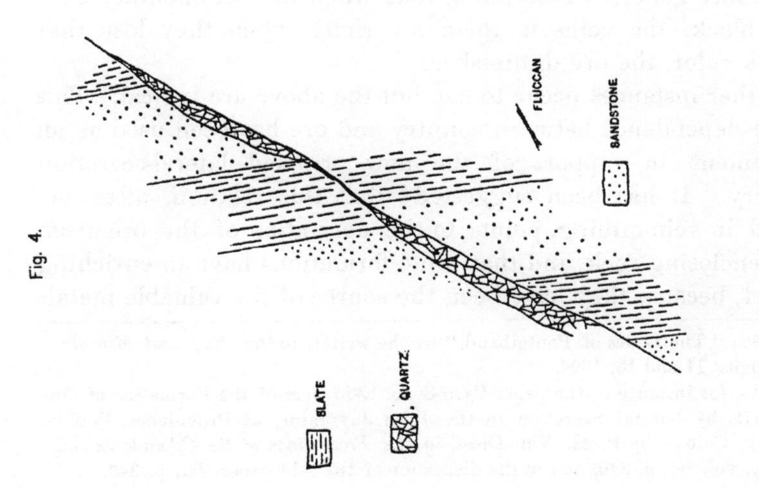

Fig. 4.

SUNDAY REEF, VICTORIA

occupied by them were produced by the movement of one of the walls of a fissure, following a line whose undulatory form was caused by the unequal texture and hardness of the beds traversed by it.

Of the change observable in the character and value of the mineral ingredients of a vein in its passage from one kind of rock into another it is hardly possible to speak in parenthesis. One of the best known examples is that of the old Dolcoath mine in Cornwall, where the vein, in leaving the clay-slate (killas) and penetrating the granite, changed from a copper-bearing into a tin-bearing lode. I might also mention the silver-lead veins of Pontgibaud,* in France, which are in a gneiss country, diversified by dikes of granulite. The ore-veins have been formed along fractures within the dikes, and on their line of contact with the gneiss. When the dike diminishes in size, the ore decreases in width; when the vein penetrates into the gneiss, the ore disappears. The best ore is associated with the kaolinization of the feldspar of the granulite; and when the latter becomes hard and unaltered in depth, the ore pinches out.

On Newman Hill, Rico, Colorado, the veins of rich gold- and silver-bearing ores are noticeably affected by the character of their rock-walls. The particular changes due to penetrating from lime into sandstone have already been mentioned in connection with the veins of the Enterprise mine, but there is also the more general observation, that when the sedimentary beds are black, the veins in them are rich; when they lose that black color, the ore diminishes.

Other instances occur to me, but the above are typical. This inter-dependence between country and ore has been used as an argument in support of the now crippled lateral-secretion theory. It has been suggested† that this relation, often noticed in vein-mining, points to the derivation of the ore from the enclosing rock, and that some formations have an enriching effect, because they have been the source of the valuable metals

---

* See "The Lodes of Pontgibaud," by the writer, in the *Eng. and Min. Jour.* of August 11 and 18, 1894.

† As, for instance in the paper "On Some Evidences of the Formation of Ore-Deposits by Lateral Secretion in the John Jay mine, at Providence, Boulder county, Colo.," by P. H. Van Diest, in the *Proceedings of the Colorado Scientific Society*, vol. iv., p. 340, and in the discussion of the said paper, *Id.*, p. 340.

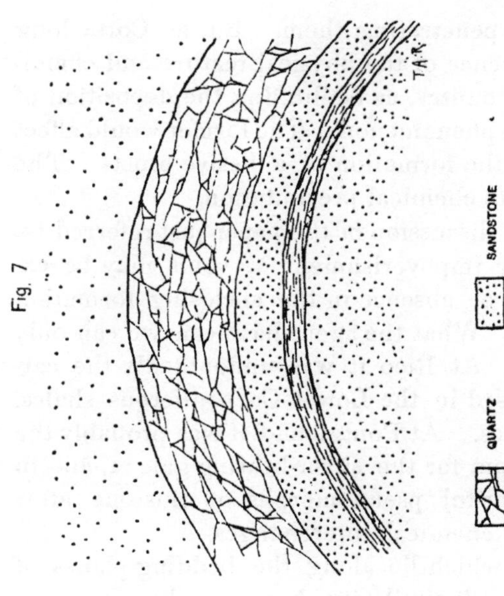

Fig. 7

JOHNSON'S MINE, VICTORIA

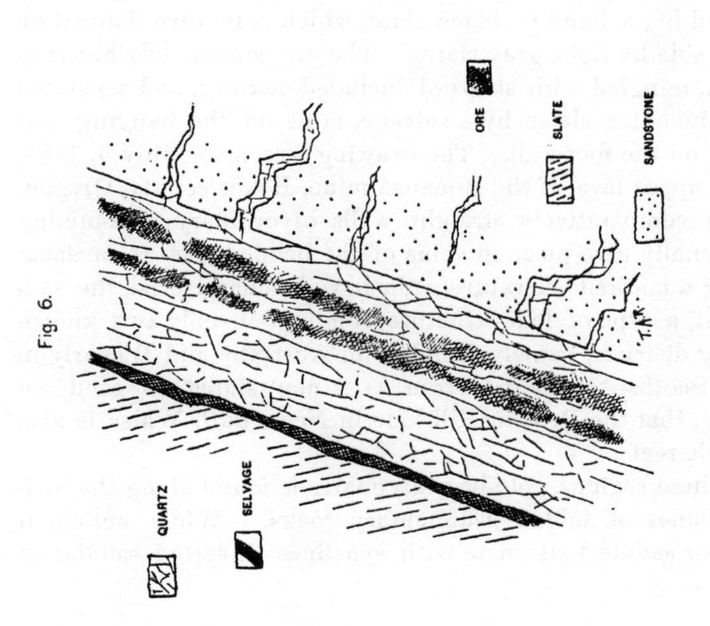

Fig. 6.

SHENANDOAH MINE, VICTORIA.

now found in the veins penetrating them. But as Cotta long ago suggested, the influence of the physical texture and chemical composition of the country, as facilitating the deposition of the ore, may explain this phenomenon. The former would affect the rate of cooling and the formation of adhesive crusts. The latter would act by direct chemical precipitation.

As I suggested in the discussion of the paper just referred to, the local enrichment or impoverishment of veins may be explained by the presence or absence in the enclosing formation of precipitating agents. What the agent has been we can only in rare instances guess. At Rico it was undoubtedly the carbonaceous matter enclosed in the Lower Carboniferous shales, limestones and sandstones. At Pontgibaud it was probably the feldspar which made room for the silver-bearing galena, and in Cornwall also the beautiful pseudomorphs of tinstone after feldspar suggest similar chemical interchanges.

In the case of veins which lie along the bedding-planes of sedimentary rocks, the dissimilarity between the enclosing walls may not go further than a slight difference in the grain of two beds of sandstone, the color of two beds of slate, etc., or it may reach the more marked diversity presented by rocks as entirely unlike as a quartzitic sandstone and a soft slate.

Fig. 5 represents a gold-vein, following the bedding of, and encased by, a band of black slate, which is in turn flanked on either side by light gray slates. The ore consists of ribbons of quartz, mingled with strips of included country, and separated from the outer slates by a selvage, faint on the hanging- but strong on the foot-wall. The drawing was made July 3, 1895, in the upper level of the Bonanza mine, Baker county, Oregon.

The comparatively straight walls of ordinary vein-mining occasionally give place in veins of the bedded class to surfaces having a marked curvature. Such walls characterize the saddle-reef, a type of lode-structure common in only two known mining districts, namely, Bendigo in Australia and Waverly in Nova Scotia—unless it be true, as is now stated on good authority, that the Broken Hill lode in New South Wales is also a saddle-reef.

In these regions, gold-bearing quartz is found along the bedding-planes of folded sedimentary rocks. While anticlinal folds (or saddles) alternate with synclines (inverted saddles or

Fig. 8

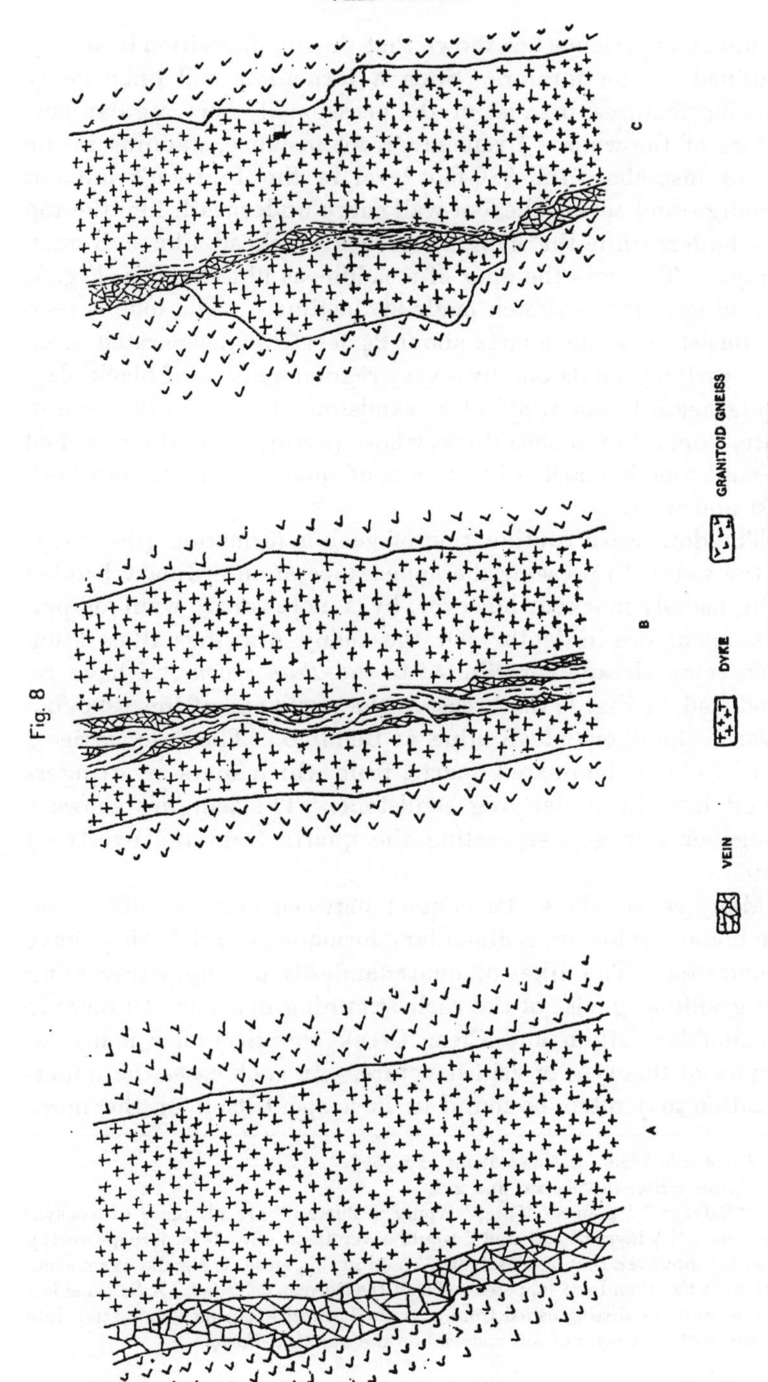

TYPES OF VEIN STRUCTURE IN GILPIN COUNTY COLORADO.

troughs), experience has shown that the ore-deposition is mainly confined to the former. Such a formation will offer many striking features, because of the occasionally very regular curvature of the walls. I remember, for instance, standing in the stopes just above the 980-foot level in the Johnson's mine at Bendigo, and seeing the foot-wall curve underneath like the top of a boiler, while the hanging arched overhead like a Roman bridge. This was the apex of a saddle, as illustrated in Fig. 7, reproduced from a sketch made at the time.* The lode is seen to consist of white quartz about 2½ feet thick, separated from the overlying sandstone by a very regular parting of black clay. Underneath is about a foot of sandstone, then a dark seam of slate, from 5 to 6 inches thick, whose parting from the next bed of sandstone is marked by streaks of quartz, thinning out both east and west.

The downward continuation of such a formation (the "legs of the saddle") presents the appearance of an ordinary bedded vein, usually marked, however, by a noteworthy want of persistence of ore in depth. Of the many drawings illustrating such veins already contributed to the *Transactions*,† I have reproduced, in Fig. 6, the breast of the north end of the 1990-foot level in the Shenandoah mine at Bendigo. The lode carries 2 feet of closely-laminated quartz, from which spurs or stringers go off into the underlying sandstone. The hanging shows a gouge or selvage,‡ separating the quartz from the overlying slate.

Many veins follow the contact between eruptive dikes and the metamorphic or sedimentary formations which they have penetrated. The dikes of quartz-andesite porphyry traversing the granitoid gneiss of the earliest mining districts of Colorado (in Boulder, Gilpin and Clear Creek counties) offer many examples of this type of vein-structure. In such cases the mineralization may often be found to have spent itself on the more

---

* October 5, 1890. See also *Trans.*, xx., 506.

† By the writer in vols. xx. and xxi.

‡ "Selvage," "gouge," "dig," "pug," "fluccan" are all more or less synonymous. "A layer of soft stuff" would cover them all. It is perhaps worthy of notice, however, that our "selvage," used in this sense, is not the exact synonym, as it has often been supposed to be, of the German *Saalband*. A *Saalband* is a definite wall, as distinguished from a gradual transition from vein-matter into country-rock. A layer of soft material on the wall is a *Besteg*.

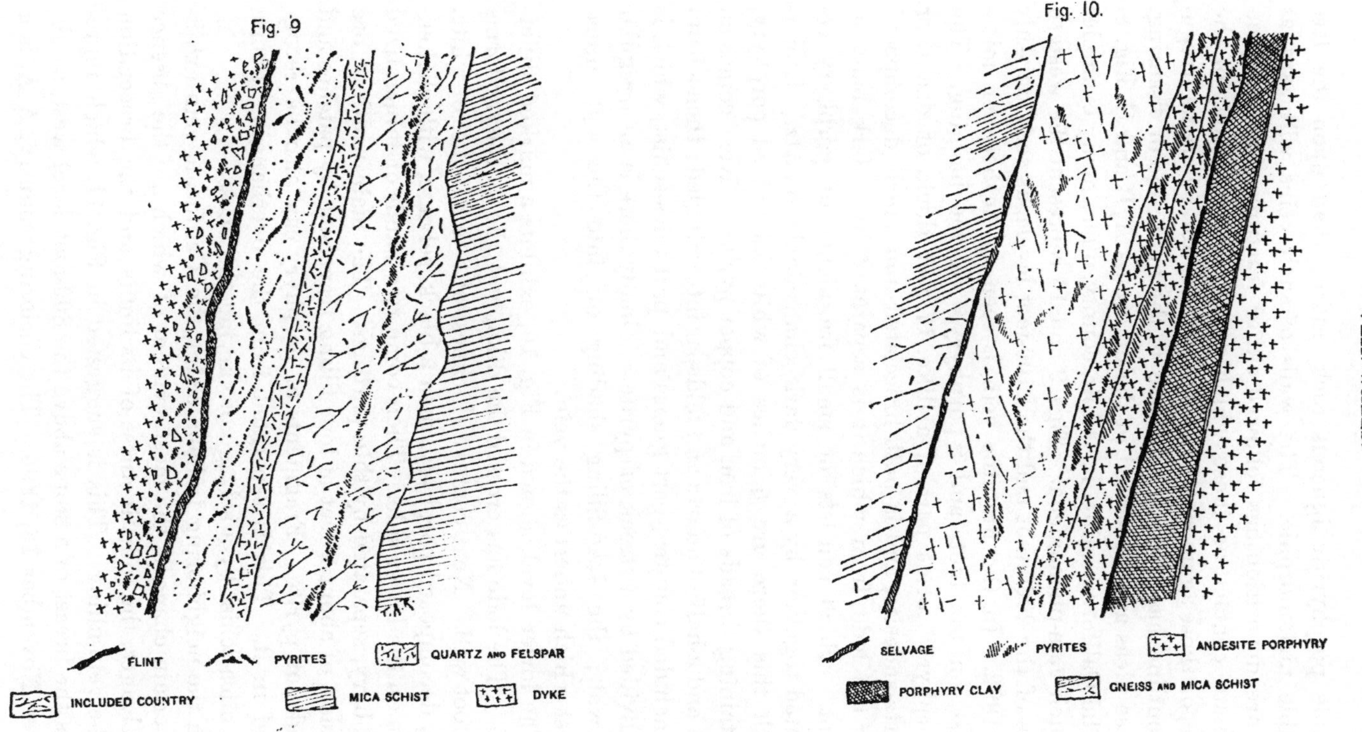

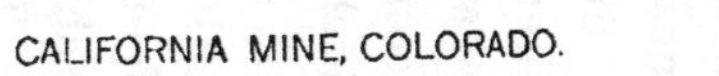
CALIFORNIA MINE, COLORADO.

soluble porphyritic igneous rock, rather than upon the less soluble metamorphic. The walls of such veins will vary, as the ore deposition has followed either fractures along the immediate contact, or those which ramify into the body of the dike, or those again which cut across the latter, where its irregular outline has been an obstacle to the main line of fissuring. These ideas are illustrated in the diagrams A, B and C, Fig. 8.

The California mine, in Gilpin county, offers many examples of such vein-phenomena. Figs. 9 and 10 represent the western ends of the 2000-foot and the 2100-foot levels, as seen on July 13, 1892. In the first the vein is seen to lie between mica-schist, on the foot, and "porphyry," on the hanging. The "porphyry" forms part of a dike, 17 feet thick, of dacite or quartz-andesite, and is both brecciated and much decomposed near the lode, from which it is separated by a dark band of "flint," which consists of small fragments of porphyry cemented together by a very dark chalcedonic quartz. Underneath this there are 5 inches of white kaolinized porphyry, containing threads of iron and copper pyrites. Next comes an inch and a-half of quartz and feldspar intermingled; then a band of included country, part gneiss and part mica-schist, which is subdivided by a streak of pyrite. Finally there is an irregular foot-wall; the load-filling shading off into the soft mica-schist which underlies the vein.

The lower level, shown in Fig. 10, exhibits a marked difference. The lode has crossed the dike, and the porphyry forms the foot-wall. Next comes a thickness of 6 to 8 inches of white, soft, decomposed porphyry, then a black selvage, with slickensides on the lower side. Then come two bands of mineralized porphyry, separated by thin partings. The main width of ore consists of about 2 feet of lode-filling traversed by patches and streaks of pyrite. Fragments of porphyry can also be recognized in it. This is separated from the overhanging gneiss and mica-schist by a selvage of varying thickness.

In the neighboring Indiana claim, the California vein exhibits certain changes, the most evident of which are the absence of selvages, the indistinctness of its limits and the brecciation of the vein-filling. This is suggested in Fig. 11, which represents the breast of a stope above the 800-foot level west, as observed November 13, 1895. The enclosing country, A A, is a granite almost destitute of mica. The part B is bespattered with pyrite. The best ore is a seam, C C, of black zinc-blende lining the hanging-wall. D is evidently brecciated. The larger part of the section consists of slightly altered country (E E) reticulated with seams of blende, following joint-fractures. The foot-wall of the vein is considered to be under the bands of zinc-blende and copper pyrites occurring along F F. The en-

Fig. 11

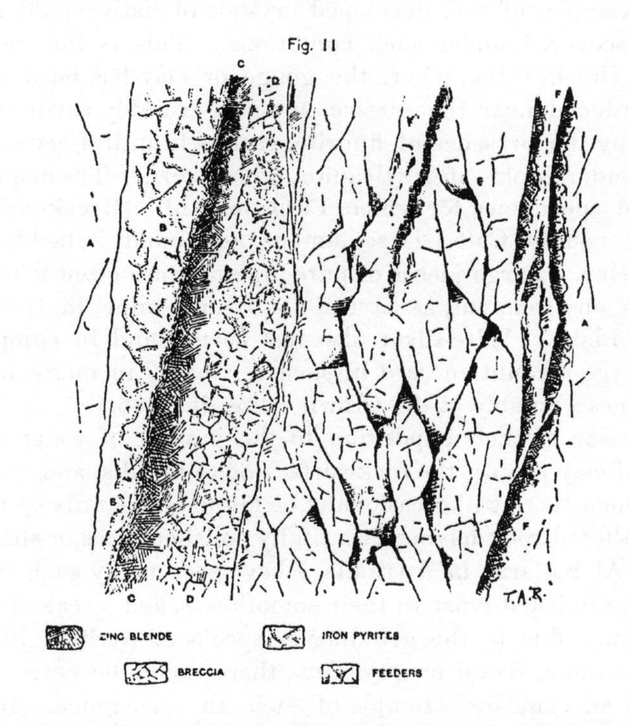

tire width is about 4 feet. The lode has departed from the dike, with which it is so closely associated in the neighboring mine; but the workings show that it meets this dike at intervals, and is benefited by the intersection.

That the vein follows the line of a fault can be seen by examining the walls of the 2000-foot level in the California mine, more particularly at points between 350 and 450 feet west of the shaft, where the lode has left the dike entirely, and is encased in the gneiss and mica-schist. The country-rock on the two sides of the drift is not the same. The extent of the throw of the fault, however, could not be measured.

In the course of the foregoing descriptions of lode-structures, mention has been repeatedly made of the occurrence of clay selvage, following sometimes one, sometimes both, of the walls of a vein. This "clay" may occasionally be material precipitated from solution; ordinarily it is only crushed rock. It frequently encloses exquisite mineral specimens, because its soft consistency has permitted untrammeled crystalline growth. Most examples of well-developed crystals of native gold have been discovered under such conditions. This is the case at Cripple Creek, Colo., where the gouge or clay has been dried and hardened near the surface, and as a crumbly earth, made purple by the presence of fluorite, carries beautiful crystals of gold pseudomorphic after sylvanite and calaverite. The exquisite leaf-gold specimens, for which Farncomb hill (Breckenridge, Summit county, Colo.) is so famous, are found imbedded in talcose clay. Large pieces of pure argentite are often found in such an environment, as at the De Lamar mine, in Owyhee county, Idaho. Wire-silver also has been found in comparatively large amount encased in such a "mud" in many Leadville mines; notably at the Crown Point, in 1886.

By reason of their opposition to the passage of water such seams of clay protect the rock-surface of vein-walls, and underneath them there will occasionally be found comparatively fresh and unaltered rock having beautifully polished faces or slickensides. At Ballarat, in Australia, I have seen many such rock-faces like finished ivory in their smoothness, and streaked with black lines, due to the grinding of specks of pyrite. In the Bonanza mine, Baker county, Ore., there could be seen quite recently an exquisite example of such an occurrence. In an upper drift there was at one place a surface of a few feet square (on one of the walls of a gold-bearing quartz-vein) covered by a thin layer of black clay, under which lay what seemed a white enamel of very remarkable delicacy. It could not be removed without breaking, because it was very friable, consisting essentially of crushed quartz partially recemented, probably by pressure.

"The handwriting on the wall" is not always easy to decipher. The lines or striæ occasionally to be seen upon its surface have been held to indicate the direction of that movement (or succession of movements) of the opposing rock-planes to which the deposit of ore primarily owed the opportunity for its existence. These lines, however, sometimes have opposite di-

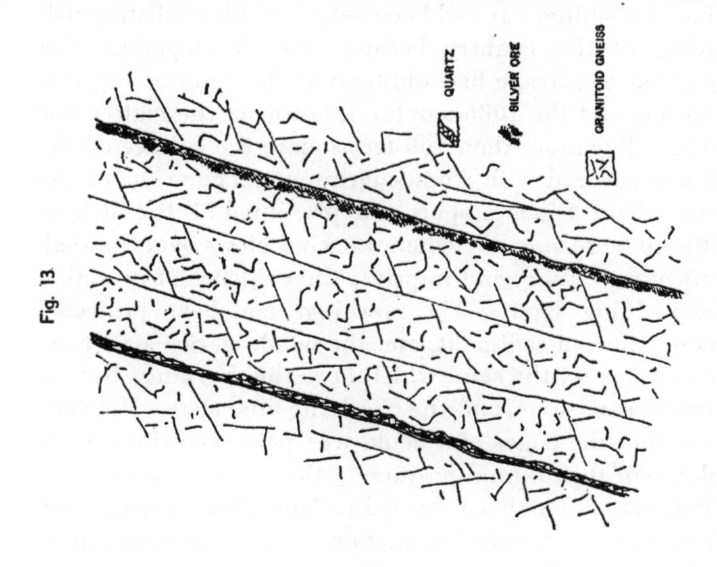

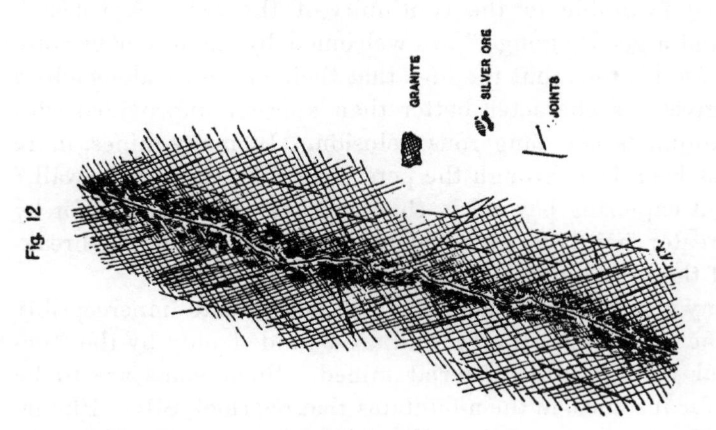

SEVEN-THIRTY MINE, COLORADO.

rections within a short distance and offer conflicting evidence hard to explain.

Rarely is a story told more clearly than in the ripple-marked foot-wall which was to be seen in October, 1891, in the Johnson's mine at Bendigo. It had been very difficult to distinguish the bedding of the country because the development of a strongly-marked cleavage had obliterated the lines of original sedimentation. At the 1065-foot level, however, the matter was made plain. For more than 100 feet square the surface of the foot-wall was covered with ripple-markings. The crests of the waves were about 3 inches apart and presented all the little irregularities to be seen to-day when the wind blows over the shallow waters of an estuary and imprints the evidence of its action upon the yielding sand. The markings had been protected by layers of Silurian sediment, and the whole series had been indurated into rock, the sand which bore the markings becoming quartzitic sandstone, and the overlying mud slate. Between them, as within the pages of a book, was preserved the conclusive evidence of the original position of the beds of rock enclosing the reef, which had been formed in later times, when fissuring had made room for the circulation of underground waters and the deposition of the gold-bearing quartz.

In the above interesting case the corrugation of the foot-wall, due to the ripple-markings, rendered difficult the detachment of the ore. Distinct walls, especially when accompanied by selvage, are very useful in actual mining; but they are not by any means necessarily indicative of a productive vein, or particularly favorable to the continuity of the ore. A "clean" wall and a good "gouge" are welcomed by the miner because they ease his toil; but the idea that their presence alongside a lode gives it a character better than another unprovided with such adjuncts is a dangerous delusion. In many mines, more ore has been lost through the persistent following of a "wall," without exploring beyond it, than was ever compensated for by the greater facility given by such a parting-plane for the breaking of the ore found.

Many veins have no defined walls, but gradate imperceptibly into the enclosing country, and are bounded only by the commercial value of the material mined. Such veins are to be seen, for instance, in the mountains that overlook Silver Plume, Clear Creek county, Colo. Fig. 12 represents a sketch made May 27, 1892, from the 300-foot level of the Seven-Thirty mine.

A fracture penetrating the metamorphic granite carries ore on both sides, which diminishes in richness as it spreads into the encasing country. The joints in the granite are evident.

In this mine the so-called "walls" are often simply two parallel veins (rich, but very small), separated by clean, hard country. This is illustrated in Fig. 13, which was obtained

Fig. 14.

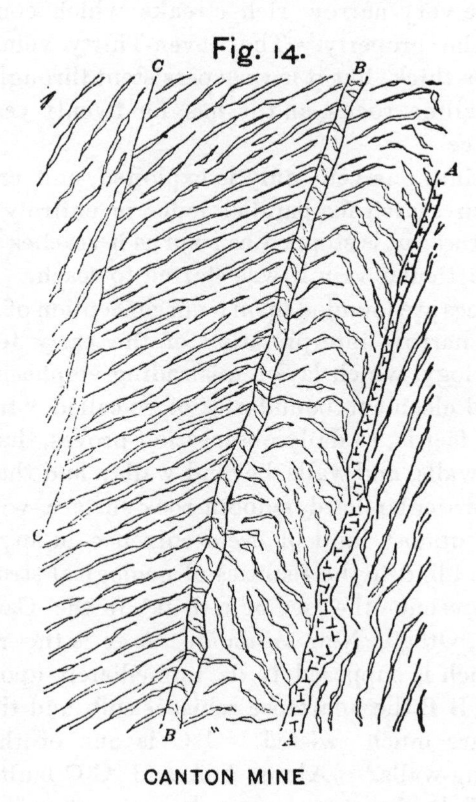

CANTON MINE

from the same level about 1000 feet further east. The granitoid gneiss is traversed by two streaks of ore, of which the one to the right is much the richer. Between them there are at least two well-marked parallel fractures devoid of ore. The vein to the left has a thin selvage, under which there is a streak of quartz carrying a little silver-ore; but the companion-vein to the right follows a fracture, unaccompanied by any selvage, whose upper side is impregnated with about 3 inches of tetrahedrite, galena, and polybasite.

Where ore is absent in the Seventy-Thirty mine, the walls are apt to be particularly well-defined; and when there is any thickness of rich silver-bearing mineral present, the walls are scarcely to be distinguished, and the rock is hard to break, because it is destitute of convenient partings. The large veins carrying gouge are found to be uniformly poor, except where they meet the very narrow rich streaks which constitute the resource of the property. The Seven-Thirty vein proper is only 2½ inches thick, but it is very persistent through the midst of hard crystalline rocks, and it has, for twenty years, proved very productive.

In many mines one vein only is exploited, and cross-cutting the country in search for parallel lodes is entirely neglected. In others, a cross-cut is stopped as soon as it reaches the further wall of the particular vein it was started to reach. Both these unwise practices are founded upon a misconception of lode-structure, due to a narrow interpretation of the early teachings of economic geology, which lays a misleading emphasis upon the definition and clean-cut boundaries of so-called "true fissure-veins." The fact is, as daily observation proves, that there are walls within walls, and walls beyond walls; and that to follow closely any particular hard, smooth rock-surface, with the idea that it is the utmost limit of ore-occurrence in any particular mine, is to be blind to the realities of geological structure.

Fig. 14 represents the face of a drift* in the Canton mine, near Waipori, Otago, New Zealand. A A is the reef, a vein of quartz which is supposed to lie immediately upon the foot-wall. Along B B the quartzose schist is soft, and the included quartz-folia are much twisted. C C is one of the so-called "false hanging-walls." Along A A and C C faulting is evident, along B B distortion only. It was not possible to say where the lode ended, or where it began. The whole width from A to C was known to be gold-bearing, although A A served as a guide in following the gold-bearing channel. Nevertheless those who were working the mine had little comprehension of the formation, particularly of its essential lack of definition, and, while admitting that there were several "false hanging-walls," insisted that there was only one foot-wall (underneath A A) which was stated to be of a different

* On November 15, 1890. See also *Trans.*, xxi., 415.

kind of rock, and exceptionally hard. On examination I found that the rock of the supposed foot-wall was similar to that of of the rest of the gold-bearing country forming the lode, and on a sample of it being crushed and tested in a prospector's

Fig. 15.

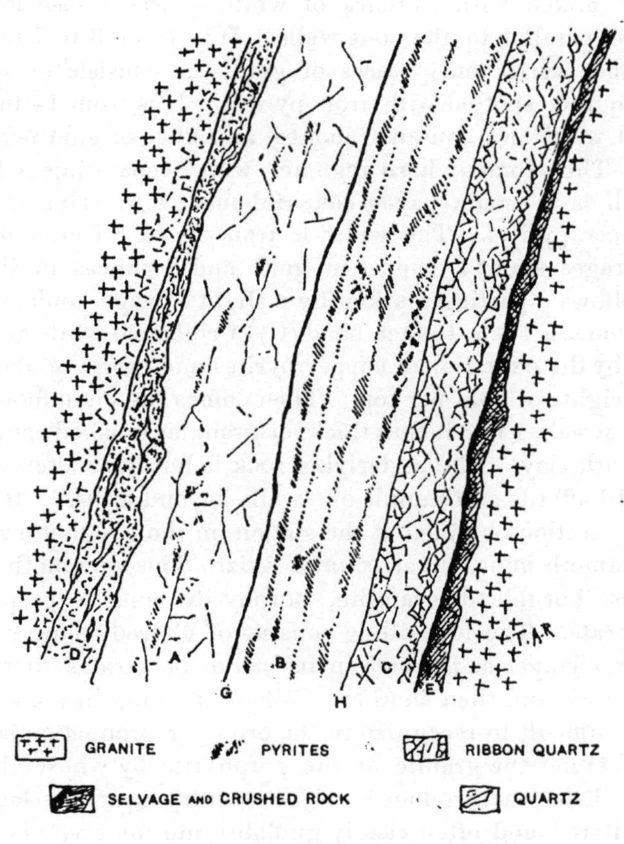

UNION AND COMPANION MINE, OREGON.

pan, it was discovered to be richer than that which was being actually mined. It was scarcely necessary after that to insist that a cross-cut should be made into the foot-wall.

Fig. 15 represents the north breast* of the lower level on the main lode in the Union and Companion mine, Union

* On June 26, 1895.

county, Oregon. It illustrates the occurrence of "walls within walls," for while the lode may be limited by the main boundaries along E and D, there are at least two partings (G and H) equally well-defined, subdividing the enclosed width of ore. The country is a fine-grained granite, which, near the hanging, is decomposed and ore-bearing. D is a streak of granular crushed country, mixed with lenticles of white quartz whose longer axes are parallel to the lode-walls. D is from 3 to 7 inches wide, and carries only traces of gold. A consists of white hackly quartz spotted with iron pyrites. It is from 14 inches to 2 feet wide, and contains about ½ an ounce of gold per ton of ore. Then comes a hard regular "wall," separating A from B, which is the main pay-streak, ribboned with veins of iron and copper pyrites. The width is from 2½ to 3 feet, and the ore averages about 2 ounces in gold and 8 ounces in silver. Then follows a parting marked by a slight selvage, underneath which comes a 10- to 15-inch band (C) of ribboned white quartz, stained by the oxidation of copper pyrites and carrying about 5 pennyweights of gold per ton. Then comes the main foot-wall with its streak, 1 to 3 inches thick, of granular crushed country, mixed with clay. The underlying rock is but little altered.

Fig. 16 affords an example of "walls beyond walls." It represents a section obtained at the station on the 500-foot level in the Mammoth mine, Pinal county, Ariz. The Mammoth lode traverses hornblende-granitite, porphyrite and a porphyry agglomerate. The lode-filling consists of altered country, and therefore changes as the lode in its strike penetrates first one kind of rock and then another. When standing in the stopes it is not difficult to recognize in the ore the reproduction of the habit of either the granite or the porphyrite by whose alteration the lode was produced. The country near the lode is much altered and often visibly gradates into the ore, while, as the lode is receded from, these effects diminish until they become confined to the faces of the rock lining the fractures. The granitite carries two feldspars, of which the pink orthoclase is evidently more stable and succumbs to decomposition less quickly than the green plagioclase. The ore is both gold- and silver-bearing, but chiefly valuable for gold. The great variety of associated minerals includes some uncommon species, such as wulfenite (usually colored by vanadic acid), vanadinite, descloizite, ecdemite, dechenite, linarite, besides the commoner anglesite, pyromorphite, cerussite, malachite, dioptase, azurite, and a little galena and pyrite. Referring to the drawing (made March 17, 1893), the edge of the ore is shown at A; it becomes

Fig. 16.

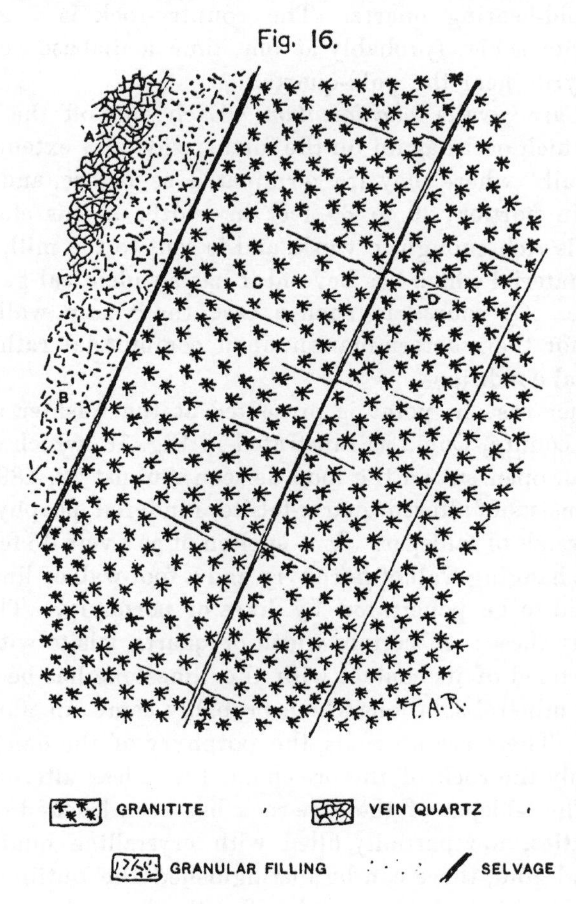

MAMMOTH MINE, ARIZONA.

mixed with altered granular country (along B) in approaching the "main footwall." This is followed by the granitite itself, in which there are well-defined walls (or fractures parallel to the lode-channel) and cross-joints, often lined with exquisite crystals of vanadinite and wulfenite.

Going to the Pacific coast, Fig. 17 represents a part of the west side of the so-called "mother lode" of California. The drawing, made May 21, 1891, is a portion of the face of a large open cut at the Gold Cliff mine, Angel's Camp, Calaveras county, Cal., near the now well-known Utica mine. The ore-channel consists of a country-rock traversed by cross-veins of white gold-bearing quartz. The country-rock is a greenish gray augite schist (probably at one time a diabase), carrying coarse pyrite near the gold-quartz.

There are "walls" *ad infinitum.* Each cuts off the quartz-seams, which occur again on the further side and extend to the next "wall," where they are terminated as before, and so on. A certain portion, 20 to 30 feet in width, of this channel of country is rich enough to work, and is sent to the mill, but the poorer material which lies beyond it has an identical geological structure. Of course, in such a case the "main walls" will depend for their determination upon commercial rather than geological conditions.

Another case in point is presented at the Cashier mine in Summit county, Colo., as illustrated in Fig. 18, which shows a part of an open cut on the lode, as seen August 22, 1895. The latter consists of altered quartz-felsite, rendered porphyritic by large crystals of feldspar. It is spoken of as a vein 45 feet wide, having a hanging-wall of porphyry and a foot-wall of lime. The ore is said to be penetrated by dikes of porphyry. The facts are really these: A certain width of quartz felsite within the neighborhood of its contact with the limestone has been acted upon by mineral solutions which probably came up along that contact. There are no walls, the porphyry of the hanging being simply the rock of the ore-channel in a less altered condition. The feldspar of the lode-rock has been leached out. In the cavities, now partially filled with crystalline quartz, iron oxide and gold, there can be distinguished the outlines of the large (½ to 1½ inches) crystals of orthoclase whose removal made the rock porous to circulating waters. The mineralization is indicated by the softening and reddening of the porphyry and is most marked along the joints, especially where they intersect. There are occasional portions of the rock comparatively unaffected by the leaching agencies, and therefore appearing as hard, unstained nuclei amid a mass of softer reddish ore. It is these that are locally termed "horses" and "dikes" of porphyry.

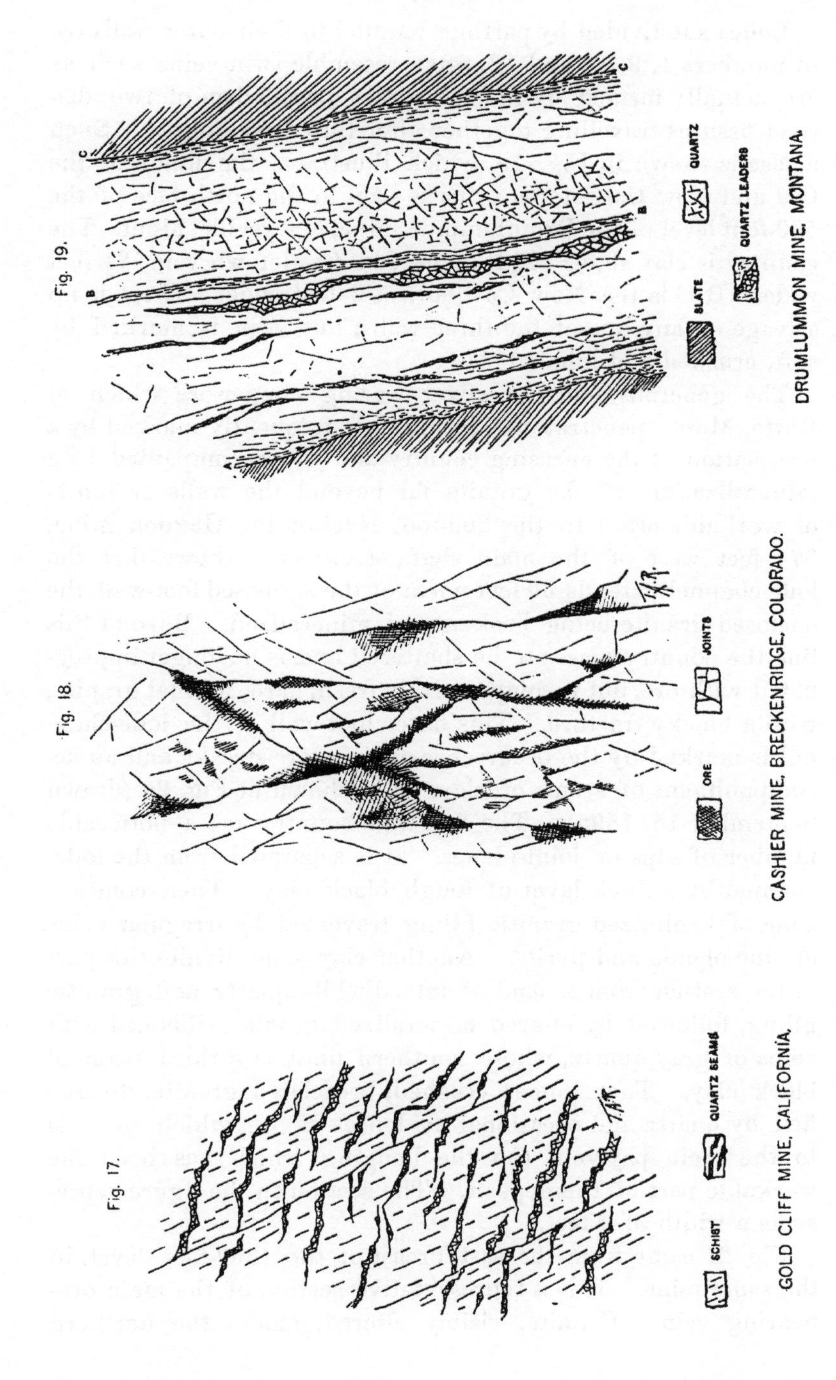

GOLD CLIFF MINE, CALIFORNIA.

CASHIER MINE, BRECKENRIDGE, COLORADO.

DRUMLUMMON MINE, MONTANA.

Lodes subdivided by partings parallel to their outer walls (as in numbers 1, 9, 10 and 12) often resemble twin veins such as are actually formed by the temporary parallelism of two distinct fissures travelling together after they have united. Such a case is shown in Fig. 19, which illustrates the union of the Old and New Castletown veins as seen in the north face of the 500-foot level of the Drumlummon mine, Marysville, Mont. The country is clay slate. A B is the Old Castletown vein, 2½ feet wide. B C is the New Castletown, 2 feet wide. There is no selvage on any one of the three walls, but each is marked by soft, crushed and foliated slate.

The generous lodes of silver-bearing copper-ore which at Butte, Mont., penetrate the granite are frequently marked by a brecciation of the encasing country and are accompanied by a mineralization of the granite far beyond the walls or limits of workable ore. In the 300-foot level of the Gagnon mine, 374 feet west of the main shaft, a cross-cut shows that the lode-channel extends 30 feet north of the supposed foot-wall, the enclosed granite being broken and mineralized. Beyond this line the country ceases to be shattered and is no longer impregnated with ore, but is comparatively fresh, hard, normal granite, with a blocky fracture. This outer foot-wall of the lode-channel is marked by the occurrence of some ore-streaks and an accompaniment of seams of clay, as is shown in Fig. 20 (drawn September 15, 1895). The foot-wall country has a noticeable number of slips or joint-planes. It is separated from the lode-channel by a thick layer of tough black clay. Then comes a zone of kaolinized granitic filling traversed by irregular veins of zinc-blende and pyrite. Another clay-seam divides this part of the section from a band of mixed white quartz and granitic filling, followed by altered mineralized granite, ribboned with veins of gray quartz, whose southern limit is a third seam of black clay. Then comes crushed, brecciated granite, diversified by quartz and occasional evidences of ore, which extends to the main pay-vein (on the hanging) which has been the workable part of the deposit. The section in the figure represents a width of 6 feet.

Fig. 21 came from the east breast of the 1300-foot level, in the same mine. It is a representative section of the main ore-bearing vein. Granite, visibly altered, marks the northern

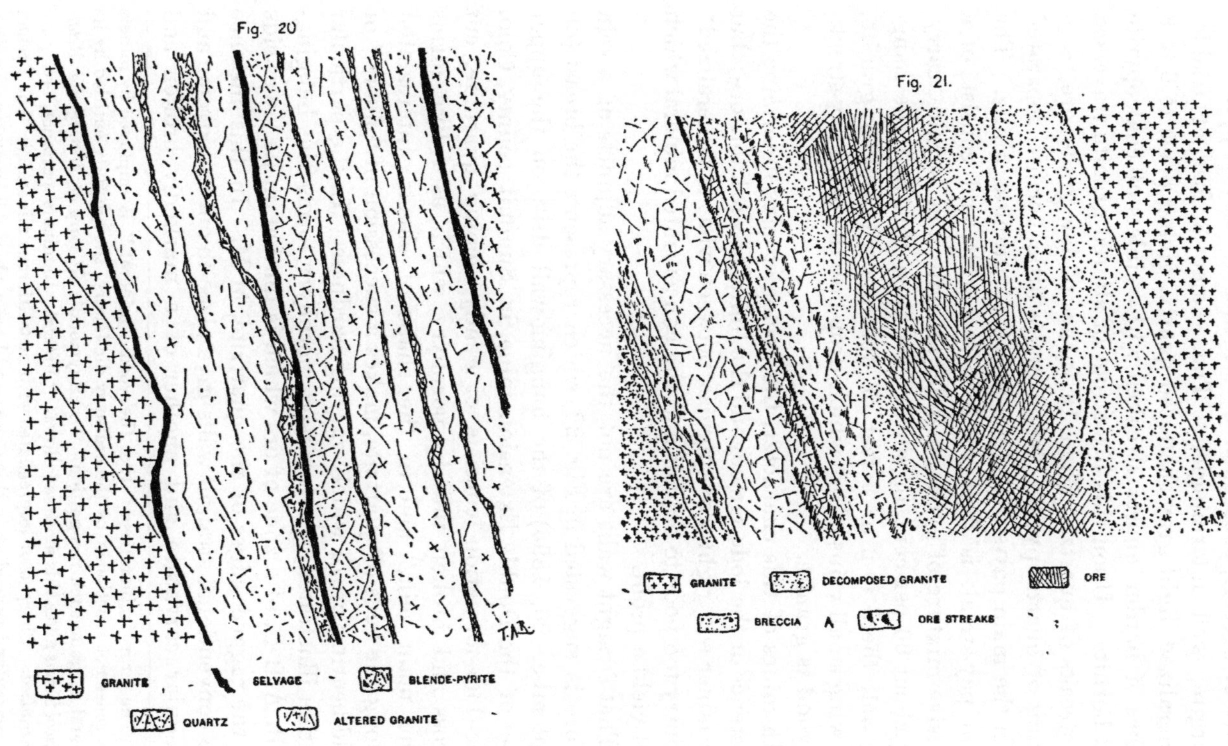

THE GAGNON MINE, BUTTE CITY, MONTANA

edge of the section, which is the foot-wall. Upon it lie a few inches of breccia, succeeded by 8 inches of blende, pyrite, and enargite, well intermingled. Then come 3 feet of friable, mineralized, light gray shattered quartz, giving place to 6 to 8 inches of harder quartz streaked with veinlets of chalcopyrite and bornite. Upon this lies a foot of altered granite traversed by streaks of quartz. Then 18 inches of low-grade ore, consisting of quartz, pyrite, blende and a little bornite, separated from the main pay-streak by 6 inches of granitic filling. The main pay-streak is from 5½ to 6 feet wide, and consists of a massive mixture of pyrite, blende, enargite and bornite, carrying about 60 per cent. of silica. Between this and the hanging-wall there is 2 or 3 feet of decomposed broken granite, showing small veins of ore which drop into the main pay-streak. Beyond is granite.

In mines of this character, the geologist may determine the existence of the lode far beyond the limits of workable ore; but the miner will rightly distinguish between what is mineralized* country too poor to exploit and the concentrated mineral which will yield a profit.

That straight walls are not the necessary adjuncts of a vein of ore is suggested in Fig. 22, which represents the breast (on September 26, 1895) of the hanging-wall drift on the upper level of the Double Extension mine, in Summit county, Colo. The lode-formation consists of gently sloping quartzite, cut across and broken into by porphyry, which, as a dike, forms "the main vein," and in the shape of sheets, intercalated among the beds of quartzite, makes a succession of "floors" of gold-bearing ore of widely varying hardness. In the particular section illustrated, the intrusive porphyry forms the hanging-wall, A B C, of a zone of ore which is limited on its lower side by the ragged edges of the quartzite, M M. In this instance the conventional straight walls give place to one of extreme and irregular curvature and another of a markedly broken and

* The term "mineralized," like the word "mineral," is employed by miners in a sense not sanctioned by the ordinary dictionary, though fully entitled by its general usage to such recognition. Dr. Raymond's *Glossary of Mining and Metallurgical Terms* gives this sense as follows: "*Mineral.* In miners' parlance, ore. . . . *Mineralized.* Charged or impregnated with metalliferous mineral." The French use *minerai* and *mineraliser* in this sense; and I have adopted it because no English equivalent occurs to me.

jagged outline; but they are nevertheless walls, as truly as would be the most perfectly straight, smooth rock faces.

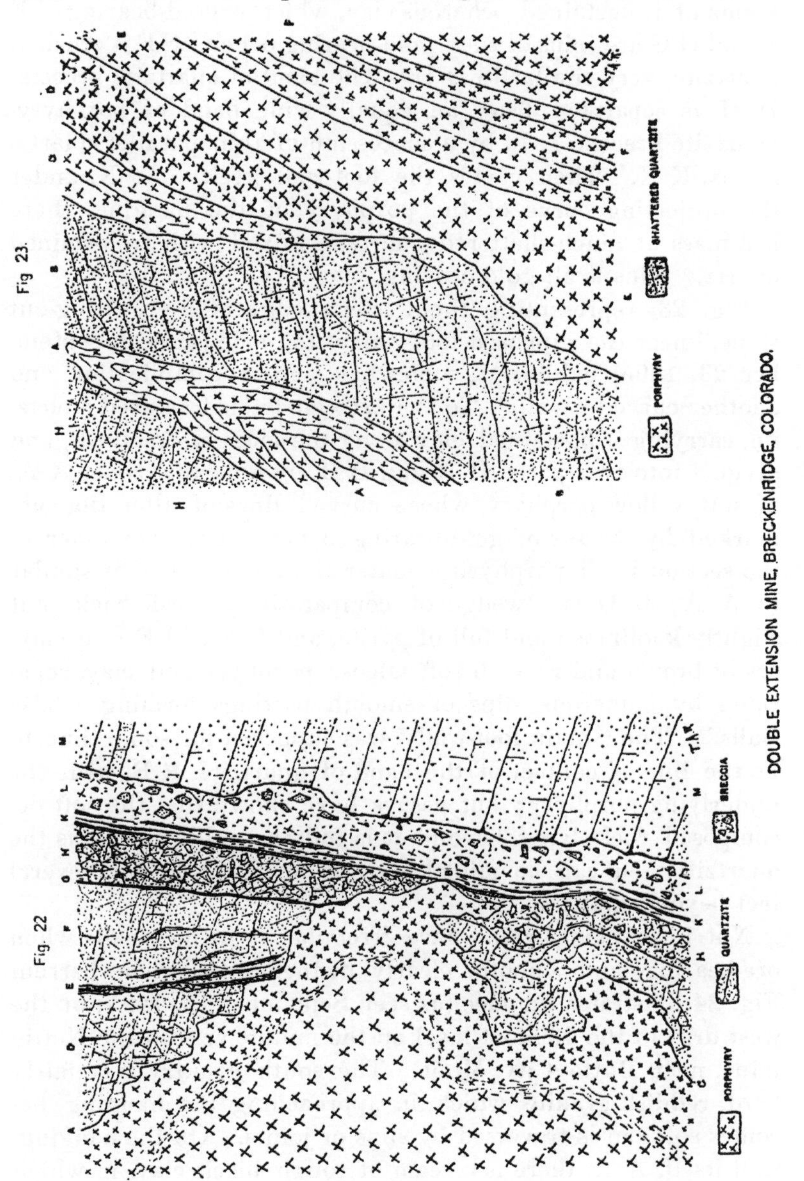

DOUBLE EXTENSION MINE, BRECKENRIDGE, COLORADO.

The porphyry, A C, a quartz-felsite, is bleached to a yellow white and softened to a granular clay, as it approaches its contact with the broken quartzite, D D. The latter is dark bluish-gray, and carries, along its joints and other fractures, minute seams of iron-stained ocherous clay, which is gold-bearing. E E and G G are veins of such gold-bearing ocher. F F is crushed quartzite, very similar to D D. The band of quartzite breccia, H H, is separated from an equally wide band of porphyry-quartzite breccia, L L, by a succession of thin parallel quartz-seams, K K. Then comes the foot-wall itself. At N, under the projecting curve of the porphyry of the hanging, there is a mass of much-shattered quartzite mixed with iron-stained quartz. This is all gold-bearing.

Fig. 23, representing the western edge of a "cutting-out stope," near the supposed foot-wall of the lode (as seen September 23, 1895) exhibits a somewhat similar complication and another curved "wall." To the extreme left is fractured quartzite, carrying iron-stained clay along the faces of fractures, and divided into two parts, H H and B B, by a narrow zone, A A, of soft yellow porphyry, whose curved lines of alteration are marked by streaks of gold-bearing ocher. The remainder of the section is all porphyritic material, of which C C is similar to A A; D D is a wedge of comparatively fresh rock, but slightly kaolinized and full of pyrite, and E E and F F are layers of brown and reddish soft talcose porphyry and clay, separated by numerous slips or smooth partings forming "false walls." The "main foot-wall" was supposed for some time to be the line of contact of the band of quartzite, B B, with the underlying porphyry; but assays have shown that the soft decomposed rock lying beyond it is fully as gold-bearing as the quartzite, and can be mined with as much profit for several feet beyond that line.

Not infrequently veins have irregular indistinct walls when ore-bearing, and smooth, clearly defined ones when barren. Fig. 24 illustrates the face (as seen September 15, 1895) of the west drift of the 430-foot level on the middle vein in the Nettie mine, near Butte City, Mont. The south country is a fairly hard, reddish granite, which, in approaching the hanging, becomes soft and is traversed by slips or joints. On the hanging-wall itself, A A, there is a seam of tough black clay, in which can be seen frequent films of minutely crystalline blende and galena, and small imbedded shots of ore and rock. This overlies a filling of white decomposed granitic material, full of partings and seams of black clay, such as C C. The foot-wall, B B, is also marked by a black selvage. Underneath it comes comparatively fresh "Blue-bird" granite.

Fig. 24

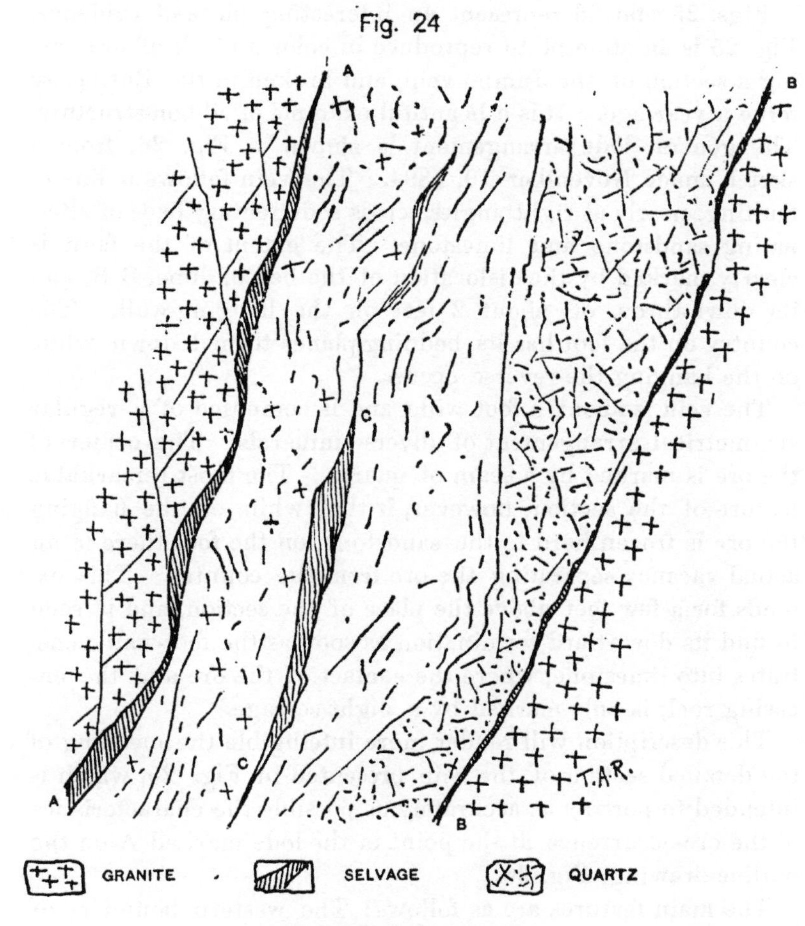

NETTIE MINE, MONTANA.

A few feet further east this level carried an ore-body, A B in Fig. 24 being the zone so transformed. The brecciated quartz upon the foot-wall was the part of the vein which first became ore. The sides of the drift are now coated with a delicate efflorescence of goslarite (sulphate of zinc). In this connection it may be of interest to state that in the Gagnon mine, at Butte, three miles away, the apparently clean country, at some distance from the lode, was found* to carry 3 per cent. of zinc, indicating the extent to which the mineralizing action had penetrated.

Figs. 25 and 26 represent an interesting piece of evidence. Fig. 25 is an attempt to reproduce in color a block of ore giving a section of the Jumbo vein, and broken in the Enterprise mine a year ago. It is a beautiful example of ribbon-structure. The general lode arrangement is shown in Fig. 26, from a sketch made November 19, 1894. The vein follows a line of faulting, nearly at right angles across sedimentary beds of alternating sandstone and limestone. The extent of the fault is clearly marked by the dislocation of the bed of lime, B B, and its down-throw of about 2 feet on the hanging-wall. The country on the foot has its bedding-planes turned down, while on the hanging the reverse occurs.

The vein is about a foot wide and is composed of a regular symmetrical arrangement of diverse minerals. The center of the ore is marked by a seam of quartz. The most remarkable feature of the section, however, is that while on the hanging the ore is frozen hard to the sandstone, on the foot there is an actual vacancy separating the ore from the country. This extends for a few feet above the place of the section, and is seen to find its downward termination as soon as the foot-wall penetrates into limestone, where the contact of the ore and the encasing rock is only marked by a slight selvage.

This description will render more intelligible the meaning of the detailed section of the vein presented in Fig. 25, which is intended to portray as accurately as possible the characteristics of the ore-occurrence at the point in the lode marked A on the outline-drawing, Fig. 26.

The main features are as follows: The western boundary of the vein is fairly straight, dipping, as the vein does, eastward. It is separated from the country-rock, a light-gray medium-grained sandstone, A A, by an actual vacant space, B B, of about half an inch. The edge of the ore nearest the foot-wall consists of an irregular band, C C, of about $\frac{3}{4}$ of an inch of quartz, speckled with pyrite and chalcopyrite. Toward the bot-

* According to Mr. C. W. Goodale, the manager.

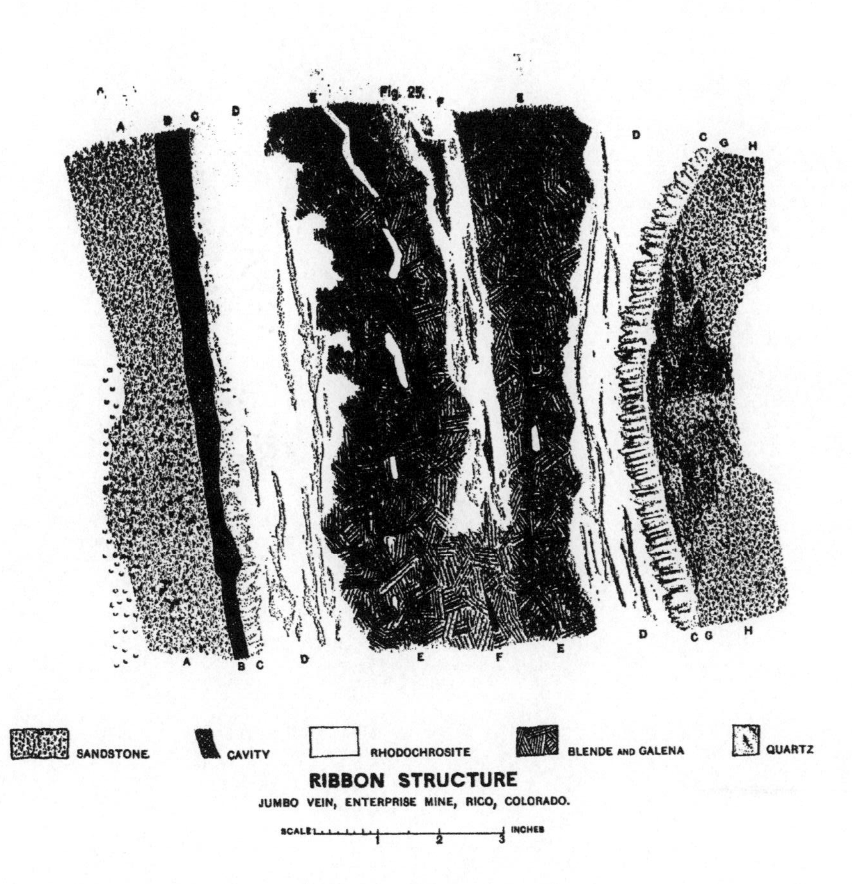

RIBBON STRUCTURE

JUMBO VEIN, ENTERPRISE MINE, RICO, COLORADO.

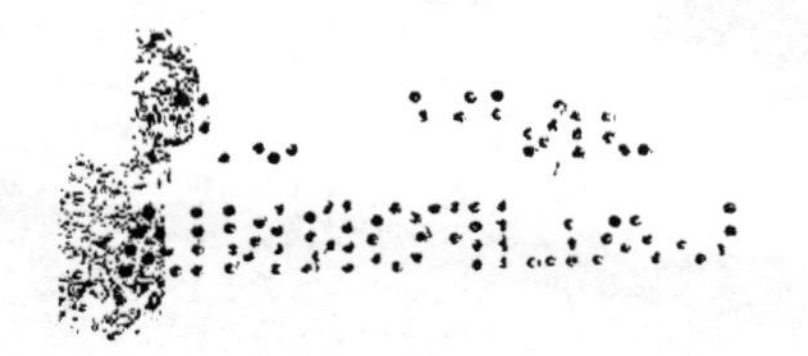

tom this quartz narrows, but becomes clearly defined into a crystalline comb, with teeth at right angles to the vein.

Then comes a zone, D D, averaging 1½ inches, of pink rhodochrosite. This band is broken into by veinlets of quartz, some of which are only branches from the outer seam, C C, while others traverse the rhodochrosite in bluish-gray streaks parallel

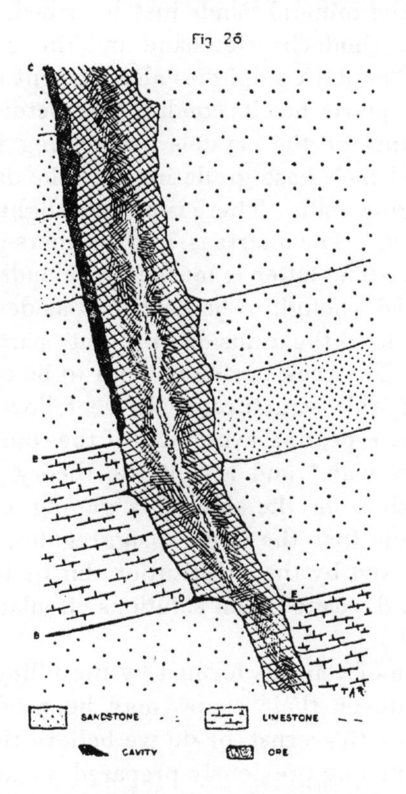

to the general structure, and are peppered over with particles of pyrites.

The rhodochrosite band is broken on the right by the irregular outline of the blende and galena, E E, which is about 2 inches wide. There are blotches of yellowish "resin-blende" and patches of bluish-black galena distributed throughout a dark and intricate mixture of these minerals. They shade out into the white quartz, F F, which makes a bilaterally symmetrical division in the ore. The dark mass of the sulphides encloses shreds of rhodochrosite having distinct outlines.

Along the median line of the central quartz-seam, F F, occurs a succession of geodes, lined not only with crystals of the quartz itself but also with beautiful crystals of stephanite. The latter are seen in irregularly distributed clusters. In the outer quartz there are numerous specks of pyrite.

The eastern half of the vein presents, in reverse order, an exact repetition of the mineral bands just described. The separation between the rhodochrosite band and the outer quartz is more distinct. The dark sulphides also present a cleaner outline. The outer quartz has its comb-like structure strongly developed, the points of the crystals penetrating into the pink rhodochrosite and their base gradating into the dark-gray sandstone of the hanging-wall. There is not the slightest selvage or parting of any sort. The quartz is, as the miners say, "frozen" to the sandstone. The latter is marked by clouds of dark mineral hardly defined enough to be described as dendritic. This feature is traceable to the diffusion of minute particles of pyrite and stephanite. The rock is rich enough to be classed as ore.

In interpreting this structure, shall we follow the explanations given for the repeating symmetry of the comb-structure of the Drei Prinzen vein,* and accept the theory of successive crystalline growth from the sides of a gaping crevasse? Or are we to conclude that the mineral aggregates, now forming the ore, were derived by the substitution, bit by bit, of rock in place by material deposited from solutions circulating along the line of fissuring?

Do we conceive of veins as formed by the filling of pre-existing cavities, whatever their shape may be, produced by the rupturing of the earth's crust, or do we believe that lodes can be formed without any previously prepared vacant space and simply by the chemical interchange vaguely covered by the term metasomasis? or, again, do both these explanations find corroboration in the daily observations of the mine?

---

* As drawn by Von Weisenbach in his book published at Leipzig in 1836. Other notable examples of this structure are C. Le Neve Foster's drawing of the Huel Mary Ann lode (in the *Transactions* of the Royal Geological Society of Cornwall, vol. xix., 1875), and that of the Carn Marth lode by J. H. Collins (in the *Proceedings* of the Institute of Mechanical Engineers, 1873). Reference may also be permitted to the writer's four colored drawings of the Eureka, Songbird, Jumbo and Kitchen veins, accompanying a paper entitled "Vein Structure in the Enterprise Mine," in the *Proceedings* of the Colorado Scientific Society for 1895.

Walking recently along the railroad grade between Anaconda and Cripple Creek, in El Paso county, Colorado, I found

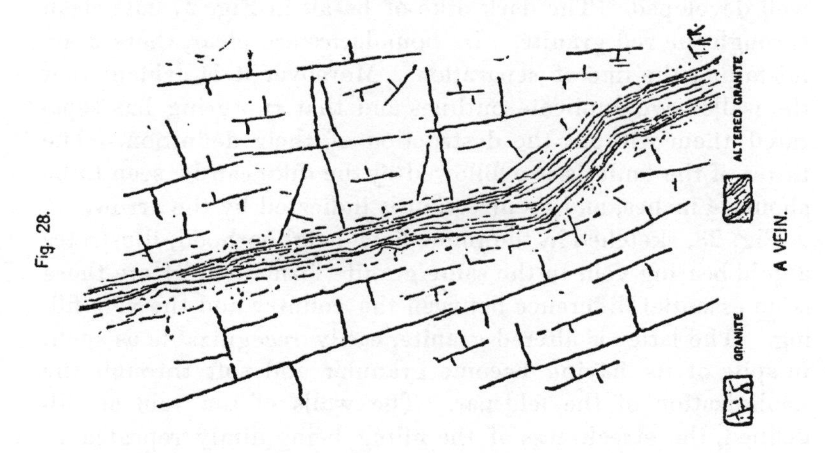

Fig. 28. A VEIN.

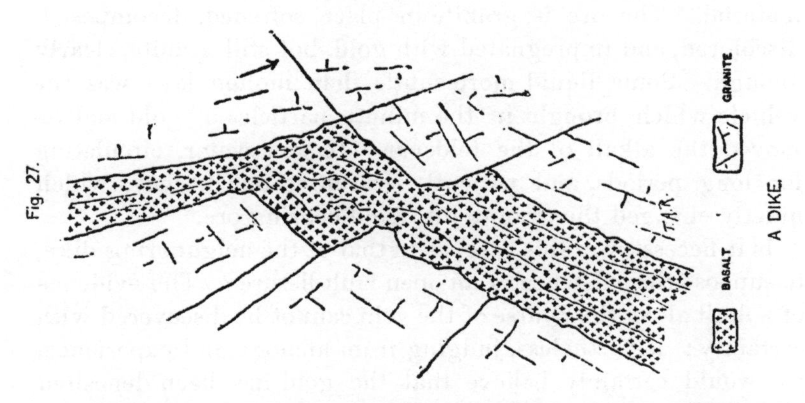

Fig. 27. A DIKE.

in the sides of two open cuts the testimony transcribed in Figs. 27 and 28, one representing a typical dike and the other a typical ore-vein. In both cases the country is the coarse-grained, red granite of the Pike's Peak region. In both the jointing is well developed. The dark dike of basalt in Fig. 27 cuts clean through the red granite. Its boundaries are clear, there is no mistaking the line of separation. Moreover it is evident that the walls have duplicate outlines and that rupturing has separated them without the destruction of their definition. The throw of the fault-fissure followed by the dike can be seen to be about 14 inches, and its direction is indicated by the arrow.

Fig. 28, sketched in the immediate neighborhood, illustrates a gold-bearing vein in the same granite formation. Here there is no essential difference between the country and the vein-filling. The latter is altered granite, easily recognizable as such, in spite of its having become granular and soft through the kaolinization of the feldspar. The walls of the vein are ill-defined, the streakiness of the filling being dimly repeated in the encasing rock. The vein-filling assays $2.60 gold per ton at this place, but is richer, without other material change of character, a few rods distant.

The dike, Fig. 27, is composed of foreign matter filling an evident fissure; the vein, Fig. 28, is rock in place changed into ore by the removal of some of its constituents and the substitution of new ones. In the former case liquid material rose into the fissure, probably *pari passu* with its formation. On the other hand, the vein of gold-ore traversing the granite gives no evidence of the occupation of a fissure by the incoming of new material. The ore is granite in place, softened, decomposed, discolored, and impregnated with gold, but still granite, clearly enough. Some liquid more subtle than molten lava was the vehicle which brought in the minute particles of gold and removed the alkali of the feldspar. It was water, circulating for long periods, and patiently searching out its way, which quietly changed the granite into gold-bearing ore.

Is it necessary in this case, as in that of the neighboring dike, to suppose the existence of an open fault-fissure? The evidence of a fault along the course of the vein cannot be discovered with certainty; nevertheless, judging from analogy and experience, we would certainly believe that the gold has been deposited along a line of displacement. It seems difficult to conceive that any fracturing, such as marks the beginnings of vein-formation, can take place without some displacement, however slight, of the two opposing rock-faces. Without such dislocation, though it be comparatively insignificant in amount, the fracture is only latent, and can hardly be said to exist. The possibility of a simple rupture, without any shearing movement or relative displacement, cannot be denied;* but observation underground indicates that, so far as the deposition of ore is concerned, we

Fig. 29.

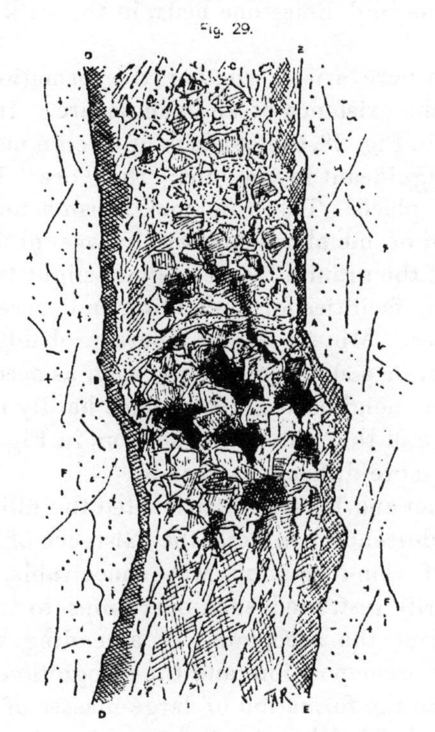

INDIANA VEIN, GILPIN COUNTY, COLORADO

have invariably to deal with rupturing accompanied by a relative displacement of the rock-walls. In other words, veins are generally built on fault-lines. The absence of evidence of such movement in a section on one particular plane is not conclusive, since the displacement may have been at right-angles to the section.

* In this connection I would refer the reader to the suggestive paper of Mr. William Glenn on "The Form of Fissure-Walls, as Affected by Sub-fissuring, and by the Flow of Rocks," read at the Atlanta Meeting of the Institute, October, 1895, and printed in *Trans.*, xxv., 499.

Where a vein cuts across sedimentary rocks, the dislocation may be looked for along the bedding-planes. Such is the case at Rico, in the Enterprise mine already referred to, where the breast of a stope will show a vein traversed by a fracture at right-angles to its walls, and apparently unaccompanied by any dislocation, but further examination will frequently show that there has been a displacement of the country along the dip of the sandstone and limestone beds, in the strike of the vein itself.

The question here arises, whether the formation of the ore-vein required the existence of an open fissure. In the particular case shown in Fig. 28, the quantity of foreign material within the vein is insignificant in amount; the "ore" being simply altered rock in place. That this rock became mineralized by the penetration of metal-bearing waters was probably due to the crushing of the granite by an original slight faulting movement, presenting facilities for circulation and consequent chemical interchanges. Minute spaces there probably were; but a clear opening, or a slow crevassing, such as accompanied the formation of the neighboring dike, seems hardly needed. The ribbon-structure of the Enterprise section, in Fig. 25, presents features much more difficult to explain.

When Werner and his school attributed the filling of veins to the agency of descending waters, the existence of open fissures at the time of vein-formation was conceivable, because the theory necessarily restricted such operations to the vicinity of the surface. But the acceptance of ascending waters as the main agents of ore-deposition, and the recognition of the conditions possible to the formation of large masses of sulphides, at once transferred the laboratory of ore-formation to a deeper horizon; and the suggestion that veins were filled by the deposition of layers of mineral precipitated from waters passing upward along fissures which were kept wide open during such time as was required for crystalline growth to choke them with ore, was immediately ridiculed by the miner, because his daily experience taught him that the vein once deprived of its filling did not remain open, but was inevitably closed by the pressure of the surrounding rock. In many cases, in the absence of artificial means of support, his mine-workings collapsed, so that where there was once a level wide enough for a man to walk through, there came to be only a seam of mud enclosed in shattered rock.

Despite the miner's objection, however, there is evidence that fissures do sometimes occur, which have been sufficiently open to permit the tumbling in of large pieces of rock. Such

Fig. 30.

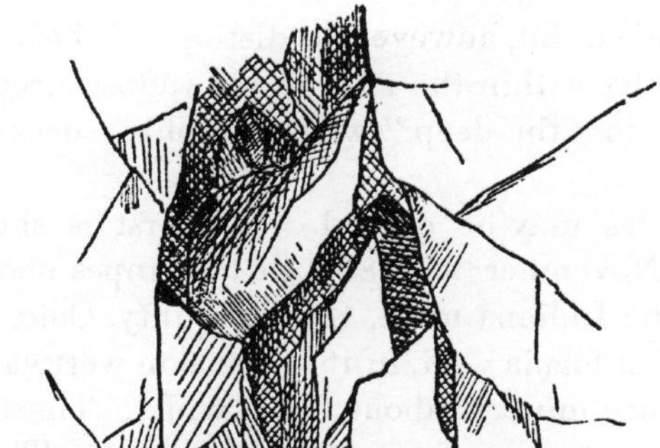

MAMMOTH MINE, ARIZONA.

an occurrence was observed in connection with certain faults which disturb the Virginie lode at Roure, near Pontgibaud, in France,* where, at a depth of 164 feet from the surface, a fault-fissure encloses a mass of clayey material containing boulders of a black, soft, and porous rock, which can be identified as

* See *Étude sur les gîtes metallifères de Pontgibaud* par M. Lodin, Ingénieur-en-chef des Mines. *Annales des Mines*, April, 1892, and "The Lodes of Pontgibaud," by the writer, in *Eng. and Min. Jour.*, August 11 and 18, 1894.

pieces of scoriaceous lava. No such rock occurred elsewhere underground; and the boulders must have been portions of the Quarternary alluvium which covered the outcrop of the lode, and fell into it at the time of its intersection by an open fissure, which long post-dated the formation of the ore-vein itself. The mines are in a district which has frequently been subjected to earthquakes, and in the heart of a region formerly the scene of great volcanic activity.

We must be careful, however, to distinguish between the formation of cavities within the zone of the vadose circulation, and their existence in "the deep," where sulphide ores have their origin.

Two examples may be quoted. The first is shown in Fig. 29, sketched November 25, 1895, in the stopes above the 800-foot level in the Indiana mine, Gilpin county, Colo. The lode, which is the California vein, in its extension westward from the Hidden Treasure mine, is about 2 feet wide. There is no parting or selvage separating it from the country. The latter is a quartz-feldspar rock, best described as granulite. Near the lode it is seamed and sprinkled with pyrite, and sufficiently gold-bearing to be sent to the stamp-mill. The main pay-streak is almost entirely composed of black zinc-blende which, by candle-light underground, contrasts strongly with its encasement of light gray country. The upper part of the vein, in this particular stope, consists of a breccia of zinc-blende, with an occasional spattering of wall-rock, the latter so disintegrated as to resemble gravel. At one point, A B, there is a shred of wall-rock lying across the vein. Lower down there are a number of cavities or vugs scattered among angular fragments of ore. It all looks loose, like an old stope filled with ore that has been mined, but the material is hard and difficult to detach without explosives. Lower again, the vein loses both its cavernous and its brecciated character, and consists of a compact body of blende. It may be added that, even where the brecciation is most evident, both walls are lined with a few inches of ore unbroken and firmly attached to the wall-rock into which it gradates. The vugs, when first found, were full of gas (CO_2, probably) and the miners suffered from bad air when working in ground of this character. The pieces of blende are held together by a siliceous cement, which also covers each fragment in the form of a gray-blue chalcedonic coating. It is almost certain that the cavities above described contained water, previous to the drainage of the ground by the penetration of the level underneath.

Fig 31

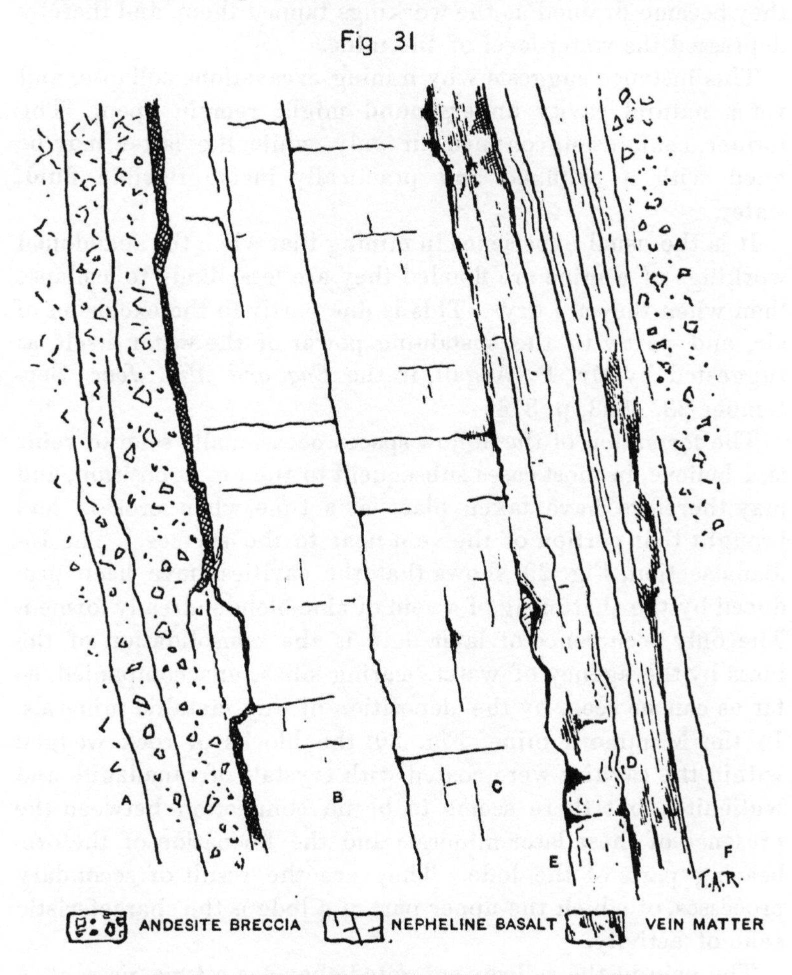

THE MOOSE VEIN, CRIPPLE CREEK, COLORADO.

Another instance is suggestive. In the Mammoth mine, Pinal county, Ariz., already described, the granite in the east cross-cut at the 300-foot level, north, has an extraordinary number of fissures partially occupied by broken pieces of rock, so wedged in as to leave open spaces. The pieces are not of any foreign rock, but are identical with the enclosing granite. Fig.

30 is a reproduction from a sketch made on the spot, March 15, 1893. The elongated cavities, such as that illustrated, were found full of water when first reached by the cross-cut; but they became drained as the workings tapped them, and thereby depressed the water-level of the mine.

This instance suggests why mining excavations collapse, and yet a natural cavity underground might remain open. The former contains unconfined air only, while the latter may be filled with a confined and practically incompressible fluid, water.

It is the usual experience in mining that when the abandoned workings of a mine are flooded they are less likely to collapse than when they are dry. This is due partly to the exclusion of air, and partly to the sustaining power of the water itself, as suggested by Mr. P. Argall, in the *Eng. and Min. Jour.*, September 23, 1893, p. 314.

The formation of the hollow spaces occasionally seen in veins is, I believe, in most cases subsequent to the ore deposition, and may therefore have taken place at a time when erosion had brought that portion of the vein near to the surface. The Indiana section, Fig. 29, shows that the cavities have been produced by the shattering of a vein of zinc-blende already formed. The only occurrence of later date is the consolidation of the mass by the agency of water bearing silica, unaccompanied, so far as can be seen, by the deposition of any metallic minerals. In the Mammoth mine, Fig. 30, the blocks of rock wedged within the cavities were coated with crystals of vanadanite and wulfenite; but there seems to be no connection between the presence of these later minerals and the formation of the ore-bearing parts of the lode. They are the result of secondary processes, of which the upper part of a lode is the characteristic zone of activity.

The vein in the railway-cut, cited above as a type, presents a filling readily recognizable as simply altered rock containing only an insignificant percentage of material foreign to the composition of the original granite. Nor is this an abnormal type of vein-structure. The rich gold mines on the adjacent hills afford numerous examples of this very kind of lode-formation. (And incidentally I would say that I know of no mining district which illustrates modern views on ore-deposition so clearly as does Cripple Creek.) Of such mines I would quote the Independence vein, whose richness is such as to cause its commercial value to obscure its scientific interest. It does illustrate very aptly, however, this part of our enquiry, because the ore is so very evidently only altered country-rock. In 1893, when the workings had not penetrated far from the surface, the car-loads of ore sent from this mine to the Denver smelters gave the impression that some one had blundered and either shipped waste from a cross-cut or else switched cars of ballast into the place of loads of ore. One could see that it was the normal Pike's Peak granite with its big pink feldspar, but

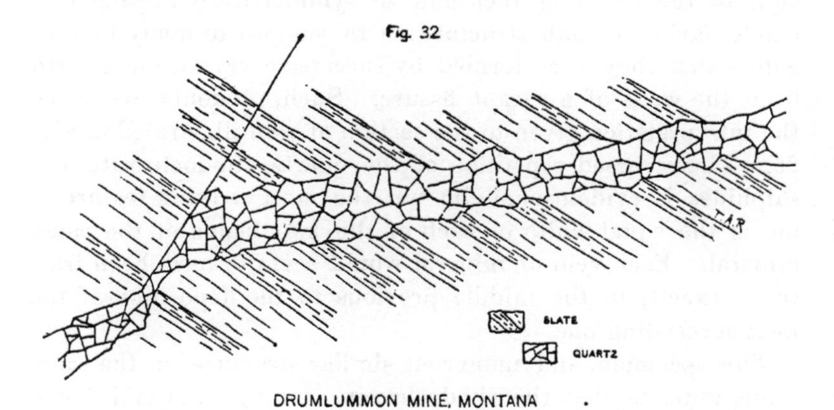

DRUMLUMMON MINE, MONTANA

it required a trained eye to note that the mica had been largely removed, leaving small iron-stained patches. It was ore by ourtesy, because there was enough gold present to give it as certain commercial value; but to the petrographer it was clearly granite, not much altered and but slightly mineralized.

The vein leaves the granite and, going northward, penetrates into andesite breccia. Its character remains the same; the ore is still altered country-rock; only now it exactly reproduces the structure of its new encasement, and the habit of the andesite breccia is quite evident, although blotches of sylvanite and fluorite may occasionally try to obscure it. The strike of the vein, its width, its richness, all appear unaffected by the passage from one formation into the other, while the change in the structure of the ore is so marked as to render it easy for the observer to know what is the enclosing rock without looking at the walls.

In a case such as this—and it is not abnormal—it is not necessary to suppose the original existence of an open cavernous fissure since the material of the vein is the material of the rock which was there before vein formation began. The vein follows a line which became a path for metal-bearing waters. Minute interspaces there probably existed, such as would be produced by the crushing and slight dislocation of particles of rock lying along a line of fracture; but a clear opening, a crevassing, such as accompanied the origination of the dike, seems hardly needed.

Occasionally, it is true, we do find veins full of minerals foreign to the encasing rock and so symmetrically arranged in bands having a comb structure as to suggest to many investigators that they were formed by successive crystalline growth from the walls of a vacant fissure. Such, no doubt, would be the interpretation given to the section of vein illustrated in Fig. 25. The reversed repetition of the quartz, rhodochrosite and sulphides is evident enough; but the most striking feature to me is the equal width of each of the two bands of the same mineral. Each vein of mineral would seem to have been fractured exactly in the middle previous to the deposition of the next succeeding one.

This specimen, and numerous similar structures in the same mine, indicate that the rhodochrosite was the first laid down, replacing, in part at least, the crushed rock which encased an original line of fault-fissuring. Subsequently another fracturing occurred, and this time the line of least resistance was the rhodochrosite itself, which, being homogeneous, broke down its center. The shattered carbonate offered an easy prey to the sulphide-bearing waters which laid down the blende and galena. The presence of bits of rhodochrosite within the sulphide band indicates that the substitution was irregular. Later, new disturbing forces were at play and the vein was fractured not only along its middle, as heretofore, but also along the lines of its contact with the encasing rock. These fractures were healed by the deposition of quartz, accompanied first by iron and copper pyrite, and then by rich silver-bearing minerals, such as the stephanite. The corrosion of the sandstone on the hanging had on that side irregularly widened the vein so as to give it greater strength; therefore the next movement, the last, took place along the foot-wall. This apparently resulted in nothing save the crushing of some of the encasing rock and the formation of a selvage whose removal produced the cavity which was so striking a feature of the stope.

Another typical illustration of this structure is presented by the Amethyst—Last Chance vein (at Creede, Colo.) which is certainly a magnificent example of an ore-break.* The country-rock, trachyte, has undergone multiple fracturing and ore has

Fig. 33.

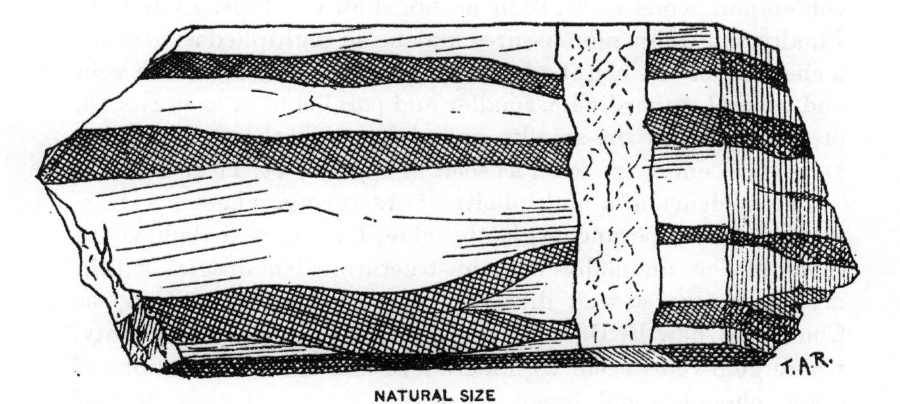

NATURAL SIZE

DRUMLUMMON MINE, MONTANA.

been deposited along the division-planes so that there are walls *ad libitum*. The regular ribbon-structure produced by the deposition of agatized quartz in a sheeted rock is very beautifully marked, and the same process of silicification is further evidenced in those places where the lode consists of breccia composed of pieces of country covered by concentric layers of agate. The lode itself is much wider than the pay-streak of silver-ore, which usually follows the foot-wall. On the hanging the boundary between vein and country is fairly discernible; on the foot less so, because for several feet beyond the ore there is a red jasperoid which gradates into country.

* At Red Mountain, in Ouray county, Colo., it has been the practice to speak of the veins (the Guston, Yankee Girl, and other celebrated lodes) as "ore-breaks," a break in the rock accompanied by ore—a term, it seems to me, much preferable to "fissure vein."

In the Enterprise example, Figs. 25 and 26, each succeeding fracture occurred in the mineral deposit which had healed the previous fracture. In other instances the mineral deposit appears to have proved harder than the encasing rock and the second fracturing took place near the original one, but in the soft rock rather than in the hard vein, thereby producing a new break parallel to the first one, and close to it, causing a repetition of vein-walls such as have already been described in connection with the sections given in Figs. 1, 9, 10, 15 and 21. Or there may be the production of companion-fissures forming contemporaneous veins, such as are shown in Figs. 13 and 19. Finally, the companion-fissures may be so multiplied as to cause a sheeting of the country and the formation either of one vein and several, subordinate, smaller and parallel to it, as in No. 15, or of a series of ore-streaks united by mineralized country so as to form one large lode, as seen in Figs. 5, 14, 17 and 31.

The evidence of a multiplicity of fracturing, whether successive or contemporaneous, is the clue, I venture to believe, to many of the anomalies of vein-structure. No district within my knowledge so well illustrates this aspect of the inquiry as Colorado's new El Dorado, Cripple Creek, in El Paso county, where gold-veins occur as mineralized and enriched portions of dikes, phonolite and basalt, traversing masses of andesite tuff and breccia. Other types are observable, but these are to-day the most characteristic. The mineralized rock forming the vein and that less distinctly gold-bearing country which encloses it, have been subjected to such multiple fissuring as to produce a very marked division of the rock into parallel bands or sheets which may be a fraction of an inch apart or several yards. This structure can be seen no less in hand-specimens than in blocks an acre big. The Moose vein, on Raven hill, is a fair example. It is illustrated in Fig. 31, as seen October 27, 1895, in the back of the sixth (or 350-foot) level. A is andesite tuff and breccia, B C D is a dike of dark, blue-gray nepheline basalt, subdivided into two barren parts, B and C, and one ore-bearing portion, D. Native gold and telluride compounds (sylvanite and calaverite) occur along the seams in the basalt where it is decomposed and iron-stained. The pay-streak extends from E to F, about 10 inches.

This sheeting or multiple fissuring was probably the result of shrinkage accompanying the cooling of the volcanic rock. The fractures have a contemporaneity of origin quite distinct from the successive ruptures discussed in connection with the ribbon-structure of the Enterprise section. The latter were marked by the precipitation of diverse minerals, while those of a Cripple Creek vein are characterized by a similarity of mineral deposition.

Cases also occur where there can be discerned a combination of both these types of multiple fissuring.

A line of weakness, or even a region of weakness, once developed in the earth's crust is apt to continue to be a line of least resistance available for future fracturing. Even when a quartz-vein is formed along a line of rupture, healing the break and strengthening it with a substance harder than the rock-walls themselves, we may suppose that the next break will take place along the line of weakness presented by the imperfect cohesion existing along the plane of contact between the hard quartz and the less resisting rock.

The gradual penetration of mineral solutions into the immediately encasing country may finally obliberate the divisions due to multiple fissuring. The sheets of rock separating one from the other would be replaced by ore, and nothing might remain of the original structure save faint partings in the lode, such as are less evident to the eye than to the hand of the miner who instinctively uses them to assist him in breaking ore.

Thus, I believe, the collection of observations in various mining districts tends to the modification of that idea of clean-cut definition which accompanied the early ideas of vein-structure. The evident contact between two dissimilar rocks, such as is seen along the walls of a dike, will be often found in veins to be replaced by an indistinct gradation from mineralized to unmineralized rock, originally the same but now rendered unlike by the selecting action of chemical solutions.

We are justified, however, in putting some limit to the depth of possible ore-formation, since that formation is dependent on the presence of water. The record of the largest number of careful observations has shown that as we sink into the earth the increment of temperature is 1° F. per each 48 feet of descent. At this rate the critical point of water would be reached at 34,704 feet or 6½ miles from the surface. Where the temperature is that of the critical point (773° F.) water cannot exist as a liquid no matter how great the pressure, but becomes dissociated into its gaseous elements. Moreover we are warranted in believing that the thermometrical gradient becomes more rapid at depths beyond those reached by human observation because of a decreased conductivity in the rocks, or as Professor Prestwich, the best authority on these matters, puts it:*

"Taking into consideration the probable limitation of the percolation of water, and the possible diminution of conductivity with increase of depth, if there should be any alteration in the thermometric gradient, at great depth, it will be more likely to be in the direction influenced by these more or less certain factors."

Therefore, taking these conditions into consideration, we may expect the circulation of water to cease at 20,000 feet or thereabout. But at the maximum depth the maxima of temperature and pressure must obtain. It must necessarily be a horizon of solution. Precipitation would hardly begin until a lowering of the temperature and a lessening of pressure permitted it. The deposition of ore is the direct result of precipitation, therefore actual ore-formation is likely to be limited to a depth often of 15,000 feet.

It is not difficult to surmise why clean-cut fractures are not necessarily most favorable to ore occurrence. In the Drummond mine, Montana, the distribution of the ore appears to be connected with the change in the angle of intersection between the course of the veins and the strike of the slate country. Most of the ore-bodies have been found where the course of the veins (N. 15° E.) cuts the slates at an oblique angle and the levels run out of ore when their direction is either at right angles to, or conforms with, the strike of the country (N. 17° W.).

Fig. 32 is a sketch made in one of the surface-workings of that mine which illustrates in miniature the fact above noted It represents a small quartz-seam 2 inches wide, traversing the slates, whose structure is very clearly marked by the color bands following lines of original sedimentation. Near the left of the sketch the quartz follows a joint and becomes narrowed, while where it crosses (along a line of slight dislocation) the country it has irregularities and bulges which answer to the alternating slate-bands. A rough, ragged fracture, when continuous, may

* *Controverted Questions of Geology*, by Joseph Prestwich, D.C.L., F.R.S., etc., Macmillan, 1895, p. 247.

be expected to offer more surface to solvent action and more, but not too many, obstacles to a rapid circulation of underground waters. Its structure also means more opposition to the closing in of the walls, because the irregular faces of the fracture, when they come together, leave openings which, if not along one section then along another, have intercommunication, and so permit of a passage which would be badly impeded, if not absolutely stopped, by the closing in of smooth walls.

Fig. 33 represents, to actual scale, a piece of slate enclosing a quartz-vein, which came from near the end of the 700-foot level, also in the Drumlummon mine. It so happens that this is a true illustration in miniature of what the lode itself is doing at this point. The New Castletown lode, on which the level is driven, is at this point cutting at right-angles across the bedding of the slates and is barren of ore. In the hand specimen, reproduced in the drawing, a quartz-vein, not quite half an inch wide, cuts perpendicularly across the slate whose bedding is rendered beautifully marked by dark bands. The vein has a uniform width, it has regular well-defined walls guiltless of the projections and bulges noticed in the previous illustration. It may be only a convenient coincidence, but it is a fact that the quartz in Fig. 25 was opalescent and destitute of other minerals while that in Fig. 24 was true ferruginous vein-quartz.

Thus underground work bears daily testimony to the close dependence of ore-occurrence upon the geological structure of the enclosing country, a relation, the importance of which Mr. S. F. Emmons has done invaluable service by clearly stating in more than one of his contributions to the *Transactions*. Wanting a proper understanding of the structure of the rock encasing his vein, the miner gropes but blindly in a maze of tangled phenomena until the geologist, by their proper elucidation, gives him a light which dissipates much of the darkness obscuring his progress underground.

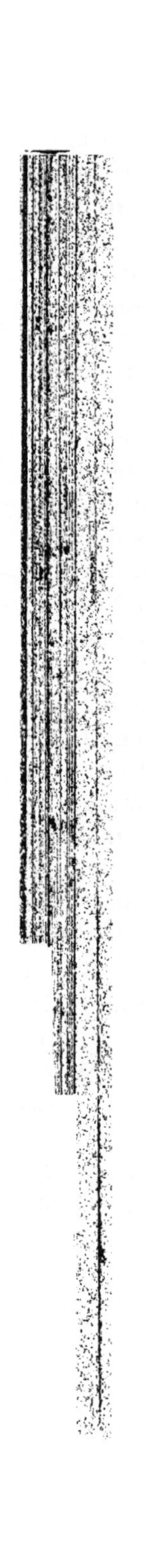

Subject to Revision.

[TRANSACTIONS OF THE AMERICAN INSTITUTE OF MINING ENGINEERS.]

THE BENDIGO GOLD-FIELD.

BY T. A. RICKARD, ALLEMONT, ISERE, FRANCE.

(Glen Summit Meeting, October, 1891.)

AMONG the names which won a world-wide fame during the golden age of the early fifties, Bendigo and Ballarat were to Australia, what the Yuba and Grass Valley were to California. The map of Victoria did not for a long time show the name of old Bendigo;* for this first and more distinctive name was replaced with the more English Sandhurst, just as the alluvial diggings gave place to quartz mining. Towards the close of last year, steps were taken to give back the old name, as associated with the early days of rich alluvium, and more suggestive than the application taken second-hand from an English military academy.

HISTORY.

The first discovery of gold† in this district, was made in the autumn of 1851; but there has never been any certainty as to the day or the man. At that time the country around Bendigo Creek was a part of the Ravenswood sheep-run, and its resemblance to the Forest Creek district (now Castlemaine), induced the first prospecting. It was late in November when the "rush" broke out; the shepherds left their flocks; the sylvan solitudes were disturbed by

* The name of Bendigo is said to have been derived from a hut-keeper on the Ravenswood sheep-run, who on account of his fondness for "fisticuffs" was nick named Bendigo, after the prize-fighter of that name. It is not aboriginal, as is often supposed, but Spanish; and equivalent to our Benedict. There must always be some confusion between the two names of Sandhurst and Bendigo; the town, and with it the gold-field, having been three times named. Several other old familiar names have been likewise unfortunately replaced by second-hand English ones; so that an old digger talks of Forest Creek when he means the modern Castlemaine, Mt. Ida for Heathcote, Growler's Creek for Bright, etc. I shall use the names Sandhurst and Bendigo interchangeably.

† The first discovery of gold in Australia, was made by E. H. Hargraves, February 12, 1851, near Bathurst in New South Wales. In August of the same year the discovery at Buninyong, near Ballarat, inaugurated the first of a series of rushes to the Victoria gold-fields.

the voices of the first "diggers;" the green glades of Bendigo Creek were ruthlessly uptorn by pick and spade; and the busy energetic life of a young and thriving mining camp replaced the sleepy idle routine of an out-of-the way sheep-station.

Since that date the gold-field has yielded over fifteen million ounces of the precious metal, valued at more than sixty millions sterling. Like most mining districts, it has passed through days of severe depression and extreme inflation. Until 1854 the alluvium only was worked. The pan and the cradle early gave way to the paddling-tub, a machine more suited to the clayey character of the wash. By reason of the limited supply of water, the long tom was never very widely used. The highest output on record was reached in 1853, when 661,729 ounces, valued at £2,646,800, were obtained from the alluvium only. Long before the diggers were aware of the real value of gold-quartz, they used to amuse themselves by breaking off specimens from the outcrops for decorative purposes.* Diamond Hill owes its name to the beautiful specimens which its surface yielded in the early days. Vein-mining or "quartz-reefing" had its inception with the discoveries made by Ballestedt on Victoria Hill, which has now been pierced for a depth of over half a mile, and is honeycombed with deep workings. From 1854 to 1862, the quartz-mining industry had a chequered career; as will be readily understood when the peculiar character of the ore-deposits has been passed in review. In 1859 the first regular registration of a quartz-mine† took place. About this period the shallow alluvial deposits became exhausted and, in the absence of deeper channels, the whole energy of a very enterprising community was concentrated upon the exploitation of the quartz-lodes. In 1862 the first limited-liability company was brought out; and in the immediately succeeding years the steady development of the field and the increasing knowledge of the lode-structure were accompanied by an increasing gold-output, which rose in 1870 to 241,380 ounces. The formation of numerous companies which marked the early seventies, introduced fresh capital into the district, and thus led to great increase in the work done in the mines, the output from which averaged at this time over 300,000 ounces per annum. But company-promotion soon devel-

* So ignorant, it is reported, were the pioneers of gold-mining in those days, that many immigrants went to the rushes or mining stampedes with the idea that "gold in quartz" meant gold in *quarts* or pailfuls!

† This was the Johnson's gold-mine, which is still a rich producer, having headed the dividend list for 1890 with £30,800.

oped into a mania which rapidly did its evil work. Paper mining replaced honest work, as it has done in many another mining district before and since. A reaction set in; the gold-field saw its darkest days; and a collapse took place which ended the wild speculation in bogus companies and marked the commencement of a new era of steady progress. The opening up of deeper ground and a renewal of

Fig. 1.

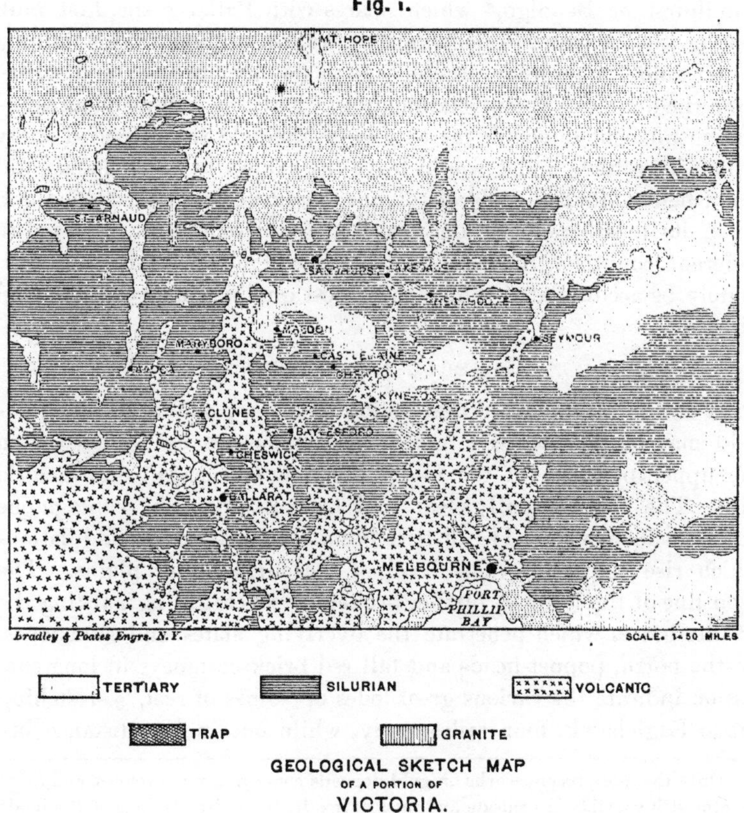

GEOLOGICAL SKETCH MAP OF A PORTION OF VICTORIA.

rich discoveries, brought about a revival in the first years of the last decade, which, accompanied as it has been by great enterprise, and the intelligent development of the mines with a more accurate knowledge of their ore-deposits, has enabled the Bendigo gold-field to hold the first place among the quartz-mining districts of Australasia, and to make its record unique in the statistics of this industry.

To the members of the Institute, probably few mining districts are less known, considering their importance, than those of the colony of Victoria. The steamship, the railway, and the telegraph, have linked together the most distant mining centers, Broken Hill and Virginia City, Johannesburg and Grass Valley; but the older gold-fields of Australia have received but a superficial and passing notice from the pens of standard writers. More particularly is this true of Sandhurst or Bendigo,* which shares with Ballarat the first rank among the Australian localities of quartz-mining. This mining center and its peculiar ore-deposits it is my purpose to describe, from the notes of a recent examination. The time—three months†—was insufficient to enable me to give to the field the extended study which it invited; and I would leave the subject to better hands, did I not believe that the objects of our Institute are best carried out, when its members place on record the different facts observed in different countries, in order that a mass of observations may ultimately be accumulated, which those best qualified may combine and discuss.

General Description.

At the outset a general description of the appearance of the gold-field may be of interest. A very good view can be obtained from the upper platform of the Old Chum poppet-heads,‡ on Victoria hill. The air is usually clear, and from this point of vantage one can see a long way. The district lies among a series of undulations which rise above the tame level of the surrounding plains. The even line of the latter is further broken by several hills, marking the granite bosses which penetrate the overlying slates and sandstones. To the north, poppet-heads and tall red brick chimneys in long succession indicate the various great lodes or "lines of reef,"§ stretching out to Eaglehawk, four miles away, while out in the distance be-

* Only the more recent works on gold-deposits contain any references, and these are almost invariably inaccurate and misleading in their descriptions of the mode of occurrence of the quartz. Lock's *Gold* which is, generally speaking, a very complete compilation on the subject, contains only one reference to Sandhurst, comprised within a paragraph of a dozen lines.

† April, September, and part of October, 1890, also a part of February, 1891.

‡ "Poppet heads" is the English and colonial equivalent for the Western "gallas" or gallow's frame. "Three legs" and "heapsteads" are the names used in the north of England.

§ A "line of reef" is a reef or lode taken as a whole. The mines of Amador, Calaveras, and Tuolumne counties would be said to be on "the same line of reef," viz., "the Mother-lode" of California.

yond is the dark blue sea of the trackless bush,* broken in front by the hills at Kerang and to the right by the promontory of the White hills,† a name familiar to every Australian digger. White hills they are no longer, for the "cement" which covered the rich alluvium has become oxidized, and patches of red gravel-heaps have replaced what was once glistening white. Between these and the spectator lies the modern city of Sandhurst, its well laid out streets lined with the English trees which have replaced the gums that once covered the site. A wavering, blue, irregular line marks Bendigo creek, once flowing through the forest glades, but now meandering past tailing-heaps and back yards. Out beyond the town we

Fig. 2.

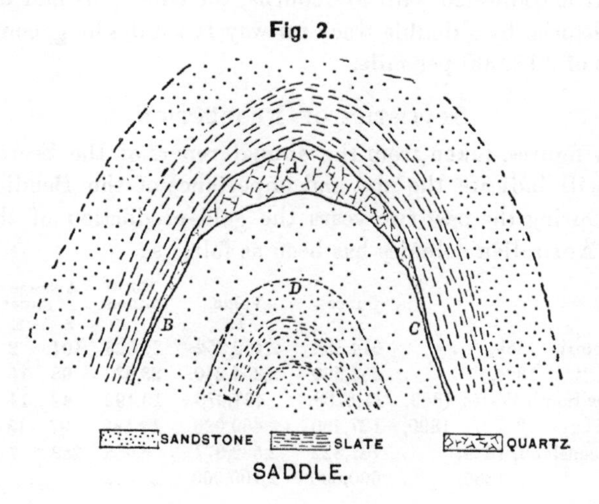

SADDLE.

can dimly see Mt. Ida, where Heathcote lies. Southward the succession of poppet-heads marks the auriferous belt. The engine-houses of the South Bellevue and Eureka Extended mines indicate the New Chum reef, on which there is operated to-day perhaps the greatest series of deep gold-quartz mines which the world has yet seen. A new church rises above Golden Square, the scene of some of the first and richest of the alluvial diggings. The view is further diversified by gray heaps of tailings, bluish piles of waste rock, gar-

* The forest of *Eucalyptus*, which at one time covered the whole of the habitable portion of the Australian continent.

† The White Hills of Bendigo and the Black Hill at Ballarat are in the history of Australian gold discovery what Mokelumne Hill in old Calaveras, and Table Mountain in Tuolumne were to California, or what Ruby Hill at Eureka, and Mt. Davidson at the Comstock were to Nevada.

dens and houses, among which is seen the sinuous curve of the railway along which the Melbourne express is now coming. Further south, in the distance, are the granitic slopes of Mt. Alexander, beyond which lies Castlemaine. To the west, clouds of sulphurous smoke indicate the various pyrites-works, and a black line against the blue sky marks the water supply flume, while immediately below us are the gardens and house of one of the mine-owners,* an oasis of pleasant green among the stern practical surroundings of mines and mills.

Sandhurst or Bendigo has a population of over 30,000 inhabitants, several fine public buildings, and some very beautiful public gardens. It is connected with Melbourne, the chief port and metropolis of Victoria, by a double-track railway 101 miles long, constructed at a cost of £18,000 per mile.

Statistics of Production.

A few figures, taken from the annual report of the Secretary for Mines, will indicate the size and importance of the Bendigo gold-field. During the past two years the gold-production of the three leading Australian colonies has been as follows:

	Ounces.	Value.	No. of miners.	Average per miner.		
		£.		£.	s.	d.
Victoria, 1889,	614,838	2,459,352	24,323	101	2	2
" 1890,	588,560	2,354,240	23,833	98	15	7
New South Wales, 1889,	119,758	434,070	10,192	42	11	9
" " 1890,	127,460	459,086	12,182	37	13	8
Queensland, 1889,	737,822	2,582,377	8,955	288	7	5
" 1890,	600,000	2,100,000	....	....		

The year 1889 was the first in which Victoria did not lead the colonies, being surpassed by Queensland through the Mt. Morgan mine, whose output alone was over 300,000 ounces. The marked decrease in the yield of Queensland for 1890 is similarly due to a decline in the output from Mt. Morgan. The explanation of the Queensland average of £288 per miner, in 1889, is explained by the fact that the most of the value of the product of Mt. Morgan in that year found its way into the pockets of half a dozen men.

The yield of 588,560 ounces (1890) is the lowest on record in the history of gold-mining in Victoria. There has been a gradual decrease during the past decade, mainly due to the exhaustion of the more readily accessible alluvial deposits.

* Fortuna Villa, the residence of M. George Lansell.

From 1851 to the end of 1890 the yield of gold from Victoria amounted to 56,870,574 ounces, valued at £227,482,296, an average per year of 1,452,761 ounces, the highest for any single year being 3,053,744 ounces, in 1856. Of this total, Bendigo has contributed 11,168,414 ounces, valued at £44,673,656. In addition, it is estimated that the gold taken away privately to Melbourne and the neighboring colonies, and not included in the government returns for the district, will amount to 4,000,000 ounces, which would give a round total of over 15,000,000 ounces, valued at over £60,000,000 sterling.

Victoria is divided into seven mining districts, of which the

Fig. 3.

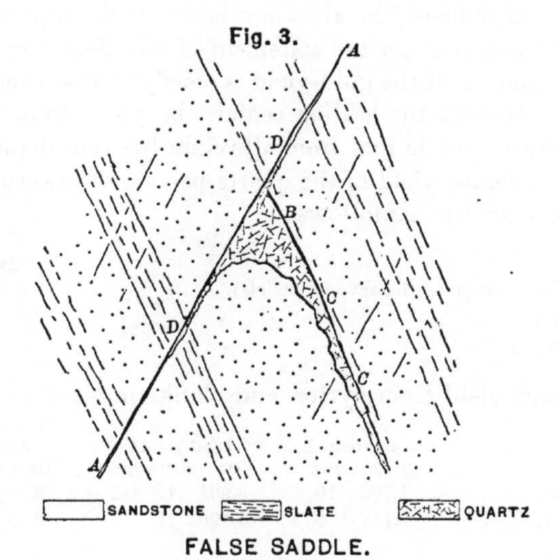

FALSE SADDLE.

most important are Ballarat and Bendigo, the former being the chief alluvial center, while the latter leads the quartz-mining districts. The total area of the Victoria gold-fields, 86,760 square miles, is worked and prospected by a force of 23,833 miners; and the proportion belonging to the two principal gold-fields is as follows:

	Area in sq. miles.	Miners.
Bendigo, ...	5,870	4,420
Ballarat, ...	5,180	6,249

While the mining operations in the Ballarat district are distributed over an actual area of 40 square miles, and the boundaries include such important centers as Clunes and Creswick, those of the Bendigo district are practically concentrated upon an area of 21 square miles within the adjacent townships of Eaglehawk and Sandhurst. During the past 2 years the yield of gold has been:

	Alluvial.			Quartz.			Total.		
	Oz.	dwts.	grs.	Oz.	dwts.	grs.	Oz.	dwts.	grs.
Bendigo, 1889,	6,973	12	10	134,547	8	21	141,521	1	7
" 1890,	3,293	3	18	134,671	10	11	137,964	14	5
Ballarat, 1889,	98,342	6	13	117,321	18	0	215,664	4	13
" 1890,	92,836	2	10	117,597	0	8	210,433	2	18

It is seen that, while Ballarat is the leading alluvial district, it produces also a large proportion of gold from quartz,* while at Bendigo, on the other hand, the alluvium is relatively important. The figures just given bear out the statement of the Secretary for Mines that the diminution of the gold-yield is chiefly in that from the alluvial mines. In both the leading centers the yield from quartz has slightly increased, while that from alluvium has considerably diminished. The average yield of the quartz per ton of ore crushed, during the past year, was as follows:

	dwts.	grs.
Victoria (average of the seven districts),	9	4
Bendigo,	9	5
Ballarat,	7	21

The average yield from pyrites and blanketings was:

	Total treated.		Total yield.			Average.		
	Tons.	cwt.	Oz.	dwts.	grs.	Oz.	dwts.	grs.
Bendigo,	1,766	10	3,901	16	12	2	4	4
Ballarat,	2,148	0	5,196	19	7	2	8	9

The highest and lowest prices paid for the gold per oz. were, at Bendigo, £3 17*s.* and £3 19*s.*, and at Ballarat, £3 17*s.* 6*d.* and £4 3*s.*

While in the colony as a whole the number of miners is pretty equally distributed between alluvial and quartz mines—11,470 in the former to 12,363 in the latter—there is, in the two leading districts, a greater disproportion; Bendigo having 949 alluvial and 3375 quartz-miners, while Ballarat has 2440 alluvial and 3677 quartz-miners. At Sandhurst there are 270 Chinamen, nearly all engaged in surface-washing, while at Ballarat there are 736 Mongolians to 5508 Europeans.

* At Ballarat, as the deep leads (alluvial) were becoming worked out, the croppings of the quartz-veins were found in the bed-rock of the alluvium. Several claims which have been rich in alluvium are now good quartz-mines.

The Bendigo gold-field includes several scattered subdivisions; but it consists practically of Eaglehawk and Sandhurst (or Bendigo), two distinct municipalities, forming one long straggling township, somewhat after the manner* of Gold Hill and Virginia City, or Blackhawk and Central City.

In 1890 the machinery employed in quartz-mining was thus distributed:

	Steam-engines.	Stamps.	Machine drills.	Concentrators.	Arrastras.
Sandhurst, ...	175	625	70	65	25
Eaglehawk, ...	105	501	40	40	6

For the Bendigo gold-field as a whole, the numbers are as fol-

Fig. 4.

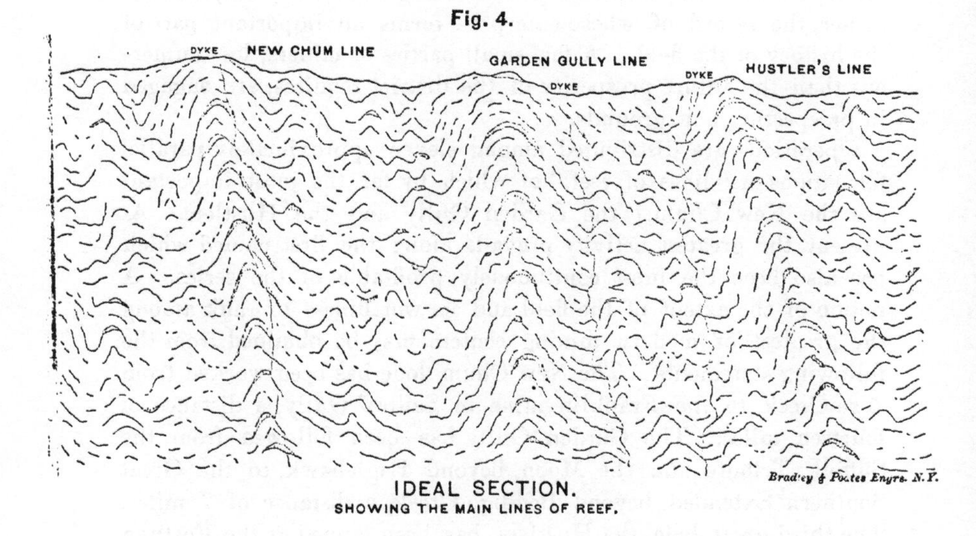

IDEAL SECTION.
SHOWING THE MAIN LINES OF REEF.

lows: Steam engines, 316; aggregate nominal, H.P. 6873; stamp-heads, 1328; whims, 76; whips, 127; machine-drills, 128; concentrators, 105; arrastras, 82. The total value is estimated at £478,611. Considering the size and yield of the district, the value of the machinery is small indeed, when compared to that of a Western American mining-camp, with its much larger and more costly hoisting-engines, mills, etc.

* But the mining townships of Victoria are very different from those of the great West. Ballarat, Sandhurst, Clunes, and Creswick are pretty towns, situated in rich agricultural districts, and presenting none of that bare ruggedness which is characteristic of most mining camps of Colorado, Nevada, and California.

The quantity and cost of the timber consumed for mining purposes during the year were as follows:

Firewood,*	146,628 tons.
Props and cap-pieces, ...	105,533 pieces.
Laths and slabs, ...	80,820 "
Sawn timber, ...	821,232 feet.

The value of this is set down at £63,856 6*s.* 6*d.*

Mining is carried on for the most part by companies, whose share capital is invariably small and their reserve capital usually *nil.* The tribute-system is largely in use, and, here as elsewhere, often leads to the discovery of valuable ore-bodies. There are few private mines, the most important of which belong to Mr. Lansell, a mine-owner, the record of whose enterprise forms an important part of the history of the field. A few small parties of miners, less numerous than the future prosperity of the district requires, are engaged in prospecting new grounds.

Operations are distributed among eleven approximately parallel formations or "lines of reef," of which by far the most important are the New Chum,† the Garden Gully and the Hustlers. At present the greatest activity prevails along the first-named which has also been the most continuously profitable of the series. A notion of the extent of the field and its worthiness to rank among the greatest of modern mining centers may be obtained from the following statements. The New Chum lode has been worked from Axe Creek to the Franklin mine in Sailors' Gully, a distance of fourteen miles. The Garden Gully has been followed from the Suffolk Tribute and the Moon, beyond Eaglehawk, to the Great Southern Extended, beyond Bendigo Creek, a distance of 7 miles. The third great lode, the Hustlers, has been tapped at the Fortuna Hustlers in the city of Sandhurst, and from there has been worked as far as the King of Prussia, in Opossum Gully, which is 5 miles away. So much for the extent of working along the strike. Downwards, the New Chum has been followed for a vertical depth of over half a mile (2641 feet), the Garden Gully for over 2300 feet and the Hustlers 2000 feet. There are in Sandhurst 18 shafts exceeding in depth 2000 feet. Several of these are still going down; and of the

* Fifty feet of firewood equal one ton of 2240 pounds. Laths and slabs are the same as what the American calls "lagging."

† "New Chum" is the colonial equivalent of the Western "tenderfoot," *i.e.*, a fresh arrival in the country.

20 companies among which these 18 shafts are distributed, 3 are in rich ore, 7 are breaking pay-ore, 9 are prospecting and 1 is idle.

Though the output has suffered during the past two years by reason of the greater attractions presented by gambling in silver shares at Broken Hill, Sandhurst contained during the past year 28 dividend-paying companies which produced 101,879 ounces 10 dwts. (not including 7 tribute parties which were dividend paying), enabling the payment of £149,381 17*s.* in dividends. The work of the past year has shown an improvement as indicated by the following comparison. In 1889 the calls amounted to £137,489, the

Fig. 5.

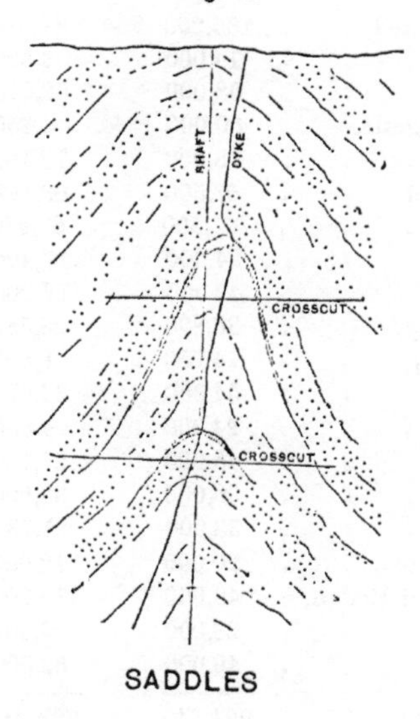

SADDLES

dividends to £118,473, a balance on the wrong side of £19,016. In 1880 the calls were diminished to £111,142, while the dividends increased to £149,381, leaving a profit of £38,239. During the last 14 years the dividends have exceeded the calls by £1,101,836, the year 1889 being the only unprofitable period. Roughly speaking, during 1890 each ounce of gold was obtained at a cost of £2 10*s.* 8*d.*, leaving £1 9*s.* 4*d.* as profit for the sharehoulders.

Out of the 28 companies which appeared on the dividend-list at the beginning of 1891, the 20 whose records are obtainable show that only four have failed to pay back the capital expended upon them, and of these four, one is not yet a year old. The average total dividends of each of these 20 properties* amounts to £81,947, the average nominal capital of each is £49,742, while the average paid up capital is only £34,167. It is seen, therefore, that these 20 companies have returned in dividends nearly twice the amount of capital called up for their equipment and development. Further, it should be stated that the number under consideration does not include several of the mines† which have the

* The following list gives the detailed figures of the twenty companies which were dividend-paying at the beginning of this year (1891). The results are given up to the end of 1890:

Company.	Nominal capital. £.	Paid up capital. £.	Dividends. £.
Catherine Reef United,	135,200	82,810	50,755
Fortuna Hustlers,	14,000	3,500	2,800
Gt. Ex. Hustlers,	68,000	62,050	419,200
Hercules and Energetic,	60,000	8,250	74,625
Johnson's Reef,	78,000	72,100	223,950
Lazarus New Chum,	67,500	63,187	67,500
Lazarus No. 1,	67,500	63,950	95,062
Lady Barkly,	24,000	17,400	58,315
New Chum Con,	42,000	18,200	132,300
New Chum Railway,	36,890	28,589	52,078
New Chum United,	14,750	8,850	66,375
North Old Chum,	54,000	33,075	86,495
Rose of Denmark,	24,000	4,800	63,600
Shamrock,	45,000	27,750	15,750
Shenandoah,	96,000	33,600	59,600
South New Chum,	32,000	21,735	800
Specimen Hill United,	20,000	12,000	23,500
United Hustlers and Redan,	48,000	19,200	114,000
United Devonshire,	28,000	9,800	21,700
Young Chum,	40,000	32,500	10,550
Totals,	994,840	623,346	1,638,955
Averages,	49,742	31,167	81,947

† The greatest producers have been the following mines:

Name.	Called up capital. £.	Dividends. £.
Garden Gully United,	21,642	667,796
Gt. Extended Hustlers,	62,050	419,200
Johnson's Reef,	72,100	223,950
Gt. Hustler's Tribute,	61,200	620,200
North Johnson's,	31,850	148,625
United Devonshire,	8,244	224,000
New Chum Con.,	18,200	132,300

There are many others, the totals of whose dividends is more than £100,000.

greatest records, but have temporarily dropped out of the dividend-list.

The instances of individual productiveness here have been surpassed in other parts of the world; but it is doubtful whether in the history of gold-mining there can be shown a better record in the proportion of dividends paid to capital expended. The product of the mines from tribute-parties is not included in the above list unless especially mentioned. It is a matter of record, for instance, that, including the product of the tributers, the Garden Gully United has yielded over £1,000,000. The Kentish mine of the late I. B. Watson is said to have yielded over £2,500,000; and another private mine, G. Lansell's "180," has given magnificent profits to its owner, as have also several tribute-workings, the records of which are not accessible now.

To the Australian the name of Sandhurst is always associated with

Fig. 6.

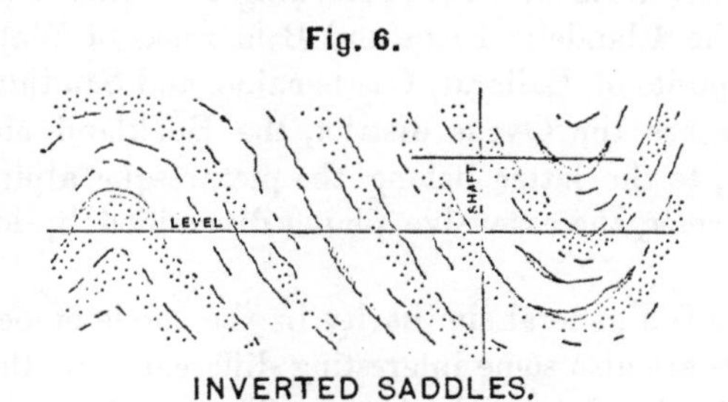

INVERTED SADDLES.

the "saddle-reefs." It was the frequent mention of these which led me to visit the mines; and it is the peculiar interest which they must have for all geologists and mining engineers which has induced the publication of these notes. Before discussing, however, these and other matters of detail, a general account of the geology of the district should be given.

Geology.

As will be seen by a reference to the map* of Victoria (Fig. 1), the gold-field is situated near the northern edges of the exposure of the Lower Silurian, which, a short distance further north, is overlain by Pliocene shales. The boss of granite at Mt. Hope suggests

* The geological map of a portion of Victoria which accompanies this paper, was copied by me from the geological sketch-map in *The Geology and Physical Geography of Victoria*, by R. A. F. Murray, the geologist of the Victorian government.

the deeper masses of crystalline rocks which also form the Mt. Alexander ranges, dividing the Sandhurst and Castlemaine gold-fields. The gold-mining districts of Victoria are almost entirely confined to beds of Upper and Lower Silurian age, of which Mr. Murray says:* "As surface or underlying rocks, they occupy the greater part of Victoria from the sea-coast to elevations exceeding 6000 feet." They form the bed-rock of the alluvium the yield of which astonished the world in 1851, and the country-rock of the quartz-lodes from which that alluvial gold was derived. Selwyn† computed their total thickness to be not less than 35,000 feet. While the line of division between the two horizons has not been found, certain differences in lithological character and fossil remains have led the Victorian Geological Survey to refer to the Upper Silurian that portion of the Silurian rocks lying east of a line drawn from Melbourne to Heathcote, while the Silurian west of that line is regarded as Lower. It is in the lower horizon, which, according to Professor McCoy, corresponds to the Llandeilo Flags and Bala rocks of Wales, that the auriferous deposits of Ballarat, Castlemaine, and Sandhurst‡ occur. In the upper are the Ovens district, the Buckland and Harrietville. Again, to the latter belong the picturesque alpine districts, and to the former, the extensive plains diversified by low rounded hills.

While there is a general similiarity in the mode of occurrence of the gold, there are also some interesting differences in the habits of the quartz-lodes in the two horizons. The auriferous deposits are found traversing certain defined belts, which have a general strike 20° to 30° west of north. These parallel belts contain veins of quartz which conform to the general strike, and which, like the belts themselves, are separated by barren portions of country. My observations lead me to believe that the gold-veins of the Lower Silurian are more often conformable to the stratification, while those of the Upper are more frequently true fissures, traversing the country at an angle to the bedding. R. A. F. Murray, the government geologist, has noted that while the quartz-reefs in the Upper Silurian

* Page 33 of the book mentioned in the preceding note.

† A. R. C. Selwyn, formerly head of the Geological Survey of Victoria, and now occupying a similar position under the Canadian government.

‡ The fossils most common in the rocks at Sandhurst, are graptolites, particularly *stellatus*, *extensus* and *tripedes*, together with *Sertularia magna* and *vergata*, *Didymograpsus*, *fruticosus* and *Phyllograpsus folium*. In the weathered slate-beds they are easily found, but underground the dark color of the rock, and the development of fine cleavage renders them difficult of recognition.

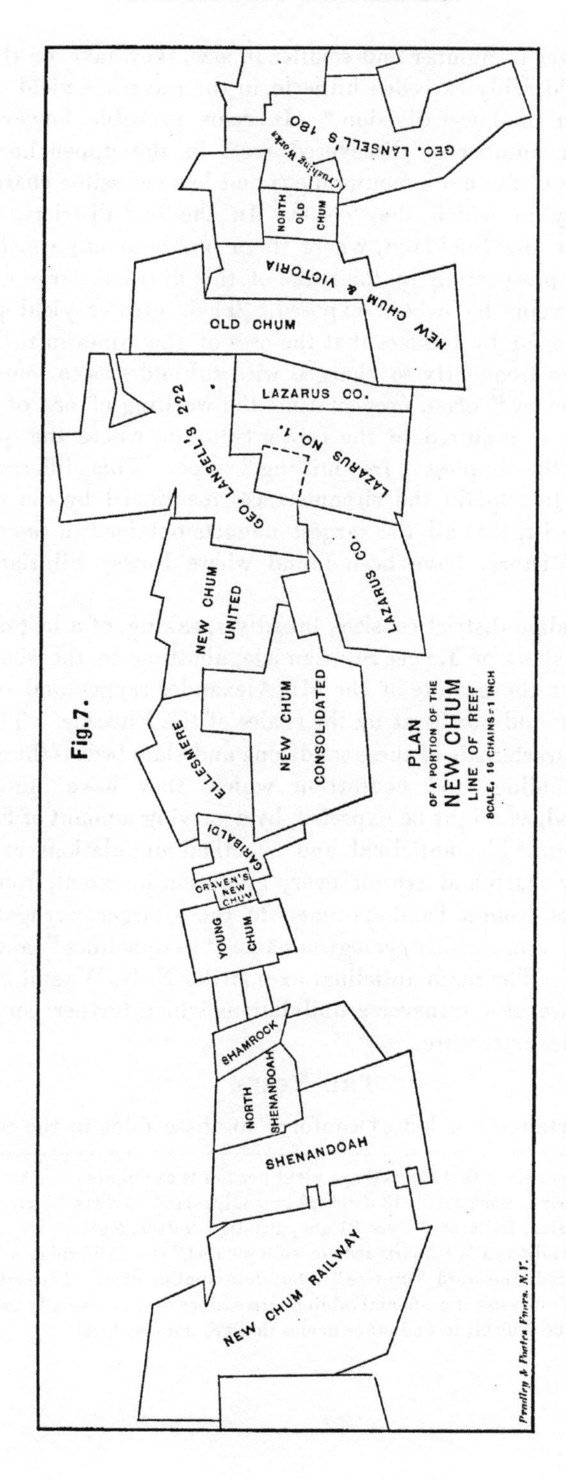

Fig. 7.

PLAN OF A PORTION OF THE NEW CHUM LINE OF REEF
SCALE, 16 CHAINS = 1 INCH

may be fewer in number and smaller in size, they have, on the other hand, considerably exceeded hitherto, in their average yield of gold, the reefs in the lower division.* It seems probable, however, that the smaller number of discovered reefs in the upper horizon is largely due to the more mountainous and less accessible character of the country in which they occur. In the few districts, such as Bright and the Buckland, where there has been any considerable amount of prospecting in the rocks of this division, large numbers of parallel veins have been exposed. Their greater yield per ton is accounted for by the fact that the ores of the mines in the Upper Silurian are frequently so charged with sulphides as to come under the "refractory" class, necessitating the working of ore of higher tenor than is required in the Lower Silurian, where the quartz is usually of the simplest "free-milling" type. This difference will also serve to explain the circumstance, mentioned by the government geologist, that all the largest nuggets obtained in the alluvial mines of Victoria have been found where Lower Silurian rocks prevail.

The Bendigo district consists, broadly speaking, of a belt of sandstones and slates of Lower Silurian age, abutting to the south and west against the granite of the Mt. Alexander ranges and overlain to the north and northeast by the shales of the Pliocene. The most marked characteristic of these sandstone and slate beds is the extreme bending, folding, and contortion which they have undergone, accompanied, as might be expected, by a varying amount of fissuring and faulting. The anticlinal and synclinal undulations are often remarkably sharp and exhibit every gradation in extent, from a few feet to miles, from a hand-specimen to those larger corrugations of the earth's crust which geologists name "ge-synclines" and "ge-anticlines." The main anticlinal axes strike N. N. W. and S. S. E., but there are also transverse undulations which further complicate the geological structure.

The Reefs.

The quartz reefs or lodes† conform to these folds in the country-

* The returns show that the average yield per ton is as follows:
Upper Silurian, Buckworth, 13 dwts. 23 grs., Gippsland 15 dwts. 21 grs.
Lower Silurian, Ballarat, 7 dwts. 21 grs., Bendigo, 9 dwts. 5 grs.

† The Australian calls a quartz lode or vein a "reef," the Californian a "ledge," while in Colorado the word "crevice" often does similar duty. Throughout this article I shall often use the colonial mining terms, since they are usually expressive and it would be difficult to find other names that are not also local.

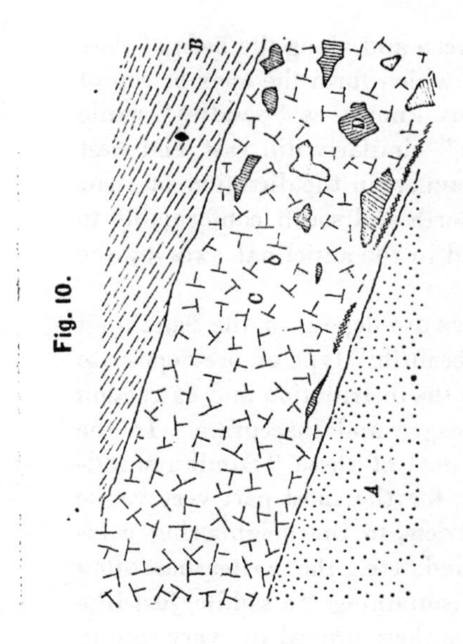

Fig. 10.

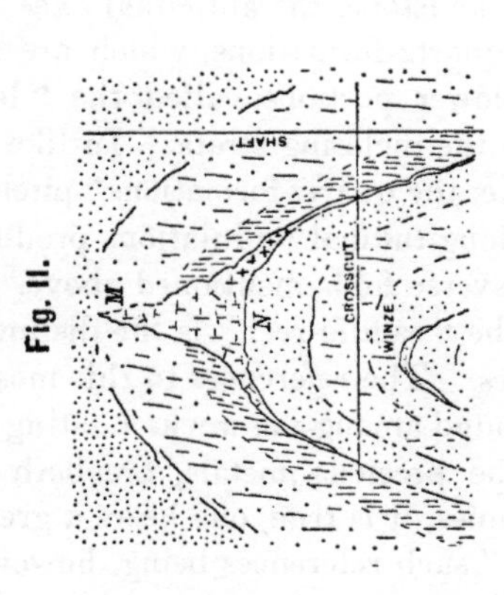

Fig. 11.

SHENANDOAH MINE.

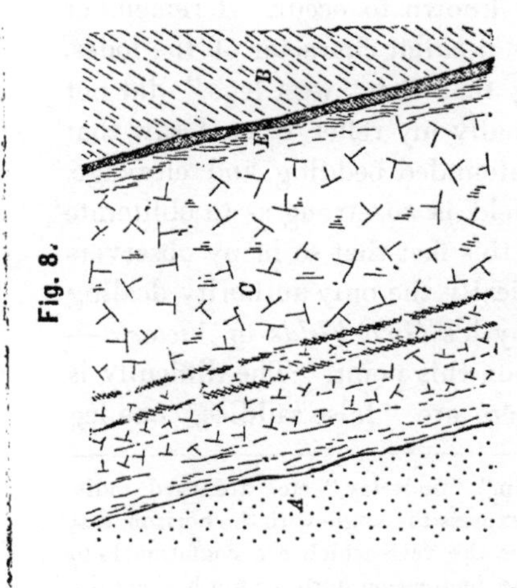

Fig. 8.

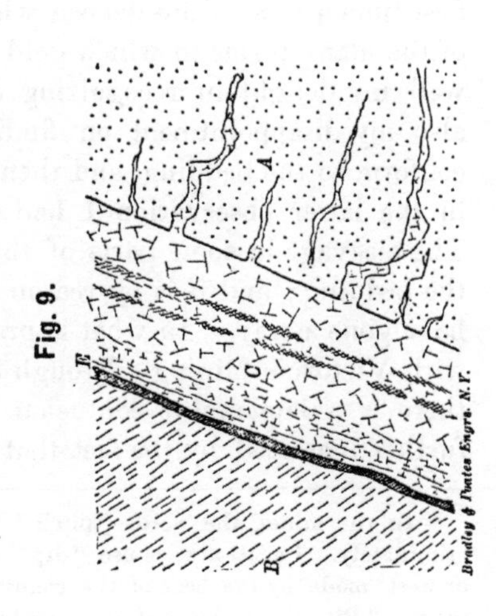

Fig. 9.

Bradley & Poates Engrs. N.Y.

rock, that is, the ore-deposits lie between and along the beds of slate and sandstone, the anticlinal axes of which form the apex or cap of the quartz-formations, which are thus known as "saddles," while the lower portions, called the "legs," similarly dip east and west with the inclosing strata. In like manner, in the direction of their strike, the quartz-formations* pitch north and south conformably to the longitudinal undulations produced in the anticlinal axes by the transverse folds mentioned above.

The "saddle-reef" is the distinctive ore-deposit of the Sandhurst mines. The references to this most beautiful type of ore-deposit to be found among the works relating to the distribution and extraction of the precious metals, are both meager and inaccurate. In the Colonies, it is true, one hears a great deal of these "Sandhurst saddles," such references being, however, for the most part very vague and incorrect. In other mining districts, in the neighboring colonies of New South Wales, Queensland, etc., the writer has often heard this or that mine spoken of as containing "a saddle, just like those of Sandhurst." On examination these proved in every case to be different forms of the ordinary junction of two lodes, not true anticlines, producing however bodies of quartz which the old Bendigo digger would promptly label as "saddles," for the sake, perhaps, of *auld lang syne*. I was beginning to fear that the saddle-reef as a distinct formation did not exist at all, even at Sandhurst, until underground, at the New Chum and Victoria mine, I saw for the first time a type of ore-deposit which is perhaps the most interesting of the many forms in which gold is known to occur. I remember well my delight in recognizing the peculiar character of the lodes, also my disappointment on finding that the "west legs" did not conform to the bedding, and then finally my relief when I saw that in the latter observation I had confounded bedding and cleavage. The cleavage in some parts of the field is so strong as to obliterate the bedding; and it is by reason of this fact that so many observers have gone astray In what is practically the only authority dealing with Victorian mining—Brough Smyth's *Gold-Fields of Victoria*—there is a lamentable confusion upon this point. The difficulty is further increased by the fact that there are "false saddles," one leg

---

* In the mines, the words "pitch," "dip," "underlay," etc., are used indiscriminately. I shall always use "dip" to express the angle with the horizon, east or west, made by the beds of the country or the reefs which are conformable to them. "Pitch" will be used to describe the inclination north or south along the strike of the quartz-formations.

of which conforms to the stratification, while the other follows a joint or some other cross-fracture in the country. Figs. 2 and 3 serve to illustrate the simplest type of true and false saddles. In both cases the bedding-planes are indicated, while, to avoid confusion, the lines of cleavage are omitted. In Fig. 2, A would be called the cap or apex, B the west leg, and C the east leg. A cross-cut passing through D would be said to be in "center-country;" as soon as the dip of the beds became distinctly east or west the cross-

Fig. 12.

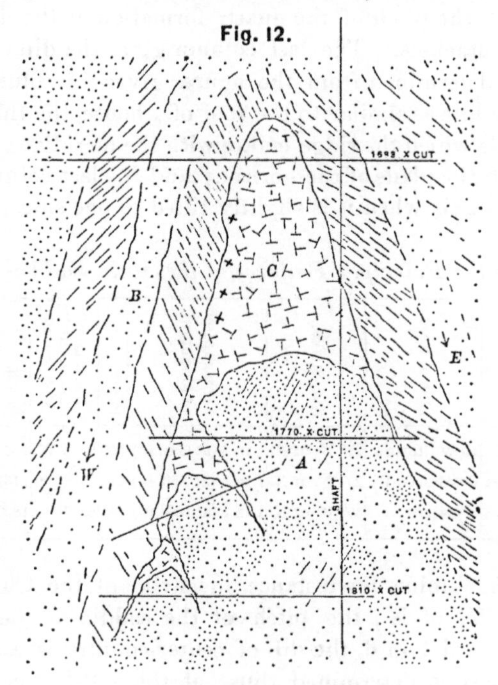

NEW CHUM CONED.

cut would have penetrated into "east" or "west country." In Fig. 3, A A illustrates a fissure, sometimes a fault, sometimes only a joint; D D is the ore formed along this line of fissure, while B is the body of quartz formed at its junction with the bedding-plane C C, which carries another vein of quartz. When a formation like this is further complicated by a few minor faults and the development of a strong slaty cleavage at varying angles, it is very difficult to determine correctly the true facts of the case; and the quartz-body may be mistaken for a true saddle, that is, an anticline of quartz.

The mines are located along the various "lines of reef," which are coincident with the surface-exposure of the quartz formed along the anticlinal axes. The general dip of the country (as distinguished from that of the individual beds comprising it) is eastward; or, as the miners put it, "center-country dips east." The following statement illustrates this point. Three mines (they happen to be three famous producers) are taken approximately opposite each other on the three great lines of reef—the New Chum, the Garden Gully, and the Hustlers. In the accompanying table, the third column gives the average strike of the anticline which forms the "line of reef," and the fourth the pitch of the quartz-formation at the depths indicated in the brackets. The last column gives the dip of "center-country" as determined from the figures given in columns 5 to 8. Taking the saddle as similar to the roof of a house, the third column gives the angle which the ridge makes with the meridian; the fourth, the angle with the plane of the horizon; and the last column, the inclination of its axis when it is slightly tilted to one side.

Table showing the Principal Features of the Main Saddle-Formations.

SHAFT.	Line of Reef.	Ave. Strike.	Average pitch of formation.	Depth.	Distance of center-country from Shaft.	Depth.	Distance of center-country from Shaft.	Average dip.
"180" Mine........	New Chum...	21° W. of N.	1 in 6 (at 2500 ft)	560	65 ft W.	2500	70 ft. E.	1 in 14
Victory and Pandora	Garden Gully	25° W. of N.	1 in 6 (at 650 ft.)	140	9 " "	2160	130 " "	1 in 15½
Gt. Ex. Hustlers	Hustlers	35° W of N.	1 in 5 (at 1800 ft)	200	80 " "	1800	160 " "	1 in 6½

At the "180" mine, for instance, the strike of the Chum formation is 21° W. of N, the pitch of the saddle at the 2500-foot level is northward 1 in 6, the dip of center-country is east about 1 in 14. The last is determined thus: at the 560-foot cross-cut the center country is 65 feet *west* of the shaft, while at the 2500 it is 70 feet *east* of the shaft, so that it has travelled eastward 135 feet in about 1940 feet, the general dip of the country having taken from one side to the other of this particular shaft the series of beds intersected by it. The ideal section, Fig. 4, represents the structure, and shows how, the general trend of the country being to the east, any one of the deep mine-workings will intercept, in succession, many saddles, some of which will prove gold-bearing and some barren. It will be understood that only the gold-bearing ones of the series are developed in a manner permitting their proper examination. One mine may have only two gold-bearing saddles, while another may have exposed a dozen or more.* As soon as one formation has been worked out, when the legs intercepted by the deeper cross-cuts are found to become too small or too poor for profit, prospecting is renewed until another gold-bearing saddle is cut. Its apex may occur between the lower portions of the legs of the last saddle or it may not be found for 200 or 300 feet deeper. *Deeper*, not further east or west on the same horizon; for the working of the mine has

Fig. 13

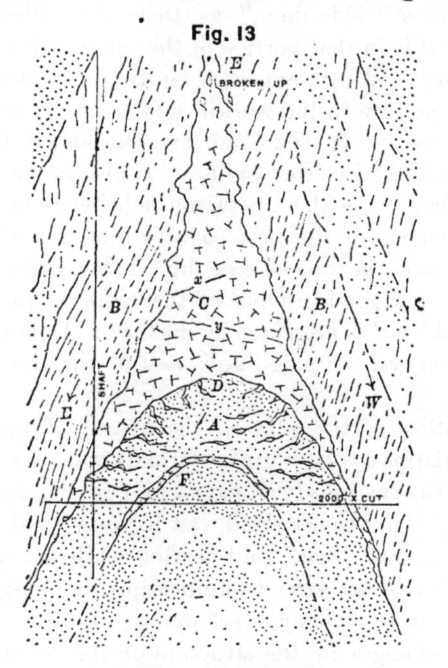

222 AND LAZARUS.

shown that a gold-bearing succession of saddles will go down in step-like gradations, each succeeding one being slightly eastward of the one above it (see Fig. 5). This simplicity of arrangement is, of course, much complicated, sometimes utterly destroyed, by faulting, and by the formation of irregular bodies of quartz that may be recognized occasionally as imperfect saddles.

* In the New Chum and Victoria mine, for instance, as many as 30 saddles have been passed through from the surface to 2300 feet. In the "180" mine, five have been discovered and explored, of which number three have proved profitably auriferous.

As would be expected by the geologist, these anticlines alternate with synclinal undulations; but this fact has been recognized by few of those engaged in the development of the field, though it is a point of paramount importance to the proper conception of the mode of occurrence of the quartz. The explanation of this neglect is that exploration has been confined, particularly in the deep mines on the New Chum reef, to a narrow strip of country in the immediate vicinity of the great anticlinal axes; and, though it is known that quartz lodes occur, for instance, between the New Chum and Garden Gully lines, these "side-lines," as they are called, have been neglected,* and it is in that portion of the country that the synclines would be situated. Where extensive faulting has taken place, these synclines are to be seen in the workings of some of the mines on the main "lines of reef." In the Confidence Extended, for instance, I observed some small "inverted saddles," as also in the Hercules and Energetic. In both cases there is reason to believe that a dislocation has brought a portion of the west country into line with one of the chief anticlinal axes, in the one case the Garden Gully, in the other the New Chum. A good instance of the more frequent occurrence of "inverted saddles" is presented by the Great Britain mine, located on the Carshalton reef, a "side-line" west of the New Chum. Fig. 6 illustrates it.

The distribution of the gold in the quartz of the saddle-reefs shows great variations. In a given formation in one mine the cap or a portion of the cap only will pay, while on the same formation in a neighboring claim the cap and west leg may yield good returns, or again both legs may prove gold-bearing while the cap is barren. The quartz-bodies do not extend uninterruptedly along their strike; their continuity is broken by overlappings or disturbed by faults. Different local changes in the structure of the country give very different shapes to the enclosed formations.

The most extensive ore-body and the greatest yield of gold from any one formation is claimed by the Garden Gully line from a saddle which traversed a group of mines (of which the Garden Gully United and the Victory and Pandora were the chief producers), at

* Since writing the above, and during a later visit to Sandhurst, I learn that the recent legislation as to mining on private property, together with the notable success of the New Red, White and Blue Consolidated, which is on a "side line" (the Sheepshead), has caused the beginning of active exploration on these subsidiary formations, this exploration being due, as has often been the case in this gold-field to the enterprise of Mr. Lansell.

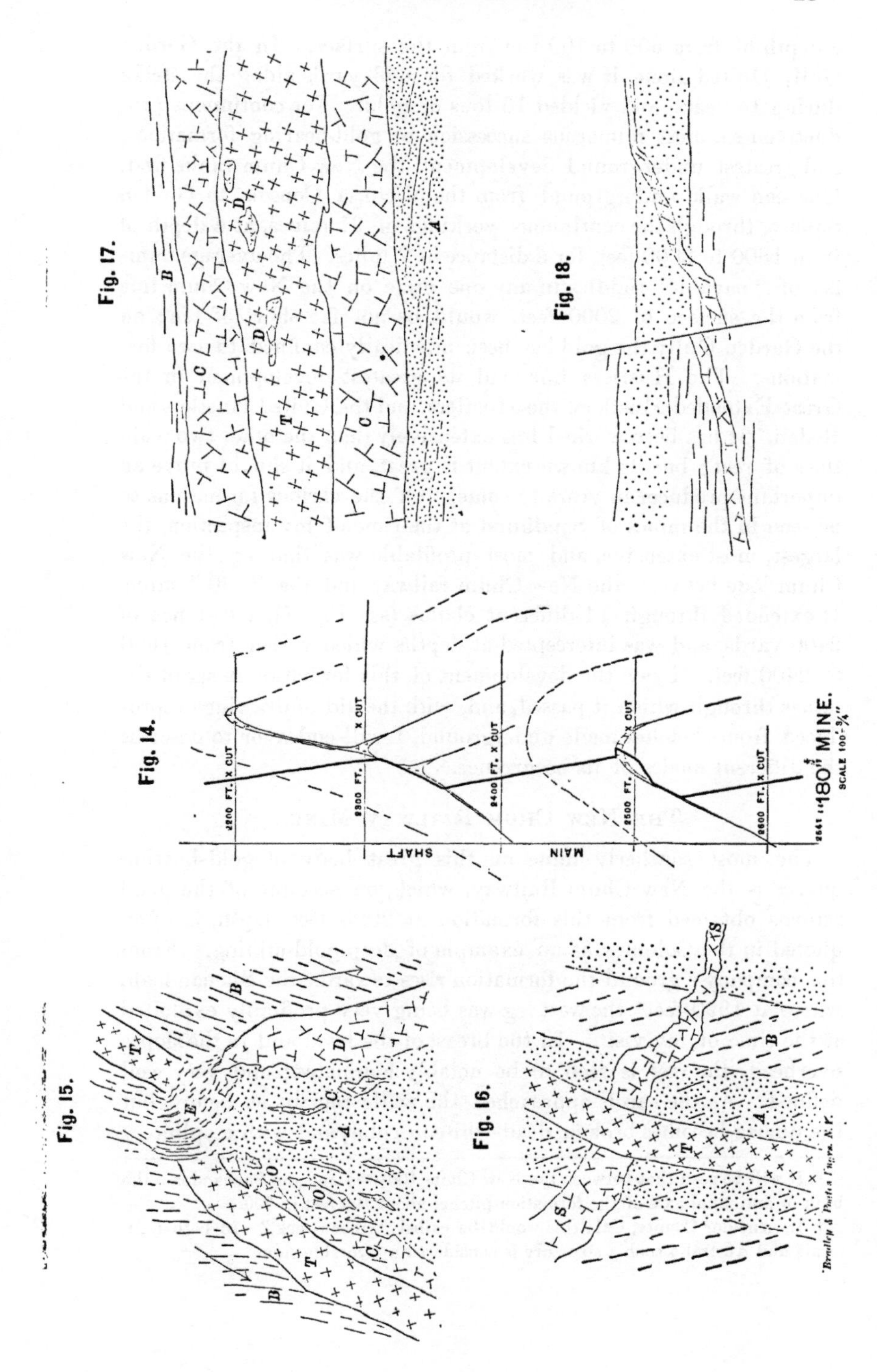
Fig. 17.
Fig. 18.
Fig. 14.
2200 FT. X CUT
2300 FT. X CUT
2400 FT. X CUT
2500 FT. X CUT
2600 FT. X CUT
MAIN
SHAFT
180" MINE.
SCALE 100'-3/4"
Fig. 15.
Fig. 16.
Bradley & Poates Engr's. N.Y.

a depth of from 600 to 700 feet from the surface. In the Garden Gully United alone, it was worked for 646 yards along the strike during 14 years, and yielded 13 tons of gold. For continuous productiveness, most numerous succession of gold-bearing formations, and greatest underground development, the New Chum ranks first. One can walk underground from the Victoria Consols to Golden Square, through the continuous workings of 21 mines, at a depth of from 1800 to 2000 feet, for a distance of 2 miles. The average number of "payable" saddles in any one mine on the New Chum line from the surface to 2000 feet, would be not far short of ten; on the Garden Gully the gold has been practically confined to two formations. The Hustlers line had its greatest development in the Great Extended Hustlers, the Hustlers, and the United Hustlers and Redan. It has been worked less extensively than the other two main lines of reef; but its known extent is great, and it should prove an important producer in years to come. Of the various formations to be seen in the mines of Sandhurst at the time of my inspection, the largest, most extensive, and most profitable was that on the New Chum lode between the New Chum railway and the "180" mine. It extended through 14 different claims (see Fig. 7), a distance of 2400 yards, and was intercepted at depths which varied from 1600 to 2400 feet. I saw the development of this formation in six of the mines through which it passed, and, with the aid of drawings reproduced from sketches made underground, I will endeavor to describe the different modes of its occurrence.

THE NEW CHUM RAILWAY MINE.

The most southerly mine on this great body of gold-bearing quartz is the New Chum Railway, which, on account of the good returns obtained from this formation at 2025 feet depth, is often quoted in the Colonies as an example of deep gold-mining. From this company's ground the formation rises towards the Shenandoah, where at 1990 feet* the west leg was being very profitably exploited at the time of my visit. In the breast of the level and in the slopes overhead, the reef is seen to be notably laminated† and very well defined. As the cap is approached the walls remain as before, but the quartz becomes broken, dead-white in color, and with a splintery

* It will be seen that between the New Chum Railway and the Shenandoah, the back or anticlinal axis of the formation pitches about 100 feet in 600.

† In Amador County, Cal., this would be called "ribbon-rock." In both California and Australia such a structure is considered a favorable sign.

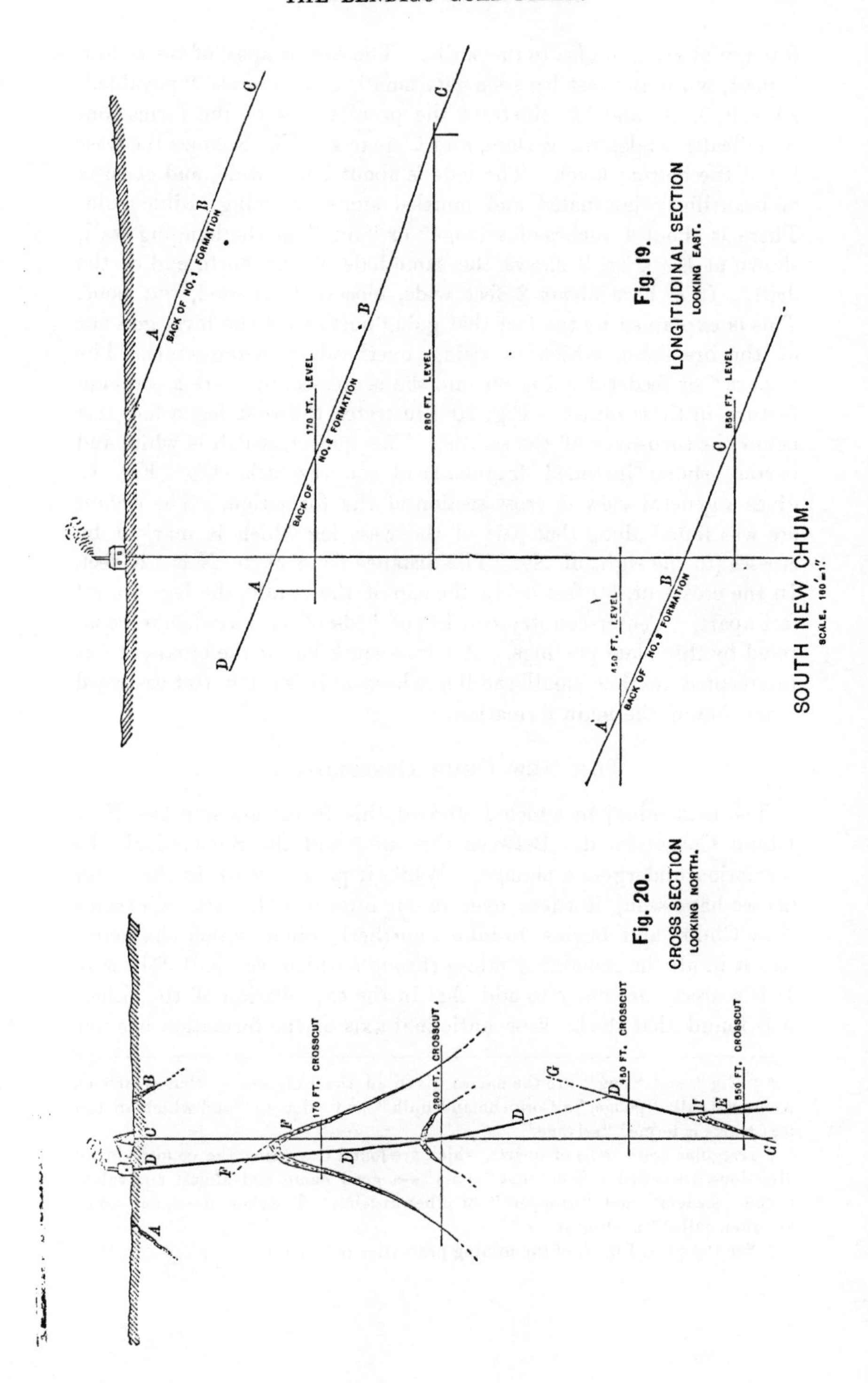

Fig. 19.
LONGITUDINAL SECTION
LOOKING EAST.

Fig. 20.
CROSS SECTION
LOOKING NORTH.

SOUTH NEW CHUM.

fracture at right angles to the walls. The cap or apex of the saddle is poor, while the east leg soon gets small, and is never "payable." Figs. 8, 9, 10, and 11 illustrate the peculiarities of the formation. A indicates sandstone, B slate, and C quartz. Fig. 8 shows the west leg at the bottom level. The lode is about 7 feet wide, and consists of beautifully laminated and mottled stone, carrying visible gold. There is about 4 inches of selvage* or "dig" on the hanging-wall, shown at E. Fig. 9 shows the same lode at the north end of the drift. It is here about 2 feet wide, closely laminated, but poor. This is explained by the fact that going northward the level gets out of the ore-shoot, which is rising overhead, pitching south. The "spurs" or feeders† going off into the center-country are a common feature in these mines. Fig. 10 illustrates the west leg a few feet below the turn-over of the saddle. The quartz, which is white and barren, shows included fragments of country-rock (D). Fig. 11 gives a general view in cross-section of the formation. The richest ore was found along that part of the west leg which is marked by crosses (to the right of N). The distance from M to N is 120 feet. In the cross-cut, 60 feet below the cap of the saddle, the legs are 90 feet apart. Center-country consists of beds of hard sandstone, separated by thin slate partings. A winze sunk below the cross-cut has intercepted another small saddle, whose axis is a few feet eastward from that of the main formation.

The New Chum Consolidated.

The next mine‡ in which I studied this formation was the New Chum Consolidated. Between this mine and the Shenandoah the formation undergoes a change. While it pitches south in the latter (as we have seen), it turns over in an intermediate mine, Craven's New Chum, and begins to take a northerly pitch, which characterizes it in all the remaining mines through which we shall follow it. It is scarcely necessary to add that in the exploitation of the mines it is found that the back or anticlinal axis of the formation has not

* "Dig" and "pug" are the names given in the Colonies to that which an American calls "gouge," a Cornishman "hulk" and "fluccan," and which in the text-books is termed "selvage."

† Irregular cross-veins of quartz, which are found traversing the country in all directions, are called at Sandhurst "spurs"—a good name, and almost equivalent to the "feeders" and "droppers" of other districts. The short-lived, flat seams are often called "floating spurs"

‡ See the plan, Fig. 7, of the mining properties referred to.

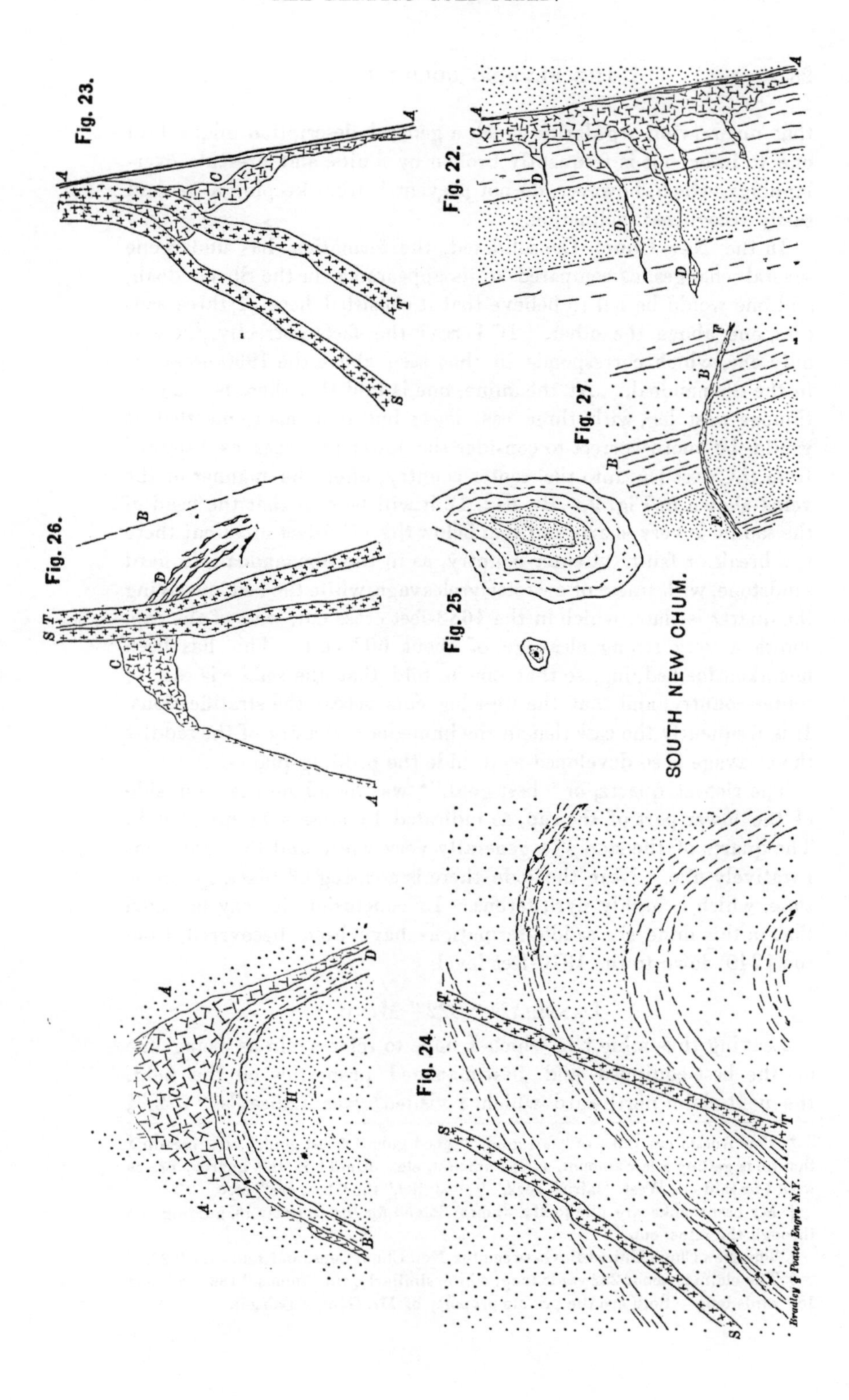
Fig. 23.
Fig. 26.
Fig. 22.
Fig. 27.
Fig. 25
Fig. 24.
SOUTH NEW CHUM.
Bradley & Poates Engrs. N.Y.

that unswerving regularity which a general description might lead one to infer, but is frequently broken by faults, and presents overlappings, which, however, do not prevent it from keeping a general easily-recognized inclination.

In the New Chum Consolidated, the formation has undergone several changes, as compared to its appearance in the Shenandoah, and one would be led to believe that it consisted here of three saddles, one above the other. If I read the facts correctly, there is only one, which corresponds to that seen above the 1990 cross-cut in the Shenandoah. At the mine, one is told that there is one continuous west leg, with three east legs; but it seems to me that it would be more correct to consider the lower two legs as "spurs" from the west leg into the center-country, after the manner of the veins shown in Fig. 9. By Fig. 12 it will be seen that the bend of the saddle is very sharp, and that below the 1770-feet cross-cut there is a break or fault. Center-country, as in the Shenandoah, is hard sandstone, with traces of a westerly cleavage, while the bed overlying the quartz is slate, which in the 1683-feet cross-cut, west of the cap, shows a very strong cleavage of about 60° east. This has been mistaken for bedding, so that one is told that the saddle is east of center-country, and that the west leg cuts across the stratifications. It is frequently the case that in the immediate vicinity of the saddles the cleavage is so developed as to hide the bedding-planes.

The richest quartz, or "best gold,"* was found on the west side of the lower part of the cap, as indicated by crosses in my sketch. The quartz of the stopes is generally very white and the gold comparatively fine. Near the lode there is a casing of black graphitic slate, which is highly pyritiferous. In conclusion, it may be noted that in this mine six saddle-formations have been discovered, from the 1240 down to the 1810-feet level.

Lansell's "222" Mine.

Leaving this mine,† to which I hope to refer again as being one of the best-equipped and best-managed properties on the field, the next claim northward which I visited‡ was Lansell's "222,"§

* The Australian talks of "fair gold," "good gold," "best gold," etc., meaning that he is getting a fair amount, a good amount, etc., of gold in the quartz; or, as would be said out West, "fairish rock," "good dirt," the "best ore," etc.

† My best thanks are due to Mr. James Boland for his courtesy in guiding me through the mine, etc.

‡ The New Chum United lies between the New Chum Cons. and Lansell's "222."

§ The claim is about 222 yards long, while, similarly, the famous "180" mine is 180 yards long. Both are the private property of Mr. George Lansell.

Fig. 28.

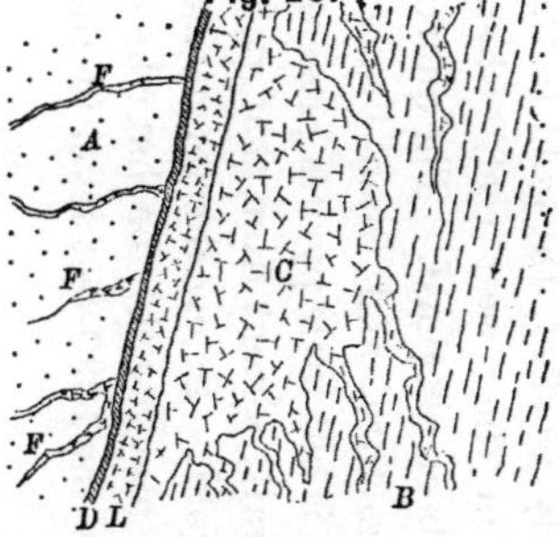

F
A
F
C
F
D L
B

Fig. 33.

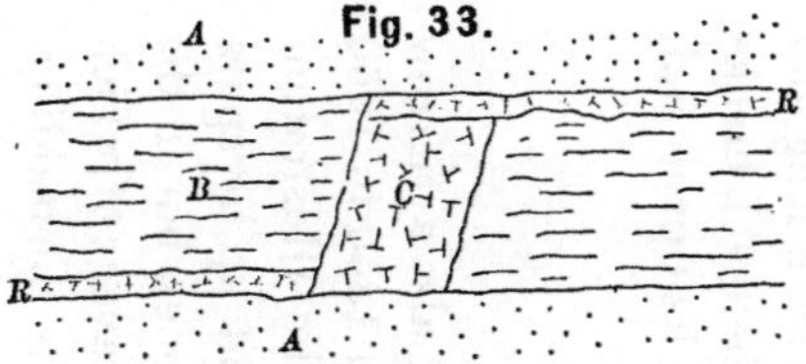

A
R
B
C
R
A

Fig. 31.

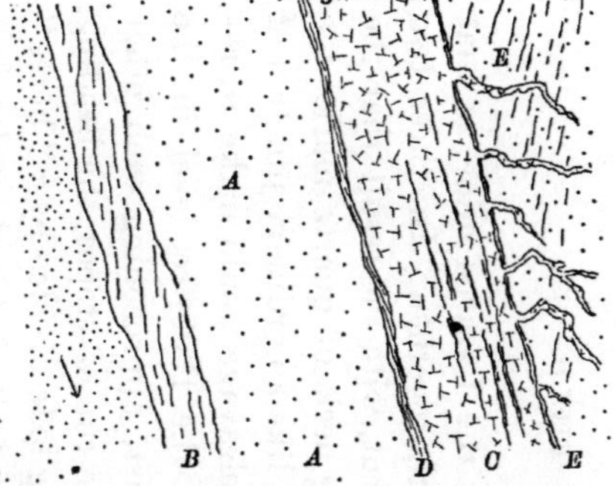

E
A
B
A
D
C
E

Fig. 30.

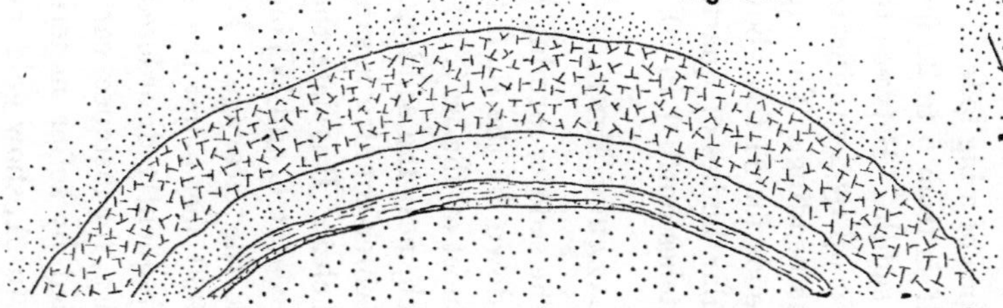

Fig. 29.

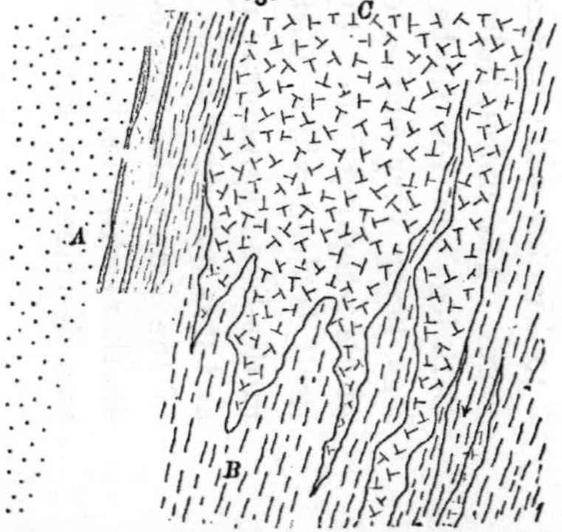

C
A
B

THE JOHNSONS MINE.

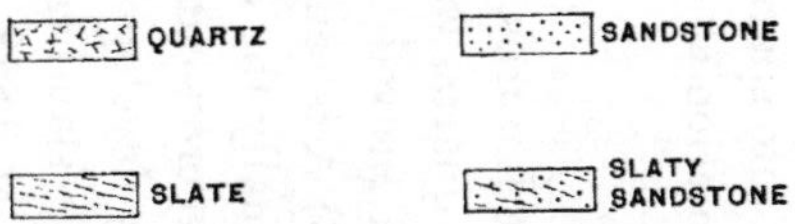

QUARTZ

SANDSTONE

SLATE

SLATY SANDSTONE

Fig. 32.

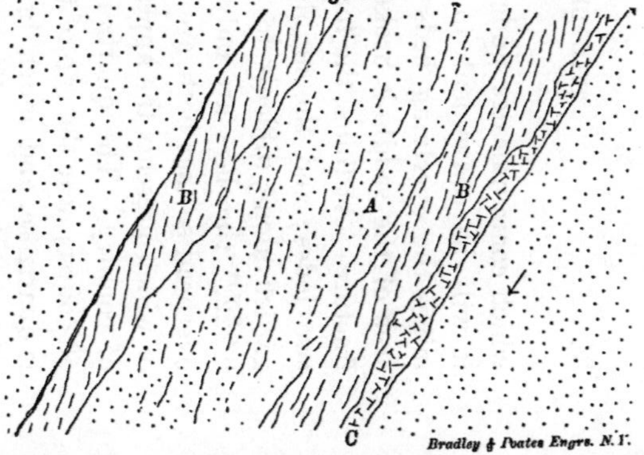

B
A
B
C

Bradley & Poates Engrs. N.Y.

which communicates underground with the Lazarus mine, so that it will be convenient to consider both together. In these two mines the overlying beds have been bent, with the result that the arch of the saddle is broken, and a long neck of quartz marks the fracture. At times, the formation looks like an ordinary junction of two lodes; and the highly developed cleavage would make it, regarded alone, difficult to unravel. At the 1900-feet level in the "222," the legs are only 20 feet apart, approaching each other overhead so that both are worked in one run of stopes, the heading of which presents the appearance of a big lode divided by a " horse " of country. The quartz itself is from 5 to 20 feet wide, enlarging, of course, near the cap of the saddle. Between the legs the country is quartzose, and much broken up by a ramification of spurs which with the cleavage obliterate the bedding. The cap forms a long and narrow neck of ore, which extends for 81 feet above the level, breaking up into spurs and irregular bodies of quartz at its upper end. The formation at this level pitches north 1 in 10.

The Lazarus Mine.

In the Lazarus, the cap of the saddle, forming here also a narrow neck of quartz, was followed down by the shaft from 1850 to 1960 feet, and proved " payable " most of the way. In the illustration (Fig. 13) A is sandstone, which, near the saddle, is cut up by spurs (as is usually the case where center-country is sandstone), but in the 2000-feet cross-cut is hard and clean. B is slate, the cleavage of which is highly developed. The arrows (at E and W) indicate the bedding, while the slanting lines mark the cleavage. E (at the top of the Fig.) is at 1889, and D at 1960 feet below the surface. C is the main body of ore, which is frequently disordered by minor faults, such as x and y. That marked y, for instance, carries crushed quartz 3 to 4 inches wide, accompanied by softer materials (fluccan or selvage). The spurs, indicated in A, show gold near the main lode. Both legs and the cap have been profitably worked, the gold in the legs being very coarse, more especially in the " west reef," where pieces weighing from 1 to 3 ounces have been obtained. The quartz itself is dead-white, with a splintery fracture, and when broken rattles like fragments of porcelain. There is but little pyrites present, and it is noteworthy that pyrites is generally absent from that part of the lode which carries the coarsest gold. Below the formation another saddle has been discovered, which may possibly be that which was described as cut by the winze below the 1990-feet level in the

Shenandoah. In Fig. 11 this "make of stone" is shown. It is separated by 20 feet of sandstone from the main saddle; it is more regular than the upper one, and its west leg carries "fair gold," while the eastern is poor.

Only two paying formations have been intersected in this mine in the last 1000 feet of sinking (that is, below the 1000-feet level), namely, the saddle at the 1750, and that at the 2000-feet level above described. In that portion of the mine below 850 feet, twelve for-

Fig. 34.

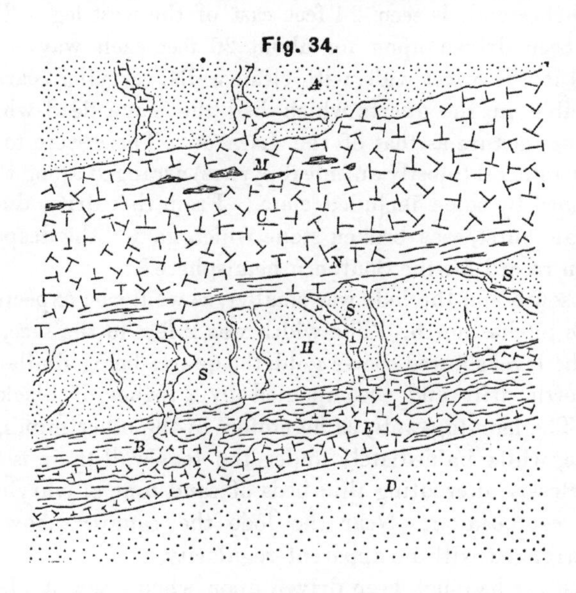

mations have been cut, most of which showed some gold; but the two above mentioned have been much the most productive.

The following list may be of interest:

Saddle cut at	Remarks.
850 feet,	"payable."
930 "	no good.
1040 "	no good.
1160 "	no good.
1240 "	7 dwts. first crushing, followed by a 4 dwt. yield.
1263 "	no good.
1435 "	no good.
1630 "	small, not payable.
1750 "	good.
1790 "	poor.
1837 "	broken, poor.
1889 to 2000 feet,	very good.

THE NORTH OLD CHUM MINE.

Next to the Lazarus,* is the Old Chum, which I did not visit, and then follows the New Chum and Victoria, which was not deep enough† to cut the formation which we have been following through the different mines. Then follows the North Old Chum, where, at 2290 feet, the main formation has been profitably developed. A cross-cut at that depth cut the west leg at 70 feet from the shaft, and 45 feet further intercepted the east leg. A lava dyke about 4 inches wide is seen 24 feet east of the west leg. The west reef has been driven upon for about 20 feet each way. In the north end it was 8 feet wide, with clean walls, carrying coarse‡ gold, easily visible in the quartz which had the same dead-white color and curious fracture as that of the Lazarus. There were to be seen patches of mineral (mostly arsenical pyrites) arranged along the black lines formed by some included slate. From this drift, during my stay in Sandhurst, was broken stone which gave "the deepest dividends" on record for the Southern hemisphere.§

In the south end the lode was smaller, as would be expected, since the saddle pitches north. The width was 2 feet in the face, and the back of the level showed coarse gold, the quartz being similar to that of the north drift and accompanied by 2 inches of black clay or "dig." The hanging-wall, as heretofore in the other mines, is seen to be slate, while immediately under the foot-wall there is "corduroy" or closely alternating thin beds of sandstone of varying hardness and composition. Near the lode the country shows coarse pyrites, arranged with no apparent regularity.

The east leg had not been driven upon when I saw it; but since then it has been developed and has given good stone.

When the north drift from the 2290-feet level in the North Old Chum has been driven 300 feet further, it will connect with the 2300 feet level in Lansell's "180" mine,|| the deepest in Australia.

---

* I take this opportunity to thank the mine-manager, Mr. T. Whitford, for his courtesy in taking me through the workings.

† Since my visit the shaft has been sunk and the reef cut at 2260 feet.

‡ The coarse character of the gold is indicated by the yield of 223 oz. of gold from only 341 oz. of amalgam.

§ For the week ending September 13, 1890, it yielded 223 oz. 9 dwts. from 141 tons, which result enabled a sixpenny dividend to be paid. The depth was 2290 feet.

|| The opportunity is offered me in this connection to express my great obligations to Mr. George Lansell, the man who has done most to develop the Sandhurst gold-field, for assistance given me in visiting the mines, as well as for valuable information regarding mining and milling in the gold-field.

In this mine the crown or cap of the saddle is cut by the 2200-feet cross-cut, 80 feet east from the shaft, showing a width of 12 feet of quartz. At 43 feet a dike is cut. In the 2300-feet cross-cut the west leg is cut at 58 feet, while the east leg is seen, broken up and irregular, at 130 feet. The dike is cut at 56 feet. In the 2400-feet cross-cut the west leg is 8 feet east of the shaft and the dike 46 feet. In the 2500-feet cross-cut this formation has not been cut, but the west leg of *another* saddle is formed at 48 feet east, the corresponding east leg being at 81 feet. The dike is at 58 feet. In the 2600-feet cross-cut the same dike occurs at 56 feet. The cross-section (Fig. 14) illustrates these formations. The dikes, one of which

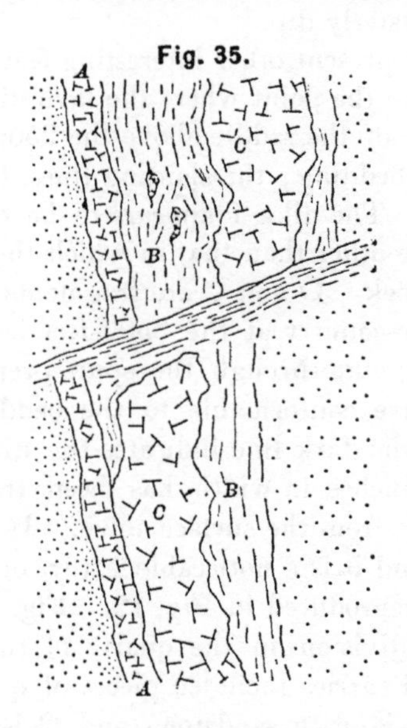

Fig. 35.

is shown in the section, are a most important and interesting feature of the mining at Sandhurst, and will be referred to frequently.

THE "180" MINE.

In the "180" mine we see the furthest northern development of the great saddle which we have followed through several mines and which is first cut by the New Chum Railway. In the "180" mine it has not been opened up much; but at 2400 feet, in the south level, it forms the deepest gold-bearing quartz now worked in the colonies. The west leg, the only portion of the formation which is worked at present, shows about 1 foot of very beautifully laminated quartz. The quartz is brittle and flaky, the gold coarse and associated with a good percentage of iron pyrites and zinc-blende. On the hanging-wall there is a black graphitic slate which carries a noteworthy amount of mundic. The foot-wall shows cross-joints about 30 inches apart and nearly at right angles to the strike. The country is "corduroy," similar to that noted in the North Old Chum, and carries pyrites near the reef. The cleavage is highly developed, and to the east (about 60 feet) in the cross-cut, it obliterates the bedding, though the latter is to be distinguished with difficulty to the east near the east leg and to the west near the west leg. Between them the country is broken up and penetrated by the dike, which has a slight easterly dip.

The illustrations represent other interesting features of the "180" mine. Fig. 15 shows the point where the lava dike, about 8 inches wide, strikes the top of the saddle above the 2500-feet level. The arrows indicate the bedding; the slanting lines, the cleavage; E is disordered country. The dike T appears to be cut off, but it finds a passage in a plane other than that in which the section is taken. A is broken sandy rock. O and C are fragments of the cap. Fig. 16 is taken in center-country at the 2500-feet level. The dike T (here 10 inches wide) cuts through the spurs preparatory to taking a more regular course conformable to the bedding. In Fig. 14, already referred to, the dark line indicates the dike, which, though it rarely exceeds 9 inches in width, has been traced through the half-mile of workings from the surface down. It can be seen in the 2600-feet cross-cut and it is a noticeable feature of a surface cutting, a sketch of which is reproduced in Fig. 12. Fig. 17 shows the dike T where it forms a division in the quartz of the west leg (of the 2500-feet saddle) and carries included pieces of quartz, D. In this figure B is slate, A is slaty sandstone, and C is quartz. Fig. 18 shows a fault, the throw of which is about 7 inches, as seen in the 2400-feet level. The sketch explains itself.

The underground workings of this very interesting mine* have uncovered five saddles, of which the one at the 2400-feet level is the third "payable" one. The other formations were cut at 560, 1560, 2000 and 2200 feet respectively. The east leg of the 2400-feet formation is the longest east leg in the mine, 40 feet. The longest west leg was worked from the 1600 to the 1870-feet level. The cap of the

---

* Mr James Northcote, the mine-manager, gave me a great deal of this information and in many other ways assisted me in my examination of the gold-field.

560-feet saddle was the most payable portion of that particular formation. The east leg was short-lived, the quartz pinching, while the wall or "back" continued. Generally speaking, the west legs are the strongest and most auriferous in this, as in most of the saddles of the New Chum line of reef. No considerable body of gold-bearing stone has been worked below 1870 feet.

Barren quartz is rarely met with, but some is not of paying grade. From 1560 to 1870 feet was worked the great saddle-reef of the "180" mine, there being a large body of stone or ore of extraordinary richness on the west leg of the formation, which was first cut

Fig. 36.

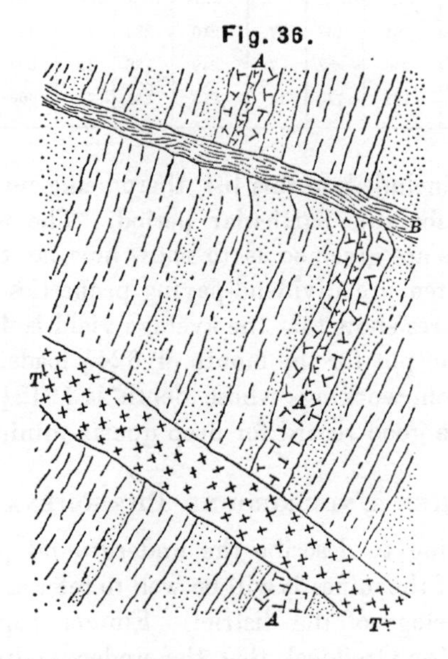

at 1560 feet. The deeper levels show no change in the character of the country, and no large influx of water* hinders further sinking.

RECAPITULATION.

This concludes that portion of my notes which refers to the series of mines in which is to be seen the main saddle-formation of the Sandhurst of to-day. It may be taken as typical of the great runs of pay-quartz from which the field has obtained most of its gold. During the time (the four weeks preceding October 13th, 1890) in which I inspected the larger number of this group of mines, the returns were as shown in the accompanying table.

* The quantity of water hoisted is 3000 gals. per 24 hours.

Mine.	1st Fortnight, ending Sept. 27, '90.			2d Fortnight, ending Oct. 11, '90.			For the Month.			Depth.
	Loads.	Ozs.	Dwts	Loads.	Ozs.	Dwts.	L'ads	Ozs.	Dwts.	Feet.
New Chum Railway....	527	474	1	512	643	1	1039	1117	2	2025
Shenandoah..............	385	479	3	365	379	1	750	858	4	1990
North Shenandoah	209	95	10	145	100	4	354	195	14	1900
Young Chum..............	108	45	5	160	56	14	268	101	19	1400
New Chum Con	736	532	12	757	378	2	1493	910	14	1850
New Chum United.......	416	117	11	420	72	8	836	189	19	1850
Lansell's "222".........	......	... ..	......	......	......	......	98	88	0	1900
Lazarus Co.................	231	163	17	210	336	11	441	400	8	1960
Lazarus No. 1..............	256	52	7	249	138	13	505	191	0	1960
North Old Chum..........	181	743	11	279	241	10	460	395	1	2290

Ten mines are included in this list, there being no crushing from the "180" mine for this particular period. The returns for the separate fortnights are given so as to show how far the yields vary. Nine out of the ten are dividend-paying properties at the present time, and the figures show that the average yield is 444 oz. 16 dwt. from an average output for the month of 624½ loads,* these results being obtained from workings whose depth is 1912½ feet from the surface. This is a good record for deep quartz-mining.

Other Underground Phenomena.

Before proceeding to describe the underground phenomena observed in others of the mines it will be well to refer again to the general structural geelogy of the district. Emmons† quotes a saying of the geologist Von Groddeck that the understanding of the true character of veins can only be arrived at by study of the structure of the region in which they occur. Of no mining district can this be said more forcibly than of Sandhurst. As we have already seen, the gold-field is situated among the highly contorted folds of the Lower Silurian slates and standstones. The fissuring which is the accompaniment and result of extreme contortion has led to the production of very diverse forms of ore-deposition. We have seen something of the saddles and synclines (troughs or inverted saddles).

* A load is equal to 25 cwt. (avoirdupois, at 112 pounds) of ordinary ore, or 30 to 35 cwt. of concentrates.

† "The Structural Relations of Ore-deposits," *Trans.*, xvi., 804.

Fig. 37.

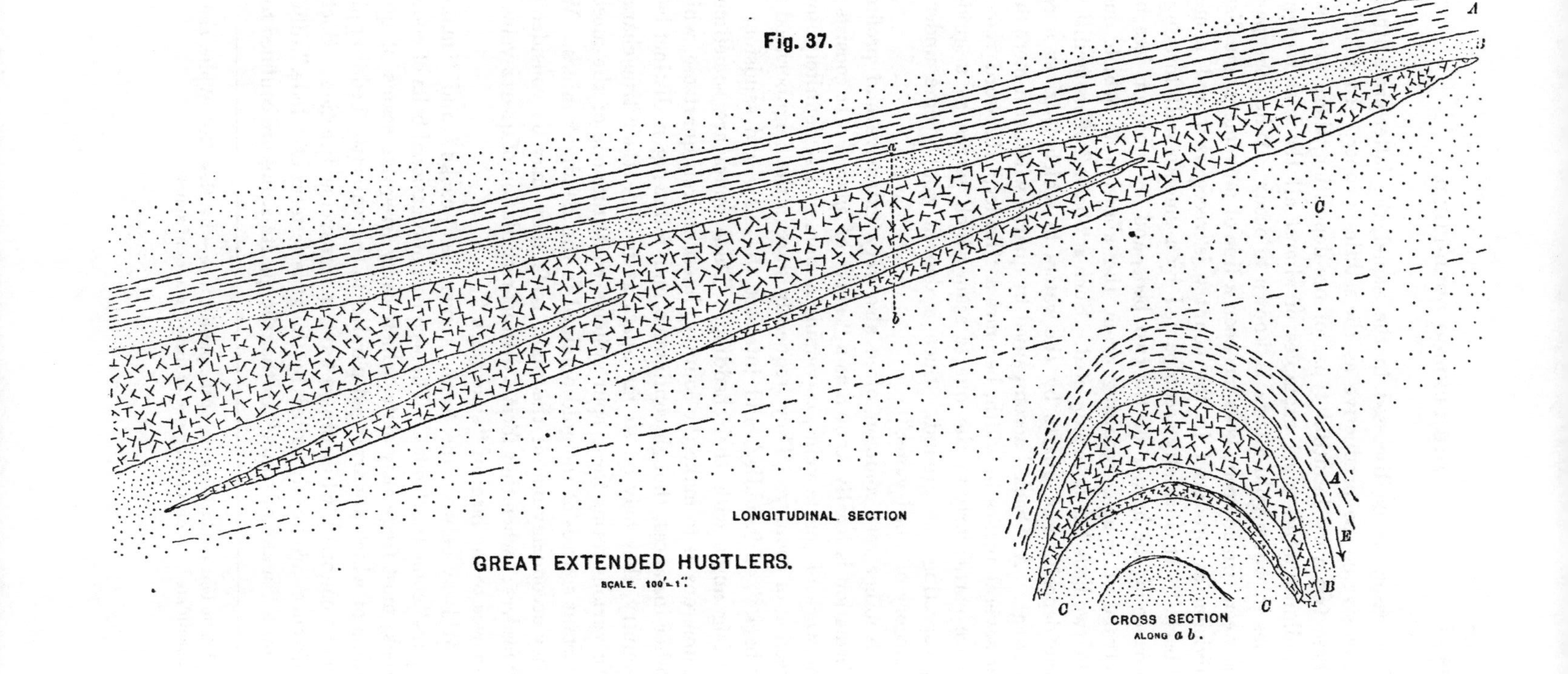

GREAT EXTENDED HUSTLERS.

SCALE. 100' = 1".

For reasons to be discussed in the sequel, the latter are neither so extensive nor so productive as the saddles proper. In addition to these two very beautiful types of ore-deposition, there are in the Sandhurst mines good examples of almost all the best recognized forms in which gold-quartz is known to occur. When the parting between two adjacent beds has been a line of movement, the distinct division as produced in the country is recognized and named a "back." It is usually accompanied by a smooth wall forming the surface of the harder of the two beds, and by more or less black selvage or gouge, resulting from the abrasion to which one of the two beds has been subjected. Such a "back" offers facilities for the deposition of quartz by the waters to which it gives a ready passage. A "back" accompanied by quartz, whether a mere thread or several inches in width, becomes a "leader," the word referring more particularly to the quartz seam which serves as a guide in prospecting the ground. Such a deposit would come under the category of "bed-veins."

Another very common, often also very extensive and profitable, formation is locally known as a "*make of spurs.*" It consists of a network of quartz-veins, traversing a more or less definitely limited section of country. They frequently start in the neighborhood of a "back" or "saddle," and intersect the slate and sandstone at a strong angle with their bedding. These spurs are sometimes so numerous as to make the country almost entirely quartzose, while in other instances, though continuous and following a distinct belt of country, they may be so far apart as to hide their true character. In various forms these spur-systems constitute one of the most important sources of the gold obtained in the Bendigo* mines. When they accompany one of the main reefs they may be considered as "feeders;" when they form a distinct reticulation of quartz-veins they answer to the term of "stockwork."

Midway between the forms known as "backs" and "makes of spurs" come the "lodes." These consist of occasional belts of country-rock, most frequently slate, which carry irregular seams of quartz, some of which are arranged parallel to the walls (the bedding planes of the country), while others have a transverse direction. Such occurrences are locally called by the generic term of "lode," differing from a "make of spurs" in that the quartz-seams are confined to one

* The best-producing mine at the present time—the New Red, White, and Blue Consolidated—is developing one of these "makes of spurs."

bed alone, and from the backs and leaders in being larger and less defined, and in carrying a greater proportion of country-rock.

Fissures containing quartz and traversing the country unconformably to the bedding, are often faults and would answer to the "true fissure-vein,"* a type which is of frequent occurrence in this field, though not so important a repository of the precious metal as the saddles or spurs. "Bulges," "blocks," etc., are names given to irregular bodies of quartz, presenting features more or less in common with the preceding types. At Sandhurst, as elsewhere, the different forms of ore-deposition cannot be arbitrarily labelled, exhibiting as they do frequent gradations from the one to the other. In the paragraphs which follow, I will describe instances of the type known as "saddles," leaving the other forms for subsequent consideration.

Fig. 38.

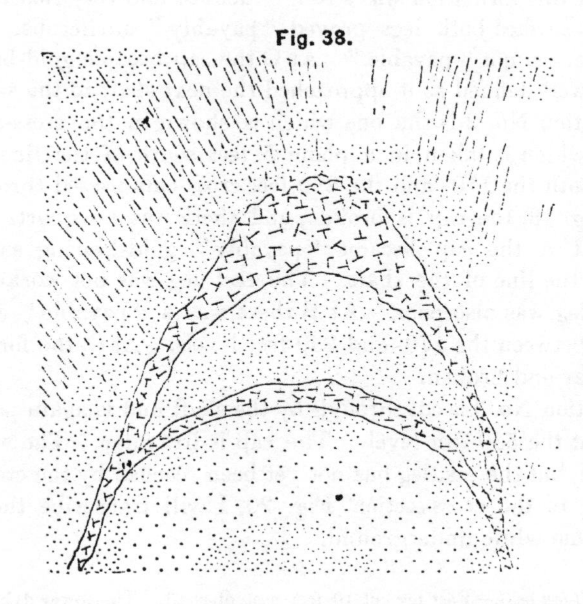

The South New Chum Mine.

At the southern end of the district—about 3½ miles from the group of big mines on Victoria Hill, some of which have been passed in review—there is a small mine called the South New Chum, which

---

* This form of ore-deposit was at one time erroneously considered as particularly favorable to the occurrence of continuous shoots of gold-bearing stone.

admirably illustrates the particular formation most characteristic of Sandhurst.*

This mine is working in regular "saddle-country." The longitudinal and cross-sections given in Figs. 19 and 20 will serve to explain it. The underground developments of this mine and its northern neighbor, have proved the existence of three formations, all pitching strongly to the south.

Formation No. 1 came to the surface 120 feet from the South New New Chum shaft. The line drawn in the section, Fig. 19, is the anticlinal axis of the saddle. Looking along the strike, it would appear like the ridge of a roof, sloping rapidly southward. At A the cap of this formation was strongly defined and very rich in gold. Near the surface both legs proved "payably" auriferous. At B, the saddle was not "payable." At C the cap became gold-bearing, and was well defined as it approached the next shaft to the south.

Formation No. 2 is the one which is shown in the cross-section, Fig. 20, which is taken in a plane 5 feet south of the line of the shaft. Both the 170- and the 280-feet cross-cuts passed through it. At D, Fig. 19, the cap is broken, and there was no quartz in the legs. At A the cap became "payably" gold-bearing as it approached the line of the shaft. The east leg was not worked, and the west leg was also poor. At B there was a "payable" cap and west leg, between the 170- and 280-feet levels. At C, the formation is irregular and broken.

Formation No. 3 is cut through by the shaft and is again partially exposed at the 550-feet level. The cap is irregular. The west leg is worked, but the east leg has not yet been reached by the cross-cut. Referring to the cross-section, Fig. 20, I will transcribe the notes made by me while underground.

At the 170-*feet level.*—East leg cut 10 feet east of shaft. The lower dyke 12 feet and west leg 34 feet, both west of shaft. The lower portion of the cap is passed through in sinking at 125 fee , the top of the cap being 5 feet west of the shaft. The lava dike is 4 feet wide, divided by 2 to 3 inches of slate, which, throughout the upper workings, separates it into nearly equal portions. The cleavage of the country is with the east leg, slightly over the cap and stronger a short distance above. On the other side of the saddle the cleavage is decidedly *across* the west leg. The formation pitches strongly southward, 1 foot in 6, and at times even 1 in 5.

---

* I owe thanks to Mr. L. A. Samuels, the mine manager, who conducted me through the workings during my visits at the mine and gave much interesting information.

At the 280 *feet level.*—East leg 74 feet east of shaft. Lava dike in the plat.* West leg 84 feet west of shaft. The dike is increased in size, the two portions being divided by two feet of slate. The cross-cut shows a subordinate saddle underneath the main formation. This secondary saddle carries a little quartz (4 to 5 inches), which in the shaft showed "nice gold." The west leg (of the main saddle) consists of two parts, each about 5 inches in width, the quartz of which is very beautifully laminated or ribboned. The hanging-wall is clean and defined, inclining at a less angle than in the level above. The cleavage west of the center-country is slightly west, nearly upright, and cuts across the bedding, which is readily recognizable. The country east of the west leg is slate (2 feet thick) overlying sandstone, the thin bedding of the latter being very marked. The east leg as cut in the cross-cut is small and poor, but is overlain by country corresponding to that which covers the west leg.

Fig. 39.

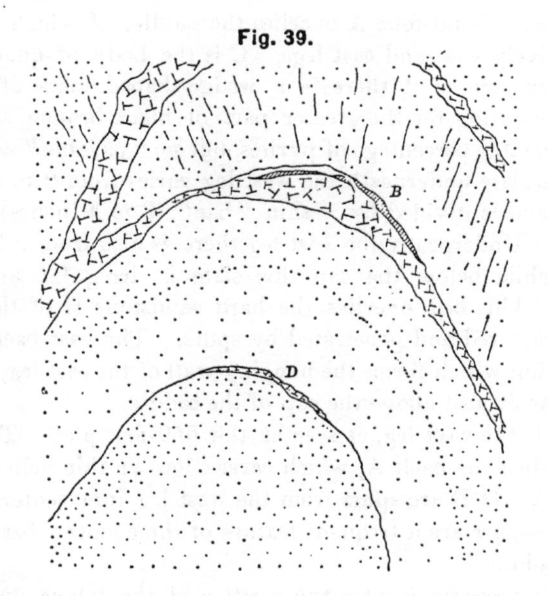

At the 450-*feet level.*—The legs of the upper formation have not been sought for, as they both became small and poor below the 280-feet level. The east leg would be about 150 and the west leg 200 feet from the shaft. At 35 feet east of the shaft a western "back" is struck, this proving to be the continuation of a wall coming from a saddle below the level. The lava is cut at 36 feet from the plat; it here shows only a fine black line dividing the two portions.

At the 550-*feet level.* In the plat the sandstone of the west country shows ripple markings. The west leg of the third formation is cut close to the shaft, and has been driven upon for 118 feet southward.

The east leg has not yet been cut, though the cross-cut has been put out in

* The station at the shaft, often called the "station." Literally, "plat" means a flat place cut out. The Cornish talk of a "tip plat," the flat place where the ore is accumulated, with facilities for discharge below the level.

that direction for 69 feet. It is estimated that 20 feet more would find it. The cross-cut very clearly shows the bedding of the country which is pierced by the lava dikes, which here are 16 feet apart. The south drift is on the west leg. While the wall (the "west back" of the 450-foot cross-cut) continues its course, accompanied by a seam of quartz ("the leader"), the west "reef" turns off eastward and strikes against the lava dikes, but has not yet been crossed by them. The quartz which here is highly auriferous is found to occur both between the two dikes and west of them. The end of the drift is in lava.

Figs. 21 to 27 inclusive illustrate some of the more interesting of the underground features already referred to. Fig. 21 shows the second formation as disclosed in the workings above the 170-feet level. There is no doubt as to the continuity of the beds forming this anticline. Sandstone A overlies the saddle, of which B and D are respectively west and east legs. C is the body of quartz forming the apex; above it there is a wedge-shaped mass of crushed country, consisting for the greater part of black broken slate containing a notable percentage of pyrites, but no "payable" amount of gold. The slate underneath the west leg varies from 2 to 3 feet in thickness, and is divided by a thin "bar" (3 to 4 inches) of hard sandstone. Underneath the east leg there is also from 2 to 2½ feet of slate, while below the cap this slate is irregular and forms "bulges." This bed overlies the hard sandstone H of the center-country, fractured and penetrated by spurs. The east back F, that is, the parting which forms the hanging-wall of the east leg, has been followed for 30 feet above the cap of the saddle.

Fig. 22 is the west leg, as seen at the 550-foot plat. The reef C is hard against the back A, which carries its own thin vein of quartz (the leader). D D are spurs from the west leg into center-country. These cross-spurs are a frequent feature of the country between the legs of a saddle.

Fig. 23 represents *in plan* the position of the "lava streaks" in the south drift of the 550-feet level. S and T represent the dikes; C is the quartz; and A, A the "back."

Fig. 24 shows the relative position of the dikes, as disclosed in the 550-foot east cross-cut. The bedding of the country is very evident. T, which is from 18 to 20 inches thick, is just east of center-country, and has cut through it. S is 20 inches thick, and has just begun to shape its course with the bedding.

Fig. 25 represents the markings, natural size, in the decomposing lava of the west dyke in the 550 cross-cut.

Fig. 26 shows the "west back," which, 40 feet above the 550-ft. level, runs into the west leg of an incomplete saddle (the third formation). It will be understood that a "back" is such only so long as it follows the line of parting between two beds of the country; that when it follows a quartz seam it often becomes the hanging-wall of a reef; and finally that, before striking the quartz or after doing so, it may change its character and be a cross-fissure. In Fig 26, A is the back; C is quartz; S and T are the two portions of the lava dike which cuts through the formation. There is a fault, the throw of which is from 8 to 10 inches. The dike does not fault the quartz, but follows the line of an earlier dislocation. D is broken ground, traversed by spurs. B is an east back.

Fig. 27 is taken from the cross-cut at 190 feet, and indicates a local fault which throws the country about 3 feet east. B, B is a

Fig. 40.

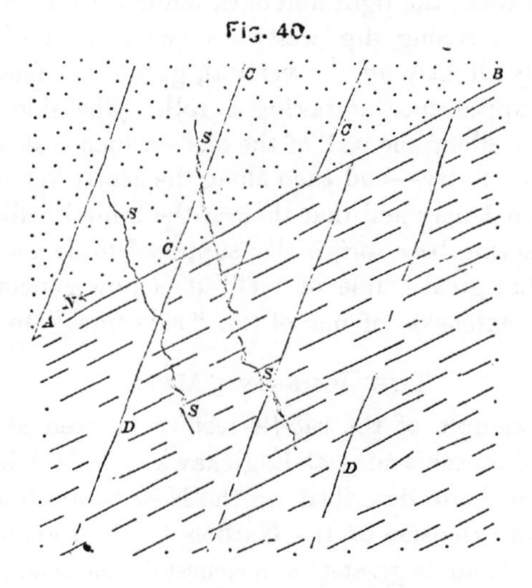

bed of slate; F, F is the line of fault, carrying a couple of inches of selvage and, on the under side, a little quartz.

This mine illustrates the general character of the saddle-reefs. To any one desirous of obtaining the key to the geological structure of the district I recommend the examination of formation No. 2 as cut by the 170 and 280-ft. cross-cuts, where the different features of cleavage and bedding of the enclosing country, the pitch and underlay of the quartz bodies, are all more distinctly marked than I observed them in any other mine. On the other hand the lower or No. 3 formation, as cut by the 450- and 550-ft. cross-cuts well illustrates the frequency with which the simpler type of ore-deposits becomes complicated almost beyond recognition. That this is due to the action of faults, is evidenced by the slickensides on the walls and the crushed condition of portions of the country, such as the wedge of slate above the No. 2 saddle.

In this mine there is none of the perplexity occasioned in the deep mines on Victoria Hill by the difficulty in distinguishing the bedding from the cleavage, where the latter is highly developed. The cleavage which elsewhere so often obliterates the bedding, especially of the west country, is clearly distinguishable in the slate and sandstone beds disclosed by the different cross-cuts.

The 280-ft. level, where it is driven upon the western leg particularly, well illustrates this feature. The sandstone there seen overlying the reef is composed of a number of thin beds of different shades of gray rock, the light and dark laminæ are beautifully regular, and have a strong dip westward (with the reef), while the cleavage is only slightly off the vertical, giving the foot-wall in the "drive" the appearance of taking a roll. The side of the level had broken first along the wall of the reef—which is also the bedding of the enclosed country—and then along the nearly vertical cleavage.

In conclusion I may add that though the mine is called the South New Chum, having been originally supposed to be a southern continuation of that great "line of reef," it is now generally accepted as being on an extension of one of the "side-lines" to the west.

The Johnson's Mine.

The finest example of the saddle-reef to be seen at the present time is in the Johnson's mine at Eaglehawk. This claim is not, like most of those hitherto described, on the New Chum line of reef, but on the northern extension of the Garden Gully, so called after the mine in which it had its greatest and richest development, the Garden Gully United. There has been but one saddle worked in the Johnson's mine, but that one has been exceptionally regular and rich. The apex is 980 feet from the surface; the west leg has been but little developed as yet, but the east leg still continues profitable beyond the 1340-ft. level, which is the deepest in the mine.

All the gold obtained in the workings came from "backs" and "spurs," as well as other irregular bodies of stone or quartz.* Descending No. 2 shaft a thousand feet, one finds that the level is driven upon "the main east back," as it is called. It may be considered as an east leg which has no corresponding west leg; nor, so

* The Colonial talks of "stone," where the American speaks of "rock," and the Frenchman "mineral." Gold-quartz in Australia is "golden stone."

far as it has been followed, does it turn over to form a saddle. This "back" has been worked successfully for 300 feet along its strike, the quartz making in bulges, sometimes 100 feet in height, but of very variable longitudinal extent. The gold-contents varied from 3 to 10 dwt. per ton. These bodies of quartz pitch north. Before coming to an end they usually break up into a number of spurs, and they are generally to be found at those points where a number of small veins—feeders or spurs—join the main back. Such a "make of stone"* is to be seen above this 1000-ft. level. The

Fig. 41.

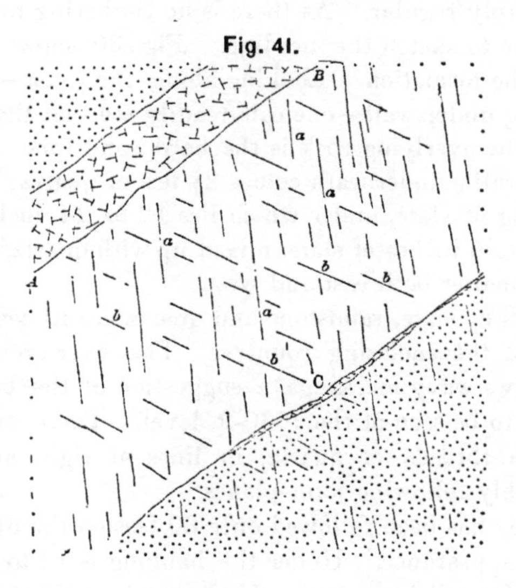

quartz, 12 feet wide, is crystalline and white, but well mineralized, containing vugs or geodes and occasionally patches of included country. This block of quartz pitches north, but 60 feet further is another which pitches in the contrary direction. A small regular seam usually accompanies the back. It is known as the "leader," and is distinct from the other irregular bodies of quartz which may be found along the line of the back. The terms "back" and "leader" are used interchangeably. The wall of the back is very well defined, and carries a couple of inches of black clay—the "dig"—which continues long after the quartz has died out. Figs. 28 and 29 will illustrate this. A is sandstone; C is massive quartz; B is slate; L is the leader; D is the dig or selvage; F, F are small spurs. A cross-cut 85 feet east from this back brings us to the east

* A "make of stone" is the equivalent of a "shoot of ore" or an "ore-body."

leg of the big saddle. The cross-cut shows that on the hanging-wall there are 14 feet of sandstone, divided by three thin partings of slate. Both walls are very clean and distinct, the hanging particularly.

Above the stopes the saddle itself is seen. The appearance of the workings is very striking. The removal of the quartz has left a large vaulted chamber, over which the curving hanging-wall extends with the regularity of an arch of masonry, while the foot-wall underneath looks like the back of a boiler, the curve of the saddle being remarkably regular. As there is no timbering to obstruct the vein, it is easy to sketch the heading. Fig. 30 shows the general structure of the formation. Looking along the crest—the summit of the curving under-wall—one can readily see that the pitch is to the north. The overlying rock is the hard sandstone which forms the hanging-wall; underneath comes 2½ feet of quartz; then follows a dark parting of slate, under which lies 12 to 14 inches of sandstone, then 5 to 6 inches of slate, mixed up with quartz, which latter gets rapidly smaller both east and west.

These layers of slate, sandstone and quartz are all conformable to the bedding of the enclosing country. The over-arching wall of sandstone shows wavy markings, a suggestion of the beautiful ripple-markings to be seen at the 1060-ft. level. There are also to be observed thin threads of quartz, in lines at right angles to the strike—probably following joint-planes.

The west leg has been followed only 30 feet on the dip. Fig. 31 illustrates its appearance. Under the hanging is 12 to 14 inches of clean quartz, C, well laminated. Underneath comes 2 feet of sandstone, A, separated by a black slate parting. Between this sandstone and the main foot-wall there is 5 inches of slate, B. The hanging is broken by feeders which carry gold. The cleavage of the country-rock is to the east, as indicated on the right of the drawing.

The east leg has supplied the major portion of the output of the mine, and has proved highly auriferous. Fig. 34 shows its appearance just below the cap of the saddle. The reef at this point is divided into two parts by about 12 inches of sandstone, H. The upper portion, C, is 2½ feet wide, mottled by the inclusion of bits of country, M, near the hanging-wall, and well laminated or ribboned (N) near the included sandstone, H, which last is cut up by spurs, S, S.

The underlying slate, B, is black and graphitic. The lower portion, E, of the reef is less regular, and lies directly upon the hard sandstone, D, of the main foot-wall. At about 40 feet below the cap the east leg, as seen in Fig. 32, pinches to 3 inches at C, with 4 feet of slate, B, and sandstone, A, between it and the hanging. The decrease in size is compensated for economically by an increase in the gold-contents.

At the 1065-foot level the lode has been worked out. At the 1280-foot level a cross course throws the east leg, which is here from 9 to 10 inches wide, eastward 6 feet, which is equivalent to its being moved from one wall to the other of the lode-channel in which it is situated. This is shown in Fig. 33. At this point a body of quartz, C, about 6 feet by 6 feet, is formed. It is very white, and has been

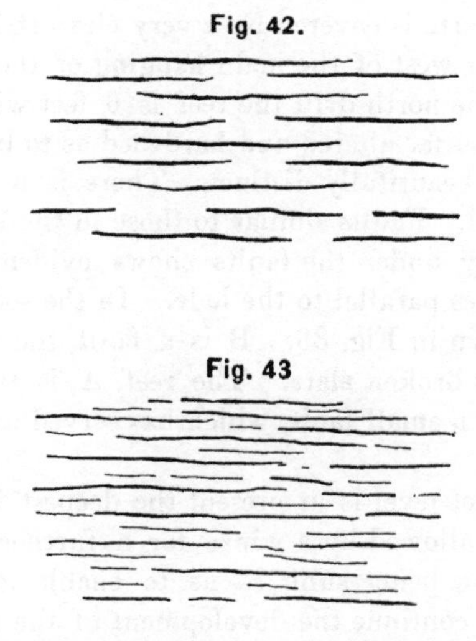

known to carry coarse gold. A is sandstone; B, slate; R, the reef. The north end of the 1280-foot drive shows two faults* in the lode. See Fig. 35. The larger has a throw of 8 inches. Both dip westward, but at different angles. The quartz, A, on the foot-wall side, is much the best in appearance and in yield. The foot-wall itself is well defined and unbroken. B is broken, slaty rock, containing pieces of quartz. C is a body of quartz, white, and containing but little gold.

In the stopes the same faults are found closer together. The spurs

* In the mine they call them "slides." Elsewhere in Australia this term alternates with "heads" and "breaks." In the New Zealand coal-mines the synonym is "troubles."

in the foot-wall going west from the reef pay to follow for 20 or 25 feet. The quartz of which they are formed is white and the gold is coarse, not varying in this regard whether near or distant from the main lode. The tenor of the ore is such that the reef will pay to work when only 3 inches thick, the miners breaking with it feeders, spurs, mullock,* everything as it comes. The average width, however, is from 12 to 16 inches, at which size the lode is most gold-bearing,—better than when it increases to its maximum width of 3 feet.

At the 1340-feet level the east leg is still rich in gold. It is becoming more inclined, having travelled east 10 feet in the last 60, while between the 1000- and 1280-feet levels it moved 90 feet. There is not so much country-rock between the walls as in the levels above. The quartz is covered by a very clean striated hanging-wall, which is 4½ feet west of the main hanging of the lode-channel. In the breast of the north drift the reef is 6 feet wide. The foot-wall is a sandstone so far altered and hardened as to be a quartzite. Ripple-marks are beautifully distinct. There is a fine line of stopes above this level. Faults similar to those in the 1280 are seen. The broken country under the faults shows evidence of having been arranged in lines parallel to the lode. In the south drift is seen the lava dike, shown in Fig. 36. B is a fault, the course of which is marked by soft broken slate. The reef, A, is twice dislocated, the second time by a small fault, which has served as a passage-way for the dike, T.

This 1340-feet level is at present the deepest in the mine. The reef has been followed by a winze for a further depth of 35 feet. The shaft is now being sunk so as to enable deeper levels to be opened, and so continue the development of the mine. A cross-cut is being driven to intercept the west leg.

The richest ore which this formation yielded was obtained just below the turn-over of the saddle, but very rich shoots were found irregularly distributed through the reef, having the same pitch as the formation itself. The quartz, speaking generally, is patchy, and sometimes extremely rich bunches are found. One of these lately gave 150 ounces of gold from a half-sack of quartz.

---

* Mullock is waste rock. A "mullock-tip" is brother to the western "dump." Originally, mullock was the name given to the basalt which covered the deep leads of Ballarat; but since then the alluvial diggers have introduced the term into quartz-mining; and at Sandhurst it refers to the slate and sandstone which is broken with the quartz of the reef.

As we have seen, most of the workings are on the east leg, which, as observed in the levels underneath the cap of the saddles, presents all the appearance of an ordinary quartz-lode with walls parallel to the bedding of the enclosing country. The cross-cuts show the bedding. The alternating slate and sandstone are called "bars" by the miners. Thus, one hears of a cross-cut having just passed through a "tight bar of sandstone." Even where the difference in the nature of the rock forming adjacent beds is very slight, there are lines of parting (usually seams of slate) which indicate the stratification. The cleavage is distinct and uniformly to the east at an angle of 60°.

Fig. 44.

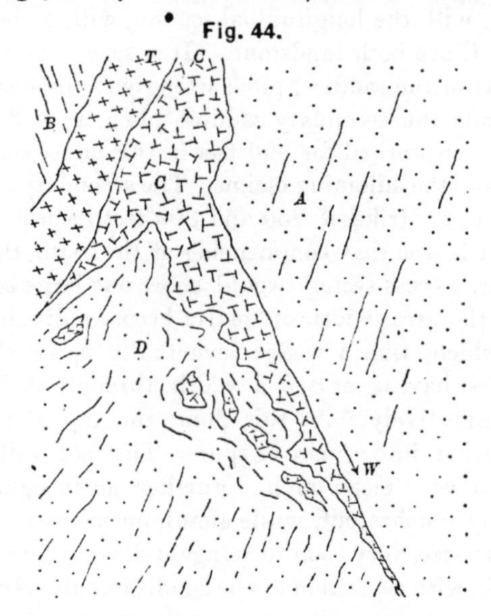

The Johnson's mine for the half-year ending June 30, 1890, paid dividends amounting to £19,441 2*s.* The yield of 12,367 loads of stone was 9146 oz. 3 dwt., worth £38,639 5*s.* 3*d.*, and averaging 14 dwt. 19 gr. per load. The following figures will be of value as indicating the tenor of the ore at the different levels during the half-year:

Part of the Mine.	Loads. (About 2800 lbs. each.)	Yield. Oz.	dwts.
Above the 550-feet level, ...	121	13	10
" 1000-feet level, . . .	1886	991	3
" 1130-feet level, . . .	289	206	16
" 1200- and 1210-feet levels, .	2156	1274	17
" 1280-feet level, . . .	5145	3276	17
" 1340-feet level, . . .	2113	2803	19

The average yield from the stone at the bottom of the mine was 1 oz. 7 dwt. per load.*

The Johnson's mine is on the Garden Gully line of reef, but the next to be described is on the third of the three great anticlinal axes—the Hustlers. The Great Extended Hustlers is one of the "record-mines" of the district,† and at the present time exhibits a formation of subsidiary saddles which is both peculiar and interesting. Fig. 37 illustrates the formation as it is seen in the workings between the 1700- and 1800-feet levels. The cross-section is taken along *a b*, and, with the longitudinal section, will explain itself. A is slate, B and C are both sandstone. It is seen that the anticlinal axes (the ridges of the saddles) pitch strongly to the north, the main axis 1 in 5, while the secondary saddles pitch 1 in 3. The main formation has been worked for 800 feet in this mine, and extends for some distance into the adjoining claims. The secondary saddles‡ have a length along their strike of 250 and 300 feet respectively.

In Fig. 37 it is seen that, commencing at the south, the right hand of the drawing, a cross-section would show one large saddle; as we proceed north the great width of quartz becomes divided by a horse of sandstone which, like a wedge, eventually splits the formation into two saddles, having at *a–b* a width through their caps of 40 and 12 feet respectively. At this point the top of the main cap is irregular, broken into a mass of spurs. The foot-wall of the lower saddle forms a very clean arch. Further north again, the lower saddle gradually pinches out, while simultaneously the main formation increases in size, very soon to be again divided by another wedge of country-rock, with repetition of the conditions already noted in the previous case.

Before the two saddles become divided by the country-sandstone,

---

* My thanks are due to Mr. Williams, the mine-manager, as also to the underground foreman, for the details here given.

† It is a noteworthy fact that the three mines which perhaps rank first in the returns of gold which they have made, should be on the three main "lines of reef," and opposite each other, illustrating the old mining aphorism, "ore against ore." I refer to the "180" on the New Chum, the Garden Gully on the line of the same name, and the great Extended Hustlers on the Hustlers line.

‡ During my later visit to Sandhurst the manager informed me that he had come across a third of these subsidiary saddles. I had not the time to inspect it. I have here an opportunity of thanking the mine-manager, Mr. Thomas Heckley, for the valuable assistance he gave me in visiting this mine, as well as for his guidance in examining the old workings of the Unity mine.

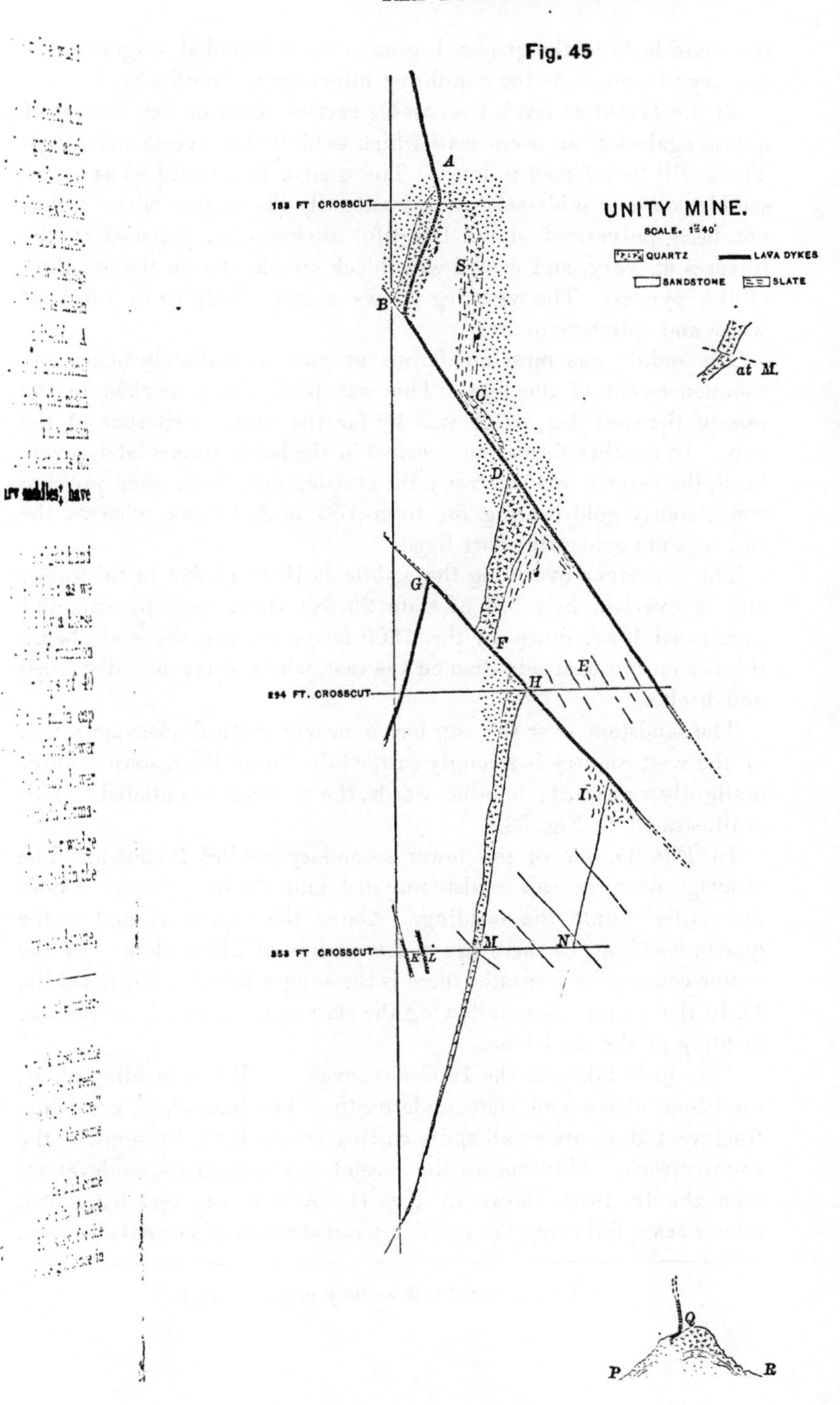
UNITY MINE.
SCALE. 1"=40'
QUARTZ
LAVA DYKES
SANDSTONE
SLATE
153 FT CROSSCUT
294 FT. CROSSCUT
353 FT CROSSCUT
A
S
B
C
D
G
F
E
H
I
K
L
M
N
at M.
Q
P
R

Fig. 45

the main body of the quartz begins to show included fragments of country; becomes, as the Sandhurst miners say, "mullocky."*

At the 1750-feet level, the east leg carries three or four inches of quartz against a sandstone wall which exhibits transverse markings. These will be referred to later. The quartz is crushed so as to resemble common table-salt, and occasionally the surface where it has not been pulverized shows beautiful slickensides, polished to the likeness of ivory, and veined with black streaks due to the grinding of the pyrites. The west leg shows a large body (5 to 6 feet) of white and splintery quartz.

The saddle was most auriferous at and immediately below the commencement of the legs. This was particularly marked in the case of the west leg, which was by far the more auriferous of the two. In another formation, worked in the levels immediately overhead, the reverse was the case; the east leg paid best, often proving continuously gold-bearing for from 150 to 200 feet; whereas the east legs are generally short-lived.

The sandstone overlying the saddle is 10 to 15 feet in thickness, and is overlain by a bed of slate 25 feet thick over the cap, and penetrated lower down by the 1800-feet cross-cut, the slate being thinner on the west side than on the east, where it is much disturbed and broken.

The sandstone over the cap has a nearly vertical cleavage; that of the west country is strongly east, while that of the eastern country is slightly westward; in other words, the cleavage is radiated. This is illustrated in Fig. 38.

In Fig. 39, one of the lower secondary saddles is shown. The cleavage over the cap is also radiated and distinct, though it does not entirely hide the bedding. Above the cap, at B, and in the quartz itself, at C, there are a few inches of black slate. In the center-country underneath, there is the suggestion of a third saddle, D, in the quartz seam following the slate-parting which marks the bedding of the sandstone.

Fig. 40 is taken in the 1805-feet level. A B is a bedding-plane, sandstone above and slate underneath. The lines C, C, are joint-fractures; S, S, are small spurs cutting across both divisions in the country-rock. Underneath the second saddle in the same level are seen the fractures shown in Fig. 41; A B is the east leg, C is a quartz seam following the bedding; *a a* are lines of normal cleavage.

* That is, mixed with mullock or country-rock.

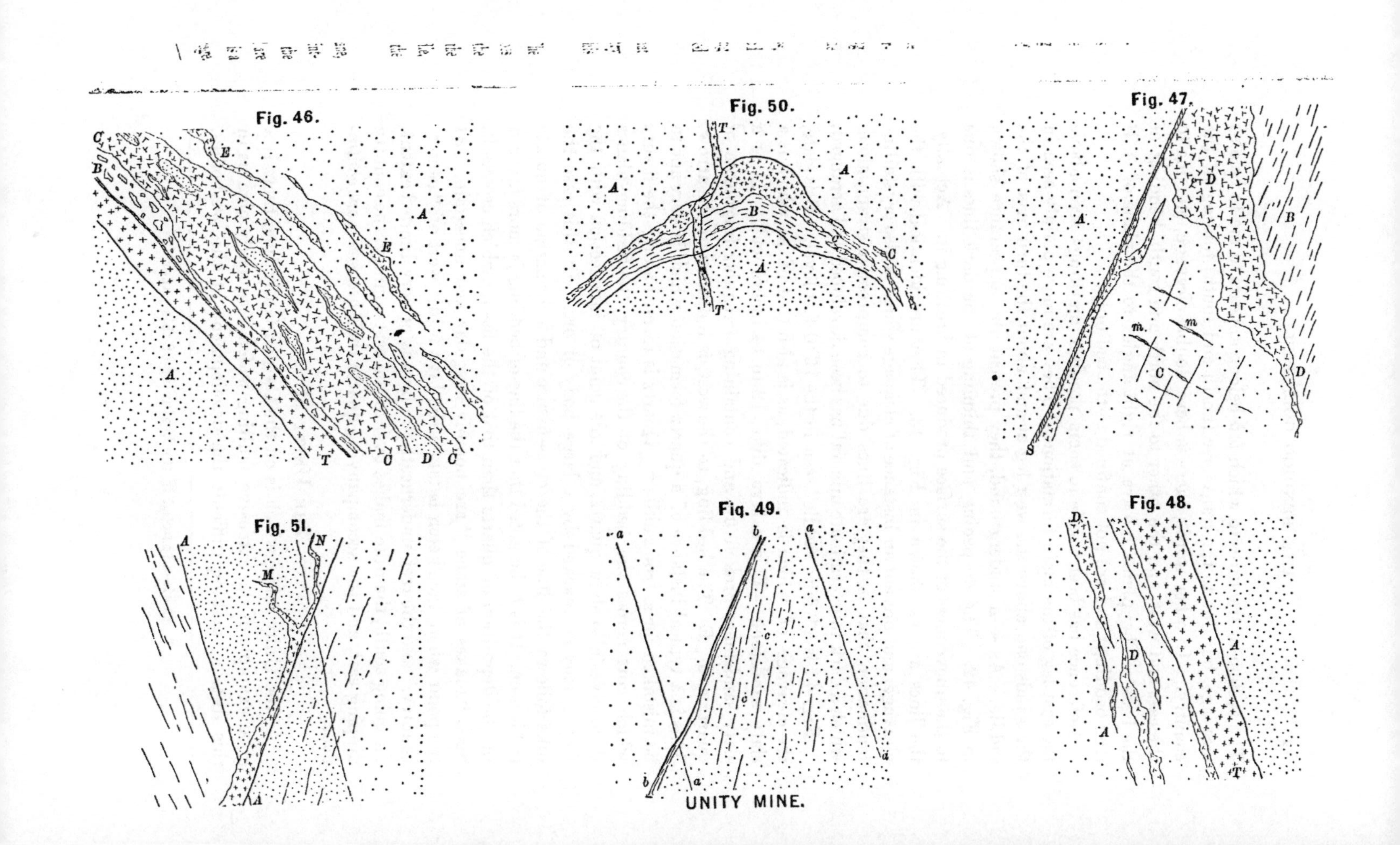

Fig. 46.
C
B
E
A
E
A
T
C
D
C
Fig. 50.
T
A
A
B
C
A
T
Fig. 47.
D
B
A
m
m
C
D
S
Fig. 51.
A
N
M
A
Fig. 49.
a
b
a
c
c
b
a
a
UNITY MINE.
Fig. 48.
D
D
A
A
T

The system of fractures to which *b b* belong appears to me to be also cleavage, but due to the transverse (north and south) folding of the country. In other words, there is here double cleavage, one due to the east and west, and the other to the north and south, bending of the beds. The planes *b b* are at right angles to the plane of A B (the bedding) and are not confined to the bed only.

Reference has been made to some peculiar markings seen above the east leg of the main formations. They are more plainly seen in the sandstone above the west leg at and about the turn-over of the saddle. As seen underground, they present the appearance shown in Fig. 42. The deepening and thinning of the dark lines is due to the unevenness of the surface produced in breaking it. Actually, the lines are as shown in Fig. 43. They are identical with the cleavage, and present an instance of cleavage which is also crystalline lamination, that is, they are lines due to compressive strain, along which a dark indeterminate mineral has been developed and arranged.

Fig. 44 is a "false saddle" seen in the 1670-feet level. B is slate, the cleavage of which is indicated, as is also that of A, which is a slaty sandstone. T is a lava dike, 12 to 14 inches wide. D is a slaty sandstone, broken up and containing irregular fragments of quartz at E, E. The bedding, to the west, is indicated by the arrow. This is a typical instance of a quartz formation which is very apt to be mistaken for a true saddle.* Quartz is formed along the fissure which cuts across the bedding of the country, the bedding-planes themselves also show quartz, and the point of intersection with the fissure may be marked by a large body of ore. In this case the dike follows the line of the cross-fissure and a formation of quartz (C) is seen. It is to be noted that the line of bedding is more favored in the deposition of quartz than that of the fissure which crosses it. Such "makes of stone" are notably irregular and uncertain. At the point where the sketch is taken the lava is just west of "center-country," and the reef underneath the lava is of very limited extent. In going south, the dike makes over to the east country, the country dipping east, and the accompanying quartz then *becomes more continuous.*

The Unity Mine.

At the Unity mine, which is on the Garden Gully line, there is a good example of the extensive faulting which frequently obtains in this district, making difficult the development of the ore-bodies.

* As already noted in the beginning of this paper.

Fig. 45 is a section, 40 feet to the inch, which will serve to explain the occurrence. B A C is an imperfect but true saddle, since the country dips east and west with the two legs. There is no true east leg, but rudiments of it are seen in the spurs which traverse the slate at S. Sandstone overlies the cap and the west leg, as well as the bed of slate in which are the spurs which replace the east leg. At 28 feet below the cap, the country is faulted, the throw bringing the west leg, the only part of the formation which continues well-

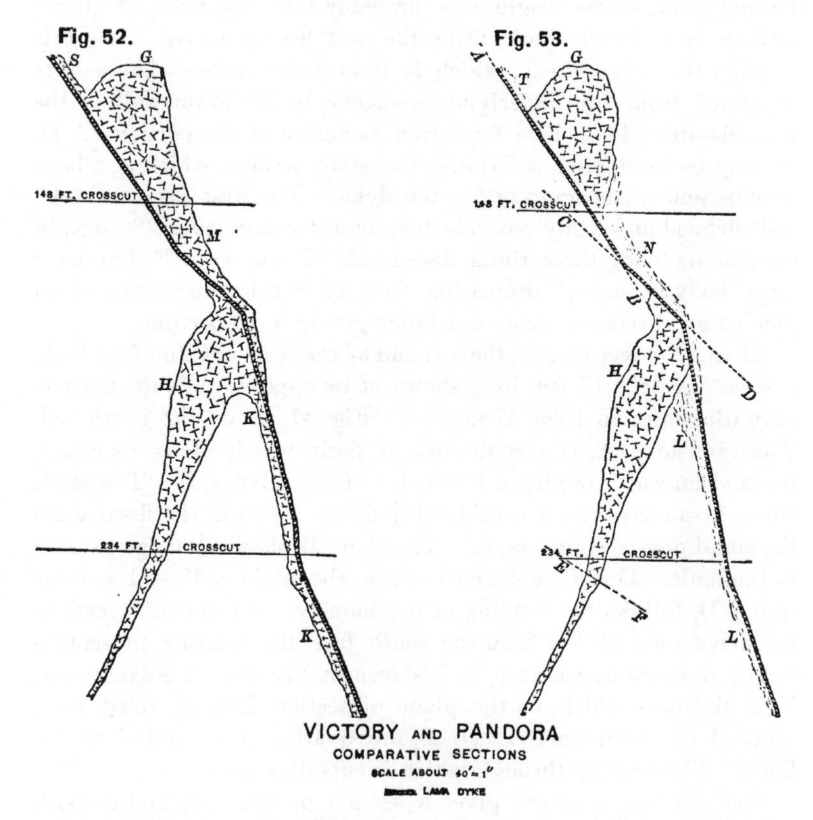

defined, from B to D, a distance of 35 feet. Another slide at a further depth of 38 feet (along the dip) causes another break, this time of 8 feet only. Again at M there is a dislocation, which has not gone further than to cause a derangement in the straight course of the lode. We will follow the formation through the different changes which it undergoes. At the 183-feet cross-cut, where it is first seen, the top of the cap has been stoped away, offering to view the partial formation of the arched structure, which is so beautifully distinct in some of the mines, the Johnson's for instance. The west leg is strongly defined, and yielded well at the time it was stoped. The spurs which replace the east leg were worked for a depth of 25 feet, and eventually merge into the formation shown at C in Figs. 45 and 46. In the latter, A is the sandstone of the overlying country; E E are spurs; C is quartz, which includes several fragments of country at D D; B is a layer of brecciated quartzose material containing gold, whose origin was probably the fragments of quartz broken away by the fault from the west leg up above. All this overlies the lava dike T, which is from 6 to 8 inches wide, and is separated from the underlying sandstone by the lower wall of the fault-fissure. The whole formation, inclusive of the portions C, D, B, may be considered as forming the slide or fault which at a later date became a passage-way for the dyke. The west leg continues, well-defined and easily recognizable, for a depth of over 200 feet, in spite of its being three times dislocated. From D to F it forms a large body of stone,* decreasing from 12 to 5 feet in width, which yielded good returns, many crushings giving 5 oz. per ton.

At the 290-feet level at the east end of the cross-cut, the first fault is seen. A drift 15 feet long shows at its opposite ends the appearance illustrated in Figs. 47 and 48. Fig. 47 shows the south end. The sandstone, A, is over the line of fault, which along its course has a clean wall carrying a few inches of soft selvage, S. The sandstone, C, underneath, is considerably fissured, and in the fissures are the small quartz-seams, *m*, *m*. The slate, B, shows cleavage parallel to the fault. Quartz is formed along the main wall, and a large spur,† D, follows the bedding of the country. At the other end of the drive, only 15 feet from the south face, the heading presents a totally different appearance, as is shown in Fig. 48. A is sandstone; T is the dike which, in the plane of section, Fig. 47, must have pinched out temporarily. D, D are quartz-veins parallel to the fault. The country through which it cuts dips east.

The 291-feet cross-cut gives a section of the lode and of both slides. The lower fault is 28, and the first, or upper, 57 feet east of the shaft. The second fault has more dip, but, like the first, it shows a bulge of lava at one end and none at all at the other end, 6

* That is, "mill-stuff," "quartz," or "dirt," as it is variously termed by the miners in various parts.

† Were it to continue for any distance it would be called a "back," since it follows the line of bedding.

feet distant, of a raise which has been put up to cut the lode. The country between the two slides is much disturbed, and shows lines of fissure in sympathy with the faults. It is full of flat spurs, having a slight dip eastward. The country west of the faults contains no spurs, but it is hard sandstone and much fissured. At 15 feet west of the second fault or slide, the country begins to show a westerly dip. In Fig. 49, taken at this point, *a*, *a* are fissures parallel to the faults; *b*, *b* is the bedding, indicated by a slate parting between two layers of sandstone; *c*, *c* is cleavage, nearly vertical.

At the 353-feet level (see Fig. 45) the quartz has decreased to 2½ feet, but it is still a strong lode. Neither of the faults is to be seen, as the cross-cut is not far enough east. At M there is seen an attempt at faulting which has gone so far only as partially to dislocate

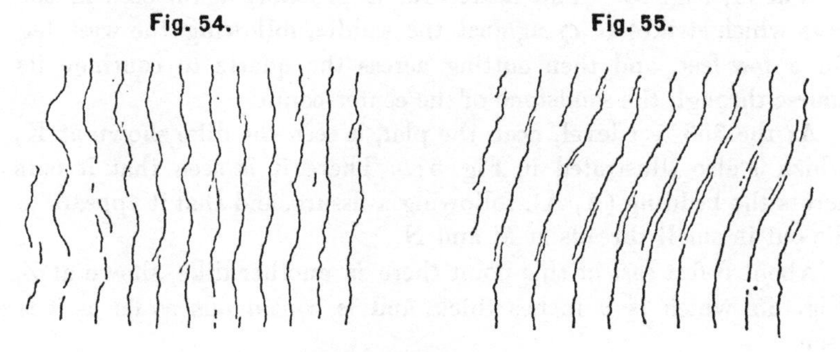

RIPPLE MARKS.
JOHNSON'S MINE. 1065 FT. LEVEL.

the reef. At N there is a "back," a line of parting between two beds, parallel to the lode, which goes up as far as the second fault, where a formation of spurs is found on both sides of it. This is the last we see of the reef, which decreases rapidly in size and becomes unprofitable, the dip taking it across the shaft before reaching the level of the 451-feet cross-cut.

This cross-cut intercepts at O, Fig. 45, the back which we saw at N, in the 353-feet cross-cut. It is well-defined and continues under foot. The cross-cut ends in a very well-formed saddle, P, Q, R, which is shown in greater detail in Fig. 50. The outline is suggestive of a double contortion. The cap makes a very clean curve. Sandstone, A, overlies the quartz, both being penetrated by the dike, T. Underneath the quartz is slate, B, in which are the spurs at C, which replace the east leg. The west leg is divided into two portions, separated by slate. Going south, they are wider apart. The slate under the cap shows lines of bedding, and its division from the underlying sandstone forms a clean, very beautiful arch.

The lava dikes, which form so marked a feature of the section, Fig. 45, have many points of interest. A dike (varying from 12 to 20 inches in size) cuts through the cap at the 183-feet level. It has a slight eastern dip which, on striking the quartz, is changed for the dip of the west leg, through which it passes (it is now 6 to 8 inches wide) until meeting the fault, which it then follows, as the line of least resistance. It underlies the formation at C, and is last seen in the 294-feet cross-cut.

Another dike (4 inches only in width) comes down, keeping company with a "leader," cut near the shaft at the 294-foot level. It is probably a branch from that which follows the No. 2 fault, as seen at H, Fig. 45. This latter dike is probably again seen in the lava which strikes at Q against the saddle, following the west leg for a few feet, and then cutting across the quartz to continue its course through the sandstone of the center-country.

At the 353-feet level, near the plat, is seen the dike shown at K, which is also illustrated in Fig. 51. There it is seen that it cuts across the bedding (A, A), following a fissure, and that it appears to die out in small threads at M and N.

About 5 feet east of this point there is another dike, shown at Z, Fig. 45, which is 9 inches thick, and is continuous so far as it is seen.

The two faults encountered in this mine are found again in the neighboring mines. They get wider apart to the north. In the Carlisle they are divided by 40 feet of country, while southward they approach each other, and in the Victory and Pandora Company's ground they are seen to cross. I saw a section of the upper workings of the Pandora in which these same faults are readily recognizable, though they had been misinterpreted by the draughtsman. In Figs. 52 and 53 are given two comparative sections. Fig. 52 is a sketch reproducing on a smaller scale the old drawing to be seen in the office of the company. Fig. 53 indicates what appears to me to be the real situation. In Fig. 52 S is called the leader, G the cap of the saddle, H the west leg, and K the east leg of another more complete formation. The lava dike follows a slide which has an irregular course through the two bodies of quartz. Just below the 234-feet cross-cut the legs are bent.

Compare this with the section of the Unity. The two mines adjoin, and we know that southward the two faults or slides are ap-

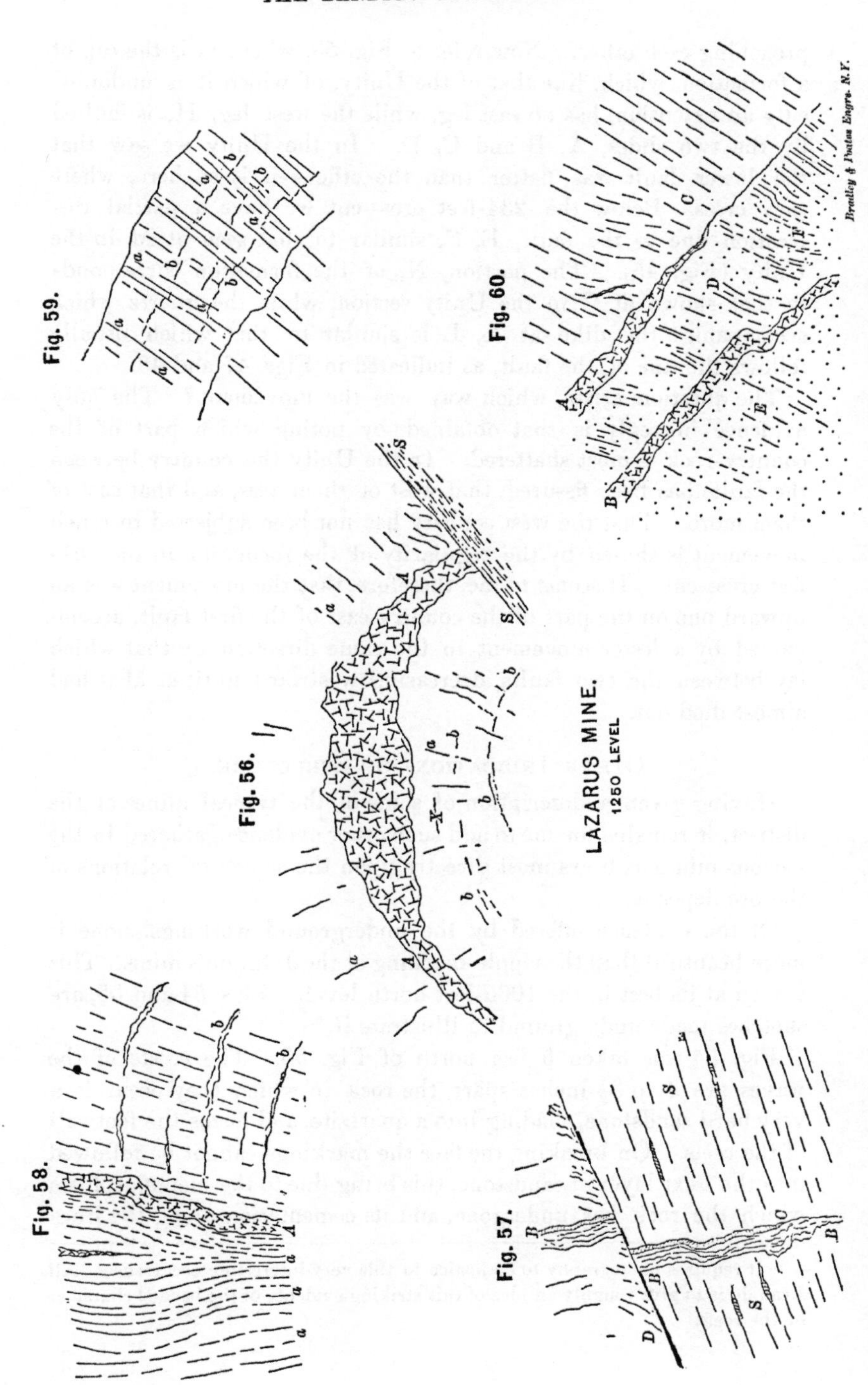
Fig. 58.
Fig. 57.
Fig. 56.
LAZARUS MINE.
1250 LEVEL
Fig. 59.
Fig. 60.
Bradley & Poates Engrs. N.Y.

proaching each other. Now refer to Fig. 53, where G is the cap of a formation, which, like that of the Unity, of which it is undoubtedly an extension, has no east leg, while the west leg, H, is faulted by the two slides, A, B and C, D. In the Unity we saw that the lower fault was flatter than the other—it is so here, where they cross. Below the 234-feet cross-cut we have a partial dislocation due to the fault, E, F, similar to that seen at M in the Unity (Fig. 45). The portion, N, of the formation corresponds to that shown at C in the Unity section, while the quartz which accompanies the dike at L, L is similar to that which usually follows the line of the fault, as indicated in Figs. 47 and 48.

The question arises, which way was the movement? The only evidence in reply is that obtained by noting which part of the country-rock is most shattered. In the Unity the country between the faults has been fissured, that west of them less, and that east of them more. That the west country has not been subjected to much movement is shown by the regularity of the formation in the 451-feet cross-cut. It seems to me, therefore, that the movement was an upward one on the part of the country east of the first fault, accompanied by a lesser movement in the same direction by that which lay between the two faults, decreasing westward until at M it had almost died out.

Other Indications of Structure.

Having given a description of some of the typical mines of the district, it remains for me to add such other evidence gathered in the various mines as bears most directly upon the structural relations of the ore-deposits.

Of the evidence offered by the underground workings, none is more beautiful than the ripple-marking at the Johnson's mine. This is seen at its best in the 1065-feet north level. Figs. 54 and 55, are sketches made underground to illustrate it.*

Fig. 54 was taken 5 feet north of Fig. 55. The crests of the waves are 3 to 3½ inches apart, the rock in which they occur is a very hard sandstone, shading into a quartzite, and forms the footwall of the reefs. On breaking the face the markings cannot be followed into the next layer of sandstone, this being due to the metamorphism which the rock has undergone, and its cementation along the lines

* It requires photography to do justice to this very interesting occurrence. All I can do is to give roughly an idea of this striking evidence of the bedded character of the reefs.

of original deposition. The preservation of the ripple-marks is due to the fact that the sandstone face which shows them was overlain by a soft slate, which has since in part been replaced by the quartz of the lode. The markings can be followed for 100 feet in height, and for more than 200 feet in length. The reef of which this is the foot-wall dips east about 70° ; it was in this part of the mine richly gold-bearing, and it was found in breaking the quartz* that it broke with difficulty, owing to the resistance offered by the projections of the corrugations which were so striking a feature of the stopes.†

Fig. 61.

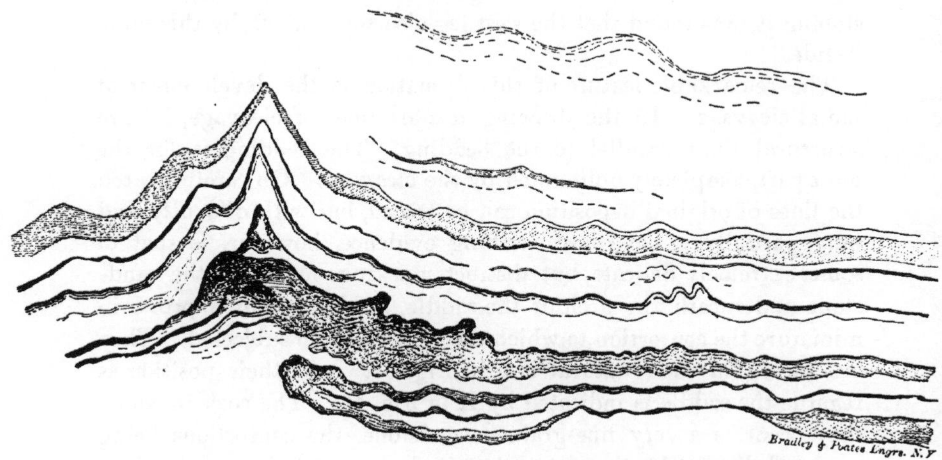

CONTORTIONS IN SANDSTONE.

LAZARUS. 1250 LEVEL

¾ NATURAL SIZE.

These ripple-markings were in east country. At the 550-feet plat in the latter mine, immediately behind the "west reef," there is a

---

* † The ripple-marked wall formed "the shooting-wall," that is, it was utilized in blasting as the line along which the quartz detached itself most readily.

‡ Soon after my third visit to Sandhurst, I saw an almost exact duplication of these markings in the sands of to-day. It was in New Zealand. We were trawling by moonlight; and while pulling in the net I noticed that the sand forming the bottom of the shallow water in which we stood was covered by a series of ripple-marks, whose distance apart and little irregularities resembled those which I had seen in the Johnson's mine; but these on which I stood were of to-day, those were laid down in the estuaries of the Silurian seas; these were soft and yielding, while those had been consolidated into a quartzitic sandstone, now one thousand, and at one time many thousand feet below the surface.

bed of sandstone whose *under* side shows a *cast* of ripple-markings, in other words instead of getting ridges, you get hollows, the true ripples on the underlying bed are not to be seen, but the sandstone which was deposited upon it still retains the impression, the negative of the face of sand which it overspread.

At the Lazarus mine is found the formation shown in Figs. 56 to 61, reproduced from sketches taken in some old stopes at the bottom of a winze leading from the 1250-feet east level. Fig. 56 gives a general view of the saddle. A A is the cap, which is not at all regular in outline, and both above and below breaks up into spurs. The east leg is cut off by a fault, on which the winze was sunk. The manager informed me (the ground is filled up now), that in stoping it, was found that the west leg also was cut off by this same "slide."

The remarkable feature of this formation is the development of radial cleavage. In the drawing, *a a* are lines of cleavage, *b b* are structural lines parallel to the bedding. The bedding is for the most part completely obliterated by the cleavage. On careful search the lines of original deposition can be traced, but with difficulty and only partially. The most striking evidence, however, is that of some extremely delicate but distinct markings found in the sandstone underneath the arch of the saddle, markings which prove in miniature the contortion to which the rock has been subjected. They are shown in their natural size in Fig. 61,* and their position as regards the saddle is indicated by X in Fig. 56. The rock in which they occur is a very fine-grained sandstone, the contortions being rendered distinct by the alternation of dark- and light-gray laminæ.

Fig. 57 was taken in the winze. The slide S is 5 feet wide, and the slaty materials of which it is composed are arranged parallel to the bounding walls. A small quartz-seam B is faulted,† the throw being about four inches. It is accompanied by a narrow band of black slaty filling. The edges of the beds of the country are bent and broken at their contact with the walls of the "slide"—at D.

Fig. 58 is taken in center-country, under the saddle. A is a vein of quartz, a spur, filling one of the many fractures produced in the bending of the beds; *a a* are lines of cleavage, at this point very marked; *b b* are quartz-seams arranged along the structural lines,

* This might be taken as the cross-section of a mountain range; and certainly these foldings, though only covering a few inches, are typical of the plication which has produced mountain ranges.

† A fault within a fault.

which are parallel to the bedding. This may be called "secret bedding," since the fractures are doubtless in sympathy with the bedding and along lines of original deposition, only now made visible by the fissuring through which relief was obtained at the time of the plication of the rocks.

Fig. 59 is in east-country, just above the slide or fault. C is slate; D, slaty sandstone; E, sandstone. A and B are quartz-spurs thrown off by the east leg. At F the cleavage is twisted.

This formation, the several portions of which have been briefly described, throws a good deal of light upon the structure characteristic of saddle-reefs.

Evidence of the structure of the district can be obtained without

Fig. 62.

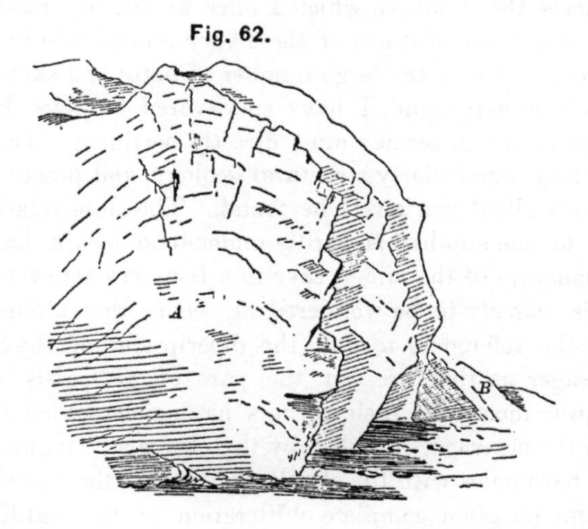

going underground, since large open cuts are numerous along the main lines of reef. Several sketches made in the surface-excavations are shown in Figs. 62 to 66, inclusive. Figs. 62 and 63 were both taken behind the engine-house of the "180" mine on Victoria Hill. Fig. 63 shows where the quartz has been for the most part removed, leaving an arch overhead. C is *débris;* D, slaty sandstone; while A is a lava dike, the same dike which is repeatedly seen underground and which forms one of the most striking features of the deepest workings, as shown in Fig. 14.

Fig. 62 is about 100 yards south of Fig. 63, and shows the southern continuation of the same anticline. At B, the bedding is very distinctly to the west, the rock being "corduroy," which weathers very slowly. The country about the apex of the saddle has undergone considerable alteration, and only traces of the original bedding are now to be seen.

Fig. 64 is taken in the vicinity of the New Red, White and Blue mine. The saddle is very sharp and distinct. C is an old drift, while D consists of material which has been filled in.

Figs. 65 and 66 will appear curious. They are sections of the same anticline, taken from the opposite sides of a narrow cutting. The space covered is only 7 to 8 feet in width.* The wavy lines are marked by small seams of quartz which follow the contorted bedding in the country-rock. The nearly vertical veins are interesting as indicating the fractures which took place along the axis of the fold.

This closes the evidence which I offer as bearing most directly upon the structural relations of these very remarkable ore-deposits of Sandhurst. Out of the large number of notes and sketches made by me while underground, I have endeavored to place before the Institute those which seemed most directly pertinent. Instances in which geology, particularly structural geology, and practical mining are so closely allied can rarely be found. The close relation which they bear to one another is hardly understood by the busy underground managers of the mines, save in a few very noteworthy cases; and this is scarcely to be wondered at, seeing that almost without exception the references to and the descriptions of the Sandhurst mines, meager as they are, on the part of geologists, engineers, etc., are quite incorrect; their authors having confounded the bedding with the cleavage. Misled by the cleavage, "its mysteriously deceptive harmonies with the stratification "† on the east side of the saddles, and its often complete obliteration of the bedding on the west side, they have considered that the east leg did conform to the bedding, but that the western cut across it, that is, the ore-deposits were "saddles" in name only.

There is one noteworthy exception, namely, a short note which appeared in the quarterly report of the Mining Department of Victoria, for December, 1888, by E. T. Dunn, formerly government geologist of the Cape Colony. In that note attention is directed to

* There are several large open cuts behind the Victoria quartz mine, and having noticed that in different parts of the workings the country dipped east and west, I hunted for the turn-over of the saddle. It was only a ter a second search that, in leaving the old excavations by a narrow road-cutting, I found what is represented in my sketches. Recent rain had rendered the structure beautifully distinct.

† Ruskin, in *Schisma Montium, Deucalion.*

this point. Had it come under my notice previous to going to Sandhurst, I should have been spared much of the difficulty which I experienced before the true nature of the deposits became manifest.*

That the typical ore-deposit of the Sandhurst gold-field is a true anticline there can be no doubt. As already pointed out, the saddle is not the only formation in which the gold-quartz is formed, since the spurs, backs, lodes, etc., have at all times been responsible for a very large share of the output; but the saddle is the distinctive form to which the others are for the most part accessory and subsidiary.

The ripple-marks frequently observed underground are incon-

Fig. 63.

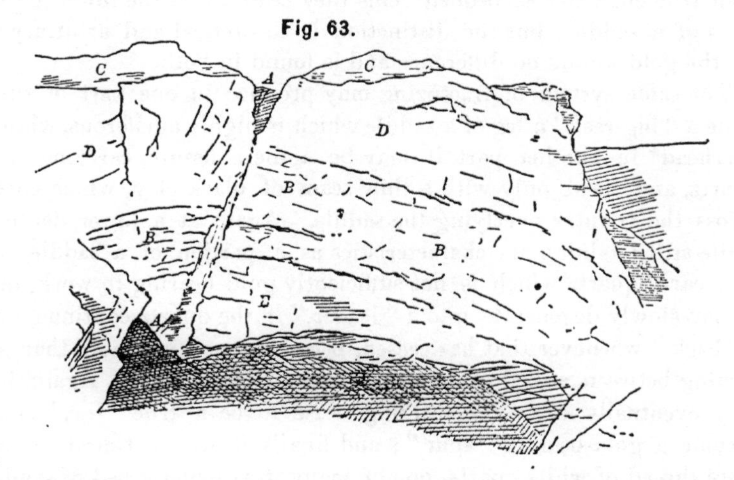

trovertible evidence of the original position of the beds. That, for instance, the east leg of the big saddle in the Johnson's follows the true line of stratification, is amply proved by the very beautiful markings seen in the 1060-feet level. Similarly, ripple-marks are seen on the faces of beds dipping west. Other no less conclusive though less striking evidence has been given in the descriptions of the various mines.

Mining is a commercial business and not a scientific pursuit. The following of a lode on its dip ceases whenever the quartz which it

* In visiting Sandhurst one usually visits first the more famous mines on Victoria Hilt, such as the celebrated "180" mine and others near it. It so happens that in these mines that the structure of the country-rock is most hidden, because the cleavage is most highly developed. The smaller mines at the other end of the field show the true character of the ore-deposits far more clearly. Among the latter the South New Chum has been described here.

contains is no longer auriferous; and as a consequence the mine-workings rarely disclose the ultimate limits of the lower portions of a saddle-formation. Lower cross-cuts will intercept a thin but pronounced parting in the country, a "back" which could be traced upward until it formed the boundary of a big lode or "reef;" or they may cut through a "spur," which if followed would be found to lose itself in an east or west leg, the distinction in the two cases being solely that of conformity with the bedding of the country-rock. Where the legs, as parts of a saddle-formation, end, and where they become one or other of the various types of cross-fissure, no man can say. It may be considered that whenever they lose their true character as bedded veins they cease to be the lower portions of a saddle; but the distinction is theoretical and arbitrary; for the gold knows no difference and is found in both.

The same system of fracturing may produce in one part of the mine a "big reef," a leg of a saddle which is highly auriferous, while overhead* in another part it may be a mere fissure, carrying no quartz, and lined only with a thin seam of black clay, which cuts across the country overlying the saddle. Again at a lower depth, while still retaining its characteristics as a portion of a saddle, it may carry quartz which is not sufficiently gold-bearing to work, or it may slowly degenerate into a "leader," if the quartz continue, or a "back" whenever that has ceased, becoming nothing more than a parting between a bed of sandstone and a bed of slate. Again, it may eventually cross the bedding, cease to be a true "leg," and become a gold-bearing "spur"; and finally it may be traced into a mere thread of white quartz, one of many, traversing a bed of sandstone.

To make nice distinctions would be arbitrary and out of accord with the facts; the same fissure in one mine plays many parts.

While, therefore, I would not say that the legs of a saddle-formation never go astray from their correct course between the bedding planes, such an occurrence would be found only when they are beginning to die out, and does in no way modify our view of the true nature of the "saddle-reef," which, notwithstanding interesting local variations of structure, is a distinct type of ore-deposit—essentially traversing the Silurian slates and sandstones in conformity with the bedding, and forming along the anticlinal undulations of the enclosing country those bodies of gold-bearing quartz which are the most pecu-

* See Fig 21, where an instance of this kind is shown in the South New Chum.

liar feature of the mining industry and the most valuable depository of the wealth of the Bendigo gold-field.

Mine-Management.

Fig. 67 is a view of the New Chum line of reef from Eaglehawk, copied from the quarterly mining report of the Victorian government.

There are at present engaged in the exploitation of the mines 143 companies and 21 tribute-parties, the former including four private mines. The gold-field has been developed by the energy of its own people, the amount of outside—chiefly Melbourne—money invested being comparatively small. Sandhurst has profited greatly by the enterprise and success of one or two mine-owners, such as the late I. B Watson and the present owner of the "180" mine; but its development has been retarded, on the other hand, by the prevalence

Fig. 64.

of one of the worst systems of company-financiering. The actual management of the mines, both above and under ground, is in the hands of capable and careful men; but they are hampered by the fact that reserve capital is almost unknown. Of the 28 dividend-paying companies of 1890, four appear also on the call list. That is to say, these four companies both levied calls and disbursed dividends during the year. A series of fortnightly sixpenny dividends may be followed by a number of fortnightly calls, supplemented by a bank-overdraft—a state of affairs which may swell the volume of business and keep a number of clerks employed, but which, to say the least, is quite unbusinesslike and highly prejudicial to the proper development of mines which require extensive and systematic exploration. The fortnightly dividend is the bane of Bendigo. Whenever a rich saddle-formation is discovered, all haste is made to open it up; as soon as the cross-cuts and drifts have intersected the ore, stoping is at once commenced, the quartz being broken away as rapidly as possible, while all development-work is for the time practically suspended. The results of a few crushings enable the directors to pay off the bank-overdraft and enter upon the regular announcement of dividend after dividend, at intervals of a week if a fortnight seems too long! No part of the profit is laid aside as a reserve-fund, and, as a consequence, when the ore-body is worked out, the dividends abruptly cease, and the bank-account is overdrawn to pay running expenses, until it is necessary to call upon the shareholders for fresh assessments. This foolish round of alternating surplus and deficit, of squandering, borrowing and taxing, has gone on year after year, notwithstanding some of the more thoughtful managers are fully aware of its harmful character. The result may be seen in the cases of mines which divided in profits a few years ago sums varying from £100,000 to over £250,000, and are now almost idle for want of funds to open up new ground. The immediate distribution of profits leads to a continual change in the *personnel* of the shareholders, so that a mine when poorest may be owned by men who had no share in the property when it was most productive. Mining is thus brought down to the level of mere speculation—an evil not unknown, under different forms, in other districts, and certain everywhere to cripple the legitimate industry of mining.

The management of the mines is controlled by the directors, the "legal manager" and the "mine-manager." Since, as a rule, these are all resident in the vicinity, and exercise more or less direct supervision, it is largely a question of personal influence which is in the ascendant. A mine-manager, as known to English companies, that is, a manager of all the mining work, or a superintendent, as known in the United States, is not found in this gold-field. The responsibility of superintendence is much scattered. The "mine-manager" receives the pay of a foreman—about £20 or $100 per month. The "legal manager" is the business man and book-keeper. The directors control the policy of the company, and also do much of the work often delegated elsewhere to a consulting engineer. Such a thing as a trained mining engineer is not to be found upon the staff of any of the companies; and the work done, as a consequence, lacks progressive character. The actual mining work under ground leaves nothing to be desired. It is carried out under the eyes of men whose superiors as miners cannot be found; but a want of system is to be observed in the development of the mines.

Treatment of the Ore.

The handling of the ore in the mills is far from satisfactory. The explanation is simple. Of the managers of 27 mines which I examined, the names of 25 indicated the fact that they hailed from the west of England, while the other two, though native-born, were of the same stock. The men of Doon and Cornwall are miners second to none; but they make the most unprogressive of mill-men.

The gold-ore of this field belongs to the simplest type of the free-milling class. As a rule, it consists of white quartz carrying a very small proportion (from 0.25 to 1 per cent.) of arsenical pyrites. Iron and copper pyrites, blende and galena are also frequently present, but in very inconsiderable amount. The gold varies greatly in its mode of occurrence, but is usually coarse, in the rich ore

Fig. 65.

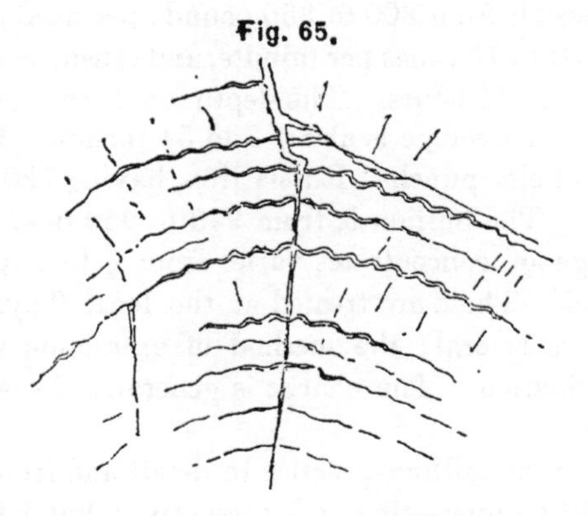

almost invariably visible, and of uniformly high purity, being worth £3 17*s.* 0*d.* to £3 19*s.* 0*d.* per ounce—pure gold being worth at the London mint £4 4*s.* 11½*d.*

The quartz itself is usually dead-white, with an appearance which, in California, would be considered most unfavorable to the presence of gold. It has a splintery fracture, rendering it easy to break. The mill-stuff consists of the quartz mixed with from one-third to over one-half of vein-filling and country-rock, which varies. The ratio in amount of slate to sandstone and the consequent hardness of the material to be crushed is constantly changing. It is the fragments of excessively hard quarzitic sandstone which cause most wear and tear to the milling-machinery.

Stamp-milling of the simplest kind is the method employed to extract the gold. The mills are constructed on the spot by excellent local foundries, at a price usually far below that which is in the western United States. A first-class stamp-mill of 40 heads, engine and shed included, with shaking-tables but without rock-breaker or automatic feeder, can be contracted for at from £6000 to £7000 or about $30,000. The battery-frame is generally of iron. Whether this mode of construction is advantageous to the work of the mill is open to doubt. Both the breaking and the feeding of the ore is done by hand, and in this respect the mills of Sandhurst, in common with most of the colonial batteries, are a standing disgrace to the modern mining industry. Inside plates are not used. Plain, not electro-silver-plated, copper plates are used to arrest the gold on amalgamating tables. Blankets precede the shaking-tables, which have the end-shock, and are a variation of the much-used Rittinger type.

The stamps weigh from 800 to 950 pounds per head; drop 8 to 9 inches, usually 70 to 75 times per minute, and crush, as a rule, about 2 tons per head per 24 hours. The depth of discharge or issue varies widely, but the average is about 3 to 3½ inches. The screen or grating is made of slot-punched Russia iron, having 120 to 180 holes per square inch. The bullion is from 940 to 960 fine.

The percentage of concentrates varies from ½ to 2 per cent. per ton of ore milled. These are treated at the local "pyrites works," of which there are several; the method of extraction varying from arrastra to chlorination. The charge is generally £3 per load of 25 to 30 cwt.

A discussion of the milling-practice in detail and from a technical standpoint would be interesting and instructive; but I must content myself in the present paper with the above indication of its general character.

Government Assistance.

The serious decline in the output of gold in Victoria has led to the introduction of government assistance to mining. Whether this course has done good or harm is a vexed question. In the hope of giving renewed vigor to the mining of the colony, the Victorian government has distributed for several years sums amounting to £80,000 per annum, the intention being to encourage exploration. A miner, a group of miners, or a company needing funds to proceed with a cross-cut, a level or a shaft to cut a known lode, or to explore for a suspected one, makes application before a local committee, elected by the miners, and called the Prospecting Board. This body reports it favorably or otherwise to the Secretary for Mines. Out of a large number of applications submitted, the most promising and deserving are selected. The terms of the grant are "pound for pound;" that is, for a grant, say of £500, there must be another £500 furnished by the applicants, making £1000 available for the particular work (and that only) designated in the application.

Such prospecting-grants are now generally condemned. Beyond question the system has completely failed to bring about fresh discoveries by prospecting. I have seen its operation in different parts of the colony, particularly in the mountainous districts which have been, as yet, least prospected; and the result appears to have been only to pauperize mining. It is always understood, that should the subsidized party succeed in placing the assisted mine in a dividend-

Fig. 66.

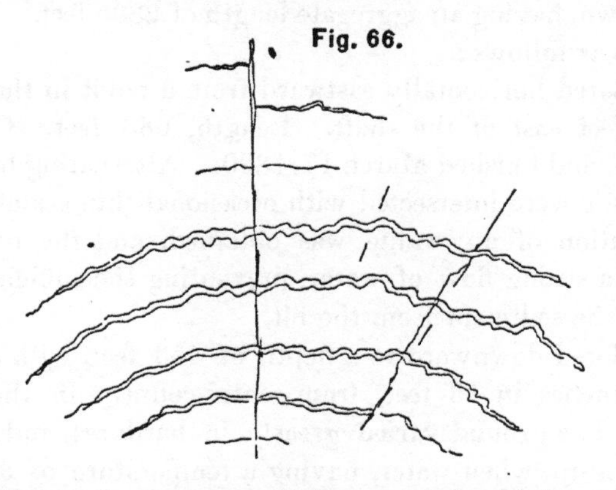

paying condition, the amount of the grant should be repaid to the government from the first profits. This has been done in instances so few as to be remarkable.

On the other hand, this annual dole has done harm by encouraging that improvidence, whether of companies or of individuals, which is the chief hindrance to systematic mining. It has long been a bad feature of colonial mining, as I have shown above, that no portion of the profits is put aside as a reserve to carry on the work of development during lean years; and this prospecting-grant encourages that evil. Having been too eager for dividends to provide for the time when the reef might "pinch out," the mine-owners fall back upon the fatherly assistance of a benevolent government. Mining is a business; and if a mine has sound prospects of success it is very rarely indeed that the money is not forthcoming to carry on the work of development. The money thus distributed could be far better employed in the endowment of a good central mining school, to replace the second-rate technical institutions which at present usurp the name.

Prospecting with the aid of the diamond-drill may be said to form a part of the scheme of governmental assistance.

During the past year, some interesting work in this direction has been carried out at Sandhurst. At the invitation of Mr. George Lansell, who offered to provide the necessary compressed air and other facilities, the Victorian mining department decided, toward the end of 1889, to put down a series of bore-holes from the bottom of the celebrated "180" mine, which had then reached a depth of half a mile. The machine used was Leschot's patent, manufactured in San Francisco, the core being 1 inch in diameter. Four holes were put down, having an aggregate length of 2335 feet. The record of each was as follows:

No. 1. Bored horizontally eastward from a point in the 1300-feet level 197 feet east of the shaft. Length, 683 feet. Commenced January 16, and finished March 17, 1890. Alternating beds of slate and sandstone were intersected with occasional thin seams of quartz. No information of any value was obtained, and the drilling was stopped by a strong flow of water, preventing the sufficiently rapid removal of the sediment from the bit.

No. 2. Bored downward to a depth of 453 feet, with an inclination of 9 inches in 16 feet, from center-country in the 2600-feet cross-cut. The ground varied greatly in hardness, and finally became broken up, when water, having a temperature of 97° F., and under a pressure of 110 pounds per square inch, was struck.*

* While the diamond-drill was at work, a good deal of notice was drawn to the heat of the water encountered in boring at the bottom of the "180" mine. The thermometer showed an average of 97° F., and at one time a maximum of 107° was reached, as recorded by three instruments. This was at a depth of 450 feet below the 2600-feet level or 3050 from surface. This temperature is by no means surprisingly high. The average annual mean temperature at Sandhurst is 60.2° F., and allowing for an increment of 1° for every 55 feet, we should expect at 450 feet a temperature of about 115° F, which agrees fairly well with the facts as disclosed in the "180" mine. However, the matter was considered of sufficient interest to induce a thorough examination by the government analyst, Cosmo Newberry, of the contents of the water so obtained. To this end, two and a half tons of the water was evaporated down to 10 gallons, care being taken to keep out impurities, and 3125 grains of solid matter remained, mostly carbonate of iron. Assays were made, but no gold was obtained, though a trace of a metal of the platinum group was given by the precipitated material. The results were disappointing, in failing to throw light on the occurrence of the gold in the quartz formation. The annual mining report of the department says that a further (optical) examination is to be made.

Fig. 67.

View of New Chum Line of Reef, Bendigo.

No. 3. Bored at a slight inclination below horizontal for 636 feet westward from the plat at the 2500-feet level. It intercepted several veins of quartz, the largest of which, however, was only 9 inches thick.

No. 4. Bored nearly vertical (inclining only slightly to the west) at a point 30 feet west of No. 2, in the 2600-feet cross-cut, to the depth of 562 feet below the level, or 3162 feet below the surface. It passed through about 200 feet of sandstone, as well as several beds of quartzitic sandstone, and cuts several series of small quartz-veins. Hot water under considerable pressure interfered with progress, as in the case of No. 2, though the temperature in this instance was lower.

The result of these borings was practically *nil.* No discoveries of importance were made; and it is doubtful whether the drilling had even a negative value in proving the absence of auriferous quartz in that portion of the country which it was attempted to test. If relied upon as a prospecting instrument under such conditions, the diamond-drill is likely to do serious harm by causing the condemnation of large areas of what may be good mining ground. In strata possessing regularity and continuity, such as the coal-seams; in the determination of the thickness of certain beds overlying known deposits, such as the sheets of basalt over the "deep leads" of California; in the tapping of old workings which are under water; in the measurement of large ore-bodies (as was done in the case of the great bonanza at Virginia City); and in many other instances, no doubt, the diamond-drill, if judiciously employed, is a most useful aid to the miner; but in quartz-mining in general, and in such districts as Bendigo in particular, it is often likely to do irretrievable harm by fostering delusive hopes on one hand or unnecessary discouragement on the other.

To support this view, it is not necessary to go outside Victoria. In the discovery of gold-seams and the exploration of deep beds of auriferous alluvium, the diamond-drill, under the judicious direction of the government geologist, has been very successful; but it has proved a very mixed and doubtful blessing indeed to quartz-mining. In the Sandhurst district, the peculiar structure of the "saddle-reefs," and the comparatively short extent of most of the bodies of gold-bearing quartz, are extremely unfavorable for tests of this character. This is particularly true of horizontal holes, which might penetrate the immediate neighborhood of rich formations without discovering them. If the drill is to be used at all, the dangers attendant upon its use should be minimized by planning a series of holes close together, so as to test thoroughly and satisfactorily, at least a small portion of ground.

Dimensions of Claims.

Among the interesting features of this gold-field is the small area of the claims. The following table gives the areas of eleven contiguous properties on the New Chum "line of reef." The claims are of very irregular shape (see Fig. 7), and in some cases the length of the tract is much greater than the figures given, but for purposes of comparison the measurement is taken along the supposed line of the New Chum reef.

Areas and Depths of a Group of Mines, Victoria Hill, Bendigo.

Name of Mine.	Area of the Property.			Length along strike of reef.	Depth of Shaft.
	Acres	R.	P.	Yards.	Feet.
Ironbark Quartz	4	1	36	220	2140
Victoria Consols	22	3	15	259.6	2162
Gt. Central Victoria	10	1	17	250.8	1970
Victoria Reef Quartz	7	0	25	140.8	2302
Lansell's "180"	14	1	12	171.6	2641
North Old Chum	2	1	25	88	2310
New Chum and Victoria	13	3	12	710	2300
Old Chum	11	2	32	710	2208
Lazarus Co. Lazarus No 1	22	1	7	83.6 764.6	2173
Lansell's "222"	16	0	19	99.4	2100

It will be seen that the average area of each property is but a little over 11 acres; that one of the claims, having an area of only two acres, has workings over 2300 feet deep; and that the ten shafts belonging to eleven companies have an average depth of 2230 feet, and are distributed over a total distance of less than a mile along the strike of the lode. The small area of these properties is primarily a relic of the early days when the claims consisted of twenty yards "along the line of reef" for two men. A certain amount of consolidation followed on the initiation of deep mining; but this was limited to the immediate needs of the time, and has long since been outgrown by the very extensive exploration of more recent years. A circumstance permitting the working of these small areas is the absence of any large quantity of water in the mines. In the particular group of properties cited above there is not a single pump. The little water that finds its way into the workings is raised with tanks alone. At the "180," the deepest mine in Australia, 3000 gallons are hoisted per 24 hours.

The largest areas held by mining companies are as follows:

	Acres.	Rods.	Perches.
Hercules and Energetic,	85	3	17
Johnson's Reef,	71	2	21
Catherine Reef United,	65	3	12
New Red, White and Blue Cons.,	41	3	27
Garden Gully United,	41	3	0

It is seen that even the largest properties are of a size which would be considered small in many districts. It must be admitted, however, that the holding of an extensive acreage does not, in Sandhurst, necessarily mean extensive explorations; since, by reason of the want of working-capital, most of the companies confine their operations to such portions of their holdings as are in the immediate vicinity of the main shaft. This bad feature of the mining work will be slowly overcome, no doubt, by further consolidation. The twin Lazarus Companies have set a good example by sharing one working-shaft; and a recent attempt to consolidate the North Old Chum, New Chum, and Victoria, though unsuccessful, may be taken as an indication of the growth of a proper appreciation of the necessity for such measures.

Deep Mining.

The list of deep mines just given suggests the general subject of deep sinking, a feature of mining at Sandhurst to which reference is often made in the colonies. This practice owes its beginning to the "180" mine. When the mine was purchased in 1873, the shaft was only 400 feet deep, but the neighboring North Old Chum, New Chum, and Victoria, Old Chum, and Lazarus, were doing well on a reef which traversed the mines from 600 to 700 feet below the surface. The sinking of the shaft of the "180" mine was commenced. At about 600 feet £120,000 of gold was won. Sinking was continued in the face of much ridicule, and slowly the mine attained a depth which in those days was considered very geat. Much money was expended without any return, until, in 1883, after ten years of steady development, the top of the saddle was cut in the 1548-feet level. An ore-body of extraordinary richness was uncovered, and stoping began; but the sinking of the shaft continued. The Melbourne *Argus* of that day said: "The success of this venture in deep mining decides the prosperity of Sandhurst for a further decade at least." The mine is now 2641 feet deep. The success of deep prospecting having been proved in this trial-instance, imitators were not wanting; and there were soon several mines competing for the record of greatest depth. Deep sinking developed into a craze which seemed at one time likely to do serious injury to the proper exploitation of the mines. The result is seen to-day in the 18 shafts, each of which has a depth of over 2000 feet.

Among the causes of this state of things the structure of the ore-deposits may be named first. The Bendigo miner has an *embarrass de richesse* in the matter of quartz-formation. Should a saddle be intercepted, the first crushing from which yields scarcely enough gold to pay expenses, it is soon left on one side in the search for the next formation, which is usually not far below. *Below*, not *ahead*, for, instead of longitudinal extension, the miner in this gold-field looks for a vertical repetition of the auriferous formations. Did each mine uncover one lode only, there would be more extensive and patient prospecting at any one given level; but, as matters stand, the exploration is not so thorough as it should be, for the reason that it is divided among several levels.

The most potent cause is, however, seen in the success of the deep mines. Deep sinking has not drawn blanks alone, but many prizes, such as the *bonanzas* of the Great Extended Hustlers, at 1800 feet; the New Chum Consolidated, at 1810; the Shenandoah, at 1900; the Lazarus, at 2000; Lansell's "222," at 2000; the New Chum Railway, at 2025; and the North Old Chum, at 2290 feet. As I have said, there are 18 shafts over 2000 feet in depth. They belong to 20 companies, there being two instances—the two in Shenandoah and the twin Lazarus companies—where one main working-shaft is used in common. Out of these 20 companies, 7 are upon the dividend-list; or, to inquire further, 4 are working quartz which is highly auriferous and 6 are breaking ore which is yielding profits, while of the remaining ten, nine are in that process of prospecting and development which here succeeds (though it ought to accompany) a period of stoping, and one is idle. This is a record above the average of even shallow mining.

The actual yield, during 1890, of the deepest-producing mines is given as follows in the quarterly report of the Victorian Mining Department:

Name of Mine.	Quantity crushed.	Yield.			Average.			Depth of Stopes.	Depth of Shaft.
	Tons.	Oz.	Dwt.	Gr.	Oz.	Dwt.	Gr.	Feet.	Feet.
Gt. Extended Hustlers...	16,612	3,780	4	0	0	4	13	1,800	2,040
Lazarus Company.........	6,232	5,059	0	0	0	16	6	2,000	2,110
Lazarus No. 1............	7,282	3,665	16	0	0	10	1	2,000	2,110
Lansell's "222"..........	3,454	2,510	2	0	0	14	12	1,950	2,105
New Chum Con............	18,721	5,153	14	0	0	5	12	1,800	1,850
New Chum United........	9,797	2,243	10	0	0	4	14	1,900	1,940
New Chum Railway......	7,325	10,371	17	12	1	8	8	2,025	2,078
Shenandoah	7,474	5,039	3	0	0	13	11	1,990	2,010
North Shenandoah........	1,048	948	3	0	0	18	2	1,990	2,010
Shamrock..................	2,496	1,446	19	21	0	11	5	1,800	1,840
North Old Chum..........	1,752	1,169	13	0	0	13	8	2,290	2,310
Kentish	3,117	1,183	13	0	0	7	14	1,800	2,113
Totals....................	85,310	42,571	5	9	...	......	......	23,345	24,516
Averages..................	......	......	...	...	...	9	23	1,945	2,043

Leaving on one side numerous small lots of ore, and taking into account only the results obtained from 12 of the deepest mines which were working regularly, it is found that during the year these dozen claims yielded 85,310 tons of ore, giving 42,571 ounces or an average of 10 pennyweights per ton, the average depth at which the ore was broken being 1945 feet.

All but one of these mines are on the dividend-list. We must inquire further in order to arrive at a just appreciation of the significance of this analysis. Of these 12 mines, 11 are paying dividends on ores broken at an average depth of about 2000 feet. There are a similar number, working at an equal average depth, which are not at present making profits. It is most probable that, during an interval of a year fully a third of the former will step into the ranks of the latter; that is to say, they will cease to break highly auriferous quartz and become prospectors for a new run of ore-ground, while, on the other hand, an approximately similar number will replace them from the list of mines which at present are in process of development. We see, then, that at the present time half of the mines working at depths between 1800 and 2600 feet are profit-paying. The average yield of their ore is 10 dwt., while that of the district as a whole is 9 dwt. 5 gr. per ton. The best returns at present come from the New Chum Railway, which, at 2025 feet depth, is breaking quartz averaging nearly 1½ ounces per ton. During the year this mine paid dividends amounting to £21,672 on a paid-up capital of £28,589 and a nominal capital of £36,890. Its total dividends to date have been £52,078.

During the second half-year the Shenandoah, working at 1990 feet depth, produced 4555 tons of quartz, yielding 3717 oz. 17 dwt., or an average of 16 dwt. per ton. Dividends of £8400 were paid during the 6 months, on a called-up capital of £12,800. Altogether this company has returned £59,600 in dividends.

The Lazarus Company, working at 2000 feet, on 76-dwt. ore, paid out during 1890 dividends of £15,750 on a paid-up capital of £63,-187. The total dividends of this mine to date have been exactly the full amount of its nominal capital, £67,500.

The results obtained in some cases appear to leave a very slight margin of profit. The average of the Great Extended Hustlers, the New Chum United, and the New Chum Consolidated mines is very low. It speaks well for the size of the lodes and the handling of the ore, that dividends are possible at a depth of nearly 2000 feet on 5-dwt. mill-stuff. The New Chum Consolidated is an instance of very successful gold-mining. The return of 5 dwt. 12 gr. shown in the list given above is higher than the ordinary average of the mine, since richer quartz than usual was broken in the second half year. For the year ending 30th June, 1890, 9586 tons were sent up from the 1800-foot level, yielding 1722 oz. 19 dwt. of gold at the mill and 278 oz. 8 dwt. (from 106 loads) at the pyrites works. The total value of the yield was £7943 5*s.* 9*d.* The average of the ore was 3 dwt. 14 gr. only, but this gave a profit of £1510 6*s.* 9*d.*, equivalent to a shilling dividend on each of the 28,000 shares. The yield was 16*s.* 9*d.* (say $4) per ton, and the cost 13*s.* 5*d.* (say $2.70) per ton. Even on this low return the company, having a paid-up capital of only £18,200, was working at a profit of 11 per cent. per annum. The New Chum Consolidated has paid £132,300 on its nominal capital of £42,000 and its paid-up capital of £18,200. The other mines will show scarcely less striking results so far as concerns the ratio of money returned to that expended.*

* As stated in the first part of this paper, out of the 28 dividend-paying mines of

The favorable record of deep mining in this gold-field, as outlined in the above paragraphs, has led many to quote the district in proof of the richness of the deeper portions of quartz lodes. This argument ignores the unique character of the geological structure of the district. Bendigo does not furnish proof of the continuance of pay-ore to great depths in ordinary veins. If we look upon each saddle-formation as a distinct lode (which it no doubt is), we perceive that the vertical extent of the auriferous quartz is very limited. As a rule, in this district, a "reef" which can be worked to profit in three levels or stoped in two lifts may be considered above the average. The different saddles form distinct lode-formations, traversing beds of the country-rock which are different from those either above or below.

PROSPECTING.

It is the fact last mentioned which necessitates the most patient prospecting. Systematic development is always one of the most important features of mine-work. In such ore-deposits as these it is of paramount importance. Owing to lack of reserve-capital, and the haste made to sink the shafts deeper, there has not been that careful and patient prospecting, which the peculiar structure of the district demands, of the ground rendered accessible by each level.

Several of the richest saddle-formations have passed longitudinally through a succession of adjacent claims. Such was the main Garden Gully saddle, and such is the formation now so successfully worked from the New Chum Railway to the "180" mine. As a rule, however, the ore-shoots, particularly the richest, are small in extent, and, unless the ground is carefully cut up by drifts and crosscuts, they are easily missed. Two instances will illustrate this. In the Old Hustlers mine, after the workings had reached a depth of over 1500 feet, there was accidentally found, last February, a good body of auriferous quartz, between the 730- and 830-foot levels, in ground which had been considered thoroughly tested. In the Hercules and Energetic mine there was found, in December, 1889, a body of quartz 8 feet square, which gave £9000 worth of gold, 46 tons yielding at the rate of 46½ ounces per ton. This was exposed in making a connection with a winze, and was as unexpected as it was valuable.

1890, I obtained the results given by 20, and found that all save four had more than returned their paid-up capital. The average of the 20 showed an expenditure of £31,167 per mine on a nominal capital of £18,742 per mine.

At present, the work of development is confined almost entirely to the three great "lines of reef,"—the New Chum, the Garden Gully, and the Hustlers,—and more particularly to the immediate vicinity of the big mines. This is not surprising. New mines are opened as continuations of already proved properties, and there is a strong tendency to remain in the neighborhood of ground which is known to have been rich. But at Sandhurst this has been overdone, and the result is that the north and south extensions of the main "lines of reef" have been but little tested, while similarly the intermediate lodes or "side-lines" have been much neglected. Since it has been demonstrated by the Geological Survey, as well as by actual mining, that the gold-belt continues for several miles on either side of the towns of Eaglehawk and Bendigo, it would be well to distribute upon shallow ground, which is as yet unworked, a little of that energy and capital which is now expended on deep sinking; for here, as in most mining districts,* the first few hundred feet are generally richer than any succeeding horizon.

As to the exploration of the "side-lines" or quartz formations which occur between the main anticlinal axes, attention has been drawn to these portions of the district by the marked success of the New Red, White, and Blue Consolidated, which is working a run of rich spurs on the Sheepshead lode or reef, intermediate between the Garden Gully and the New Chum. In this connection it is a curious fact that, notwithstanding the system of government assistance, no cross-cut has been driven between the main "lines of reef" in that portion of the district (in the town itself) where they have been proved to be richest. Such an exploration is recommended by the structure of the country, the richness of this section of the gold-field, and the experience of other mining districts.

Outlook for the Future.

The prospect of future deep mining on this gold-field is most encouraging. The depths yet reached are far within the capabilities

* There are exceptions, but they are not numerous. I recall the history of some of the mines on the "Mother-lode" in Amador county, Cal., where no profitable returns were obtained until several hundred feet had been sunk. As a rule, however, the increasing richness of mines with depth is a fallacy too often found among the paragraphs of a mine-report. In most districts there are few gold-mines, the first 50 to 200 feet of which did not exceed in richness any succeeding horizon.

of modern mining machinery.* There is no reason to expect any increase of water; on the contrary, the experience of deep mining in other places points to a diminution in the quantity of water as depth is gained. No chemical causes have been observed as active in this district which would tend to make the underground temperature increase other than with the usual increment of 1° for every 50 or 60 feet of additional depth. Moreover, there is every probability, from a geological standpoint, that the saddle-reef formations will be repeated, as in the 2000 feet or more already pierced; for the only cause likely to bring them to an end, the granite contact, is at a depth far beyond human reach. The cost of exploitation will increase with depth, but, as the mines are comparatively dry and the encasing rock does not become harder, it need not be expected that this increase will be material between 2000 and 3000 feet.† Below 3000 feet, increased power and efficiency of the winding-engines, for which there is yet plenty of room at Sandhurst, will compensate for the increasing distance from which the ore will be lifted.

At present, generally speaking, 5 dwt. of gold per ton will, with a 3-feet or 4-feet reef, pay the expenses of the mine and mill. The New Chum Consolidated has shown how, at the end of the second thousand feet from the surface, 3½-dwt. ore may be made to pay. The width of the lode in the New Chum Railway, the Shenandoah, the New Chum Consolidated, the New Chum United, etc., averages about 6 to 8 feet; and it is the width of the quartz in the reefs which will determine the economical results of its extraction. In this respect there is no reason to foresee a diminution. Nor is there any apparent reason to expect a decline in the gold-contents of the quartz. In most gold-mines the first 200 or 300 feet are richer than any succeeding horizon of similar thickness; but having once passed out of the region of surface-waters there is nothing to cause one to expect a marked change. Poor zones and rich ones will alternate as hereto-

* The shafts are invariably vertical, the winding being done by steam-power with round wire-ropes and cages. Double-decked cages are not in use. Under the government regulations and the supervision of the mine-inspector, the winding-apparatus is always in good order, and the safety-appliances are frequently and regularly tested.

† The speed of sinking varies. The country traversed by the New Chum lode is less hard than that of the Garden Gully; hence, while at a depth of over 2000 feet in the "180" mine, the speed of sinking the last 200 feet averaged 20 feet per fortnight (the distance from 2440 to 2640 feet being accomplished in exactly ten fortnights); the average rate in the Victory and Pandora, on the Garden Gully line, is 12½ feet per fortnight. In both cases timbering, etc., is included.

fore; but, owing to the folding of the country in which the reefs occur and the recurrence of the beds, the mine-workings will, at greater depths, pass through portions of country already met with nearer the surface. The gold does not appear to have become more refractory with increasing distance from daylight; the percentage of pyrites has varied but little, while the fineness of the gold has not changed.*

It will be primarily a question of the economical handling of the ore, and in this respect the underground exploitation is at present as good as the treatment at the surface is bad. With mills of enlarged capacity, properly equipped with modern improvements (particularly with automatic feeders and with rock-breakers), the Bendigo gold-field should long continue to be the home of successful gold quartz-mining and the pioneer of deep sinking.

NOTE BY THE SECRETARY.—Comments or criticisms upon all papers, whether private corrections of typographical or other errors, or communications for publication as "Discussion," or independent papers on the same or a related subject, are earnestly invited.

---

* In the Lazarus, at 2000 feet, I saw white quartz containing beautiful coarse gold in pieces weighing several pennyweights. The gold at the greatest depths eached is still 22¾ to 23 carats or 945 to 960 fine.

THE BENDIGO GOLD-FIELD (SECOND PAPER): ORE-DEPOSITS OTHER THAN SADDLES.

BY T. A. RICKARD, DENVER, COLORADO.

(Schuylkill Valley Meeting, Reading, October, 1892.)

THE earlier paper (*Trans.*, xx., 463) describing this Victorian mining district, to which the present is supplementary, was mainly confined to the consideration of the "saddle-reefs," as scientifically the most interesting and economically the most important of the ore-deposits of Bendigo. Incidentally, however, references were also made to other gold-bearing quartz-formations worked in the district; and of these a brief account will now be given. Among them the most productive is that which is locally named a "make of spurs" or "spur-formation," that is, a network of quartz-veins bearing some resemblance to the type known as *Stockwerk.*

THE CATHERINE REEF UNITED.

This mine, situated at Eaglehawk, is a representative of the class characterized by this form of ore-deposit. The underground workings disclose a strikingly regular succession of alternating slate and sandstone beds with easterly dip. The different members of the series are named according to their thickness, which in any one bed varies between narrow limits only. Fig. 1 is a cross-section, illustrating the structure of that portion of the Lower Silurian country-rock which contains most of the mine-workings. The scale of the drawing is too small, and it would otherwise be confusing, to indicate the quartz-spurs themselves; but it will be understood that certain members of this series of slates and sandstones are traversed by comparatively small and generally irregular veins of quartz called "spurs," and confined either to one bed or to a certain succession of beds.

The "ore-shoots" of the mine are the "runs" or "makes" of spurs, which in one plane (that of the strike of the enclosing rock) pitch north 22 inches in 6 feet, and in the other plane (that of its dip) cross the bedding of the country nearly at right angles, thus presenting a contrary or westward dip to the extent of the width of the belt. This width, measured upon the westward dip, varies for the "runs" in this mine from 150 to 200 feet. Along the strike they have been followed north and south for several hundred feet, but the extent of their development necessarily depends upon economic considerations, the amount of gold they carry and the hardness of the rock they traverse. In the Catherine Reef United, the two shafts are 947 feet apart, and one run of spurs has been traced throughout this distance and 300 feet beyond. Such a network may have its regular course disturbed by faults, and is subject to the irregularities of strike and dip which usually characterize such formations.

In this mine the bed richest in spurs and most productive in gold is the "forty-foot sandstone" (see Fig. 1), so called from its thickness. It is overlain by the ten-foot slate, the two-foot sandstone and the five-foot slate. The two-foot sandstone is so hard, and forms so marked a feature of the underground workings, that the miners have named it "The Devil's Back," and by this term it is generally known. Underneath and west of the forty-foot sandstone comes a regular series, including, in the order named, the eighteen-foot slate, the thirty-foot sandstone, the six-foot slate, the eighteen-foot sandstone and the thirty-five-foot slate. The wider beds of sandstone are often subdivided by slate partings, as the thirty-foot slate, for instance, is cut up by several thin seams of sandstone. The divisions into slate and sandstone are frequently somewhat arbitrary, since there are rocks of intermediate composition, sandy slate and slaty sandstone, recognized by the miners as "bastard slate" and "bastard sandstone." The main divisions into the "bars" or beds of stated thickness are determined at the time of driving the cross-cuts, when the drill decides the question of relative hardness and fracture.

The spurs worked in the forty-foot sandstone, in the stopes above the 676-foot level, will serve as illustrations. Figs. 2 and 3 are reproduced from the sketches made underground. In Fig. 2, A is the bed of sandstone dipping east and penetrated by a series of quartz-veins inclining flatly to the west. These do not penetrate the overlying slate B, but they are found in the underlying bed C, where their position is almost invariably changed to a steeper angle. The fractures filled with quartz have many of the features of joints, as shown in Fig. 3, which illustrates the same bed of sandstone in another part of the mine, where it is barren of quartz but fractured in lines which are almost at right angles to the plane of bedding.

The width of a breast (4 feet) in the stopes ordinarily shows an average of three spurs, the largest about 8 inches and the smallest 1½ inches in thickness, giving an average width of 15 inches of quartz in the 4 feet of sandstone rock broken for the mill.

In Fig. 4 a longitudinal section is presented, as seen in the north stopes of the 676-foot level. It will be noted that the spurs pitch to the north, their dip being (as we have seen) to the west. Lines of cross-fracture or jointage are indicated. The upper spur A carries an inclusion of country C. This is considered favorable to the occurrence of gold, which, as in ordinary vein-mining, often "makes at the point of a horse." The lower spur B shows two splices.

Fig 5 illustrates a behavior on the part of the spurs, which is characteristic of them all over the gold-field. A quartz-seam, upon leaving a bed of sandstone and entering one of slate, invariably turns its dip in sympathy with the cleavage of the slate. Frequently a big "blow" or irregular body of quartz is formed in the slate separating two beds of sandstone. This is very marked in the case of the ten-foot slate, as shown in Fig. 5, where S S are spurs, A is the forty-foot sandstone, B the ten-foot slate, C the "Devil's Back," and D the five-foot slate.

The "leaders," the position of which is marked on the general cross-section (Fig. 1), form a marked feature, and are useful guides in the underground workings. They are essentially thin seams of quartz and clay, following the line of parting between certain members of the series of slate and sandstone beds. The thickness of quartz which they carry, though not uniform, is usually limited to 4 or 5 inches, and they are accompanied by a variable amount of graphitic clay, derived from the softer of the two bounding walls. It will be noted that in every case save one the overlying bed is sandstone and the underlying slate. In these cases the leaders carry a very clean hanging-wall, often showing smooth, polished surfaces, sometimes also marked by *striæ*. The foot-wall country exhibits a variable amount of distortion, and is frequently in a crushed condition for 2 or 3 feet below the black clay or mud of the leader into which it gradually merges.

Following the stratification as they do, the leaders may be considered true bed-veins. There is abundant evidence that the plane of parting which they follow has also been a plane of movement of the enclosing rocks, and that the black clay which is one of their chief characteristics is the result of the attrition which has crushed the softer slate constituting one—usually the lower—of the two surfaces in contact.

In the development of the mine, such a leader is often found to be the apparent starting-line of a series of spurs. It is also found that the latter are frequently faulted by the former. The second fact explains the first. Occasionally, the displacement is so small as to be of no importance to the miner, though interesting to the geologist.

Fig. 6 illustrates "the leader," so-called to distinguish it from the "east," "west," and "black" leaders, shown in Fig. 1. The sketch was made in a short cross-cut from the 676-foot level. The leader D separates the forty-foot sandstone A from the eighteen-foot slate B. It consists of a thin but regular vein of quartz E, accompanied with black clay or fluccan, varying from one to three inches in thickness. The spur S in the forty-foot sandstone is seen to reappear in the eighteen-foot slate; but in crossing the line of the leader it has been dislocated and broken. The throw is two feet. C C are fragments of quartz broken off the spur by the movement along the plane of the leader. The greater resistance offered at this particular point by the quartz of the spur has caused a local increase of the thickness of crushed material; and consequently there is a width of 4 or 5 inches of black clay between the two faulted portions of the spur.

Fig. 7 shows the "black leader," so-called because in this instance quartz is frequently absent, and entirely replaced by graphitic clay. The unusual thickness of the latter is explained by the fact that the overlying bed A, the eighteen-foot sandstone, has more than ordinary hardness, while B, the underlying thirty-five foot slate is comparatively soft. The sandstone forms a straight, clean, polished hanging-wall. The leader C carries imbedded in the clay small pieces of quartz, probably the shattered remnants of quartz-veins broken through by the movements along the plane of the leader. The width is about 4 inches. The underlying slate is much crushed and disturbed near the leader and is full of distorted threads and irregular pieces of quartz, which become regular spurs at no great distance from the leader.

In Fig. 8 the "east leader" is illustrated. The sketch was made in the stopes, called the "No. 13 backs," above the 870-foot level, main shaft. This leader follows the line of parting between the thirty-foot sandstone and the six-foot slate. A and B indicate the thirty-foot sandstone, of which the part B is beginning to be marked by slaty cleavage. C and D are the six-foot slate which, near the leader, is sandy. The country on the two sides of the parting has a nearly equal hardness, which explains the fact that this leader is no-

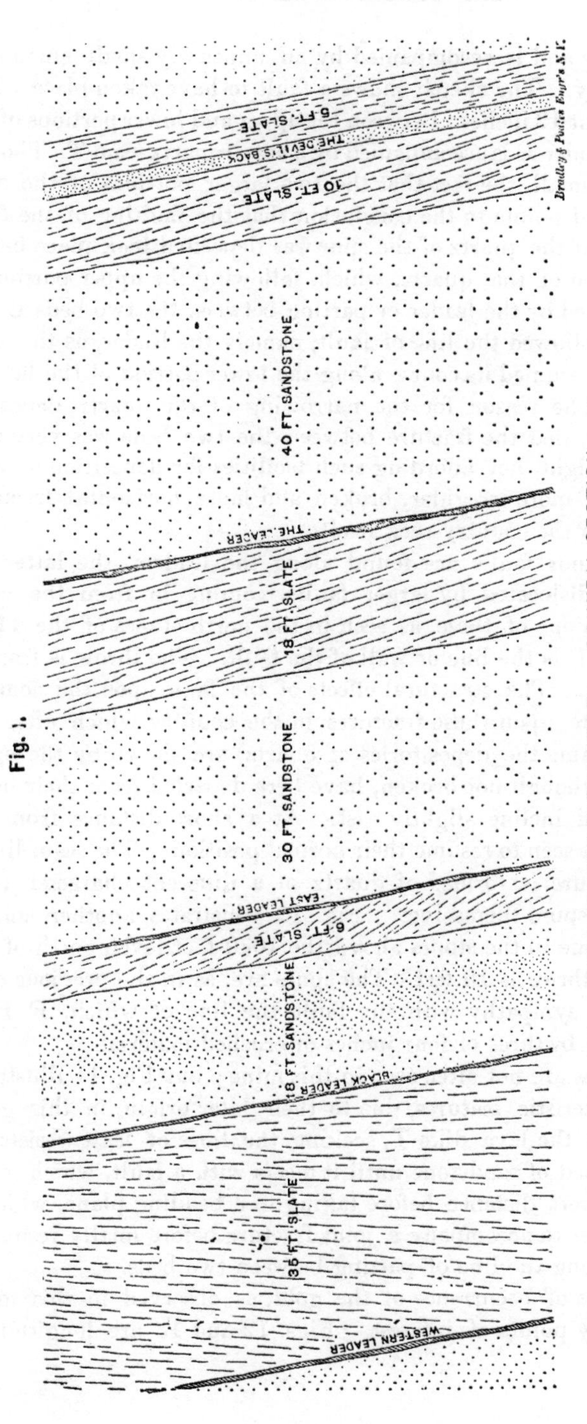

Fig. 1.

CATHERINE REEF UNITED. GENERAL SECTION.

SCALE 1 INCH 24 FT.

tably narrow and is accompanied by an unusually small amount of graphitic clay. The sketch shows a fault to have taken place. The throw is about 12 inches. Between the upper and lower portions of the spur S the leader L is accompanied by a small quartz seam T. Though this is very small, the fact that the two larger portions of the spur are connected points to the conclusion that the faulting of the fracture in which the quartz of the spur was deposited took place before the deposition of that quartz, which, following the upper portion of S, was arrested by the leader or parting between the two beds B and C; that it followed the line of fault, namely the leader, as the seam T, and then resumed its course along the lower portion of the faulted fracture. The reason for the narrowing of the quartz deposited along T was, that the fracture between the two beds was very narrow and straight, not affording such facilities for deposition as were given by the more irregular, broken and more horizontal fractures across and in the country-rock itself.

While minor faults are found along the leaders, the latter are themselves dislocated by larger faults, coming in from the west. Fig. 9 shows one of these, as seen in the north stopes of the 1140-foot level. F is the line or wall of the fault. The throw is from A to B, 12 feet. The structural effects of the fault upon the country (and therefore upon those fractures in the country which after the faulting became the depositories of quartz) are shown by the spurs S S, which, though not broken, have here departed from their usual west dip, and incline slightly east. At a short distance from the fault they are seen to resume their normal position. The seam lining the fault-fissure is formed of quartz of a different character from that of the spurs themselves. Fig. 10 illustrates another similar fault, this time in the stopes above the 1070-foot level, north of the shaft. The throw is 15 feet. The spurs are, as in the previous case, distorted in sympathy with the fault, the line of which, F F, is accompanied by three or four inches of crushed material.

Lava dikes are not often seen in this mine; but Fig. 11 illustrates their characteristic features, due to their later origin, in this gold-field. Here the lava dike T, seeking the lines of least resistance, traverses a bed of sandstone until it meets with a fault, which it follows for a short distance before taking to a bedding plane, which it in turn leaves to accompany a joint fracture before finally resuming its course along the line of parting between two beds.

The modes of occurrence of the gold, as observed in this mine, present many points of interest. Figs. 12 and 13 are longitudinal

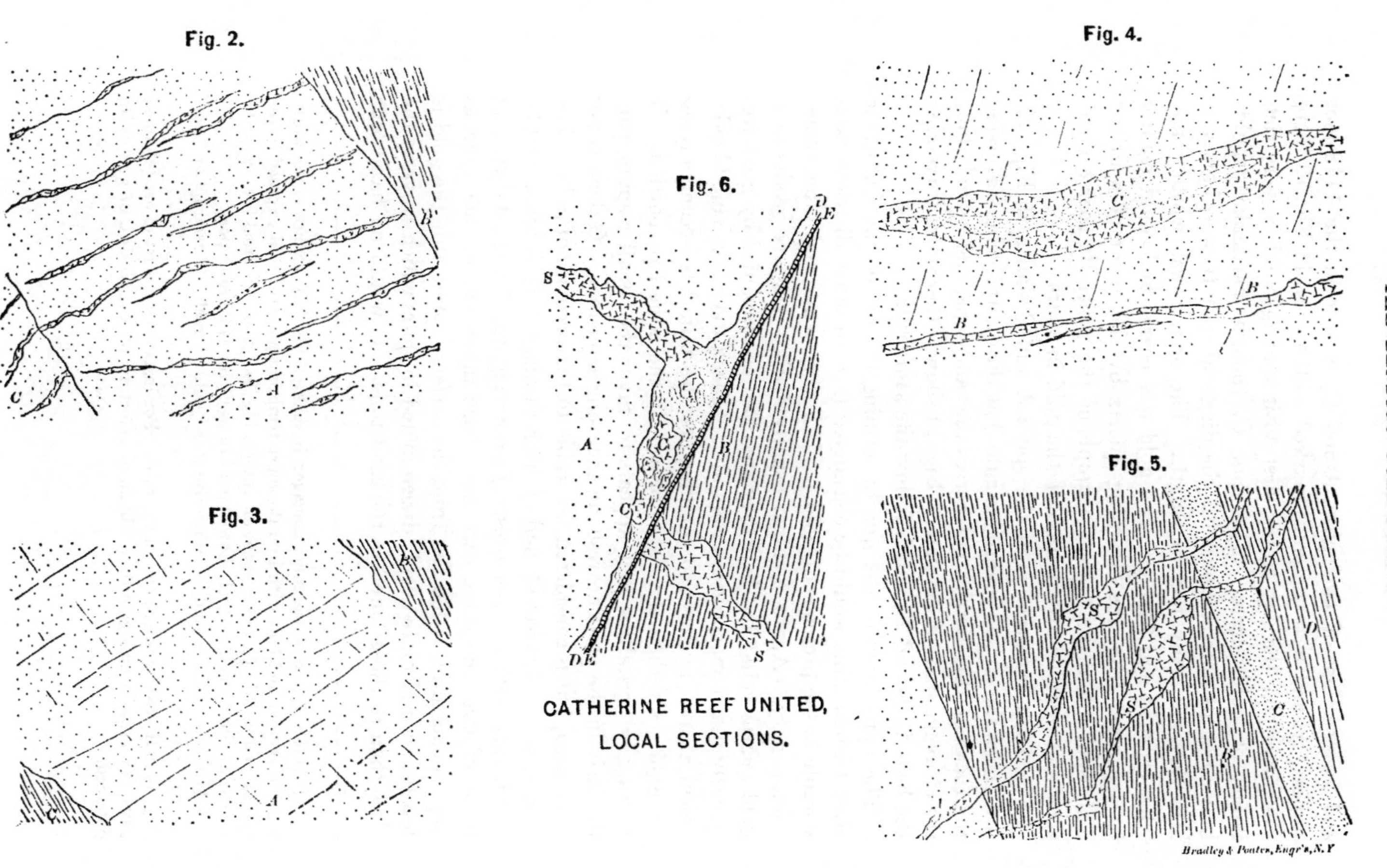

CATHERINE REEF UNITED,
LOCAL SECTIONS.

and cross-sections of two spurs, A and C, as seen in the 1118-foot stopes. At and near the point marked with a cross in Fig. 12 there was found, on the day of my first visit underground, a very rich bunch of gold-quartz. The spur C, though particularly gold-bearing at its intersection with the diagonal spur B, was, however, also rich for many feet in length. The quartz was white, somewhat crushed and friable. The gold was coarse and could readily be separated from the enclosing quartz by the point of a drill.* This spot afforded a typical example of the generally coarse character and iregular distribution of the gold in spurs.

Fig. 13, a cross-section of the spurs A and B shown in Fig. 12, illustrates the almost invariable rule that the quartz-veins traversing the rocks of this district cut across the sandstone, and travel along the cleavage lines of the slate. Fig. 14 shows other spurs traversing the beds C, D and E, which follow the same rule.

The appearance of the quartz forming the ore of the Catherine Reef United mine would be considered in most other districts unfavorable to the presence of gold.† It is very white and often somewhat sugary. Again, a miner usually dislikes vugs or geodes in a gold-quartz vein; but here they are often accompanied by gold, frequently pseudomorphic after mundic. "Black Jack" or zinc-blende, arsenical iron pyrites and mundic, are the minerals accompanying the gold. Of these the last is present in the largest proportion. It is coarsely crystalline and appears to favor the smaller spurs, particularly in the parts nearest the enclosing sandstone. Galena is not often seen, but is considered favorable to the presence of gold. The gold is generally coarse‡ and of high caratage. It is often visible, and invariably so in rich ores. Pieces weighing 7 dwt., 4 dwt., and many of two and three dwt. have been taken out of one spur, at depths exceeding 1000 feet from the surface. As a rule the gold is less coarse and more evenly disseminated in spurs which traverse the slate than in those found in the sandstone. It would be hazardous

---

* It is the custom of the mine manager, in making his daily rounds of the workings, to remove and take with him the more tempting pieces of gold exposed in the spurs; and at this place we spent an interesting half-hour in breaking away the richest of the "stone," which showed gold in particles as large as peas.

† Miners elsewhere would characterize it as "wild," "rash," "bastard," or "bull quartz."

‡ As indicated by the retorts in the mill. For instance a test lot of ore (46 tons) gave 42 ounces of amalgam from which, on retorting, 34 oz., 15 dwt. of gold were obtained.

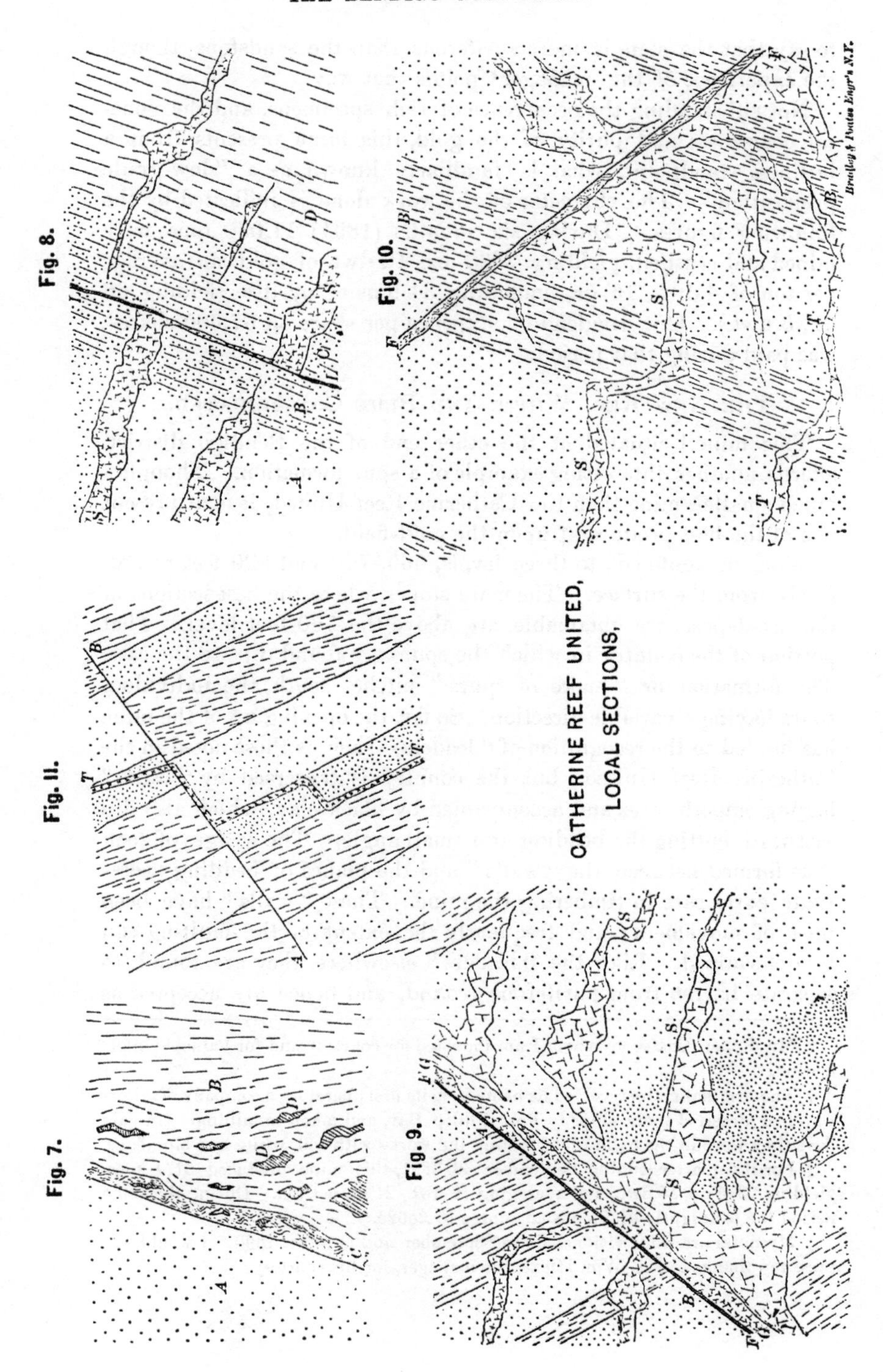

CATHERINE REEF UNITED,
LOCAL SECTIONS.

to say that the slate is more auriferous than the sandstone, though the experience of the manager* points that way.

Notwithstanding the occurrence of rich specimens, and the general coarseness and purity of the gold, this mine presents, from a business standpoint, what is familiarly known as a "low-grade proposition." The character of the work done is indicated by the following figures. During six months (1891) 11,080 tons were mined and crushed, yielding 2583 oz., 12 dwt. of gold, an average per ton of 4 dwt., 16 gr. (including 34 tons of pyrites yielding 66 oz., 5 dwt.). One dividend of sixpence per share on 67,600 shares, was paid during that period.

The New Red, White and Blue Consolidated.

This mine,† situated at the other end of the Bendigo district, affords another interesting example of a spur formation. Though a much smaller mine than the Catherine Reef United, it is at present one of the best producers‡ upon the gold-field.

Work is confined§ to three levels, 660, 760 and 820 feet respectively from the surface. The main stopes, where the best sections of the ore-deposit are obtainable, are above the 760-foot level. That portion of the country in which the spurs are found dips to the east. The formation or "make of spurs" pitches south, the individual spurs having a variable direction. So far, the opening up of the mine has not led to the recognition of "leaders," such as those noted in the Catherine Reef United; but the country is traversed by "walls" having smooth faces and accompanied by black clay, which also dip eastward, cutting the bedding at a small angle. The wedges of rock thus formed between the "walls" and the planes of bedding necessitate extra care in timbering the mine. These "walls" have been lines of movement; and frequently the extent of the faulting can be determined. Like the "leaders" elsewhere, they are found to separate barren from auriferous ground, and hence are accepted as

---

* Mr. Robert Coates, to whom I am indebted for courtesy and for valuable information.

† The original claim received its name from its first operators, a company of sailors who had deserted their ships in Port Philip Bay, and taken to mining. On gala days they were in the habit of decorating the works with red, white and blue flags.

‡ For the quarter ending September 30, 1891, this mine produced 5174 tons, yielding 2815 oz., 15 dwt., and averaging 10 dwt., 21 gr. per ton. During the same period the fortnightly dividends amounted to £5625.

§ My notes are from visits made in September and October, 1890. I would express my thanks to Mr. Wm. Hicks, the manager, for his courtesy.

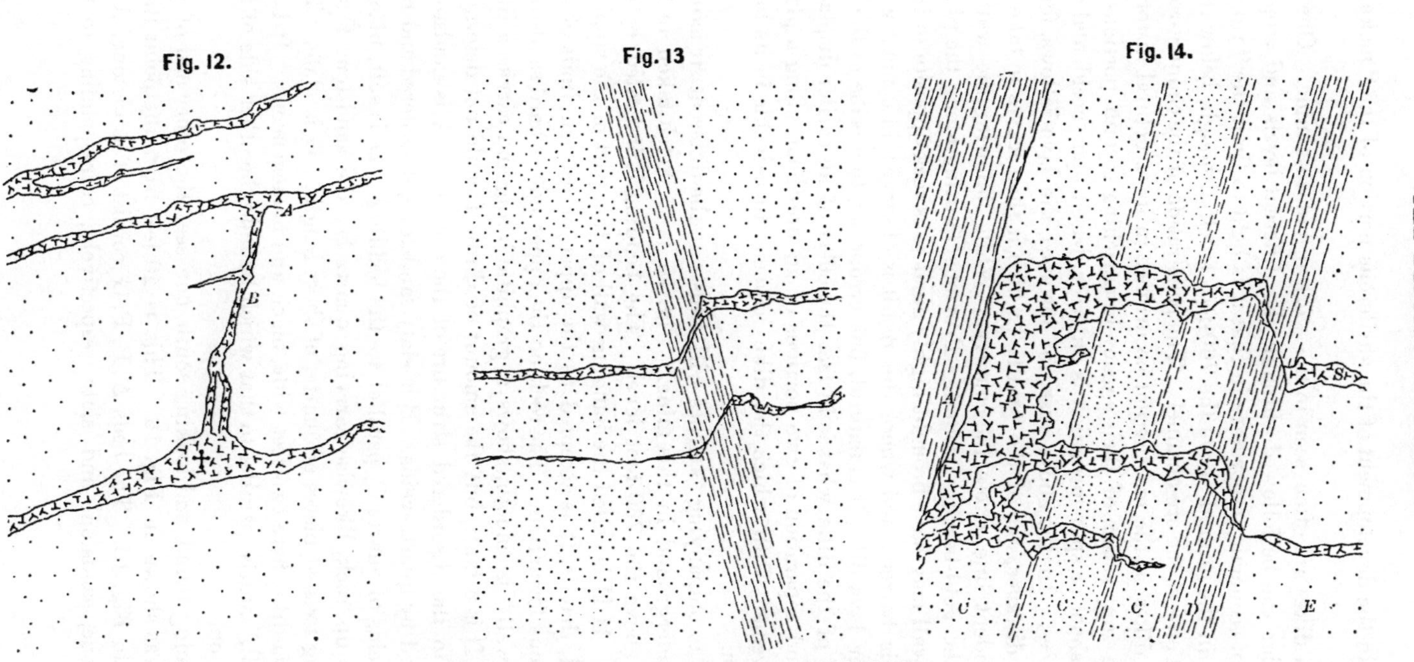

CATHERINE REEF UNITED, LOCAL SECTIONS

guides in the development of the auriferous portion of the "make of spurs."

Lava dikes are often seen in the workings of this mine. One in particular can be followed through the different levels and stopes, and is of economic importance in that it is held to limit the "run of spurs" in the ground above the 760-foot level, although, below that level, it crosses the ore-channel. The penetration of the country-rock by the dikes is the most recent occurrence in its structural history; and the supposition of their agency in enriching certain portions of an ore-shoot or spur-formation is an error, based upon accidental coincidences. In the above case, for instance, the lava follows, for a certain distance, one of the "walls," which it then leaves to take an independent line across the country. While following the wall it limits the gold-bearing ground. Previous faulting along the plane of the wall may have brought highly auriferous rock opposite to that which is barren: and where this result is observed in contact with the later lava-dike, it is natural, but erroneous, to ascribe to the influence of the dike what is in fact the effect of an earlier displacement and separation of two portions of an ore-shoot, along a plane of movement with which the dike happens to coincide in its local position.

The accompanying sketches will illustrate the different structures seen underground in this mine. Fig. 15, is a sketch taken in the stopes above the 760-foot level. The bedding is indicated by the arrow. H H is called the hanging-wall of this part of the stoping-ground, that is, it is accepted as the upper or eastern limit of the auriferous formation. The rock on the upper side is hard sandstone. The "wall" is smooth, shows a graphitic lining and carries a little clay. It is a fault, but the amount of throw I could not determine owing to the disordered structure of the country. A is sandstone, threaded by quartz-veins. B is slaty sandstone, disordered and carying veins of quartz C, parallel to the bedding. D is soft, black, broken-up rock, likewise carrying quartz E; the sandstone F also showing twisted pieces of quartz, of short length. G is slate. T is the lava-dike, here two feet wide, black and homogeneous. It hugs a "wall," nearly parallel to that which bounds the other side of the formation.

Turning round and looking south, one sees the other end of the stopes, as shown in Fig. 16. This is 40 feet from the point illustrated in Fig. 15. The beds A, B, F, G consist of sandstone, slaty sandstone, sandstone and slate respectively, corresponding to the

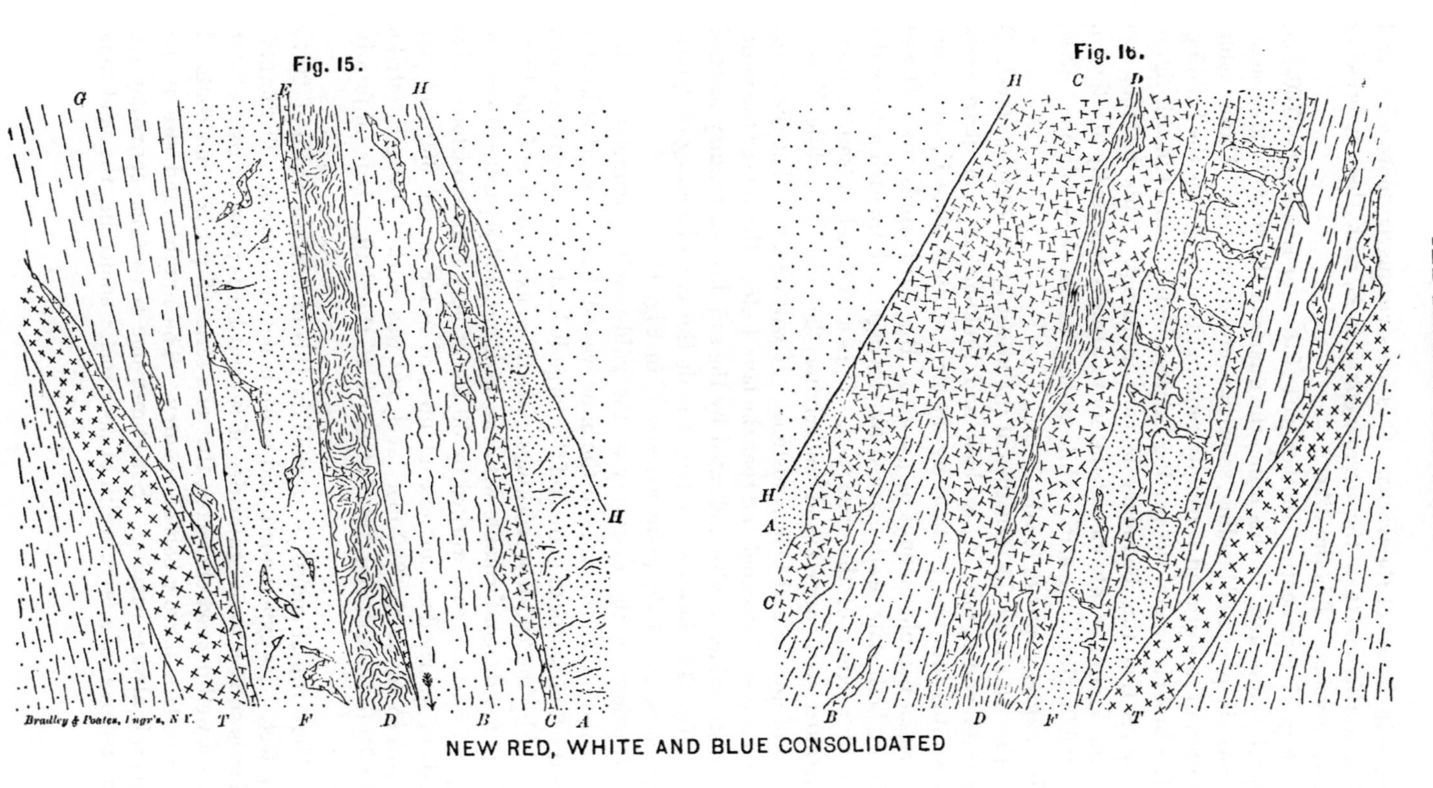

NEW RED, WHITE AND BLUE CONSOLIDATED

opposite breast of the works. C is massive quartz, more white and less auriferous than the narrow spurs E E, with their interlacing quartz-veins. T is the dike.

Fig. 17 representing another part of the stopes above the 760-foot level, shows how, in this mine, the quartz of the spurs follows sometimes the planes of bedding, sometimes cross-fractures. Close examination of the ground is usually required to distinguish the bedding. The broken line on the parting between beds C and D in this figure denotes that this boundary is very indistinct. This instance exhibits again the behavior of the spurs, already repeatedly mentioned, in "following" the slate, while cutting across the sandstone. (See, for example, spur *a*, *b*, *c*.)

Fig. 18 shows the structure of the breast of the 760-foot level. D is slate which, near A, is black and broken. A is a quartz-vein separated into two portions by the strip of black slate E, which last carries a notable amount of pyrites. B is hard sandstone. G and H are both slate, well marked by cleavage and appearing to be divided portions of the same bed. C is a channel of soft country, streaked with quartz and underlying the dike T; while F is a vein of quartz, and O is sandstone. It will be noted that the country has been dislocated and that the lava follows the line of the fault-fracture, which is also indicated by the soft broken country marked C. Fig. 19 shows on a larger scale the cross-hatching of quartz which occurs at the point marked K in Fig. 18.

Previous to the discovery of the gold-bearing character of this spur-formation, the work of the mine was confined to what is known as the "Sheepshead main reef," a large lode of massive quartz which is cut by the 760-foot workings. As there exposed, it is from 12 to 14 feet wide, conformable with the bedding, and separated from the spur-formation now worked by 30 feet of very hard sandstone. This lode of quartz does not extend upward so far as the 660-foot, nor down so far as the 820-foot level, but was worked along the strike for 70 feet north and 169 feet south of the shaft. It is uniformly poor.

The "run of spurs" is worked in this mine for a width of 20 to 25 feet. Its extent north and south is not yet known. The formation is subject to irregularities of several kinds; but occasionally it has a well-defined character, that of a series of quartz seams confined to a certain narrow belt of country. The number and size of the spurs is always varying. Sometimes one or two rich spurs pay for the removal of a large width of poor rock; sometimes the sandstone

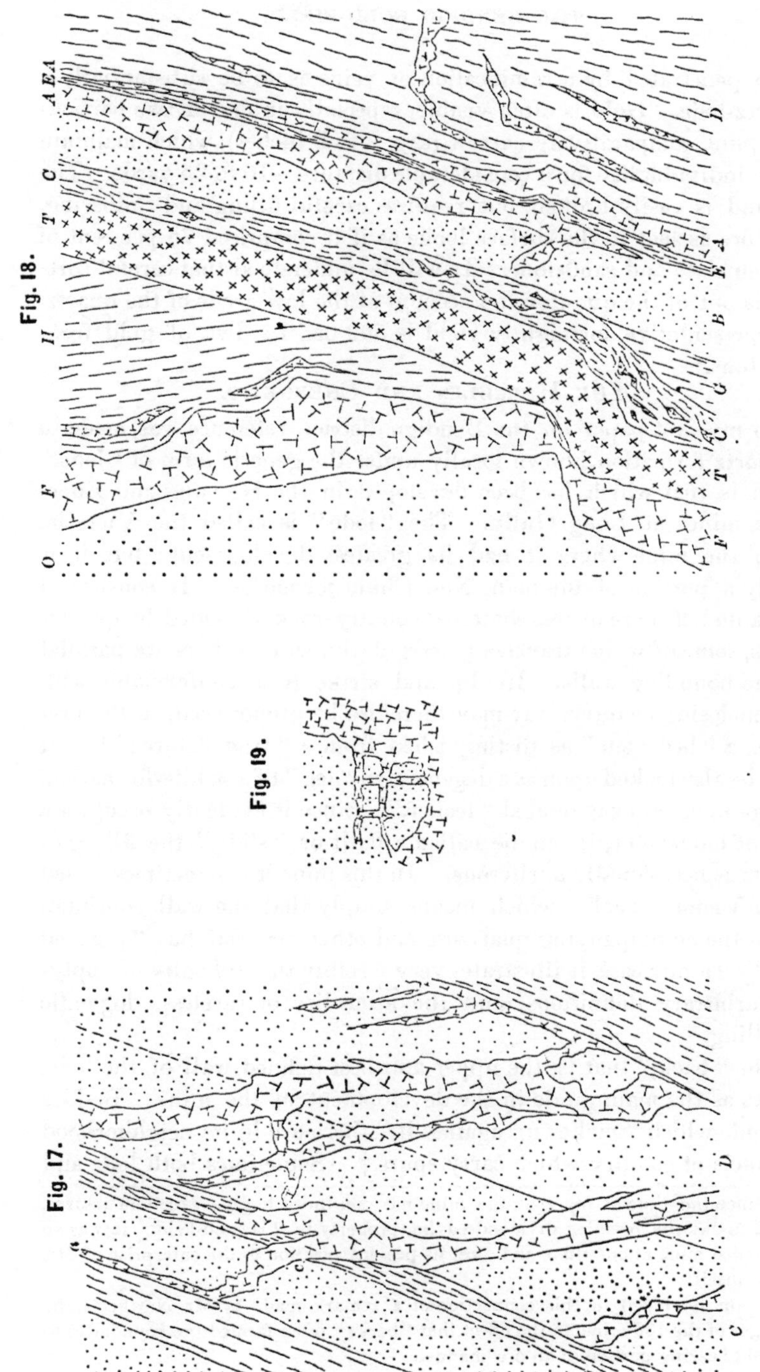

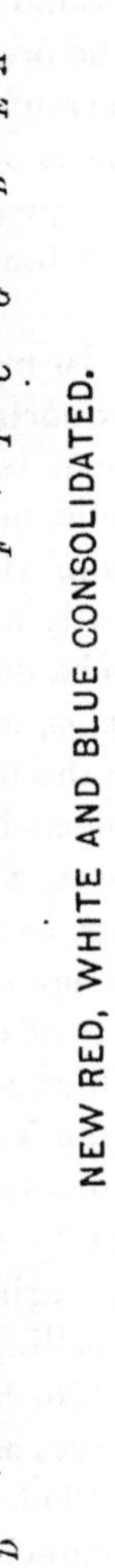
NEW RED, WHITE AND BLUE CONSOLIDATED.

is so penetrated by a ramification of veins as to be substantially a quartz-lode. Gold is often visible, especially in one or two particular spurs, comparatively narrow (say, 2 to 3 inches), which maintain their individuality for a considerable distance—10 to 25 yards. The ground is easily broken, but requires careful filling and timbering. The ore as sent to the surface looks as if it comprised 75 per cent of barren slate and sandstone. It will be understood that careful sorting is out of the question, on account of the low grade of the quartz. A representative fortnightly yield is 559 oz., 14 dwt. of gold from 641 tons.*

THE HERCULES AND ENERGETIC.

In my first paper on the Bendigo district, reference was made to a quartz formation known locally under the generic term of "lode." Such is that which has been developed in the Hercules and Energetic mines at Long Gully. The "lode" is called the Victoria, from the mine where it had its greatest development; but it is really a portion of the main New Chum formation. It consists of a channel of more or less shattered country-rock, threaded by quartz-veins, some of which traverse it irregularly, while others are parallel to the bounding walls. In dip and strike it is conformable with the enclosing country. It may therefore be pronounced, in the first place, a "bed-vein," as distinguished from a "true fissure;" but it may be also looked upon as a degenerated "leg" of a saddle-formation, or again, as an exaggerated "leader." Since it evidently occupies a line of movement, it can be called a fault or "slide," the filling of which is occasionally auriferous. In this mine it is sometimes called "the Victoria back," which means simply that the wall continues when the accompanying quartzose and other material has "pinched out." In any case, it illustrates very forcibly the difficulty of applying arbitrary definitions, and the absurdity of careless, dogmatic labelling.

The "back," that is, the upper and best defined wall of the lode, serves as the main guide in the development of the mine. Besides the lode which "makes up against it," there are in its neighborhood channels of country which carry spurs.† Sometimes both lode and

* Since the time of my visit, the mine has continued to give excellent returns, and is to-day the premier dividend-paying property in this gold-field. It was an abandoned mine at one time, and owes its present position to the enterprise of Mr. Geo. Lansell.

† I am reminded in this connection of a remark made by Mr. Northcote, the manager of the 180 mine, to the effect that "backs" often formed the boundaries to the gold-bearing portion of an ore-channel.

spurs are sufficiently gold-bearing to warrant exploitation, sometimes the lode but not the spurs, or *vice versa*.

Nearly parallel to the "main back," there are other divisions in the country, which are called "leaders." These are simply lines of movement in sympathy with a main fault, and are marked by small clay seams. They tend to meet the main lode on its dip; and at the junctions so formed, large bodies of ore have been found and profit-

Fig. 20.

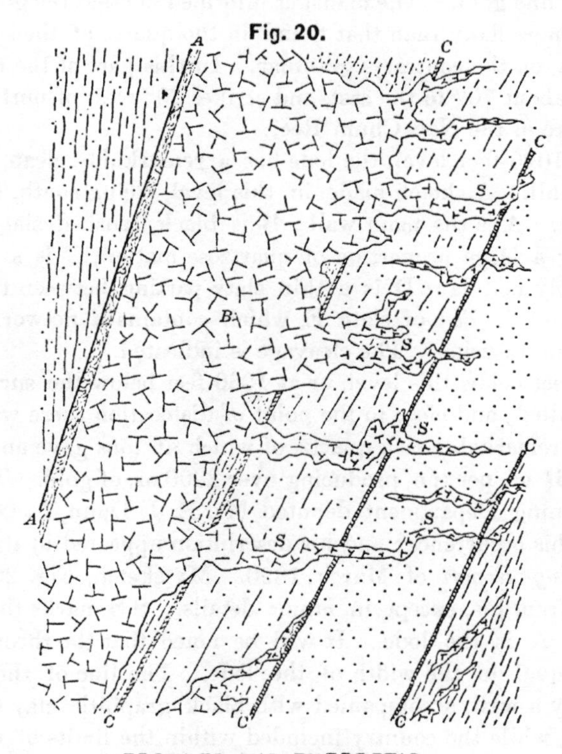

HERCULES AND ENERGETIC

ably worked. Bodies of quartz as much as 40 feet wide have been discovered under these conditions, bounded to the west by the "back," but breaking into spurs towards the east. These spurs cut through the so-called "leaders" and are found richest at the intersection. An example may be seen in Fig. 20, where A A indicates the main back, and B is a large body of quartz breaking eastward into the spurs S S, while C C are the "leaders." The spurs are more flat, and there are more good spurs in the sandstone than in the slate; although if a good (*i.e.*, a highly gold-bearing) spur traverses more than one bed of the country, it is not found notably poorer in the slate. The gold is coarsest in the sandstone. It is also noted that frequently the heaviest gold is on the foot-wall or lower part of any given spur.

At the 600-foot level, the "Victoria lode" consists of a bed of black slaty material, containing threads of quartz which carry both coarse and fine gold. The manager informed me that the gold in the slate was more flaky than that found in the quartz of the sandstone which was, on the contrary, granular. In this part of the mine the lode dips about 70° to the east, and strikes 20° west of north, which is the strike of the New Chum Reef.

At the 1020-foot level, the lode has a remarkably clean, straight hanging-wall. A sketch made in this level, facing south, is shown in Fig. 21. A is the main wall; B is black, crushed slaty filling, containing a large proportion of quartzose matter; C is sandstone, here notably nodular; D is a thin slaty parting between the sandstone C and the bed of slate E, which contains a network, 9 feet wide, of quartz-veins. The cleavage is indicated.

Forty feet below this level, or at 1060 feet below the surface, the lode is faulted; and close to the point of dislocation there was found a body of remarkably rich quartz, of which 46 tons gave an average yield of 46½ oz. per ton, producing over 2000 oz. of gold. The Victorian Mining Department deputed Mr. E. J. Dunn, F. G. S.,* to examine this occurrence; and his description appeared in the *Quarterly Mining Report* of March, 1890. My sketch, Fig. 22, differs but little from his, except in minor details. S S marks the line of the fault; A is the lode. It will be noted that the throw of the fault is equal to the width of the lode. The line of the fault is marked by a wall, accompanied with black graphitic clay of varying width, while the country included within the limits of the lode is shattered and traversed by a number of quartz-veins, of varying size and direction. There is every gradation from clean black slate to clean white quartz. The broken portions of included country have become so interpenetrated by minute quartz-veins that it is impossible to say where the quartz begins and where the slate ends. Locally it would be termed an extreme case of a "mullocky" reef. The cleavage cuts through the slaty filling of the lode.

* This gentleman, one of the leading colonial geologists, has recently been engaged in the preparation for the Victorian Mining Department, of a monograph upon the Bendigo gold-field, which is to be published shortly.

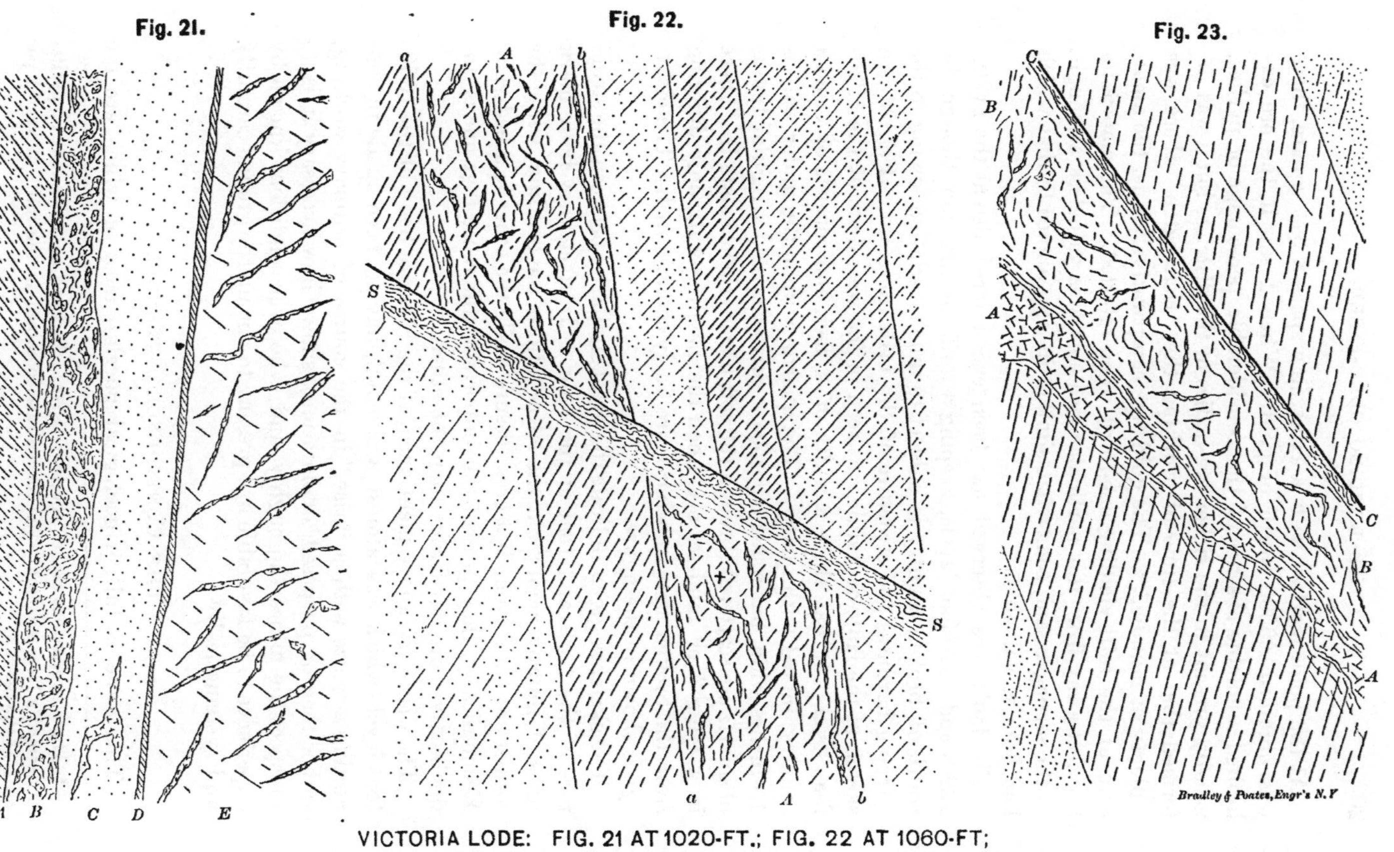

VICTORIA LODE: FIG. 21 AT 1020-FT.; FIG. 22 AT 1060-FT; FIG. 23 AT 1220-FT. LEVEL, IN HERCULES AND ENERGETIC MINE

At the point marked with a cross, in a space 8 feet square, there was obtained £9000 worth of gold. When I saw it, the ore exhibited a notable absence of the considerable percentage of pyrites, usually present in the richer ore-bodies of the district; but I was informed that at the time of working the "rich patch" there was "plenty of mundic."* The gold was comparatively fine, the amalgam at the mill yielding one-third when retorted. The material a few feet away from this particular rich spot was valueless, although, as the manager said, somewhat similar in appearance to that which had been so valuable.

The fault here observed has been encountered also at the 1120-, 1220- and 1320-foot levels, keeping a uniform direction, though the width of broken country is subject to much variation, owing to the varying hardness and structure of the beds traversed. Fig. 23 is a sketch of the fault or "slide," as seen in the west drift of the 1220-foot level. It has there a width of 4 feet, and consists of broken country, mixed with quartz. The quartz A upon the under wall is 10 inches wide, but non-auriferous. Stringers and irregular lumps of quartz are scattered through the soft black crushed slate B. There is a "dig" or mud-seam C upon the hanging-wall, only differing from the broken slate in being more finely comminuted. The cleavage is shown in the figure.

The sketches given indicate the great similarity between the structure of the fault and that of the lode. In fact, they both mark lines of improvement; but the latter follows the bedding, while the former crosses it; one is a cross-fissure while the other is a bed-vein. Economically, they present the important difference that one is undoubtedly more auriferous than the other.

The discovery above mentioned, of the rich ore-body at the 1060-foot level (which was almost accidental, having been made in connecting a winze with a "rise," in the course of ordinary working, rather than of purposed exploration), shows how necessary it is to prospect the mines thoroughly, and, wherever conditions known by experience to be favorable are present, to cut up the ground carefully and systematically.

THE CONFIDENCE EXTENDED.

This mine is full of geological puzzles, most of which are too

* These expressions are of course relative. At Bendigo, 2 per cent. of pyrites is deemed a large amount. The concentrates obtained in the mills average 1 per cent. of the ore.

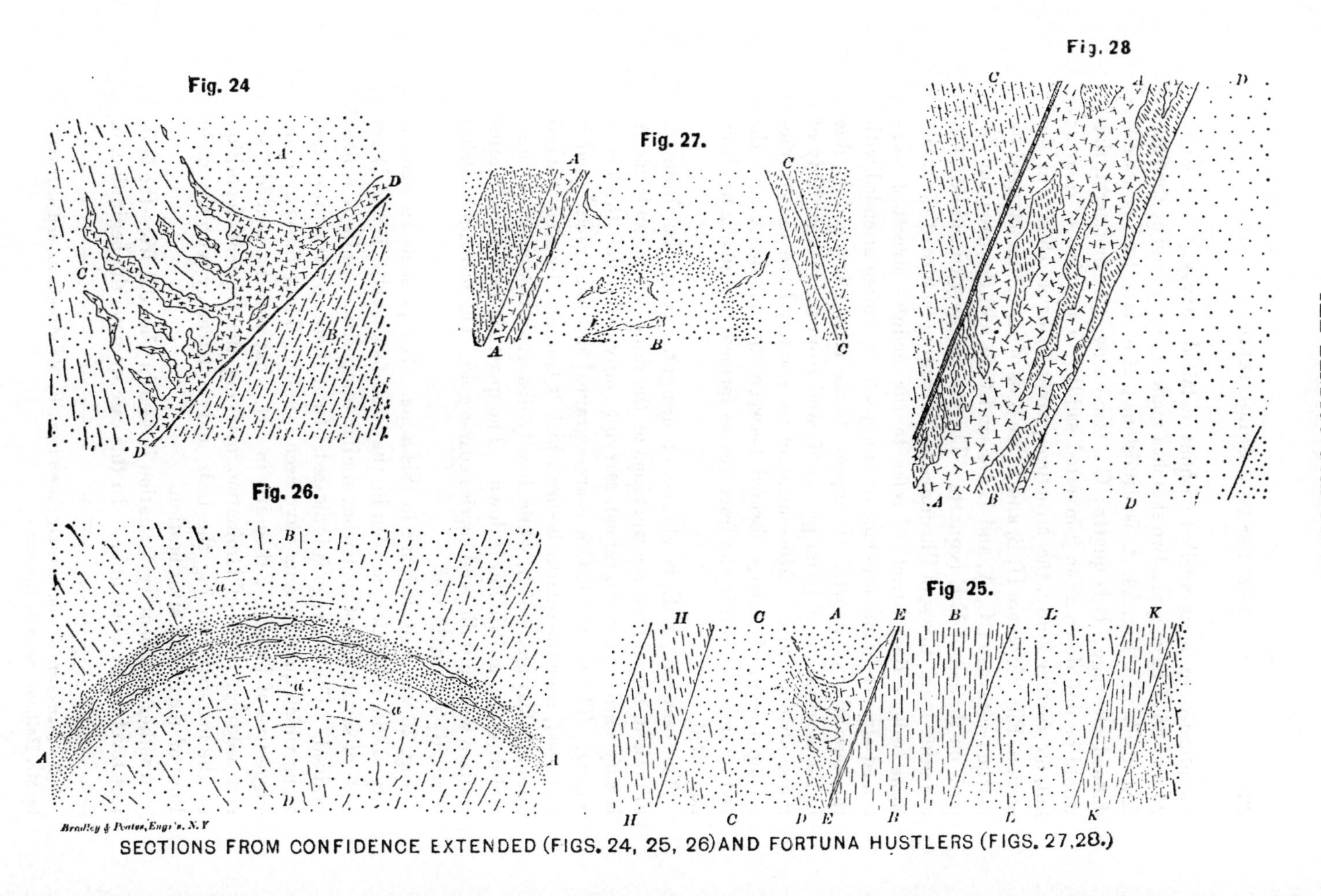

SECTIONS FROM CONFIDENCE EXTENDED (FIGS. 24, 25, 26) AND FORTUNA HUSTLERS (FIGS. 27, 28.)

complicated to be unravelled without detailed surveys.* In Figs. 24 and 25, I have given two sketches made underground at the 980-foot level. In Fig. 24, A is sandstone; B is slate, having a well-defined cleavage; D is quartz; C is sandstone. At first sight, this would be put down as an "inverted saddle," a syncline of quartz; but, on further study, the structure proves to be as shown in the section, Fig. 25, where H, B, and K are beds of slate, characterized by marked cleavage; C, A, and L are sandstone, and the irregular deposit on the left of E E is quartz. There is no inverted saddle. On the other hand, this spot illustrates an interesting feature of the sandstone beds of the district. I refer to the nodular structure mentioned already in one previous instance, which is often attended with the deposition of quartz. It appears to me that A is a nodule, due to the segregation of purer grains of sand from the general body of the sandstone bed C. Movements of the country have caused fracturing around the wedge formed between the nodule A and the bed of slate B; and in the fractures so formed the quartz has been deposited.

The east wall, E E, is clean and straight. It can be followed, unbroken, throughout the workings of the mine. The sandstone A is very hard and fine-grained, carrying only a few fine threads of quartz. The sandstone C is coarser grained and not so hard, and is traversed, near the quartz bodies which it contains, by well-marked lines of cleavage, which, while locally distorted, resume their normal direction a little lower down. The quartz is very heavily mineralized with mundic, and carries coarse gold. The formation pitches north, 6¾ feet in 33 feet.

Fig. 26 was also sketched in this mine. It represents an elementary saddle formation, as seen in the 1250-foot level. B and D are beds of sandstone, of somewhat similar composition, while A is finer grained. The cleavage is indicated. In A there are structure-lines, along which quartz seams have been formed. These lines, *a*, *a*, are suggestive of bedding; they are, in fact, what may be termed "secret-bedding," lines of lamination marking the original deposition of the material forming the beds, and in harmony with the main lines of parting or stratification.

Another rudimentary formation is seen in Fig. 32, from the 500-foot level of the same mine. In this case we have the beginnings of

---

* My thanks are due for courtesy shown to me by Mr. Abrahams, and afterwards by Mr. Hall, managers of the mine.

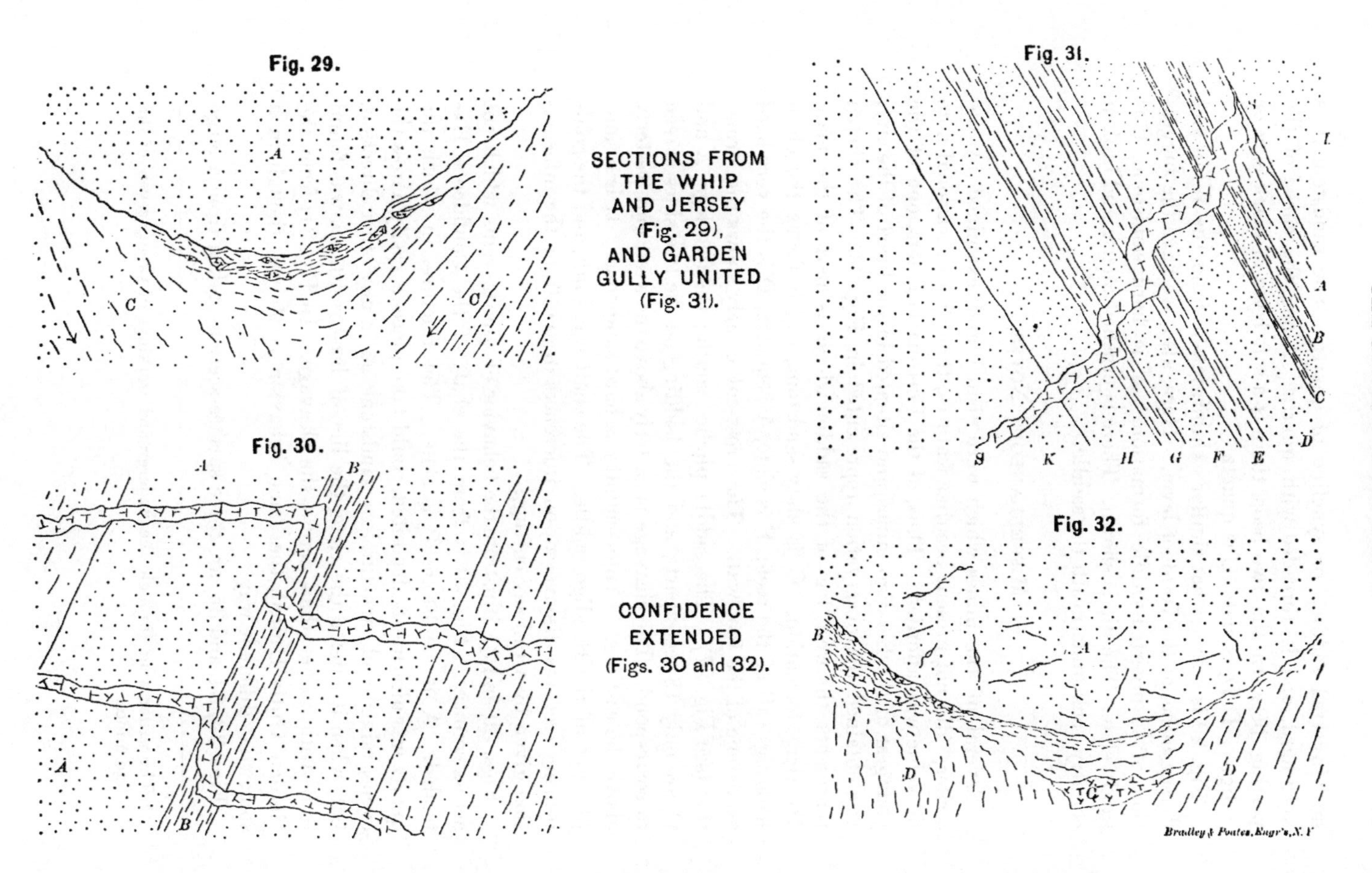
Fig. 29.
A
C
C
SECTIONS FROM THE WHIP AND JERSEY (Fig. 29), AND GARDEN GULLY UNITED (Fig. 31).
Fig. 31.
I
A
B
C
D
S
K
H
G
F
E
Fig. 30.
A
B
A
B
CONFIDENCE EXTENDED (Figs. 30 and 32).
Fig. 32.
A
B
D
D
Bradley & Poates, Engr's, N.Y.

an "inverted saddle," or syncline of quartz. The sandstone A is very hard, and is threaded with quartz. Along the curve of the parting, between the two beds, there is a soft layer of black clay or "dig." D D, is slate, C is quartz.

Fig. 29 is a somewhat similar exhibit, taken from the Whip and Jersey mine, at the 700-foot level. Here, also, we have the suggestion of an inverted-saddle formation. A is hard sandstone, C is sandy slate. The dip is shown. There is crushed quartz and broken country-rock underneath the saddle.

Miscellaneous Notes.

Several of the mines which were visited and examined by the writer,* did not present peculiar features, though an occasional note of value was afforded. Thus, at the Fortuna Hustlers, one of the youngest of the Bendigo mines, and situated in the heart of the city, was obtained the observation represented in Fig. 28. The sketch represents the west leg of the saddle, which is part of the main Hustler's formation. C is slaty sandstone, which forms the clean hanging-wall of the reef; B is slate, A is quartz, D is the very hard sandstone of the foot-wall. The cross-cut clearly shows the formation (see Fig. 27). The saddle pitches south; the two legs, A and C, are only 18 feet apart; and the bedding on the two sides is seen to correspond. The cleavage is easterly above the west, and westerly above the east leg. Center-country is hard sandstone. It is a sample type of the Bendigo saddle. The shaft was sunk, and the gold-bearing reef was discovered, on inferences drawn from the information given by a railway cutting.

Figs. 30 and 31 illustrate the behavior of quartz-veins which cut across country. Fig. 30 is from the shaft of the Confidence Extended. A is sandstone, B is slate. The fracture, in which the quartz was afterwards deposited, could take place as easily across the even-grained and structureless sandstone as in any other direction, but, when it entered the slate, the line of least resistance was found to be with, rather than across, the cleavage. In Fig. 31, from the Garden Gully United, the same behavior is shown, though in a somewhat different way.

Figs. 33, 34, and 35, are three surface-sections, taken from a large

* The writer went through the underground workings of twenty-seven of the Bendigo mines.

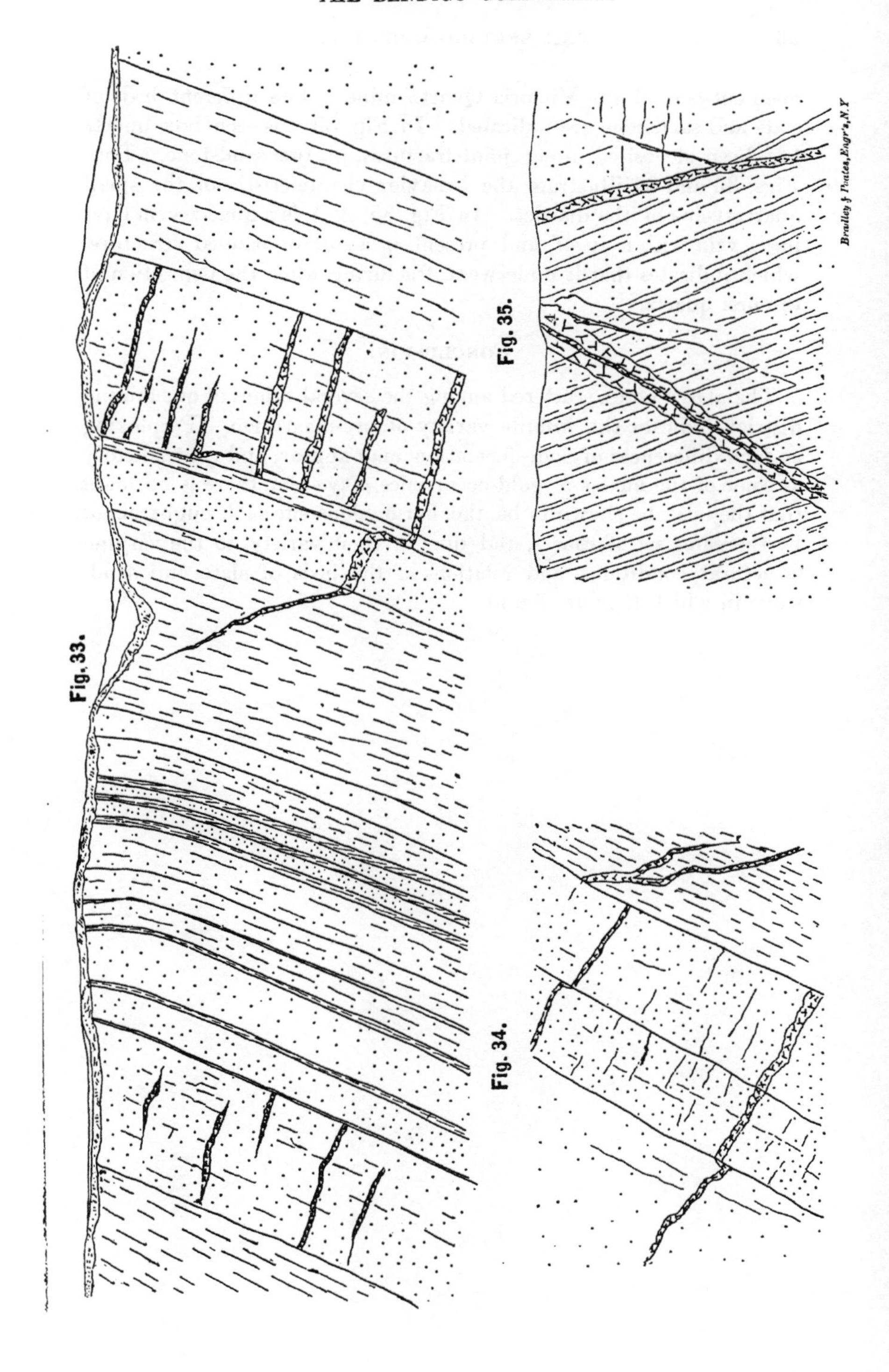

Fig. 33.

Fig. 34.

Fig. 35.

open cut behind the Victoria Quartz mine. The different beds of slate and sandstone are indicated. In Fig. 33 it is seen how quartz has been deposited along joint-fractures in the sandstone. Both Figs. 33 and 34 illustrate the behavior characteristic of the small quartz-veins of the district. In Fig. 35, A A is a quartz-vein having a vein within itself, and presenting a sort of banded structure, which indicates that it underwent fracturing after the deposition of the first quartz.

CONCLUSION.

The above notes, scattered among the several mines of one mining district, indicate the infinite variety of ore-deposition. Perplexing as such differences of lode-formation may appear, and contradictory as some of the modes of gold-occurrence may seem, there is no doubt that they can be shown to be the harmonious effects brought about by the same set of causes, and due, in their variety, to the varying structural conditions and relations of the beds of slate and sandstone in which they are found.

[TRANSACTIONS OF THE AMERICAN INSTITUTE OF MINING ENGINEERS.]

THE ORIGIN OF THE GOLD-BEARING QUARTZ OF THE BENDIGO REEFS, AUSTRALIA.

BY T. A. RICKARD, DENVER, COLORADO.

(Chicago Meeting, being part of the International Engineering Congress, August, 1893.)

THE lode-formation of the Bendigo gold-field was described in a former paper.* It presents a striking identity of arrangement with the general geological structure of the region, which is one of comparative simplicity. The alternating beds of slate and sandstone which constitute the prevailing country-rock have been extremely contorted; yet notwithstanding their highly developed cleavage it is possible to discern in their structure the evidences of original sedimentation. Along the crests of the waves which are the axes of anticlinal folds occur bodies of gold-bearing quartz, of great economical importance, which the miners have appropriately termed "saddles." They are more extensive in strike than in dip, are found to occur in a recognized succession, and have been followed by a very complete system of mine-workings reaching from the surface to more than half a mile below.

I. THE ROCK-FORMATIONS.

At the commencement of our inquiry into the origin of this gold-bearing quartz, it will be necessary to consider the relative ages of the different formations of the region. Numerous graptolites have made it easy to label the slates and sandstones in which the reefs occur as Lower Silurian. Any further subdivision, however, of this great thickness of rocks has been rendered almost impossible by reason of the striking lack of variety in the fossil remains, the general confinement of their occurrence to the slates; the marked scarcity of conglomerates and breccias; and finally, the notable similarity of texture and composition which characterizes the successive beds of slate and sandstone.

The enormous thickness of this series has been referred to. Dr. A. R. C. Selwyn, at one time head of the geological survey of Vic-

---

* "The Bendigo Gold-Field," *Trans.* xx., 463.

toria,* estimated it at 35,000 feet. The same authority is in accord with the present government geologist, R. A. F. Murray, in considering that a thickness of 7500 feet of superincumbent rock at one time covered the beds which now form the surface.

The geological map of the colony† here reproduced (Fig. 1) indi-

Fig. I.

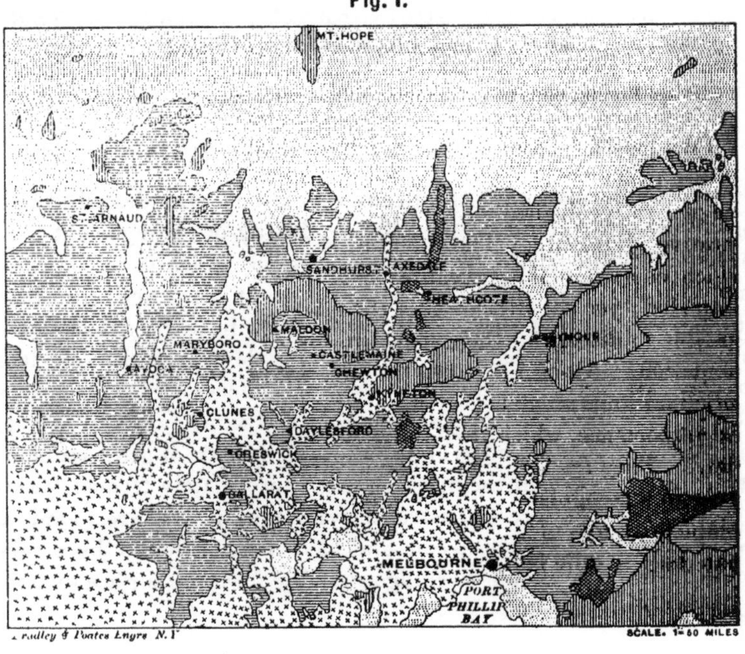

TERTIARY SILURIAN VOLCANIC

TRAP GRANITE

GEOLOGICAL SKETCH MAP
OF A PORTION OF
VICTORIA.

cates that the Silurian rocks of the Bendigo‡ gold-field are overlain to the north by Tertiary shales, while south they abut against the granite mass of Mount Alexander, which, in a horseshoe form, separates this mining district from that of Castlemaine.

The contact between the two older formations is best seen in cer-

* Now director of the Geological Survey of Canada.

† Taken from "Geological and Physical Geography of Victoria," by R. A. F. Murray.

‡ The old name "Sandhurst" appears on the map in place of "Bendigo."

tain road-cuttings at Big Hill, 7 miles from the town. The accompanying map* (Fig. 2) shows, on a scale larger than that of the general map, the principal features of the surface geology of this part of the district. It will be noticed that the crest of the ridge is formed, not by the granite but by the Silurian rocks where hardened by their contact with the granite. The railway passes through the ridge; but the tunnel indicated on the map is rendered useless for the purpose of obtaining a geological section by reason of the

FIG. 2.

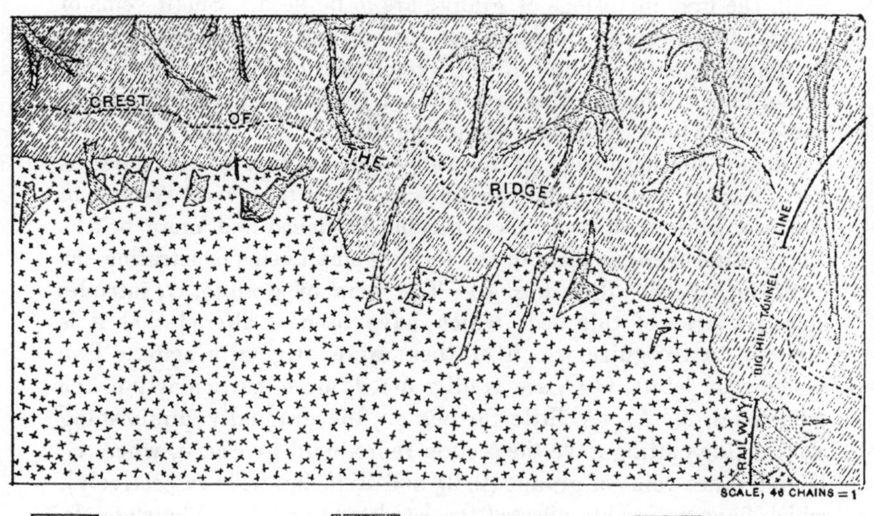

GEOLOGICAL MAP
BIG HILL, BENDIGO.

brickwork which effectually hides the rock-structure. A short distance to the west, however, the main wagon-road, in crossing the hill, affords several interesting sections.

The northern edge of the cutting shows the commencement of the physical and chemical changes produced in the Silurian rocks by contact-metamorphism.† The slate graduates into a crystalline schist and the sandstone becomes a quartzite. More mica has been

---

* A reproduction of a small part of a general map of the gold-field, published by the department of mines, Melbourne.

† Also termed by Daubrée, "the metamorphism of juxtaposition," page 133, *Études Synthetiques de Géologie Expérimentale.*

developed. At this point the structure of the country shows no marked disturbance; but in approaching the highest point of the road, attention is drawn to a distortion of the bedding by the occurrence of a large black lava dike, which evidently follows a line of fracture.

Further southward the rocks begin to exhibit more marked alteration, being hardened, somewhat bleached, and very much jointed. At the crest of the ridge the granite itself is first noticed. (See Fig. 3.) On the west the cutting gives a partial syncline, the members of which show joints. Underneath, and upon a level with the road itself, the first intrusions of granite are to be seen. Small veins of it extend halfway up the height of the cutting. The rocks forming the immediate foreground of the sketch are cut up by joints and cross-joints. The bedding is indicated by the contour of the embankment which forms the side of the road.

On the southern slope of the ridge the cleavage of the slates is very marked, and the sandstones are traversed by several systems of fracture, while both rocks have become hardened and more brittle. The granite again appears in the form of small veins penetrating the overlying rocks. (See Fig. 4.) The vein of granite, G, intrudes among a series of thin beds which have been so altered that it is difficult to recognize which of them were originally slate. Two systems of joints, C C, are readily noted, more particularly in the sandstones, A, A. A little further on, the gradually diminishing sides of the cutting gave the section reproduced in Fig. 5, where G, H and K are veins of granite (2, 2½ and 15 inches wide respectively), which have noticeably affected the sandstone bed A. The slates B, B, are also baked, and their ordinary cleavage (60° to 65°) is largely obliterated by cross-fracturing.

The wagon road does not show the actual contact of the two formations; but the railway, in approaching the tunnel from the south, gave me the sketch seen in Fig. 6. Here the main mass of the granite throws out small branches which intrude between the bedding-planes of the slates and sandstones, penetrating them for a considerable distance. Some of these veins, C C, are not more than a couple of inches thick. The darker lines, B B, indicate segregrations of ironstone.

Approaching the tunnel-entrance a crossing of two granite veins is observed (Fig. 7).

To the facts above given, the following observations may be added. At 500 yards from the contact the effects produced are only faintly

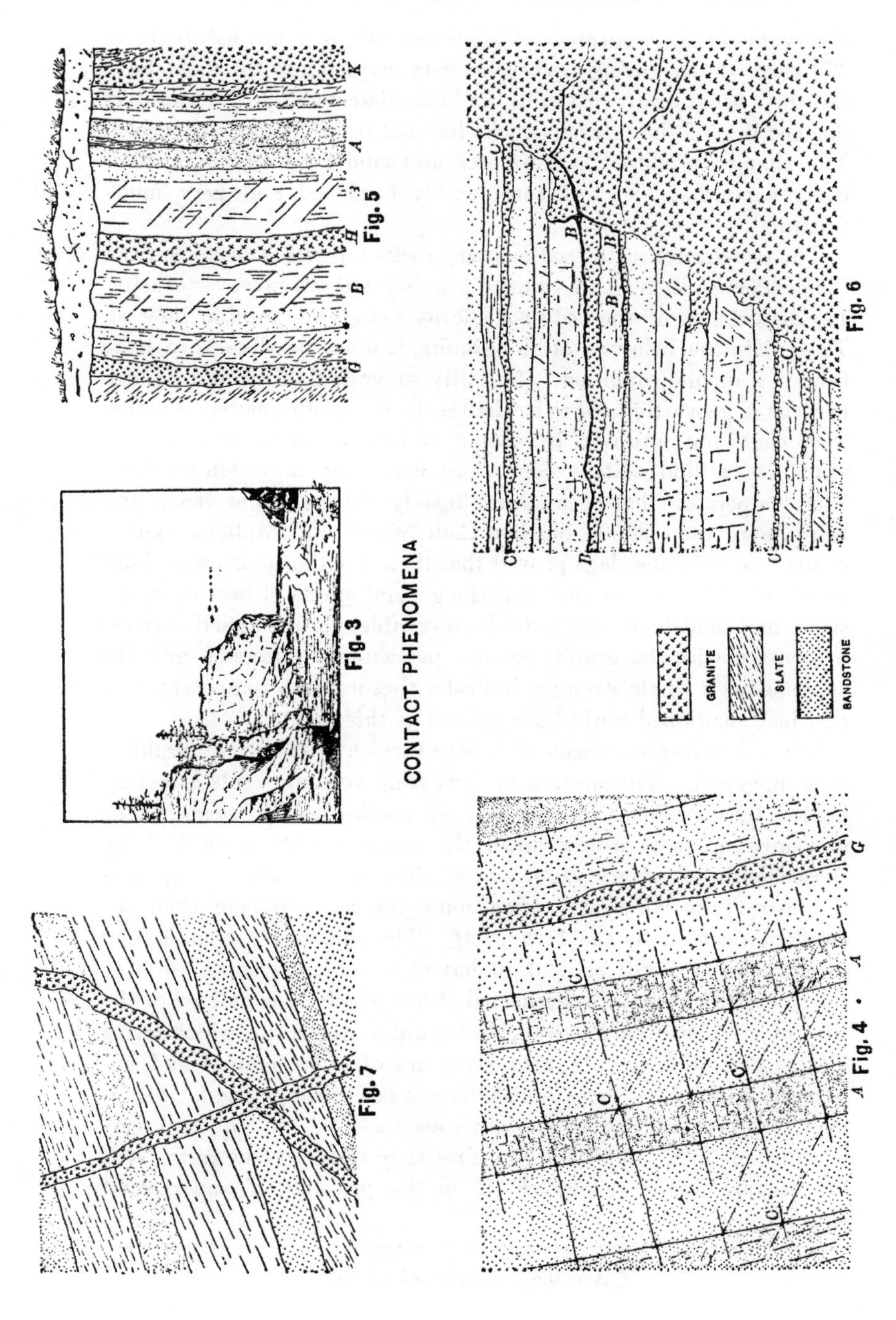

Fig. 5

Fig. 3

Fig. 6

Fig. 7

Fig. 4

CONTACT PHENOMENA

observable in the occurrence of joints recognized as not habitual; at 300 yards the fracturing has become very marked and the rocks have lost their usual color. Even in the immediate vicinity of the contact the bedding of the Silurian rocks has not been notably disturbed. The changes produced in the slates and sandstones consist in their being somewhat bleached, considerably hardened, and very much fractured.

The penetration of the sedimentary rocks for such a distance from the contact by such small veins of granite, without however causing any disturbance of the bedding, affords a study of peculiar interest. To explain the behavior of the granite, it must be allowed to have been in a mobile condition. Mobility suggests fluidity, and fluidity in rocks is generally held to be the result of a molten state. An acid rock could not, however, by reason of heat alone, be in a plastic, much less a fluid state, except at an extremely high temperature. That temperature has been approximately determined as being probably more than 2500° but less than 3000° F. Mallet's experiments * on siliceous slags proved that their melting point was about 3000° F. Though we find that the general effects of heat upon the slates and sandstones are indeed observable, yet the actual surfaces in contact with the granite are not particularly affected; and the evidence as a whole strongly indicates that no such temperature as that just mentioned could have existed at this point.

The volcanic phenomena of to-day furnish a clue to the problem here suggested. The ejection of lava from volcanic vents is accompanied by enormous volumes, not of smoke but of steam. The movement of a lava-flow down the mountain-side is marked by those clouds of watery vapor whose after condensation precipitates the heavy rains which are often more injurious to man than the desolation caused by the lava itself. The subsequent examination of the cold lava will show it to have a vesicular structure due to the myriad bubbles of superheated steam which it contained at the time of ejection. In this way, by the aid of the steam which forms a large part of their bulk, acid lavas are often as mobile and apparently as fluid as those which have a basic composition. Moreover, Daubrée has proved that siliceous rocks which require a temperature of 2500° to 3000° F. before they undergo true fusion will yet become liquid at 800° F. if in the presence of superheated steam.

* Also those of Sir Lowthian Bell.

It is suggested, therefore, that the penetration of these slates and sandstones for some distance from the contact by small veins of granite was due to a condition of suspended solidification caused by the presence in the igneous rock of imprisoned steam, taken up by the intrusive mass of granite when at lower depths.* The granite, it is true, does not now exhibit a vesicular structure such as would ordinarily characterize a rock whose mass has been interpenetrated by steam; but it must be remembered that it cooled under the pressure of a great thickness of overlying formations and was extruded at so early a period in geological time that any such structure would long ago have been obliterated by those infinitely slow physical and chemical changes which are forever bringing about the "decay and repair" of rocks.

To the petrographer the granite presents no particular feature of interest. It is of the normal type, consisting of mica, quartz and orthoclastic feldspar.

It was extruded at a period previous to the Devonian, but later than the Lower Silurian. The following facts warrant this statement. In the sections afforded by the road-cutting and railway-embankments at Big Hill, we have seen that it intrudes between and across the bedding-planes of the slates and sandstones. It must necessarily, therefore, be of later origin. Further, while the great series of the Lower Silurian rocks has been much folded and contorted, yet this has not altogether obliterated the original lines of their sedimentation, and we find that when the strike of the slates and sandstones makes an acute angle with the locally irregular line of the contact, their dip abuts against the granite. Viewed as a whole, however, the general line of the contact is at right angles to the strike, and, therefore, to the axes of the folds of the Silurian rocks. This fact proves not only that the granite was extruded at a period following the deposition of the Silurian sediments, but also subsequently to that folding which is one of their most noteworthy features. The stratigraphical position of the granite is further indicated by the fact that elsewhere in the Colony its eroded edges are overlain by rocks known to be Devonian, which are neither penetrated nor altered at the surface of contact.

Fig. 8 † represents a section at Mt. Hump Creek.

---

* Since the above was written I have read Rev. Osmond Fisher's "Physics of the Earth's Crust," in which the author argues in favor of the existence beneath the earth's crust of a liquid substratum of rock holding water-vapor in solution.

† From *Geology and Physical Geography of Victoria*, by R. A. F. Murray.

Having determined the relative ages of the slates and sandstones in which the quartz-reefs occur, and of the granite with which the former come in contact, we have one other formation to consider. I refer to the lava dikes, incidental references to which have been frequently made in my previous papers on this subject. They form a marked and instructive feature of the Bendigo mine-workings. A few additional sketches will serve to illustrate further the characteristic behavior of these "lava-streaks," as the miners call them. This is

FIG. 8.

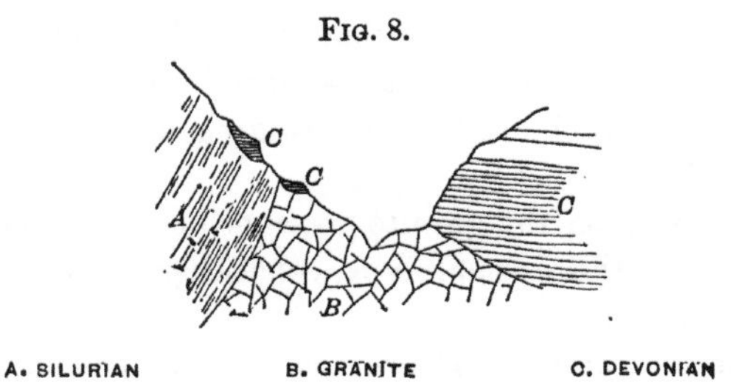

well shown in the cross-section of the deep levels of the 180 mine. (Fig. 9.) The course of the dike, whose width averages about 9 inches, can be readily followed as it cuts through the successive saddle-formations. In Fig. 10 a dike (in another mine) is seen to traverse a sandstone, cutting through several small quartz-veins, until it reaches a bedding-plane, which it then follows. In Fig. 11 another lava-streak, following the structural lines of the country-rock, becomes "pinched out" in the plane of the section along which the sketch is taken. Doubtless it found an easier passage elsewhere. In Fig. 12 a larger dike, 14 inches wide, exhibits centers of decomposition marked by concentric formations of white zeolites. There is a dark band of less altered lava following the center of the dike. In Fig. 13 a lava-streak separates into two small branches before finally dying out.

It requires but little observation underground to determine the recent origin of these dikes, which evidently followed a passage along those lines in the country-rock offering the least resistance.

The lava is of Tertiary age, either Pliocene or post-Pliocene. It is identical in lithological character with the basalt, successive sheets of which, in a manner much resembling that to be observed in California, overlie the Miocene and Early Pliocene gravel of the "deep leads" of the alluvial mines of Victoria. The gold-field of Bendigo was probably also, at one time, covered by the lava extruded from the dikes; and while all vestiges of such an outflow have since been removed by erosion, yet by a comparison of its structure, composition and behavior it is easy to see the similarity existing between the rock which forms the "lava-streaks" of the deep workings of

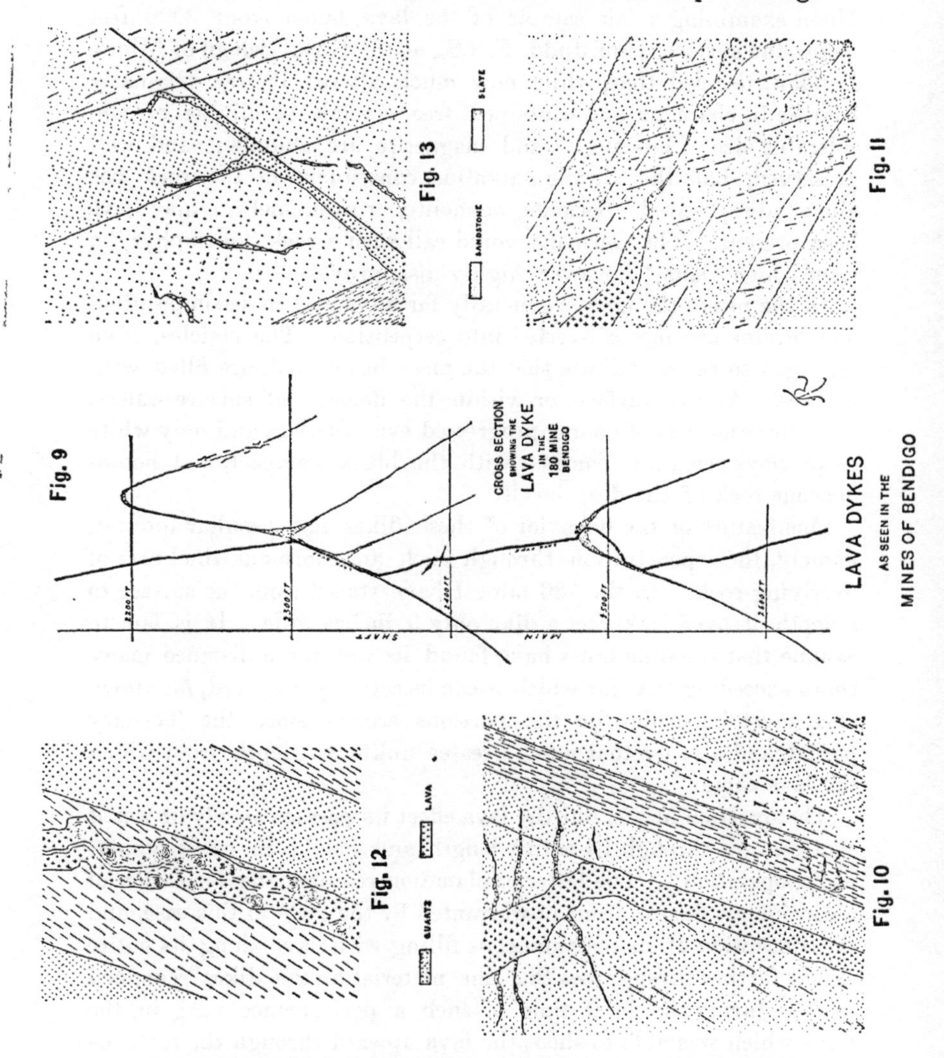

the mines and that of the basaltic plateaux overlying the rich alluvium of the neighboring districts of Clunes and Heathcote, and the more distant ones of Ballarat and Ararat.

It is not easy to determine accurately such a rock as the Bendigo lava. Igneous rocks in the vicinity of ore-deposits are almost invariably much altered, and therefore very difficult of examination under the microscope. In the case of a basic rock, such as the one under consideration, these difficulties are very much increased. Upon examining a fair sample of the lava, taken from 2000 feet below the surface, Prof. Judd, F.R.S., at the Royal School of Mines, London, decided that, while now much altered, it was originally basaltic in character and contained free crystals of olivine of considerable size. The augite and magnetite are abundant and well preserved, but the alumino-alkaline constituent of the rock has almost disappeared. Leucite, nephelite, or even melilite may have been present. The Germans would call such a rock "melaphyr," a term which Dana,* however, highly disapproves.

In the mines the lava is generally far advanced in decomposition. The olivine becomes converted into serpentine. The vesicles, often arranged so as to indicate that the mass has flowed, are filled with zeolites. At the surface, or within the domain of surface-waters, this alteration has of course progressed even further, and only white soapy clays remain to contrast with the black, compact, and homogeneous rock of the deep levels.

One feature of the behavior of these dikes has peculiar interest, namely, their penetration through such an enormous thickness of overlying rock. In the 180 mine I have traced from the surface to a depth of over 2600 feet a dike only 9 inches wide. It is fair to assume that the dike must have found its way for a distance many times exceeding that for which it can be actually observed, for allowance must be made for the portions eroded since the Tertiary period, and also for that much greater unknown depth from which the lava came.

The question arises, did the lava effect its passage slowly or was it shot upward through its entire length and height in one instantaneous operation? The latter explanation requires the formation of a continuous fissure, itself unwarranted by the facts in this case, and the conception of its instantaneous filling with lava along its entire extent further presupposes that the material of the dikes remained molten during the brief time of such a performance. Again, the force which was able to shoot the lava upward through the tortuous fractures, penetrating many thousands of feet of overlying rock, would also cause it, when it reached the surface, to be ejected with great violence and to a great height into the air.

---

* *Manual of Mineralogy and Petrography*, p. 485.

Such an explanation is in accord with the catastrophic theories of the past, but it is opposed by the modern study of volcanic action, and does not harmonize with the facts as observed in the mines of Bendigo.

The accurate observation of volcanic phenomena by Scrope and Judd* has shown that the molten material issuing from the vent of a volcano usually wells up slowly. The violent paroxysmic outbursts which do occasionally take place are essentially surface-occurrences, and are due to a sudden relief from pressure obtained by the escape of accumulated superheated steam.

The course of the dikes and the effect of the lava upon the rock-surfaces in contact must be now considered. Nothing in their mode of occurrence is more remarkable than their tortuous and very irregular passage through the overlying rocks. They do not fill a clean-cut, continuous fissure; they rarely preserve a straight line for any great distance, but follow bedding-plane, joint and cross-fracture, as each in turn presents itself. Where their line of passage takes them across the quartz-reefs, the crossing is generally effected near the apex of the saddle. The conclusion arrived at is, therefore, that the penetration of the basalt through the slates and sandstones was not the work of a few seconds, during which it was shot upward instantaneously through its entire height and length, but rather that it required a long time, and took place during a period when the Silurian rocks were subject to a tangential strain which caused their fracturing and thereby, little by little, offered a passage to the basalt, which being at the time under pressure, was seeking its way upward.

The effects produced upon the Silurian rocks, through which at Bendigo the basalt penetrates, and upon the granite and gravel, upon which in the neighboring districts it lies, are in each instance so slight as to be barely observable. The feeble changes produced upon the rock-surfaces in contact suggest themselves as due rather to the action of water than to the effects of heat.

It is, indeed, true that rocks having the composition of these basalts are fluid at a temperature low in comparison with that required to melt an acid lava; but observation of the behavior of the dikes, and of the effects produced upon the beds through which they pass, impresses one with the conviction that the temperature of the material filling them did not reach that which even the more basic of basalts require for their fusion. The conclusion forces itself upon us that

---

* As described by the former in his book entitled *Volcanos*, and by the latter in his *Volcanoes*.

the basalt of the "lava-streaks" did not owe its mobility to a molten condition arising from intense heat. The government geologist, Mr. R. A. F. Murray, states the belief that the dikes were "in great measure the product of hydro-thermal action."* This is my view also, which, more definitely expressed, is as follows: The material of the dikes, when forcing its passage through the Silurian rocks, was more in the condition of a boiling mud than what we ordinarily imagine as the state of a liquid basalt; its mobility was due not so much to the fact that it was basic in chemical composition, and therefore more readily fused, but because of the superheated steam of which it was full. Such imprisoned steam was superheated because originating at a great depth. It has been found that within the very restricted limits of human observation heat increases with depth. As we sink through the earth's crust the average increment of heat is at the rate of 1° F. per 47½ feet of descent.† Water cannot remain a liquid, in spite of increasing pressure, at a temperature above that of its critical point. This was first determined by Caigniard de la Tour to be 773° F. More recently, however, Battelli proved‡ that it was 364° C., equivalent to 687° F. Such a temperature, taking the mean surface temperature to be 50° F., would be reached at a depth of about 30,000 feet. In areas under disturbance and in volcanic regions local conditions would tend to raise the temperature and consequently to diminish the depth required to reach the horizon where water can no longer remain liquid. A depth of less than 25,000 feet might then suffice. In the gold-field of Bendigo we have to deal with a series of rocks over 35,000 feet thick. The basalt of the dikes has come up through that thickness, and consequently must have had its origin at a depth greater than that required to reach the temperature of the critical point of water. Such water as it contained must have existed, therefore, as water-vapor, which, as it approaches the surface, we may call superheated steam. This steam was in a state of compression, seeking to be relieved of its load, and striving to go where only that load could be lightened, that is, upward. The expansive force of such imprisoned steam rendered it a

---

* *Op. cit.*, p. 137.

† Sir William Thompson (Lord Kelvin) estimated it at 1° per 51 feet. Prof. Prestwich collected 530 observations made in 248 localities, and selecting those only in which the necessary precautions had been taken, found the average to be as above stated, 1° per 47½ feet.

‡ At Turin, in 1890. Strauss gives the critical point as 370° C. and A. Najedin as 358° C. This information I owe to Prof. Hallock and to Mr. Carl Barus.

powerful agent in assisting the basalt to force its slow way through the crevices and fractures of the overlying rocks until relief was obtained at the surface. The lava would have lost some of its heat and moisture by contact with the rock-faces along which it passed, but having a temperature much above the lowest temperature of hydro-thermal fusion, it could lose much heat without solidification, and was therefore enabled to arrive at the surface where it probably welled forth water, steam and mud, overspreading the older rocks and the later gravel with one of those sheets of basalt which form so marked a feature of the surface geology of the Colony of Victoria.

The Silurian rocks form the Ultima Thule of all research into the geological history of the region. Of the rock-masses by the disintegration of which they were formed, no vestige remains. Their uniformity of structure and composition, their thin bedding and general regularity, their wide extent and great thickness, all prove that they were the sediments deposited during enormous periods of time from a comparatively shallow ocean. The frequent ripple-marks now observed at a depth of many hundred feet below the surface tell of estuarine seas and shallow reaches of water. The absence of any large bodies of conglomerate and breccia indicates a continuity of uniform conditions.

If the members of this great series of beds were deposited in water of a generally uniform depth then it follows that the bed of the Silurian ocean must have undergone a subsidence which kept pace with the rate of sedimentation. The subsidence occurred along a certain line of weakness in the earth's crust, which later became the longer axis of a trough-like depression. The gradual deepening of that depression corresponded with the slow rate at which it was being filled.

As each layer of material became covered by another it began to undergo a change; as the superincumbent mass became thicker and heavier, the thin slime and the fine sand were consolidated by pressure and commenced to take to themselves the form and structure of the rocks which we term slate and sandstone.

When the area of depression had been loaded with a thickness of seven miles of sediment, subsidence ceased and elevation began. The existence of so great an accumulation of material determined the choice of this part of the earth's crust as the place of an elevatory movement. The lowermost layers of sediment were subject to an enormous pressure whose effect was to cause condensation followed by contraction. This was the local cause which started the action of the tangential strains due to the general compression of the earth's crust. This compressive strain brought about the folding of the Silurian sediments, it caused them to be gradually elevated and compelled the sea-waters covering them to recede slowly. Their upper portions became dry land, and erosion at once commenced.

The forces still at work continued to exert a lateral compression upon the sediments which now had become rock-masses. The effects are seen in their highly developed cleavage and in the flexures and undulations which are to-day their most marked feature. Their corrugation made them stronger but less pliant.

Though bent and folded, the Silurian beds were not at this time broken by extensive fissures. This was prevented by their great thickness and by their comparative flexibility. That flexibility was due in the first place to the water which they still held over from the period of their sedimentation, and secondly to their fine grain and thin bedding.

The neighboring rock-masses, forming the rim of the original area of depression, were, however, unprotected by any such depth of overlying deposits; they were older and more rigid; and they became consequently broken by fractures and fissures through which volcanic agencies found a vent. It was then that the plutonic granite, perhaps, absorbing into its mass the lowermost portions of the Silurian beds,* welled up toward the surface, giving out its heat as it advanced. The granite as it was extruded must have exerted a certain pressure against the slates and sandstones. It may have been the agent which caused those minor transverse undulations whose axes are at right angles to the main ore-bearing anticlinal axes of the Bendigo gold-field.

A period of comparative quiet followed. The Silurian rocks un-

* At Mt. Tarrangower there is an inlier of sandstone which suggests this. For the following particulars I am indebted to Prof. Ulrich, of New Zealand: "The Silurian inlier lies about 16 chains away from the Granite-Silurian boundary, forming a small hillock about one acre in extent. It consists of a nearly black, dense, metamorphic sandstone, and its boundary with the granite can, in several places round its circumference, plainly be seen, but there is *no* change in the rock other than that feldspar particles make their appearance near the boundary. However, in a water-course which cuts across the contact the dark Silurian rock can in one place be seen to run down into the granite; and there it becomes gradually quite light-colored, *i e.*, strongly feldspathic and also micaceous; in fact, so altered that any geologist seeing a hand-specimen of it would no doubt call it a fine-grained granite."

derwent denudation; their upper portions were disintegrated; from their *débris* other formations were built up, to be in turn eroded and washed into the waters of the ocean. While the upper earth was undergoing change, the under world was also the scene of silent chemical and physical processes, removing matter here and laying it down there, destroying and upbuilding, ever shifting their centers of activity but never at rest. In the latter part of the Tertiary period the basaltic lava found its passage through the fractures traversing the Silurian rocks and covered the golden alluvium lying upon their eroded edges.

II. The Auriferous Quartz.

The origin of the deposits of gold-bearing quartz, belongs to no particular period. The agencies which brought the ore of the reefs to its present position began to operate when first the sediments of the Silurian seas were laid down, and have continued until now.

In the reefs of Bendigo the two most important substances are, the metal, gold, and the matrix quartz. The other mineral constituents found in association with these two are relatively unimportant. Their origin is a matter of much interest but it can be discussed apart.

1. *The Quartz.*

The Silurian sediments contained a large proportion of sand, and the resulting rocks are very siliceous. Silica occurs in the rocks partly in a free state, as quartz, but a proportionally larger amount is found combined as a silicate. Free quartz is soluble in heated waters. The quartz combined in the more complex form of silicates has been demonstrated by Daubrée and others to be readily dissolved in hot water, more particularly superheated steam.

The formation of the quartz lodes has taken place in two stages, namely, leaching followed by precipitation. In both operations water was a necessary factor. It is the all-powerful disintegrating and transporting agent of the land-surface, where its chemical activity is intensified by the presence of dissolved carbonic acid; it is also, though in a different way, no less efficient underground where by reason of the prevalence of a higher temperature it becomes more energetic both physically as a force and chemically as a reagent; and finally it is increasingly powerful at great depths where even the tremendous pressure of superincumbent rock-masses will not prevent it from being transformed into steam and developing a chemical activity to which the mineral constituents of the rocks can offer but feeble resistance. It was the agent which, from a state of general distribution, collected the quartz and brought it into the fractures and fissures, crevices and cavities where the miner now finds it.

All rocks, though compressed and dried as they appear, contain moisture.* All rocks are pervious to water† to a varying extent by reason of capillary action, and all rock-masses are permeable because of the fractures and joints which traverse them. It has been shown by Daubrée, Bischof and others, that only a very small quantity of water such as we find the rocks actually to contain is required to produce the most pronounced changes in their chemical constitution, particularly when aided by pressure, accompanied by high temperature.

The heat, in most instances accompanied by pressure, required to make the water in the rocks intensely active, was afforded at various times and for long periods. At the earliest stage of their history they were subjected to high temperature. From the moment that one layer of sediment was deposited on the ocean bed and became covered by another it began to acquire a more elevated temperature in consequence of its increased distance from the surface of radiation.‡ As the thickness of the sediments kept pace with the sinking of the bottom of the depression in which they were laid, the successive beds acquired a temperature proportioned to their depth. The lowermost members attained a distance of over 35,000 feet from the surface. Though the increment of temperature be not constant with increasing depth,§ yet the heat which obtained at a horizon seven miles from daylight must have been extremely high.‖

The great thickness to which they attained caused an enormous pressure to be exerted by the upper upon the lower members of the series. Pressure does not develop heat unless motion also occurs. When, however, at the close of the period of their deposition certain elevatory movements took place, their great thickness must have helped to develop an energy which became converted into an exceed-

---

* The French call it *eau de constitution*. They also have a term *eau de carrière*, which is the equivalent of our "quarry-water."

† The amount of water in the rocks and their porosity has been measured by Sterry Hunt, Prestwich, Delesse, and others. See *Chemical and Geological Essays* by the first and *Geology, Chemical, Physical and Stratigraphical*, by the second.

‡ An observation due to Babbage, and quoted by Daubrée.

§ See Rev. Osmond Fisher's *Physics of the Earth's Crust*, 1889, chap. i.

‖ Le Conte, *Elements of Geology*, p. 93, says, "the lower portion of sediments 10,000 feet thick would be raised to a temperature of about 260°, and of 40,000 feet thick to that of 860°." In the instance under discussion the eroded portions of the Silurian sediments are estimated to have been 7500 feet thick, and their total original thickness to have been 35,000 feet.

ing heat. This period of elevation and corrugation was long; and during its prevalence the slates and sandstones must have been subject to the full play of those slow chemical and physical forces which, though less striking than violent volcanic outbursts and sudden earth-movements, are yet the most powerful of the agencies which modify the rocks.

Afterwards, when the granite was extruded, further heat was diffused through the adjoining slates and sandstones. In the long interval which separated the time of the extrusion of the granite from that which marked the ejection of the basalt, the Silurian rocks were subjected to slow movements of elevation and depression which, while not so energetic as those which took place at the beginning of their life-history, were yet sufficient to develop a rise of temperature. In later times, the period during which the lava penetrated the older rocks must have been marked by an increase of their heat. Though the extrusion of both granite and basalt have left more striking evidence of their occurrence than the slower movements referred to, which took place at other periods, yet the heat they afforded was not so widely diffused, and the effects produced were comparatively local. The heat given out by both granite and basalt was however subsequently widely distributed.

This distribution was effected through their contact with underground waters which, becoming converted into steam, transferred some of their heat to other circulating waters, and these in turn injected their heat and steam through the pores and crevices of the slates and sandstones.

The hot water and superheated steam * thus produced during long continuing periods were the restless agents in leaching the rocks, in dissolving out the silica from its state of combination and in afterwards transferring it to underground currents which bore it away until changed conditions compelled them to deliver it up by precipitation as quartz.

Thus, in this and other ways, the quartz became separated out through the pores of the rocks by a kind of sweating process,† to be segregated along the joints and fractures traversing them.

---

* One feels inclined to speak of "superheated" water, but there is no such thing. Superheated water, without reference to pressure, is steam. Geikie, though his writing is generally a model of accurate expression, slips into the use of this term on page 284, *Text-book of Geology*.

† Or, as Daubrée puts it: "Le quartz a été fourni aux veines par une sorte d'exsudation de la roche encaissante."

The experiments of Daubrée proved the action of superheated steam in dissolving the silicates and the subsequent precipitation, upon the lowering of the temperature, of crystalline quartz of a character similar to that found in association with the gold and other metals of ore-deposits.

The changed conditions compelling precipitation were very various in kind and due partly to physical, partly to direct chemical causes. In traversing the minute underground passages of the rocks, the hot solutions would meet with portions of loose or crushed rock, giving larger space, increased surface, and diminished pressure. This would favor precipitation. The workings of mines often afford illustrations of ore-deposits which owed their existence to such conditions. Similarly hot waters meeting and mingling with colder currents or passing between comparatively cool rock-surfaces would have their solvent power diminished, leading directly to the deposition of the material in solution. The hot springs of the present day in many parts of the world give us familiar examples of the precipitation of silica resulting from the lowering of the temperature of the issuing waters. A third cause may be quoted. The reducing agency of organic and other matter is able to precipitate silica from its state of solution as a silicate. Of this, silicified wood is a familiar illustration.

The seas, the rivers, the thermal springs, and the underground waters of to-day carry notable quantities of silica in solution. Forchammer found sea-water to contain silica, and in certain samples he determined the quantity to be as much as 3 parts in 1,000,000 parts of water. A cubic mile of the ocean would, at this rate, contain 13,500 tons of silica. Deville showed that the river Loire, at Orleans, contained in 100,000 parts 13.46 of solid matter, of which 30 per cent. was silica. The geysers of the Yellowstone Park, those of California, Iceland, and the north island of New Zealand, all deposit silica from the waters ejected by them. Steamboat Springs, in Nevada, and Sulphur Bank, in California, may also be instanced. The occurrence in New Mexico and Arizona of extensive areas which have undergone submergence, and whose forests have become petrified by the action of percolating siliceous solutions, affords a striking instance of the transference of silica by underground waters and its precipitation under favorable conditions. That the waters of mines contain silica in solution has been proved by the silicification*

* This, as pointed out to me by Prof. Le Conte, is a double process, consisting of the filling of the interstices by the precipitation of silica and the actual replacement of the woody fiber itself.

of drift-wood, beautiful specimens of which have been found in the auriferous gravels of the "deep leads" of California and Australia. Such silicified wood has also been shown to be gold-bearing.

The concentration of quartz in one place rather than in another, its deposition between certain beds and along certain fractures rather than elsewhere, are the results more of simple physical conditions than of complicated chemical reactions, and are due to causes to be discussed in considering the structural geology of this district.

2. *The Gold.*

The waters of the ocean contain gold. In 1851 Malaguti and Durocher determined the occurrence of silver, but did not extend their inquiries into the question of the presence of gold in sea-water. This fact was first accurately determined by Sonstadt in 1872.* His experiments were not quantitative, but he stated, in parenthesis, that the amount was "certainly less than 1 grain in the ton."† More recently, however, Münster found an average of 5 milligrammes per ton.‡ In endeavoring to arrive at an approximate estimate it must be remembered that local conditions, such as the temperature of the water, will affect the amount in solution. Sonstadt's researches were made with water obtained near Ramsey, in the Isle of Man, while Münster got his from the Kristiania Fjord. In each case the sea-water was that of a northern latitude. In warmer regions it is probable that precipitation, due to the presence of putrescent organic matter, may diminish the amount of gold held in solution.§ Let us, however, take 5 milligrammes (equivalent to $\frac{1}{13}$ of a grain) as an approximation. This, though in itself a minute quantity, will be found to represent an enormous total amount of gold in the waters of the ocean. From the results obtained from the careful soundings carried out by the Challenger and similar scientific expeditions, it has been computed that the ocean has an average depth of 2500 fathoms, and that it contains 400 million cubic miles of water.|| This is equivalent to about 1,837,030,272,000

---

* "On the Presence of Gold in Sea-Water," E. Sonstadt, *Chemical News*, Oct. 4, 1872, xxvi., p. 159.

† Sonstadt is often incorrectly quoted as having shown that sea-water contains 1, grain of gold per ton.

‡ The amount of silver was determined to be 20 mg. per ton. See "On The Possibility of Extracting the Precious Metals from Sea-Water," *Journal of the Society of Chemical Industry*, April 30, 1892, xi., p. 351; abstract from *Norsk Tekniak Tidsskrift*, vol. 10, No. 1.

§ See note in the *Chemical News*, Oct. 4, 1872, xxvi., p. 161.

|| Geikie's *Text-book of Geology*, 2d ed., p. 33.

million tons, which, upon the basis of 5 milligrammes per ton, would represent 10,250 million tons of gold. By way of contrast, it may be added that, according to Soetbeer, Leech, and others, the gold-production of the world, from the beginning of 1493 to the end of 1892, a period of exactly four centuries, has amounted to only 5020 tons. The present output is equal to about 200 tons per annum.*

The gold in sea-water is kept in solution as an iodide.† The amount of free iodine present in the ocean is very minute;‡ but a large proportion of that element occurs combined as an iodate of calcium.§ From the results of a series of six experiments, Sonstadt found that a cubic mile of sea-water contains about 17,000 tons of iodate of calcium or 11,072 tons of iodine.|| This represents the occurrence in the entire ocean of no less than 4,428,800 million tons of iodine.

The iodine which maintains the gold in solution is obtained from the iodate of calcium. Gold is soluble in extremely dilute solutions of iodine, which under ordinary conditions are in turn readily reduced by organic matter. That the gold in the sea is not precipitated is due to the presence of the iodate of calcium in which it is not soluble but which, being readily decomposed by *putrescible* organic matter, liberates the iodine required to keep the gold in solution.

There is reason to believe that the sea-waters of to-day contain much less iodine than those of former geological periods.¶ That there is so little free iodine in the ocean is due to causes parallel to those which bring about the noteworthy absence of carbonate of lime. Marine animals abstract the latter while marine plants absorb the former. How great is the work done in this way, is evidenced by the dimensions of the coral reefs and by the extent of the foraminiferous and other marine limestones.

The abstraction of iodine is no less striking. Sea-weeds, and more particularly those which grow at great depths, are the chief source of the iodine of commerce.** When, after a storm, such sea-

* The colonies of Australasia, since 1851, have produced 2830 tons, of which Bendigo contributed 375 tons.

† *The Chemical News*, October 4, 1872, xxvi, p. 159.

‡ *Ibid.*, April 26, 1872, xxv., p. 196.

§ *Ibid.*, April 26, 1872, p. 197.

|| *Ibid.*, May 24, 1872, p. 242.

¶ Thomas Sterry Hunt, *Chemical and Geological Essays*, p. 142.

** It is also recovered from the nitrate of Chili.

weeds are cast upon the shores of Great Britain,* France and Sweden, they are collected and burnt, and from their fused ashes, termed "kelp," the iodine is subsequently extracted by a simple chemical process. From 13,000 kilos of kelp about 100 kilos of sodium carbonate and 15 kilos of iodine are obtained.†

That iodine is not now so plentiful in the sea as during former geological periods has been suggested by chemical investigations into the composition of rocks. Certain sedimentary formations contain notable quantities of it. It has been found in some aluminous shales in Sweden, and also in certain varieties of coal and turf.‡ The saline waters of several springs contain large amounts of it. Even rain-water has been known to give a recognizable iodine reaction when tested, such iodine having been obtained by the agency of winds which have been blowing over certain areas of the sea where it was being liberated by the action of organic matter upon the iodate of calcium.§

Let us now return to Bendigo. The conditions which prevailed during the time when the Silurian seas washed the earth's surface, were no less favorable to the solution of gold than those which obtain to-day. It is indeed probable that the waters of the palæozoic ocean contained more gold than those of the present geological period. When the sediments sunk to the bottom they carried with them, entangled amid the silt and sand, a large proportion of sea-water, most of which was subsequently yielded up as they were pressed down by the weight of the overlying deposits. The sea-water as it was rejected by the solidifying strata, left behind it a residuum which contained in a highly concentrated form the original constituents of the Silurian ocean. The sediments which fell to the ocean floor also contained portions and fragments of vegetable life which in their decay served to decompose the iodate of calcium contained in the residual sea-water and so set free the iodine. This is a very active element and would at once form a fresh combination. The excess of iodine thus obtained may have served as a solvent for any particles of metallic gold which by mechanical means accompanied the sands laid down beneath the sea and which thus became added to that already derived from original chemical solution in the waters of the ocean.

* Particularly the western isles of Scotland, and the west coasts of Wales and Cornwall.

† *The Principles of Chemistry*, D. Mendeljeff, vol. i., p. 490.

‡ Fownes's *Chemistry*, p. 185. § *The Chemical News*, May 17, 1872, xxv., p. 231.

The great body of slates and sandstones, not less than seven miles in thickness, was thus slightly but distinctly gold-bearing. It is so still; the amount of the precious metal has remained practically the same and the changes which have occurred have merely brought about its less uniform distribution.

As the sediments were further consolidated, the vegetation which they originally contained became disintegrated into its elements, and the iodine which it had abstracted from the sea-water when alive was now yielded up. The iodine thus derived became another factor in the solution of the gold which was disseminated through the solidifying strata.

From a state of even dissemination through great rock-masses to its concentration along certain lines of fracture which we call reefs, the gold arrived through the agency of the water still retained by those rock-masses. The geological history of the district has shown that from various causes the heat required to make those underground waters intensely active was at hand for long periods. It is probable that the time of the extrusion of the granite was in this respect one of unwonted chemical activity. So also was the later period, when the lava dikes penetrated the overlying strata.

As an iodide, the gold readily circulated through the medium of the waters occurring in the rocks. We know gold to be readily soluble, even in extremely dilute iodide solutions. It would be wrong, however, to suppose that this or any other one chemical agency was the universal solvent for the precious metal. There is reason to suppose that in different formations and under diverse conditions gold is soluble in varying combinations, all having a common want of stability. It has been suggested by Le Conte and others, and some recent discoveries* in the mines have tended to confirm the supposition, that the gold of certain lodes has been deposited from the solutions of the persulphate of iron. Alkaline sulphides have long been mentioned in the text-books as the chief agents in the dissolving of the metallic sulphides, and very pretty chemical formulæ have been evolved to explain the resulting complicated reactions.†

* "Gold Deposits in the Quartzite Formation of Battle Mountain," F. Guiterman, *Proceedings of the Colorado Scientific Society*, vol. iii., part iii.

† The supposition that gold is insoluble in alkaline sulphides requires modification. At the June, 1893, meeting of the Colorado Scientific Society, Mr. L. G. Eakins reported the results of recent experiments, which show that fine gold, digested for four to five days in a cold, strong yellow solution of ammonium sulphide, has a solubility equal to 2 per cent. Weak solutions, with seven days' digestion, gave confirmatory results. Sodium sulphide also was found to be a solvent,

Occasionally gold is found free from association with metallic sulphides; then Bischof's experiments are quoted, and its origin is put down to a silicate of gold soluble in alkaline waters. Again, gold readily combines with free chlorine, forming a salt easily soluble in water. The chloride of gold has, however, not been considered as likely to exist in nature because of the rarity of the occurrence of free chlorine and because of the instability of the compound which it forms with gold. The latter quality is not peculiar, but characteristic of all the salts of gold as we know them. Free chlorine *does* exist in nature, and though not frequently detected, this is not to be wondered at, seeing that its occurrence is usually associated with that of hot water or steam. Volcanoes, especially during times of temporary quiescence, emit large quantities of hydrochloric acid,* and the rims of the volcanic vents, then called "fumaroles," show incrustations of chloride of iron, chloride of ammonium, etc. The comparatively large amounts of such salts as exist in sea-water, which were found at Vesuvius† was one of the facts used in support of the theory that the steam of volcanic eruptions came from the ocean. That these and other compounds, including hydrochloric acid, owe their origin to free chlorine, and were formed in the presence of hot water and steam, is most probable. Therefore it would seem well to add the chloride to the other probable combinations in which gold may circulate underground.

In the great laboratory of nature, the chemical changes which take place in the long periods of time with which we have to deal must be of an infinite variety, and therefore no number of chemical formulæ can represent the reactions which have followed each other from the moment when the gold was liberated by the disintegration of some pre-Silurian rock to this later day, when the miner finds it enclosed in the quartz of the reefs. Therefore, while it is suggested that as an iodide it may have travelled through the waterways of the Silurian rocks, it is not meant that in its long journeyings before it arrived at the place where it is now found, it did not pass through many changes of chemical combination.

The precipitation of gold from solution can be brought about in many ways. More than twenty-five years ago certain members of the Geological Survey of Victoria carried out a series of experiments

---

but to a very slight degree only. Mr. Eakins is still prosecuting this series of interesting experiments.

* Judd's *Volcanoes*, p. 213.

† Palmieri's *Vesuvius*, p. 121.

which yet remain our main source of information on this subject. In 1864, an accidental discovery by Richard Daintree led to the determination by him and his colleague, Prof. George H. F. Ulrich, of certain conditions under which the precious metal is deposited from solution.* It was demonstrated subsequently by a number of careful experiments made by Wilkinson,† that organic matter, in any of the ordinary forms in which it is found to occur in the alluvial drifts, is readily capable of precipitating the gold of even the weakest solutions, and of depositing it as a thin metallic coating upon other particles of gold or upon any of the sulphide minerals, such as pyrite, galena, or blende, with which its occurrence is usually associated.

Several years later, in 1871, Daintree was in London and at Dr. Percy's laboratory, at the Royal School of Mines, commenced another series of experiments. In different bottles he placed solutions of chloride of gold in strong solutions of chloride of sodium, containing a piece of pure rock-salt, and to each he added a crystal of one of the several metallic sulphides commonly found in gold-ores. Four years later, and before any results had been obtained, he died. The chemical reactions which he had set to work were, however, proceeding slowly but surely. I now quote Mr. Richard Pearce, who was an eye-witness of both the beginning and the consummation of these experiments.

"When Daintree died, one of the bottles, namely, that containing a gold solution and a crystal of common pyrite, was removed to Dr. Percy's laboratory in Gloucester Crescent, and there, in 1886, the work which Daintree had begun came to fruition. On the smooth surface of the crystal of pyrite there had been deposited a cluster of crystals of gold."

---

* The following interesting account of what has become an historical event was given to me by my distinguished friend, Prof. Ulrich, now of the University of Otago, New Zealand, and with his permission, I reproduce it here:

"I was engaged in the laboratory analyzing a zeolitic mineral, whilst Daintree himself was busy with some photographs. Suddenly he made ejaculations of astonishment, and on my asking the reason, he showed me a small, common medicine-bottle, which contained (more than half full) a water-clear fluid in which floated a part of the cork. The cork remaining in the neck of the bottle was acid-eaten. At the bottom of the bottle there was a large speck, or rather little nugget, of bright gold. He explained that he had made a concentrated solution of chloride of gold for the purpose of toning his photographic prints. He had placed in it a small speck of gold. 'And now see (turning the bottle upside down), the speck has grown to such a size that it won't go through the neck of the bottle.' We both agreed that the organic matter of the cork had been the cause of the decomposition of the solution and the growth of the gold."

† The late Mr. Charles Wilkinson, Government Geologist of New South Wales, at that time junior assistant in the Geological Survey of Victoria.

It had taken fifteen years to obtain the result, and the man who started the investigation did not live to see his work fulfilled; but what, it may well be asked, is the brief time of even a generation when compared to those vast æons during which Nature, in her greater laboratory, is carrying out operations similar to these?

It is interesting to be able to add as a sequel to the above that in the ore of the St. Louis mine, in Gilpin county, Colorado, crystals of pyrite have been found which, while in themselves non-auriferous, were yet dotted over with clusters of crystalline gold remarkably similar to those obtained by Daintree's experiment.*

The persulphate of iron, which is also a vehicle for the removal of gold from one place to another, is reduced by organic matter, with the formation of gold-bearing pyrite. Under like conditions silicate of gold solutions would precipitate their gold. Again, metallic sulphates or alkaline sulphide waters containing metallic sulphides in solution would, on meeting alkaline carbonates in the presence of organic matter, be compelled, by reduction and neutralization respectively, to form a deposit of metallic sulphides.† Finally, iodide solutions are readily reduced by organic matter or by ferrous sulphate, while the presence of both together would lead to the formation of an auriferous pyrite.

We need now to recall the fact that the original sediments contained imbedded with them numerous remains of the organic life of the ancient seas, many of whose forms were subsequently preserved in the rocks as the fossils which now enable us to determine the stratigraphical position of the slates and sandstones. Other remnants of the more minute organisms such as abounded in the waters of the ocean, though they left behind them no recognizable forms, yet were doubtless mixed among the silt and mud which fell to the bottom of the Silurian seas. Such organic matter in process of time became decomposed into its constituent elements, those which were soluble being removed by underground waters while the insoluble remained to undergo further decomposition. In this case, as often happens in the laboratory of Nature, the solvent and the precipitant formed part of the same substance, whose disintegration liberated the one from the other, both, however, to meet again and to react upon each other at a later period. Thus the iodine, which is a solvent for gold, was set free while the carbon, which is a precipitant,

* My authority is again Mr. Richard Pearce.

† Le Conte's *Elements of Geology*, p. 245.

remained in the residue and became converted into the graphitic material which darkens the rock encasing the quartz lodes.

This graphitic material was the precipitant for the gold solutions. Its occurrence in certain lodes has been pointed out by others* but its very frequent association with gold quartz has not been fairly recognized. In the main auriferous belt of California, passing through the counties of Amador, Calaveras, Tuolumne and Mariposa, an encasing formation of black slate is known by experience to be a favorable indication, while a gray or greenish-gray country-rock is considered less encouraging. At Amador City, in the county of the same name, I have seen men coming up from underground with faces and hands all sooty black by reason of contact with the graphitic slate which there and elsewhere in that region encloses the quartz veins. In New Zealand, more particularly in the province of Otago, the gold-bearing lodes are similarly characterized by a black selvage, the somber tint of which is due to carbonaceous material. In the mining districts of the continent of Australia, as well as in the gold-fields of the island of Tasmania, the slates and metamorphic schists, which so generally form the prevailing country-rock, are dark, and the clay which lines the walls of the reefs is black and graphitic. Such is the case also in the mines of Bendigo. It is not, therefore, necessary to go far to find an agent capable of precipitating the gold from its state of solution in the underground waters of the rocks.†

That gold does occur in solution in the underground waters of to-day has been shown by the evaporation of large quantities of ordinary mine-water and the finding of gold in the residuum,‡ also by its occurrence in the incrustation of boilers using mine-water.§ It has been proved by the examination of old mine timbers left in abandoned workings and the discovery in the decayed and often silicified wood of crystals of pyrite which were distinctly gold-bearing.|| Furthermore, it may be added that in the gravel of the deep leads, at Ballarat, for example, there has been found driftwood

* As, for instance, by Sandberger, in connection with the veins of the Erzgebirge.

† How far electro-chemical and electro-magnetic forces may have aided these reactions cannot be estimated. That they took some part is highly probable.

‡ First proved by Daintree.

§ Also by Daintree, at the mines at Maryborough.

|| *Notes on the Physical Geography, Geology and Mineralogy of Victoria*, by A. R. C. Selwyn and G. H. F. Ulrich. Melbourne, 1866.

yielding assays of from a few pennyweights to several ounces of gold per ton.*

The occurrence of the metallic sulphides, such as arsenical and ordinary iron pyrites, galena, and blende, the minerals most commonly found at Bendigo in association with the gold, brings up a large field of conjecture. Sandberger† showed that iron, zinc, lead, copper and other metals occurring in lodes can also be found in certain silicates common to the crystalline rocks, such as olivine, augite, hornblende and mica. His deductions have not been, however, entirely accepted. At Bendigo the lode-formation, though evidencing metamorphic action, is yet sedimentary. It may, perhaps, be suggested that the granite contributed the metallic sulphides associated with the gold and quartz, but there is no evidence to support such an explanation. While the frequency of the occurrence of ore-deposits in association with eruptive rocks is a fact now widely recognized, yet it has given rise to generalizations not altogether warranted. In this particular district the contact is not a place of ore-deposition, and the main series of producing mines is seven miles distant. There are, it is true, certain anticlinal formations of quartz not far from the granite; but they appear to have no relation to the contact, and they are not economically of any importance. The geological evidence of the region has suggested that the granite was a factor in the process of ore-deposition, but it does not indicate that this rock was the source of the gold or of the associated sulphides.

In this connection I would hazard the remark that the near neighborhood of igneous rocks is favorable to the occurrence of ore-bodies, not always or necessarily because such rocks were the origin of the precious metals which were leached out from them, but often because the extrusion of such eruptive rocks afforded the heat and steam which gave an intensified chemical activity to percolating solutions.

The Silurian sediments were obtained from the erosion of pre-existing rocks, whose disintegrated particles probably contained the material required to form the sulphide minerals of the reefs. Whether the metals were dissolved in the waters of the palæozoic

---

* Investigated by the above-mentioned, and also by J. Cosmo Newberry, analyst to the Geol. Survey of Victoria.

† *Engineering and Mining Journal*, March 22 and 29, 1884, xxxvii., pp. 218 and 232. Translation of first chapter of "Untersuchungen über Erzgänge," by Fridolin Sandberger. Wiesbaden, 1882.

seas and were subsequently chemically precipitated, or whether they were deposited by mechanical agency among the silt and sand which fell to the ocean bed, is a question not now to be determined.

In the reefs of the Bendigo gold-field the metallic sulphides are present in an unusually small proportion. Iron pyrites is much the most abundant. A great deal of the richest ore of the deepest workings of the mines is a clean white quartz, almost entirely free from any accessory minerals. Such quartz frequently contains the gold in an extremely coarse form, in pieces weighing many penny-weights.* The presence of pyrite is not necessarily an evidence of the poverty or richness of the quartz. Rich ore-bodies often contain a good deal of it, just as poor ones sometimes do. Other mining-districts afford a similar experience. We must recognize that the sulphates of iron act very differently with an increase or decrease of the oxygen they carry. Ferrous sulphate, FeO, SO_3, is a precipitant† for gold solutions, while ferric sulphate, $Fe_2O_3, 3 SO_3$,‡ is a solvent. May not this fact help to explain the irregularity of the phenomena attending the presence of pyrite in gold-ores?

The limitations to the length of this contribution must prevent the discussion of theories explanatory of the process by which the gold and quartz were actually brought to the place where we now find them. Modern teaching and experience has advanced that theory of ore-deposition which has been, not at all happily, called "lateral secretion." It is to the effect that the origin of the precious metals as found in lodes is to be ascribed to the immediately encasing rock, out of which they have been leached by solutions, which afterwards by endosmotic flow, penetrated the walls of the fissure and there deposited the gold and silver. Such an explanation is, I respectfully submit, more in harmony with the teachings of a professor's laboratory than with the testimony of underground observation. No theory so narrow as that framed on a phenomenon as specific as endosmosis can live in the air of the mines where the modes of occurrence of the ore have an infinitude of variety which even the most general of explanations can hardly hope to cover. As our stock of ascertained facts slowly accumulates, as each dis-

---

* At the Lazarus Mine, at a depth of over 2000 feet, I saw pieces of gold exceeding an ounce in weight in a large reef of clean, white, splintery quartz.

† It is the precipitant employed in the chlorination-works of California and elsewhere. Charcoal is also used for the same purpose.

‡ The formula is not rigid. The so-called sesqui-sulphates of iron are of variable composition.

tant mining district sends in its quota of recorded observations, we shall, I believe, find that "lateral secretion" in its narrowest meaning is rarely tenable, that is, that while the material of ore-deposits may have been and probably was derived from the leaching of the rocks, it did not necessarily or often come only from those which are immediately adjacent to the walls of the lodes. In the case of the Bendigo "saddles" we are led to believe that the gold and the quartz were derived from the mass of the surrounding formation rather than from that small portion only which immediately adjoins the reefs. In the process of ore-deposition and lode-formation this gold-field was a part of a greater area, in which similar phenomena of segregation* took place, an area which included nearly all of the numerous and productive gold-mining districts of the colony of Victoria. Bendigo was especially favored because of the very peculiar structure of its beds of slate and sandstone.

It may be objected to the general explanation which has been offered with regard to the origin of the gold and the quartz that if it be accepted as true, all sedimentary rock formations should be equally capable of profitable exploitation. The answer is obvious, namely, that to the miner the mere dissemination of gold in rocks is a fact economically of no importance, since it is only by its concentration in certain quantities and in certain forms of ore-deposit that it can repay him for the toil and expense of its extraction. In the Silurian rocks of the Bendigo district conditions obtained and agencies were at work which were particularly effectual in collecting the gold from its wide and even dissemination to its concentration in the reefs.

The richness of the gold-field is owing primarily to two causes, the structure of the country-rock whose extreme and very regular folding gave rise to unusual facilities for the percolation of underground waters, and the occurrence in the slates and sandstones of a precipitant able to compel the deposition of the gold.

It is a striking fact that while the anticlines necessarily alternate with synclines, the latter are not the places of ore-deposition. "Inverted saddles," as the miners call them, do indeed occasionally occur; but they are not economically of any importance. The explanation is a simple one. By the mechanical principle of the arch the anticlinal structure tends to preserve a passage for mineral solu-

* This word "segregation"—the separating out of material and its regathering together elsewhere—best covers the process of lode-formation in this particular region.

FIG. 14.

Drawing of Bendigo made in 1851.

tions, while on the contrary the basin or trough, formed by the synclinal arrangement of the beds, tends by the action of gravity to become closed. It is not probable that the apex of the anticlines was marked by an open way; but we are justified in supposing that along the anticlinal axes there were portions of rock more loose and more permeable than the country surrounding them. These we may liken to arched canals through whose long passages mineral solutions have circulated from pre-Devonian times till now, bearing with them that golden freight which by reason of the reducing action of the carbonaceous matter lining their walls they were compelled little by little to lay down.

Thus we arrive at a stage when from wide and uniform dissemination through enormous rock-masses the gold and the quartz have become concentrated along certain lines and in certain localities, but nature is never at rest; they are no sooner laid down than they become again wanderers through the underground waterways. The causes operating to remove them from one place and to concentrate them in another are forever at work. We speak of secondary deposition in cases where we think we clearly recognize the removal of metallic ore from one place to another; but as a matter of fact all the ore-deposits of the mines are concentrations, and in their nature secondary, from the period when their constituent parts formed a portion, relatively large or infinitely small, of the first sediments laid down upon the floor of the ocean to that time long afterward when the pick of the miner disturbs that which for ages has been going through a process of continual change and evolution.

Thus the silica which as fine sand fell to the bed of the sea, in process of time united with other elements and became a part of a complex mineral which we call a silicate. That silicate, by the reversal of the reactions which had brought it into existence, became subsequently disintegrated, the silica was set free, and as quartz later on became the matrix enclosing the gold. From that state of admixture it was separated by the contrivances of man or by the less noisy and more powerful agencies of heat and cold, wind and rain, to be swept into the running stream which carried it into the waters of the ocean in whose silent depths it was destined again to "sow the dust of continents to be."

Similarly the Silurian slates and sandstones may have been themselves derived from the erosion and disintegration of the granites, other portions of which afterwards intruded among them. Or, again, there is reason to believe that the granite may have been formed by the extreme metamorphism of the lowermost members of the Silurian series. Upbuilding and disintegration in the mineral creation like life and death in the organic world, are but correlative parts of one continuous process. They are different aspects of that indestructibility of matter and transmutation of energy which are taught no less by the formation of a quartz reef than by the unfolding of a flower.

At that point in the long sequence which marks the present time, the eye lingers on rolling woodland and winding road, grassy meadows and fleecy flocks, glancing from the busy activity of the railway to the peaceful quietness of the farm, to be finally arrested as it catches the gleam of the Bendigo mines, white islands in the dark blue sea of the lovely Australian bush, that vast forest of *Eucalyptus* whose leafy waves have replaced the watery wastes of palæozoic times.

Fig. 14 is a reproduction of a drawing of the Bendigo field as it appeared in 1851. The upright mass of quartz in the foreground is the "west leg" of a saddle-formation, while just beyond the two pools of water (now ornamental lakes in the grounds of Mr. George Lansell's residence) there is seen an actual saddle, outcropping at the surface.

LA GARDETTE: THE HISTORY OF A FRENCH GOLD MINE.

BY T. A. RICKARD, ALLEMONT, ISÈRE, FRANCE.

(Baltimore Meeting, February, 1892.)

THE mountains of the picturesque Dauphiné, in southeastern France, have long been known to collectors as the source of many minerals of rare occurrence; but they contain also several mines, which, though less known, date from the time of the Saracens, many centuries back, to that of the graduate from the *École des Mines*.

In the department of the Isère, upon the cliffs of Villard Eymond and overlooking the lovely valley of the Oisans, is the historic gold-mine of France, a mine interesting as much for its record as for its geological features.

The gold-bearing character of the quartz of La Gardette was known to the Greeks, the Romans, and later to the Saracens, all of whom at various periods extracted the gold of the outcrop which ribs the mountain-side. Mention of this mine is found in certain rare manuscripts of the seventeenth century, but it was in 1733 that the first serious working was inaugurated. In that year the exploration of the vein was undertaken by order of the king, much to the disgust of the inhabitants of Bourg d'Oisans, to whom it has been, both in earlier and later times, a source of revenue, either directly by the extraction of small quantities of gold-ore or indirectly by the attraction which it has presented to passing travellers. The working of the mine by the State had no results of importance and shortly ceased. In 1765, the mine was again given over to the peasantry. In 1776, the king conceded to his brother, the Comte de Provence, afterwards Louis XVIII., a large number of mines in this district. La Gardette was included; but work under this grant was directed to the mines of Les Chalanches, near Allemont, in the neighboring commune. Here a foundry was erected under the supervision of M. Binelli, considered one of the most celebrated mineralogists of his time. In 1781, Binelli was replaced by Schreiber, a Saxon engineer, whose name is interwoven with that of nearly all the old mines of this part of Europe. At that time a peasant named Laurent Gardent came to Allemont to see M. Schreiber and to show him certain specimens of gold-quartz recently extracted from the lode at La Gardette. Schreiber made an examination, was much impressed with what he saw, and forthwith started for Paris, where, upon his arrival, it was decided to commence at once the exploitation of the mine. Systematic working was then started under the personal supervision of the best mining engineers of that period.

This discovery of a gold-mine is said to have produced a great sensation in Paris. All the literature of that time makes mention of the matter, and the arrival in 1783 of the first ingots of gold aroused great enthusiasm at the Court.

In 1786, the Comte de Provence caused a medal to be struck, the gold of which was the product of this mine. The matter was placed in the hands of *L'Académie des Inscriptions et Belles Lettres,* which chose Dupré to execute the designs. One side of the medal represented Louis XVI., and the other the Comte de Provence, in the act of offering to the king the first product of a French gold-mine.

The cost of this medal was put down at 8000 livres, and it has been sarcastically said that this represented all the results thus far obtained by the enterprise. It certainly marked the most brilliant period of the history of the mine, for soon after, in 1788, work was discontinued. It is reported that gold was found chiefly in the upper levels, but that the irregular distribution of the precious metal, together with other circumstances, caused the abandonment of active exploitation. Among the "other circumstances" it is safe to include the insufficient quantity of gold in the ore; for in those days, as now, there were men who knew how to avoid laying unnecessary stress upon this somewhat important feature of gold-mining.

After 1788, followed a blank period, during which the mine was the property of the commune. The extraction of specimens, both of gold and of rock-crystal, was actively carried on, and helped to beautify collections in many of the less distant towns. More particularly it was then that the basis was laid for the fine collection now to be seen at the museum in Grenoble.

Under the first Empire no work was done on the mine; but the mining engineer-in-chief at that time, Hericart de Thury, was instructed to write a report, which appeared in the *Journal des Mines,* in 1806. The times were troubled, and the complications of politics allowed the mine to remain in oblivion until 1837, when a fitful re-opening took place. In 1839, the property was acquired by M. Van de Velde for a Parisian company, which, under the management of M. Graff, caused the mine to be re-opened in 1841. Some very rich pockets of specimen-gold were found, but no serious prospecting was done, and no important results were obtained.

The mine then passed through the hands of several proprietors, under whose management far more gold was expended than was returned, and more than was returned was stolen.

In 1862, La Gardette came into the possession of two Englishmen, Messrs. Fisher and Watson; but, upon the departure of the latter for the more productive mines of Australia, the former became discouraged and ceased work. Since then nothing has been done to forward its development, save the writing of reports by the successive mining engineers-in-chief, to whom it seems always to have furnished the subject of a sort of graduation-essay.

Recently the re-opening of La Gardette was considered by the English company now working the mines at Allemont; and the writer was called upon to examine it. The results of that examination, as stated below, bring the history of La Gardette up to date.

The history of the mine, as outlined above, finds a ready explanation in the study of the conditions under which the gold occurs. An obtuse disregard of the facts as disclosed in the workings and a belief in wild theories of ore-deposition have been the causes of the repeated failures to make it profitable.

The gold of La Gardette occurs in a quartz-vein traversing the gneiss, which in this district has a great variety of hardness and composition.* The lode is of extraordinary regularity in strike, dip and parallelism of walls. It runs nearly due east and west,† and makes an angle of 75° with the horizon.

The gangue is almost entirely quartz, with a highly developed banded or ribbon-structure. Slickensides characterize the walls. The minerals accompanying the quartz are galena, iron and copper pyrites and (rarely) gray copper. It has been noted that the coarse galena ‡ is favorable to the presence of gold. The richest quartz

* The French engineers variously apply to it the terms gneiss, protogine and crystalline schists

† Or, as the French simply put it—"*vers sept à huit heures.*"

‡ "*À grandes facettes.*"

is that which has a bluish-black color,* an experience in harmony with that of other gold mining districts.

This is not the only known occurrence of gold in the district, though it is the only one which has risen to the dignity of a mine. M. de Thury, mining engineer-in-chief under the first empire, has left the record† that there are several gold-deposits in the department of the Isère :

1. At Portraut, at the foot of the glaciers of the Grandes Rousses, where it occurs in association with argentiferous lead ore

2. At Auris, on the southern slope of the same range, and nearly opposite La Gardette, in a complex lead-copper ore, associated with antimony.

3. At Mollard, near Allemont, in argentiferous galena; also in the copper pyrites of La Corchette and in that of Les Chalanches, Allemont.

In later times I have found it in noteworthy quantities in the quartz of the Mine des Arabes, at Allemont, where it is associated with galena, blende and pyrites, but occurs more particularly in thin seams of maroon-colored earth. It is found in the argentiferous galena of Grand Clos, near La Grave, as well as in certain quartz-veins above La Villette. Nearly all the ores of this district which I have had cause to assay have shown traces of gold.

Returning to La Gardette, we may consider briefly the geological conditions. The gneiss, traversed as already remarked by this lode, is overlain by the dolomite of the Trias and that in turn by the lime-shale of the Lias, both of which formations have been identified by their fossil remains. The Trias is but slightly represented, and here varies in thickness from 5 to 20 meters. The Lias is represented by lime-shale, but its upper horizon is diversified by the slate, the contorted bedding of which is a striking feature of the cliffs above Bourg d'Oisans. In it occur the slate-quarries of Allemont and Oz.

The vein does not penetrate the overlying secondary formations, but ceases at the upper eroded edges of the gneiss. Where the vein ceases, two stringers or small seams of quartz extend a little way up into the Trias. These stringers contain galena which is gold-bearing, also copper and iron pyrites; in fact, they have much the same mineralization as the main lode itself. They do not reach the Lias. The accompanying sketch (Fig. 1) illustrates these conditions.

---

* *"Quartz enfumé, d'un bleu noirâtre."*

† *Journal des Mines*, 1806.

The Trias is a metalliferous horizon containing a notable proportion of iron pyrites, usually in fine grains, arranged in small seams, which ramify through the mass of the formation.

The lode of La Gardette answers to the term, much misused, and often but little understood, "true fissure-vein." In this instance we have a "gold-quartz-vein," and it is necessary to recognize separately each one of the words which make up the name. In practice it is too little recognized that a vein may (as "true fissure-veins" are popularly supposed to do) continue indefinitely downward through the crust of the earth, but unless it carries gold-

Fig. I.

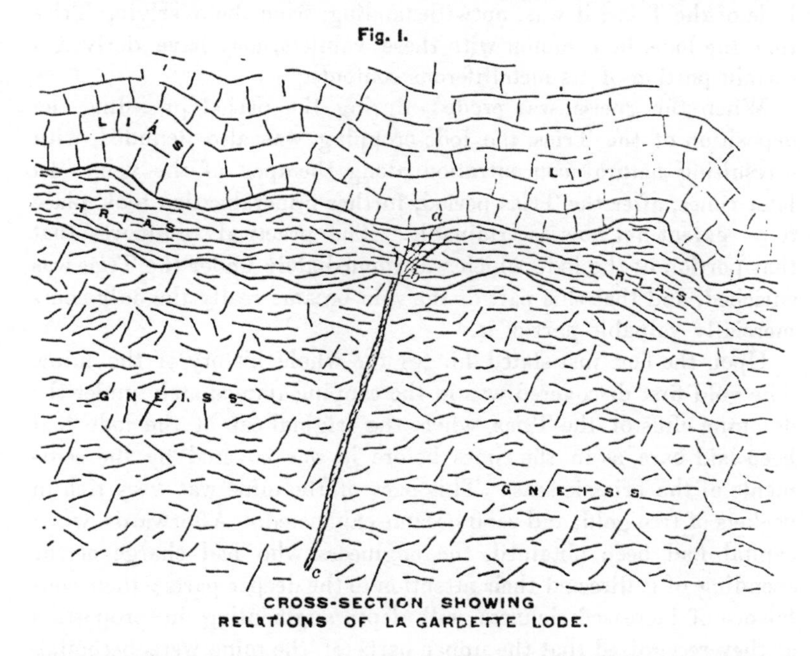

CROSS-SECTION SHOWING RELATIONS OF LA GARDETTE LODE.

quartz it is commercially of no value, however interesting scientifically it may be. Further, the gold-quartz may be continuous, but unless the precious metal is present in a certain proportion it ceases to be valuable to the miner. Here at La Gardette we have a very beautiful* specimen of this type of ore-deposit; and in attempting to arrive at its probable history it is necessary to distinguish between the vein and the lode, between the fissure and the mineral matter deposited in or along that fissure.

It is evident that the vein is anterior to the deposition of the

* By reason of its striking regularity of structure.

Trias, and consequently much older than the Lias. The two veinlets which start from the upper edge of the vein and penetrate the Trias owe their origin to movements which caused fractures to be formed in the Trias, and at this point are in line with the course of the main fissure, because the latter was a line of least resistance.

The fact that the mineralization of the veinlets in the Trias is identical with that of the upper part of the main lode in the gneiss points to a common origin. It appears to me, therefore, that while the formation of the fissure of La Gardette, as well as the deposition of the quartz of the lode, antedated the laying down of the beds of the Trias, it was, notwithstanding, from the overlying Trias that the lode, in common with these veinlets, may have derived a certain portion of its metalliferous contents.

When the gneiss was eroded during the period preceding the deposition of the Trias, the lode-cropping was also denuded, with a resulting natural concentration along the apex of the vein. In later times, after the Trias period, further mineralization took place to an extent not now ascertainable. The practical result was that that portion of the lode which was immediately under the Trias was enriched, and that this part of the vein became really the only commercially valuable part of it.

Upon the fact just stated hinges the whole history of the mine. The gold first discovered was in the capping immediately under the dolomite cliffs of the Trias, where the original cap of the lode had been laid bare, as in the times before it was covered by the sediments of the Triassic seas. This part of the mine was very rich in pockets of free gold, and soon became exhausted. Afterwards, when capital had been obtained, the engineers who had charge of the spending of it directed their attention to the deeper parts; their confidence of increased richness with depth augmenting in proportion as they recognized that the upper parts of the mine were becoming exhausted.*

Successive administrations essayed and failed to open up a productive mine at lower levels; while, during the intervals, the peasants scratched about in the upper workings and found occasional specimens.

To-day the history of the development of the mine can be read in the underground workings. Owing to the comparative smallness

* Many of the reports of the French engineers declare that increasing richness may be expected with increasing depth. This is a fallacy too frequently repeated in later times by men who have seen far more of gold-mining.

of the lode, its great regularity and the hardness of the enclosing rock, the workings are all readily accessible. The upper development consists of adits alone; the lower, of shafts connecting a series of levels. The total vertical depth explored exceeds 400 feet. It is seen that while all the quartz has been stoped away from the first level up to the sloping floor of the gneiss-dolomite contact above, and a winze follows that contact for a short distance under the level,* in the lower workings, on the other hand, hundreds of feet of drifts have been extended without the stoping of a fathom of ground. In the uppermost part of the lode the quartz is bluish and ribboned, containing galena and pyrites; but lower down, while it still carries occasional patches of pyrites, it is white and barren. The upper portions of the lode have a width of 20 to 25 centimeters, but in the lower levels this is increased to 30 and 40; above, it is compact; below, it is full of geodes lined with crystals. While the upper part has been a gold-mine, the lower part has been, and is still to-day, valuable† for rock-crystals alone. Assays made by me of the quartz standing in the lower level gave traces of gold only.

The vein is remarkable for the extreme development of two features common to many gold mines—slickensides and ribbon-structure.

The slickensides‡ extend continuously over large portions of the surface of the walls, for a length of 400 meters and a height of over 80. The *striae* are nearly horizontal, an evidence of a very great change of position, indicating a movement through a large angle. That this is not impossible, or indeed improbable, is shown by the evidence of similar movements to be found in the vicinity; at Allemont, for instance. This change of position was probably contemporaneous with the period of the contortion of the schists (or gneiss); and both were in all likelihood due to the uplifting action of the granite, a boss of which peeps from under the limestones of the Grandes Rousses.

The very beautiful lamination or ribbon-structure which characterizes the vein has received frequent mention from visitors. M. Charles Levy records that in one band alone, 9 centimeters thick,

* This has much to remind one of the "contact" and "verticals" in which the ore occurs at Rico, Colorado.

† M. Napoleon Albertas, who kindly guided me over the mine, exploits the lower workings for the rock-crystals found there, and has lately shipped groups to Paris which brought him 600 to 1000 francs per lot.

‡ Called "*miroirs*" by the French and "*harnische*" by the Saxons.

he counted 38 *laminæ* or ribbons. To some of those who have described the mine, it has suggested successive reopenings and fillings by crystallization. To others, it has indicated merely intermittent crystallization.

This structure of gold-quartz-veins, frequently observed in widely separated mining districts, is generally recognized by miners as a favorable indication. Allied to it is the observation that milk-white quartz is usually less likely to be gold-bearing than that which has a dark or bluish tinge.

The explanation of the ribbon-structure does not require the fanciful imaginings of the "successive reopening and refilling" theory. That the lode is usually an altered form of the country which the gold-vein traverses, is a fact now generally recognized. All underground observation leads one to look upon the dark, nearly parallel lines which cause the ribboning as the last remnants of included fragments of country-rock. In the gold-mines of California and Australia, one can frequently note the transition from the large, irregular pieces broken from the lode-walls to the small, more regularly arranged fragments which serve to subdivide the width of a lode. In certain cases one may follow the slow gradation from quartz filled with portions of "country" to dark quartz in which the traces of included country are entirely unrecognizable.* At La Gardette it is to be noted (the fact is recorded by M. Graff) that when the vein includes fragments of the inclosing country, such inclusions are always smaller than the width of the particular band of quartz in which they are arranged.

It is not possible to imagine that, under ordinary circumstances, a cavity could long remain empty under the weight of the overlying rocks. Any fracture or fissure formed must immediately tend to close up, and where it is a line of movement, a fault, as many veins are, the effect would be to choke it up with fragments torn from that wall which was of inferior hardness. Later movements along the same line would tend to shear off additional portions of the enclosing country. Further, it has been shown (by Becker at the Comstock, for instance), that the action of a fault is to cause a sheeting of the country by fractures in sympathy with the main line of fault. This structural effect might also come into play within a lode; for, as the mineral solutions percolating along the main fissure or

* In Australia this would be the transition from "mullocky" to "mottled quartz.

line of fracture dissolved portions of the included broken country and in turn deposited quartz or other lode-filling, they would also follow the other lines of fracture lying latent in the adjoining country. The result would be one of those cases where one or more "false walls" are found, and where it is only the proportion of gold present which determines where the lode may be practically considered to cease.*

In thus insensibly passing from a discussion of ribbon-structure to that of lode-walls I have but followed the transition which occurs underground. But to return: When the lode filling is less hard than the enclosing country, there is always a likelihood of its becoming the line of movements similar to those which brought it into existence. Later faulting would take place along the lines of included country rather than along that of the hard quartz, and would result in the crushing of those portions of included country and their arrangement along parallel lines. Occasionally the included country is harder than the portions of the vein more or less incompletely filled with quartz, and then we get the crushed sugar-like quartz which characterizes several well-known lodes.

Experience shows that ribbon-structure is a desirable feature of a gold-vein. In the mines of California and Australia these black lines of lamination greatly facilitate the actual breaking of the quartz; but besides this they are generally found to characterize gold-bearing rock. This fact, common to districts otherwise very dissimilar, is something more than a coincidence. These black lines are often graphitic,† the evidence left by metamorphism of the organic remains deposited in the original sedimentary rocks. Such carbonaceous matter would act as a reducing agent upon mineral solutions, leading to the precipitation, for instance, of gold from a chloride solution. In some of the mines of Victoria, the thin, black seams of slate dividing the quartz are found to be covered with a mosaic of fine gold. In California similar instances are not unknown.

In this connection, one is reminded of the fact that dark and bluish quartz are found by experience more likely to be gold-bearing than white quartz. There is a distinction to be made, however, be-

---

* It is a bad tendency which some miners have, to seek for "walls," and then, having found one, to desist from any further cross-cutting through the lode.

† The miners of Amador county, California, when they come up from underground, look like coal miners, on account of the black slate having discolored their clothes.

tween the dark and the bluish-tinted varieties; for while the former appearance is probably due to the presence of minute fragments of black slate or other country-rock, now almost entirely replaced by quartz, the latter is due to the presence of sulphate of iron, which may be either original or a secondary result of the decomposition of iron pyrites. Such quartz becomes rusty by oxidation within the zone of surface-decomposition. In this case, ferrous sulphate may play the part of the graphitic matter of the black slate.

In leaving the subject it is well to add that, in mining, generalizations are always dangerous if followed too far. An instance is suggested by the above paragraph. White or dull whitish quartz is invariably considered by the California miner an evidence of the poverty of the vein; and yet I have seen at Bendigo quartz of this character, at a depth of over 2000 feet, which carried gold freely, not in an isolated pocket, but as characterizing a lode-formation over a great distance.

NOTE BY THE SECRETARY.—Comments or criticisms upon all papers, whether private corrections of typographical or other errors, or communications for publication as "Discussion," or independent papers on the same or a related subject, are earnestly invited.

Subject to Revision.

[TRANSACTIONS OF THE AMERICAN INSTITUTE OF MINING ENGINEERS.]

THE MINES OF THE CHALANCHES, FRANCE.

BY T. A. RICKARD, DENVER, COLORADO.

(Bridgeport Meeting, October, 1894.)

IN southeastern France, among the magnificent alpine masses of the Dauphiné, there is a group of celebrated mines of silver-, nickel- and cobalt-ores, the deposits of which present many features of interest. In 1891 I directed the work done at the Chalanches, and at that time made the notes upon which the present paper is based.

The workings consist of a complex series of adit-levels entering the heart of a mountain, the summit of which is one of the lower peaks of the Belledonne, in the commune of Allemont, and overlooking the valley of the Romanche and its tributary, the Eau d'Olle. The lower adits are 3700 feet above the village of Allemont, or 7250 feet above sea-level.

HISTORICAL.

The discovery of these, as of many other notable mines, was accidental. In 1767, Marie Payen, a shepherdess (bergère) of Allemont, found an outcrop of silver-ore, and brought away, in ignorant curiosity, a lump of heavy stone, which she handed to the village smith. When tested on his forge, the molten silver trickled from it. The shepherdess received 600 francs upon her wedding-day as a reward for the discovery.* Thereupon some of the peasants of the commune began to dig and to smelt the silver-chloride ores, the croppings of which they traced down the mountain-side. The exploitation thus inaugurated continued with but little interruption for more than a century.

The first mining done was digging.† A loss of life through a crush in the excavations caused the authorities at Grenoble to take

* There is an entry on the pay-sheet for September, 1768, made in the handwriting of M. Schreiber. "Paid to *bergère* Marie Payen, at her marriage with Jean Roux, master charcoal-burner at the *fonderie* d'Allemont, the sum of 600 livres as recompense for discovery of the mine."

† "Gophering" or "coyoting."

official cognizance of these operations; and M. Lemonnier, a member of the Academy of Sciences, was sent by order of the king to investigate the discoveries of silver reported to have been made by the peasants.

In 1769, systematic work was commenced under the direction of a Piedmontese engineer, M. Binelli.*

On the 10th of June, 1776, the king granted a concession of these mines, as well as those of Allemont and La Gardette,† to his brother, the Comte de Provence, afterwards Louis XVIII. Smelting-furnaces were erected at the base of the mountain near Allemont.

In 1781, Binelli gave place to Schreiber, who assumed the direction of the smelting establishment and of the various mines. Schreiber's name is interwoven with the history of most of the mines in this part of Europe. He was a Saxon engineer of much ability, and was the father of the *École des Mines*, which was first established at Moutiers in Savoy.

On the 2d of August, 1792, as a consequence of the Revolution, the mines became national, and passed into the hands of the new government.

The years intervening between 1776 and 1791 cover the most prosperous period in the history of the mines. In 1791, the amount of development-work was decreased, and the profits dwindled away under bad administration. The silver produced was sent away, but the amount of money necessary to pay for the work was not forwarded to the mines. Accounts were liquidated with *billets* or promissory notes instead of cash. Matters went slowly from bad to worse until 1807, when the State abandoned operations and made a concession of the mines to a public company.

From the first discovery up to 1801, a period of thirty-two years, the mines produced 9453 kilos, or 303,914 Troy ounces of silver. According to Schreiber, the receipts were 2,098,421 francs, or 65,577 francs per annum. The expenses, 1,890,096 francs, equal to 59,090 francs per annum. During this period there were times when the lodes became very much impoverished, and the profits were further diminished by the erection of reduction-works, buildings and other outlays. There remained, nevertheless, a net balance of 207,585

---

* For many of these data I am indebted to an *Extrait des Mémoires de la Société des Ingénieurs Civils* by M. Alfred Caillaux.

† See "La Gardette: The History of a French Gold-Mine," by the writer. *Trans.*, vol. xxi., p. 79.

francs. The best years were 1784 and 1785, when the profits amounted to 55,000 and 54,000 francs respectively.

In 1808, Schreiber became director of the *École des Mines* at Moutiers,* and ceased to manage the Chalanches. In 1809, a new company obtained control of the mines through an imperial decree. It was short-lived, and was succeeded by a series of other companies, which did more or less work in an unsystematic manner up to 1873. In 1889, application was made by Pierre Manin for a concession of the then abandoned mines. A company, *La Société Savoysienne*,† was formed to operate mines in Savoy and in the Dauphiné, including those of the Chalanches. This ambitious enterprise exhausted its energy in making large promises, and, after a brief existence, was sold out for the benefit of its creditors. The Countess de Grailly became proprietress of the Chalanches, and of Grand Clos, the mines of which have always been worked in conjunction with the former.

In 1890 the mines passed into the hands of an English Company, the "French Mines, Ltd.," and vigorous work was commenced at the Chalanches, as well as at Les Arabes, Villaret and Grand Clos. Owing to the fact that the other mines of this extensive group offered better inducements for the investment of capital in their development, the Chalanches was operated to a limited extent only. The exploratory work led to the discovery of several rich pockets of silver-ore and small patches of nickel- and cobalt-ores. On the whole, however, the results were not such as to encourage further work, and in September, 1891, operations ceased. These interesting old mines are therefore again abandoned.

Resumé of the History of the Mines.—The record of the Chalanches presents a story similar to that which is told of mines in more modern mining districts. The inaccessibility of the mines in winter, the richness of the ore, its great fusibility and the consequent systematic robbery of the silver are local commonplaces. Circumstances all worked together to make the Chalanches mines the prey of the most barefaced plunder. With the aid of a common forge-fire, even without the intervention of a crucible, and with little knowl-

---

* Sometimes spoken of as the "*École des Mines de Pesey.*" The mines of Pesey and Macot, in the hills of old Savoy, were the cause of the choice of Moutiers, the nearest town, as the locality for the first school of mines.

† That Gallic imagination by no means falls short of the Anglo-Saxon faculty when it is applied to prospectus-making, is proved by the flamboyant style in which the promoters of the *Société Savoysienne* addressed the public.

edge or skill, lumps of silver could be produced from the very rich chlorides, ruby silver and black sulphides which constituted in the main the soft earthy ores or *terres* found in the crevices of the outcrop. Aged inhabitants still talk sportively of the theft like old smugglers, and point out nooks in the woods which the remaining ruins of the little furnaces dug out by the miners, show to have been the scenes of former illicit silver-ore smelting. In these furnaces, no larger than an ordinary fire-place, dug in the earth and smeared with clay, with charcoal, or, failing that, clods of dung for fuel, and two or three little urchins to blow, like cherubs on the old maps, out trickled the white metal. Clergy and people joined cheerfully in these moonlighting operations without in any degree shocking local ethics. The priest at Allemont, who lately restored the parish church, says that the old church had a room adjoining the sacristry, in which a former reverend father used to melt down the silver-ore brought to him by the faithful. The slags were concealed in an excavation under the floor, where a large accumulation of them was found when the church was restored.

During the earliest period of mining at the Chalanches, some bodies of extremely rich ore were found near the surface. It is said* that two shots produced sufficient silver to pay for the two buildings known as the pavilions at Allemont, with their various ornamentations, including the *fleurs-de-lis* which still adorn the roof. As 200 to 300 kilos of silver would at that time be worth from $10,000 to $15,000, this statement does not seem incredible.†

A "pockety" mine is notoriously apt to be loosely and extravagantly managed. The uncertainty of the work is prejudicial to the maintenance of system. The various companies that operated these mines from 1808 to 1873, did so at a loss, due largely to inexperienced engineering and loose financial management.

M. Geymard, an engineer of repute, says of this interval of sixty-five years:

"Explored, exploited, abandoned and resumed by new companies, the mine paid dividends or levied assessments according to the ability of the men sent to take charge

* For this information, and many notes, I am indebted to my father, Mr. Thomas Rickard.

† An idea of the value of the ore, outside the patches of extremely rich matetial, may be gathered from a statement made by M. Schreiber, in a report sent to the *Académie* and afterwards reproduced in the *Annales des Mines*, that the average richness of the Chalanches ore treated in the furnaces at Allemont, up to that time, was 750 grammes per 100 kilos, equivalent to about 219 Troy ounces per ton of 2000 pounds avoirdupois.

of the works and according to the amount of the capital placed at their disposal. To sum up, the mining was restricted to the workings opened up by Schreiber; for all his successors confined themselves to a few meters of development and exploration among the old drifts."

M. Caillaux, summarizing the history of this period, adds:

"The mines of the Chalanches were never worked on a proper scale during the present century and, as has been well expressed by M. Gruner, the want of success marking the various attempts made during this lapse of time does not in any way prove the sterility of the ore-deposits, or the impossibility of its being operated to advantage."

The latter part of this statement is open to discussion.

It is not a little remarkable that although the silver is always associated in the lodes with rich nickel- and cobalt-ores, often with bunches of stibnite, and more rarely and erratically with gold, the government engineers took no note of any metal other than silver. None of the valuable metals mentioned figure in the old accounts. The speiss containing nickel and cobalt was rejected with the slags, and went to fill the swamps and to form the road-beds which, in later times, were furrowed and turned over to recover their valuable contents.*

The possibility of utilizing three metals instead of one seems to

---

* It has not been found possible to ascertain the value of the old mattes with which roads were made and marshes filled, but the following notes will assist: Herr O. F. Köttig, writing from Oberschlema, August 7, 1889, with regard to the speiss, picked up in the old smelter-dump and shipped by Pierre Manin, says that it contained 3.6 per cent. metallic cobalt, 8.9 per cent. metallic nickel, and 0.2 per cent. silver. Another lot gave 3 Co, 11.8 Ni, and 0.3 Ag. This speiss was neglected during Schreiber's time.

In August, 1863, the Viscount de Talon sent 7 barrels of ore to Vivian & Sons, Swansea. The ore in these barrels ranged in value (net) from £3 to £31 per ton, not including the silver. The average was about £16. The nickel was valued at 1*s*. 9*d*. per pound, the cobalt at 7*s*. per pound. The smelting charge was £1.10.0 per ton. The contents of the barrels were:

No.	Cwt.	Qr.	Lbs.	Nickel. Per cent.	Cobalt. Per cent.	Silver. Oz. per ton.
1, ...	16	0	9	.6	.07	11
2, ...	2	0	6	2.6	1.30	47
2, ...	3	0	6	1.1	0.60	19
3, ...	6	0	10	3.2	2.00	81
4, ...	17	3	26	3.0	1.60	201
5, ...	1	0	8	2.3	1.6	86
5, ...	3	0	7	3.8	1.9	192
6, ...	23	3	9	3.2	1.9	27
7, ...	1	1	8	4.3	3.3	39

have dawned upon the later engineers quite as a discovery; and this fact stimulated the repeated spasmodic attempts to rehabilitate the old mine. The arsenides of nickel and cobalt were sold in England and in Germany.* More recently, a German chemist was employed at Allemont in an experiment to manufacture cobalt pigments for the arts. He was not successful, and the attempt was abandoned.

In 1891 the gold-value was first recognized.† Its importance proved greater from a scientific than from a commercial point of view. The old mine-workings, aggregating 20 kilometers in length, showed that a great deal of unsuccessful exploration had been carried out. Search among these galleries, particularly near the surface, resulted in the finding of certain rich bunches of ore, which were soon exhausted. An attempt to introduce the tribute- or lease-system was made, with partial success. The necessity for concentrating the operations of the company led to more active work at Grand Clos and Les Arabes, and, at the same time, made it advisable to abandon the Chalanches.

The Ore-Deposits.

The geological formation is simple. A network of veins traverses crystalline schists of very variable character. The country forms a part of the great crystalline formation usually referred to as the Archaic schists of the Alps, though in point of fact they probably include rocks from the granite up to the Carboniferous. Lithologically, certain sections suggest the Huronian and Laurentian. These schists lie immediately upon the granite; they are extremely variable in character, so that at different places they can be described as

---

* The Chalanches mines were always worked in conjunction with those of Grand Clos, near La Grave. The lead-ores of the latter were brought about 20 miles to Allemont, to be smelted with the products of the Chalanches. The smelting was simple. The lead-sulphides and the arsenides of nickel and cobalt were submitted to stall-roasting, and then passed through a low blast-furnace (*four à manche*), giving as products: (*a*) work-lead, containing the silver, and, of course, the gold, if present; (*b*) speiss, containing the nickel and cobalt; (*c*) slags. The work-lead went in succession to *Pattinsonage* and cupellation for the extraction of the silver. The speiss and slag went over the dump. It was only in later years that the nickel and cobalt were taken account of.

† Small samples of the earthy ore gave occasionally as much as 1 to 1½ ounces per ton. At Les Arabes, a mine just above the village of Allemont, there occurred narrow streaks of maroon-colored earthy stuff, which assayed from 2 to as high as 7 ounces of gold per ton.

gneissose, granitoid, talcose, micaceous, graphitic, or amphibolic.* At the base of the slope leading to the mines there are superb blocks of rock, containing crystalline epidote.

Chalanches, like the mines of the Alps generally, in France, in Savoy, or in Switzerland, is far up toward the summit of the complicated schist-region. Looking across the valley of the lovely Romanche, one can see La Gardette, the historic gold-mine of France, perched upon the cliffs overlooking Bourg d'Oisans. Across the winding Eau d'Olle rises the imposing mass of mountains known as "les Grandes Rousses."† One can distinguish their structure, which more immediate examination proves to be that of crystalline schist overlain patch-wise by the dolomite of the Trias, in turn succeeded by the shales and slates of the Lias.

Immediately above the Chalanches mines and in rocks of Carboniferous age there is a deposit of anthracite. To the anthracite succeeds, in fragmentary deposits, the dolomite limestone, which, though less constant than the Carboniferous sandstone and shale, almost always accompanies the anthracite basins, and in the Hautes Alpes‡ is associated with what appears to be Permian sandstone and the *grès bigarré*.§ The latter fact gives rise to the suggestion that the dolomite belongs rather to the Permian than to the Trias.

The country which more immediately holds the lode-channel of the Chalanches mines, and which I may term the encasing rock, has the character of gneiss. In actual contact with the veins it is amphibolic and contains a notable amount of pyrite. This last characteristic causes it to emit sparks when struck with steel and indicates the origin of its local name, *la roche martiale*.

The oxidation of the pyrite is the reason of the red bands which seam the steep bluff crowning the upper precipitous summit of the mountain. These bands of pyritic schist dip into the hill just as the main series of veins does, a fact which led M. Graff, a distin-

---

* On the government map the whole mass of the Chalanches, as also the Corneillion, on the opposite side of the river, is marked "amphibolites." Above and beyond Chalanches, there is shown a peak of "euphotide amphibolique," flanked by serpentine. There is no doubt that these rocks, on either side of the Romanche, are in places highly amphibolic, but it is an error to describe the entire mass of the mountain as such.

† *I.e.*, the great "roughs," ruffians, or rugged ones.

‡ At Argentière and Valgaudemar.

§ The *grès bigarré* is the Bunter sandstone of the Vosges, where it rests comformably on the red Permian sandstone. (See Geikie's *Textbook of Geology*, 3d ed., p. 870.)

guished engineer, to conclude that they played the part attributed to the fahlbands of Kongsberg, in Norway.

The maps of the mine exhibit a wonderful network of galleries, spreading like a cobweb over an area of about 600 by 300 meters.

It is computed that the workings aggregate in length not less than 12 miles, an extent in remarkable contrast to the relatively small quantity of ore produced.

The principal veins are six in number, four of which, the Cobalt, Prince, Simeon and Hercule, lying about 15 meters apart, dip rather flatly into the mountain, in conformity with the bedding of the enclosing rock-formation, while the two others, the Directoire, a nearly

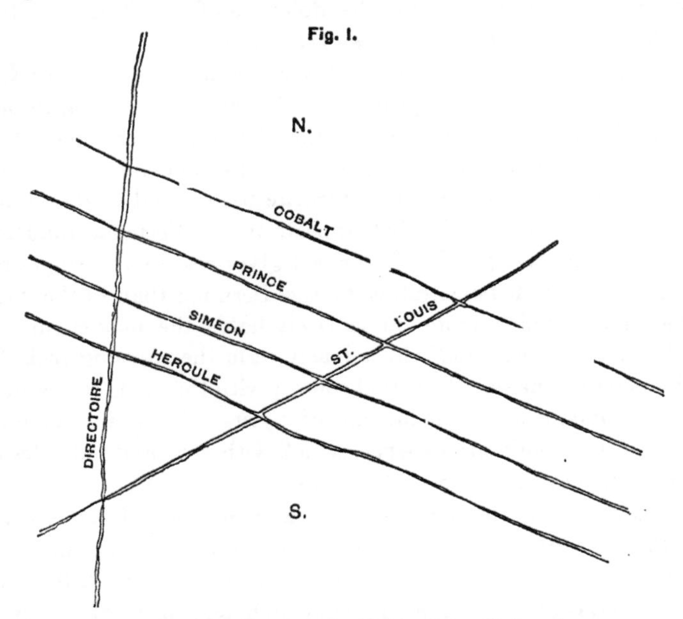

Cross-Section, Showing the Veins of the Chalanches.

vertical vein, and the St. Louis, which cuts it diagonally, may be considered as cross-veins or counter-lodes to the main-series. The accompanying cross-section (Fig. 1) will explain these relations.

In addition to the six lodes mentioned there are several minor veins which may be considered as branches or feeders of the main series, the whole forming a complicated web of ore-bearing fissures. Schreiber is quoted as having said that "the veins have as many directions as there are points of the compass." Gruner speaks of Chalanches as "a mountain radiated and fractured in every direction, the largest fissures being filled with fragments of the encasing rock and the smaller fissures with metalliferous ores of very varied character." Geymard considered the lode-structure to be a *stockwerk;* but Graff distinguished two principal groups of veins, namely (1) those which have a north-south strike and dipped either east or west, apparently parallel to the lamination of the encasing schist, and (2) those which strike east-west and dip uniformly to the north. According to the same authority, the veins north-south were the most regular and continuous. Both series, however, were stated to throw off numerous branches and thus gave rise to an apparently inextricable confusion.

Of the engineers quoted, Graff was the most trustworthy observer. His description of the veins is good. Others, even Schreiber, seem to have entirely failed in mastering the difficulties of the deposit. In the more fissured ground near the surface the lode-formation may bear the description of a *stockwerk*, or, more properly, little separated patches of *stockwerk;* but in depth this character is not apparent. Observation at every point at which the numerous veins occur along the southeastern flanks of the Belledonne *massif* goes to prove that by far the greater number of the veins which have been mined conform to the bedding and, therefore, dip into the hill, as the Chalanches lodes do. They follow the flexures and folds of the schist, and naturally present all kinds of variations of direction when exposed in the mine workings. This is also observable at Les Arabes, Villaret and Oulles. The cross-lodes appear to be relatively unproductive, the St. Louis and Directoire being exceptions. It may be said, speaking generally, that these bedded veins in the Dauphiné* are weak, inconstant and difficult to follow.† They are frequently altogether obliterated, leaving to the miner the bedding of the country as his only guide. Hence the apparent extravagance of drifts.

Towards the center of the mine workings of the Chalanches there are three dikes of diabase, respectively 23, 3 and 30 meters in thickness. Furthermore, all those who have at various periods directed the mines take note of large barren fissures which traverse the mountain and dislocate the ore-bearing veins. They are filled with fragments of country and with clay, both sandy and micaceous. Schreiber considered them as barren lodes and termed them "*filons*

---

* The more modern departments of the Isère and Hautes Alpes.

† These bedded veins in the schists remind one often of the lodes in the quartzose schists of Otago. (See *Trans.*, xxi., 411 *et seq.*)

sauvages privés de substances métalliques." A Cornishman would call them "flookan."

It has been thought by several observers that the lodes were more numerous near the surface than in the interior of the mine. This is due to the fact that any single fissure, in approaching the surface, spreads itself out into a number of subordinate fractures. It has also appeared that the lodes gained in regularity as they penetrated the mountain. Caillaux therefore adds that this fact seems to indicate the probable occurrence in depth of only a small number of lodes, but that those surviving will have a regularity greater than those which have been hitherto exploited. Regularity of structure would be a poor compensation to the miner for the fact that the enclosing rock is much harder, and the thickness of ore smaller, than in the ground nearer to daylight.

The veins vary in width from a knife-blade to 80 centimeters (31.5 inches); their usual thickness lies between 3 and 30 centimeters (0.1 to 1 foot). The following data from my note-book will indicate, in a general way, the size and nature of the veins:

July 14, 1891.

Première Hercule.—12 cm. in two parts: the upper, quartz and fine-grained galena; the lower, efflorescence of nickel (annabergite) in earthy ore.

Troisième Hercule.—30 cm. Antimonial ore (stibnite and oxide) accompanied by cobalt-bloom (erythrine).

Cinquième Hercule.—Vein crossed by cross-course. Broken; barren.

St. Nicholas.—35 cm. Brown earthy ore, containing arsenates of nickel and cobalt, with native silver.

Galerie d' Argent.—5 to 7 cm. Wire silver, with black sulphides of silver, in an earthy gangue.

August 11, 1891.

Première Hercule.—10 cm. Chiefly calcite, a few spots of galena.

Troisième Hercule.—35 cm. Calcite, with threads of galena.

Prince.—20 cm. Red and black earthy ore (found afterwards to contain 35 ozs. Ag) stained with copper.

Cinquième Hercule.—7 to 10 cm. Calcspar, with splashes of stibnite.

St. Nicholas.—12 cm. Black earthy ore, with stones of fahlerz and kupfernickel.

Galerie d'Argent.—5 cm. Black earthy ore, full of native silver.

General Conclusions Regarding the Ore-Occurrence.—Some of the conclusions of several accomplished French engineers have been quoted. They differ according to the condition of the mine at various periods, depending upon whether the development-work was being vigorously pushed ahead, new ground opened up and rich ore extracted, or whether the work of exploration was restricted and the operations confined to the search for the pockets of silver which were found from time to time irregularly distributed amid the complex of veins. At the time when I directed the work, attention was mainly confined to the exploration of blocks of ground not previously intersected by drifts or cross-cuts. No work was done in the ends of galleries farthest advanced into the mountain. My experience of the mine and its ore-deposits led me to the following conclusions:

The formation of crystalline schists has been subjected to fissuring at more than one epoch, an earlier one being marked by the fractures filled with ore and now forming the lodes, and a later one characterized by the formation of cross-fractures which broke across the previously formed lodes, and are themselves non-metal-bearing.

Near the surface the lodes are soft, especially where they are richest in silver. Though they agree in the main as to their mineral impregnation, the daily advance of drifts will exhibit an extraordinary variety of vein-filling. They all contain nickel, cobalt, silver, and in places also antimony and gold, as their commercially valuable elements. With these are associated the numerous mineral species for which the mines are famous, notably acicular crystals of epidote. The minerals of the Chalanches are to be found in most of the important collections of Europe.*

Examination of the old workings proves clearly that with increasing distance from the surface the country gets harder, the veinstuff loses its soft character, the veins become fewer in number, more regular, less wide and less ore-bearing. Approaching the surface, on the contrary, the schists are fractured in a multiplicity of directions, the veins become larger, their filling is generally earthy, and they throw off branches, at the intersections of which ore-bodies are found. In general, mineralization becomes more pronounced with approach to daylight; this being due, not merely to the oxidation of the sulphides, but to an actual relative increase of "orey" matter.

The outer portion of the mountain is jointed and otherwise frac-

* Its contribution to the Museum of Natural History at Grenoble would itself repay a voyage across the Atlantic. All the mines of Leadville, for instance, would hardly furnish a richer collection of minerals than the Chalanches alone.

tured to a very remarkable degree. The seams of ore which follow such joints and fractures are often composed of earths rich in nickel and cobalt, or of wire-silver mud. An illustration of such an occurrence is afforded by the pocket of native silver found within a few

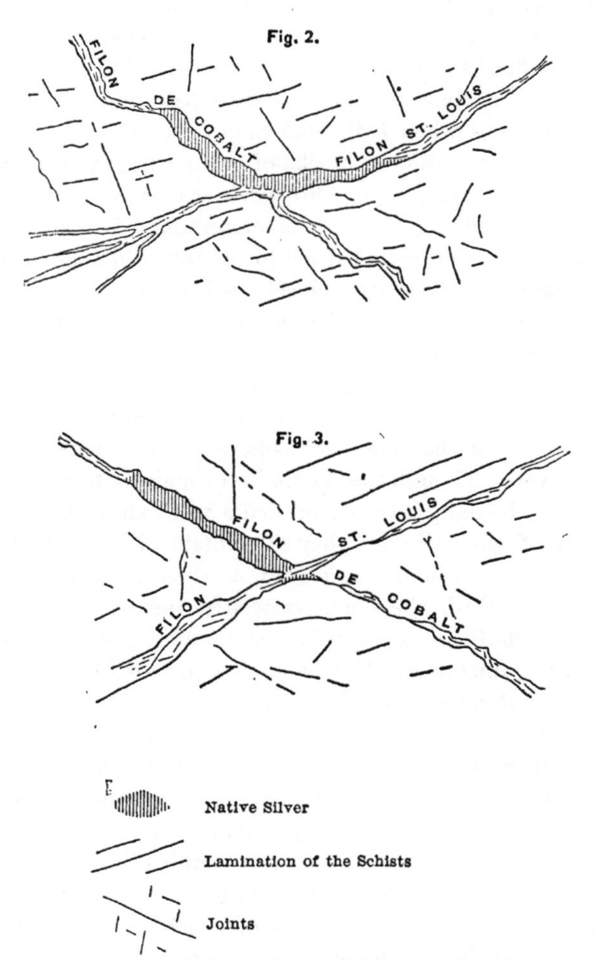

Sections Showing Pocket of Native Silver.

feet of the surface, at the intersection of the St. Louis and Cobalt veins, in July, 1891. The accompanying sketches, Figs. 2 and 3, show the change in the lode-structure as the working progressed, the time between the two drawings being four days. In Fig. 2 the native silver, enclosed in a black mud, and accompanied by ochreous earth, extends from one vein, across the point of intersection, into the other. In Fig. 3 the silver is confined to the upper part of the Cobalt vein. Red iron-earth formed the filling of the St. Louis. The structure of the enclosing country is merely suggested in the drawings.

The observations made from day to day led me to conclude that the richest part of the mine was that which was within the influence of oxidation, and that both chemical agencies and structural conditions favored an enrichment of ore near the surface. This statement is particularly applicable to the silver contents. It also holds true of the gold, but it is less accurate with respect to the nickel and cobalt. The richness in silver of the oxidized ores suggests secondary precipitation. This is confirmed by the fact that the silver appears to be thrown down upon the nickel and cobalt arsenides, and often envelopes them in such a way as to impart to them the rudiments of a nodular structure. The hard, undecomposed arsenides contain only small amounts of silver. The gold, only occasionally present, is associated invariably with soft, maroon-colored, earthy, iron-bearing vein-stuff. The nickel and cobalt minerals appear to be primary ores, and are more persistent than those of silver and gold.

The dependence of the occurrence of large amounts of rich ore upon the broken and fractured character of the country would seem to me to indicate that that particular concentration of metallic minerals which renders the deposit economically important is of comparatively recent geological date; for, only near the surface (the surface of any given time, not necessarily only that of to-day) were there the conditions favoring such a concentration. A study of the vein-structure of the surrounding region shows that the ore-bearing veins are younger than the Jurassic age. This, of course, applies only to the fissures in which the ore is now found. The actual deposition of ore could not have commenced before the fracture took place, but it has probably been going on ever since. At the Chalanches, the change in the nature of the ore-deposits, both in structure and in mineral contents, is measured, not from any imaginary nearly horizontal surface of a former unknown epoch, but from the steep slope of the hillside of to-day.

The occurrence of deposits of nickel-ores in close association with basic eruptives, and more particularly magnesian rocks, has been frequently noted. The country-rock of the Chalanches lodes consists of the crystalline magnesian schistose rocks, which have been already mentioned. They are overlain by dolomitic limestone, and are intruded upon by a mass of altered gabbro or euphotide, which is, in turn, flanked by serpentine. The serpentine may have been derived by the metamorphism of amphibolic schists, euphotide, or limestone.* That the origin of nickel-deposits is traceable to the leaching of basic eruptives; that the metal and its ores occur in a finely disseminated condition in such rocks, and have been by them brought within the reach of circulating waters, is, at present, a strongly favored theory.†

As bearing on this part of the subject, the following additional facts are pertinent. At the Chalanches, in addition to the magnesian silicate schists, forming the encasing rock of the lodes, there are, at least, three dikes of diabase. Above Bourg d'Oisans—six miles distant—there is an amygdaloidal melaphyre (the *spilite* of the French geologists), which carries nodules of calcite accompanied by sulphide of nickel.‡ Between Allemont and Vizille there are, according to the government map, several outcrops of *spilite*. They are usually associated with the upper Trias, and occasionally appear to belong to the Jurassic. Certain of the ore-bearing veins of the district penetrate from the crystalline schists into the anthracite beds, and even into the Lias, the last being the youngest formation of the lode-mining portion of the region.

The *roche martiale*, or pyrites-bearing bands of schist, which immediately contain the most productive veins, illustrate that association of nickel and iron pyrites which has been often remarked by geologists.

The origin of the metals which enrich the veins of the Chalanches is a matter which, owing to the limited data bearing upon it, cannot be discussed at great length with profit. The silver and gold may be supposed to have been derived, as elsewhere, from ascending solutions which, in approaching the surface deposited their precious contents according as the structural conditions of the rock or the chemical composition of the casing of the fissures may have regulated that deposition. The nickel and cobalt will be considered,

* That serpentine can be derived by metamorphism from magnesian silicate rocks, or from limestone, has often been pointed out—quite recently by Mr. S. F. Emmons, in "Geological Distribution of the Useful Metals," *Trans.*, xxii., 71.

† It is advocated, for instance, by Mr. P. Argall, in a contribution to the Colorado Scientific Society, entitled, "Nickel: the Occurrence, Geological Distribution, and Genesis of its Ore-Deposits."

‡ A similar occurrence is that of millerite, or sulphide of nickel, recently noted by me in certain hornblende-schists at the Gipsey Queen Mine, 3½ miles east of Salida, Colo.

by many, to have had a more definite and immediate origin in the magnesian silicates of the diabase and schists, out of which they will be supposed to have been leached. These explanations of the origin of the four metals mentioned would, in that case, be a compromise between the contending views of the two sides in the controversy between the extremes of lateral secretion and ascension.

It will be claimed, however, that, if the nickel and cobalt were obtained from out of the encasing (the wall-rock), or the enclosing (the remoter country) rock, then, there should be some difference in the schists penetrated by those parts of the lodes carrying a notable amount of nickel and cobalt, as compared to the schists in which the lodes are barren. It cannot be said that this is the case. The bands of pyritic schist are parallel to the lodes, and enclose both their rich and their barren portions. That the lode-channel is marked by the presence of schist rich in pyrite is true; but the fact points not to the pyrite as the source of the metal, but simply to a probable identity of source, and contemporaneity of deposition. Within the reach of oxidizing agencies, both the nickel and cobalt arsenides and the iron pyrite are found decomposed in some places, unaltered in others. The interior workings of the mine show veins carrying hard unoxidized ores encased in harder schists, the pyrite of which is unaltered. While the rock which has been most affected by oxidation contains veins richer in silver and gold than that which does not show the action of such agencies, there is no noteworthy difference in the nickel- and cobalt-contents.

In these mines, as in others in widely separated regions, I have observed that it often happens that a very narrow but very rich streak of ore may occur encased in hard undecomposed country, while, on the contrary, a large width of poor veinstuff may be enclosed by highly altered and mineralized rock. This, which is, I believe, a common observation to those who spend much time underground, is a fact forever opposed to the narrower* views of any lateral-secretion theory. In such cases it is evident that the encasing rock has been mineralized and enriched through the agency of solutions which travelled in the lode-channel; that the mineralization took place from the lodes to the country and not from the outside country toward the interior of the lode-channel.†

If we accept the current theory that the nickel and cobalt came

* As distinguished from the wider interpretation given to that theory, not by Prof. Sandberger, but by Mr. Emmons and others in this country.

† This view was advocated by Mr. Pearce thirty years ago. (See *Trans.* xxii., 740.)

from the leaching of magnesian silicates (and facts are numerous pointing that way), then, we must conclude that the origin of the nickel and cobalt of the Chalanches was not the immediately enclosing country, but rocks similar to it, which underlie it at a greater depth. The silver and gold, it may be suggested, were precipitated from other solutions, and at a period other than that which saw the deposition of the nickel and cobalt. The precious metals were probably derived from a deeper-seated source; and may have been leached from the granite which underlies the schists and is penetrated by the basic eruptives. In both cases, the various metals must have come from a depth where leaching action was powerful, and from which ascending currents brought the metallic constituents, the subsequent precipitation of which produced valuable ore-deposits.

NOTE BY THE SECRETARY.—Comments or criticisms upon all papers, whether private corrections of typographical or other errors or communications for publication as "Discussion," or independent papers on the same or a related subject, are earnestly invited.

The Enterprise Mine, Rico, Colorado.

BY T. A. RICKARD, STATE GEOLOGIST, DENVER, COLO.

(Colorado Meeting, September, 1896.)

I.—HISTORICAL.

RICO, in the southwestern corner of Colorado, is one of the productive mining centers of the San Juan region, so-called because its waters drain into the river of that name, which is tributary to the Colorado. The San Juan region includes the counties of Ouray, Hinsdale, San Miguel, Dolores, San Juan and Montezuma. It is traversed by a network of picturesque mountain ranges on whose lofty summits there rests perpetual snow. The region is peculiarly rugged, and, in the early days of its development, tested to the full the hardihood of the adventurers who first explored its cañons in search for gold and silver.

A prospecting party, guided by Jim Baker, scout and trapper, penetrated in 1861 this part of the territory of Kansas. At that time the country was in the possession of the Ute Indians. In October, 1873, by the Brunot treaty, they ceded to the United States Government the richest mineral-bearing portion of their domain. But in the interval much prospecting had already been done, in defiance of difficulties among which snowslides and redskins were the most noteworthy. The mountains bordering the Animas and its tributaries were first explored by the pioneers, but the gathering wave of immigration soon swept further westward, and in 1864 a guide named Robert Darling brought a party of United States army officers and Mexicans from Santa Fé to the croppings of certain lodes which he had found on the Dolores* river. This party erected

* If the Spaniard devastated the countries he conquered, he at least left a poetic nomenclature in his wake. The river Animas was called Rio de las Animas perdidas—"the river of lost souls," and the gloomy magnificence of its tumultuous way renders the name appropriate. Dolores, Durango, San Miguel, San Juan, Ignacio, Dulce, Juniata, etc., compare well with Cripple Creek, Leadville, Central City, Corkscrew, Coke Ovens, etc.

FIG. A.

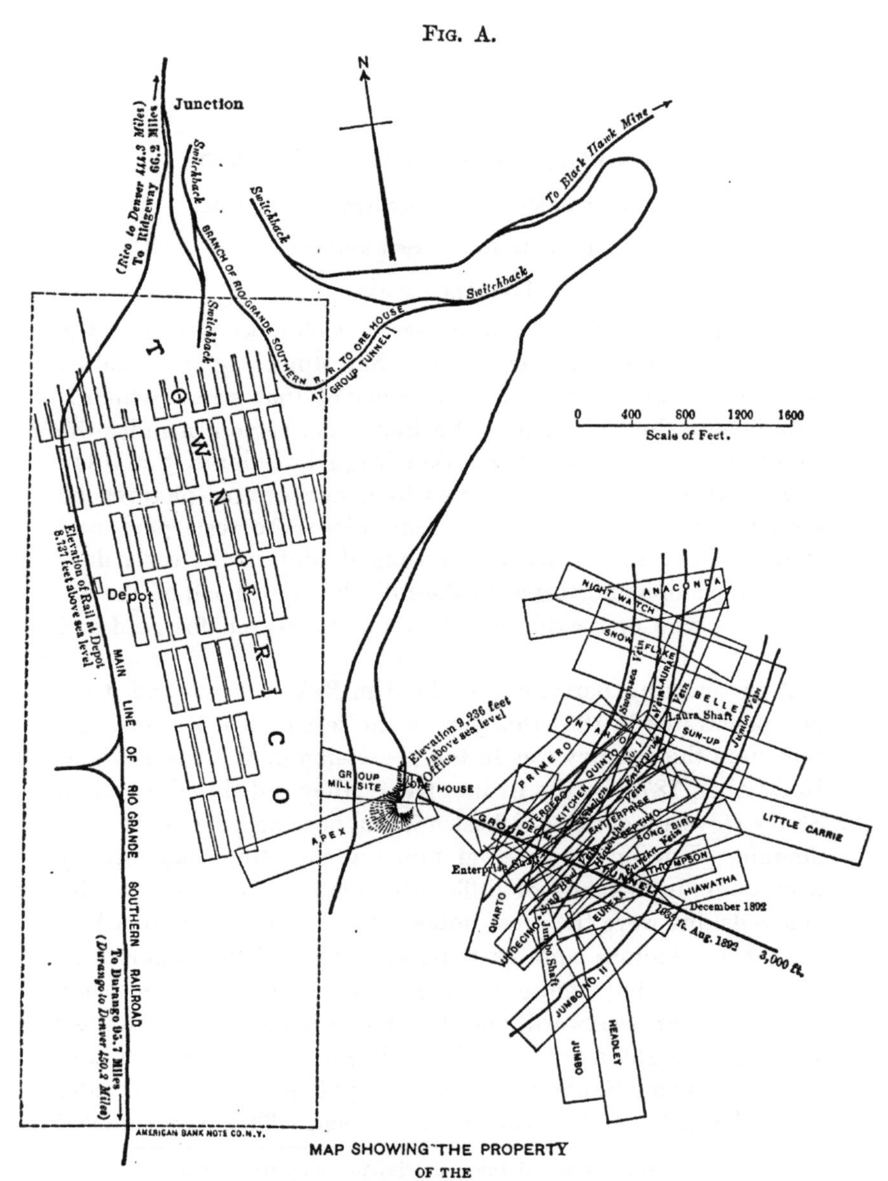

MAP SHOWING THE PROPERTY
OF THE
ENTERPRISE MINING COMPANY
AT RICO, COLORADO.

an adobe furnace and spent an entire summer in an abortive attempt to smelt the ores, the outcrop of which can still be seen at the north end of the main street of the town of Rico, upon claims now owned by the Atlantic Cable Company. In the autumn they returned to Santa Fé, and the valley was given up to the trappers and hunters, who found beaver along the stream and bear and deer on the hillsides.

In 1869 another expedition arrived. It consisted of John Eckels, William Hill, Pony Whitmore and two others, all of whom had made their way from the Moreno mines, a district near Elisabethtown, in what is now New Mexico. They discovered several large lodes near the site of the present settlement of Dolores. In the following year, Gus Begole came across the range from Silverton, and brought an assay-outfit with him. He and his partner, Eckels, discovered and located the Nigger Baby (now Yellow Jacket) and Dolores (now Aztec) mines. They sank several shafts and ran several drifts, but the ore proved too low in value to meet the costs of treatment and transportation, and they abandoned their claims. Others, who came from time to time, had a like experience.

In 1878, John Glasgow, Charles Hummiston and Sandy Campbell found their way northward from La Plata City. They spent the summer in active work, and located the Atlantic Cable, Grand View, Phœnix, Yellow Jacket and other claims. During the succeeding winter and in the early spring of 1879, the news went out that "carbonates" had been found at Rico, and a second Leadville uncovered. A "rush" set in. In the fall of that year Messrs. Jones and Mackay, of Comstock fame, visited the camp and purchased the Grand View group of mines. Next year, 1880, the boom continued, and the erection of a smelter* was begun. The material required for construction all came on mule-back over the ranges from Alamosa at a cost of 16 cents per pound. In the fall the furnaces were blown in under the superintendance of Messrs. Endlich and Arnold.

All the early discoveries of this district centered around Nigger Baby hill and the valley at its base. In 1879, however,

* That smelter still exists. It has afforded many well-known metallurgists their early and hard-bought experience. Its history would present an amusing commentary on the struggles of ill-digested enterprises.

Fig. B.

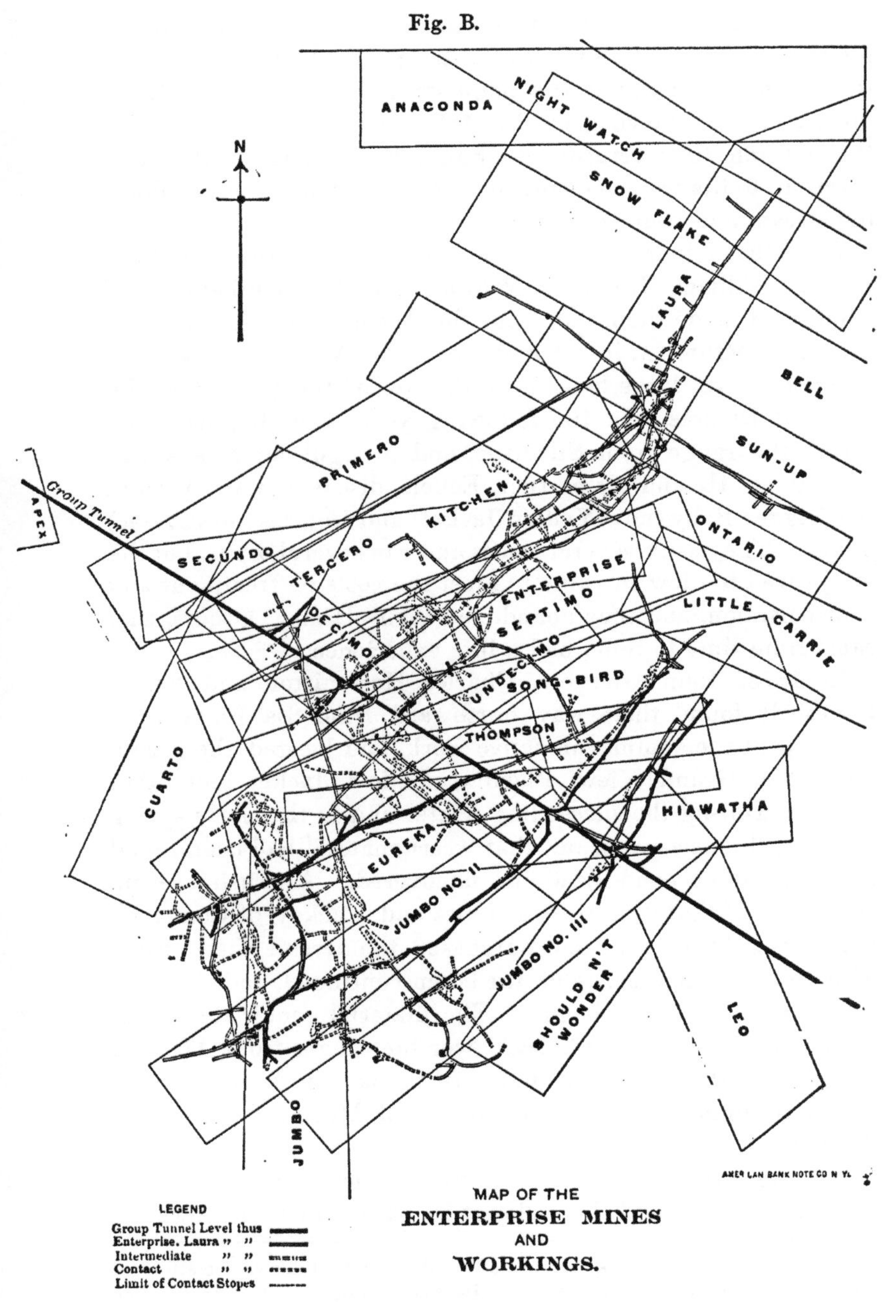

E. W. Hunt, Surveyor.

a shipment was made to Swansea from a discovery by Harry Irving on a claim located further south, on Newman hill, which is like a footstool to Mount Dolores. This event, unimportant as it seemed at the time, marked the beginning of the development which more than ten years later led to the prolific production of gold and silver out of the workings of the Enterprise and Rico-Aspen mines.

In the spring of 1881 David Swickhimer, Patrick Cain and John Gault sunk a shaft 35 feet deep upon their Enterprise claim on Newman hill. This work was undertaken not upon the evidence of ore, but in the expectation of cutting the continuation of the veins successfully worked in certain claims further south, owned by the Swansea Gold and Silver Mining Company. Without entering into a detailed description of the geological structure of Newman hill it is necessary, in order to make the early story of discovery clear to the reader, to say that the true rock (sandstone and limestone) is overlain by drift, through which shafts must penetrate before reaching the ore-bearing formation. The veins do not reach the present surface, save in the face of the landslip where Harry Irving first detected them. The three owners above mentioned traded their claim to George S. Barlow for $300 worth of lumber. Barlow continued the sinking of the shaft to a depth of 146 feet. On an adjoining claim, named the Songbird, another miner, A. A. Waggener, sank a shaft to the depth of 203 feet. The latter penetrated through the drift into lime shale; but the Enterprise shaft did not at that time reach the true rock. Both shafts got into very wet ground. In the meantime the Swansea workings were reported to be impoverished and, finally, exhausted of ore. It was also said that the veins did not extend northward, but the real fact was that cross-veins had faulted the ore-bearing veins in a manner to be rendered clear later on in this account. Newman hill was discredited, and early in 1883 the Enterprise and Songbird shafts were abandoned.

A year later, Larned and Hackett resumed work in the Swansea levels, and, by mere accident, discovered that the veins had not come to an end, but were simply dislocated. They prosecuted development, proved the continuity of the ore and made large shipments. Their success induced Waggener and Barlow to relocate their abandoned claims late in 1886. But neither of them had any capital, and they were unable to overcome the heavy flow of water. In December, 1886, David Swickhimer bought out Waggener's interest, acting on knowledge obtained while working in the Swansea mine, which had satisfied him that the veins must extend into the Enterprise and Songbird claims. In March, 1887, he recommenced the sinking of the Enterprise shaft. In May he acquired one-half of Barlow's interest. In July the windlass was replaced with a steam-engine and a pump. All this time Larned and Hackett were drifting rapidly northward and threatened soon to reach the boundary separating their territory from that of Swickimer and Barlow. Unless the two latter succeeded soon in finding a vein in place, so as to permit a valid location, the claims could be successfully disputed.* They therefore hurried the sinking, and in spite of bad luck, floods of water and a general lack of experience, they struck ore on the 6th of October at a depth of 262 feet. The first assay gave 2.1 ounces of gold and 519.4 ounces of silver per ton.

This ore was one foot thick and formed part of a "flat lode." In the light of later developments, this discovery is known to have been a piece ot particular good fortune, for the maps of to-day prove that it was the edge of the biggest ore-body ever found on Newman hill, and that a shaft put down 20 feet further east would have missed it. This was the first evidence of the existence of a flat ore-deposit. Swickhimer thought at first that it was merely a roll in the Enterprise, an almost vertical vein. It was, however, soon proved by the workings to be a bedded formation, conformable to the enclosing country. The shaft was sunk 60 feet below this "contact," and a drift was run westward until the increased seepage of water, in the following spring, proved too much for the pump, and caused work to be confined to the contact. In July the water diminished, drifting was resumed, and in August, at a distance of 118 feet southwest of the shaft, the Enterprise vein was at last intercepted. The ore was 20 inches thick and assayed 3.2 ounces of gold and 285.5 ounces of silver per ton.

In May, 1890, the Songbird and Enterprise mines, together with much adjoining property, were acquired by the Enterprise

---

* A good example of the iniquitous operation of our absurd mining law.

Mining Company, the operations of which were directed by the writer from March 1, 1894, to February 28, 1895.

Fig. A shows the group of claims forming the property, and the railway connecting them with the town of Rico. The mine map (Fig. B) indicates the complex of drifts and crosscuts which follows the ramification of veins. The workings aggregate 8 miles in length. They have yielded ore whose gross value exceeds $3,500,000. Of this nearly one-quarter has been gold, the remainder silver.*

Entrance to the mine is made by the Group tunnel which has a course S. 56° 58′ E., and consequently cuts the ore-bearing veins almost at right angles. Moreover, since its line corresponds closely to the strike of the country, it intercepts the veins at an approximately equal depth below the contact. It would do so without variation, but for the step-faulting which accompanies the vein-structure.

The tunnel or adit is 2920 feet long. Near the entrance the contact is 210 feet overhead, at the breast it has approached to within 35 feet. The largest drop is due to a down throw on the so-called Leo cross-vein.

II.—The Country-Rock.

The Dolores, as it flows southward from the town of Rico, is overlooked on the west by Mount Expectation, and on the east by Newman hill. The river has eroded its own way and does not follow the line of a fault. The Lower Carboniferous beds, which form both its bed and the immediately flanking hillsides, can be traced across the valley. On Newman hill they are for the most part hidden by a deposit of Quaternary drift, the maximum thickness of which is about 400 feet, diminishing

* The analysis of representative lots of ore gave the following results:

	First class.	Second class.
SiO_2,	29.2 per cent.	50 to 55 per cent.
Mn,	2.0 "	6 to 10 "
Fe,	11.8 "	6 to 10 "
Zn,	12.0	5 to 7
Pb,	10.2	2 to 3
S,	11.6 "	5 to 8 "
Au,	0.87 oz. per ton.	0.3 to 0.5 oz. per ton.
Ag,	221.50 " "	45 to 75 " "

The first class was mostly contact-ore, while the second class consisted of the bulk of the product of the verticals.

southwestward. The underlying shales, limestones and sandstones contain fossils which determine their stratigraphical place. The intrusions of porphyrite,* both plentiful and irregular in form, particularly at the northern end of Newman hill, afford an explanation of the metamorphism of the sedimentary rocks.

The country enclosing the ore-deposits consists of these shales, limestones and sandstones, having a strike N. 20° W. and an average dip of 10°. They are thinly bedded. Single beds are not extensive, one layer dwindling in thickness until it dovetails into another. Without necessary variation in width, the composition may change so that lime graduates into sandstone. These facts indicate that the sediments were laid down in estuaries and in such shallow reaches of water as permitted of swift changes in the conditions of sedimentation. The fossil remnants are of a kind that accords with this view.

The foregoing description applies especially to that portion (about 200 feet thick) of the formation to which the mine-workings are practically confined. The veins do not penetrate upwards beyond the horizon known as the "contact," and they become barren at an average depth of about 150 feet below that horizon. For this reason the overlying rock has been merely penetrated in sinking to the ore-bearing horizon, and, similarly, the underlying beds have only been pierced by one or two unsuccessful shafts and bore-holes.

The beds above the contact consist, in ascending order, of:

2 to 5 ft. of lime breccia;

A thin bed of soft, crushed sandstone, which rarely reaches a thickness of 9 ft., averages less than 1ft., and is occasionally entirely absent;

---

* The following notes on a thin section of this rock, a hornblende-augite porphyrite, were given by Mr. R. C. Hills, geologist of the Colorado Fuel and Iron Company.

Macroscopic Character.—The rock is grayish in color, and shows white, opaque feldspar (plagioclase), evidently much kaolinized; also small, partly-altered green hornblendes. Apatites are occasionally visible under the lens.

Microscopic Character.—Under the microscope the feldspars are seen to be largely altered to kaolin. So far as determinable in the only section available, they are plagioclase. The green hornblendes are largely altered to chlorite. Small pale-green augites and stout, relatively large apatites are rather numerous, together with ore particles (magnetite). The granular groundmass is much kaolinized, and abundantly distributed through it are grains and microlithic crystals of feldspar, also kaolinized.

6 to 8 ft. of black shale;
30 to 40 ft. of sandstone beds, and
40 to 50 ft. of black lime-shales.

The last graduate into a series of blocky limestones, the escarpments of which appear on the face of Mount Dolores. In none of the beds of this series have profitable ore-deposits been found, although large veins of calc-spar traverse them at intervals.

The contact is not an ore-measure lying between two persistent beds of shale and limestone, as has been stated.* The composition of the encasing rocks is variable because of the comparatively brief persistence of individual members of the sedimentary series. It may be said, however, that the ore of the contact is invariably found in rock which has undergone shattering. An appearance of undisturbed solidity is occasionally given by later cementation. Several raises put up to the contact from the upper main level of the mine afford sections of the formation. Two are quoted.

No. 1.

10 in. of compact pulverulent, lime,
14 in. of lime breccia,
2 in. black shale,
19 in. limstone,
12 in. sandy limestone,
10 in. sandstone,
1 in. parting of black mud,
2 ft. lime shale,
8 in. crushed lime,
25 in. blocky limestone,
10 in. soft sandstone,
1 in. parting of shale,
Then a series of thin beds of sandstone aggregating 21 ft.
18 in. light-colored limestone,
10 ft. sandstone,
Then a further series of sandstones.

No. 2.

2 ft. of brown lime breccia,
6 in. of fine-grained sandstone.
21 in. dark blocky limestone,
3 in. black soft shale,
28 in. blocky lime,
5 in. laminated sandstone,
18 in. black shale,
1 in. parting black mud,
15 in. light gray limestone,
1 in. parting,
58 ft. sandstone,
$2\frac{1}{2}$ ft. shale,
7 in. sandstone,
3 in. shale,
$4\frac{1}{2}$ ft. sandstone,
2 in. lime shale,
3 ft. coarse sandstone.

---

* J. B. Farish, *Proceedings Colorado Scientific Society*, vol. iv., p. 154.

The section given in the first column came from the first raise on the Jumbo No. 3 upper level, the second from a raise on the Kitchen vein from the Enterprise level at the end of the cross-cut between raises 7 and 8, Songbird. The linear distance between the two sections is only 900 feet. Beyond the similarity of the breccia, which marks the "contact," these two sections are entirely dissimilar, and it seems impossible to recognize any continuity in the stratification. The comparison serves to explain the statement already made that the beds of the series are notably non-persistent. The sketch reproduced in Fig. 1 represents the face of a cross-cut where a bed of sandstone has been caught in the act, as it were, of merging into a bed of lime-shale.

FIG. 1.

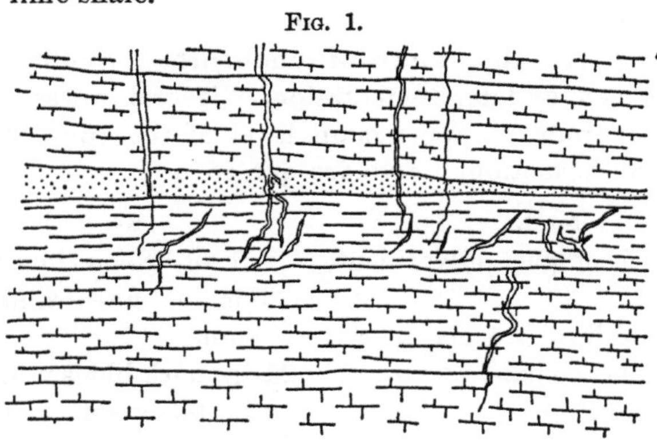

Non-persistence of Beds.

Further sections of the contact-horizon will be given when we come to consider the ore-distribution at that level.

It has been seen that below the contact comes a series of very thin lime and shale beds, interrupted by occasional sandstone divisions. These beds are all dark in color, graduating from coal-blackness at the contact to dark grays at a distance of 100 or 150 feet from it. As the contact is left, the sandstone beds become more frequent, their grain is notably coarser, the limestones become less shaly and more blocky, the black shale is absent, and soon the workings penetrate into thick beds of a coarse, light-colored quartzitic sandstone.

One or two shafts have been sunk, but the records which have been kept are unfortunately so vague as to be useless for the purposes of a geological section. The Jumbo shaft penetrated 524 feet below the contact, without any discovery of importance. The Skeptical shaft, just north of the Enterprise property, was sunk 365 feet in porphyrite and then penetrated 15 to 20 feet of shales and limestone. A bore-hole snbsequently put down in the bottom of this shaft went through 200 feet of shales and limestone before entering quartzite, where operations ceased. The Lexington tunnel, which is 400 feet below the contact, penetrates the Newman hill formation for a distance of 2740 feet, and is in coarse, light-colored, hard sandstone for most of its length.

There has always been much surmise regarding a certain undiscovered "second contact." It has been the cause of much nonsensical mining. The Enterprise Company, in 1893, sunk the Jumbo shaft in search of this lower ore-measure, not realizing that, owing to the position of the shaft and the dip of the formation, they would have to go down 300 feet below their

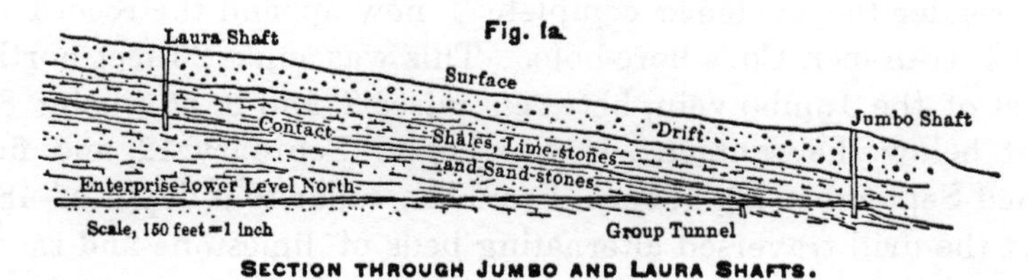

SECTION THROUGH JUMBO AND LAURA SHAFTS.

main adit before they would be even level, in a geological sense, with existing northern workings.

This piece of exploratory work was badly planned and proved without result. The neighboring company, the Rico-Aspen Consolidated Mining Co. put down a bore-hole, which was as barren of encouragement. The "second contact" of Newman hill is a vain imagination. A hazy idea of the geology of Leadville, blended with a misconception of that of Rico, has caused the growth of an idea having no facts for its support. It was suggested by the occurrence, on the hills north of the town (above the village of Piedmont), of a series of at least three ore-bearing contacts, and it seemed to be indicated by the developments in the Atlantic Cable mine. But deductions from this evidence are vitiated by reason of the fact that the valley of the Dolores, near the north of the town, is crossed by a large dike of porphyrite, marking a fault which breaks the continuity of the country on either side.

The Atlantic Cable Co.'s bore-hole gave the following downward section:

	Feet.
Limestone,	7
Lead- and zinc-ore,	4
Limestone,	5½
Lead and zinc-ore,	5
Limestone,	13
White marble,	20
Zinc-blende ore,	3
Specular iron-ore	18
Limestone,	43
Porphyrite,	1
Limestone,	25
Porphyrite,	2
Limestone,	3
Mineralized porphyrite,	3
Porphyrite,.	21

The remaining 170 feet of the hole continued in quartzite. To render the evidence complete, I now append the record of the Rico-Aspen Co.'s bore-hole. This was sunk 20 feet northwest of the Jumbo vein, between raises 7 and 8, at a point 85 feet below the contact. It was begun February 12, and finished September 10, 1895. From the collar to a depth of 481 feet the drill traversed alternating beds of limestone and sandstone; the latter becoming coarser as depth was attained. From 481 to 541 feet the drill traversed porphyrite. Between 541 and 573 feet the rock was quartzite. Then porphyrite continued to the bottom of the hole, at 706 feet. The evidence afforded by these borings will be referred to after other matters have been passed in review.

III.—The Ore-Occurrence.

In the investigation of the relation between the ore-occurrence and the rock-structure it is found that there are two distinct systems of vein-fissuring. One series of veins has a N.E.–S.W. strike and a nearly vertical dip; and this series is crossed and faulted by a second system, having an approximately N. and S. trend and a flat dip. The former are ore-bearing and are called "verticals" or "pay-veins;" the latter are barren of valuable ore and are termed "cross-veins." Both series fail to reach the present surface, save where deep erosion has occurred; because in coming up through the Carboniferous formation they are abruptly terminated in their near approach to a certain horizon marked by the occurrence of black shale and beds of crushed lime. This, the contact, is disturbed by the cross-veins.

The verticals are not productive immediately under the contact. On the contrary, when within a distance varying from 5 to 15 feet from that contact they split up into stringers, which scatter the ore so as to render exploitation unprofitable. Apart from this dispersion of the vein, the total amount of ore which it carries is also decidedly lessened. The plane of the contact is itself ore-bearing; the bodies occurring in the form of narrow channels ramifying through the crushed rock, in directions which correspond exactly to the strike of the veins underneath. It is a notable fact, moreover, that the cross-veins, barren as they are, are yet related to ore-bodies on the contact as rich as, if not richer than, those above the line of the verticals.

The knowledge of the relationships just outlined has proved vital to the intelligent exploration of the mines; and the theoretical consideration of them is, to the student of ore-deposition, highly suggestive. The structure of the formation, on account of the narrowness of the veins and the thinness of the beds through which they pass, affords within the space of a few square feet sections which ordinarily it requires acres to encompass. The coloring, moreover, of the minerals accompanying the ore and of the rocks enclosing it, is so marked as to assist the ready intepretation of structure, and enable the observer to portray them by pen and pencil. For these reasons the writer has endeavored to give the testimony collected by him in the form of a series of drawings, rendering much comment unnecessary.

IV.—The "Verticals" or Ore-Bearing Veins.

These belong to the simplest type of ore-deposit. They are fractures cutting across the sedimentary rocks almost at right angles to the bedding-planes. They have a simple structure. Their width averages less than a foot; they are built along fault-lines, are sensitive to the changes in the encasing rock, and are themselves faulted by veins of later formation.

About a dozen veins have undergone noteworthy development, and of these, five have yielded the bulk of the ore-production of the mine. The Enterprise, Jumbo No. 2, Jumbo No. 3, and Hiawatha all dip to the northwest at angles varying from 5° to 15° from the vertical; the Eureka is practically vertical; while the Kitchen, Swansea and Songbird veins have an opposite (southeast) dip and a flatter angle, viz., from 12° to 22° from the vertical. Their strike varies between 50° and

FIG. 2.

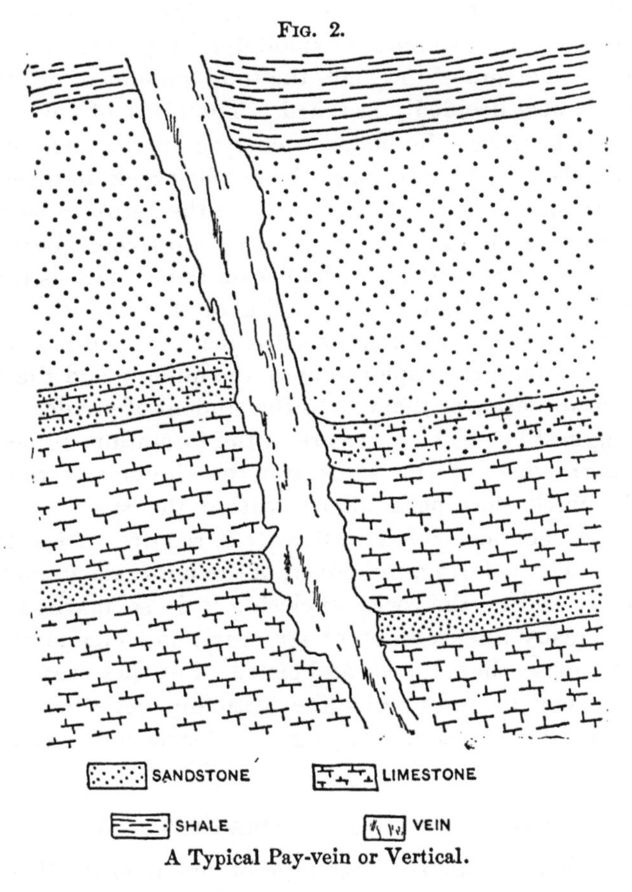

A Typical Pay-vein or Vertical.

65° east of north, the Eureka being conspicuous for its regularity.

Fig. 2 is a typical illustration obtained from the end of one of the levels. The vein is from 5 to 6 inches wide and cuts the country-bedding at a right angle. It will be noted that the ore occupies the line of a fault, the throw of which is 7 inches; the direction of the movement which caused it being indicated

Fig. 3.

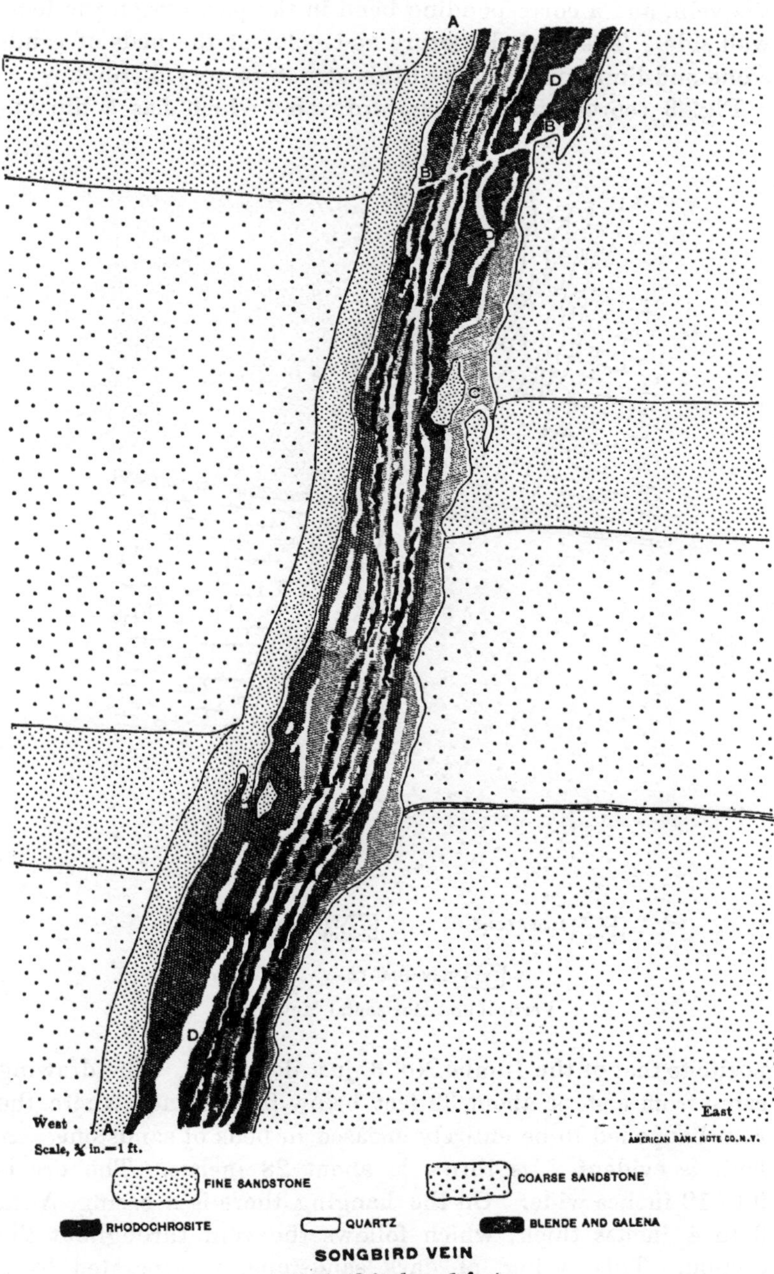

Scale, ¾-inch = 1 foot.

by the turning up of the bedding-planes on the hanging-wall of the vein, and a corresponding bend in the partings on the foot-wall side. The varied texture and color of the beds of sandstone and limestone rendered this structure very distinct.

Fig. 3 represents the Songbird vein, which has a dip op-

FIG. 4.

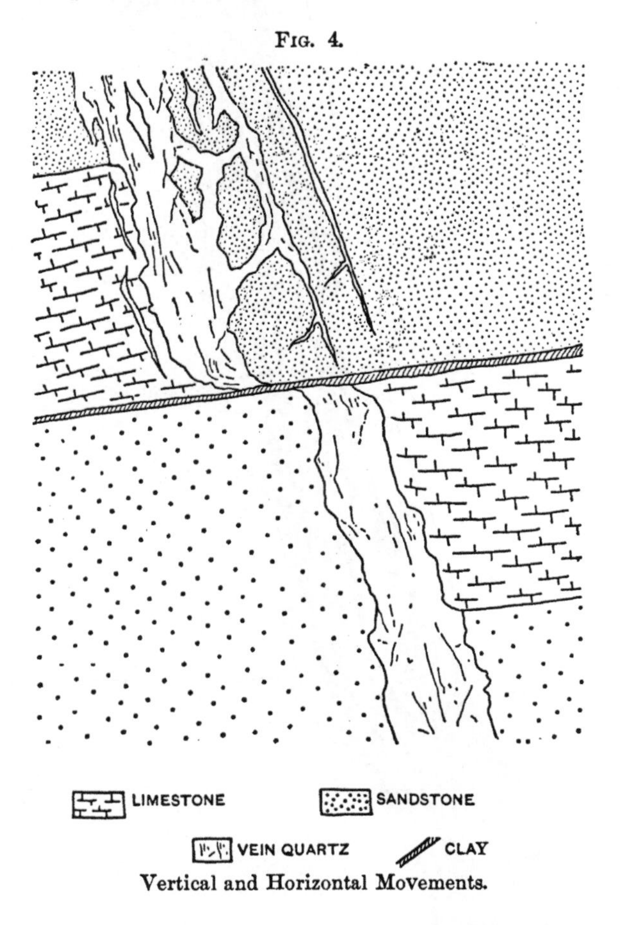

Vertical and Horizontal Movements.

posite to that of the Jumbo No. 3, just described, The drawing came from a stope about 30 feet below the contact, where the vein happened to be entirely encased in beds of sandstone. A fault is evident. Its throw is about 28 inches. The ore is 9 to 12 inches wide. On the hanging there is a casing, A A, 3 to 4 inches thick, which follows the vein throughout the section. This casing, of dark sandstone, is separated by a slight selvage from the outer country, but graduates gently into the vein-stuff adjoining it on the east. On the foot-wall there is a marked selvage, accompanied by crushed rock. Two inclusions, C C, of sandstone occur within the ore. The latter is banded by streaks of zinc-blende and by ribbons of rhodochrosite within the quartz. D D is quartz. At B B the ore has been slightly dislocated.

In Fig. 4 more complex conditions are represented. Two dislocations of the country are evident. The vertical break followed by the ore-formation has a throw of about 2 feet, while the lateral fault, evidently of later occurrence, has caused a disturbance measured by 9 inches only. It will be noticed that the parting separating the upper fine-grained sandstone from the bed of limestone has been brought into line with that dividing the same bed of lime from the underlying coarser sandstone. This coincidence must have facilitated the subsequent lateral shifting of the rocks. Such occurrences are frequently observable in the mine.

It may be questioned whether the movement along the bedding took place before or after the ore had been laid down. There is evidence elsewhere in the mine that such movements have both preceded and succeeded the vein-formation. In this case it preceded, because the ore is seen to be not abruptly broken off, but shaped to the structural conditions created at this point previous to its precipitation.

A different state of things is disclosed in Fig. 5, which represents the Jumbo No. 2 vein, as seen in the end of the lower level. Here, as usual, the ore occupies a fault-fracture, which has dislocated the bedding to the amount of 20 inches; but, in addition, a later movement along coincident partings has broken the vein and thrown it about a foot. The ore has been shattered, and in the clay accompanying the line of fault there are pieces of quartz and rhodochrosite, evidently due to this shattering. The quartz-veins, unaccompanied by ore, observable to the left of the vein, are of later origin. The dike of porphyrite will be referred to elsewhere.

In Fig. 6 a similar later movement, but this time in a vertical direction, is illustrated. The Jumbo No. 3 vein, here shown, follows a fault the throw of which is about 2 feet. Since the ore was laid down a later shifting of the country has been accompanied by the formation of a fracture, which approximately follows the line of weakness of the older movement, and breaks across the ore lying in its path. The amount of this dislocation cannot be measured with certainty; it is probably slight.

Fig. 5.

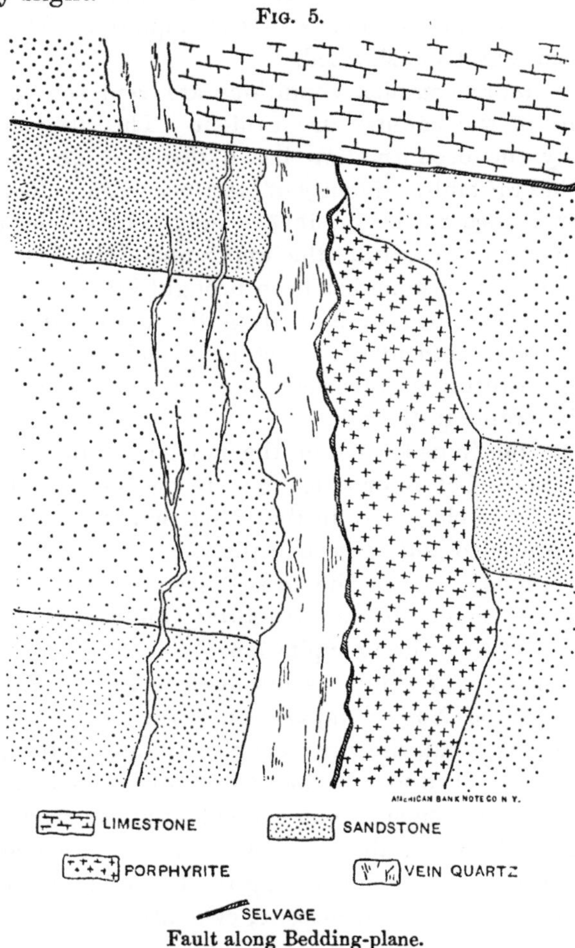

Fault along Bedding-plane.

The veins are, as to their size, behavior and ore-bearing character, very sensitive to the structure of the enclosing rock. They flatten when traversing lime, the increased deviation from the vertical being accompanied by a diminution of ore. Even in those cases where the actual width may not decrease the percentage of valuable minerals does. In sandstone they usually straighten up, and are marked by an enrichment. When crossing a parting between the beds an offset of ore is often formed underneath the parting. Some of these characteristics are illustrated in Figs. 7 and 8, both representing the Enterprise vein. In both the faulting along the ore-break can be measured, since the dislocated portions of the same bed are similarly lettered.

FIG. 6.

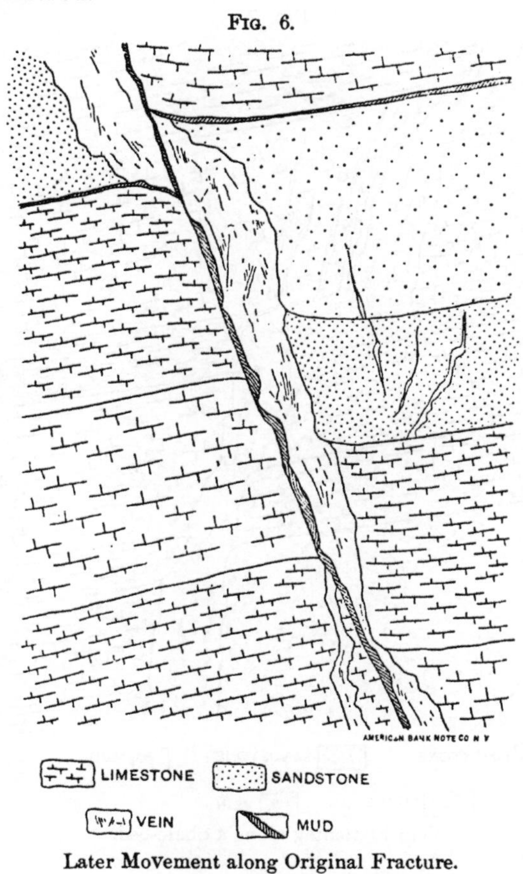

Later Movement along Original Fracture.

Among the miners it was the common saying that "the vein makes ore in sandstone," but my observations did not quite confirm this generalization. When traversing lime the veins tend to split up into stringers, and this is the case, to a lesser degree perhaps, in sandstone. On the whole, my experience was that unlike walls give the best environment for rich ore, and that a foot-wall of sandstone with a hanging of lime is a particularly favorable combination.

In some cases comparatively modern shiftings of the country are evidenced. Thus in Fig. 7 the flattened part of the vein, E F, formed along the shale band, is crushed, and it is my belief that this was caused by a movement along the bedding

FIG. 7.

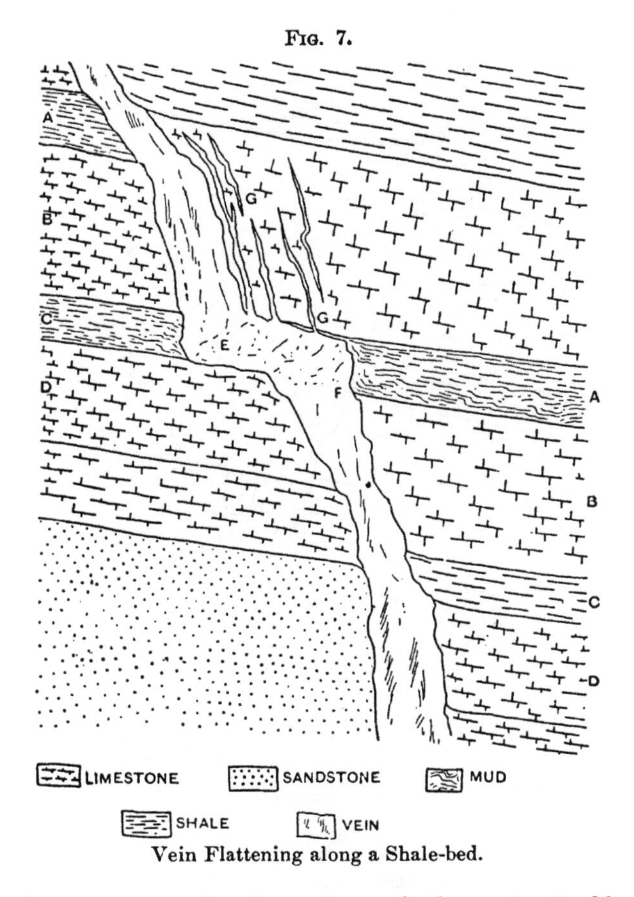

Vein Flattening along a Shale-bed.

long subsequent to the formation of the vein itself. The stringers, G G, are also of late origin, and are composed of barren quartz dissimilar to the gangue. In Fig. 8 there is a similar jog in the vein, but in this case no crushing or disturbance is suggested.

Pronounced selvages are not characteristic of these veins. When noticeable they are usually in lime-beds. In sandstone they are comparatively infrequent, and the absence of a parting causes the ore to separate with difficulty from the sandstone, into which it merges in such a manner as to cause the miners to say that it is "frozen" to it. A casing of sandstone is occasionally seen when the vein is entirely enclosed by lime, proving that the sandstone must have been shorn off upper beds in

FIG. 8.

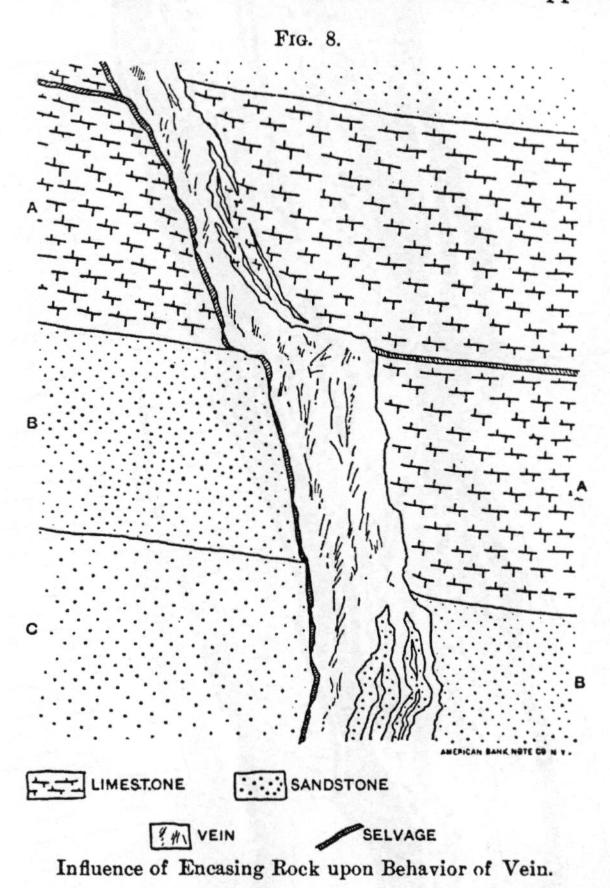

Influence of Encasing Rock upon Behavior of Vein.

the course of that movement which determined the existence of the vein. This feature is illustrated in Figs. 9 and 10. The first of these represents the Jumbo No. 3 vein, which in this instance is identified with a fault of about 3 feet throw. On the foot-wall the ore is divided from the country by a clay selvage, but on the hanging there is no such parting. Toward the bottom of the section the vein exhibits, on its hanging-wall,

FIG. 9.

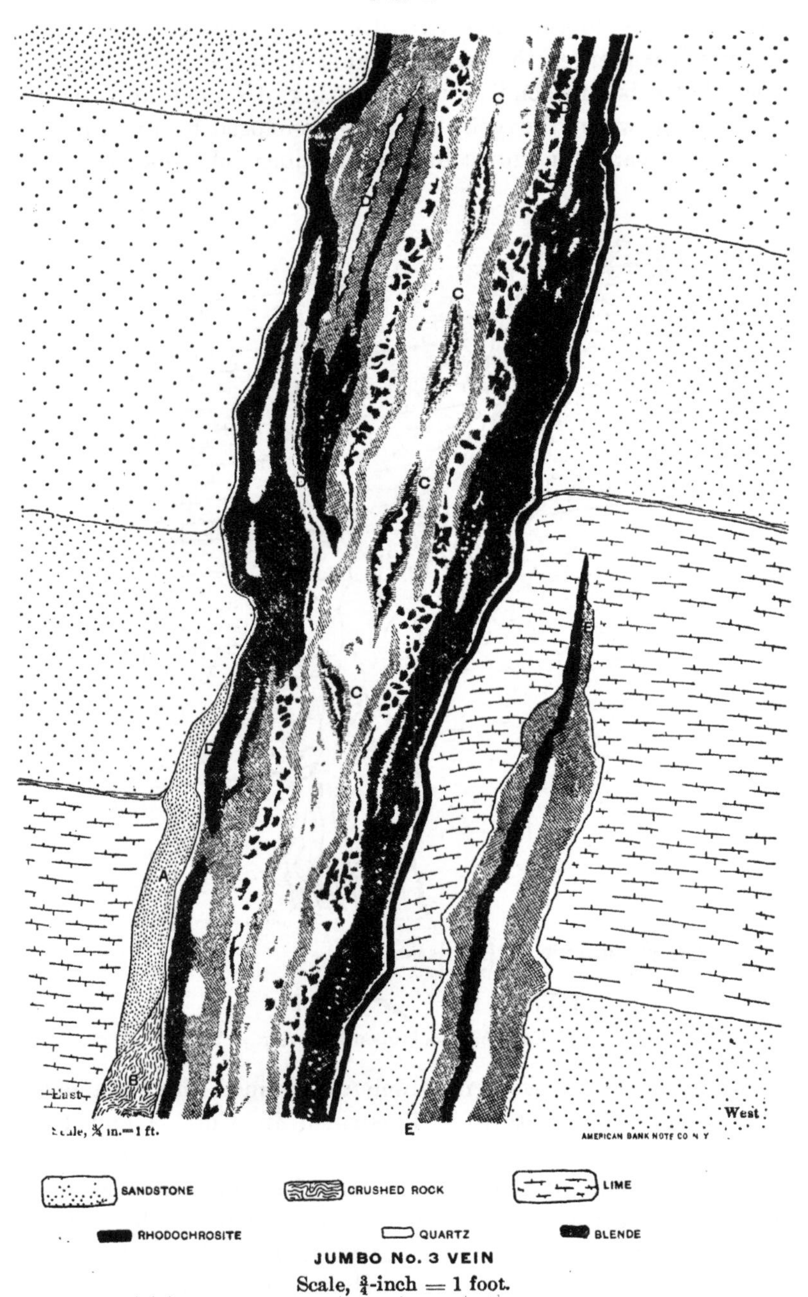

JUMBO No. 3 VEIN

Scale, $\frac{3}{4}$-inch = 1 foot.

FIG. 10.

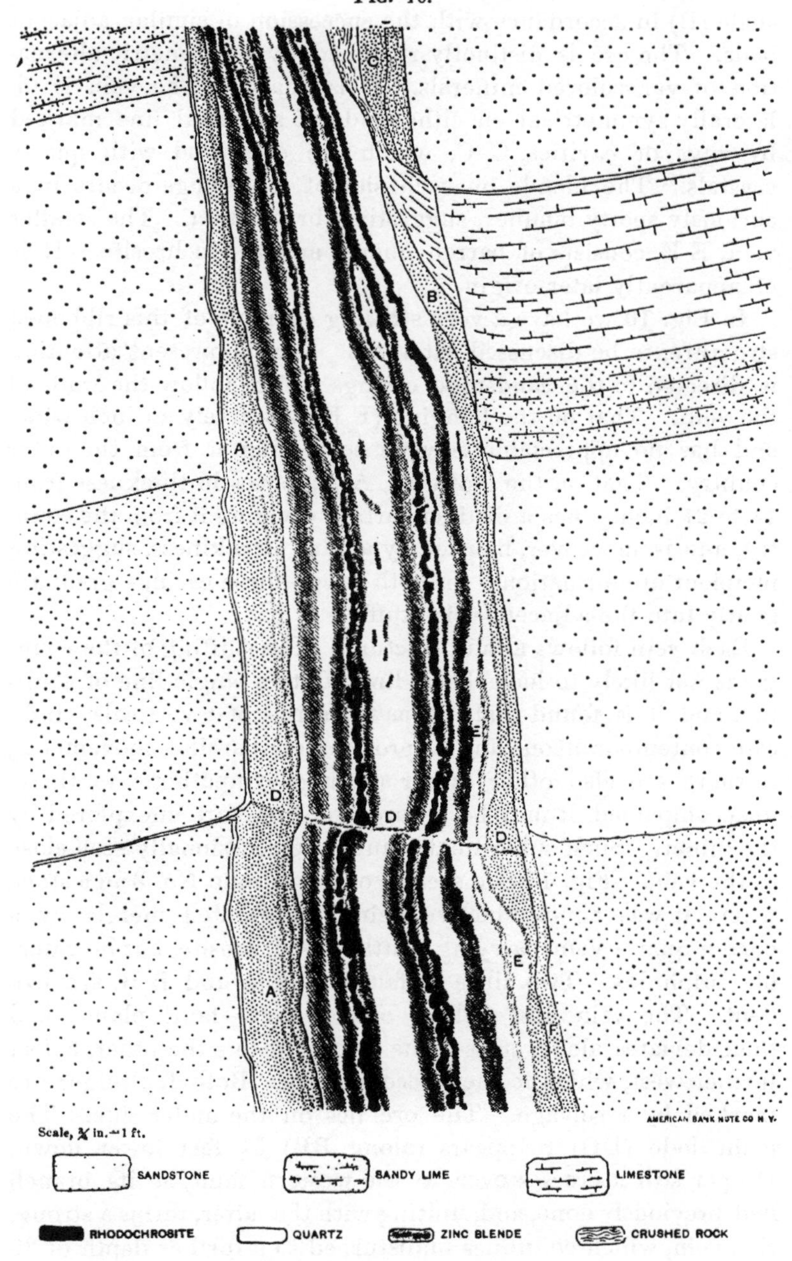

EUREKA VEIN
Scale, ¾-inch = 1 foot.

a casing of country which changes from sandstone (A) to lime shale (B) in accordance with the succession of similar adjacent beds. The ore is distinctly ribboned by a symmetric alternation of vari-colored minerals. D is quartz. The vein is bilaterally symmetrical on either side of a central line marked by vugs or cavities, C C, which are encrusted with quartz crystals. The blende on each side of these vugs occurs in a curiously spotty manner, suggesting brecciation. The smaller vein, E E, consists of barren quartz and rhodochrosite. It is of apparently later origin.

In Fig. 10 we have a very striking example of this ribboned structure, to be discussed later on. For the present attention is directed to the sandstone casings which follow the walls of the vein. That on the hanging (F F) is scarcely an inch wide, and has no appreciable selvage separating it from the outer country. That on the foot-wall, A A, varies in thickness from 1½ to 2¾ inches, has a distinct parting dividing it from the country, and is, moreover, marked by a dark streakiness, suggesting incipient ore-formation. In both cases these casings graduate gently into the adjacent vein-stuff.

Each vein follows a fault-fracture. The shifting of the country is not likely to have been limited to a single line of faulting, and it is found that just as the series of "verticals" indicate contemporaneous and approximately parallel movements, so there are also other minor shiftings sympathetic to these, unaccompanied, it may be, by ore, and therefore unexplored by the miner. Such subordinate faults are occasionally seen close to the vein. Fig. 11 is a section of the Jumbo No. 3 in a stope where it was non-productive, being represented merely by a barren quartz vein, carrying a little rhodochrosite, but no valuable sulphides. Two lines of faulting, A A and B B, are evident. The vein first follows one of these fault planes and then deviates along cross-joints until it meets the other, 5 feet further west, which it then accompanies. Both fault-lines are marked by a selvage. The ore lies on the under side. The main lode (DD) reappears (along BB) 5½ feet lower down. Deeper still it crosses over to the western fault, as its branch had previously done, and, uniting with the latter, forms a strong, rich vein, which continues undisturbed to a further depth of 30 feet, when another irregularity breaks its continuity.

Fig. 12 affords further evidence. In this case, as in the last, the vein is seen where it is small and poor. Under such conditions its structure is more readily discernible, because enrichment and enlargement generally produce confused outlines and are accompanied by a generous mineralization, destructive of

Fig. 11.

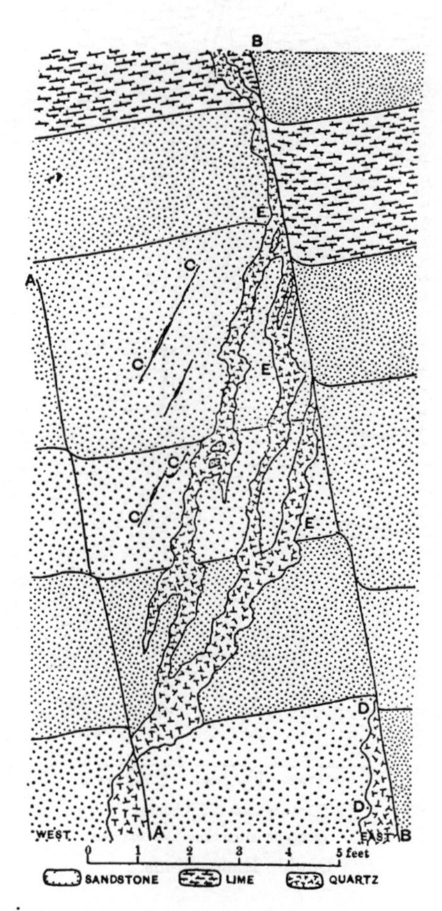

Vein Changing from one Fault-plane to Another.

definition. Hence the stopes most instructive to the scientific investigator are least pleasing to the mine manager. In this section there are two veins, both small, of which the western may be regarded as a mere off-shoot. Two dislocations and several minor disturbances of the country are noticeable. The western quartz-seam occupies a fault of a few inches, which dies out into a mere distortion of the bedding; the larger vein is identified with a fault having a throw of 2¼ feet. The bending

FIG. 12.

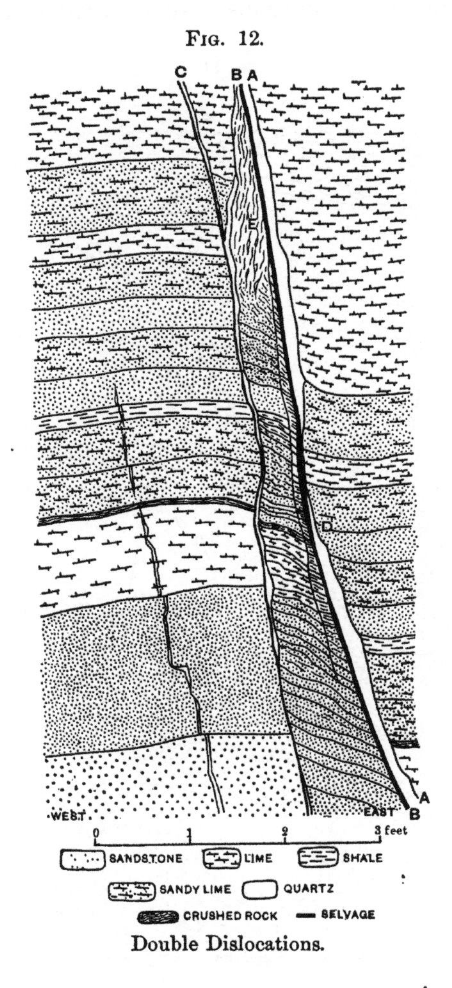

Double Dislocations.

of the edges of the beds as they abut against the vertical quartz-veins is very marked. Slight shiftings along bedding-planes are indicated by the behavior of the small stringers traversing the country.

The veins are built up of many-colored minerals, which give them a rare beauty, and serve also to accentuate their structure.

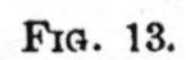

FIG. 13.

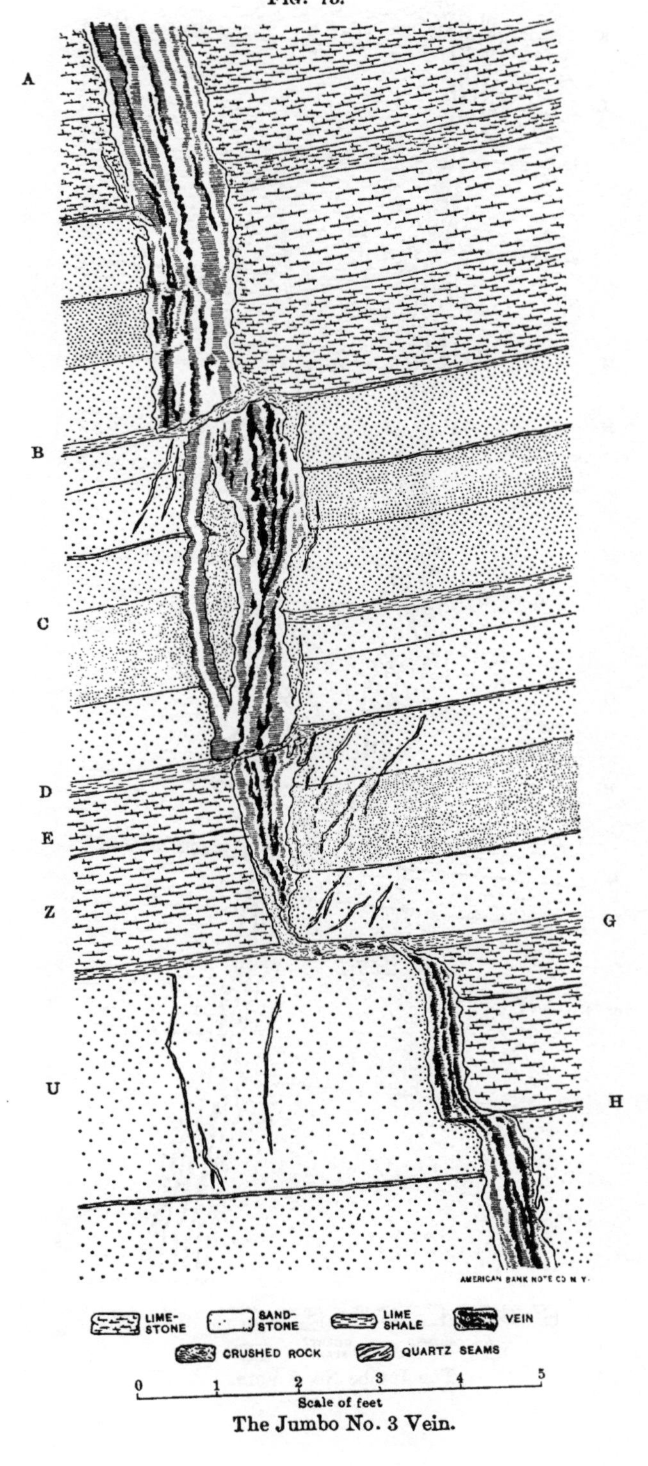

The Jumbo No. 3 Vein.

Fig. 14.

The Jumbo No. 3 Vein.

Rhodochrosite and quartz enclose the sulphides of zinc, lead, iron, copper and silver in the form of galena, blende, iron and copper pyrites, argentite and stephanite. Native gold and native silver both occasionally accompany the argentite. A banded- or ribbon-structure is frequently brought about by the alternation of quartz, rhodochrosite and the sulphides. This structure assists the breaking of the ore, which will often part in ribbons within itself more readily than it can be detached from the encasing country. Fig. 25 illustrates a typical piece of vein-stone.* The gentle graduation of ore into country is noteworthy. The banding due to the rhodochrosite is a distinguishing feature, while the inclusion of portions of sandstone and the comb-structure of the quartz are additional testimony as to the origin of the ore, which will be discussed under another heading.

Much more might be said concerning the behavior of these veins; but sketches of the actual occurrences are better than verbal description. Fig. 13 shows the Jumbo, No. 3 lode, as seen in the stopes of the mine. The vein structure is illustrated for a height of sixteen feet. It will be observed that the vein follows a line of fracture which has faulted the country. The vertical dislocation measures 2½ feet, and is rendered easily evident by the partings of shale which separate alternating sandstone and limestone-beds. Following the section downward, the ore is about 14 inches wide at the top, opposite A, and is distinctly ribboned with bands of quartz, rhodochrosite and zinc-blende, the last being also mixed with galena. The vein continues fairly uniform for five feet, and is then interrupted by a break opposite B, which indicates that the country has been shifted to the right for a distance of six inches. This movement took place at a point where a coincidence occurs between the partings between two sandstone-beds on one side and lime-sandstone-beds on the other. Below this point the vein opposite C is less regular, and divides into two branches, of which the eastern carries all the pay ore, the western being merely rhodochrosite and quartz. The latter, from its composition and structure, suggested to me, at the time, that it was of later origin than the ore-bearing vein. Opposite D, four feet

* For a discussion of the evidence afforded by this drawing the reader is referred to the author's paper entitled "Vein-Wall," *Trans.*, xxvi., 193.

Fig. 15.

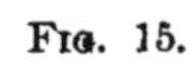

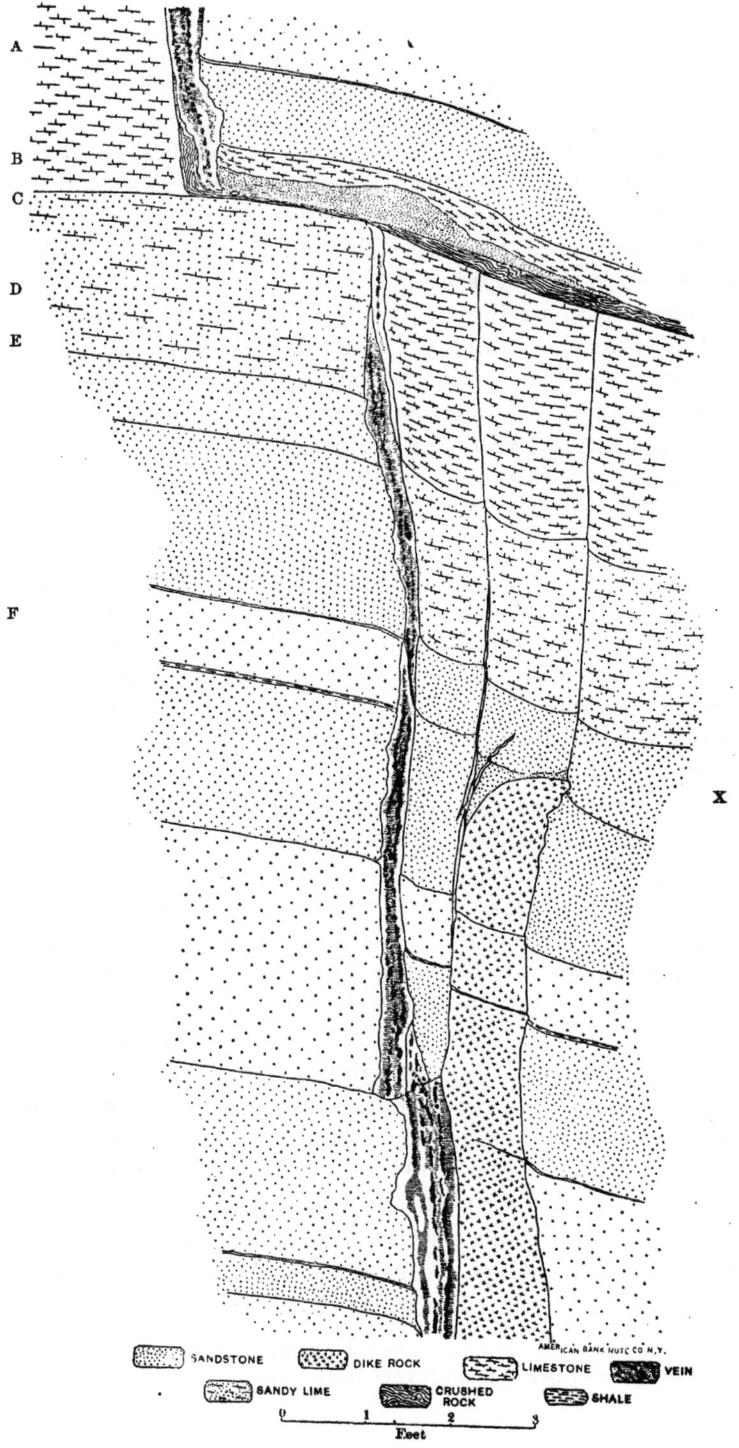
A
B
C
D
E
F
X
SANDSTONE
DIKE ROCK
LIMESTONE
VEIN
SANDY LIME
CRUSHED ROCK
SHALE
0 1 2 3
Feet
AMERICAN BANK NOTE CO N.Y.

FIG. 16.

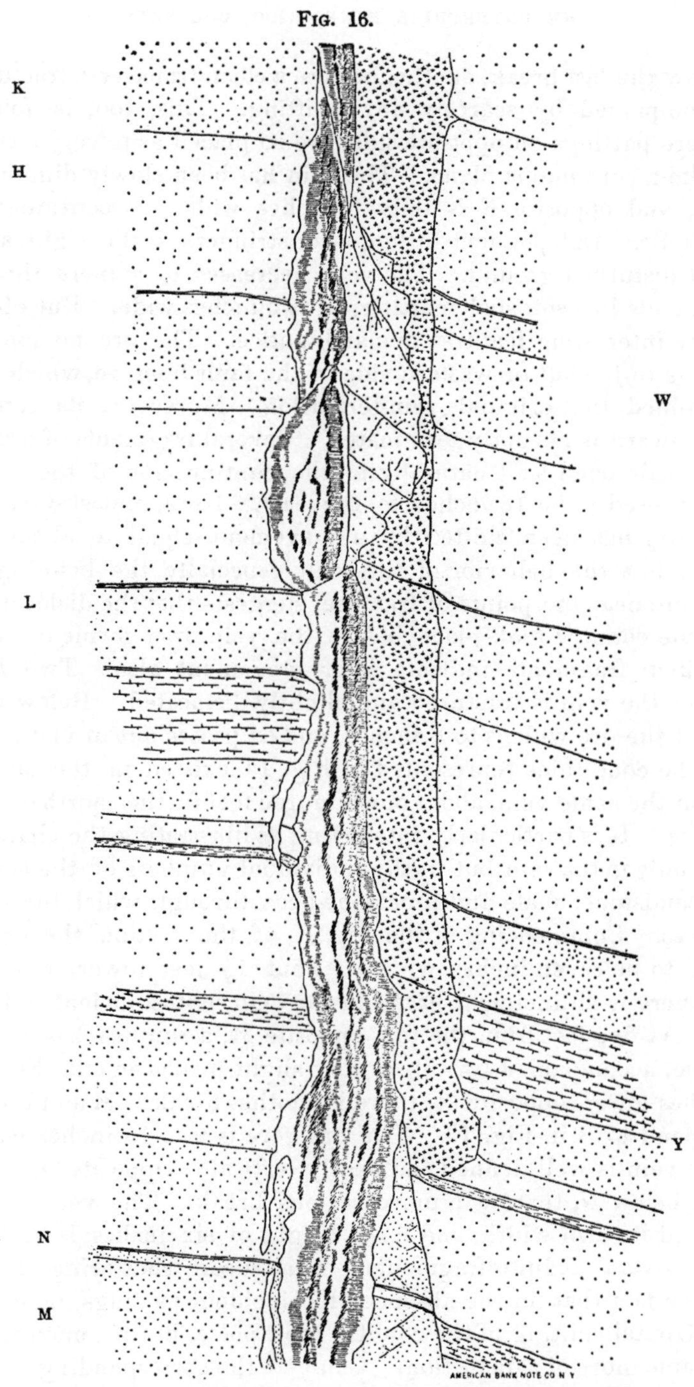

Downward Continuation of Fig. 15, Constituting Section of Jumbo No. 3 Vein through Successive Stopes.

below the last break, another occurs, a clean transverse fracture, accompanied by scarcely any shifting. This, too, is found where partings on opposite sides are so placed as to very nearly make a continuous plane. The vein has been slowly diminishing, and opposite E is only 8 inches wide. It continues to dwindle. Independent little quartz stringers to the right suggest disturbed ground. The vein decreases to a mere thread and ends in a soft shale parting, G, two inches wide. But other more interesting features demand notice. The ore no longer clings to the left or western side of the fault-fracture, which determined the original vein-formation. Moreover, its course downward is abruptly terminated. Several fragments of ore in the shale band lead naturally to the continuation of the vein, discovered to be 18 inches to the east. Here, opposite G, the country has been shifted by a movement equal to about 1½ feet; but the behavior of the vein, especially the bending of the ore near the points of fracture, suggests that the dislocation of the country took place before the vein was formed. The broken fragments indicate later movement also. Two feet lower the vein is fractured and slightly dislocated. Below this point the ore widens to 8 inches, and so passes out of view.

The conditions represented in Fig. 14 existed in the stopes upon the same vein, about one hundred feet further north on the strike. It is particularly interesting as illustrating the changes not only in the vein but in the individual numbers of the series of sandstone, shale and limestone-beds through which the vein passes. Commencing at the top, K, of the section, the ore is seen to be seven inches wide. About 1½ feet lower, opposite L, there is an incipient break in the vein, not sufficient to part it. At five feet from the top, opposite M, a rupture has taken place, accompanied by only a very slight displacement. Eleven inches lower, opposite N, there is another break. M and N correspond to B in Fig. 13. The ore here is 9 to 10 inches wide. The vein is well defined and nearly vertical. Opposite O, three feet below N, there is a more serious break. The vein is displaced its own width, and for a height of six inches is broken into several definite fragments. This severe shattering is due to the fact that, in the absence of coincident partings, no clean horizontal shifting of the country took place, but the movement became more of a distortion. Compare the corresponding break in Fig. 13, opposite D. The vein now decreases, as in Fig. 13, and leaves, opposite P, the foot-wall of the main fault-fracture. Opposite Q the ore is about 1½ inches wide and two partings exactly coincide, so that the horizontal shifting of the country is expressed by a clean-cut fault in the vein amounting to a displacement of 1½ feet. A piece of ore occurs in the clay-seam, midway between the divided parts of the vein. Below this the ore increases. Opposite R there is an incipient fracture, corresponding to H in Fig. 13. The ore then widens steadily, and, as it leaves the section at S, has a width of about one foot, neatly ribboned. In this section the vertical throw along the fault-fracture followed by the vein is a little less than in Fig. 13.

In Figs. 15 and 16 there is a representation of the same vein as seen in a series of six stopes covering a vertical height of 33 feet. These stopes were at least two hundred feet north of the place illustrated in Fig. 14, but in the same horizon of country-rock. The lime and sandstone at the top of Fig. 15 can be identified with the beds P, Q and W in Fig. 14, and with Z and U in Fig. 13.

At the top of Fig. 15, opposite A, the vein is only 5 inches wide, but is built up of rich sulphides, zinc-blende and galena. A little lower, opposite B, the ore leaves the foot-wall of the fault-fracture and then undergoes a dislocation of over two feet. This movement along the bedding of the country occurs where partings coincide. On the west the plane of movement, C, is not accompanied by much selvage, but eastward there is a seam of clay widening to 3 or 4 inches in thickness. Incidentally the want of correspondence in the composition of the country on the two sides of the vein attracts attention to the fact that it follows a fault-fracture, the vertical throw of which is three feet. Opposite D, one foot below the break, C, the ore exhibits what miners call a "splice," that is, one distinct band of ore thins out and another at the same time commences to appear. The vein opposite E is about six inches wide, and rich. Just above it was very poor and quartzose. Nothing noteworthy occurs until, three feet lower (F), another splice is seen. The vein is still small but fairly rich, and continues thus very uniformly for another five feet, when (G) several features attract attention. The vein appears fractured, although not separated, and abuts against a dike of porphyrite nearly one foot wide, which forms the hanging-wall of the ore for a distance of five feet (G, Fig. 15, to H, Fig. 16). When traced up and down the series of stopes it was found to cease in both directions and to be therefore an intruding tongue. Its behavior is clearly shown in the drawing, beginning opposite X, Fig. 15, and ending opposite Y, Fig. 16. It varies little from a uniform width of slightly less than a foot, except for a short distance (H to W), where it is squeezed to a mere thread a few inches wide. The porphyrite is evidently very sensitive to the structure of the country-rock into which it has thrust itself. Opposite G the ore is about 7 inches wide. It continues so for four feet downward, when (K) another splice occurs and the vein widens. For the next six feet the ore varies from 6 to 12 inches, and is then (L) slightly dislocated. Below this the vein widens steadily and develops a beautiful ribbon-structure, passing out of the section (M) with a thickness of 14 inches. At the bottom of the section an instructive feature is presented. The ore-bearing vein leaves the main fault-fracture (here shown by a line to the right), and pitches slightly to the west. This is only temporary. Lower down, outside the section here shown, it turns back and resumes its course along the fracture which determined its existence. It will be noted that the ore in leaving the main fault-line (about opposite Y), follows a line of minor fracturing which shows a slight but evident dislocation also.

These four sections, Figs. 13 to 16, show how the behavior of the vein is determined by the structure of the enclosing rock. The fracturing through the country, which opened a possible channel for the circulation of the mineral-bearing waters, was continuous but irregular. The irregularity was due, in the first place, to the varying composition of the beds through which the fracture passed. That fracturing was accompanied, as we have seen, by a vertical displacement of from two to three feet. The direction of the throw is beautifully evidenced by the bent edges of the partings in the country on either side; on the hanging, upward; on the foot-wall, downward.

The horizontal shiftings which now break the vein occurred in part before ore-deposition, but mostly afterward. The later movements were also the parents of the quartz-stringers which now fringe the vein, as at C and F in Fig. 13, and M. and P. in Fig. 14. They represent the healing-action of later solutions, circulating along a reopening of the old passage-ways.

The three sections, Figs. 13, 14 and 15, illustrating the same horizon in the country, at considerable distances apart, serve to prove how individual members of the sedimentary series change in composition. The two beds, E and Z, both limestone, in Fig. 13, become, in going northward, the two beds, P and Q, one limestone and one sandy lime. In Fig. 14 the relative thicknesses have changed. Again, in Fig. 15, the same beds, now seen 200 feet further north, are to be identified in the single bed of limestone extending from A to C. The underlying sandstone also shows variations at the different points. It is noteworthy, at the same time, that the vein acts much in the same way when traversing identical beds at different points.

The subsidiary fracturing, seen in Fig. 15, to the east of the vein, between C and G, is instructive. That this structure usually accompanies vein-faulting I certainly believe. The breast of a slope or the face of a drift rarely exhibits it, because mining does not require that the ground should be broken for a width necessary to make it visible. This structure, the sheeting of the country, is very marked at Cripple Creek, in Colorado, and is the origin of parallel and multiple veins.

V.—The Cross-Veins.

These veins, although non-productive, play an important part in mining operations, because they dislocate the ore-bearing "verticals," and are themselves related to extensive ore-bodies on the contact. The picture they present is that of white bands of crushed quartz, cutting through everything, as distinguished from the verticals, which are like pink ribbons of rhodochrosite, traversing the sedimentaries with difficulty because of their faulting by these cross-veins.

The latter are built upon fault-lines marking movements greater in extent than those accompanying the older, ore-bearing fractures. They are essentially quartz-veins. A variable amount of crushed country accompanies the quartz. Rhodochrosite and valuable* sulphides are notably absent. When

* Many assays were made of the pyrites from cross-veins encountered in cross-cuts. Traces of gold and 4 to 8 ounces of silver were about the best results. Oc-

seen, they represent broken fragments of pay-veins traversed by the path of the cross-vein. Next to quartz, iron pyrites is their most characteristic mineral. The pyrite is in a crumbly, easily disintegrated state, very unlike the solid crystalline condition in which it appears amid the ore of the verticals. Sometimes no foreign minerals are present in notable quantity, and the cross-vein is simply a seam of crushed country, softening into mud. This the Cornishman would call "fluccan."

Fig. 17 represents a cross-vein in the breast of a drift on the Jumbo No. 3 vein, which had met the cross-vein and had been faulted by it. The Jumbo vein was intercepted by this drift 8 feet further ahead. Three breaks are noticeable, each marked by faulting. The cross-vein itself lies between two unlike beds of lime, marking a fault which evidence elsewhere along the drift showed to be 6 feet. The lime on the foot-wall is separated from a series of thin beds, exhibiting a good deal of variety, by a curving line of selvage, following a fault whose throw is 4 feet. Then come beds of shale, crystalline lime and sandstone, traversed by a dislocation of 10 inches. In each case the down-throw is on the hanging-wall. The cross-vein is almost vertical and carries fragmentary rhodochrosite torn from the Jumbo vein.

Fig. 18 illustrates another cross-vein in the act of cutting through the Jumbo vein. The line of the former is deflected in breaking through the latter. To the left the series of beds consists of blocky lime, lime shale, broken lime, limestone, black shale, closely laminated lime and light gray sandstone. On the right this series is seen to be succeeded by beds of dark sandstone. The fault along the cross-vein is 4½ feet, obliterating the displacement which must have accompanied the line of the ore-bearing vein. The latter is thrown its own width, 10 inches. It has been shattered by the cross-vein, but appears to have been repaired and reconsolidated (since its displacement) by

---

casionally good assays were obtained, but the pieces from which these samples came were invariably marked by the presence of fragments torn from the pay-veins. Nevertheless, from a scientific standpoint, pyrite containing a few ounces of silver is as much ore as blende or galena, carrying much higher values. In this connection it may be added that the pyrite of the verticals is not notably silver- or gold-bearing unless admixed with copper pyrites, blende or galena. The pyrite and quartz, whether in the cross-veins or the verticals, is not a sign of valuable ore.

healing seams of quartz, which now ramify through the previously broken rhodochrosite. This cross-vein was distant 28 feet from the one described above.

The comparatively small scale of the rock-formation affords many excellent illustrations of phenomena usually requiring

Fig. 17.

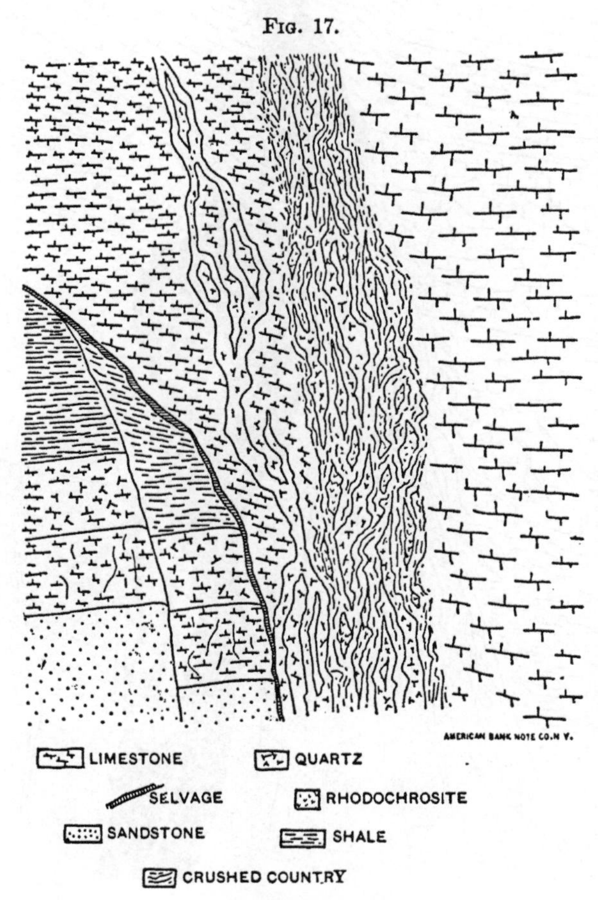

A Cross-vein in Limestone.

large areas for their exemplification. Some of the dislocations are so slight that they can be seen to die out in mere distortions. See Figs. 19 and 20. The former is particularly instructive. It shows the sides of a cross-cut which joins the workings on the Hiawatha and Enterprise veins. Near the floor of the cross-cut there is a fault of about 5 inches which follows a seam of clay nearly 1 inch thick. This dislocation diminishes upward; the thin beds break less and bend more, until finally the verti-

FIG. 18.

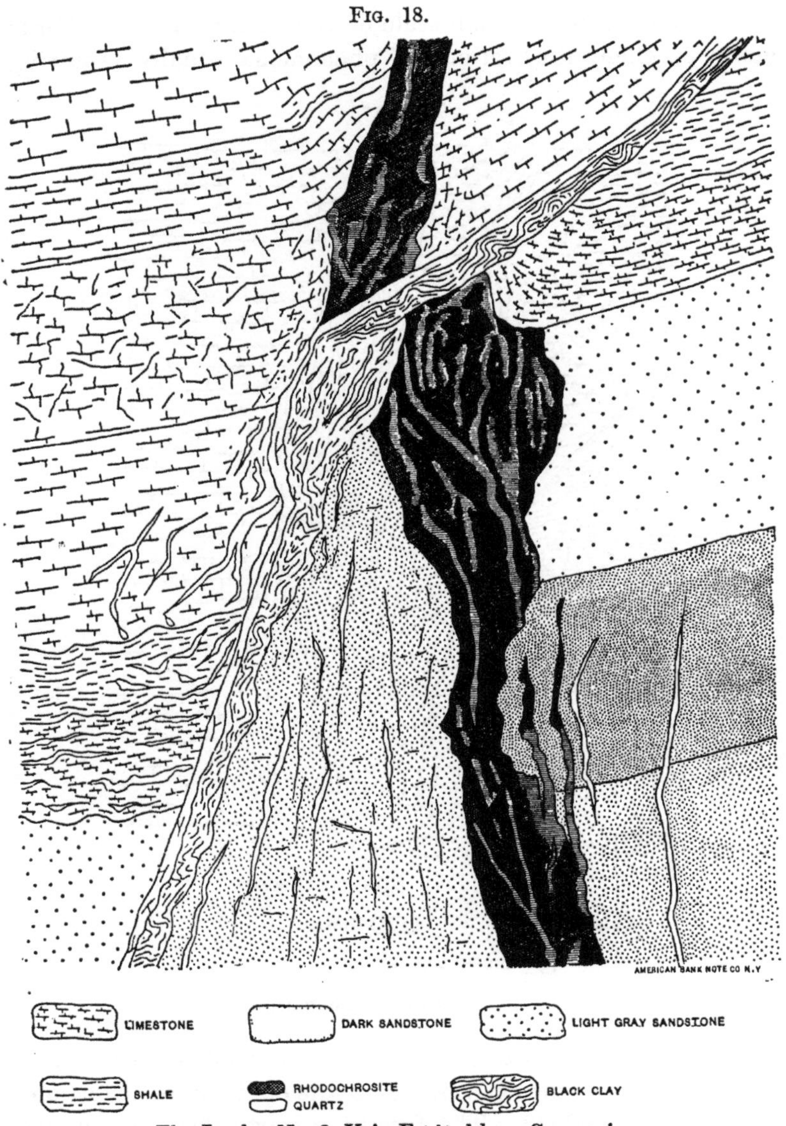

The Jumbo, No. 3, Vein Faulted by a Cross-vein.

cal displacement fades out in horizontal shifting. The latter is clearly evidenced by the minute multiple step-faulting of a

FIG. 19.

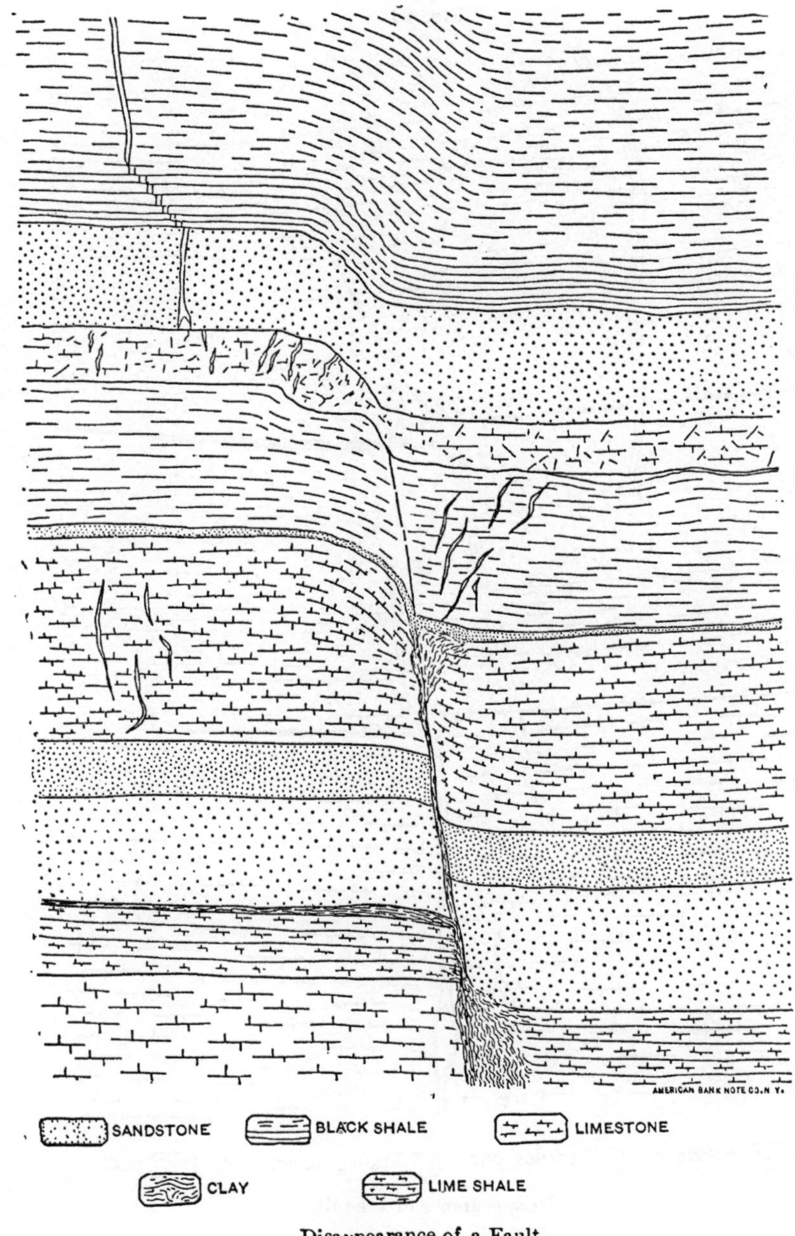

Disappearance of a Fault.

quartz-seam in the shale, near the roof of the cross-cut, which

FIG. 20.

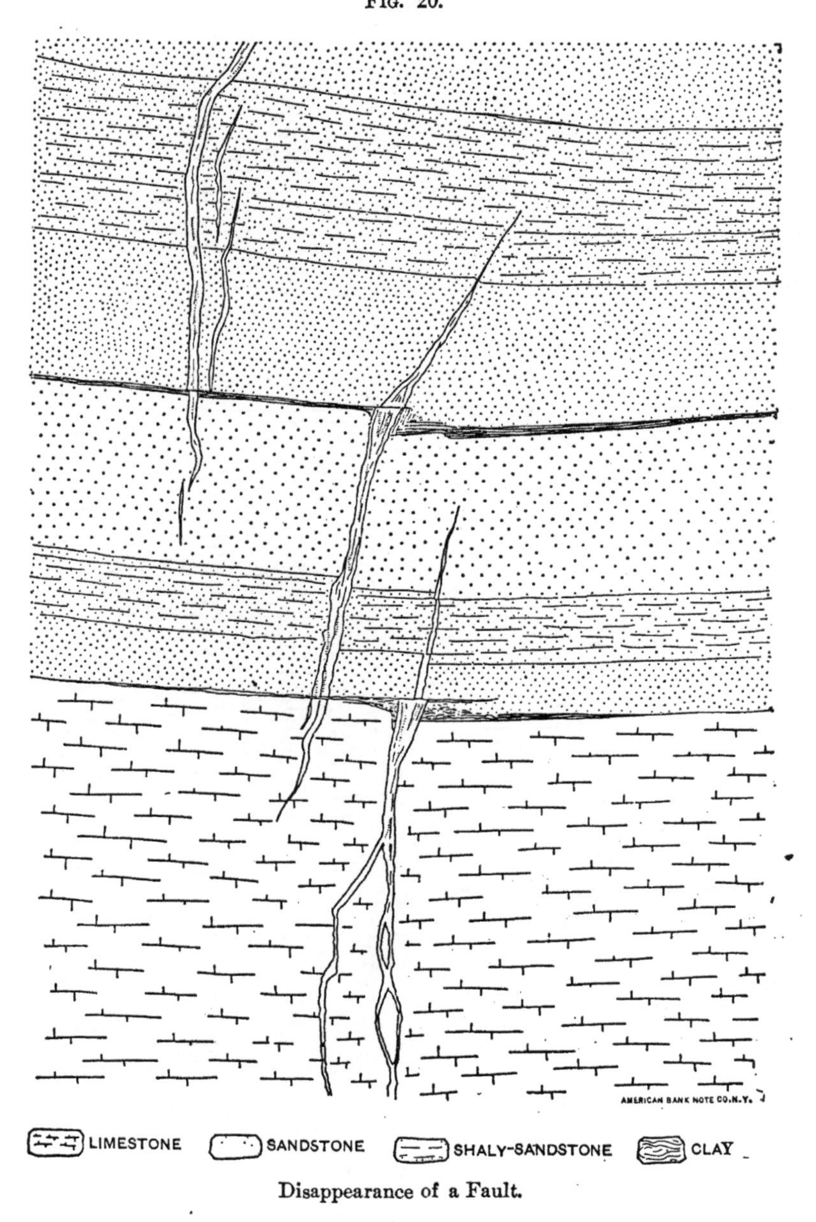

Disappearance of a Fault.

indicates that the flexibility ot the shale enabled it to resist fracturing by dissipating the force of the vertical movement in a sliding of the laminæ over one another. Fig. 20 conveys the same lesson. The three quartz-seams here shown follow lines of lessening vertical movement; the distinct faulting of the beds in the lower part of the section dwindling until in the upper portion only a slight distortion of the country is to be discerned. These examples illustrate how large faults may gradually disappear, and suggests why the vein-systems of Newman hill terminate in their approach to the horizon known as the "contact."

FIG. 21.

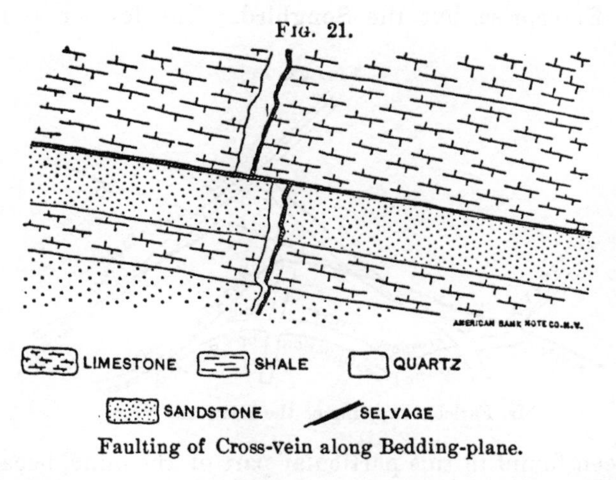

Faulting of Cross-vein along Bedding-plane.

The cross-veins are themselves displaced in some instances by slight later movements which have taken place along the bedding-planes of the country, such as indicated in Fig. 21, where a small south-dipping cross-vein is thrown 5 inches along the parting between beds of limestone and sandstone.

The dislocation of the ore-bearing veins by the cross-veins necessitates the employment of scientific methods in mining. The mine-workings of Newman hill are needlessly tortuous and complex, because most of them have been directed by men blind to the indications of geological structure. The expensive results of a bewilderment due to this cause are well illustrated in the case of the Hiawatha vein, between raises No. 9 and No. 11 on the main level. This particular instance is quoted because it appears in Mr. J. B. Farish's paper on Newman hill.* His

* *Proceedings of the Colorado Scientific Society*, vol. iv., p. 159.

drawing is here represented (in Fig. 22), together with another (Fig. 23) based on a very close sifting of the evidence, accompanied by a careful survey.

Referring to Fig 22, I quote Mr. Farish's description:

"It illustrates the occurrence of a fault in the Enterprise vein and the deflection of the Hiawatha vein by the same cross-fissure. The break in the Enterprise vein is seen to be sharp, while the Hiawatha vein is not faulted, but makes along the cross-vein for nearly 100 feet before emerging from its walls and resuming the original course."

The vein called the Enterprise in the above description is not the Enterprise, but the Songbird. The former vein has

FIG. 22.

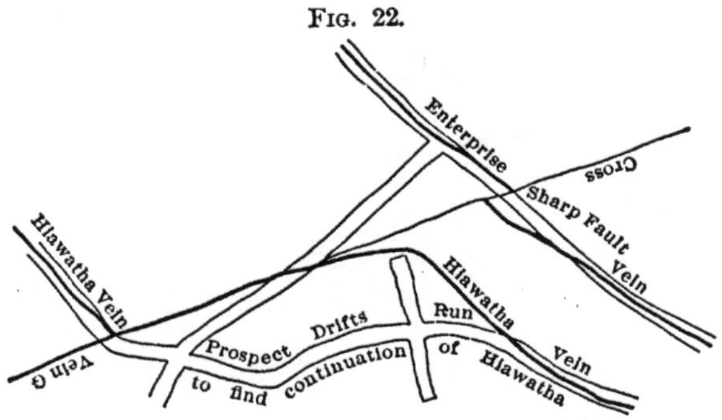

Mr. Farish's Drawing of the Hiawatha Vein.

never been found in this particular part of the mine, because it breaks up into unimportant stringers and is not definitely recognized until nearly 400 feet further north. The Enterprise dips east slightly, the Songbird dips west flatly. In their strike northward the two veins unite.

The cross-vein faults both. If Fig. 22 were correct the formation of the Hiawatha would be later than that of the cross-vein, and the latter must be later than that of the so-called Enterprise vein, and the ore-bearing veins would not be of contemporaneous origin. Such, however, are not the facts.

The cross-vein strikes the Songbird almost at right angles and throws it to the right. When it meets the Hiawatha, it throws that vein to the left. The Hiawatha and Songbird dip in opposite directions. Both faultings are in accord with Carnall and Schmidt's rule.*

---

* In *Fault-Rules*, by Francis T. Freeland, *Trans.*, xxi., 499.

The representation (in Fig. 22) of the Hiawatha as following the cross-vein is, in my judgment, incorrect; the error having been made by mistaking the "drag" for ore in place. The same error is repeated in another drawing (Fig. 24) by Mr.

FIG. 23.

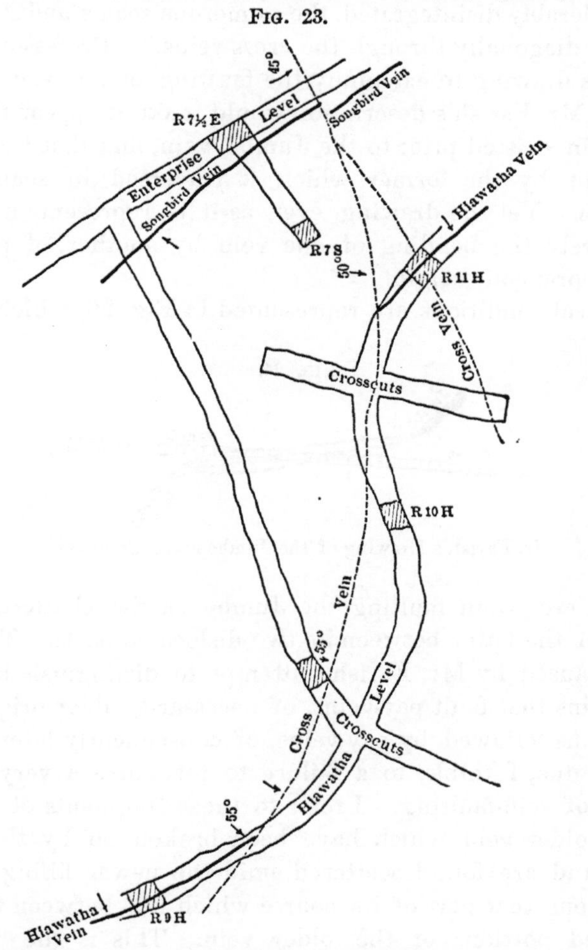

The Faulting of the Hiawatha and Songbird Veins by a Cross-vein.

Farish,* which Professor Kemp has perpetuated in his treatise on ore-deposits.† The value of the latter as a book of reference makes it imperative that the mistake should be pointed out.

* Pages 158 and 159 of the paper cited.

† *The Ore-Deposits of the United States*, by James F. Kemp, p. 20.

Mr. Farish says that the occurrence represented by him in Fig. 24 is an instance of the "bend of a vertical pay-vein as it approaches an intersecting cross-vein," and also, that "the Jumbo vein, as it approaches and departs from the cross-fissure is considerably disintegrated, the numerous seams and stringers striking diagonally through the cross-veins." Professor Kemp uses this drawing to exemplify the faulting of one vein by another. Mr. Farish's description would make it appear that the cross-vein existed prior to the Jumbo vein, and that the latter was bent by the former, which it traversed in seams and stringers. Yet the drawing, even as it is, represents no fault, but merely the bending of one vein by another, of possibly contemporaneous origin.

The real conditions are represented in Fig. 26, which shows

FIG. 24.

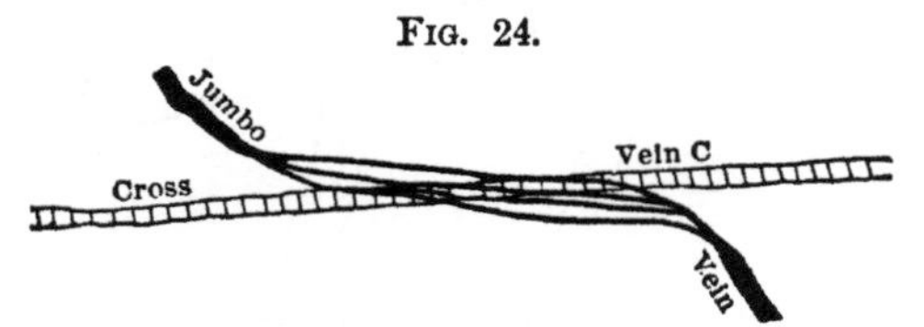

Mr. Farish's Drawing of the Jumbo and a Cross-vein.

how the cross-vein faulting the Jumbo carries shattered fragments of the latter between its two dislocated parts. The confusion caused by Mr. Farish's attempt to distinguish between cross-veins that fault pay-veins, of necessarily older origin, and cross-veins followed by pay-veins, of consequently later formation, is due, I think, to a failure to recognize a very simple feature of vein-faulting. I refer to those fragments of the ore of the older vein which have been broken off by the fault-fissure and are found scattered amid the newer filling of the latter along that part of its course which lies between the two disrupted portions of the older vein. This is the "drag," which is so valuable an aid in mining, because it enables the miner to trace the direction of the throw.

The faulting of the "verticals" by cross-veins is a prominent feature in the Enterprise mine. When half a dozen drifts were running on the several pay-veins a fault was encountered about once per month. In other words, the distance between the cross-veins averaged from 65 to 100 feet. Failure to apply the

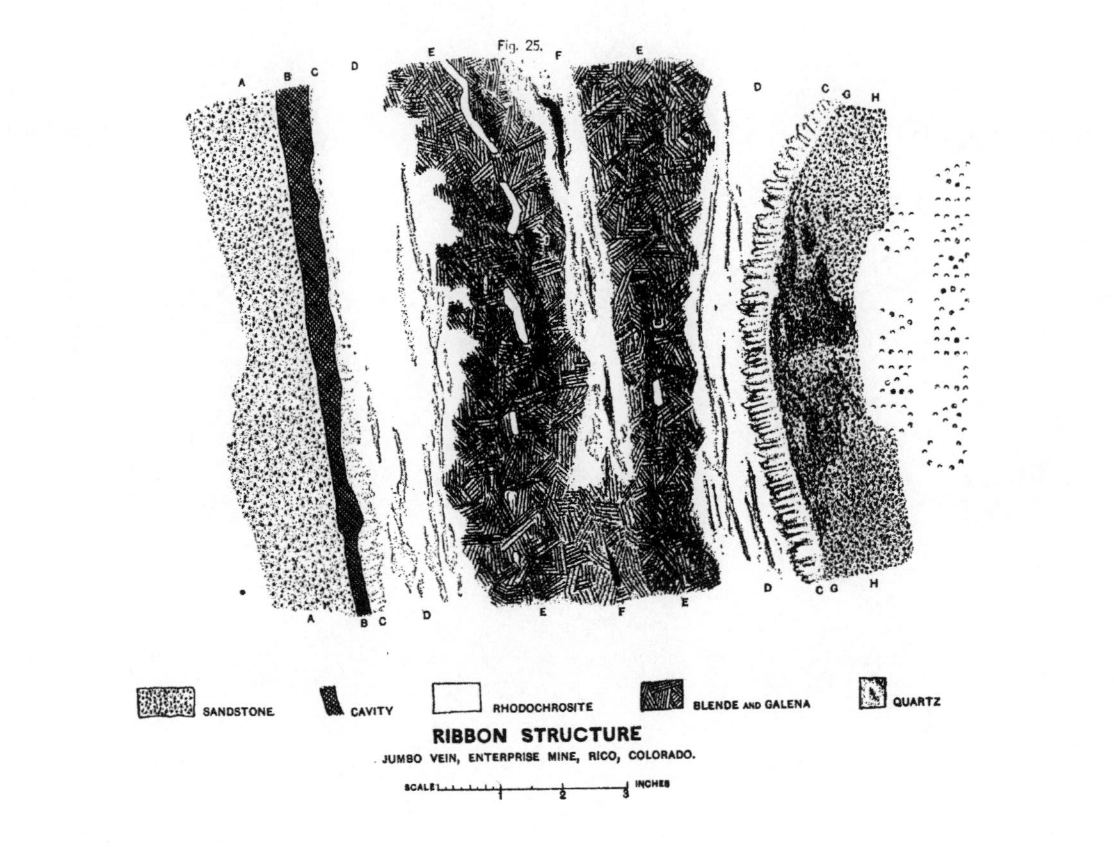

RIBBON STRUCTURE

JUMBO VEIN, ENTERPRISE MINE, RICO, COLORADO.

elementary rules of faulting and to make such observations, measurements and calculations as would serve to identify a cross-vein as it passed through the series of nearly parallel verticals, has caused much unnecessary expenditure on Newman hill. More broken ends of faulted-veins were happened upon by accident in "drifting" than were discovered by intelligent search. The practical importance in mining of the study and interpretation of geological structure is forcibly emphasized by this record of experience.

Out of over two hundred instances of faulting noted during my twelve months' direction of the Enterprise mine, I detected only one (and that an uncertain) exception to Carnall and Schmidt's rule, which is stated by Mr. Freeland as follows:*

"If the fault be encountered on its hanging-wall side, after breaking through it, prospect toward the hanging-wall side of the vein; on the contrary, if from the foot-wall side, then prospect toward the foot-wall side of the vein."

"This rule, as Mr. Freeland says, "applies only to normal faults, and is, in addition, subject to an important exception."† I do not cite it as the statement of a universal law. In other districts other rules may be found applicable. Nor is it necessarily to be expected that in any one district a single rule only will obtain. But if in a given locality the prevalence of a given fault-rule (expressing the habit of the veins in that locality) can be demonstrated, a working-hypothesis of immense value is thereby furnished to the miner. My remarks in this connection are to be taken, therefore, as applicable only to the district here under consideration, or to other districts in which similar conditions may be determined by experience.

When a cross-vein intersects a pay-vein at an acute angle it is practicable to follow the trail of the latter, as evidenced by the scattered fragments constituting the "drag." When, however the faulting occurs at a large angle there is less evidence of the direction of the throw, and a level driven on the cross-

---

* *Trans.*, xxi., 499.

† "Normal faults" are those in which the rock forming the hanging-wall of the fault has moved downward along the foot-wall of the fault. The exception to Carnall and Schmidt's rule mentioned by Mr. Freeland in the above quotation is that of an obtuse fault-angle. The rule might be so stated that this would be no exception, as the case is simply one of an acute-angled fault, *approached from the other side*.

vein until it meets the other portion of the disrupted vertical would involve turns too sharp for subsequent use in tramming. It might thus be necessary, after finding the continuation of the pay-vein, to run a new piece of level for practical mining purposes. But if the direction of the throw be known from pre-

FIG. 26.

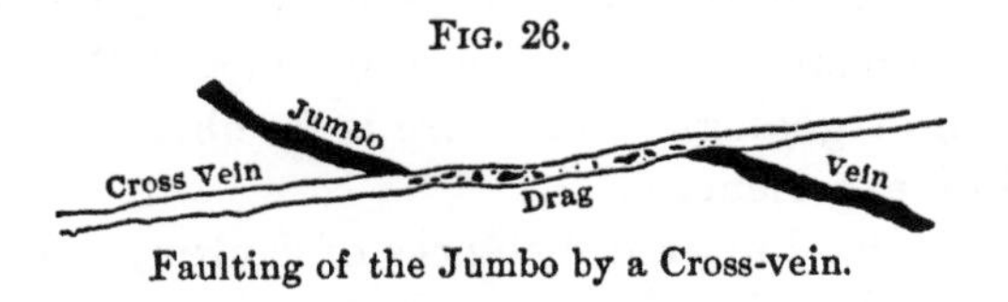

Faulting of the Jumbo by a Cross-vein.

vious experience, the heading can be so turned as to strike the vein beyond its dislocation by a course of minimum deflection, making a practicable mine-level.

When a cross-vein meets a vertical at a very small angle a

FIG. 27. FIG. 28.

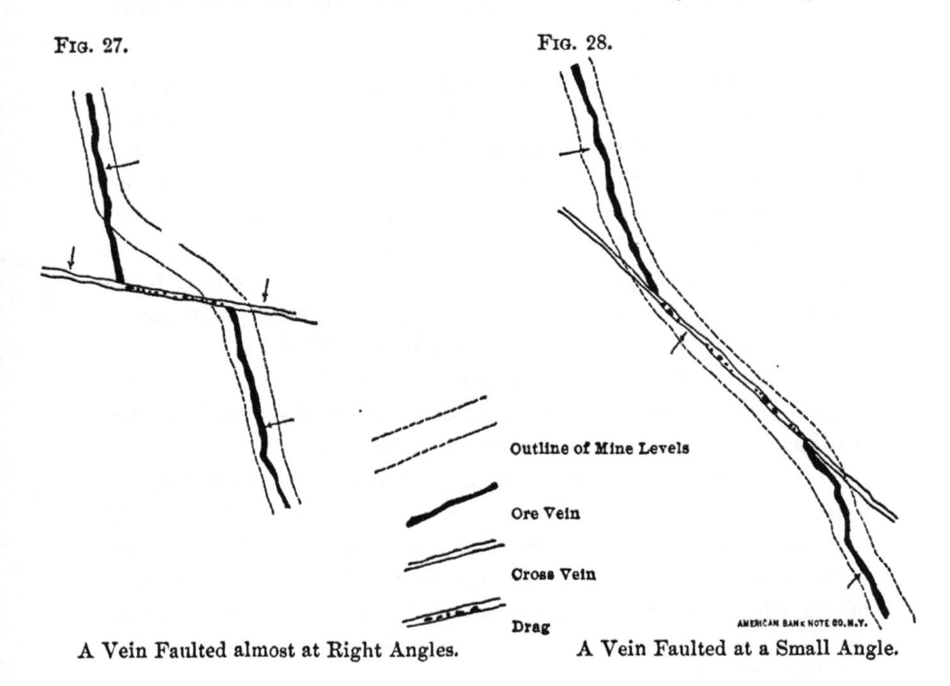

A Vein Faulted almost at Right Angles. A Vein Faulted at a Small Angle.

slight throw, as measured at right angles to the strike, becomes sufficient to produce a wide separation between the broken ends as measured along the course of the pay-vein. In such cases it is easy for the unobservant miner to be misled by the unapparent deviation in his level and to mistake a cross-vein carrying fragments of lode-matter for the pay-vein itself. (See the diagrammatic sketches in Figs. 27 and 28.)

FIG. 29.

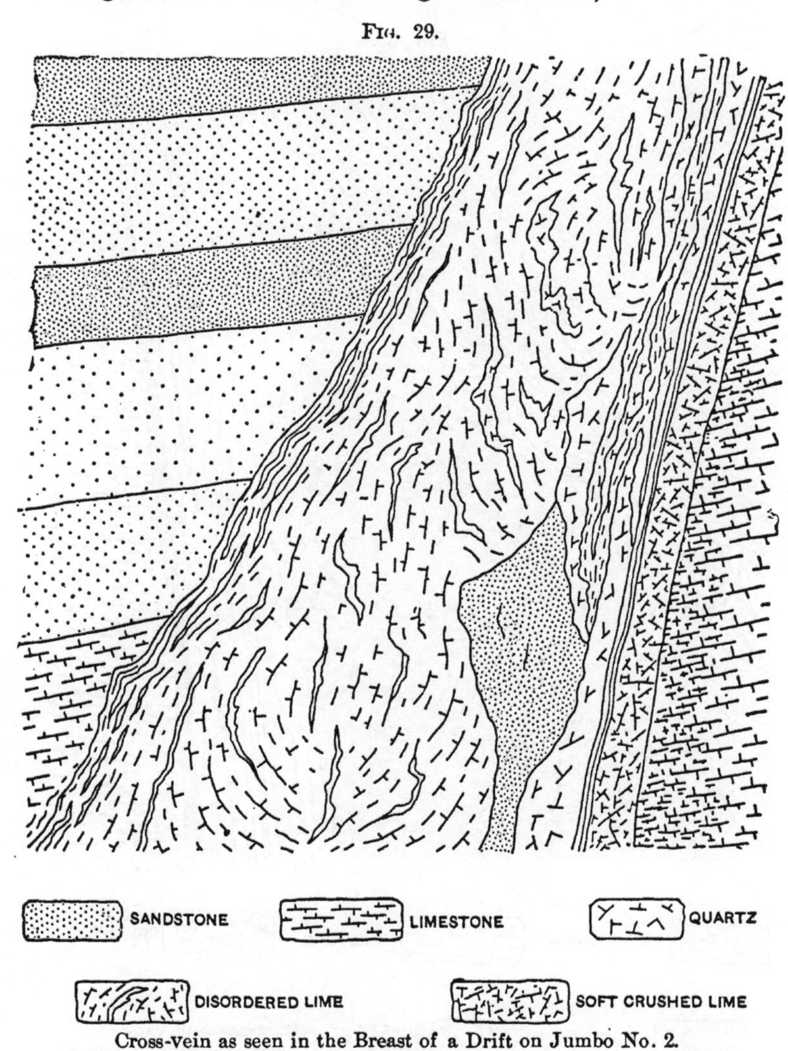

Cross-vein as seen in the Breast of a Drift on Jumbo No. 2.

The rhodochrosite is a great aid in tracing the ore-bearing veins because its bright color renders it readily distinguishable amid the white quartz or the dark shattered country of the cross-veins, and its absence distinguishes the latter just as its presence characterizes the pay-veins.

Figs. 29, 30 and 31 illustrate the behavior of two cross-veins, encountered in following the Jumbo No. 2 lode. Fig. 29 shows the cross-vein in the breast of the level. It consists of 2 to 3

FIG. 30.

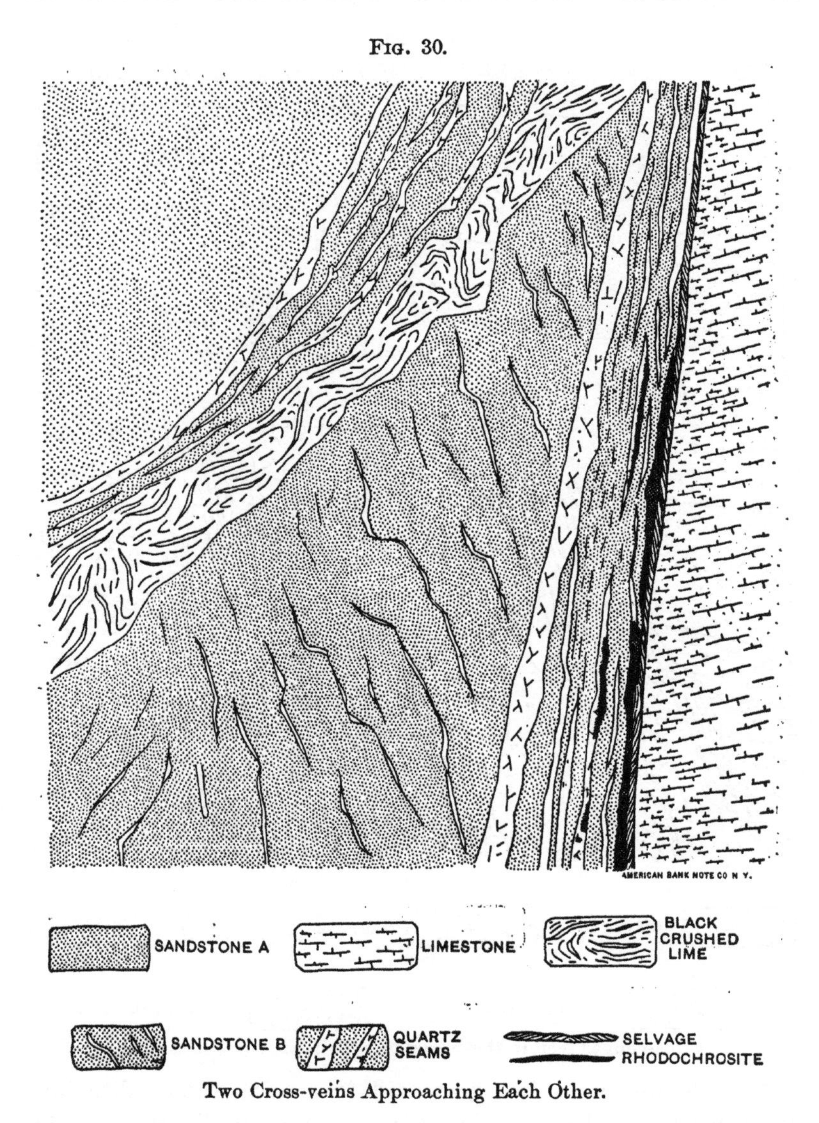

Two Cross-veins Approaching Each Other.

feet of crushed country, enclosing stringers of quartz. There is a fault, but its extent cannot be determined. The wedge of sandstone observable near the floor of the drift marks the be-

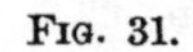

FIG. 31.

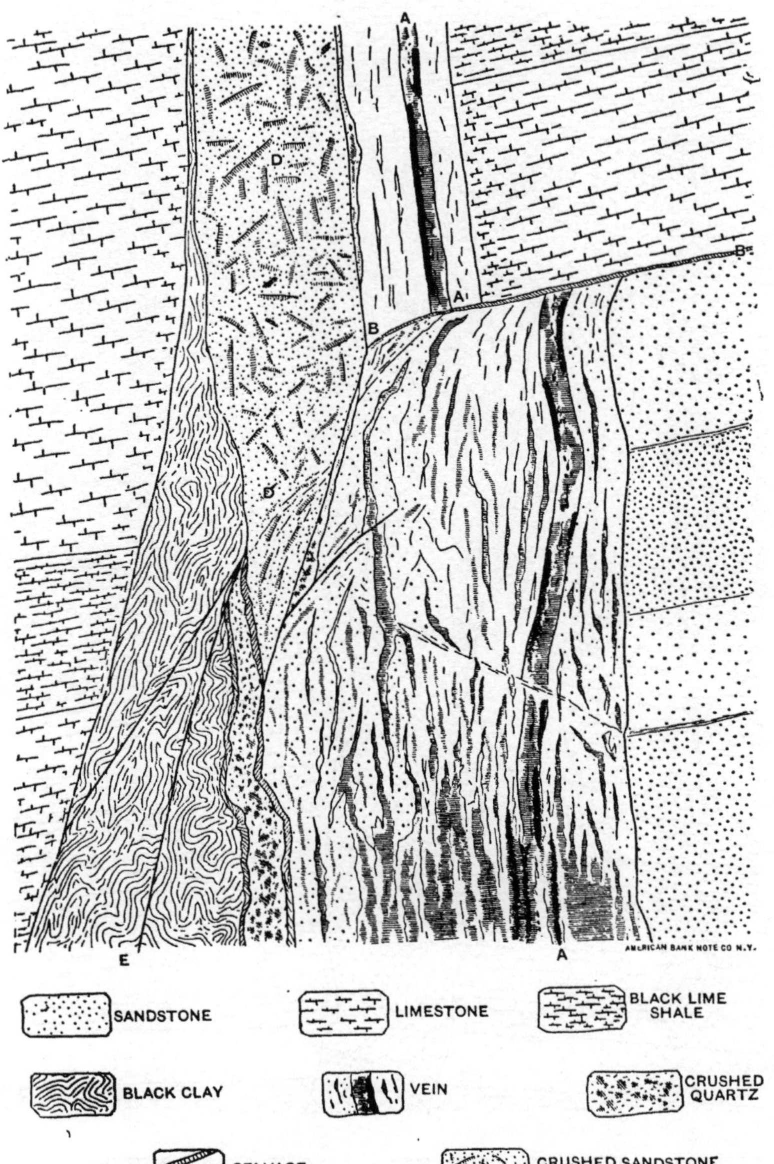

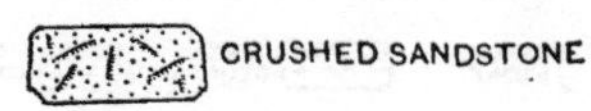

A Vertical and a Cross-vein Intersecting.

Fig. 32.

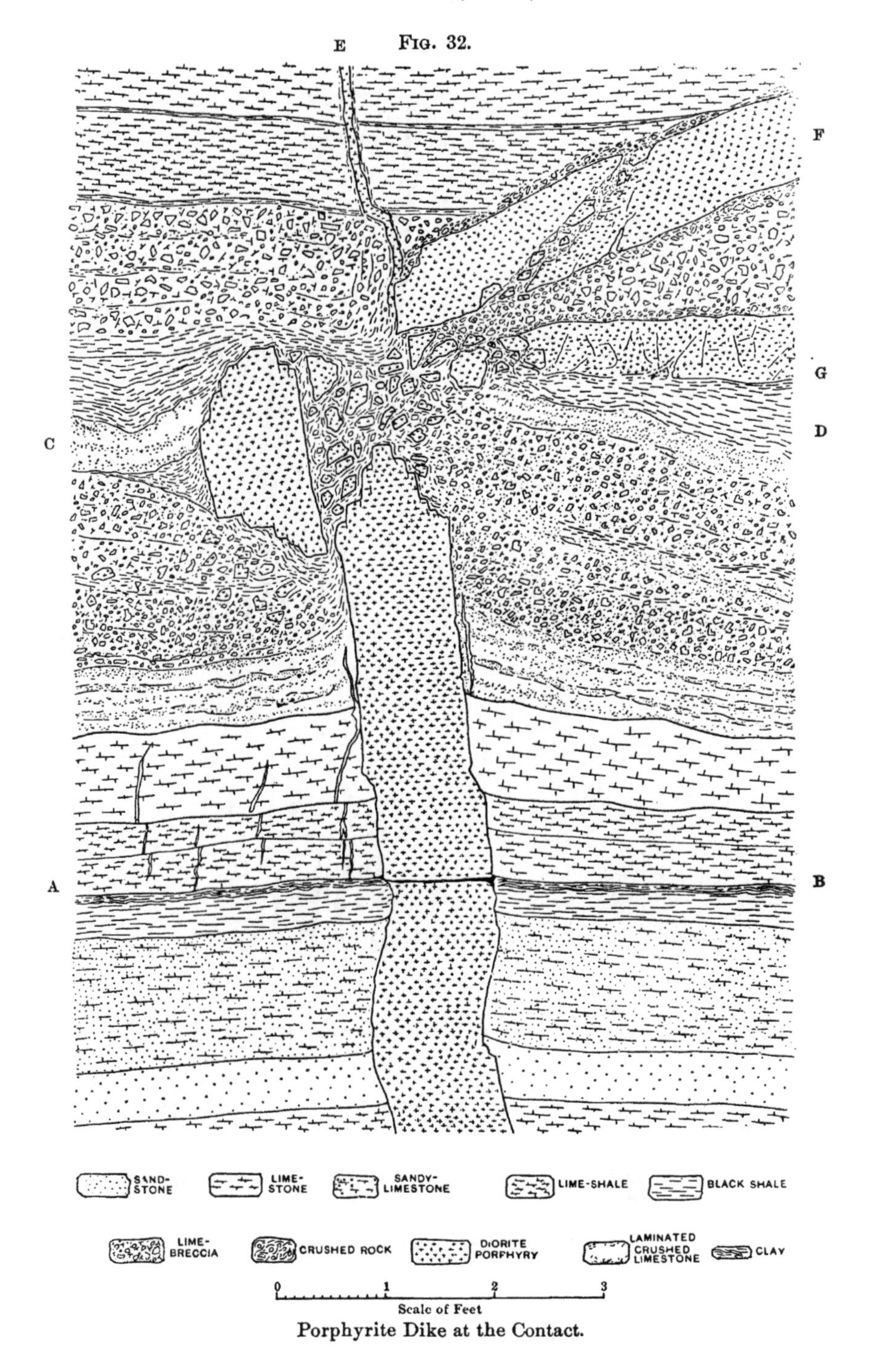

Porphyrite Dike at the Contact.

ginnings of a separation which Fig. 30 shows to be the departure of one of two consolidated cross-veins. One of these

Fig. 33.

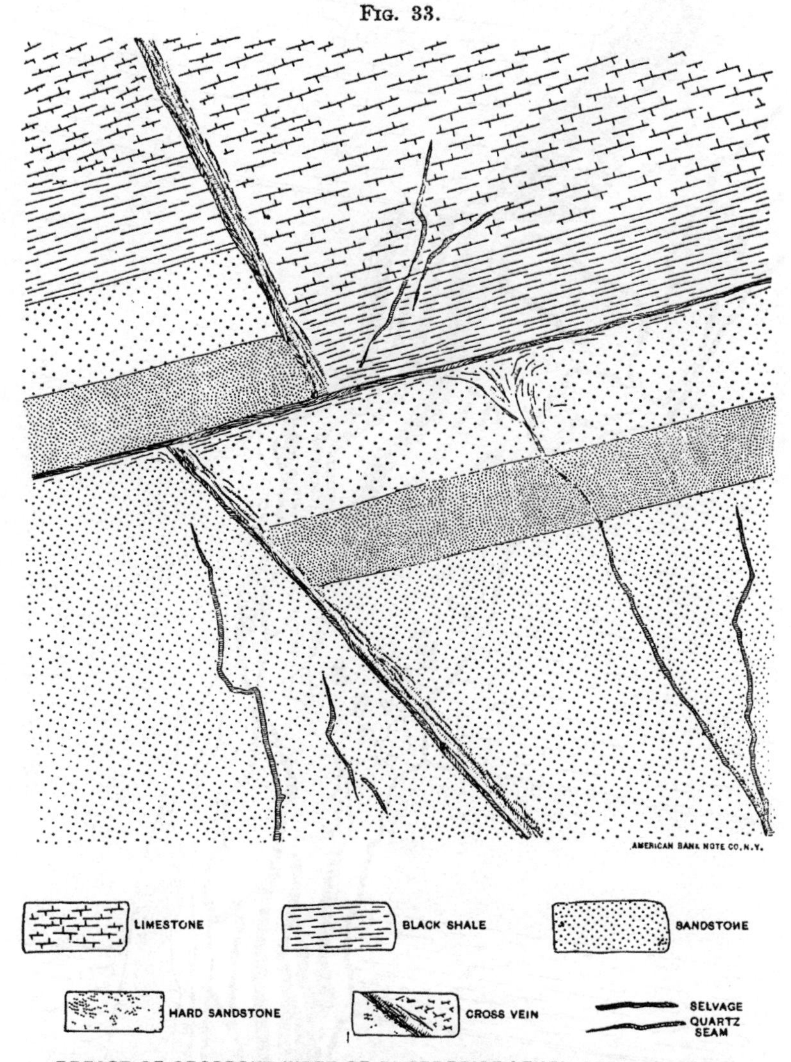

BREAST OF CROSSCUT WEST OF ENTERPRISE LEVEL, APRIL 10TH. 1894

began at this time to exhibit pieces of rhodochrosite, which soon afterwards led to the finding of the vein. Fig. 31 illustrates another cross-vein. The structure is complicated. The narrow seam, A A, and its enclosing vein-filling have been

FIG. 34.

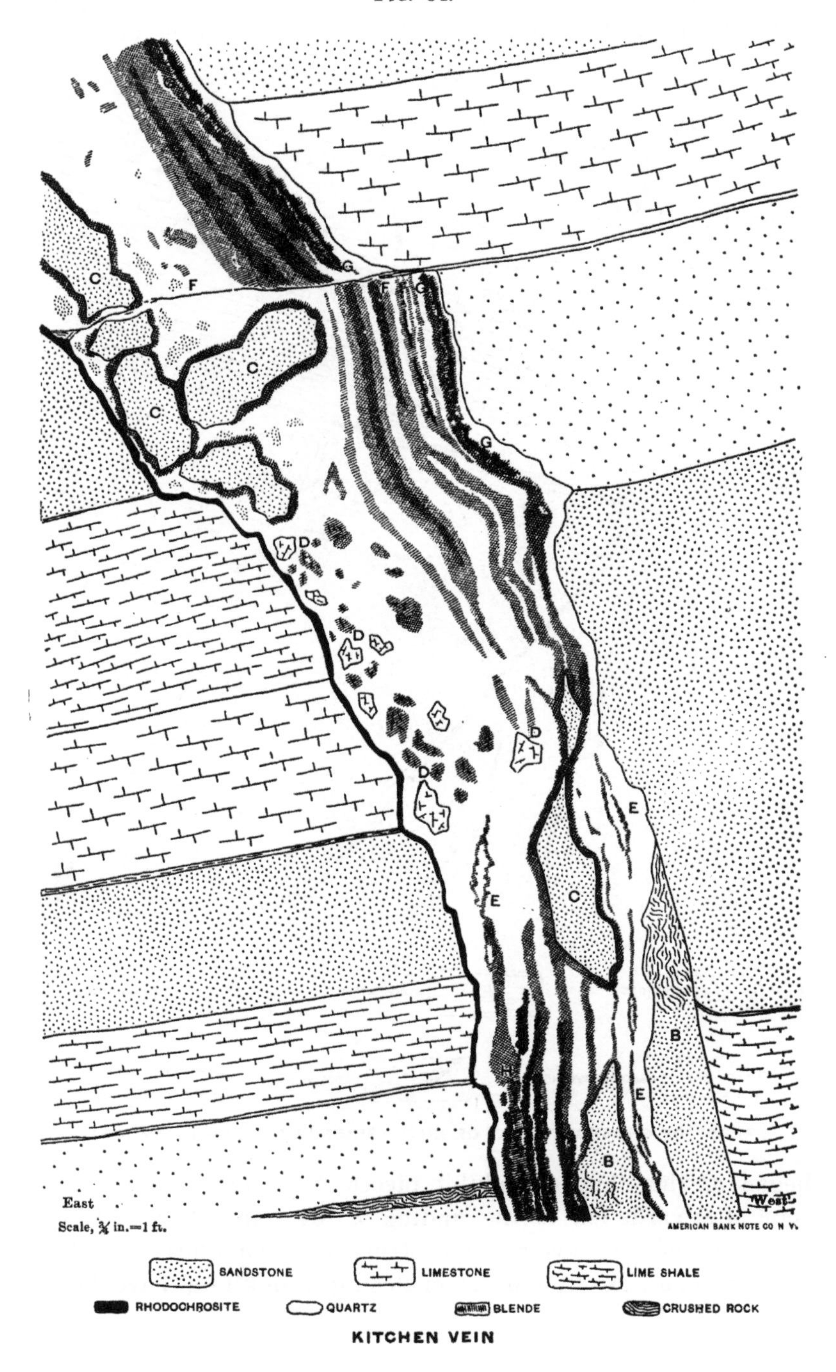

KITCHEN VEIN

faulted along a bedding plane, B B, and subsequently a movement along the hanging-wall of the cross-vein has led to another dislocation, a down-throw of the country and a distortion of the former line of movement. It is interesting to note that the secondary companion cross-vein thus formed consists of crushed sandstone, D, in the upper part of the section, and of black lime, E, which has been crushed into clay, in the lower part. This change corresponds with the alternation of rock in the enclosing country.

Fig. 33 illustrates a fine example of double faulting. The steep cross-vein has caused a displacement of a little over 2 feet, which is very clearly marked by the dark band of sandstone. Subsequently a shifting along bedding-plane has faulted the cross-vein about 20 inches.

Fig. 34 is a section of the Kitchen-vein, as seen in the face of a south drift. This vein dips westward. The example is of value because of the inclusion of fragments of country-rock. Of these, some (CC) are sandstone and others (DD) lime. Their distribution corresponds with that of the rocks on the foot-wall, and suggests that they were torn off the enclosing country. The mineralization of the rock and the substitution of ore is suggested by the fact that each of these included pieces has a well-marked rim of rhodochrosite. On the hanging-wall there is no parting from the country; on the foot-wall the ore is readily detached, because of the existence of a continuous selvage of black mud. Ribboning, due to alternations of quartz, rhodochrosite and blende, is well marked in the upper hanging-wall portion of the section. This vein follows a fault-fracture, the throw of which exceeds 5 feet, an amount unusual to the pay-veins of the Enterprise mine.

The occurrence of dikes and tongues of porphyrite has been noted incidentally. The deep shafts and bore-holes have encountered large thicknesses of it among the sedimentary rocks underlying the horizon of the existing mine-workings. The sedimentaries have been found hardened and otherwise affected in the vicinity of the eruptive rock, so that sandstone appears as quartzite and limestone is converted into marble. What the real character of this body of porphyrite may be cannot be determined from the evidence at my disposal. The more general observations of the gentlemen of the Geological Survey will doubtless throw much light on the matter. To the mining engineer one fact stands out very prominently, that whether in their strike northward, approaching the porphyrite of Silver

Fig. 35.

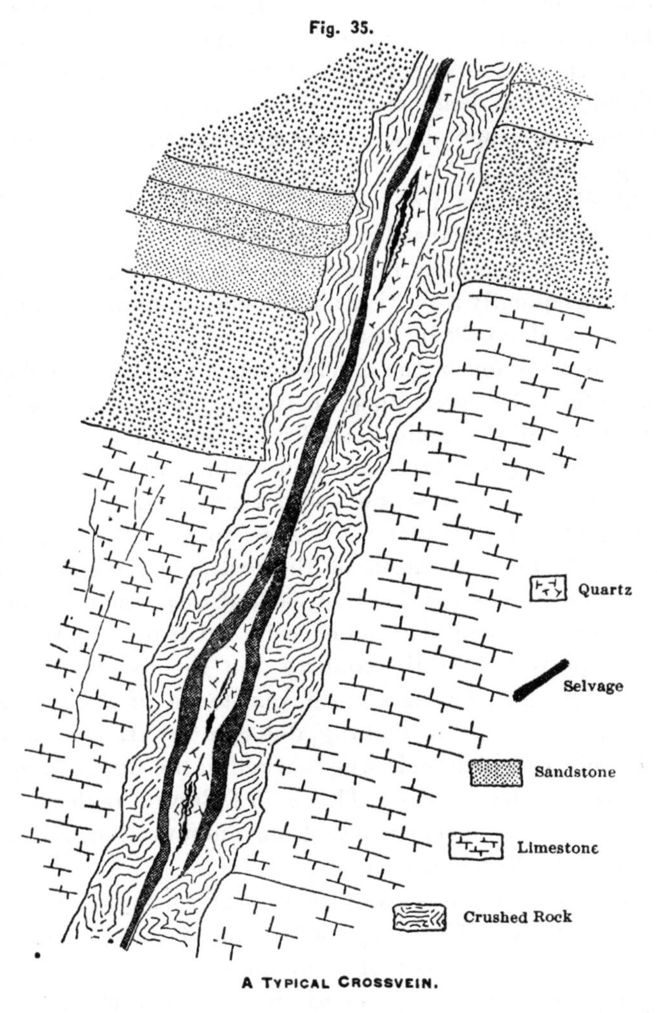

A TYPICAL CROSSVEIN.

creek, or in their extension downward, toward the lower-lying body of that rock, the veins become impoverished of pay-ore. In view of the general recognition of the beneficent association of veins with eruptive rocks, this is very instructive. The explanation suggests itself that though the shattering of the country accompanying the intrusion of the porphyrite aided vein-formation, because of the passageways thereby developed for the circulation of underground waters, yet at the same time the immediate neighborhood of hot masses of eruptive material did not favor the precipitation of ore, and that the best deposition took place at such a distance as permitted the cooling of the waters and the consequent laying down of the metals they carried in solution.

The veins penetrate the porphyry. The contact in many places carries fragments of porphyry. These facts prove, as would be readily surmised, that vein-formation succeeded the intrusion of the porphyry.

In the stopes irregular tongues of porphyrite were occasionally observable. As they only followed the vein for a short distance in its strike and dip, it was not possible to find out much about them. Such occurrences are seen in Figs. 5, 15 and 16. Fig. 32 is a drawing, which I made with particular care, showing how a dike is shattered at the contact. The porphyrite is 14 inches wide. Along the parting AB it is fractured, and a clay selvage across it suggests later movement (in the direction of the strike, so that the cross-section does not show it). At the level of CD the dike is smashed to pieces. A disturbance along the contact has shifted some of the fragments. The regular passage of the dike upward has been interfered with at the contact, and, previous to the shattering, it seems to have divided into branches, of which three (terminating in the sketch at E, F and G) still survive. The contact itself, here marked by a thickness of nearly 5 feet of crushed shale, crushed sandstone and lime, contains numerous fragments of porphyrite. Apart from its bearing on the origin of the dikes, this drawing illustrates how the contact barred the progress upward of fissuring of every kind.

VI.—The Contact.

As the ore-bearing veins are followed by the mine-workings* they are found to split up, weaken and become impoverished

* Entrance to the mines is made through long adits which aim to cut the veins at a depth of 50 to 100 feet below the contact. Hence the regular vein-structure is first seen, and the other complications come under notice as the stopes and raises are ascended. This order has been observed in describing them here.

at a certain horizon, lying from 5 to 20 feet below a zone of crushed rock, in which are enclosed ore-bodies richer than those of the veins themselves. It is rarely that even diminished seams of ore survive so as to connect with the ore on this contact. Ordinarily the highest stope on the "vertical" is separated from the nearest stope on the "contact" by a few feet of black shale

Fig. 36.

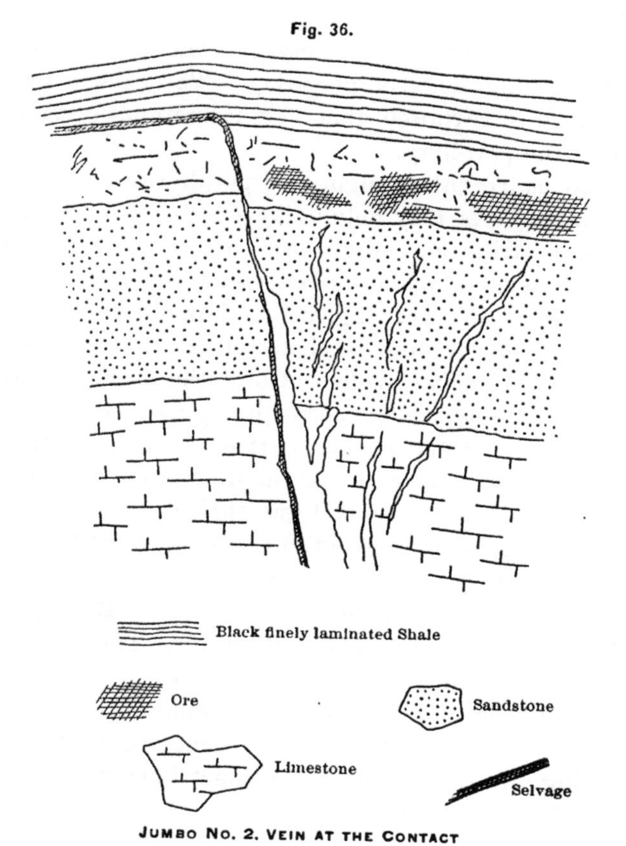

JUMBO NO. 2. VEIN AT THE CONTACT

and lime, containing, at best, only scattered remnants of the vein. Figs. 36 and 37 give simple examples. Occasional sections show no connection between the vertical and the contact overhead. Such small quartz or rhodochrosite threads as do occur have no continuity.

The cross-veins behave in relation to this contact in a manner very similar to the pay-veins. Since they are barren of ore, there are no workings which follow the cross-veins in their immediate approach to the contact, but there is scattered evidence* sufficient to confirm this statement. There is, however, a notable difference between the behavior of the two systems of vein fissuring when they reach the contact. The economically important observation has been made that the cross-veins are less regularly topped by ore-bodies along the

Fig. 37.

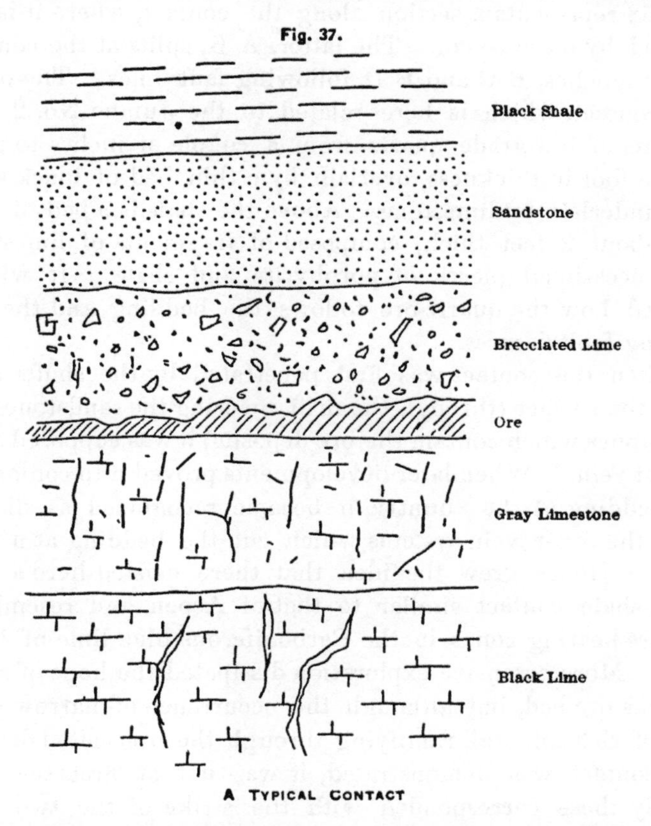

A TYPICAL CONTACT

contact; but, on the other hand, such ore-bodies as have been found over cross-veins have been usually larger and certainly richer than those apparently related to underlying verticals.

The dislocating influence of the pay-veins appears to die out in its approach to the contact-zone, so that the latter undergoes a mere undulation or roll along its bedding-planes. At the same time the pay-veins become unrecognizable, even as minute

* Obtained from stopes under the contact, in places where a cross-vein happens to cut a vertical as it is dying out.

seams or partings, in the rocks above the contact. In the case of the cross-veins, however, a more serious disturbance of the contact-zone is indicated by violent bends and occasional step-faults. The cross-vein itself, though it does not reappear above the contact as a strong quartz-seam, is yet represented by a well-developed division in the country accompanied by selvage. Fig. 38 represents a section along the contact, where it is dislocated by a cross-vein. The latter, A B, splits at the contact, both branches, B C and B D, following fault-lines. The ore of the contact, which is here related to the Jumbo No. 2 vein, consists of low-grade quartz from a couple of inches to more than a foot in thickness, overlain by a thin bed of black shale and underlain by limestone. Above the shale is a bed of breccia, about 2 feet thick, composed of fragments of limestone, with occasional pieces of porphyrite and shale. It will be noticed how the quartz-ore follows the bedding and the connecting fault-fissures.

When the contact was first penetrated by the shafts sunk from the surface (through the drift covering the sandstones and limestones which contain the ore-deposits) it was supposed to be a "flat vein." When later developments proved it to conform to the bedding of the country, it became recognized as distinct from the other vein-systems which cut the bedding at a right angle. Hence grew the idea that there existed here a limestone-shale contact similar to that of Aspen and resembling the ore-bearing zones in the Carboniferous blue lime of Leadville. More extensive exploration dissipated the hope of a continuous ore-bed, but although the occurrence of narrow channels of rich mineral, ramifying through the brecciated beds of the contact, was demonstrated, it was not at first seen how clearly these corresponded with the strike of the two vein-systems, the upward course of which was terminated by the contact. The recognition of this relationship was a key unravelling many perplexities, and a light to the intelligent exploration of a territory of complex geological structure.

The contact is not, as the term might imply, a continuous plane of division between two rock formations, nor does it mark the parting between the two adjoining beds of ore formation. The accounts which have been given of a "contact-limestone," overlain by a "drab shale" and underlain by a "finely lamina-

Fig. 38.

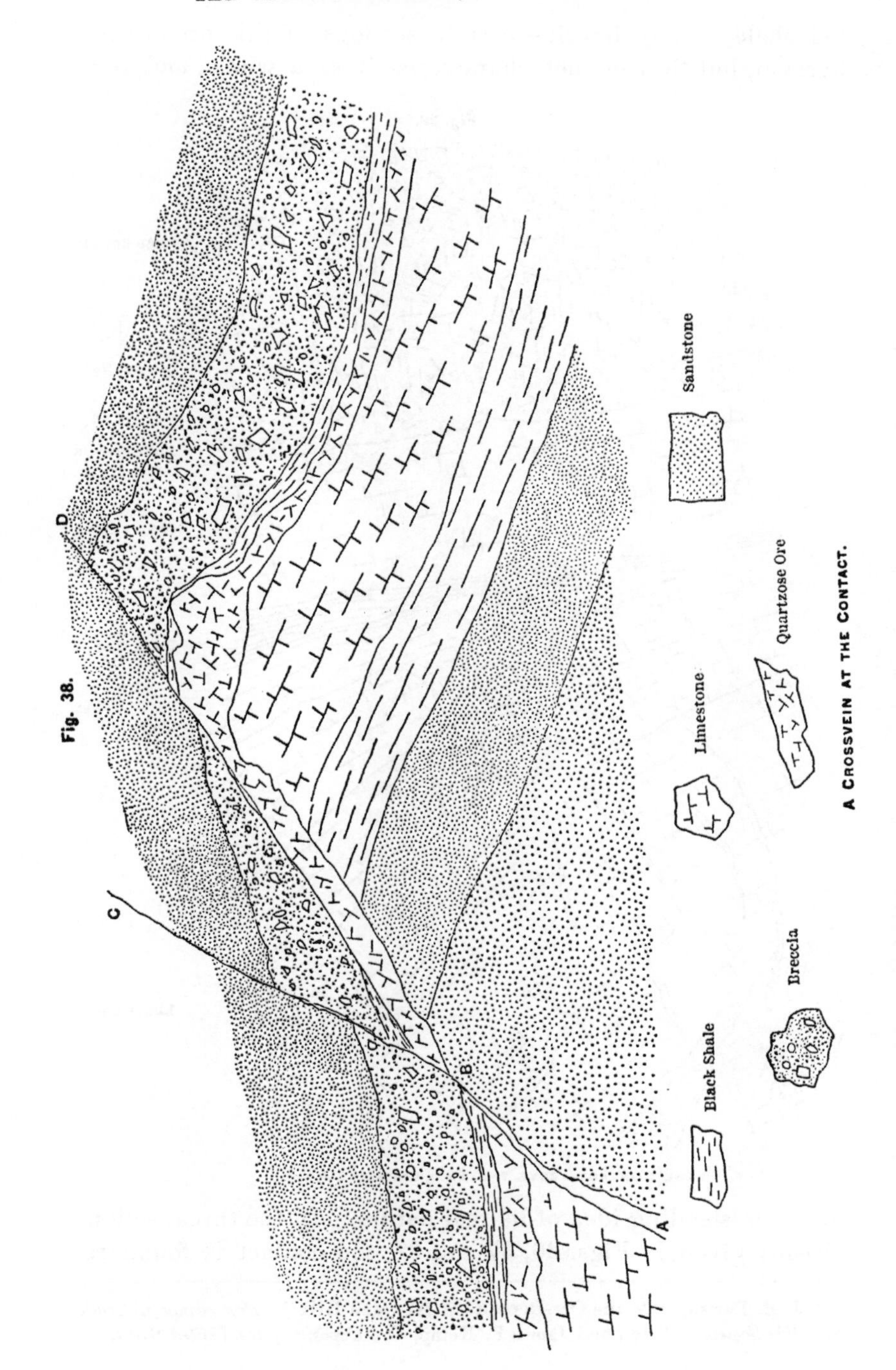

A CROSSVEIN AT THE CONTACT.

ted shale,"* may describe certain sections of this ore-bearing horizon, but they do not characterize it as a whole, and they

Fig. 39.

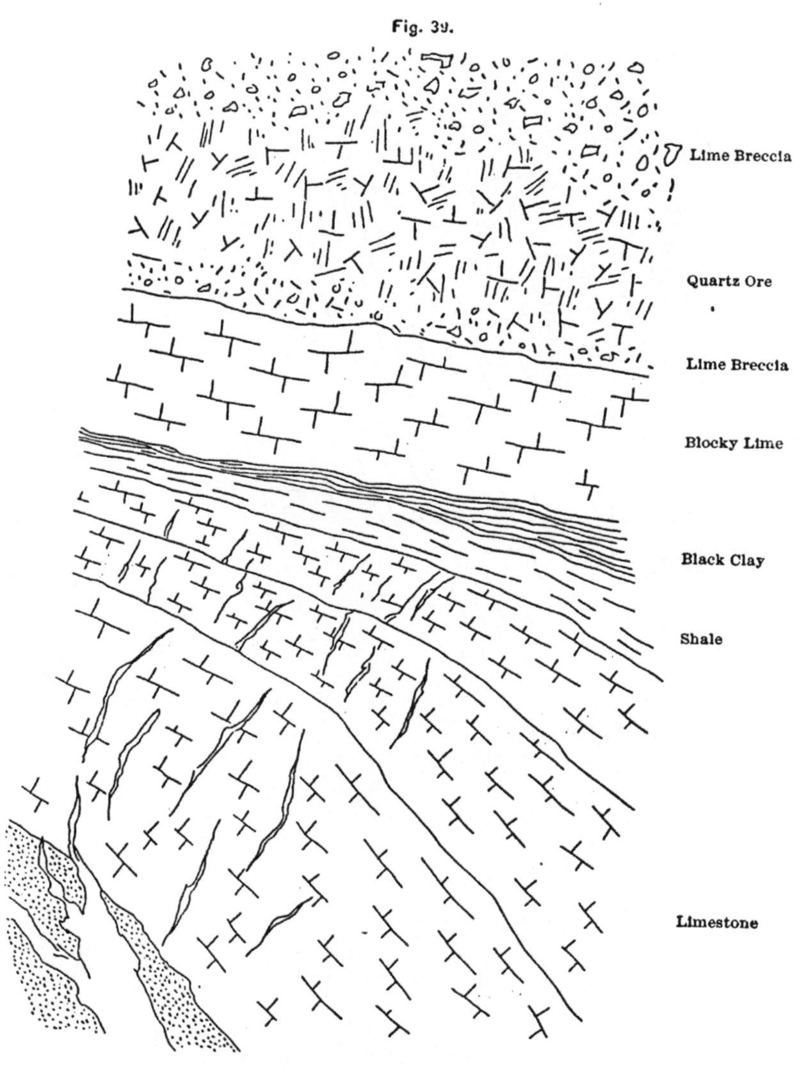

THE CONTACT ABOVE THE ENTERPRISE VEIN

give a misleading idea of its real nature. In the three sections already given, in Figs. 36, 37 and 38, the contact is found re-

* J. B. Farish, "On the Ore-Deposits of Newman Hill." *Proceedings Colorado Scientific Society*, vol. iv., and James F. Kemp, *Ore-Deposits of the United States*.

spectively in a crystalline lime, overlain by black shale and underlain by sandstone; in a lime-breccia, underlain by sandstone and underlain by a gray limestone; and in the third instance

Fig. 40.

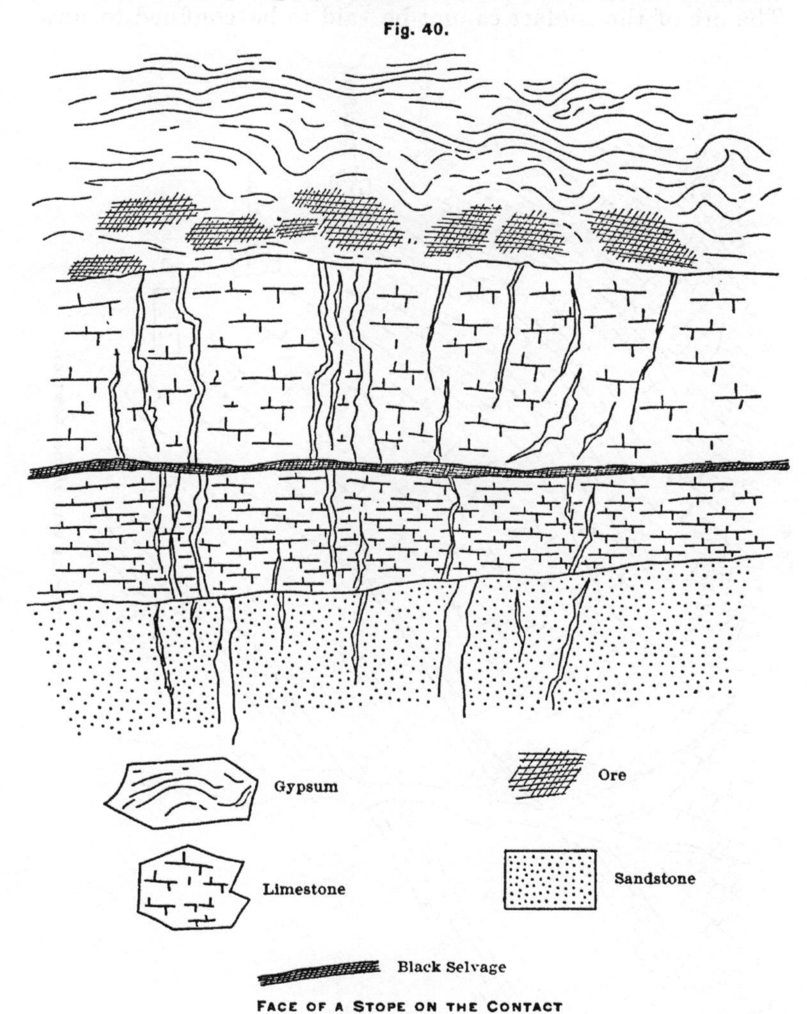

FACE OF A STOPE ON THE CONTACT

in a mass of crushed quartzose lime, covered by black shale and overlying a blocky limestone. Additional sections are now given in Figs. 39, 40 and 41. In the first of these we see the stringers thrown out by the Enterprise vein as it nears the contact, which in this case consists of brecciated lime, enclosing a low-grade quartz ore. In Fig. 41 another similar example is given. In Fig: 40 the contact-ore lies at the base of a bed of gypsum, which in turn overlies limestone, penetrated by stringers, which come from the Jumbo No. 3 vein below.

The ore of the contact cannot be said to be confined to any

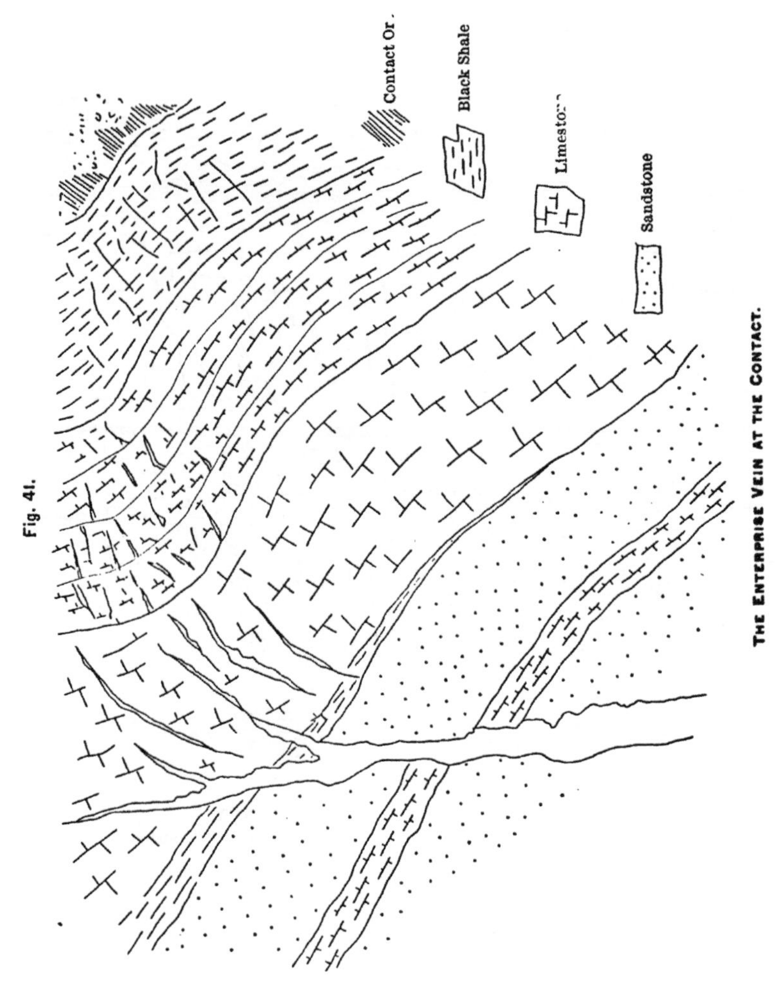

Fig. 41. THE ENTERPRISE VEIN AT THE CONTACT.

particular encasement; but one may venture the generalization that it is to be sought for in a layer of crushed rock, which occurs along a certain horizon marked by a thinly-bedded series of black limestones and shales. The parts of the contact explored during my period of management were very frequently characterized by a distinct breccia made up principally, but not solely, of lime-fragments. Pieces of shale and sandstone were recognizable as derived from adjacent beds; and fragments of porphyrite were traceable to neighboring intrusions of that rock. The contact above the Jumbo No. 2 frequently consisted of compact, pulverulent lime, graduating into breccia overhead and underlain by blocky lime; that above the Enterprise was often breccia, shading off into blocky lime overhead, and underlain

Fig. 42.

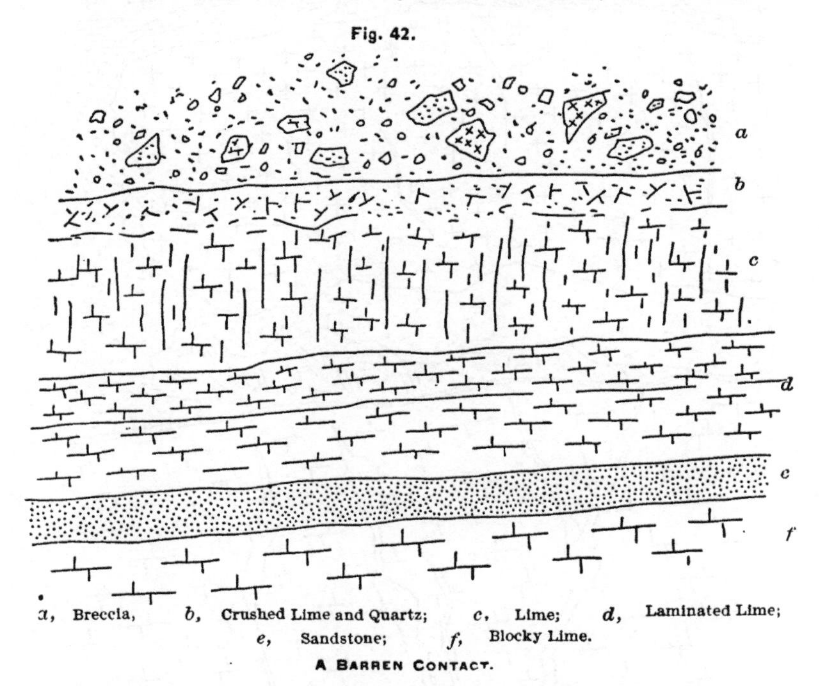

a, Breccia, *b*, Crushed Lime and Quartz; *c*, Lime; *d*, Laminated Lime; *e*, Sandstone; *f*, Blocky Lime.

A Barren Contact.

by black shale; while the ore of the Jumbo No. 3 contact was found between a powdery brown lime and a thin bed of black shale. The variability of the stratigraphical position of the ore, thus emphasized, is due to the non-persistence of individual beds.

Except when it tops the veins, the contact is barren. Fig. 42 illustrates the face of a cross-drift on the contact where no veins have enriched it. In Fig. 43 an intrusion of porphyrite is shown. The pulverulent lime indicates crushing, probably accompanying the invasion of the porphyrite.

The crystalline limestone of many sections is doubtless breccia consolidated by pressure and cemented by the underground waters which have lined the cavities with crystals of calcite. The gypsum bed to be seen in the southwestern part of the

Fig. 43.

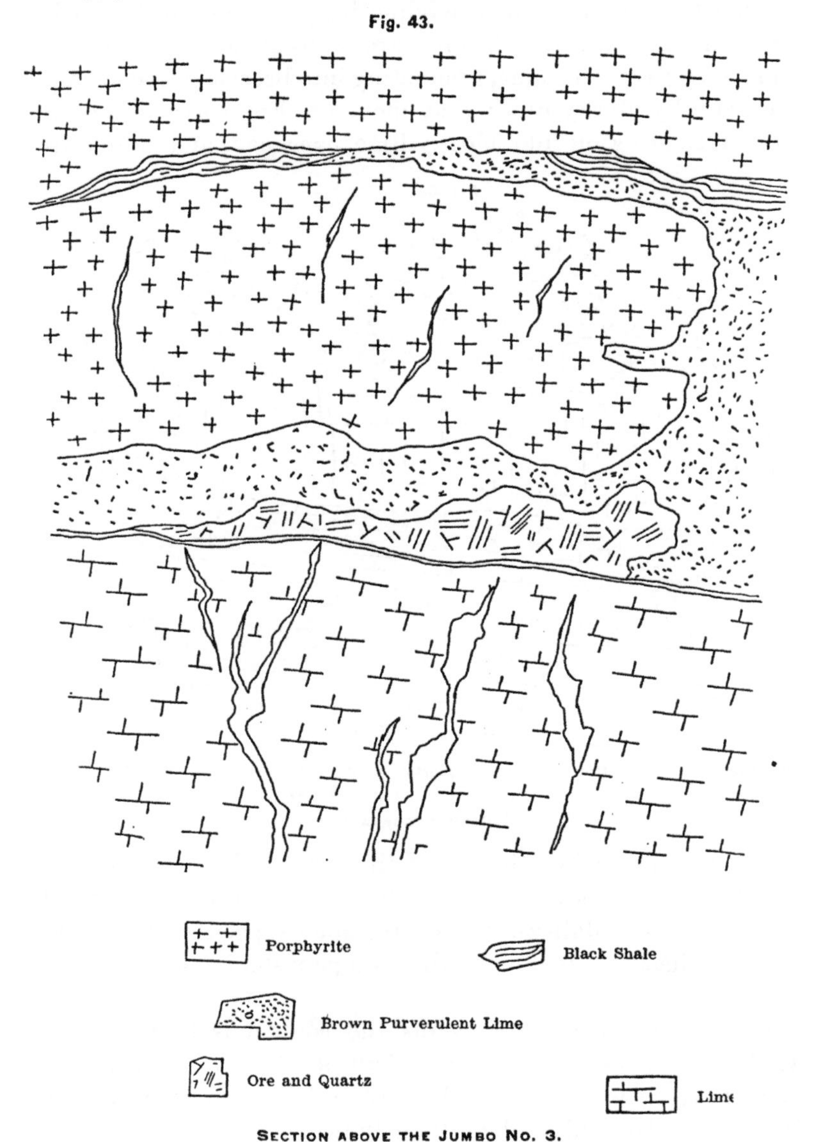

SECTION ABOVE THE JUMBO NO. 3.

mine affords a parallel instance. A section is given in Fig. 40. The vein whose shattered termination is to be seen below the ore of the contact is the Jumbo No. 3. The gypsum thins out and is a local occurrence having a maximum thickness of 15 feet. Its wavy, compact texture suggests an origin by a sulphatization of lime-breccia through the agency of solutions coming from neighboring ore-bearing measures.*

The foregoing general description will assist the reader's understanding of Fig. C, which represents a part of the workings on the main (also called the Enterprise) level, 98 feet above the Group tunnel. It covers an area about 1800 feet long by 350 feet wide, and was chosen for especial examination and study because the developments are more complete than in any other part of Newman hill, while at the same time they present features sufficiently typical of the ore-occurrence over the whole territory covered by the mine maps.

The levels are seen to follow three veins, the Enterprise, Songbird and Hiawatha, the dip of which is indicated by the arrows. Commencing at the crosscut from the Enterprise shaft the Enterprise vein is followed without difficulty as far as Raise 6,† where the vein dies out in weak stringers. Up to this point the stopes extended to the contact; but northward, stoping ceases on this vein. The vein which the level followed further on, from R8S, was once considered the Enterprise because of imperfect observation of the facts. So also the vein followed by the level further north still, from R15E to R18E, is called the Enterprise. It is a branch vein, a subordinate member of the series of many disclosed in the mine.

There is too frequent a tendency in mining to look upon veins as necessarily continuous, and to make the nomenclature correspond to the drifts. In this case, a level follows three distinct veins in different parts of its length. The crosscut between R7½E and R8S was put out to search for the Enterprise vein in case it had been faulted by the two cross-veins, G and H. It proved that this particular vein had ceased; but it led to the discovery of the Kitchen lode.

The numerous cross-veins are mapped, and their dislocating influence on the pay-veins is easily discernible.

---

* A large body of gypsum also occurs along the contact in the Vestal workings of the Rico-Aspen mine.

† The raises are numbered and initialed, so that R6E means Raise No. 6 on the Enterprise, R10H, Raise No. 10 on the Hiawatha, etc.

Returning to the crosscut from the Enterprise shaft, we will
llow the Songbird, which there appears as a small vein having

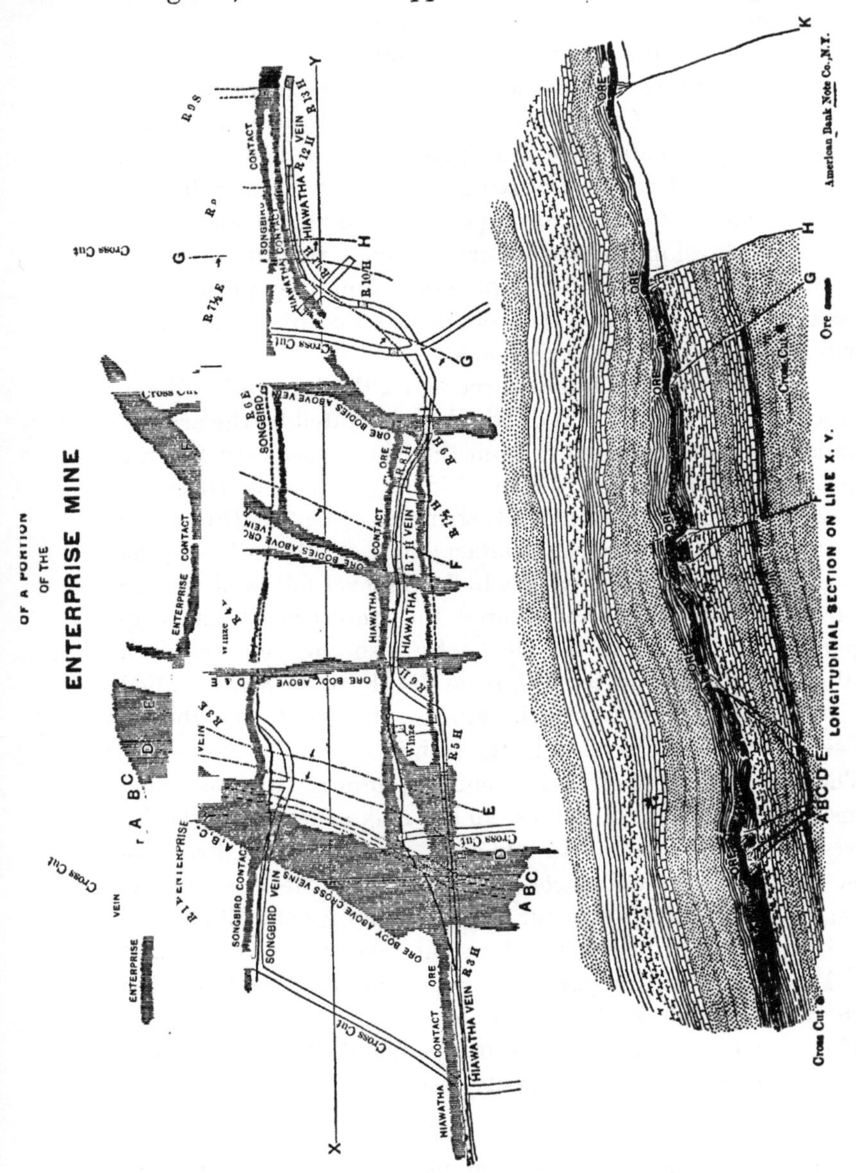

slight easterly dip. Going northward (to the right), the level
hows it to flatten. The next time it is seen, 560 feet further

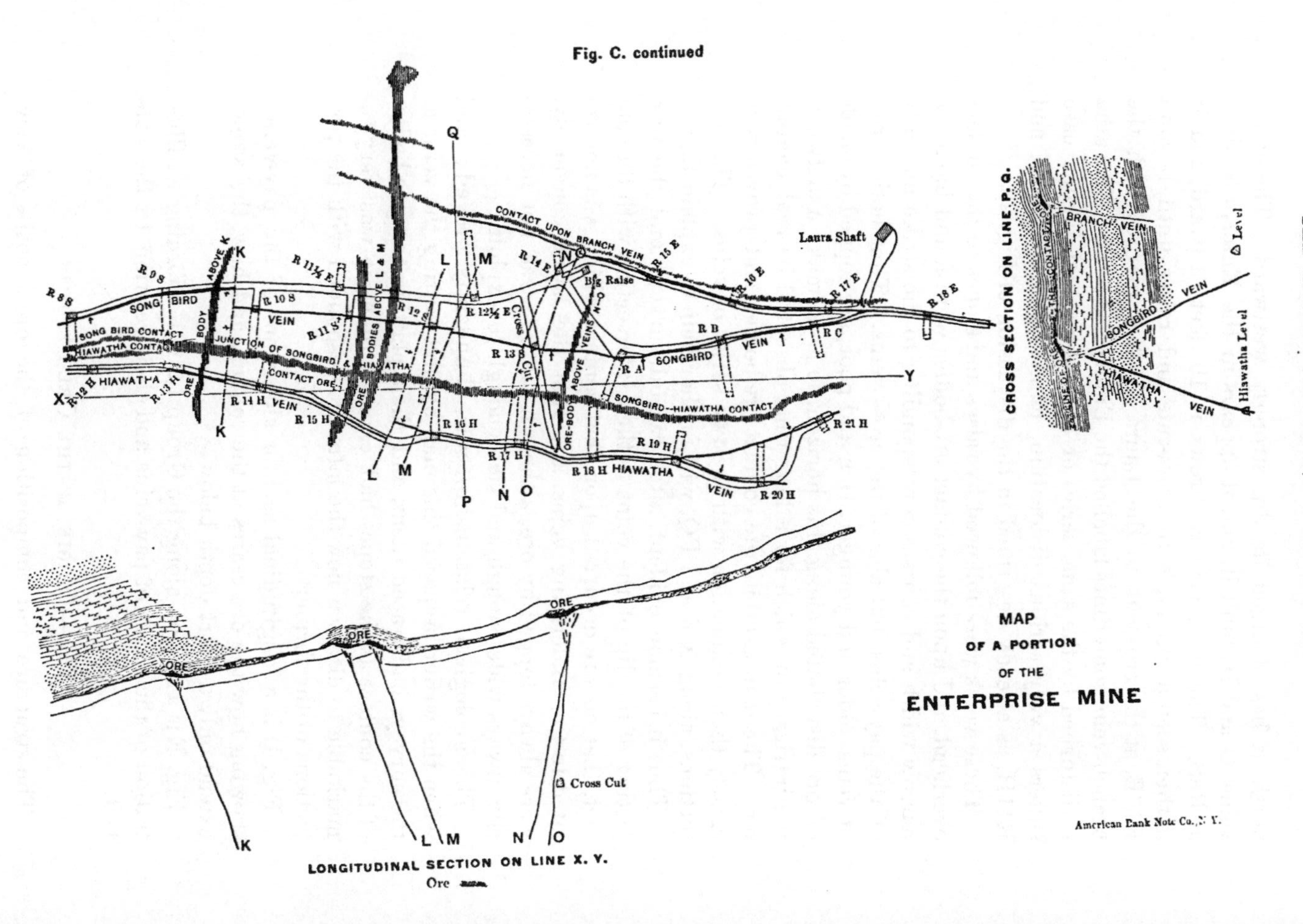
Fig. C. continued
CONTACT UPON BRANCH VEIN
Laura Shaft
Big Raise
SONG BIRD
SONG BIRD CONTACT
HIAWATHA CONTACT
JUNCTION OF SONGBIRD & HIAWATHA
CONTACT ORE
HIAWATHA
SONGBIRD VEIN
SONGBIRD--HIAWATHA CONTACT
HIAWATHA VEIN
Cross Cut
CROSS SECTION ON LINE P.Q.
BRANCH VEIN
SONGBIRD
HIAWATHA
VEIN
Level
Hiawatha Level
LONGITUDINAL SECTION ON LINE X.Y.
Ore
MAP
OF A PORTION
OF THE
ENTERPRISE MINE
American Bank Note Co., N. Y.

north, it has changed its dip strongly westward. The cross-veins, G and H, fault it, and it appears in the Enterprise drift at R8S. The cross-vein K throws it 19 feet eastward, and it is then seen at intervals in cross-cuts and raises until, beyond R17E, at the crosscut to the Laura shaft, it merges with the branch-vein, sometimes labelled the Enterprise. The Hiawatha is disturbed by the same series of cross-veins, and in one case suffers a very serious dislocation, namely, between R9H and R11H, as already described in the discussion of Fig. 23.

These veins were followed by raises and stopes to the contact overhead, and upon the contact ore-bodies were found having a narrow width and courses corresponding to the strike not only of the pay-veins, but also of the cross-veins. Thus both series of veins, older and younger, rich and poor, are topped by bands of ore distributed along this horizon, so as to make a network the intricacy of which for a long time obscured its real character. The map exhibits the contact ore-bodies and proves very clearly their connection with the two series of veins. The two sections, along XY and PQ, will further help to explain this.

Thus it became evident, after careful surveys and the projecting of the dip of the veins to their intersection with the contact, that no ore occurred upon that contact which was not related to the underlying veins, and that the latter, conversely, were always topped by ore, although that ore was not necessarily always wide enough and rich enough to exploit.

The geological relationship was abundantly proved; and when the writer prepared this map by first putting the veins as the surveys had traced them, and then platting the ore-bodies of the contact as the stopes had exposed them, it was very remarkable to discover how the latter corresponded with the projections of the former.

Fig. D is a longitudinal section along one of the pay-veins, showing how the ore ceases at the contact, and how the stopes extend only a short depth below.

Fig. E is a section along the Group tunnel, affording an illustration of the series of pay-veins and their relation to the contact.

VII.—Origin of the Ore-Deposits.

The structure and composition of the ore-deposits of New-

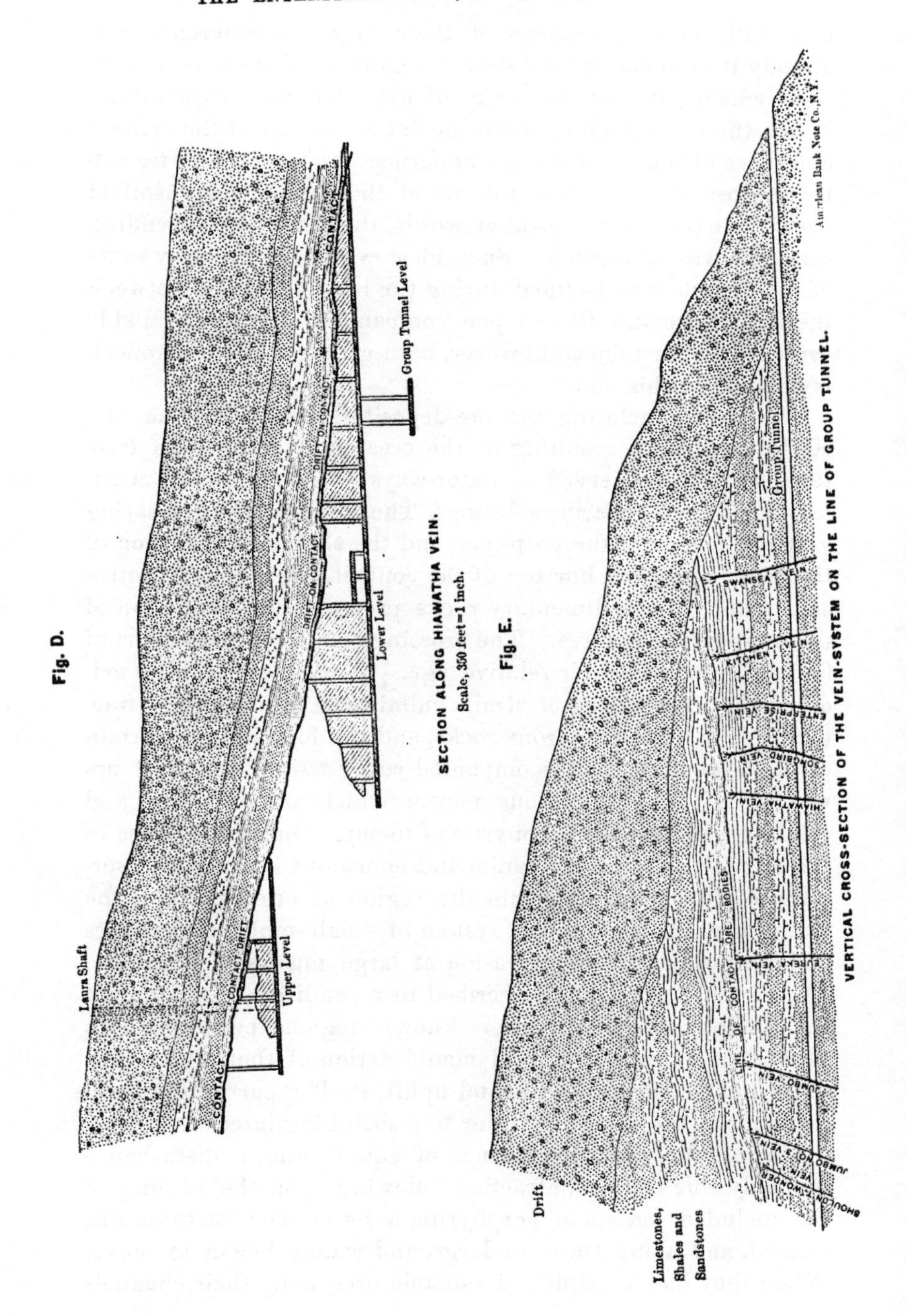

Fig. D.

SECTION ALONG HIAWATHA VEIN.

Scale, 350 feet = 1 inch.

Fig. E.

VERTICAL CROSS-SECTION OF THE VEIN-SYSTEM ON THE LINE OF GROUP TUNNEL.

man hill offer suggestions of their origin. Reference has already been made to the slow recognition of their true relations which came in the wake of extended mine-exploration. When the connection between the flat ore-bodies of the contact and those of the vertical veins underneath them was first traced- there arose the idea that the ore of the former had "spilled over" into the latter; in other words, the theory of descending solutions was advocated. Such ideas were expressed by some of the experts who testified during the long litigation between the Enterprise and Rico-Aspen companies. All the available evidence on the subject, however, both geological and chemical, is opposed to this view.

The rocks enclosing the ore-deposits have undergone successive rupturing, resulting in the creation of a series of fractures which have served as water-ways available for the circulation of mineral-bearing solutions. The fact that the ore-bearing verticals penetrate the porphyry, and the shattered condition of the latter along the horizon of the contact, prove that its intrusion among the sedimentary rocks preceded the formation of the ore-bearing fissures. The crossings of the later systems of fissuring establish their relative age. Thus, then, we have evidence that a condition of strain culminated in a multiple fracturing of the Carboniferous rocks, and the formation of certain of these fractures was accompanied *pari passu* by the slow upwelling of mobile igneous matter which, when cooled and solidified, became the porphyrite of to-day. Outside the area of the mine-workings large faults and enormous intrusions of porphyrite did occur, but within the region of ore-deposition the forces at work produced a system of small multiple fractures and did not permit the invasion of large masses of porphyry. These results are vaguely ascribed to a condition of strain, and the analogy of other better known regions permits us to ascribe this strain to a local manifestation of that wider phenomenon called the continental uplift, itself a part of the readjustment of the earth's exterior to a shrinking interior.

During the next disturbance of equilibrium, a disturbance probably due to the contraction following upon the cooling of the included masses of porphyrite, a set of new fractures was formed, and along these underground waters began to move. When they had precipitated valuable ores along their channels of circulation, later movements produced a series of cross-fractures which faulted them. Minor shiftings, which have supervened at various more recent times, have caused displacements along the bedding, affecting both of the older vein-systems.

The contact-zone has been the victim of all these disturbances. This is to be ascribed directly to its structure. A thickness of closely-laminated shales is laid upon blocky limestones and sandstones. To the formation of a fracture the latter rocks would offer no particular obstacle, because of their homogeneity; but the upward extension of a fracture would be impeded, if not stopped, by meeting a series of beds which, on account of their laminated structure, are easy to bend, but hard to break. There is nothing fanciful in this reasoning. The section given in Fig. 19 affords an illustration exactly in point. Thus, it seems to me, the structure of the rocks of the horizon now known as the "contact," was the immediate cause of the repeated shattering which that horizon underwent; it was the factor which stopped the upward extension of the vein fractures and produced the consequent limitation to the circulation of those mineral solutions which were the immediate agents of ore-deposition. Thus is explained the concentration of large masses of ore along this zone, because it became a dam, checking the circulation in an upward direction. The fractures now followed by the pay-veins were unable to break through the shales above the contact, and though the later cross-veins were stronger, they too were stopped by the elasticity of these closely-laminated beds.

In each case, therefore, the force of vertical fracturing was diverted into a horizontal displacement which soon made the zone under the shales a mass of shattered rock, peculiarly adapted to become the place of ore-deposition.

The ore is confined to the pay-veins for a depth of 100 to 175 feet below the contact. It does not occur in the cross-veins, but is found in the contact immediately above them, as above the pay-veins. This distribution cannot be satisfactorily explained with positive certainty. There are so many conditions determining the character of an ore-deposit, and ordinary mine-exploration, being a commercial enterprise and not a scientific inquiry, reveals so few of them, that the geologist, though aided by the chemist, is often at a loss. Yet, in this case, the available evidence, though in many ways inconclusive, is highly suggestive.

The ore-bodies of the contact apparently owe their existence to a combination of chemical and physical conditions. The ending of the vein-fractures at the contact-horizon put a stop to the upward flow of the metal-bearing solutions, and caused them to permeate the shattered rock and distribute themselves along the strike of the veins which had been their passage-way. At the contact they found the chemical precipitant which compelled the dissolved metals to separate out as aggregates of ore. That precipitant was probably the graphite of the black shales, as is indicated by actual experiments presently to be described.

The non-persistence of the ore of the pay-veins below a certain distance from the contact is apparently connected with the circumstance that the sedimentary rocks immediately below the contact are black, by reason of the carbonaceous residues of the vegetation imbedded amid the sand and mud on the floor of an estuary of the Carboniferous period. This carbonaceous matter probably acted as a precipitant of the metal-bearing solutions. As the depth below the contact increases, the rocks lose their blackness, and presumably, therefore, contain no precipitant carbon.

Although the pay-ore terminates at a depth fairly uniform among the different veins, it must not be supposed that the veins themselves cease at this horizon. On the contrary, the fractures maintain their course to depths far beyond the deepest mine-workings; but they become barren of valuable ore, enclosing nothing but quartz and crushed country. The rhodochrosite ceases. The Lexington tunnel, for example, cuts through the lodes which have yielded so richly in the Enterprise workings, 400 feet overhead, and discloses them as veins of white quartz, traversing light gray, coarse-grained sandstone. Both lode and country-rock have changed in character entirely.

The fact that the cross-veins are barren, and yet rich ore-bodies overlie them in the contact-zone, indicates that such bodies are due to special shattering of the ground by the cross-veins, which has furnished favorable places for ore-deposition. Moreover, barren as the cross-veins are, they appear to influence the richness of the pay-veins. It is common to find ores of more than average grade in the pay-veins, where they are broken by the cross-veins. But the mineral solutions apparently did not rise through the cross-veins. They must have circulated along the verticals; and the deposition of ore in connection with the cross-veins, as noted, may have been not only a collateral, but even a secondary process. The superior richness of these bodies in many cases may indicate that they have resulted from a re-solution and re-deposition of the contents of the verticals.

The idea of the precipitation of the ore through the agency of carbonaceous matter has been advanced in connection with ore-deposits in other regions. I may quote as instances the black Silurian slates of Bendigo, Victoria;* the Devonian slates of Gympie, Queensland; the Jurassic slates of the "mother lode" region in Calaveras and Amador counties, California; the black shale enclosing the gold-specimen ores of Farncomb Hill, Breckenridge, Summit county, Colorado; the graphitic casing occasionally seen in the ores of the Sunnyside and Mastadon veins, in San Juan county, Colorado; and the celebrated Indicator† series of Ballarat, Victoria.

In order to test this theory, I broke some pieces of black shale on the contact above Raise 12 on the Enterprise vein, and took them to the Argo smelter, where, by the kindness of Mr. Pearce, the following experiments were made. A piece of the Rico shale was put into a weak solution of sulphate of silver ($AgSO_4$) containing some free acid intended to neutralize the lime ($CaCO_3$) in the shale. The precipitation of metallic silver became visible in three days. The parallel experiment with gold was more interesting. A piece of ore (assaying 1147 ounces of gold per ton) obtained from the Prince Albert mine at Cripple Creek was taken, and its gold was extracted by a solution containing ferric sulphate ($Fe_2O_33SO_3$), common salt ($NaCl$) and a little free acid (H_4SO_4). This Cripple Creek ore carried the black oxide of manganese (MnO_2) in visible quantity, and thus the chlorine used to form the gold-solution was liberated in a manner simulating natural conditions. Of the gold in this Cripple creek ore, 99.91 per cent.

---

* Discussed by the writer in *Trans.*, xxii., 319 *et seq.* Also by Mr. Argall, same volume, p. 762.

† See also "The Indicator Veins, Ballarat," by the writer, *Eng. & Min. Jour.*, Dec. 14, 1895.

was extracted, and subsequently precipitated on the Rico shale by inserting the latter in the solution thus formed. The gilding of the black shale by the deposit of gold became visible within four hours.

The order of succession of the various minerals composing the ore is indicated in many ways. Beautiful pseudomorphs of quartz after baryta have been found. The replacement of rhodochrosite by quartz is often discernible. That the baser sulphides frequently enclose fragments of rhodochrosite (Fig. 25), establishes their relative age. Pieces of country found within the vein-matter are occasionally surrounded by a rim of rhodochrosite. (See Fig. 34.) The silver sulphides, and the native gold associated with them, occur exclusively within the geodes which are usually distributed along the center of the vein. The veins enclose shreds of country-rock; sometimes such pieces of rock are found within masses of sulphide ore, and there is an imperceptible graduation from clean sulphide to rock so impregnated with ore as to have its true character obscured. Banded structure is common.

This succession points to the following conclusions: When the fractures were first formed they consisted of lines of crushed country, afterwards healed by a deposit of carbonate of manganese. The latter (rhodochrosite) is likely to have been derived from limestones occurring at a horizon not necessarily very far below the place of the present ore-deposits. Then came a fresh fracturing, accompanied by the deposition of baryta and the sulphides of lead and zinc. Later still the earlier vein-stuff became shattered by fissuring on the old lines of movement, and along the water-way thus created there came siliceous solutions, which replaced baryta and rhodochrosites with crystalline quartz. Finally, the vein was riven along its center, and waters rich in the salts of gold and silver found their way upward to undergo precipitation through the agency of the rhodochrosite and the shattered portions of the carbonaceous country enclosed within the vein-walls.

We are driven in this case to the hypothesis of ascending solutions. A lateral flow must be a part of an upward or downward movement of the underground circulation. As a general phenomenon it is inconceivable. The deposition of ore from descending solutions is in this case chemically possible through the reduction of sulphates by carbonaceous matter. But for the hypothesis of descending sulphates there is no basis of fact. The geological evidence is all against it. The structure and environment of the ore-bodies point to their derivation from solutions which came up from below. The passage-ways open to the circulating waters cease upward and extend downward; they connect with no available origin in one direction, but lead to a possible source in the other.

Subject to Revision.

[TRANSACTIONS OF THE AMERICAN INSTITUTE OF MINING ENGINEERS.]

THE MOUNT MORGAN MINE, QUEENSLAND.

BY T. A. RICKARD, MELBOURNE, AUSTRALIA.

(Cleveland Meeting, June, 1891.)

AMONG the gold-deposits discovered in recent years none is more extraordinary in richness or interesting in structure than that of the famous mine at Mount Morgan. At a time when but few Australian mines were known to the world outside the colonies, Mount Morgan was quoted as an occurrence, unusual not only in its origin (for it was said to be due to geyser-action) but also in the purity of its gold.

The mine is situated just within the tropics, twenty-six miles southwest of Rockhampton, in central Queensland. Queensland attained in 1889 the first place among the gold-producing colonies of Australasia, a position previously always held by Victoria. In that year Queensland produced 737,822 ounces, while Victoria came second with 614,838 ounces. Of the four chief mining districts Rockhampton stood first, as the following figures show:

	Quartz crushed, Tons.	Yield, Ounces.	Average per ton, Oz.	Dwt.
Rockhampton,	81,138	340,669	4	2
Charters Towers,	109,328	165,551	1	10
Gympie,	106,625	112,847	1	1
Croydon,	29,423	51,156	1	15

It should be added that Queensland has but little alluvial mining, the total output from this source in 1889 being only 5044 ounces, none of which came from the four principal gold-fields. Leaving out Rockhampton, the output of which is practically that of the Mount Morgan mine, it will be noted that the more northerly gold-fields show the highest average (Gympie being furthest from and Croydon nearest to the equator). This is largely explained, however, by the increased cost of milling, due chiefly to the want of water for batteries and the consequent economical necessity of selecting high-grade material only for crushing.

During the year ending November 30, 1889, the Mount Morgan

mine produced 75,415 tons, yielding 323,542 ounces, 13 dwts., 13 grs., worth £1,331,484 18*s.* 5*d.* (about $6,657,424), while the expenditure was £227,769 19*s.* 8*d.* (say $1,138,849) permitting the payment of £1,100,000 ($5,500,000) in dividends. The yield per ton was 4 oz., 6 dwts., 4 grs., while the working-cost was, as is seen, only 17 per cent. of the value of the product. These figures speak for themselves.

So rich a mine would be expected to have some romance woven about the story of its discovery. Numerous and various tales are told of the first recognition of its value, but the best authenticated facts are as follows: The property consists of the original selection (No. 247) of 640 acres taken up for grazing purposes, in 1873, by Donald Gordon. Becoming acquainted with the brothers Morgan, who also held land in the district, he showed them one day a piece of gold-bearing quartz which he had picked up in Mundic creek. For a consideration, stated to have been £20 and as much whiskey as he could drink, Gordon agreed to indicate to them the locality of the find. On the hill overlooking the creek he showed them the siliceous ironstone, some of which can still be seen cropping out on the north-eastern slope. The stone carried visible gold; they found by sending samples to Sydney that it was even richer than they had imagined; so they purchased Gordon's holding at £1 per acre.

The three Morgans subsequently sold, first a part, and eventually the whole, of their interest in the mine. In 1886 a company was formed with a capital of one million shares of £1 each. These shares rose toward the end of 1888 to £17 5*s.* (about $86.25) giving the mine a market value of seventeen and a quarter millions sterling or over eighty-six million dollars. The shares are now quoted at £7; for Mount Morgan, to quote the language used by the managing director at an annual meeting, "is after all only a gold-mine and is consequently subject to the vicissitudes of all mineral formations," which is a truism too often forgotten by those who conduct mining operations.*

Description of the Mine.

The mine does not, as its name would imply, crown the summit of a mountain, properly so called, but forms a quarry at the top of a

* Of the neighboring companies (and as might be expected they are numerous, with their suggestions of an extension of the ore-deposit in such names as Mount Morgan West, Mount Morgan Extended, Mount Morgan North Consols and so on *ad nauseam*) none have proved profitable, notwithstanding the expenditure of much money in numerous and scattered trial-shafts, tunnels, etc.

hill, only 500 feet above the village at its base and 1225 feet above sea-level, surrounded by very broken hilly country and almost encircled by a small stream (the Mundic creek already mentioned), and in many respects distinct in position and geological structure from the hillocks about it. From the summit can be seen the level line of the "desert sandstone," crowning the highest ridges of the neighboring hills—spurs from the main range which under different names (Blue Mountains in New South Wales and Australian Alps in Victoria) traverses the three colonies near the east and south-eastern coast of the Australian continent. The base of the sandstone is slightly lower than the summit of Mount Morgan and overlies greywacke and quartzite. Dykes, of at least two periods, form an important feature of the structural geology.

The crest of the mount is being rapidly broken away in the quarrying operations which have supplied the great output of the mine. The removal of from 1200 to 1700 tons a week makes a big hole in a year. While the greater part of the summit has been thus removed, the northeast slopes still show the croppings which first excited Gordon's attention.

The ground is worked in terraces or benches, 30 feet high and 300 feet long. At the date of my visit five of these, started at various times, could be seen, though but little remained of the floor, 25 feet from the top, which formed the open cut first put into the ore. A central shaft passes from the floor of the second terrace and connects with the deeper tunnels, a series of which have intersected the deposit in various directions. This shaft, at 206 feet, connects with the Freehold tunnel, the main ore-way of the mine, which is 789 feet long. The next deeper tunnel, called No. 1, starting at right angles to the Freehold, from the southern face of the hill and penetrating it for 1070 feet, is 155 feet below the floor of the lowest surface-working and 320 feet below the original summit. This No. 1 tunnel, though only 33 feet deeper than the Freehold, affords a very interesting section of the mountain and indicates a great change in the form of the ore-deposit. It is the deepest of the adits (for adits and not strictly tunnels most of them are), except the "Sunbeam," which had been started at the time of my visit; but it had not been advanced sufficiently to throw any additional light upon the structure of the mountain.

The Mount Morgan ore is remarkable for its extremely heterogeneous character. The frequent alleged discoveries of "a second Mount Morgan," supported by similarity of the specimens exhibited to specimens of the Mount Morgan ore, are not to be wondered at, in view of the great difference in appearance of fragments broken in different parts of the same heading of this mine. The material quarried in the upper workings, samples of which are now to be found in collections and museums all over the globe, is generally so friable and shattered, as to render its removal with the aid of black powder only, very easy. Standing in one of the open cuts, one can see faces of bluish-gray crushed quartz very similar to Comstock ore; masses of siliceous hematite (usually considered the most typical Mount Morgan stone) resembling the croppings of ordinary gold-veins, whether in Victoria or California; heavy black iridescent iron-stone which might be the cap of a silver-lode; and light-colored reddish ore which might come from the gossan of a copper-mine. Some of the material is crushed to the consistency of sugar, while in other portions of the mine men are seen employed in breaking boulder-like masses of ore. Stalactitic forms occur in cavities, whilst a vermicular and reniform appearance is also not uncommon. Very rich returns were obtained from a body of bluish-black, beautifully iridescent ore. The Freehold tunnel-workings encountered a patch of very pure white, porous, friable "sinter," so light, owing to the imprisoned air, that it would float on water.

Mount Morgan is not a "specimen-mine," though in its early history it furnished very lovely pieces of gold-quartz, some of which I saw in Sydney. They consisted of an iron gossan, containing big "splashes" of gold, of the size of a thumb-nail. As a rule, however, the gold, even when visible, is very fine and scattered thickly through the stone; but owing to its peculiar character (the coating of oxide of iron to be referred to later on) it is not readily detected by the eye.

Examination of the many varieties of ore shows that while there may be a great difference in outward appearance, due to coatings of many-colored oxides, the ore is always substantially quartzose. As seen in the surface-workings, the deposit may be considered a mass of quartzose material of varying color and specific gravity, traversed by a series of dykes having a general N. W. and S. E. direction. Its extent is approximately indicated by the various workings which intersect the hill. In the upper part of the mine, while the north and northeastern slopes prove it to be continuous in these directions, its eastern limit has been reached in the lower floors or benches; and to the southwest it is bounded by a large felstone dyke, which forms the most marked feature of the surface-excavations. The first tunnel to pierce the hill is No. 2, which starts from the fifth or lowest floor, running N. 10° E. and penetrating 120 feet of ore before it reaches a dyke. This is probably the dyke above mentioned as the boundary of the deposit in the upper floors; but here it has evidently cut into the deposit. Beyond this dyke the tunnel is in ore for 200 feet further, before meeting another intrusion of felsitic matter, which extends for the remaining 130 feet to daylight. A branch-tunnel proves the continuity of the ore-body eastward. At this level the stone evidently holds out most satisfactorily, and is, I believe, for the most part auriferous. It is a porous siliceous material, varying between a light "sinter" (sometimes white, sometimes iron-stained) and a heavy iron-stone, often manganic, the latter in boulder-like masses.

The next tunnel is the Freehold, forming, as has been said, the main artery of the mine. It connects with the shaft (sunk from the No. 2 floor), and from the mouth of it a tramway carries the ore to the chlorination-works. The tunnel starts from the southeast face of the hill, traversing a decomposed rhyolite for the first 180 feet, and then cuts through 40 feet of pyritiferous quartzite, a rock which here first makes its appearance and which will be seen to play an important part in the geology of the locality. Leaving the quartzite, the tunnel cuts through 180 feet of a rock which I recognize* as a normal dolerite. The innermost portion is decomposed, and abuts against a much altered felspathic rock. This brings us to 390 feet from daylight; the remainder of the tunnel (397 feet) traverses the ore-deposit. The ore is a light sinter-like quartz, often iron-stained; but while it contains short rich patches the general tenor is low. Near the junction with the main shaft, almost at the end of the tunnel, a branch cross-cut runs southward for 237 feet. The first 186 feet are in the deposit, which here also is poor, though in appearance it does not differ from the richer portions.† The next 20 feet traverse the dyke which at this point, as in the upper floors, forms the limit of the ore-body. Beyond this there is only to be seen the pyritiferous quartzite, which also appears in a short cross-cut to the east.

The next deeper tunnel is No. 1, only 33 feet below the Freehold. I was denied access to this, and am, therefore, indebted to a govern-

---

* Thanks to microscopic sections shown to me at Charters Towers by Mr. Clarke.

† It is here that one sees the light, white, pumice-like material which, owing to the air in its cavities, will float on water.

ment report* for the following particulars: This adit, 1070 feet long, starts in, and traverses, the quartzite of the country for 132 feet before passing through a narrow dyke; it then passes through 67 feet more of highly pyritiferous quartzite, which continues to within 12 feet of a shaft coming down from floor No. 5. There next succeeds a large felstone dyke (the one noted as cut in the tunnel above), and the remainder of the adit continues in pyritiferous quartzite, alternating with numerous dykes, until the last 200 feet are reached, which are occupied by an altered dolerite. The auriferous material of the upper workings was represented in this adit by 25 feet only of siliceous iron-stone. Two branch cross-cuts similarly prove the absence of the deposit in this part of the mine at this depth, as they also intercept the quartzite of the country.

A still lower tunnel, the Sunbeam, starts from the west face of the mountain, passing through pyritiferous quartzite; but it was not advanced sufficiently at the time of my visit to afford any evidence as to the extent of the ore-deposit.

It will be noted that the largest section as yet obtained of the auriferous portion of the deposit is in tunnel No. 2, where its dimensions are 356 feet (26 feet of which is occupied by the big dyke) in a north direction, by 310 feet east, as proved by a branch cross-cut. The deposit has been proved to extend further eastward in the deeper levels (Freehold tunnel) than it did in the surface-excavations, while on the other hand its western limits are more restricted.

Theories of the Formation of the Ore-Deposit.

The origin of the ore-deposit has been, as might be expected, the theme of much controversy. The earliest description came from Mr. R. L. Jack, the head of the Queensland geological survey, well known as a careful observer. His report was made officially for his government and appeared in 1884. Others have contributed their opinions since; and meanwhile the rapid development of the mine, more particularly by the deeper adits, has furnished additional data for a problem, of which the following solutions have been offered:

1. That the deposit is that of a geyser (R. L. Jack).
2. That it is an auriferous zone traversed by a series of quartz veins of auriferous mundic (J. Macdonald Cameron).
3. That it is the decomposed cap of a large pyrite-lode (the view held by several local and other mining engineers).

---

* R. L. Jack's second report, which will be referred to again.

The Geyser-Theory.—The first of these explanations, known as the "geyser-theory," was promulgated by Mr. Jack in his first report, dated November 8, 1884. It has found its way into many scientific publications, and has been until recently the most widely accepted. Citing as a similar occurrence the hot springs and geysers of the Yellowstone Park, described by Dr. A. C. Peale in the United States Geological Survey reports, Mr. Jack observes in his first report: "Nothing but a thermal spring in the open air could have deposited the material under consideration." The two sections shown in Figs. 1 and 2 are taken from that report. Of Fig. 1, he says: "The above diagram represents my idea of what would take place in the case of a geyser remaining in activity for a (geologically speaking) lengthened period." The original form of the ground is shown by *aa*, the deposit of precipitated material by *bb*, the layer of solid material at *cc*, while the surface-contour is indicated by *dd*. In describing this section he concludes, "Such, I believe, is the history of Mount Morgan as we now see it."

The structure of the mountain is supposed to be shown by Fig. 2, in which Mr. Jack indicates rhyolite dykes at *d* and metamorphic rocks at *s*. He says: "*a* is the pipe of the geyser (theoretical), *b* the cup-deposit and *c* the overflow of the geyser."

These two sections have found their way into scientific publications all over the world, and have been for a long time accepted as most interesting explanations of one of the richest of modern gold-mines. But they evidently are altogether imaginary and theoretical.

The same geologist says, in his description: "After the cessation of thermal activity the powers of sub-aërial denudation would come into play. Denudation would obliterate the lateral terraces which were probably not absent from the slopes of Mount Morgan." The information obtainable in 1884 as to the nature of the deposit was totally insufficient to enable an observer, however careful and experienced, to build a theory upon. The amount of work done at that time was mostly limited to one open cut, 10 by 15 feet in size, which showed a fan-like arrangement of the ore, suggesting to Mr. Jack the structure of a geyser. It was an altogether local appearance, as was proved by the work of a few days later.

In a second report, dated December 12, 1888, Mr. Jack adds: "The evidence now to hand, in my opinion, goes far to confirm my original view that the auriferous material was deposited by a thermal spring." The deposit had been asserted in the first report to have been formed "in the open air," a condition constituting the whole

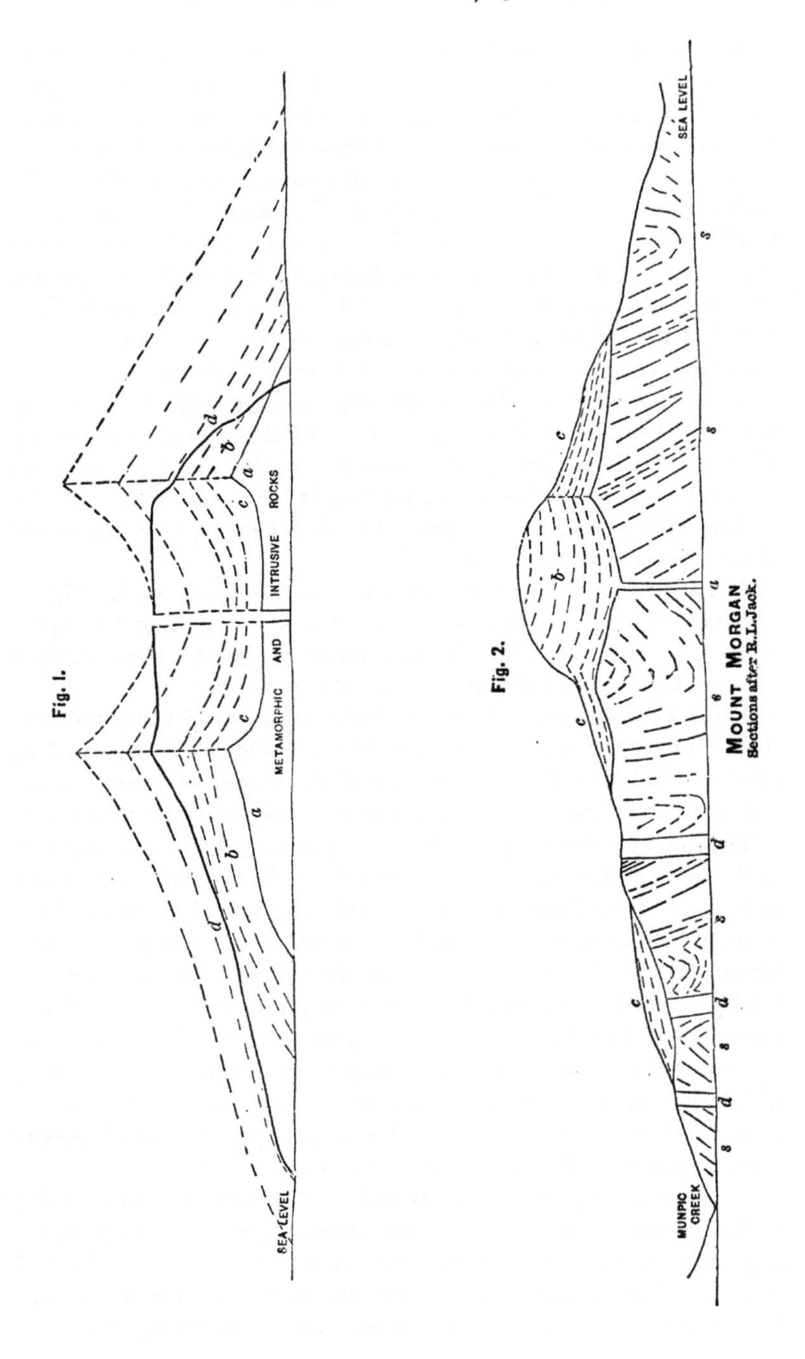

Mount Morgan
Sections after R. L. Jack.

distinctive value of the theory, which otherwise might not differ materially from the views accepted as to the aqueous origin of ore-deposits in general. In the second report the author declares his theory to have been confirmed, but produces evidence which really contradicts its essential feature.

Taking the two reports together, we find that, in support of his explanation, Mr. Jack adduces: (*a*) the "fan-like arrangement" of the material; (*b*) "the frothy and cavernous condition" of the siliceous sinter; (*c*) the hydrated condition of the silica and (*d*) the fact that one of the dykes intersecting the country does not also intersect the deposit.

(*a*) The "magazine" face, at the time of Mr. Jack's first visit, appeared to him to have a distinctly fan-like structure. But little work having been done since that time at this particular point, it should be possible to verify this statement; but the most careful observation does not confirm it. The cutting shows an arrangement of loose material such as can frequently be seen in surface-excavations, probably less marked now than when first seen by Mr. Jack; but the entire absence of any repetition of such an arrangement of material in the other workings must destroy its value as evidence, if it ever had any.

(*b*) The "frothy and cavernous" condition of the quartz is not peculiar to Mount Morgan. Those who are familiar with the pocket-mines of California will remember the very light, pumice-like quartz which often accompanies the gold in those mines. It is the light honeycombed quartz found in the gossan of gold-veins in many places. Generally speaking, the Mount Morgan gangue can be duplicated elsewhere; and the particularly light "frothy sinter," noticeable as a patch in the deeper workings, differs only from the bulk of the quartzose ore in that it is remarkably free from accessory minerals. It is the siliceous skeleton produced by surface-decomposition of more than ordinary intensity. It is possible to produce artificially in a few minutes, by the application of acid to very dissimilar iron-stone ore, a material similar to that noted, which has undergone the chemical action of underground waters during long periods.

(*c*) The hydrated condition of the silica is not unusual in quartz-lodes, and while it is valuable as evidence to disprove the suggestion that the ore is the replacement of anhydrous quartzite, it does not necessarily point to the agency of a geyser.

(*d*) The intersection of the country by a dyke which does *not* penetrate the deposit is an important fact. The dyke referred to is that cut by tunnel No. 1 at 120 feet from the shaft coming down from floor No. 5. Mr. Jack says, at the end of the second report: "This shows that the sinter and iron-stone are deposited on, and were not altered portions of, the pyritous quartzite country-rock." Unfortunately the evidence is far from conclusive. In describing No. 1 tunnel the dyke is referred to in the following sentence: "A quartzose rock full of fine pyrites is traversed for the next 37 feet, when a dolerite or rhyolite dyke is cut. The direction of this dyke is uncertain, as the tunnel is here timbered up." In his descriptions Mr. Jack always endeavors to draw a marked distinction between the "felstone" dykes which penetrate the ore and the "dolerite" and "rhyolite" which intersect the country-rock. Now, this distinction is arbitrary and misleading. I examined the rocks both *in situ* and, a few days afterwards, under the microscope. The dykes penetrating the deposit are so decomposed that it is impossible now to determine which is felstone and which rhyolite. The dykes cutting through the country are similarly, but not to so great a degree, decomposed, especially in the vicinity of the deposit. The statement that a "dolerite *or* rhyolite" dyke does not penetrate a deposit which is freely intersected by decomposed felspathic eruptives, should be supported by proof. It is not so here; for the dyke cut in No. 1 tunnel has not been followed upward, and in the south branch of the Freehold tunnel, just overhead, there is a dyke, which is possibly the one in question.

My sections, shown in Figs. 3 and 4, will illustrate this. The left half of Fig. 4 resembles one of Mr. Jack's sections, being taken along a nearly identical plane (mine is taken due north and south). His section, however, gives a curved contour to the limit of the deposit, very nicely, but unwarrantably, suggesting the shape of a geyser-basin. The right half of the section is not filled in, since this part of the mount has not been thoroughly developed. Fig. 3 is taken along a line E. 17° S., and is obtained by projecting the cross-cuts from the No. 1 and Freehold tunnels upon the plane of the No. 2 tunnel.

In further considering the conditions under which thermal springs and geysers* exist, and the similarity of such conditions to those obtaining at Mount Morgan, it may be noted:

1. Isolated geysers are unusual, nothing being more remarkable

* A geyser may be defined as a thermal spring ejecting material in the open air.

than the extended area of such phenomena. But vigorous and extensive prospecting, such prospecting as always follows great mineral discoveries, has not led in this region to the finding of anything

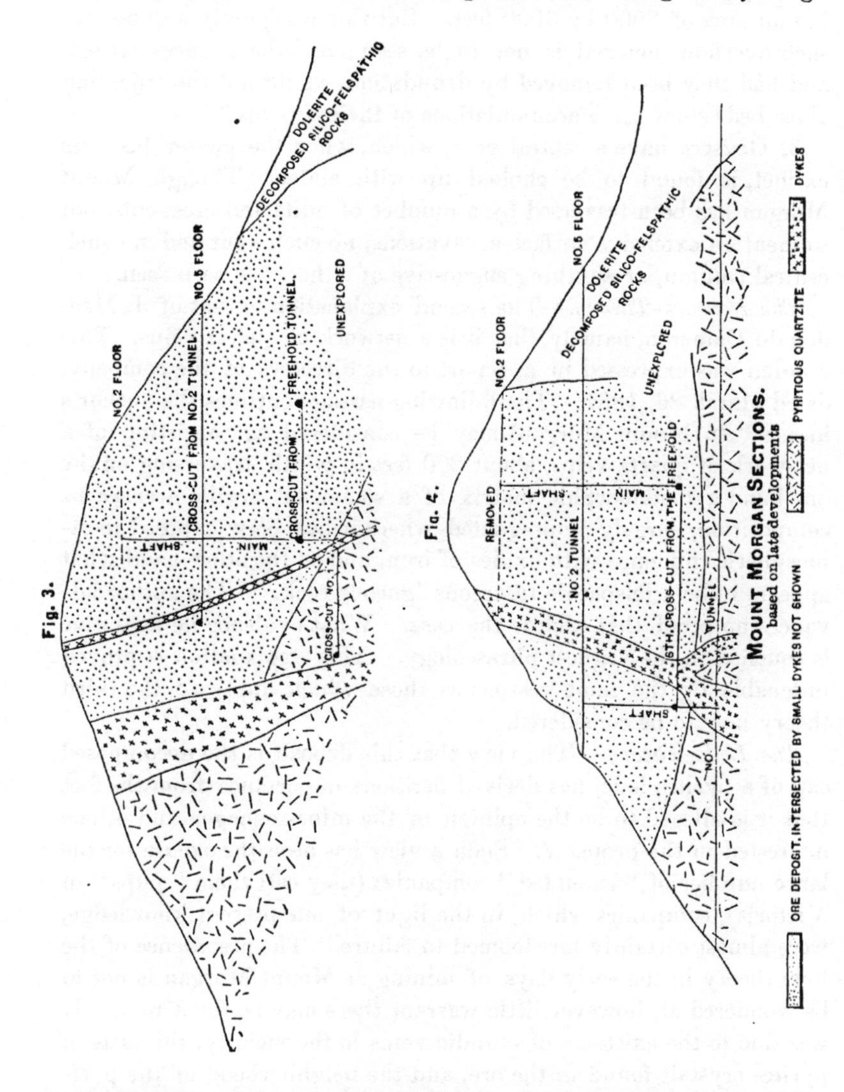

similar to Mount Morgan or to a geyser. It is true, there is a small hillock adjoining the mount to the northwest (called Callan's Knob) where somewhat similar material has been found ; but it may be considered as an offshoot or spur from the main deposit.

2. Geysers have terrace-formations due to the overflow of the siliceous or calcareous waters. The map accompanying Mr. Jack's first report indicated "the overflow from the hot spring" as covering an area of 2000 by 3000 feet. But this was purely a guess, for such overflow-material is not to be seen; had the terraces existed, and had they been removed by denudation, would not the adjoining river beds show some accumulations of their *detritus?*

3. Geysers have a central vent, which, when the geyser becomes extinct, is found to be choked up with sinter. Though Mount Morgan has been traversed by a number of adits and cross-cuts, not to mention extensive surface-excavations, no such vent and no such central column, or anything suggestive of either, has been seen.

The Network-Theory.—The second explanation is that of J. Macdonald Cameron, namely, that it is a network of quartz veins. This opinion was expressed in a report to the directors of the company, dated March 26, 1887. The following passage outlines the author's ideas: "It (Mount Morgan) may be considered as consisting of a net-work of quartz veins, about 200 feet in width, traversing on the one hand a metamorphic matrix of a somewhat argillo-arenaceous composition, largely impregnated where it has been exposed to atmospheric influence with oxides of iron, and on the other hand what appears to be a felspathic tufaceous igneous rock." The author advances no proof that such is the case. What evidence he does give is couched in the vaguest phraseology. This explanation is proved untenable by the same reasons as those which apply to the third theory now to be considered.

The Lode-Theory.—The view that this deposit is the decomposed cap of a pyrites lode, has derived fictitious importance from the fact that it is alleged to be the opinion of the mine-manager and others interested in the property. Such a view has been the excuse for the large number of "extended" companies (they call them "pups" in Victoria), companies which, in the light of our present knowledge, were almost certainly foredoomed to failure. The prevalence of the lode-theory in the early days of mining at Mount Morgan is not to be wondered at, however little warrant there may be for it now. It was due to the existence of mundic veins in the vicinity, the casts of pyrites crystals found in the ore, and the neighborhood of the pyritiferous quartzite, the true relation of which was not known, or if known, was kept very quiet. The well-marked limits of the great deposit, as outlined by the deeper developments of the last two years, completely disproves this theory and that of Cameron. No doubt, the assertion of the continuity of the gold-bearing rock, even though it should become pyritiferous, has fascinations for a shareholder which even a geologist can understand. But the deeper tunnels have proved that lode-structure is entirely absent, and that the deposit lies upon, but is not an altered portion of, the pyritiferous quartzite of the country.

The Theory of Metamorphosis and Replacement.—The more than ordinary size and value of the Mount Morgan ore-deposit, the romantic history of its discovery and its subsequent marvellous production, have all helped to place it outside the ordinary type of mineral occurrence. This is often the case with new discoveries, and it is not the first time that the geyser has been a *deus ex machina* to a perplexed geologist. The splendid work done in recent years by Sandberger, Daubreé and others in the study of the chemical change, and the distribution by underground waters, of the minerals of the older rocks has led to a gradual substitution, for theories of eruption, or direct igneous agencies in general, of the belief in the capability of aqueous agencies to bring about most of the phenomena of ore-deposition. Moreover, the action of percolating solutions on the minerals of the crystalline rocks, and subsequent interchange with the more soluble portions of the neighboring rock-masses, have, to a great extent, replaced the assumptions of the old ascension-theories. The view that the more ore-deposits are studied the more frequently they will be found to be in great part the product of the replacement of the country-rock is gaining ground, and has been much strengthened by the researches of Becker at Virginia City, Curtis at Eureka, and Emmons at Leadville, as contained in the monographs published by the U. S. Geological Survey, monographs (the last-mentioned, particularly), which are the classics of modern economic geology. This view, which underground experience is continually confirming, has explained many of the more unusual types of ore-deposition in Europe and America, and I believe it will apply to Mount Morgan.

The explanation suggested is this: that the ore-deposit of Mount Morgan represents an altered portion of shattered country-rock, which, by reason of its crushed condition, was readily acted upon by mineral solutions, and that these solutions replaced the basic and felspathic with acidic and quartzose material, which was also gold-bearing. It is its quartzose and permeable character which has saved from disintegration the mass thus affected, and has preserved it as an ore-body on the summit of the hill.

A case, in many respects very similar, occurs in the Red Mountain basin, Colorado, and the following description given by Mr. Emmons* is clearly in point:

"Instead, however, of being surface-deposits, they (the Red Mountain deposits) are simply portions of the andesitic country-rock from which acid waters have removed the basic constituents, perhaps depositing a certain amount of silica in their place; the resulting quartzose masses, offering greater resistance to the disintegrating effect of atmospheric agents and to erosion than the surrounding rocks, have been left as mound-like ridges protruding above the general surface, more or less independent of the natural drainage-channels."

This explanation, curiously enough, was also given to disprove a geyser-theory, but the deposits in question seem to possess scientific interest only.

Deposits of somewhat kindred origin are the Yankee Girl and Bassick mines in Colorado. The former, a well-known mine in the San Juan district, is perhaps the more suggestive, since in that case several fracture-planes meet to form a wedge-shaped mass of rock, which has been so acted upon by percolating waters as to become a "chimney" of very valuable silver-ore.

To return to the particular conditions which obtained in the district under discussion: The country around Mount Morgan consists, in descending series, of sandstone, greywacke, quartzite, shales and occasional bands of serpentine. Intrusive masses of dolerite interpenetrate the country, and often occupy a position between the originally stratified rocks. The sandstone is known in the geology of Queensland as "Daintree's desert sandstone," and, by virtue of a recent discovery of fossil remains, may be labelled as Cretaceous. The greywacke has not been recognized as such on the Mount itself, but it underlies the sandstone in other parts of the district. It is a mixture of silico-felspathic material, which shades off into a greywacke-slate, and is usually so much altered as to be scarcely distinguishable from a true eruptive. The quartzite has a bluish-gray color, and is highly pyritiferous, the pyrites carrying traces of gold.

The shales are usually so indurated and altered as to resemble closely the crystalline rocks. The dolerite, when fresh, is a typical rock of its class, and in various intrusive forms is cut through by most of the prospecting-shafts at the base and around the mountain. The serpentine is of very limited occurrence. Some uncertain, undetermined decomposed rocks, which have also been called dolerite,

* "Structural Relations of Ore-Deposits."—*Transactions*, xvi., 809.

are probably metamorphosed greywacke. The country is freely intersected by dykes of varying composition, especially near the ore-deposit.

The history of these rocks I read somewhat as follows: A period of dynamic disturbance is indicated by the intrusions of dolerite, which, by extreme metamorphism, might have changed a dolomite into the serpentine we now see; would have indurated the shales so that they are scarcely to be distinguished from the crystalline rocks; and would also, accompanied by chemical alteration, change a ferruginous red sandstone into a bluish-gray, highly pyritiferous quartzite. Approaching the surface, the same energy would be expended in the fracturing of the quartzite and greywacke; the intrusive dolerite would rise through the fissures in the shattered rocks, forming dykes, which, meeting a silico-felspathic granular rock (the greywacke), would give it a semi-crystalline character. The sandstone would similarly be vitrified. Later movements would result in the further intersection of this part of the district by the numerous dykes, the decomposed remains of which are now to be seen ramifying through the deposit. Those gradual chemical interchanges would take place which resulted in the alteration of the shattered country-rock, and its becoming a portion of the gangue inclosing the auriferous material, which was then, or at a later time, deposited. In process of time, subaërial denudation removed the sandstone, which now is only to be seen on the further summits of the neighboring hills. Atmospheric agency continued to carve away the less siliceous and less porous portions of the country surrounding the deposit, until Mount Morgan, owing to the pervious, quartzoze nature of its crest, remained as a low hill in an undulating country.

The district of Mount Morgan has undergone extreme metamorphism, accompanied, as such metamorphism frequently is, by those chemical reactions which more or less completely alter the rocks acted upon.

It is probable that there was more than one period of such activity, alternating with intervals of rest, during which the pressure of overlying rock-masses and the cementing action of the underground waters would slowly re-solidify the more or less shattered rocks.

The deposit occupies a shattered portion of country-rock, freely intersected by dykes. The largest dyke in the upper workings cuts through the ore at one level, while it bounds it at another. It seems in keeping with experience elsewhere, to suppose that the dykes penetrating the country formed passages along the line of contact, through which and around which, by the agency of mineral solutions, the deposition took place, sometimes on one side only, sometimes on both, the physical condition of the different portions determining, in a large measure, the extent of such deposition and replacement. It is also probable that, after the formation of the deposit, later movements caused the circulation of fresh mineral-bearing solutions, which would lead to the enrichment of the deposit at certain points, and so explain the irregular tenor of the ore.

The evidence at hand is not sufficient to enable us to determine the particular rock in which the deposit was formed, and which it almost entirely replaced; but it could not have been the underlying quartzite, the line of contact with which shows no gradation into the ore; and of the other two possible rocks, the greywacke and the sandstone, the indications point to the former.* In the neighboring prospecting-shafts, auriferous quartzose material resembling that of Mount Morgan has been found to pass into a greywacke or greywacke-slate, which, in turn, overlies the pyritiferous quartzite. Silicate of alumina is generally present throughout the deposit. The character of the ore-material is such as would result from the alteration of a silico-felspathic rock by the action of mineral-bearing solutions.

The numerous dykes which penetrate the deposit are the evidence and the result of the dynamic agency which fissured the country-rock, and in so doing prepared a ready passage for the underground waters. These waters, while they may have been "thermal" (*i. e.*, of a temperature above the mean annual temperature of the district, as are most underground waters), yet did not undergo precipitation "in the open air." The brecciated forms which the ore sometimes assumes scarcely require the explanation offered by Mr. Jack, who states that "the occasionally angular condition of the sinter and tumbled condition of the ironstone masses, would appear to indicate that explosive discharges of gases or steam occurred at intervals with sufficient violence to disturb the deposits accumulated by the thermal spring." Would it not be simpler to suppose the fracturing of the rock, of which there is abundant evidence, to have produced irregular pieces, which would subsequently become partially rounded

---

* The position of the greywacke has been shown to be between the quartzite and the sandstone. The original summit of the Mount is but little below the level of the base of the sandstone, while the deposit, as we have seen, overlies the quartzite.

by the solvent action of the waters percolating through their interstices?*

At Mount Morgan all the conditions generally considered favorable to ore-deposition were present in a marked degree. A mass of country-rock, whether greywacke or sandstone, consisting of granules of quartz, held together by a felspathic cement,† becomes fractured by the intrusion of a series of dykes, which form a ready passage for the flow of mineral solutions. Such a rock would readily lend itself to alteration, and would be well fitted to receive a mineral deposit. Later dynamic action produced a further metamorphism of the surrounding rocks, followed or accompanied by the intersection of the already shattered rocks by another series of dykes, which reopened a passage for the underground waters. In this case it is not necessary to go far for the source of the gold. The large mass of decomposed pyritic quartzite, though the pyrites contains but a trace of the precious metal, is more than sufficient to account for the wealth that has been uncovered.†

This theory explains several points which have been the source of perplexity, more particularly the want of regularity in structure, which was as much opposed to the geyser- as to the lode-theory. The great unevenness of the gold-contents, the frothy state of the quartz, so often noted, and also its hydrous condition, may be due to

* Since writing the above, I have visited and examined the equally famous Broken Hill silver-deposit. When I declared that the material of the Mount Morgan ore was not unique, that it could be duplicated elsewhere, and that it did not, *per se*, require any extraordinary explanation, I did not expect to see so soon an instance confirming my statement. The huge outcrop of the Broken Hill silver-lead lode has much to remind one of the big gossan of the Queensland gold-mine. Indeed, if one stands in the quarry between Brodribb's and Patterson's shafts one might easily imagine oneself on floor No. 3 of Mount Morgan. The face has the same singular, broken appearance, consisting of masses of ochreous earth, bands of hard, siliceous ironstone, bodies of black manganic ore, together with portions of white kaolinized material (here undoubtedly altered inclusions of country-rock, and not, as at Mount Morgan, decomposed felstone dykes). Both at the surface and underground, the sintery stalactitic structure which is found at Mount Morgan also occurs. The hydrated iron-ores produce the same "frothy," "cavernous" material, supposed to be so suggestive of a geyser—and this, too, not in isolated patches only, but widespread. I may add, that there is no doubt whatever that the Broken Hill is a "true fissure"-lode, traversing the metamorphic rocks of the district (schists, garnetiferous sandstone, and quartzite, principally the first), independent of their stratification.

† The underlying quartzite may not be the only source of the gold, for it is noteworthy that the "desert sandstone" has been known to be auriferous, such gold being waterworn and sedimentary.

later precipitation in cavities already formed by the solvent action of the waters on the shattered portions of rock. Finally, there is no need to offer any particular reason for the unusual richness of the deposit, beyond the eminently favorable physical condition of the original rock, the numerous channels offered to percolating waters and the close proximity both of the source of the gold and of the usual precipitants.

Whether this be a correct explanation of the facts, time and the further development of the mine will probably decide. It may certainly be said, without fear of contradiction, that few mines offer such an interesting field for geological speculation.*

TREATMENT OF THE ORE.

The Mount Morgan chlorination-works are among the most successful and extensive now in operation. The complete success of the treatment is largely due to the extremely friable character of the ore, which renders it pulverization easy, while its porosity assists materially in the thorough chlorination of the gold. An enthusiastic writer has spoken of the ore as "a sort of snow-drift, which melts in the chlorination-vats of the company into a golden sand, such as might be supposed to have been brought from the bed of the river Pactolus, instead of from the top of an Australian mountain."

The capacity of the present works is 1800 tons per week. The gradual alteration and enlargement of the method of treatment is shown by the stamp-mill and the two chlorination-plants. The mill (25 stamps) by the side of Dee creek, is a reminder of early attempts at the extraction of the gold by ordinary amalgamation. The rock which the battery was called upon to crush averaged 10 ounces

* Since I wrote the above there has come into my hands the report of the Mount Morgan Gold Mining Company, Limited, presented at the last annual meeting, and giving information up to July, 1890. It is accompanied with maps and sections, notably calculated to yield a minimum of information. It appears from the report that later developments have added but little to the known extent of the ore-body, the main feature of the later exploration being the driving of a cross-cut south from the Freehold, at 800 feet from its mouth, proving the extension of the deposit at this line and in this part of the mine—as was to be expected, seeing that its dip is toward the mouth of the Freehold.

Up to November 30, 1890, the mine had paid the present company £2,358,333 in dividends, the total gold obtained being 756,042 ounces, worth £3,121,741. During the last six months (ending November 30, 1890) 37,867 tons of stone were treated and yielded 113,251 ounces of gold, or an average of 2 ounces 19 dwts. per ton. Nearly all this came from the upper workings.

per ton, but the contents of the tailings proved of much greater value than the amount of amalgam obtained. This led to a critical examination of the ore at the Sydney Mint. It was found that the bullion was of a fineness hitherto unknown in nature, assaying 99.7 per cent., occasionally 99.8, of pure gold, the rest being copper, with a trace of iron. It is remarkable as being almost entirely free from silver. Its value per ounce was £4 4*s.* 8*d.*, pure gold being worth at the London Mint £4 4*s.* 11½*d.* Dr. Liebius, of the Sidney Mint, considered, as the result of numerous experiments, that the iron present was in the form of an oxide, which coated the gold and so prevented its contact with the mercury.* Parcels of the ore were then sent to the Technological Museum at Melbourne, where Messrs. Newberry and Vautin subjected it to numerous tests. It was found that when dehydrated by heat it became less refractory; and its treatment by chlorine was suggested.† The ordinary Plattner process, as worked in California, was found quite unsuitable, owing to the extremely fine crushing which it required, and which resulted in the formation of slimes, greatly hindering subsequent filtration. It was then that Messrs. Newberry and Vautin planned the method which has since, with the addition of some mechanical details, been patented under their joint names.‡

While the stamp-mill proved incapable of extracting a proper percentage of the gold, the peculiar character of the ore is remarkably suited by the process now in use. The extremely minute state of subdivision of the gold and the very friable nature of the gangue caused, in wet stamp-milling, a considerable loss in "float," carried away in the readily-formed slimes; but the character of the ore indicated that dry crushing would prove effective; and hence the early introduction of Krom rolls was the first great improvement made. The old works, which are just above the stamp-mill, receive their supply

---

* My own experience with the ores of Gilpin county, Colorado, leads me to believe that this is, more frequently than is usually supposed, the obstacle to successful amalgamation.

† It had been previously treated with chlorine by Limberner at Gympie.

‡ The mine-manager informed me that nothing of the original Newberry-Vautin plant remained save a piece of old shafting. The Mount Morgan Company bought the right to use the process, and hence does not pay any royalty. While the original plant has been entirely replaced, it cannot fairly be said that the method of treatment has been changed, or that it is any other in its main features than that known by the names of those who first adopted it. Into the question of the originality of the process it is not within the province of this paper to enter. It has been ably discussed by the New York *Engineering and Mining Journal.*

on an aërial tramway, three-fourths of a mile long, with a total fall of 400 feet. It is of the Hallidie type, uncommon in the colonies, but familiar to American mining engineers, and need not therefore be here described. The feeding of the buckets (of which there are sixty) is admirably effected by the use of a travelling supply-bucket. The old works contain two rock-breakers, one revolving ore-dryer, four sets of rolls, ten furnaces, fourteen barrels, twenty-six vats and twenty-eight filters. The power is supplied by one 40-H. P. (nominal) engine and two 30-H. P. boilers. The practice in these works is identical with that in the new works about to be described.

The new works, erected on the higher slope of the mountain, about 400 feet from the summit, while somewhat similar to the lower works, show improvements in arrangement, besides allowing the necessary fall for the tailings. The ore, coming from the open cuts and the Freehold tunnel by a series of inclines (some of which are automatic gravity-tramways) passes through two Blake-Marsden rock-breakers, and is then dried in two revolving cylindrical furnaces, self-discharging after the manner of a *trommel.* The rotation is produced by an external cog-wheel. Thence the dried material goes to Krom steel rolls, arranged in two series of three each. Each set of rolls is followed by a revolving sieve, the coarse stuff passing into an elevator which returns it for second crushing, while the fine passes on to the second rolls and thence to the third. The last pair of rolls pulverizes the ore twice before it is carried by a worm-conveyor to an elevator which sends it to a chamber above the furnaces. An 80-H. P. compound engine drives the rolls and a 20-H. P. engine the breakers and revolving dryers, while two Root blowers are employed in the removal of the dust, of which there is a great deal. Five Cornish boilers of 30-H. P. each supply the necessary steam.

The furnace-chamber is 400 by 120 feet in size, and contains 18 ordinary reverberatory furnaces, each having a capacity of 1 ton. The heat to which the material is here exposed is a cherry-red, such as will get rid of organic matter, water of crystallization, etc., but does not approach that of a thorough calcination. At the end of 3 hours the charge is removed and spread upon the cooling-floors. When it has cooled it is trucked to the 21 iron hoppers which supply 21 chlorination-barrels. The charge for chlorination is 1 ton of pulverized ore, 90 gallons of water, 40 to 50 pounds of sulphuric acid and 35 pounds of chloride of lime. It will be seen from the nature of the charge that the chlorine is generated in the barrels themselves, and attacks the gold while in a nascent state.* In order to make the charges uniform, the ore is mixed so as to yield 4 to 5 ounces per ton.†

The chlorination-barrels have been much altered from time to time. They are now made at the mine out of the native timber (*Eucalyptus*), the staves being made of "spotted gum" and the ends of "iron bark," with an inside diameter of 3 feet 6 inches. A cast-iron frame was at one time placed round the outside edges. The barrels are tarred to render them tight and are protected inside by a double lining, an innermost wooden lining protecting the lead one. They revolve on horizontal bearings, the power being supplied by a belt travelling around the barrel itself. Six revolutions per minute is the usual speed, and the charge is removed after an interval of from 1½ to 2 hours.

The barrels discharge into 84 leaching-tubs. These tubs or vats will hold 2½ to 3 tons of pulp each, and are supplied with filter-beds of sand and gravel, through which the solution passes before it is conducted by lead and antimony pipes to the charcoal-filters. These are 64 in number, **V**-shaped and packed with 2 feet of charcoal of unusual fineness, the upper layers having the size of coarse blasting-powder, and increasing in fineness toward the bottom. From 9 to 14 bushels are used in each filter.

When saturated with the precipitated gold the charcoal is removed to the assay-office, where it is burned to an ash containing 75 per cent. of metallic gold. The solution from the filters passes into tanks of concrete, whence it is pumped back to be used again in the chlorination-barrels.

The sulphuric acid used in the generation of the chlorine is manufactured upon the ground in leaden chambers, having a capacity of 20 tons per week. It is estimated that by the production of the acid on its own premises the company saves £20,000 per annum. There are also extensive work-shops, with lathes and shearing-machines, together with two assay-offices, well equipped with all the necessary appliances. The whole plant is run continuously, being lit by electricity at night, when the mountain assumes a very beautiful and animated appearance.

The mine and all its surroundings bear testimony to the great

* After the style of the processes patented by Mears in 1877 and De Lacy in 1864.

† The pleasure of producing a uniform mixture of such high grade has not been possible since the time of my visit.

energy displayed in its development, in keeping with the value and size of the great ore-deposit which has made Mount Morgan famous.

The accompanying plate, made from a photograph, gives a good

idea of the surroundings of the mine. In the foreground is the bridge crossing the Dee creek; immediately above is the old stamp-mill; behind it is a pile of the ore first broken; and behind that again are the lower assay-offices. To the left are the old chlorination-works, while the upper chimney indicates the new works. Beyond is the Mount itself, the white spot being the iron roof of the main shoot which leads from the upper workings.

NOTE BY THE SECRETARY.—Comments or criticisms upon all papers, whether private corrections of typographical or other errors, or communications for publication as "Discussion," or independent papers on the same or a related subject, are earnestly invited.

[TRANSACTIONS OF THE AMERICAN INSTITUTE OF MINING ENGINEERS.]

THE LIMITATIONS OF THE GOLD STAMP-MILL.

BY T. A. RICKARD, DENVER, COLORADO.

(Chicago Meeting, being part of the International Engineering Congress, August, 1893.)

MILLING is one of the metallurgical arts whereby the extraction of the largest possible proportion of the value in an ore is effected at the least possible expense. Stamp-milling* is that particular process in which a heavy body of iron is caused to fall upon the ore so as to disintegrate it and thereby induce a separation between what is valuable and what is worthless. The latter is usually less in specific gravity, and is therefore, by the further aid of water, removed from the former, which is then collected by the use of mercury.

Several similes have been employed to describe this process. The stamp has been likened to a hammer of which the stem is the handle and the die the anvil. The ore upon which the stamp falls has been compared to a nut awaiting the descent of the hammer whose blow is to separate the valueless shell, the quartz, from the valuable kernel, the gold.

When we begin to pursue our inquiries, however, we find that the analogy is just sufficiently true to emphasize the departures from it. The hammer falls, the anvil is fixed; so with the stamps and the mortar. The anvil is made of softer metal than the hammer; so also the die is often, and should be always, of steel or iron less hard and more tough than that of the shoe. The movement of the hammer and the drop of the stamp are both intermittent.

In regard to their intermittent action, as in many other respects, stamp-mills arrange themselves under two types, which, though apparently contradictory, have both been evolved from a common original, and are united by a great variety of intermediate modifications. The slow speed and the high drop of the mills of Gilpin county, Colorado, appear to have very little in common with the fast speed and short drop of those of the main gold belt of California; yet the practice of the one was largely derived from that of the other,

* I shall confine the discussion to simple gold stamp-milling.

and each has been adapted to the treatment of the ores of its particular region.

The first and most apparent difference is that of speed. In Colorado the drop is regulated at 30 per minute, while in California it averages from 90 to 105. The more rapid drop gives a less intermittent action, and in this respect more nearly approaches the ideal machine.

The work done by the hammer is, however, dependent not only on the rapidity of its blows but also upon its weight and the distance through which it falls. Keeping to the two types, which we have chosen as representatives of the two systems of milling, we find that in Colorado the stamp weighs 550 to 600 pounds and falls a height of 18 to 20 inches, while in California the stamp weighs from 750 to 850 pounds and drops only 4 to 6 inches. Upon multiplication of these three factors—weight, drop, and speed—we find that the theoretical work done is nearly equal and is about one horse-power.

In milling, however, the efficiency of the stamp as a crushing machine is gauged by the quantity of ore which it can reduce, and we find that this does not at all correspond to the theoretical equality of the mills. In Colorado the stamp crushes 1 ton per twenty-four hours, while in California, with an ore of similar hardness, the amount is from two and a half to three times as much. Why is this difference? To explain it we must suppose the hammer to fall not upon the dry and wide surface of an anvil, but upon a face of iron confined within a narrow box and under water. This box corresponds to the mortar or coffer of the stamps. It has no opening* save in front, where a metallic grating or screen permits the escape of only that part of the material which has been crushed sufficiently small to pass through the openings. The ore upon the die is under water. The depth of that water depends upon the level of the bottom of the aperture occupied by the screen-frame. In Colorado the depth of discharge, as measured by the distance from the bottom of the screen to the top of the die, is 14 inches, but in California it is 4 inches only.† Herein lies the key to the difference in the crushing capacity of the two mills. Though the same amount of power be expended, and though the screen used be of similar mesh, yet in the Colorado mill the stamp falls through 10

---

* The feed-hole is higher up and does not concern us here.

† In making the comparison between the two systems of milling, I have purposely chosen extreme types.

inches more of water and has to discharge the pulp at a level 10 inches higher than in the California mill. The greater depth of discharge deadens the effectiveness of the blow of the stamp and weakens the force of the splash. Another result is obtained. While the screen does not in either case succeed in sizing the material discharged through it, yet it will be found that, though provided with similar screens, the pulp issuing from the deep mortar has a fineness much greater than that discharged by the shallow one. The pulverized ore is retained by the deep mortar long after its particles have been crushed to a size permitting their passage through the screen-openings, and they therefore become repulverized to a further degree of fineness.

This touches upon one of the points in respect of which the stamp-mill is most faulty. By actual test it is found that, though using a 40-mesh screen, for instance, with the theoretical supposition of crushing to that particular size, yet in a Colorado mill fully 70 per cent. of the pulp, and in a California battery about 50 per cent., will pass through a 100-mesh sieve. The percentage varies with the character of the ore, but these figures may be considered fairly representative. Two causes are chiefly responsible for this. The most important in its effects is the pause which occurs between the successive drops of the stamp. In a Colorado mill the interval is two seconds; in California it varies from three-fifths to two-thirds of a second. Particles of ore, which have been pulverized to a fineness which would permit of their exit through the screen, are enabled to settle towards the bottom of the mortar. It would be expected that the heavy metallic minerals occurring in the ore would, because of their greater specific gravity, be most affected by this feature of the treatment. In practice this is found to be so. The fine slimes contain a large proportion of metallic sulphides, generally valuable on account of their close association with the precious metals, while the coarsest particles to be found in the tailings usually consist of quartz and other minerals forming the less heavy gangue.

The want of any proper control over the regular sizing of the pulp is also due to the unequal and irregular splash of the water in the battery and the hap-hazard way in which the particles of pulverized ore strike against the screen. In the case of any single particle, for instance, it is a question of hit-or-miss whether it be thrown against an opening or a blank. If it fail to pass through, it is thrown back by the recession of the water and undergoes a further agitation and probable pulverization.

In practice this feature of the stamp-mill is recognized by both the California and the Colorado millman. On the Pacific coast the mortars are made narrow, thereby diminishing the opportunities for the settling of the particles of ore, and, by increasing the force of the splash, adding to the chances of its exit through the screen. In late years there has also been a tendency to use wire-cloth in place of punched iron, for the reason that the former, though having openings of identical size, yet has more of them per square inch than the latter, and, therefore, presents a greater area of discharge. By giving an inclination of 10 degrees to the screen-frame, the exit of the pulp is further assisted.

In Colorado, this defect of the stamp-mill has been utilized, and has been made an assistant to the millman. The mortars of this district are wide and roomy, the splash of the water inside the battery is weak, and the pulp remains inside until pulverized to a fineness much exceeding that required for its passage through the screen. There is a reason for this apparently contradictory feature of the milling practice. To explain it we must glance at the ore. We find it to contain an average of 15 per cent. of pyrite. The gold is very fine and intimately associated with the pyrite. To separate them it is necessary not only to crush to a certain degree of fineness, but also to obtain conditions which will permit the gold when once separated to settle upon amalgamated plates placed inside. The deep discharge causes the pyrite to remain in the mortar-box long after it has been pulverized to a size smaller than the screen-openings; the long drop gives the interval of time required to allow of the settling of the fine gold, while the roomy character of the mortar aids the deep discharge in affording a chance for the gold to get out of the way of the falling stamps and to become amalgamated upon the two copper plates inserted at the back and front of the mortar. In this way about two-thirds of the total yield of amalgam is obtained inside the mortar. In California the introduction of plates is not admissible, in mortars having so shallow a discharge as 4 inches, because the more constant and more violent agitation of the pulp prevents the settlement of the gold and would cause the abrasion or "scouring" of the surface of amalgamated plates. A certain varying percentage of gold is, indeed, usually arrested inside, partly by the aid of mercury added to the ore as it is fed into the battery, but this is of such a coarseness that gravity alone would serve to keep it within the mortar.

We have now entered into the discussion of the effects produced by the action of the stamp upon the ore. In many respects it departs from the analogy of the hammer which cracks open a nut. While being lifted the stamp also turns. This is effected by the friction of the cam-surface against the under side of the tappet. In a slow-drop mill the stamp makes a complete turn each time it is lifted, but with an increased speed this action is more uncertain and from 4 to 10 drops are required to make a whole revolution.

This feature of the stamp-mill breaks the analogy to the hammer and anvil, and causes it to resemble the pestle and mortar. The turning of the stamp in rising is communicated to the ore when it falls and induces a grinding-action, which has important results. The mere impact of the stamp upon the particles of gold has the effect of hammering it, of increasing its density and of preventing its amalgamation, while the turning of the shoe upon the die causes the abrasion of the surface of the gold and the rubbing off of any film of foreign matter, which, by preventing contact between the gold and the mercury is prejudicial to amalgamation. In grinding the ore the stamp, however, also tends to convert it into slime. The hammer which cracks open the nut liberates the kernel without smashing it, but in pursuing the simile we find that the stamp not only breaks the shell, but both the kernel and the shell are further crushed, and their particles become confused together. The stamp which frees the gold from the quartz has to deal with a material in which the valuable and the valueless constituents are so uneven in size and so intermixed that the one is often crushed too much and the other too little.

I have seen auriferous quartz* which very nearly approached our simile of the nut. The gold occurred in seams and cavities in a quartz which had a honeycombed character. With such an ore there is just a certain blow which will break the brittle quartz and liberate the ductile gold. Such ideal conditions are very rare. The different parts of the same ore usually vary both in hardness and composition. The same work done on two pieces of mill-stuff will produce entirely dissimilar results. In the stamp-battery the heavy sulphide minerals are pulverized to a greater fineness than the siliceous gangue. When the gold is not too closely associated with the pyrite, coarse and rapid crushing will produce an adequate separation; but when the metal is in a finely divided condition and very

* At Clunes, Victoria, Australia.

intimately mixed with the pyrite, then fine crushing is demanded and can, unfortunately, only be obtained by the production of a very undesirable excess of slimes.

We have glanced at the results produced by the turn of the stamp upon the ore. Upon the mechanism itself the results are beneficial. The revolution of the stamp equalizes the wear upon the shoes and dies. It tends, also, by maintaining an even crushing-surface, to prevent that decrease of efficiency which occurs when either hammer or anvil has an irregular face.

Water is the vehicle used for the removal of the valueless portions of the ore from those which are valuable. Its low specific gravity as compared to both the metal and its enclosing gangue enables us to use it as a medium for their separation. A liquid having a specific gravity greater than that of water, and intermediate between that of the gold and its gangue, would be more effective if its use were practicable, which it is not.

In the mill, however, specific gravity is not the only factor we have to consider. The water discharged from a stamp-mill often transports the heavy pyrite further than the light quartz. This is due to the fact, already referred to, that the pyrite remains inside the mortar longer than the quartz and becomes pulverized to a further degree of fineness. It, therefore, presents a larger surface to the water. Again, the metallic sulphides commonly occurring in gold ores have a cleavage more highly developed than that of quartz; therefore, while the latter finds its way into the water in irregular and angular grains, the former will be found in thin plates and flakes, which readily float upon a running stream.

Water is the fluid used, but air also plays its part. During the time of its violent agitation under the falling stamp, the water entangles a certain amount of air. Such air exists in the form of small bubbles which hold the finely-pulverized ore in suspension and thus become the main agent in the floating of the slimes. Warmth causes the air to expand and the bubbles to become dissipated; therefore any rise in the temperature of the water, such, even, as is caused by the impact between the stamps and the ore upon the die, is favorable to a diminution in the amount of slime.

When the pulp is discharged from the mortar-box it runs down copper plates covering long sloping tables. The copper, whether plain or silver-plated, is provided with an amalgamated surface, and it is this amalgamated surface which is supposed to do the work of arresting the gold. Mercury unites with gold forming a heavy amalgam; but, in practice, it is found that a plate which is covered with a good coating of gold-amalgam will serve to arrest gold much more effectually than a clean surface of either amalgamated silver or copper.

The amalgamating-tables have a slope varying with the amount of water used, the heaviness of the pulp, and the rapidity of the crushing. A gradient of $\frac{7}{8}$-inch per foot is common in Australia, while in Colorado the inclination is over 2 inches per foot.* The colonial mills consume 5 gallons of water per stamp per minute, while those of Colorado use less than 2 gallons. Theoretically, the use of the least possible quantity of water, and the spreading of the pulp over the largest possible surface, will give the best separation of the gold from the gangue. In practice, the varying composition of the ore prevents a nice adjustment of the conditions. You may readily determine an inclination which will be most effective in causing a separation of the gold from the quartz, but, it may be such as to cause the pyrite to settle. On the other hand, the slope may be so adjusted that the pyrite is carried away; but, such conditions may then be obtained as will also permit of the escape of the gold.

The amalgamating-tables are attached to the frame-work of the mill. The vibration set up by the falling stamps causes a pulsation of the water flowing over the plates similar, in a way, to the action of a jig. This assists the work of gravitation. The vibration has, however, another effect, namely, that of crystallizing the iron of the working parts of the mill, making them brittle and decreasing their time of service. In this, as in other respects, the stamp-mill presents contradictory features.

At the outset, we described milling as the art of treating an ore so as to extract the maximum of value at the minimum of expense. Let us apply the description to the two types of mill to which particular reference has been made. In Colorado, a stamp crushes 1 ton of the ore of the Gilpin county mines in 24 hours, and the cost, using free water power, is 70 cents. In California, the best equipped large mills crush at the rate of rather more than 2½ tons at a cost, also using free water power, of about 35 cents per ton. The extraction in both regions will be, by amalgamation alone, about 70 per cent. We will omit the amount extracted by the concentration of any valuable pyrite, because the percentage of such material is

* An Australian mill usually crushes 2 tons per stamp per 24 hours.

very variable, and it forms a by-product, the value of which depends largely upon local conditions.

The ore of the Gilpin county mines carries about 15 per cent. of pyrite, and other heavy sulphides. The gangue is more feldspathic than quartzose, and is the product of the alteration of the country-rock—granitoid gneiss—and of the dikes* which penetrate it. The gold is not only present in a state of very fine subdivision, but it is also intimately associated with the pyrite.

On the other hand, the mill-stuff treated in Amador, Calaveras and Tuolumne carries from 1 to 2 per cent. of pyrite. The gangue is quartz, but the ore also contains a very large proportion of the country-rock, which in this case is slate, augite schist, and diabase. Of these, slate predominates. The gold is coarser than that of the Colorado ore, and it is not so closely associated with the pyrite.

Let us now consider the results to be obtained by an interchange of treatment, using California batteries on Colorado ore, and *vice versa.*† The Gilpin county ore is of medium grade, say 8 dwts., or $8 per ton. The local methods extract $5.60‡ at a cost of 70 cents. A California mill would give an extraction of only $4, but would crush such soft ore fully three times as fast, so that the cost would be, say 25 cents, giving a net yield of $3.75 as against $4.90 obtained by the methods of the district. Here, the slower mill gives the best results with a particular ore, and the Colorado millman considers the Californian very stupid because he does not use Colorado methods. Let us go to California and use the Gilpin county mill upon an ore of simpler character and of lower tenor. We will consider the treatment of an ore containing 6 dwts., or worth $6 per ton. The California mill would extract 70 per cent. at a cost of 35 cents, leaving a balance of $3.95 per ton. The Colorado battery would extract an increased percentage, say 75 per cent., but the ore being much harder than that of Gilpin county, the crushing capacity would be less and the cost per ton greater than when treating Gilpin county ore, say, therefore, $1.00 per ton, leaving a net yield of $3.50 per ton. The California mill, if crushing 100 tons of ore per day, would, therefore, show a profit $45 per day greater than that of the Colorado mill. As a matter of fact, there are other practical considerations which would render inadvisable the interchange of

* The "porphyry" of the miners; really, quartz-andesite.

† In making the comparison the cost of motive power, being very variable, is left out of the count.

‡ Amalgamation only, omitting concentration afterward, is here included.

methods, among which may be mentioned the smaller size of the ore-bodies of Gilpin as compared to those of California; while it must also be remembered that the construction of a Colorado mill of a capacity equal to that of a California plant would require twice as much capital.

The comparison just made will serve as an illustration of the fact that milling is a business for getting money, and not a scientific pursuit directed to the obtaining of a perfect metallurgical treatment.

The contrast between the methods in use in two mining districts in the same country, illustrates the first axiom of all successful ore-reduction, namely, that the treatment must be suited to the character of the ore. Colorado methods in California would probably fail just as surely as California ways have been unsuccessful in Gilpin county. This is a truism not always remembered by machinery firms, who do not desire to be bothered by the making of new patterns. Too often, the ore is required to bend to a certain treatment in a mill of a particular design, instead of the mill being modified to suit the necessities of a particular ore.

The stamp-mill has presented to us many contradictory features. It is seen to be compounded of good and ill. It may be simple, but it is clumsy; it may be crude, but it is effective. As a machine, it has undergone an evolution common to all human inventions. It was founded on the first stone implement of the prehistoric savage; it became modified into the *matate* of the Mexican and the tilt-hammer* of the Chinese; it progressed until running water was called in to aid human muscle, and in the machine of the Hungarian peasant, it reached the primitive type from which our present mills were evolved. How great has been the comparatively recent improvement can be seen by stepping from Hungary to California.

In the valleys around Verospotak, in Transylvania, the larger mills† consist of twelve stamps, in coffers holding four each. The power is derived from an overshot water-wheel 10 feet in diameter. The cam-shaft is of iron, and revolves on agate bearings, lubricated with water. The lifter, or cam, is iron-shod. The stamp weighs 250 pounds, and has an agate head. The stem, the coffer, and all

* "A Chinese System of Gold-Milling," by Henry Louis, *Trans.*, xx., 324.

† Modern American mills have been lately introduced, and can be seen working side by side with those dating back to 100 A.D. For the particulars above given, I am indebted to Mr. E. H. Liveing.

The Old and the New.

the rest, are made of beechwood. Each stamp drops 30 times per minute, and crushes about 300 pounds of soft ore per 24 hours. These machines have changed but little since the time of the Roman occupation under Trajan, when this district was a part of the province of Dacia.*

Let us now go to California, whose record is little more than the record of a generation. Among the foothills of the Sierra Nevada we find mills containing 80 stamps, weighing 750 to 850 pounds each, and dropping 95 times per minute. Those of the working parts which are not of iron are made of steel. At single mills, 200 tons of ore are crushed per day. The mill building has a height of 70 feet, and the ore is never touched by manual labor from the moment that it arrives at the top in the mine-cars to the time when it is discharged at the bottom as waste.

In Transylvania, the individual shareholder often has his own mill; in California a thousand unite to operate one, which can, in 24 hours, treat as much ore as the Hungarian mill crushes in 100 days. The little machine of the Hungarian has been tapping away like a woodpecker for eighteen centuries, and yet has not produced as much gold as has been contributed in the brief time of one generation by that completer mechanism whose muffled thunder echoes among the cañons of California.†

What has been done may serve as a measure of what can yet be done. Perfection is as unattainable in milling as in any other branch of industrial art; otherwise progress were soon ended. We can compare the old mill with the new, not only with a complacent satisfaction at the advance that has been made, but with the consciousness that where so much improvement was possible much room for improvement must remain.

It is not for me to attempt to foretell what place the stamp-mill is destined to hold in the metallurgy of the future. Let me, however, in concluding, suggest the reflection that though the appliances of to-day may show a great advance upon the older more imperfect type from which they were evolved, yet there is no mining district that possesses a mill which cannot, in some essential, be improved upon.

---

* Well-preserved gold coins of the time of Trajan have been found in the mine-dumps.

† The accompanying illustration of an old and a new mill has been engraved from a pen-and-ink drawing, made by my friend, Mr. H. R. Pridham, after a photograph taken by myself during my residence in California, some years ago. In the middle distance is an old ruined water-wheel, formerly the motor of a small ten-stamp mill, the dismantled portions of which remain in the shed on the left. In the background is a fully equipped, modern, California steam-mill.

Subject to Revision.

[TRANSACTIONS OF THE AMERICAN INSTITUTE OF MINING ENGINEERS.]

THE GOLD-FIELDS OF OTAGO.

BY T. A. RICKARD, DENVER, COLORADO.

(Plattsburgh Meeting, June, 1892.)

THE province of Otago consists, roughly speaking, of the southern half of the South Island* of New Zealand. On three sides it is washed by the Pacific Ocean and on the north it abuts against Westland and Canterbury. It covers an area of over 20,000 square miles, and for the most part has an extremely broken surface; the narrow plains of the sea-board are bounded by the rounded foothills which in turn are overlooked by range after range of the snowy summits whose varied beauty has made the island known as the Switzerland of the southern hemisphere.

The gold-fields are confined to the quartzose schists which, in a broad band, 70 to 75 miles wide, cross the district extending in a northwesterly direction from the shore of the western ocean into the northern provinces. These quartzose schists are the characteristic rocks of Otago, and to their curious structure are due the interesting differences exhibited by the mining districts in the modes of occurrence of the gold. These rocks, which are almost unbroken over their full extent, have been divided by the Provincial Geologist,† F. W. Hutton, into two series of beds, named respectively the Wanaka and Kakanui formations. The only reason given for this division is the desire "to divide such an enormous thickness of rocks in order that the map might display somewhat of the geological structure of the district."‡ As Hutton and others have pointed out there is in fact no dividing line. The changes noticeable in the ascending series of schists are very gradual, and are due to the slowly decreasing effects of metamorphism. Hutton estimates the thickness of the Wanaka series at as much as 50,000 feet, and the

* New Zealand is variously spoken of as consisting of the North, Middle and South islands or of the North, South and Stewart's Island. The last is of insignificant size and importance. Otago is in the southern of the two chief islands.

† That is, the geologist of the province of Otago. It is the term used in the government publications.

‡ Page 33 of *Geology of Otago*, 1875.

Kakanui at 52,800, or twenty miles for the two. The information obtainable is not sufficient for trustworthy estimates, but it is certain that these rocks have an enormous thickness over a very large area.

The question of their age is surrounded with much difficulty. The geologists of New Zealand are not agreed as to either nomenclature or chronological position; and the maps of the several authorities show most unfortunate divergencies of opinion. For our immediate purpose it may suffice to say that the quartzose schists of Otago overlie the syenitic gneiss and granite, which Hector and Hutton agree in regarding as pre-Silurian;* while they underlie a are overlain by those of the Matai, the geological position of which was for a long time sharply debated, but has been determined by fossils as Carboniferous. Such is the evidence upon which the quartzose schists have been labelled Silurian.

The rocks themselves have been variously described as phyllite, clay-slate, mica-schist, etc. Hornblende-schist and quartzite are also known to occur. The mineral constituents are subject to frequent variation. The name "phyllite" is often used, but since the argillaceous character of the mica-schist has been seldom more than guessed at, this usage may be taken as a mere *façon de parler*. Neither of the above names describes distinctively the rock which forms the prevailing type, and I would suggest "quartzose mica-schist," or, for brevity, "quartzose schist" as being more descriptive.

The most striking—and at times a very remarkable—feature of the schists of Otago is the extraordinary development of quartz. In many localities (for instance, in the twelve miles between Lawrence and Waipori, where the road-cuttings afford numerous sections) the quartz forms half the bulk of the rock. It is interbedded among the folia of the schist, often in very regular and continuous lines. The seams of the quartz will, over a wide area, have an average thickness of from half an inch to an inch, increasing often to bands 5 or 6 inches wide.‡ Fig. 1 illustrates an outcrop of such quartz-banded rock.

It is noteworthy that over the greater part of the gold-fields the

---

* Sir James Hector calls the granite pre-Silurian, and F. W. Hutton has called the formation by the name of Manipori, and considers it Laurentian or Cambrian.

series of rocks called Devonian and Carboniferous,† which in turn

† Je Anau series (Devonian) of Hector or the Kaikora formation (Carboniferous) of Hutton.

‡ Above La Grave, Hautes Alpes, I have seen sections in the Lias which strongly reminded me of this feature of the Otago schists.

beds preserve a very flat dip over very wide areas and are rarely, and only locally, much tilted.

That the quartz is formed along lines of bedding, and not cleavage-planes, is amply proved by the observation of the changes in the character, color and composition of the rock, which take place at right angles to the lines of foliation, whose low angle of inclination is also confirmatory, since cleavage is usually characterized by a high angle.

The quartz is most probably of secondary origin, and due to the exudation of silica from an extremely siliceous silt and segregation along the lines of lamination during the period of metamorphism.

To those acquainted with the mineral deposits of the Western United States, Europe or Australia, the almost entire absence of eruptive rocks over such a wide auriferous area will appear extraordinary. It will also largely explain the scarcity of lode-formations, the comparative want of definition which belongs to most of the veins, and the small extent of the ore-bodies, notwithstanding the fact that large portions of the country-rock are undoubtedly in themselves gold-bearing. To the latter fact is due the enormous quantity of alluvium, the yield from which is much greater in importance than that of the quartz-veins of the province.

The history of the first discovery of gold in Otago has been related by one of the pioneers—Vincent Pyke. It may be considered as dating from the 4th of June, 1861, the day upon which a Dunedin newspaper published a letter from Gabriel Reed announcing the finding of gold at Tuapeka.*

Before that date, a native of Bombay, Edward Peters, or "Black Peter," as he was more generally known, had been getting gold in what was afterwards famous as Gabriel's Gully, the locality which saw the first great "rush" of 1861. In the following year the known auriferous area was extended far into the mountains of the interior, and on August 16, 1862, Hartley and Reilly—a native-born American and a Yankee Irishman—astonished Dunedin by depositing at the treasury a bag containing 87 pounds of gold. They had been quietly working on the beaches of the Clutha river near the Dunstan gorge, and the announcement of their find caused the then small population of Otago to make a hasty stampede into the in-

* The first discovery of gold in New Zealand is said to have been made at Driving Creek, Coromandel, in 1852. The Coromandel district has, however, been overshadowed by its neighbor the Thames, which has had a most brilliant though irregular record. Neither of them has been such a regular producer as Otago.

terior. From this time immigration from the Australian colonies set in, and gold-fields were proclaimed* in rapid succession.

Between September and December, 1862, 70,000 ounces were sent by escort to Dunedin from Dunstan.

The working of the gold-deposits was rendered easy by the fact that, unlike the wash of the Victorian fields, that of Tuapeka, the first Otago gold-field, and those of the province generally, contained but little clay, and did not necessitate the "puddling" which, at old Bendigo for instance, was the treatment required by the stiff, hardened "cement" of the Australian alluvium.

Nuggets do not figure much in the history of the Otago gold-fields. According to Vincent Pyke the largest was one of 27 ounces, reported to have been found at Waipori in 1863. Nuggets were comparatively rare, and the few found were very small, and in no way comparable to those which in the early "fifties" made Ballarat and Dunolly famous.

The production of Otago is shown in the accompanying statement:

The Gold-Production of New Zealand.

	Auckland.		West Coast.		Otago.		Total.[a]	
Year.	Oz.	Value.	Oz.	Value.	Oz.	Value.	Oz.	Value.
1889	28,655	£113,191	101,696	£406,451	64,419	£256,480	203,211	£808,549
1890	31,745	125,760	89,096	356,368	63,410	255,926	193,193	773,438
Highest yield,	330,326[b]	1,188,708	552,572[c]	2,140,946	614,387[d]	2,380,750	735,376[e]	2,894,517
Total since '57,	1,639,357	6,122,473	5,093,249	20,226,540	4,783,968	18,886,928	11,818,221	46,425,629

a. Including minor districts.
b. This was in 1871, at the time of the Thames rush.
c. Due to the Hokitika discoveries.
d. When the Gabriel's Gully Diggings were at their best.
e. In 1871.

The mining industry of Otago at the present time offers a variety of interesting features. The methods by which the precious metal is won are very diverse, including dredging and hydraulic elevating in addition to the ordinary forms of alluvial and deep mining. In lode- or vein-mining—the "quartz-reefing" of the colonies—Otago is considerably behind its neighbors of the Australian continent, but

* When the auriferous character of a district has been proved the colonial government "proclaims" it a gold-field.

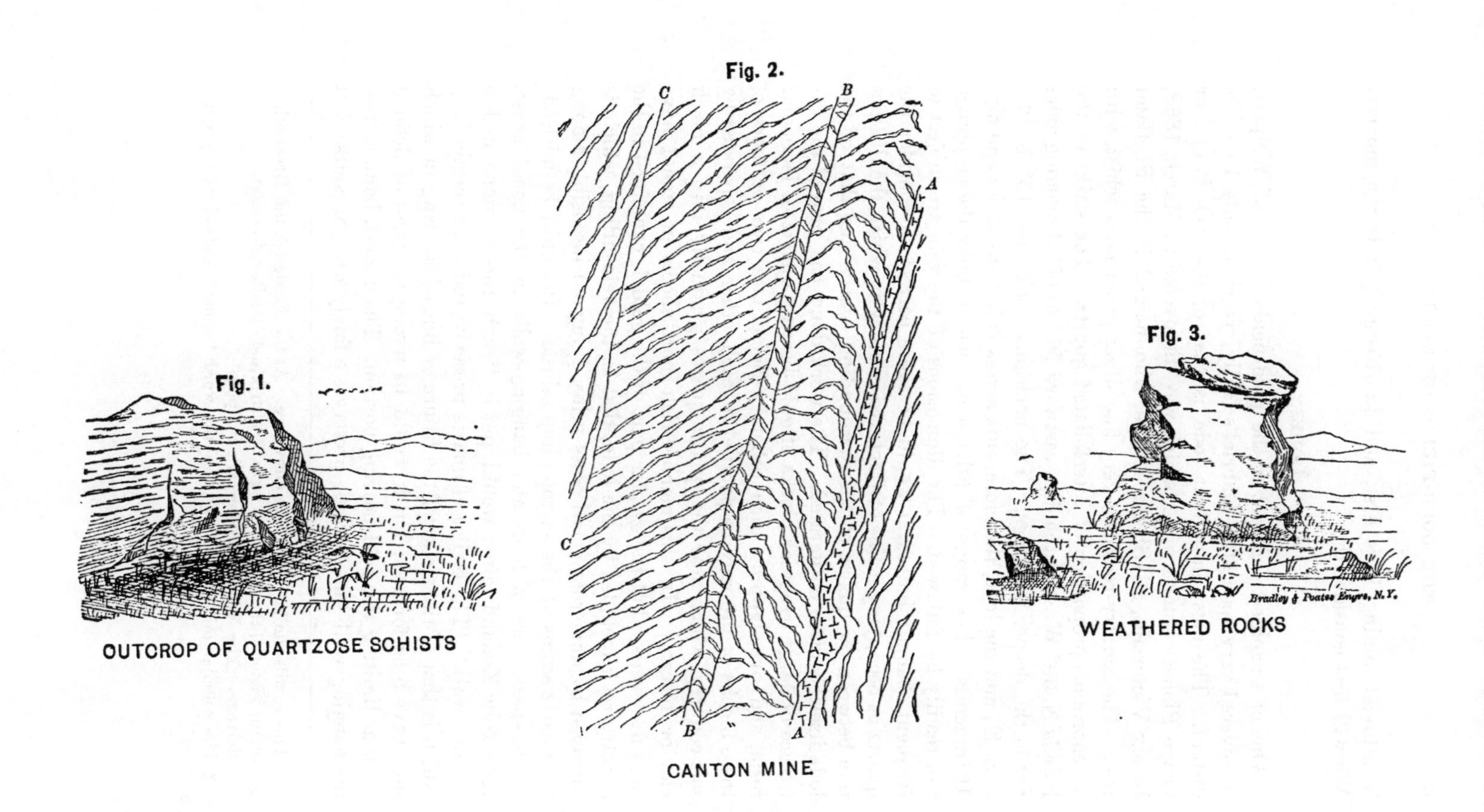

OUTCROP OF QUARTZOSE SCHISTS

CANTON MINE

WEATHERED ROCKS

in alluvial mining it is first, and in dredging it is the pioneer. We will first consider the

Lodes.

One of the oldest and best of the gold-fields is that of Waipori, discovered very soon after Gabriel's Gully, which is only 10 miles distant. The most important lode is that of the "O. P. Q." or Otago Pioneer Quartz mine, which was discovered in March, 1862, by six Victorians, all Shetland men, who named it the Shetland reef. The country-rock is a very fine silver-green mica-schist, with an enormous proportion of interfoliated quartz. The strike of the lode is N. 30° W., and the dips eastward 50° to 55°, becoming more flat in the deeper workings. The enclosing rock dips 15° to 20° S. S. E., and the lode, therefore, cuts across it in both strike and dip. It traverses a low range of hills, across whose brow the croppings can readily be followed. The dimensions of the vein are subject to frequent change; indeed, it would be more correct to consider the quartz as occupying a line of fractured country in which the distinction between the quartz of the surrounding schists and that of the lode itself is often obliterated. The work done has proved the existence of large irregular blocks of gold-bearing quartz of lenticular form, and over-lapping each other.* Sometimes two or more of these blocks of ore may be parallel to each other and separated by soft country, which near the quartz passes gradually into a black clay or fluccan.† In this manner rich short shoots of stone‡ have been discovered; but the greater part of the gold is contained in the country-rock included by the lode, that is, the vein-filling, which is threaded by numerous small cross-veins, dying out insensibly among the quartz seams of the surrounding schists. It would be misleading to speak here of foot- and hanging-walls in the usual sense. As a New Zealand miner would put it, "each run of stone makes its own wall." The bodies of quartz preserve, indeed, a certain line, which is that of the strike of the fissure or line of fissuring in which they have been formed; but it would be wrong to speak of defined walls as limiting the gold-bearing portion. The eastern boundary—the hanging-wall—of the lode preserves a fairly straight course, but

* This recalls similar formations elsewhere. At the Frederick the Great mine, Sebastian, Victoria, the ore-shoots were of a strikingly similar character.

† Selvage. The New Zealander calls it "pug."

‡ The colonial miner generally uses the word "stone" instead of "quartz" or "ore."

there are only suggestions of a foot-wall, a series of parallel or nearly parallel division-planes, separating the quartz of the enclosed from that of the enclosing schists.

On the Waipori side of the ridge, the lode traverses a portion—a bar—of country-rock more siliceous and harder than usual; and it is to be noted that within this passage the quartz is barren of gold, and contains a smaller amount of iron pyrites than in the softer portions of the hill. In washing the lode-matter, one notices a heavy white mineral, which lingers with the pyrites and gold after the rest of the gangue has been washed away. This is the somewhat rare mineral scheelite or tungstate of lime,* which in this district occurs scattered through the schists, though rarely in workable quantities. In addition, the country-rock contains occasional thin veins of cinnabar, irregular in behavior and too small to have economical importance. The presence of the mercury mineral is not observed in working the lodes, but it forms a marked feature of the alluvium with the gold of which it is mixed.

About a third of a mile westward is the parallel lode of the Canton mine, where the true character of the formation is more readily discernible. Fig. 2 illustrates the appearance of the end of the southeast drift. A A is the "reef," a vein of quartz which is supposed to lie immediately upon the "foot-wall." Along B B the country is soft, and the included quartz folia are much twisted. C C is one of the "false hanging-walls." The whole width is gold-bearing, though A A acts as a guide in following the auriferous channel. It is not possible to say where the "lode" ends or where it begins. It is a channel of country-rock which is auriferous within ill-defined limits, and in which the vein A A acts as an indicator. Along A A and C C, it has undergone faulting; along B B, distortion only. It is an extreme case of what a colonial calls a "mullocky reef,"—a lode full of mullock or waste rock—a type of ore-deposit more common than is supposed, though the usual forms are less distinct than in this case.† We shall see somewhat of the same structure in the other districts of the province.

* Specific gravity, 5.9 to 6.1; composition, CaO_4W. At Glenorchy, at the head of Lake Wakatipa, I examined a mine which had been worked for this mineral.

† In this particular case, those working the mine had little comprehension of the formation. I was informed that there were several "false hanging-walls," but that there was only one foot-wall, which was said to be of a different rock, much harder than the hanging. On examination, I found the rock of the so-called foot-wall to be similar to that of the rest of the country enclosed by the lode; and on crushing

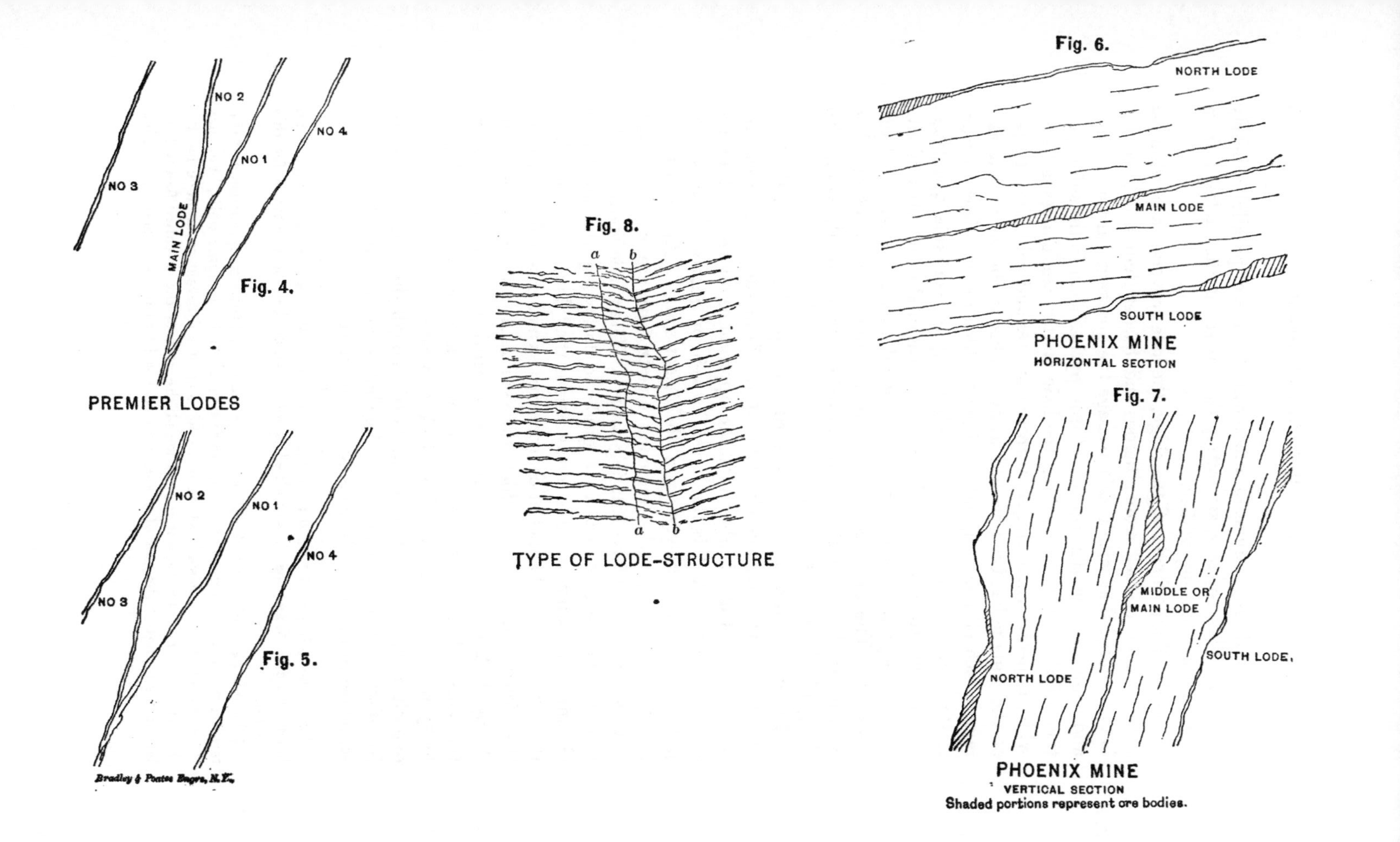

Fig. 4.

PREMIER LODES

Fig. 5.

Fig. 8.

TYPE OF LODE-STRUCTURE

Fig. 6.

PHOENIX MINE
HORIZONTAL SECTION

Fig. 7.

PHOENIX MINE
VERTICAL SECTION
Shaded portions represent ore bodies.

The chief lode-mining or "reefing" district of Otago, which lies north of Lake Wakatipu, presents a great contrast to that which surrounds Waipori. The latter consists of peculiarily monotonous rounded hills, covered with a peaty soil on which the tussock grass of the ridges and the flax in the gullies are the only vegetation. The rounded outlines of these foot-hills have been ascribed to glacial erosion in the same way that much else in Otago is put down to this little understood agency. However that may be, we have here the basal wrecks, the stumps, as it were, of what were once mountains, perhaps as rugged as those further inland. Denudation has degraded them to uninteresting hillocks, and the peaty soil which covers them now serves to protect them from rapid erosion. Farther inland, the foot-hills become more striking to the eye by reason of the very curious weathering which has left numerous isolated rocks standing above the surrounding level like an army in skirmishing order. See Fig. 3.

These have been sometimes supposed to be the rocky freight brought down by glaciers; but they are *in situ*, and the explanation of their occurrence is difficult. The peculiar weathering which originated the names Rock and Pillar, Raggedy Ridge, Rough Ridge, etc., given to some of the hills of Otago, is not of uniform occurrence, being less marked as the coast is approached. I have noticed that the schist of which these isolated rocks are composed is usually more siliceous than that of the surrounding country, and I would suggest that it is the irregular segregation of siliceous material which has induced this very unequal weathering.

At Barewood, where I examined a large number, I found that, almost without exception, each of these pillars, usually about 10 or 15 feet high, contains a cup-shaped hollow of varying size. This is on the side towards which dip the prevailing quartzose schists of which they are composed; and it is doubtless due to the formation first of a small hollow by the action of rain-water, followed by the removal, by wind and rain, of some of the softened mica-schist. In the concavity so formed, rain-water would lie, and, by the aid of frost, would shatter the interfoliated quartz. Along the banks of the mountain torrents this action has progressed more rapidly, and the rocks exhibit a honey-combed appearance.

---

it and testing it with the "dish" or pan, it was found to be richer than the portion then being mined. It was scarcely necessary after that to advise the miners to put a cross-cut into the "foot-wall."

To return to the northern mining districts. The region about Lake Wakatipu, in which they occur, is crossed by the snowy ranges of the Southern Alps, 10,000 to 12,000 feet in height, the highest summit, Mt. Cook, having the altitude of 12,350 feet. By the sculpture of frost and snow, the structure of the mountains has been laid bare; and though they are carved from the same rocky material as the lower country,* the rugged, bleak, precipitous ranges, with their coronets of snow, are in strong contrast to the rounded contours of the brown-paper-like hillocks around Waipori.

In this district, Arrowtown, Macetown, and Skippers are three of the best-known mining centers. At Macetown, near the head of the Arrow† river, are the Tipperary, Premier, and Sunrise mines.

The Premier, at the foot of Advance Peak, offers several points of interest. A series of three nearly parallel lodes is met by a counter-lode, and at the junctions ore-shoots occur. The two cross-sections, Figs. 4 and 5, indicate the relative positions of the members of this lode-system. The dip of the No. 2 or main lode brings it across the others, which are thereby deflected, but *not* cut. At the meeting-point ore-bodies occur; but so far only one of these junctions—the lower junction of the No. 2 and the No. 3—has been thoroughly developed. The ore-shoot there found has a flat inclination (about 45°) along which it was followed by a series of irregular workings. There is always a clear division-plane, or "wall," between the two lodes when after the junction they are seen to go forward in company.

The cross-cut from the lower adit to the No. 2 lode indicates the relation of the lodes to the quartzose schist of the country. See Fig. 9. The folia of quartz (dipping about 20°) which are flat in the country become more nearly vertical as the lode B is approached. They become broken by fractures A A, which can be followed until they lose themselves in the vertical lamination of the schist encasing the quartz of the "reef." Fig. 10 shows the appearance of the No. 1 reef as seen in the back of the level, the sketch being in plan or horizontal section. The width of the lode proper is 15 to 18 inches, strike N.W., dip, 75° to 80°. On the foot-wall side there are a couple of inches of "pug" or selvage, on the opposite wall

* Quartzose schist. This refers only to the mountains in the vicinity of the mining districts cited. The main ranges westward are formed in part of granitic rocks.

† So-called because not straight—*lucus a non lucendo.* In riding from Arrowtown to Macetown the track crosses the winding river twenty-five times in 10 miles.

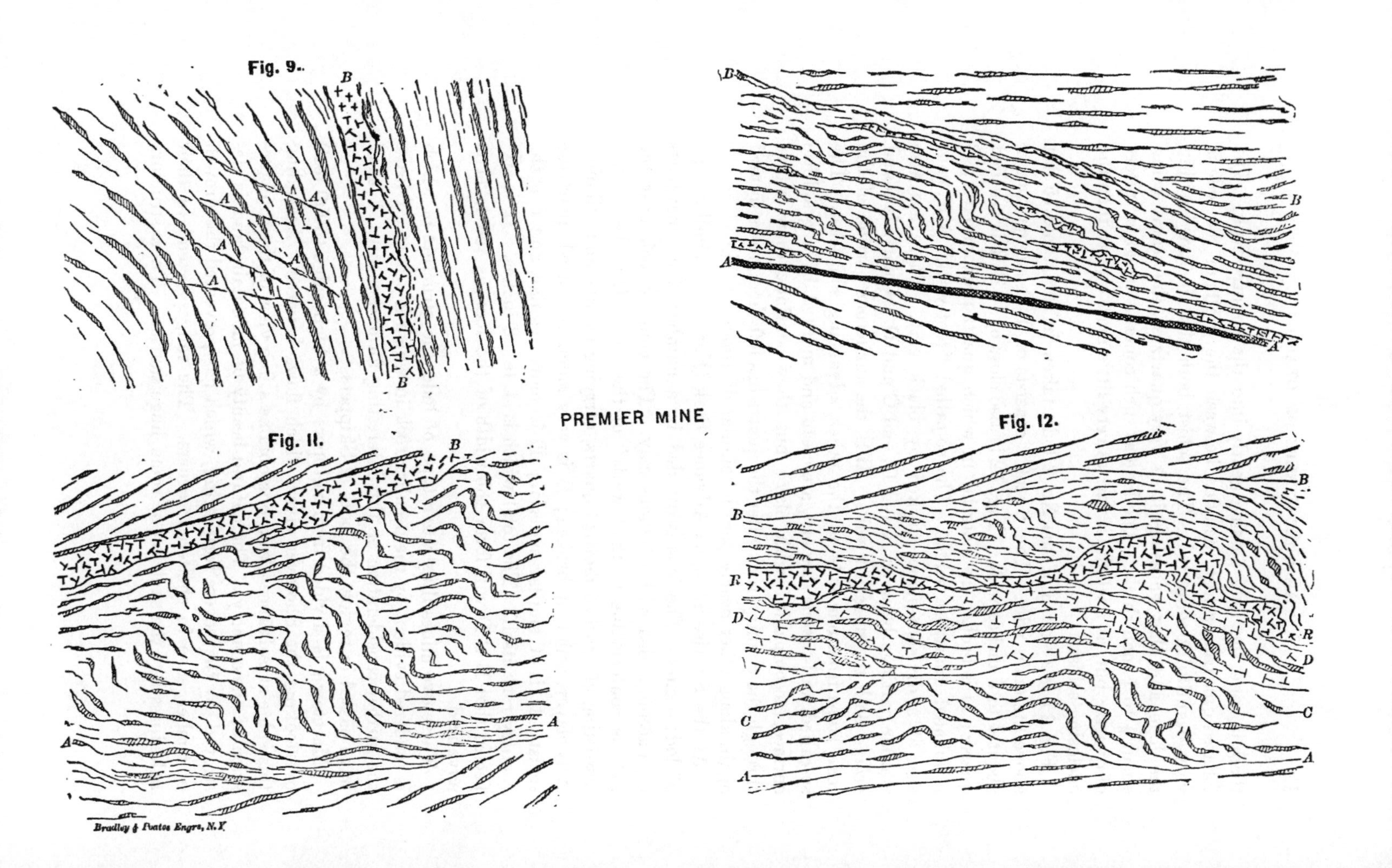
Fig. 9.
B
A
A
A
A
A
B
B
B
A
A
PREMIER MINE
Fig. 11.
B
A
A
Bradley & Poates Engrs, N.Y.
Fig. 12.
B
B
R
D
R
D
C
C
A
A

there is none, the lode passing gently into the country-rock. The foliated quartzose schist included between the limits of the lode is very beautifully contorted. The schist, both that included within the lode and that encasing it, is very quartzose, the seams being about $\frac{1}{8}$ to $\frac{1}{4}$ inch thick. The lode as such contains irregular pieces of quartz; but, as a whole, it carries very little more quartz than the enclosing schist.

Fig. 11 is a sketch of the No. 2 lode, taken in the stopes 75 feet above the main adit. The "reef" consists of B B, sugary quartz about 4 to 5 inches wide. Fig. 12 is also close by. R R is crushed quartz; D is broken-up material in which quartz predominates; C is foliated quartzose schist. The so-called "foot-wall," A A, consists of a division-plane marked by black clay. There are also similar partings between R and D, and C and D. Quartz is found *behind* both the so-called walls. B B, the auriferous portion of the lode, steps from one side to the other, advancing southeast. The so-called walls, *while they last*, are clean and marked by a graphitic surface, accompanied with clay; but they continue their course through the country-rock after the quartz has left them, to be formed again along other nearly parallel similar divisions.

At the Sunrise mine, on Advance Peak (Fig. 13), a similar type of lode occurs. The lode is divided by a number of distinct partings accompanied with unctuous clay. The country-rock near the lode is conformable to the "walls" of the lode. C is the "reef," consisting of crushed mottled quartz, carrying pyrites and stibnite; A is clayey, soft and broken; B is very much contorted quartzose schist; D is of similar material; E is more regular, and F is the country-rock itself. The total width is 4 feet 8 inches. It is expected that B C D will form the width of the lode as the level is advanced.

These two mines illustrate a type of lode, the structure of which is rendered evident by the peculiar rock in which it occurs. I leave the further consideration of this interesting formation until I shall have passed in review the lodes at Skippers.

Macetown is divided from Skippers by a high ridge (6000 feet) overlooking the Shotover river, which flows between. Ascending this ridge at Advance Peak, one obtains a fine view of the configuration of the country. The natural beauty of mountain land is destroyed by the severe weathering, which has prevented the growth of timber, save in sheltered corners. The ranges extend in their nakedness like the skeleton of some huge saurian. Their structure can be readily discovered; the dip-slopes are *débris*-strewn and incline at a low angle, while the opposite faces are precipitous and rocky in the extreme, cut up by well-marked systems of jointage. On the summits of the ridges the soft schists show the evidence of the easy and rapid degradation which they undergo when exposed to the snow and frost of these higher altitudes. Landslips on a large scale are common and temporarily dam or permanently divert the mountain streams, causing heavy floods, such as the historic one of 1878, which altogether changed the distribution of the auriferous beaches of the Shotover river.

The reefs at Skippers are perhaps the best known of the Otago lodes, on account of the Phœnix mine, locally celebrated as the first to utilize the electric transmission of power for the working of its stamp-mill. Like the other lodes we have considered, those of Skippers consist of a system of several gold-bearing quartz-veins traversing a more or less defined ore-channel. The lodes are three in number, forming a lode-channel which dips north and has a variable strike of about N. 80° W. They are known as the North, Middle and South lodes respectfully, and the Middle is the richest and strongest vein, and the main ore-producer. They are approximately parallel; but, even when close together, never lose their individuality. There is a good hanging-wall to the north of the series and a good foot to the south; but the intermediate walls of the different members of the series are irregular and indistinct. The shoots of gold-bearing stone have two marked features. The first is, that they make in step-like gradation from one lode to the other. Shoots on any two lodes are never opposite each other, whether in vertical or horizontal section. It is almost the converse of "ore against ore." See Figs. 6 and 7. The other noteworthy characteristic of this system of gold-veins is that they are particularly poor when their strike is south of west, while they are almost invariably found to become gold-bearing whenever a turn takes place to the north of west. As the result of these two features, it is found in the working of the mine that one lode is rarely rich for any great distance; that when it is gold-bearing in a given portion, the portions of the other two lodes which are immediately opposite will be poor; that when it becomes barren the gold is to be looked for at a point opposite in one of the two other lodes; and of the two, in that one whose course at that point is north of west.

The mine is worked by adits. At the level of the No. 3 adit the south lode is intersected at 96 feet from day-light, the Middle at 24 feet beyond, and the North at 60 feet further still.

Fig. 13.

Fig. 14.

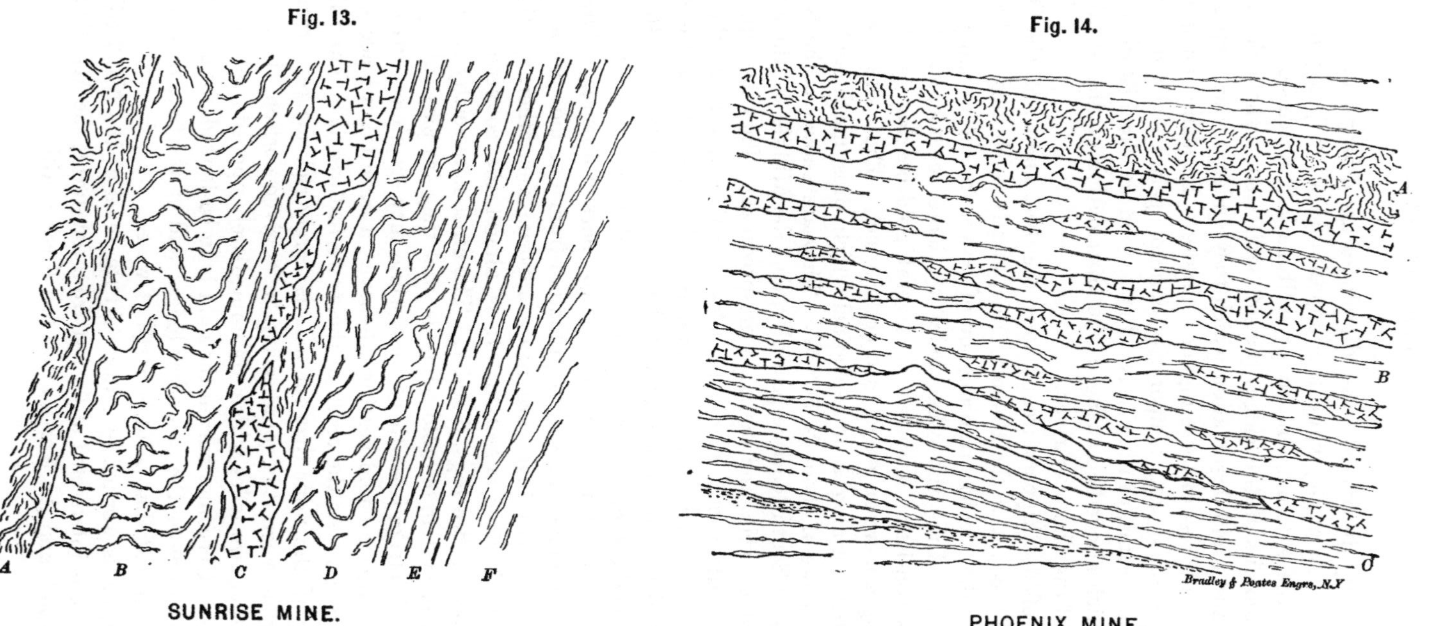

SUNRISE MINE.

PHOENIX MINE.

The shoots of gold-bearing stone are from 50 to 200 feet in length; for the most part they pitch strongly to the west, but are not continuous in depth. However, as one ore-shoot comes to an end, another is discovered below; and the deeper workings expose ore-bodies which do not extend to the surface. A cross-course cuts the lodes at a strong angle, faulting the Middle lode, which on the east side was called the "Promised Land" lode until it was recognized as a faulted portion of the Middle lode. Immediately east of the cross-course the lode was very rich. Fig. 14 illustrates the Main lode as seen in the 150-foot level stopes. A is disordered soft country lying against the hanging. B is "payable stone"—quartz and included country merging into each other; the quartz is dark and mottled, carrying a small percentage of arsenical pyrites. The gold is coarse and often visible. C is country-rock included within the lode and arranged parallel to the foot-wall which it follows. The whole width is 8 feet. A is 2½ feet. C is 3 feet.

The reefs of Skippers, Macetown and Waipori have certain features in common, due to the identity of the country-rock which they traverse, and probably also to their contemporaneous origin. In each case the structure is that of a group of lodes rather than of one single vein. The series of two or more approximately parallel reefs occupying fractures formed along a more or less defined belt represents the relief given to the compression of the rocks by a large fault-fissure such as that of the Comstock vein. The contortion observable in the country included within the limits of the ore-channel, and the frequent partings which are also lines of faulting, lead one to the belief that more than one movement of varying intensity took place. The fine-grained, closely foliated quartzose schists yielded in part, by the formation of folds, to the compression to which they were subjected, while the strain was further taken up by the production of a set of co-ordinated fissures. Along these lines, later movements crushed the intervening enclosed country, and thus by preparing a channel for the percolation of mineral solutions, created in a varying degree the *locus* of gold-deposit.

The outside members of any such system of lodes usually have walls which respectively form the boundaries of the ore-channel. Such walls are usually accompanied by a varying thickness of black clay or gouge. Generally speaking, however, the distinction between "country" and "reef" is arbitrary. The quartz of the reef, the quartzose schist of the included country, and the somewhat less quartzose schist of the enclosing country shade off imperceptibly.

Sometimes there is no line of division between the reef and the enclosing rock; at other times there is a line of parting, which is dignified by the name of "foot-wall" or "hanging-wall."

There is reason to believe that the quartz of the lodes sometimes occupies fractures formed along the axes of anticlinal folds; but, as a rule, the type of structure is that shown in miniature in Fig. 8 from a sketch made by me from a cutting on the Shotover main road. The sketch covers a width of 5 feet. Instead of forming a narrow clean-cut crack or fissure, the soft schist, by reason of its interfoliated quartz which renders it of varying hardness, is traversed by a belt of dislocated country bounded by two parallell fractures. The interfoliated quartz enables one to see clearly that there has been a movement resulting in the disarrangement of the parallelism of the quartz laminæ. In this case the fracture (from *a* to *b*) has a width of only 7 inches; but the structure is similar in origin and kind to that which we have seen forming the larger lodes of the mines. The part between the two lines of fracture forms the beginnings of the "mullocky reef" of the colonial miner, that is, a lode carrying a large proportion of included country-rock. Percolating waters deposit their quartz first along the lines of parting, and we get a twin system of veins; and if the action be continued further, the intervening filling is also silicified. It may seem a long step, from a lode like the Premier, the larger part of which is gold-bearing material which is crushed, included country-rock, to a massive vein of clean auriferous quartz; yet in truth the difference is not of origin of structure, but of degree only, being due to the variable extent to which quartz has replaced the country-rock.

Before leaving this part of the subject, I would venture the suggestion that lode-mining in Otago has a future scarcely to be inferred from the scanty results hitherto obtained. Examination of the lodes shows that they are found in channels but little divided from the main mass of the country-rock, and that the quartzose schists are auriferous in themselves, outside the boundaries of such lode-channels. The formation strongly points to the probability that there will be found certain belts of country-rock sufficiently auriferous to become mines. At present, gold-milling in Otago is in a very crude condition; but with the improvements made in the treatment of gold-quartz ores and the consequent decrease of the cost of handling them, there will come a day when large "low-grade gold-propositions" will give lode-mining in this part of New Zealand an importance now unsuspected.

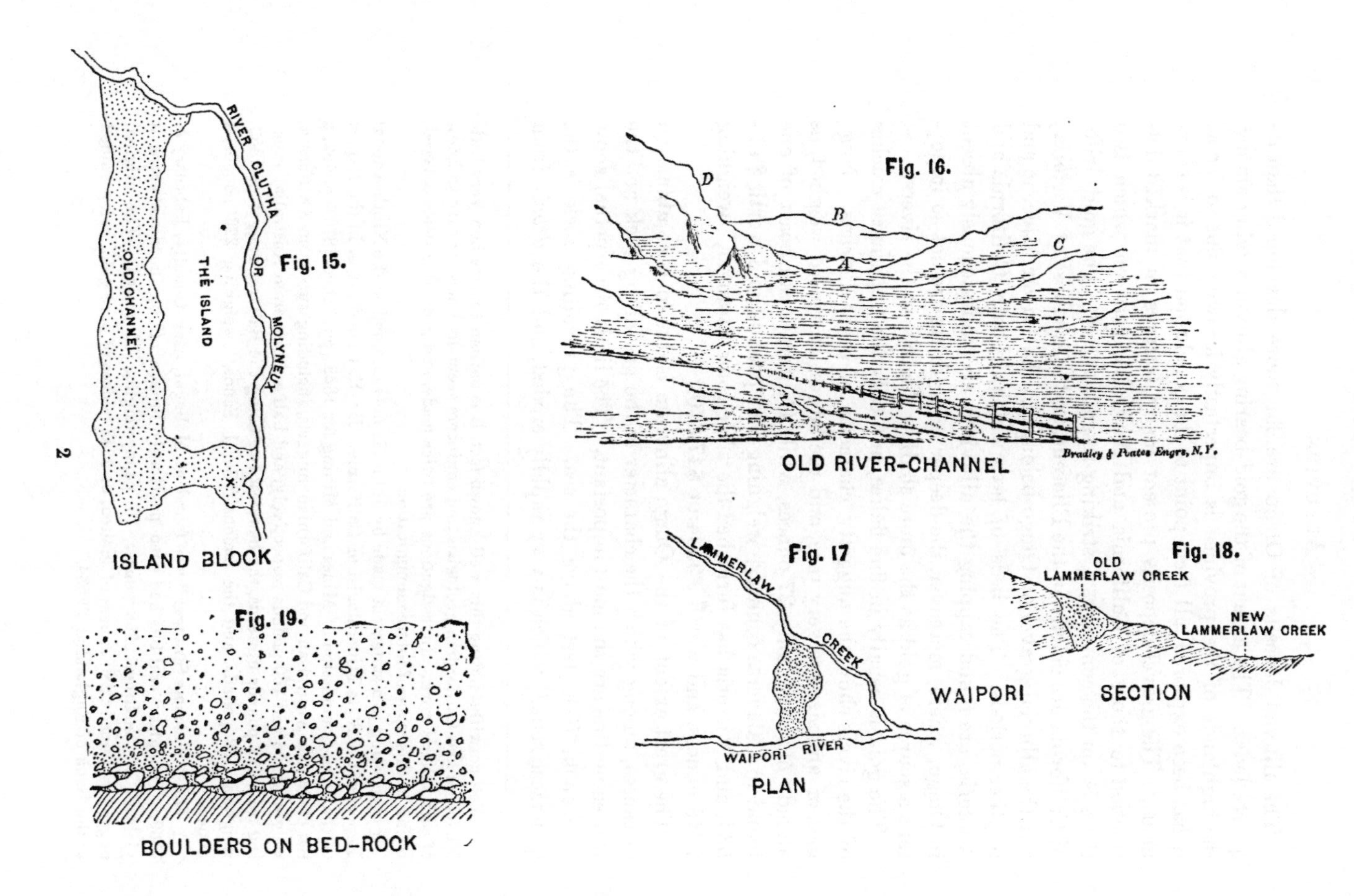
Fig. 15.
RIVER CLUTHA
OR
MOLYNEUX
THE ISLAND
OLD CHANNEL
ISLAND BLOCK
Fig. 19.
BOULDERS ON BED-ROCK
Fig. 16.
D
B
C
A
Bradley & Poates Engrs, N.Y.
OLD RIVER-CHANNEL
Fig. 17
LAMMERLAW
CREEK
WAIPORI RIVER
PLAN
Fig. 18.
OLD
LAMMERLAW CREEK
NEW
LAMMERLAW CREEK
WAIPORI
SECTION

ALLUVIUM.

The alluvial deposits of Otago are far more developed than its quartz lodes. The extent of the gold-bearing alluvium lying among the highlands of the province is imperfectly known; but so far as it has been explored, all facts point to the conclusion that it is enormous. The gravel-deposits present many features in marked distinction to those of California and Australia, between which two there is, on the contrary, a striking resemblance.* The gold-drifts of California are chiefly in the Pliocene, and so are those of Victoria; but the alluvial gravel of Otago belongs to the Lower Miocene and to other periods. The beds of lava which, in both California and Victoria, are found capping the alluvium are conspicuously absent in Otago, where, moreover, the deeper lying drifts are not so important a source of gold as the more shallow deposits of the rivers.

The gold is usually in fine flakes, and only the uppermost reaches of the rivers show the nuggetty character of other districts. Nuggets of any size are very rare, and the largest on record weighed, as already observed, only 27 ounces, while California can boast of one found in Calaveras county,† weighing 195 pounds, and worth $43,-534, and Victoria has furnished the "Welcome Stranger," weighing 2248 ounces, and worth £9534 or $47,670.‡

The great extent of the Otago alluvium is due to a combination of causes, among which the character of the prevailing rock and the low snow-line are the most important. The limit of perpetual snow is at about 7000 feet above the sea. The prevailing rock is the quartzose schist, which is very rapidly eroded, and the *débris* from

---

* Yet, notwithstanding this well-known fact, it is curious to note how very little Victorian and Californian geologists and engineers seem to know or care to know of each other's work. Investigations are twice made, and ground is twice covered, owing to this want of intercommunication.

† According to a statement made by John Hays Hammond in the Ninth Annual Report of the State Mineralogist of California. Dr. Raymond gives, in the Report of the U. S. Commission of Mines and Mining for 1869 (published 1870), p. 452, a list of famous Australian and Californian nuggets, including two from California, weighing 106 and 160 pounds respectively, and half a dozen from Australia, ranging from 112 pounds upwards, the heaviest two being the Ballarat nugget of 1853, weighing 168 pounds, and the famous "Sarah Sands," weighing 223 pounds, 4 ounces.

‡ The "Welcome Stranger" was found at Moliagul, near Dunolly, February 5, 1869. Though several bits had been previously broken off by the discoverers, it yielded, on melting, 2268 ounces, 10 pennyweights, 14 grains. The original nugget contained 2280 ounces of melted, or 2248 ounces of pure gold, being valued at the Bank of England at £9534.

which is washed down into the great natural tail-races of the Clutha, Shotover and Kawarau rivers. These rivers are subjected, by the rapid thawing of large masses of snow, to frequent flooding and the diversion of their channels by the *detritus* brought down. The soft character of the bed-rock soon makes a temporary channel a permanent one; and so it happens that, even at the higher altitudes, terraces of gravel and old river-beds are frequent and form prolific sources of the precious metal.

The most extensive alluvial deposits are found along the course of the Clutha, formerly called the Molyneux. This river is to the miners of Otago, in some respects, what the Nile* is to the Egyptian fellaheen, though it is the great artery, not of fertile lowlands, but of auriferous highlands. The anxious cultivator beside the banks of the historic northern river does not watch the rise and fall of its waters with more anxiety than does the energetic New Zealander the rapid current of the southern stream, whose quick rise or gradual fall may mean on the one hand the cessation of work and the flooding of claims, or, on the other, the steady ingathering of a golden harvest.

The Clutha affords some interesting lessons in physical geography. With its seven-knot current, it sweeps down through the easily eroded schistose rocks in which it has cut for itself a natural sluice-box, the riffles of which are the rocky bars, while its head is at the feet of the glaciers, and its lower end empties into the Pacific. Where the river now flows there was once a chain of fresh-water lakes, cut out by the glaciers which have now retired further inland. These lakes received the tribute of the upper highlands in the form of the gold-bearing gravel which was deposited on their bottom. In later times the Clutha worked it way from lake to lake, and finally, having worn down its channel deeply through the schists, it emptied the lakes, cut through the deposits which had been formed in them, and now flows unfettered to the sea.

The main features of the country through which it passes are bare mountain ridges of high elevation, diversified by narrow valleys of pastoral land. Narrow rocky gorges separate wide flats of evidently lacustrine origin. That part of the river from the Island Block to Alexandra is typical. At the former point there occurs an old channel, separated from the new by a ledge of rocks; and above the

* It has been stated that the Clutha carried a volume of water equal to that of the Nile. The estimation of the Clutha was correct, but the Nile was probably incorrectly gauged.

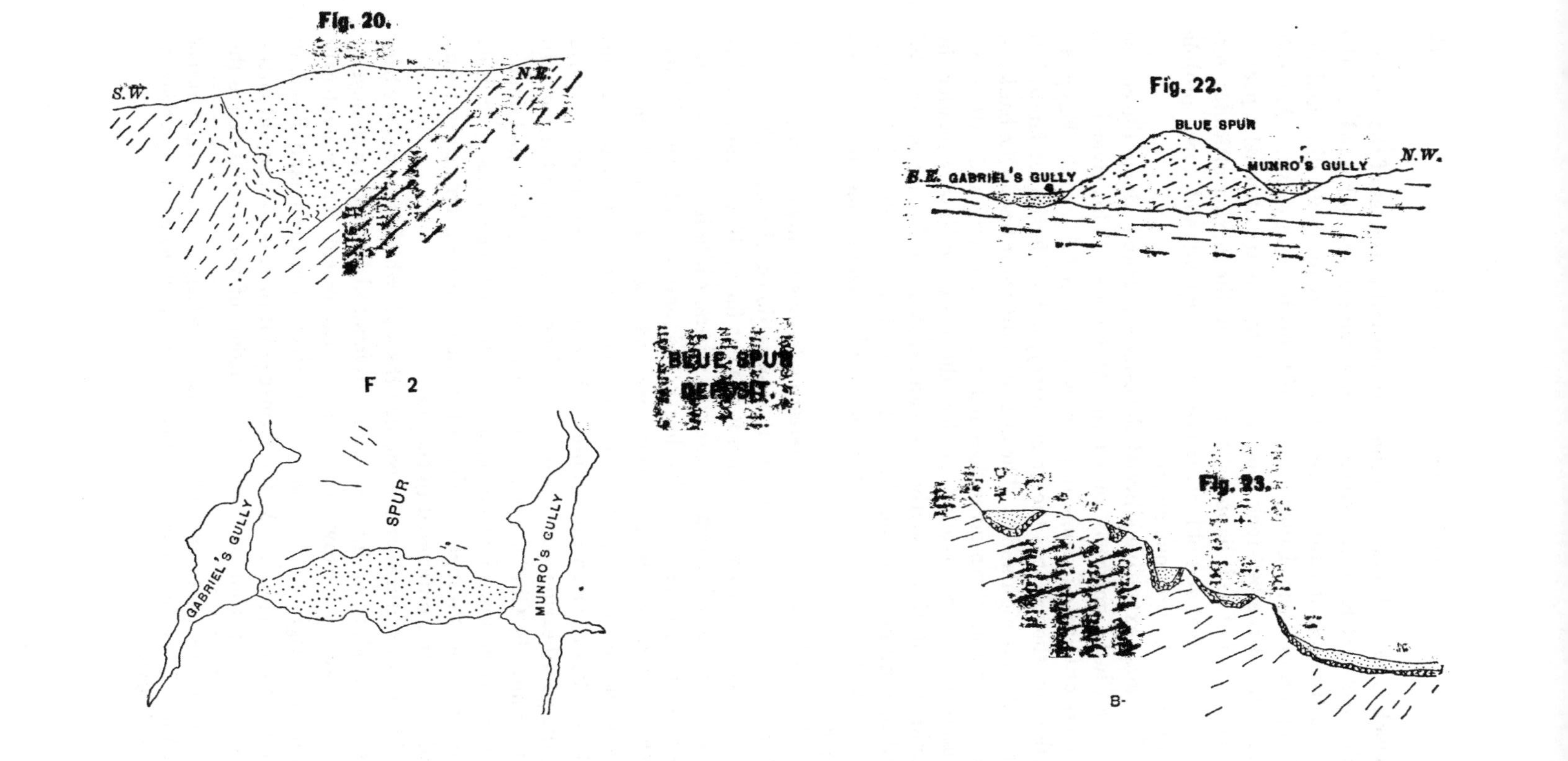

Fig. 20.
S.W.
N.E.
F 2
SPUR
GABRIEL'S GULLY
MUNRO'S GULLY
BLUE SPUR
DEPOSIT.
Fig. 22.
BLUE SPUR
S.E. GABRIEL'S GULLY
MUNRO'S GULLY
N.W.
Fig. 23.
B-

junction of the two channels there is a narrow gorge, which is succeeded by the wide flats of Ettrick, through the eastern side of which the river has cut its way. Then follows another rocky cañon, to be succeeded by the small lake-basin in which the Hercules Company is finding " pay gravel." Continuing north, below Roxburgh bridge, comes a short rocky cutting which opens out into Coal Creek flat, from which it is 6 miles to Coal Creek gorge, and then 25 miles of narrow ravines to the Alexandra terraces which mark the successive levels of an old lake.

Tne Island Block is now the scene of important mining operations. There the present river (see Fig. 15) flows on the eastern side of a rocky ridge which forms an island between the river of to-day and that of a former period. The old river-channel, about ¼ mile wide and 4½ miles long, is marked by a valley of the most valuable pastoral land in Otago. Its upper end opens out towards the banks of the present river, but forms a narrow gorge at its lower end where (at the point marked by a cross in Fig. 15) the Island Block Company is working. The configuration of the valley is further shown in Fig. 16, in which sketch C is the island, A the point where the old and new channels meet, which is overlooked in the distance by B, the Old Man Range, D indicates the lower slopes of the Spylaw Hills.

The Island Block Company commenced work at the immediate outlet of the old channel; but it was found that there was here a narrow deep rocky gorge, the overhanging sides and rapidly shelving bottom of which had prevented the deposition of the gold. Going inland, the channel widened, and the bottom became less steep. Soon however the gravel was cut off by a bar of rock. It was then decided to test the channel at a point nearer the center of the valley; but on bottoming, hard schist was encountered at 60 feet, and no pay gravel was found. The next step was to cut right across the supposed line of channel. This met with success. On the south side, after having raised 50,000 cubic feet of barren material the elevator bottomed upon "a lead" 3 to 4 feet thick, and carrying rich gravel. Mining operations have thus shown that the old channel is more irregular than would at first sight be supposed. Fig. 19 exhibits the arrangement of the large boulders which lie upon the bed rock. The gold is found in the heavy silt which has been caught among the interstices of the boulders, the arrangement of which forms a natural riffle, resembling the " pavement-riffle" of the miner.

At Waipori there is a similar deposit in the old bed of the Lammerlaw Creek. It is shown in plan and section in Figs. 17 and 18.

The deposit has a depth of 54 feet and a width varying from 25 to 30 feet. The sides are very irregular, and there are several distinct layers of gravel, making "false bottoms." The origin of this old channel is easily explained. The country rock is here traversed by a lode, the croppings of which have been found in the bed-rock. They contain gold, and have a strike which brings them in line with the "Otago Pioneer Quartz" lode worked on the opposite hill. (This is somewhat after the manner of the deep leads of Ballarat, where, when the alluvial deposits had been worked out, the mines had a second lease of life through the development of the quartz-veins, the croppings of which had been found in the bed-rock of the alluvium.) Here the lode-channel, consisting of rock softer than the average, offered an easy passage for the Lammerlaw, on its way to join the Waipori river. Having become filled up during a time of flood, the channel was diverted and formed the Lammerlaw Creek of to-day.

Near Waipori is the celebrated Blue Spur. This is perhaps the best known and most interesting alluvial deposit in New Zealand. It is situated near the head of Gabriel's Gully, about two miles east of the town of Lawrence. The name is derived from the color of the alluvium which was first discovered in the gully at the foot of the spur into which it was afterwards traced. The blue, or, more accurately greenish-blue, tint is probably due to the silicate of the protoxide of iron. The deposit consists of a mass of cement and conglomerate, occupying a cup-shaped hollow in a ridge or spur which divides two nearly parallel gullies. Figs. 20, 21 and 22 illustrate it. Fig. 20 is a cross-section. The contour-line of the ridge as here shown is obtained from an old photograph taken in 1865. The working of the deposit has entirely altered the lines. The longer axis is nearly at right angles to the present trend of the Blue Spur ridge, the depression which the deposit fills having been cut out of the Wanaka schists, the characteristic rocks of Otago. Fig. 22 gives the longitudinal section. The gully on either side is partially filled with tailings.

The plan, Fig. 21, shows the deposit to have a roughly oval shape, the longer axis having a strike N. 60° W., and a length of about 34 chains, the shorter or N.E.-S.W. axis is 22 chains in length.

For the sake of brevity, the two longer sides of the deposit, facing respectively N.E. and S.W., will be spoken of as the "Lawrence" and the "further side." On ascending the ridge on the further side, about 400 feet above the present level of Gabriel's Gully, the country

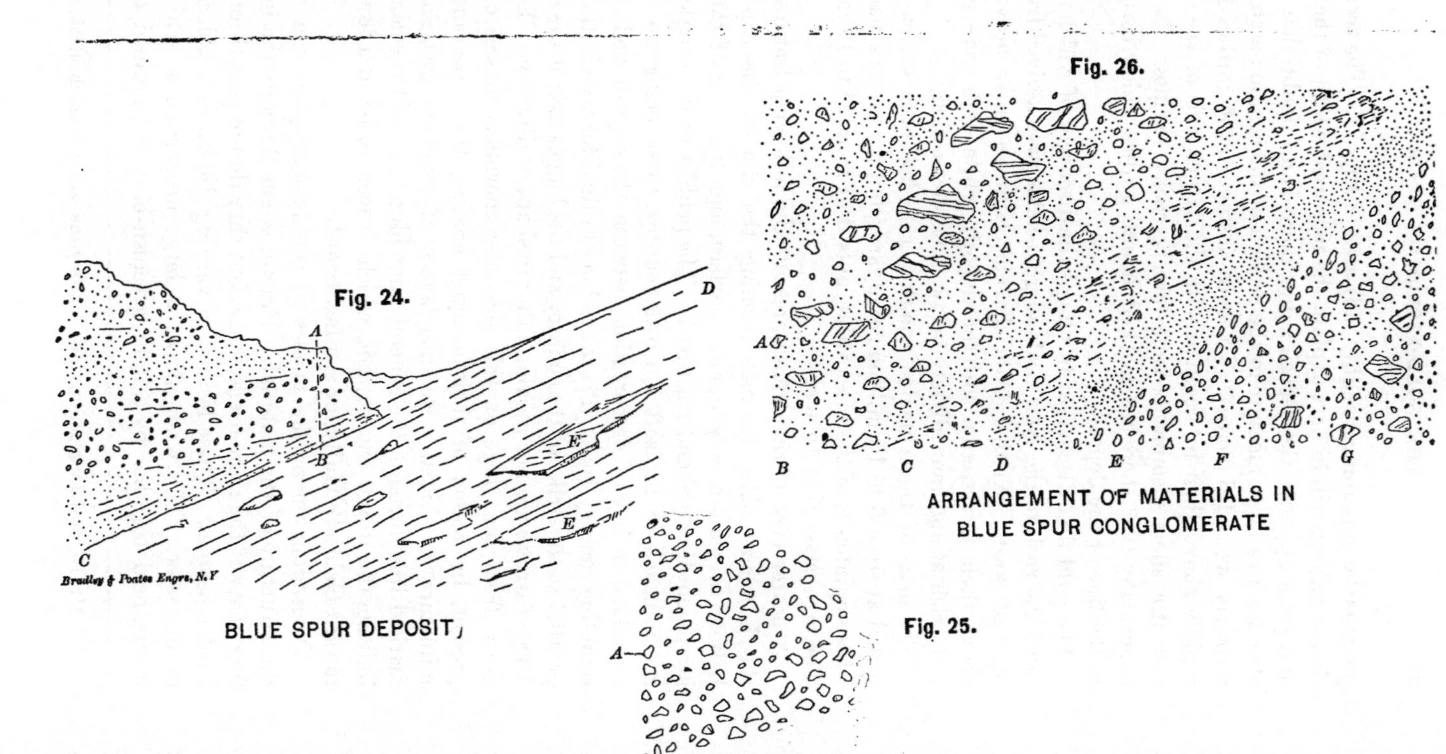

Fig. 24.

BLUE SPUR DEPOSIT

Fig. 25.

Fig. 26.

ARRANGEMENT OF MATERIALS IN BLUE SPUR CONGLOMERATE

presents the appearance roughly indicated in Fig. 21. The accumulation of tailings will in part explain the increased width of the gullies at a point opposite the deposit; but there is no doubt that this is also due in a large measure to the erosion of the conglomerate which formerly extended across. In both instances, the gullies narrow rapidly above their intersection with the larger axis of the deposit, and at a short distance become merely rocky ravines. It is also noteworthy that there are marked lateral gullies which correspond to the line of the deposit.

The gold first discovered in Otago, the discovery of which inaugurated the rush to the New Zealand gold-fields, was derived from this mass of wash. Neither Gabriel's nor Munro's gully proved rich above their intersection with the deposit, while both were particularly rich at and immediately below that intersection.

The mass of the cemented conglomerate has been variously estimated at from 8 to 10 million cubic yards, but the present manager,* whose opinion is of the most value, puts it at from 11 to 12 millions. The area covered is 45 acres.

A nearer view of the deposit presents the following features. On the Lawrence side, the rock forming the rim of the cup-shaped hollow presents a very uneven surface, and the folia of the schist are much distorted and fractured. The pebbles of the conglomerate are imbeded in the soft crushed country, there being no distinct boundary or line of demarcation between the deposit and the surrounding country-rock. There are longitudinal furrows in the schist, parallel to the strike of the country and the longer axis of the deposit. These furrows are not regular either in shape or direction. Looking down from the ridge on this side, the enormous masses of wash present, by reason of their unequal erosion, the appearance of a miniature cordillera. Reddish layers of oxidized material form marked lines through the general faint blue tinge of the cement, and indicate the dip of the deposit, which is seen to be at a low angle toward Gabriel's Gully, or southeastward.

Descending among the masses of conglomerate, one can observe the structure of the material. Fractures cut through the included fragments of the more fissile schist, but they do not pass through the harder pebbles. These last are of varying size up to boulders 2 feet in diameter. The wash shows a large proportion of schist-fragments, held together by a light greenish-blue fine-grained cement.

* Mr. Howard Jackson, to whose courtesy the writer is much indebted.

Of the remaining portions of the material, large jasperoid boulders are most prominent.

Going to the further side of the deposit, it is seen at once that the rim-rock has an altogether different appearance from that noted on the Lawrence side. The bounding wall is perfectly straight and even, with a dip of 25° 30′ to the southwest. It shows incrustations of alum. The schist of the country, which dips with the face of the rim-rock, shows no signs of having been crushed or dislocated. When first uncovered, the face of this wall is said to have been (and its present appearance suggests it) wonderfully straight and smooth. Exposure to the weather has induced the slow shaling off of its surface, which is now littered with the thin sherd-like fragments broken off by frost and rain. Originally the rim-rock was covered with a thin layer of clay which protected it; but this clay first hardened and then cracked, to be subsequently removed by the rain; and the underlying country-rock is being rapidly eroded. Fig. 24 gives a view of its present appearance. C D is the face of the rim-rock; E E are the fragments which have scaled off its surface. Along A B a section has been taken, which is indicated on a larger scale in Fig. 25. In this, F is the quartzose schist of the country; E is a chocolate brown indurated clay covering the rim-rock itself; D consists of a $\frac{1}{4}$-inch layer of hard cement; C is a $2\frac{1}{2}$ inch layer of soft brown mud; B is $2\frac{1}{2}$ feet of reddish small gravel; A is the main body of cement and wash.

Fig. 26 shows the arrangement of the material in a block of the conglomerate. A is a layer of large pieces of the quartzose schist, but slightly rounded; B contains smaller brown schistose fragments; C is schist rock, containing but little quartz; D is a band of bluish cement; E is a highly quartzose schist bed; F contains quartz pebbles, not rounded; G is mixed material, quartz fragments and schist irregularly distributed through a cement matrix.

Coming toward Gabriel's Gully, a pink layer containing lignite is seen. Near the gully, mining operations have partially uncovered the bed-rock which here presents the features shown in Fig. 27.* This is a section taken along a line N. 36° W., making an angle of 14° with the longer axis of the deposit. At the present time the edge of the rim-rock is laid bare, but the bed-rock of the gutter itself is under water. It is seen that the gutter comes abruptly

* This a rough copy of a drawing made by Mr. Howard Jackson, who was good enough to allow me to make use of it.

against a ridge over 60 feet high; from the foot of this it rises gently towards Munro's Gully, but soon meets with another interruption in the form of a wall of rock over 20 feet high,* beyond which it slopes away again before taking its regular rise. Numerous irregular crevices, found in the bed-rock of the gutter, generally contained a great deal of gold. Above this, the cement was very variable in its gold-contents, which were usually arranged along certain red bands. Some of the material was absolutely non-auriferous; while other portions would form a pudding-stone of golden wash, the gold of which, as distinguished from the fine flakey character which it usually had, was of the size of beans.

While the rim-rock rising from the gutter on the northeast side was, as we have seen, smooth and regular, that of the Lawrence side was rendered remarkable by "crab-holes," as they were called by the miners. These were irregular corrugations, not parallel or straight, but narrowing and opening out at intervals. Fig. 23 is a sketch; roughly representing some of these in section. These cavities were as much as 15 feet deep; they were covered with a casing of cement as regular as plaster; and the material which filled them was a different-looking cement from that of the main body, the line of division between the two being so marked as to be of assistance in blasting.

The northeast wall continues underneath the Lawrence rim-rock. In working along the smooth face of the northeast rim, a party of miners followed it and were working their way underneath the southwestern "reef" † before they discovered their mistake by hearing the blasting of those above them.

THE GENESIS OF THE BLUE SPUR DEPOSIT.

The above concludes the general description of a deposit the origin of which is a subject of much interest. To recapitulate its main features; it consists of a mass of cement and conglomerate, lying in a depression formed in the quartzose mica-schists of the Wanaka series and transverse to the present water-shed of the district. The country has a dip of 15° to 20° southwest, the inclination increasing near this deposit. The rock on the N.E. side conforms in dip to

* This is outside the limits covered by the section given in Fig. 27 and is therefore not shown.

† The Australian and New Zealander calls the rock rising from the bed-rock of the gutter the "reef." It must not be confounded with the same word, which is used as a synonym of lode, ledge or vein.

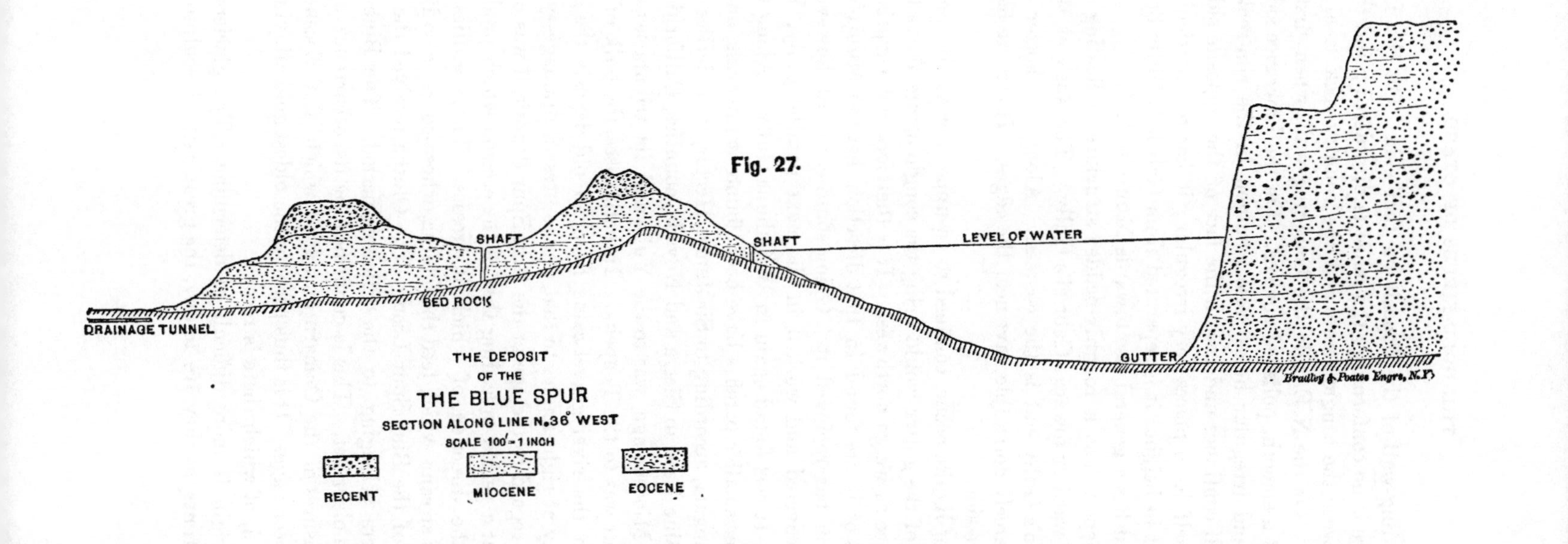
Fig. 27.
SHAFT
SHAFT
LEVEL OF WATER
BED ROCK
DRAINAGE TUNNEL
GUTTER
Bradley & Poates Engrs, N.Y.
THE DEPOSIT
OF THE
THE BLUE SPUR
SECTION ALONG LINE N.36° WEST
SCALE 100' - 1 INCH
RECENT
MIOCENE
EOCENE

the bounding-wall of the deposit on that side. On the Lawrence side, there is no conformity; the schist is disordered; and the boundary between the conglomerate and the country-rock is almost obliterated. On the N.E. side the bounding-wall, when first uncovered, had a smooth, polished surface. On the Lawrence side corrugations and irregular holes were common in the rim-rock. The N.E. wall continues underneath the face of the opposite side. The gutter itself is in places very irregular; it has one particular break of 20 feet in height; it is separated from Gabriel's Gully by a rocky ridge, and has a general rise towards Munro's Gully.

The deposit has a roughly-bedded struciure, indicating a gentle dip S.E.-ward or towards Gabriel's Gully. The wash at the N.W. or Munro's Gully end is the coarsest. Along its shorter axis, the dip is scarcely noticeable, save near the edges. It is more flat on the Lawrence side.

Beds of lignite occur, confined to an upper horizon of the deposit. The dip of the gutter would bring the conglomerate above the present surface as we go northwest. It is therefore not surprising that no traces of it are found in that direction, beyond Munro's Gully; for erosion has removed it. Going south-eastward, however, it has been uncovered and worked in the next parallel gully, Weatherstone's. It was found again in Waitahuna Gully. Along the same line, other smaller patches have been found at intervals, and finally at Kaitangata, according to Sir James Hector, it underlies the coal.

The Blue Lead in Sierra and Nevada counties, California, an old river of Miocene age, was to the Yuba and its tributaries what the Blue Spur was to the Tuapeka. In both cases, the bulk of the gold found in the river-gravel came from an old deposit, the course of which lay at right angles to that of the present drainage-system.

With regard to the age of the Blue Spur deposit, it was obviously formed at a period preceding that of the erosion which produced the gullies the direction of which it crosses. These gullies are the source of streams which feed the Clutha, whose course also lies across the line of the Blue Spur Lead. The Clutha received the drainage of glaciers belonging to the Pliocene period. The Blue Spur is therefore older yet. This is confirmed by its occurrence under the coal-measures of the Oamaru formation which is of Eocene or Cretaceo-Eocene age. It is thus perhaps the oldest gold alluvium which is worked, of which there is record.

Its origin is more difficult to determine. The glaciers were in Eocene times, as they are to-day, the great natural sculptors of the face of Otago. All investigation into the surface geology of this portion of New Zealand points to this agency as having cut out the lake-basins which form so prominent a feature of the country. It is glacial action which is generally mentioned as explaining the origin of the Blue Spur. That it is no ordinary ancient river-deposit is proved by the contour of the bed-rock; for the conglomerate lies in a pot-hole rather than in a regular channel. While the general slope is gently southeastward, the lower end at Gabriel's Gully is found to be bounded by a rocky ridge fully 75 feet in height; and midway between the two gullies another rapid rise of 20 feet is encountered. Such a configuration is best explained by considering it as a pot-hole or hollow scooped out by ice-action. Though there are no striæ or markings produced by glacier movements now to be seen, yet this is in no way remarkable; for the bed-rock is of too soft a character to have preserved any such evidence. The presence of the large jasperoid boulders is confirmatory. They do not belong to the locality, since no similar rock is found nearer than the Blue Mountains or Tapanui, 25 miles northwestward. They were carried down by the ice. The character of a large portion of the material forming the deposit similarly indicates that it was not brought down in or by a stream of water, but as the rocky freight of a glacier. A large part of the wash consists of angular fragments of quartz, as well as of pieces of the quartzose schist-rock, which are not rounded. The agency which eroded the depression in which the auriferous material lies was assisted by the structure of the rock at this particular point. I have no doubt that the N.E. bounding-wall of the deposit forms the line of a fault, and that the reason of the formation of the rocky basin at this particular spot is to be found in the fact that the schist had been crushed by the movements accompanying faulting. Of this the N.E. wall, its smooth face, and its continuation under the opposite rim-rock, together with the crushed condition of the S.W. country are ample evidence. The line of fault is not parallel to the course of the lead; the two meet between Munro's and Gabriel's Gully; and so explain the enlargement at that point of the receptacle of the ore-deposit.

This explains the natural selection of this particular place as the *locus* of the deposit. To proceed further, the glacier in its slow downward progress to the sea is temporarily arrested by the softer rock which it here encounters much in the same way as a runner is retarded in crossing a ploughed field. This arrest allowed the accumulation of a terminal moraine, which, protecting the rock on which it lay, assisted the tendency of the ice to erode the softer schist; where the terminal moraine at one time lay, we now find the rocky bar shown in Fig. 2. A hollow was scooped out. This was in early Eocene days. A little later, that subsidence took place which preceded the deposition of the Oamaru series. This caused the retirement of the glacier, or more accurately, the melting away of its lower portion. The rocky basin which had been scooped out by the ice now became a fresh-water lake, with its upper end still guarded by the glacier. The ice which broke away from the foot of the glacier bore with it the large boulders of jasperoid which had been brought down from Tapanui. This and other material was borne across the lake, to fall eventually upon its bottom as the ice-floes melted. In the meantime, up above, the glacier continued to plough through the soft quartzose schists and sent down a golden tribute, derived from the lode-formations which it cut through. The fine flakes of gold were aecompanied with micaceous mud and angular bits of quartz, all to be deposited in the capacious hollow of the lake. Thus the rocky basin became gradually filled up with confused layers of big jasperoid boulders, quartz-gravel and bluish mud, the gold sifting its way to the lower portions. The subsidence continuing, and with it the slow retirement of the glacier, and the lake being nearly full of *detritus*, it became a morass. Vegetation took root and flourished for a brief period. A time of flood due to excessive thaw brought down a volume of water bearing the sand and gravel which covered the vegetation. Being thus protected from the air, the reeds of the morass became the lignite of to-day.

A river linking a series of small lakes, of which the Blue Spur was one, now flowed along the course of the present alluvial lead. Additional material was deposited in some places, while material was removed in others. In the middle of the Eocene period, the elevation of the land culminated and changed the drainage-system of the district. In Miocene times, the Clutha and its tributaries began to flow across the line of the Blue Spur lead. That erosion then commenced which, in the cutting out of Munro's, Gabriel's and Weatherstone's gullies, left the gravel-deposit as a part of a dividing ridge.

Reference has been made by Hutton to the fact that certain of the Eocene beds contained gold in order to explain its occurrence in the Blue Spur deposit. This is not needed. The examination of the material composing the conglomerate shows that the larger part is derived from the degradation of the primary schists. The grinding by the ice of these soft rocks formed the mud which is now "cement." It was from the quartzose folia that was derived the coarse gravel, and, finally, it was the gold which elsewhere to-day is found in the lodes and even in the rocks of the Wanaka series, which made the Blue Spur not only a geological study but also a very valuable gold-mine.

Conclusion.

The consideration of the gold-deposits of Otago, whether in lodes or alluvium, is intimately connected with the study of the quartzose schists which are the characteristic country-rock of the one and the bed-rock of the other. In many respects these rocks are remarkable. Their extent and unbroken continuity have been referred to; allusion has also been made to the many interesting physiographical studies which they offer; and it now remains to briefly refer again to their auriferous character.

That the bulk of the gold of the extensive alluvium of Otago came from the degradation of the known lodes, I do not believe. The mere proportion of the known extent of the one to that of the other may be an insufficient basis for such a conviction, for in this country, as in other comparatively new regions, the mineral deposits are only in an infancy of development; yet this fact has some weight. The chief evidence comes, however, from the known relations of the lodes to the country-rock, and from the character of the alluvial gold itself.

It is not necessary to recapitulate the observations made on the lodes; how they are for the most part ore-channels, the auriferous filling of which shades off into the surrounding country, as regards both gold-contents and structure. The gold of the alluvium has not that shotty character familiar to the digger of Australia or California. An exception must be made of the small placer-deposits of some of the mountain streams, the nuggety gold of which is easily traceable to its source in the neighboring quartz-veins. Such gold is, however, quite exceptional, and altogether of a character different from that of the great fluviatile and lacustrine deposits of the Clutha and Shotover, the wealth of which is contained in the very fine flaky gold, which is like bran and notable for its uniformity of size and wide distribution. It has a character making it easy of recognition among samples coming from other districts. It is also remarkable that quartz stones showing gold are of very rare occurrence at the Blue Spur or in the gravel-deposits of the Clutha and

Shotover. The gold came, I believe, as did the gravel through which it is distributed and the cement in which it is often imbedded, not from the few comparatively insignificant quartz-lodes, but from the great mass of the quartzose schists. The quartz folia which form the characteristic feature of the schists of Otago are known in places to carry gold far outside the limits of any of the particular lode-channels which also traverse them. One could, with perfect reason, regard the whole belt of quartzose schists as one large bed-vein, in which the lodes now worked are merely small cross-veins,—a large gold-vein, through which for ages the glaciers have ploughed an easy way, cutting furrows from which the quartz and schist have been swept by wind and rain into the swift waters of the Clutha and Shotover, to be laid down by them in the form of the great banks and terraces of gold-bearing alluvium, which to-day are the chief depositories of the mineral wealth of Otago.

NOTE BY THE SECRETARY.—Comments or criticisms upon all papers, whether private corrections of typographical or other errors, or communications for publication as "Discussion," or independent papers on the same or a related subject, are earnestly invited.

[TRANSACTIONS OF THE AMERICAN INSTITUTE OF MINING ENGINEERS.]

ALLUVIAL MINING IN OTAGO.

BY T. A. RICKARD, DENVER, COLORADO.

(Lake Champlain, Plattsburgh, Meeting, June, 1892.)

In a previous contribution some description was given of the occurrence of the auriferous alluvium, extensive deposits of which lie among the highlands of the province of Otago, New Zealand. The present paper will describe some of the methods by which that alluvium is exploited.

The alluvial mining of New Zealand is not yet hindered by "débris legislation." The mines are for the most part situated in a country of little agricultural value, and large rivers with rapid currents act as tailings conveyors. In time, however, steps will have to be taken to restrict the discharge into the Clutha of the tailings which are slowly raising the river-bottom.

The methods of California have been followed in Otago with such alterations as peculiar local conditions require. The distinctive feature of the alluvial mining of Otago is the prominence given to the hydraulic elevator. The all-important factor in this form of mining is the control of an abundant water-supply at a suitable height above the level where it is to be utilized; and in this respect the gold-fields of the south island of New Zealand are most favorably situated, the low level of the snow-line insuring a splendid supply throughout the year, and under a pressure which, in practice, is only limited by the strength of the plant and the expense incurred in its installation. This is in marked contrast with the gold-fields of the Australian continent, which have, on the other hand, a magnificent supply of the best mining-timber.

Next to the hydraulic elevator in importance comes the dredge, the use of which in Otago, under several forms, is rapidly extending in the exploitation of the river-bottoms and sea-beaches.

The Hydraulic Elevator.

This machine, which is destined to have an ever-widening field of usefulness, consists essentially of an inclined pipe, at the bottom of which a jet of water under heavy pressure is allowed to play, and

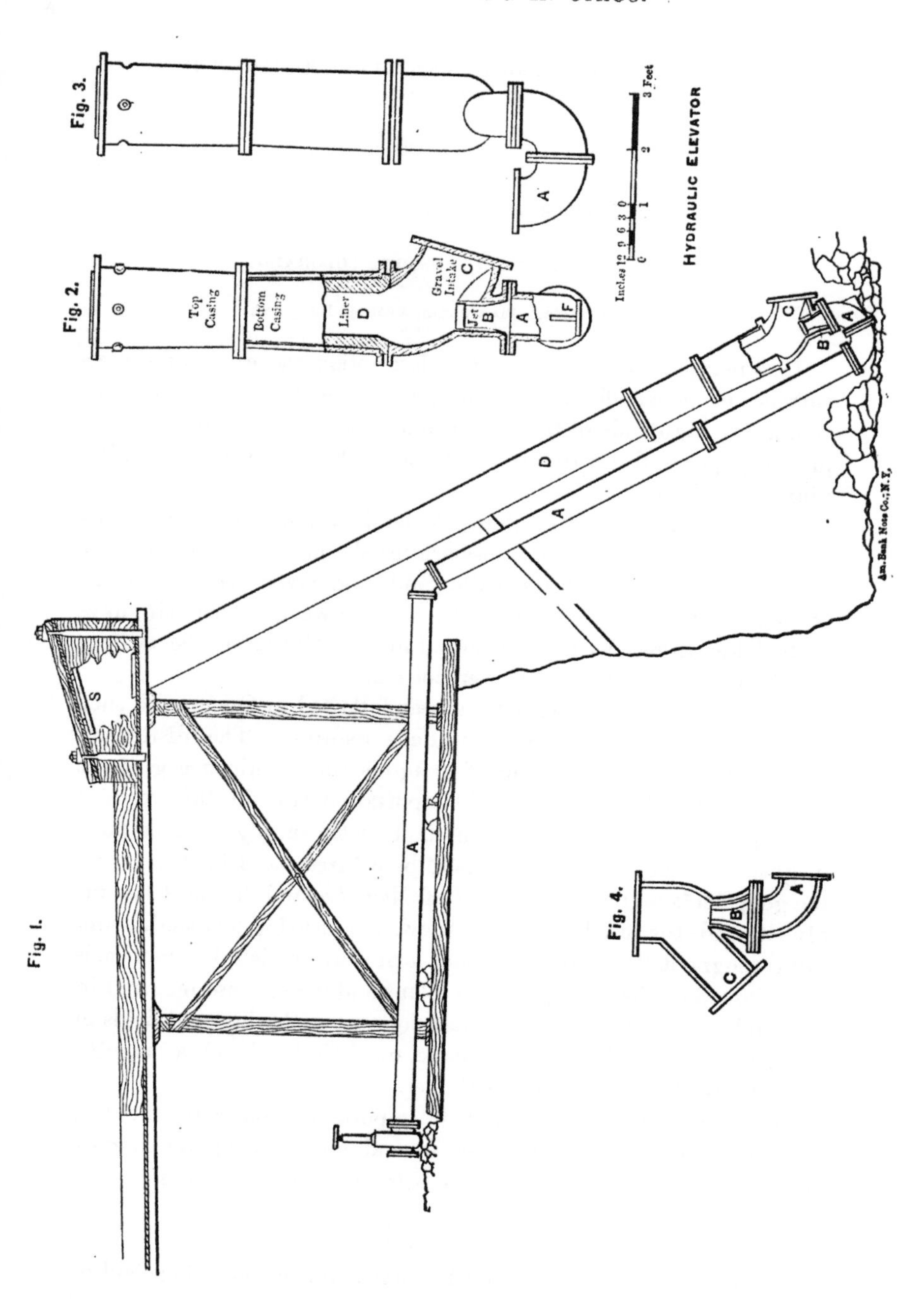

HYDRAULIC ELEVATOR

by its action propel through the pipe the gravel and water which enter at its lower opening. The action of the jet at the foot of the elevator is not limited to the direct force of the water liberated under a heavy pressure, but is also in a large measure aided by the strong suction induced by the vacuum created by the upward rush of the water. This suction is also of importance in that it draws the gravel through the opening and thereby does the feeding. Originally the pipe of the elevator—called the uptake or discharge-pipe—was separate from the jet or nozzle which was directed up it; but in the newer forms of the machine the two are combined into one compact casting. Not only has this modification made the elevator a more complete machine, but by diminishing the unnecessary distance between the jet and the opening which admits the gravel, it has allowed a greater advantage to be derived from the suction induced by the rapid passage of the water.

By reference to Figs 1 to 5 the following description will be understood. The bottom of the elevator has a bracket, called the "foot," which is placed upon the bed-rock upon which lies the gravel which it is intended to raise. The water and gravel enter at the "hopper" or intake C, being drawn in by the suction caused by a powerful jet of water which passes through the "jet" or "nozzle" B and impinging upon the entering gravel and water sweeps them up the elevator or "uptake" pipe D. At the upper end of this pipe the gravel is deflected by the "striking-plate" S (Fig. 1) and proceeds down the series of sluice-boxes arranged at a gentle incline. These sluice-boxes are provided along their bottom with riffles and other contrivances to arrest the gold carried in the gravel, from which it separates by the action of gravity.

Gravel and water hurled with such velocity as belongs to a jet under a head of several hundred feet, must produce great wear and tear upon any surface against which they impinge. The greatest amount of friction takes place at the lower part of the uptake-pipe, called the "throat" (E). This is narrowed so as to prevent the scattering of the jet and the consequent decrease of its effectiveness. A strong removable casting called the "liner," is here inserted. It increases gently in diameter from its lower end upward. See Figs. 2 and 5. A liner three inches thick lasts about six weeks.

Passing up the elevator pipe, the gravel meets the "striking-plate." This is a heavy piece of iron 2 feet square and 3 inches thick. It should last from two to three months; but a defective casting will be broken in a few hours. The force of the ejected gravel is not all expended upon the striking-plate, hence the head sluice-box, the sides of which are usually raised to 2½ feet, is lined with iron plate (No. 12 boiler-plate), to protect the wood-work.

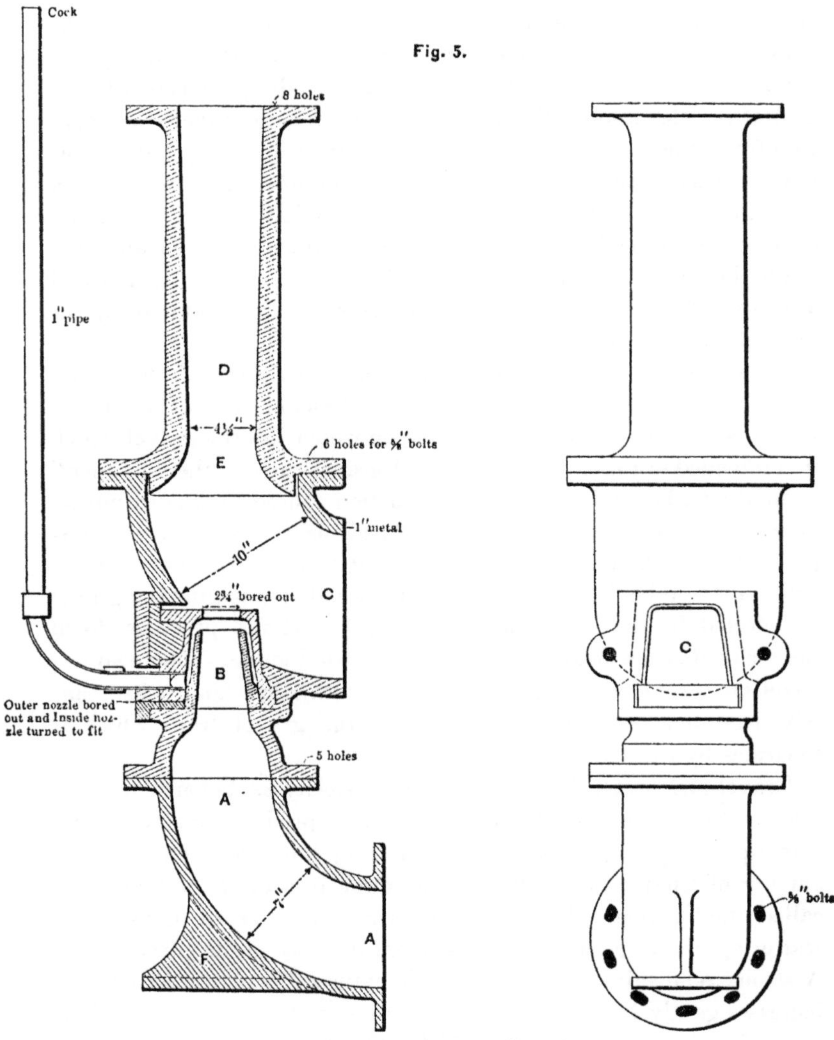

THE ATMOSPHERIC NOZZLE FITTED TO THE HYDRAULIC ELEVATOR

When properly handled, the elevator is a most effective machine, raising a cubic yard or 1½ tons of gravel per minute, together with a large volume of water. Thus it not only lifts the gravel but also acts as a pump in getting rid of the water used in breaking down the face of the wash. Its usefulness in the latter capacity is highly important to those engaged in the working of claims the bedrock of which may be below the drainage-level. It is this feature of the machine which has rendered it of special service in the alluvial mines bordering upon the rivers of Otago.

The first introduction of the hydraulic elevator* in Otago was made by Cranston and Perry in 1882. John Perry did a good deal of experimental work about that time, and in 1884 inaugurated the first successful working at the Blue Spur in Gabriel's Gully.†

The Blue Spur.—This ranks among the first of the alluvial mines of New Zealand. The English company (the Blue Spur and Gabriel's Gully Consolidated Mining Company Limited), which was formed by the amalgamation of a large number of miners' claims, is at present only working the tailings, of which there is an enormous body lying in Gabriel's Gully. These tailings have a maximum thickness in the center of the gully of 72 feet, while the face now being worked is 59 feet high. They are estimated to cover an area of 120 to 130 acres, of which the company owns 23 acres.

The original deposit (described in my former paper) was worked in the first instance by numerous small parties of miners by means of ground-sluicing supplemented with the crushing of the larger lumps of cement in small stamp-mills, the big, hard boulders being picked out previously by hand and thrown to one side. I could not find that any gold was got in the large quartz pebbles; it was almost entirely confined to the blue cement forming the bulk of the material. This cement was very hard, and the gold now obtained is due in large measure to the disintegration of the solid lumps left in the tailings.

The accompanying sketch, Fig. 6, will indicate the appearance of the workings and the method of exploitation. To the left is seen the nozzle or giant, the diameter of the brasses being 2¼ inches, supplied by a 10-inch pipe carrying water under a head of 400 feet. The water-race, fed from the Waipori river, is 31 miles long. The

* Also called "tailings elevator" at Ballarat, "ejector" and "lift" in Otago.

† Since writing the above, my attention has been drawn to a description by Dr. R. W. Raymond (*Trans.*, viii., 254), of a dredge invented by Gen. Roy Stone, in which a jet under heavy pressure is used to break the material under water (as a similar nozzle does on land), and the material so excavated is sent to the surface by the action of another, larger jet, working inside a pipe.

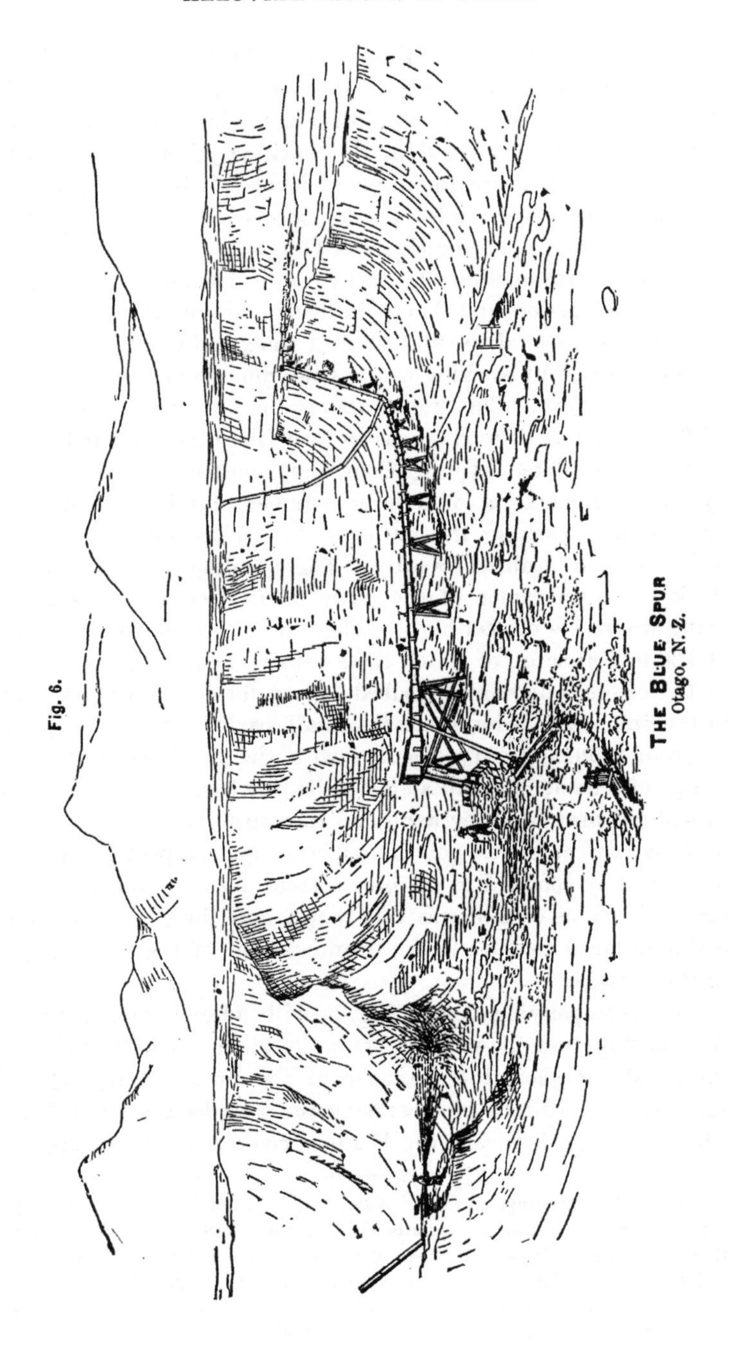

Fig. 6.

THE BLUE SPUR
Otago, N. Z.

nozzle is represented as playing upon a face of tailings about 59 feet high, separated from the bed-rock by 6 to 8 feet of black silt, in which are the layers of wash which were worked in the early days, as is evidenced by the timbers of old shafts and drifts which are occasionally displaced by the water. Immediately above the lower bed of alluvium was the site of the old township which sprung up at the time of the rush of 1861, now, however, completely buried under 50 to 60 feet of tailings. Its former position is proved by the miscellaneous assortment of articles* which the sluice-boxes collect. These occur imbedded in the silt together with a large quantity of lead in the form of shot, reminders of that eccentric fusilading which often characterizes a mining camp in its early stages. The gold obtained from this ground is usually coated with quicksilver.

The nozzle is about 50 yards from the first elevator, which is shown next to the right. The wash is driven by the water from the nozzle into the opening or hopper of the elevator, the man in charge removing the larger boulders which tend to choke it. The dimensions are: hopper, 14; jet, 2; throat, 6½, and uptake pipe 15 inches in diameter. The water which does the work comes under a pressure of 430 feet through a 10-inch pipe. The material is lifted 15½ feet, and, impinging upon the striking-plate, passes down the sluice-boxes which are 3 feet wide and in lengths of 12 feet, with a grade of 9 inches in 12 feet. Two hundred and forty feet further to the right is the second elevator, which raises the gravel again, this time through a vertical height of 56 feet through a pipe 65⅓ feet long, inclined at an angle of 58° 45′. The second elevator has dimensions similar to those of the first, and delivers the material into sluice-boxes having a grade of 9¼ inches in 12 feet, whence it is finally deposited on the tailings-heaps, most of the water finding its way into the main drain along the true bottom of the gully. One of the shafts connecting with the drain is shown to the middle right of the sketch. In the immediate foreground is the pressure-pipe of the first elevator. The smaller pipe shown near the elevator itself is only brought into use in cleaning up the sluice-boxes. The face of the bank of tailings is kept up by an arrangement of fascines, which are roughly indicated to the right, above the shaft.

All the gold saved is obtained from the run of sluice-boxes be-

---

* Coins, gold and silver, rings, copper rivets, nails, etc. Among other curiosities, the manager showed me several nuggets which had been obtained in cleaning up. These, judging by the character of the gold, were obtained in some *other* district, very probably Ballarat, and lost here.

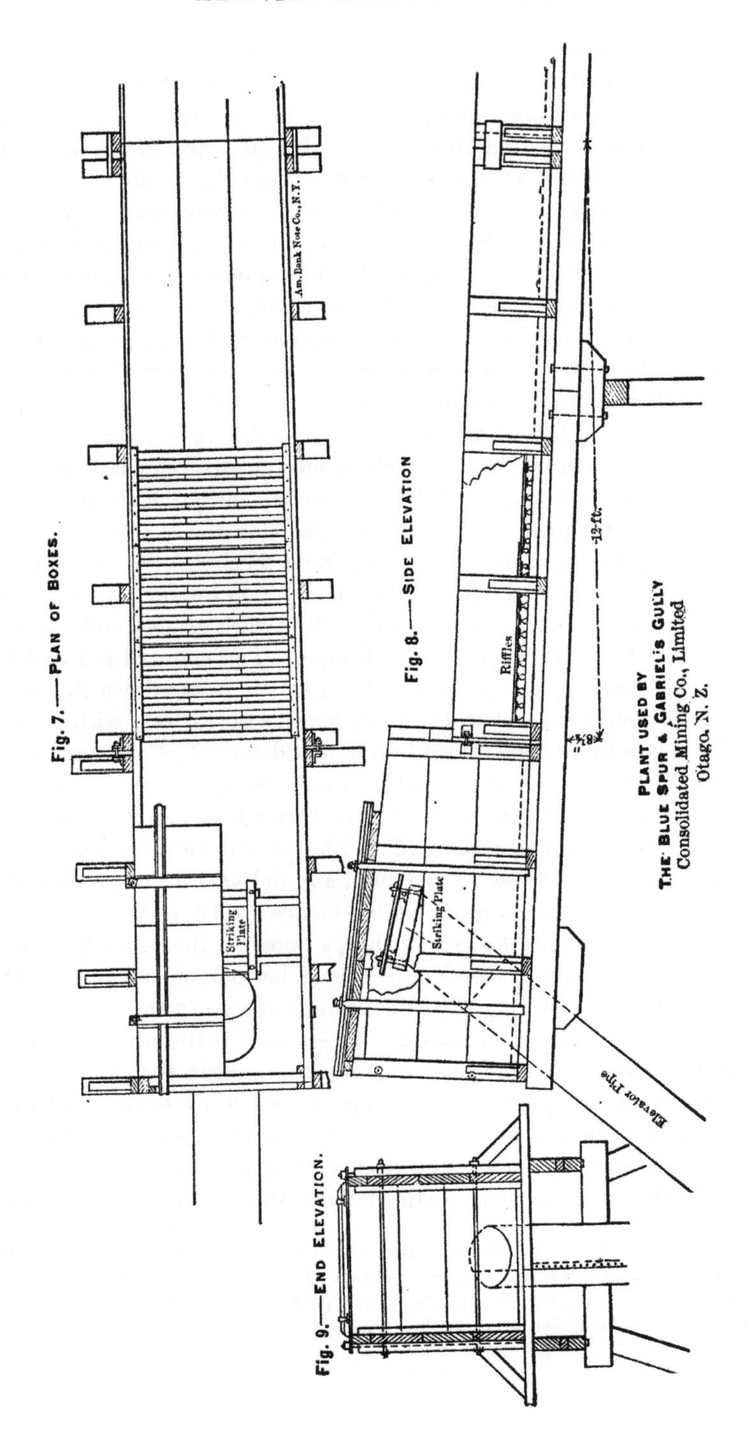

Fig. 7.——PLAN OF BOXES.

Fig. 8.——SIDE ELEVATION

Fig. 9.——END ELEVATION.

tween the two elevators. These boxes, as already stated, have a general grade of 9 inches in 12 feet, but the last two or three have 12 to 14 inches. The bottoms are covered with cocoanut matting, over which are placed riffles of angle-iron riveted to angle-iron. The end boxes are provided instead with $\frac{3}{8}$-inch perforated $\frac{1}{4}$-inch iron plate.

The details of the plant are shown in Figs. 7 to 11.* The elevator is not of the best type; the manager does not recommend it. The T-piece and blind flange mean an unnecessary resistance to the water. The best form approximates that to be described later on as in use at some of the newer mines on the Clutha river. The connection between the pressure-pipe and the elevator should take place along a greater angle, since the less the deflection of the water the greater the efficiency obtained.

The striking-plate is a casting 24 by 24 inches by 3 inches thick. The head-box is lined with No. 12 boiler-plate to protect the woodwork against the stones deflected from the striking-plate.

It is not found necessary to use deflectors with the nozzle. The usual feathers—shown in section A B, Fig. 10—for keeping the water from scattering, are provided. The ball-joint with pin permits the vertical movement of the director. Instead of the universal joint which is so apt to get jammed, a large holding-down bolt regulates the horizontal motion.

The manager of the Blue Spur has designed a new form of riffle. The angle-iron on angle-iron is a riffle which entails a great waste of material, to avoid which a reversible form would be advantageous. This is shown in Fig. 12. The eight bars are not riveted each separately, as is usual, but rest upon mortices or recesses in the two lateral retaining-bars, and are kept in position by the two bolts as indicated. In this way, when the edge of one side of the riffle bars becomes worn out, it is but the work of a few minutes to loosen the bolts and reverse the bars. With such an arrangement it is necessary that the bars should be accurately cut, otherwise they do not remain firmly in place.

While the plant at the Blue Spur is by no means the best in Otago, yet the following record, given to me by the manager, shows that very good work is done here.

In working a face of tailings, 55 feet high,† 26,929 cubic yards

* Traced from the original of Mr. Howard Jackson, the Manager, who kindly granted permission.

† The tailings have consolidated by reason of the cement of the original deposit, which they contain, and present the character of ordinary wash.

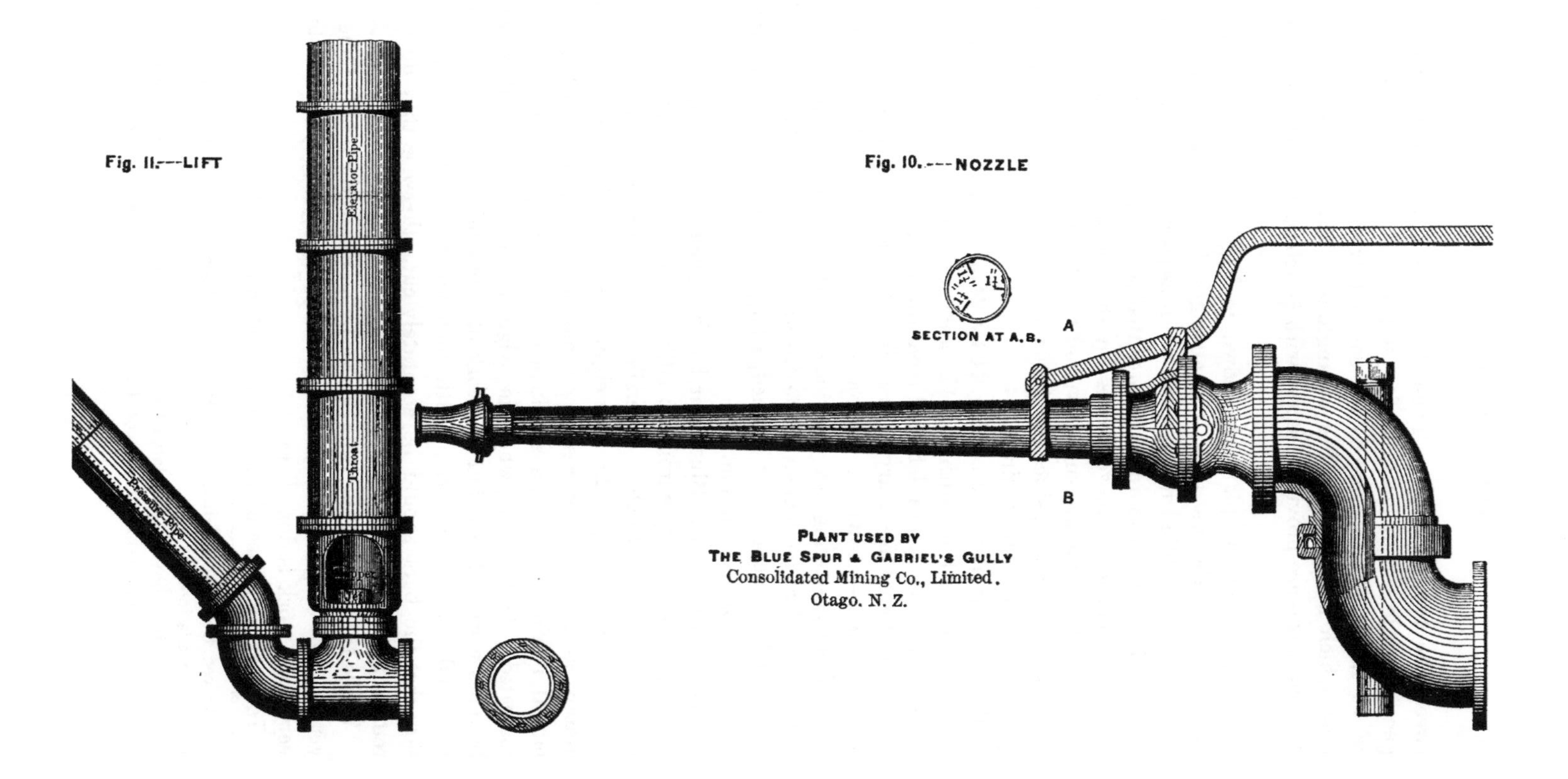

Fig. 11.---LIFT

Fig. 10.---NOZZLE

were raised and dumped a height of 87.8 feet in a period of 680 hours.

The yield obtained was 115 ounces, 11 pennyweights, 16 grains, worth		£433.8.9
The working cost included 3 men, ...	£89.5.0	
Two liners worn out, ...	10.4.0	
Add to this for depreciation of plant, management, etc., ...	99.9.0.	
The total cost was, ...		198.18.0
Leaving a profit of, ...		£234.10.9

The value of the material was 5.6 pence per yard or 3¾ pence per ton.

The cost of working was 1.77 pence per cubic yard or 7.78 pence per ton.

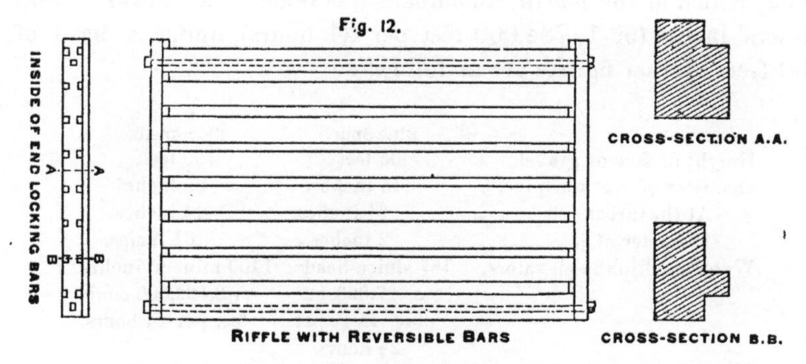

The water used with 2 elevators and 1 nozzle, was as follows:

No. 1 Elevator, 450-ft. head. Pressure 193 pounds per square inch. 8¾ sluice-heads.*

No. 2 Elevator, 375-ft. head. Pressure 164 pounds per square inch. 10 sluice-heads.

Nozzle, 375-ft. head. Pressure 164 pounds per square inch. 3¾ sluice-heads.

The gold won from the sluice-boxes was 97 ounces, 16 pennyweights, 18 grains, worth £366 17*s.* 10*d.*

Pickings from the "reef" or rim-rock, 17 ounces, 14 pennyweights, 22 grains, worth £66 10*s.* 11*d.*

The latter gold was imbedded in the crevices of the schist which forms the bed rock.

---

† A sluice-head of water equals theoretically 1 cubic foot per second; in practice it is usually calculated at 50 cubic feet per minute.

Considering that this work was done under the disadvantages of a much worn and by no means first-rate plant, it is needless to emphasize the good results obtained in so far as economy of handling is concerned.

It so happens that at the North Bloomfield claim, in California, the elevator was used under conditions which render interesting a comparison. There a more elementary type of elevator was in use, the jet not being a part of the machine itself, not even inserted inside the mouth-piece, but so placed as to shoot up the line of the elevator pipe. There one elevator lifted gravel a vertical height of 87 feet; here two elevators raised the material in two lifts of 31.8 and 56 feet respectively to a total height of 87.8 feet. At the Blue Spur a nozzle with a head of 375 feet and delivering 3¾ sluice-heads (or 5400 cubic feet per 24 hours) did the work of breaking down the bank, which at the North Bloomfield was done by a stream of 800 miners' inches (or 1, 784,000 feet per 24 hours), under a head of 450 feet. Other figures are as follows:

	Blue Spur.	North Bloomfield.
Height of face of gravel, .	55 feet	135 feet.
Diameter of elevator-pipe, .	15 inches.	22 inches.
At the throat, . .	6½ inches.	14 inches.
Diameter of jet, . .	2 inches.	6½ inches.
Water used in the elevators,	18¾ sluice-heads or 27,000 cubic feet per 24 hours.	1400 miners' inches or 3,122,000 cubic feet per 24 hours.
Under a head of . .	410 feet.	530 feet.
Capacity of the plant per 24 hours, ...	1000 cubic yards.	2400 cubic yards.

The wear and the tear at the North Bloomfield was found so excessive that the use of the elevator was discontinued. Under the conditions which obtained there, nothing less than 25 cents per cubic yard would pay.*

The Island Block.—A modern plant may be seen at the Island Block, where three elevators are at work. The nature of the deposit was described in my former paper as an old channel of the river Molyneux or Clutha. At the time of my visit, the face of wash was 40 to 51 ft. high, the top-gravel being poor but not absolutely barren, while the "pay" consisted of the lowermost 10 to 15 feet. The

* This information I owe to Mr. W. H. Radford, formerly superintendent of North Bloomfield.

gold is accompanied by a large amount of fine black iron sand, the bulk of which, on cleaning up, is collected with a magnet. The nuggets found in alluvial mining elsewhere are unknown here, for the precious metal occurs in rounded flakes, of a fairly uniform size and rarely more than $\frac{3}{16}$ inch in diameter. The wash contains numerous pebbles of what the miners call "black maori," a sort of siliceous wad,* resulting from the hydration and oxidation of rhodonite, the silicate of manganese. The transition can sometimes be observed. The quartzose schists form the bed-rock, but are not represented by any large boulders, their easy degradation having produced the small angular quartz stones, the sand and the dark blue silt which form so large a proportion of the wash.

A plant of unusual excellence has been erected here. The bed-rock being below the level of the river, the wash is raised by hydraulic elevators into sluice-boxes, a lift of 40 to 60 ft. To do this work, water is taken from a stream named the Fruid burn, and conducted through pipes of lap-welded steel, until it arrives at the elevators, where it comes under a head of 760 feet. Starting from the penstock or pressure-box with a double main, each branch having a diameter of 16½ inches, the water, after 17 chains, is delivered into a single pipe which for 27 chains has a diameter of 16½ inches, to be reduced to a pipe 15 inches diameter, which conducts it to the immediate vicinity of the heading when a final reduction to 9 inches diameter takes place. The discharge or uptake pipe of the elevator is 15 inches and the jet is 2½. The material lifted per elevator varies with the character of the gravel from 30 to 40 tons per hour. The lap-welded steel pipes are all manufactured in England. Their thickness varies from ¼ to $\frac{3}{16}$ of an inch, the greater thickness is found to make them too heavy and the lesser is far more than equal to any strain to which they may be subjected. They are said to have been tested up to 1000 pounds per square inch previous to being sent out. The price of these pipes varies from 8*s.* 2*d.* per foot for 15-inch pipes to 4*s.* 9*d.* for 11 inch; but the average has been about 6*s.* 6*d.* for the 15-inch pipes, of which the greatest length has been used. This figure should not prevent their extensive use on the part of companies desirous of erecting a first-class permanent plant.

All the joints are faced with the lathe and the pipes are in 18 feet lengths. Of course such lap-welded pipes severely test the joints, the excess of pressure on which is not relieved, as is ordinarily

* As was pointed out to me by Professor G. H. F. Ulrich, now of the University of Otago.

the case, by leakage along rivets. The pattern adopted is the Kimberly (see Fig. 13).

A is a loose angle-iron flange which is slipped on before the ring B is shrunk on to the pipe. This arrangement is very convenient, since the bolt-holes in the flanges can be made to fit without all the trouble of turning a length of pipe. . The drawback to this form of joint consists, however, in the fact that for rough country it is not well adapted, since with any movement the shoulder is apt to slip out of its recess, producing a leak. For fairly straight, permanent lines it is most desirable.

The wash is thrown by the elevators into sluice-boxes of Kauri pine, 3 feet wide, 1 foot deep, and in lengths of 12 feet. Old 40-pound rails are used to support them. (See Fig. 14.) They are

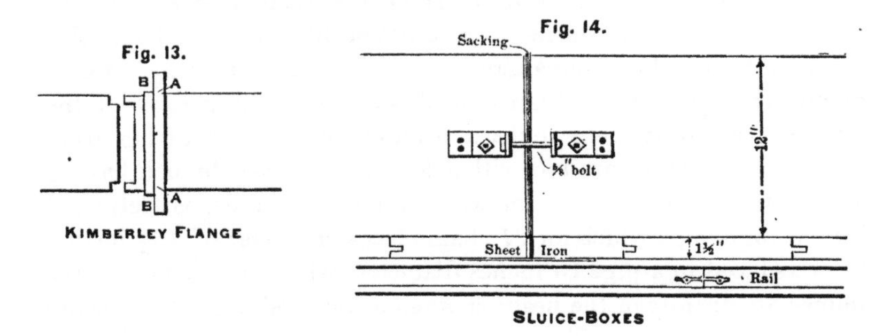

Fig. 13. KIMBERLEY FLANGE

Fig. 14. SLUICE-BOXES

lined with thin sheet-iron, and set at a grade varying with the material treated, the average being 8 inches per 12 feet. With the "stripping" or uppermost layers of "wash," this is increased to 9 inches; and, on the other hand, the last few boxes of a series are inclined at from 7 to 7½ inches only. Sacking (common bags, ripped up) is put along the bottom of the upper boxes, underneath the riffle bars. The riffles are of the usual form, angle-iron riveted to angle-iron. After a varying distance—100 to 150 feet—of ordinary riffles there succeeds a series of "under-currents" or false bottoms. These, like the "grizzly," or sizing-bars of a stamp-mill, serve to separate the larger boulders from the finer, richer wash. In their arrangement a set of iron-bar riffles follows the ordinary angle-iron riffle; that is to say, the first bar (of a set of flat iron bars, $\frac{1}{16}$ inch apart) is placed upon the last bar of angle-iron, so as to form a sieve or false bottom through which most of the fine wash finds its way.

The false bottom is placed at a less inclination than the sluice-box, and as a consequence, the depth increases gradually until, at the end of two 12 feet sluice-boxes the distance between the riffles and the bottom is 4 inches. In this length of 24 feet a succession of various forms of riffles is employed. First come the iron-bar riffles already noted, for a length of 4 feet. These are found very effective in saving the fine gold. Then come two sheets of perforated iron, followed by a "blank" or piece of plain, ordinary iron plate. This last is put in to economize the water, so much of which would otherwise pass through into the under-current that the remainder would be insufficient to carry forward the coarser material. This blank is above the lower end of the under-current and is followed by a succession of ordinary angle-iron riffles, to be again followed with another separation by a second under-current. In this way the wash is sized, the coarse passing on, while the finer, more auriferous gravel is treated apart, in the "side-runs" to be described later on.

The upper sluice-boxes are lined, as I have said, with ordinary sacking; but from the top of the first under-current, cocoanut matting is substituted, linen being placed underneath. The heavy, black iron-sand containing the gold finds its way through the matting which serves so well to arrest it and lies upon the linen which covers the wood-work. The total length of sluice-boxes varies from 200 to 300 feet, according to the changes in the coarseness and richness of the wash.

At the end of the under-currents, the finer gravel which has thus been separated, is diverted into "side-runs" or "streamers." These are tables placed by the aide of the main line of boxes in order to effect the collection of gold. The arrangement comprises two tables (See Fig. 15), each 3 feet broad by 24 feet long, placed at a grade of 10 inches in 12 feet. They are ordinarily covered with cocoanut matting; but when washing the residues obtained from the clean-up of the sluice-boxes, the uppermost 4 feet are covered with green baize. Plush was tried for this purpose but it was found that the large quantity of black sand choked it. Of such a pair of "side-runs," one is used at a time; the inlet in the delivery-box being closed with a piece of sheet-iron. The material is diverted from one table to another every two to four hours, depending upon the fineness and richness of the gravel. In cleaning up, most of the black iron-sand is removed while wet by the use of the magnet; after drying, some of the lighter impurities are eliminated by blowing. The "blowings" and the iron-sand are treated in an amalgamating barrel. This is the only use made of mercury. With the aid of electric lights at night, work goes on through the twenty-four hours uninterrupted.

The Roxburgh Amalgamated.—Twenty miles up river from the Island Block is Roxburgh, where several large hydraulic plants are at work. The Roxburgh Amalgamated Company's claim is on the east bank, facing the town, near the mouth of Teviot creek, and at the lower end of Coal Creek flat, through the western portion of which the river has cut its way. That part of the claim bordering upon the Clutha was originally worked by the diggers with good success, first by ground-sluicing and afterwards by means of a primitive dredge. Though this part of the huge deposit which forms Coal Creek flat has been considered the best, because also the only portion available for work, there is reason to believe that old river-channels may yet be found in the

Fig. 15.

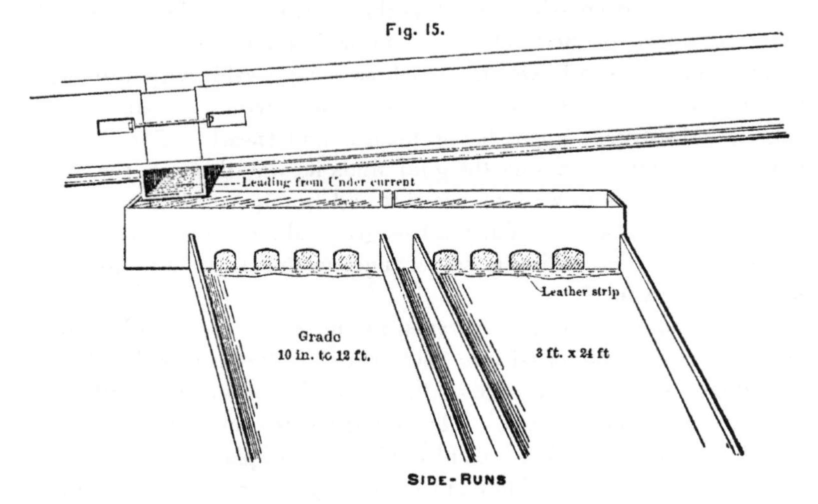

SIDE-RUNS

form of deep leads. At the time when I was in Roxburgh, the Company was engaged in opening up the high terrace; but since then the manager* has discovered a lead some distance back in the flat; and the elevators have made an average monthly output of 250 ounces. The gold is confined to the lowest 3 feet, the top-gravel being absolutely barren. This fact, only recently ascertained, is contrary to the previous general opinion that these river-terraces were more or less gold-bearing through their entire thickness.

Because of its proximity to the river, the claim† is subject to

* Mr. René Proust, to whom I am indebted for this later information, as for courtesy in every other way.

† The New Zealander calls that part of the claim which is blocked out and being worked, a "paddock."

flooding when the river is high. For this reason, at the time of my visit, only one out of three elevators was at work. The face of wash consisted of a barren over-burden of 39 feet lying upon 20 feet of auriferous material. The 59 feet of material thus necessarily to be moved was handled by three elevators, each having approximately the following dimensions :

Supply or pressure-pipe 11, uptake or elevator-pipe 17 inches in diameter. Liner narrowed at the throat to 11 inches. Jet of No. 1 has 3, each of the other two 2½ inches, diameter.

Elevator No. 1 was lifting to a vertical height of 37 feet. No. 2 to 28 feet; and No. 3 to 25 feet.

Of the directors, monitors or nozzles, which break down the face before its removal by the elevators, each has a 2¼ inch nozzle and is connected bv a 7-inch pipe to an 11-inch supply-pipe which in turn joins to the 18-inch main. There are 16 chains of 18-inch pipe and then 48 chains of 22-inch pipe, which bring the water from the penstock.* This last is 14 by 10 by 12 feet and is connected in turn with a flume 2½ miles long, leading to the supply-dams. The flumes, built of Kauri pine, is 3 feet deep and 3 feet wide, having a gradient of 9 inches to the chain.

The head of water available at the claim is 600 feet and 22 sluice-heads were in use at the time of my visit out of the total of 75 heads, said to be the largest water-right in New Zealand, and the equivalent of 3750 cubic feet per minute.

At the present time, as already stated, the pay-gravel is confined to 3 feet only; and to remove the barren over-burden, the elevators have 10-inch liners, 18-inch uptake-pipes and 3-inch jets. This, the "stripping elevator," is shown in position in Fig. 1.† The side and back-views (Figs 2 and 3) explain themselves. The face of wash is broken by 3-inch nozzles. This plant is removing 590 tons per elevator per 8-hour shift. Three men per shift per elevator are employed, the wages for the three amount to 22*s.*—so that the actual handling of the gravel costs less than one-half penny or 1 cent per ton. On reaching the "pay dirt" a smaller elevator is used, having a 2-inch jet, 8-inch liner and 15-inch uptake pipe. By this form of machine the gold-bearing material is more quietly and carefully handled than the over-burden of barren gravel.

At the present time 45 sluice-heads are in use, so that each eleva-

* Called in California the "bulk-head" or "pressure-box."

† These drawings I owe to the courtesy of Mr. René Proust, a member of this Institute, and engineer of the company.

tor, with its driving-nozzle, uses about 15 cubic feet per second, under a pressure of 250 pounds per square inch.

The sluice-boxes 3 feet wide, 18 inches deep and 12 feet long, are supplied with ordinary angle-iron riffles. The grade is 6 inches per 12 feet for elevator No. 1, 8½ inches for No. 2, where the gravel is notably coarser, and 7 inches for No. 3. 6 inches is found to be best adapted for the general run of the material treated.

It is intended to erect "side-runs," the tables to be 20 feet broad and supplied with wells or traps to hold mercury. The manager is fully alive to the difficulty of arresting the extremely fine, flaky gold by purely mechanical means. The gold occurs here, as at the Island Block, in fine flakes mixed with black iron-sand. On cleaning up, a great deal of lead in the form of shot is found.*

At this mine the hydraulic elevator has been adapted to various uses in a very ingenious manner. A common difficulty in placing an elevator in position is that the sinking of a hole to receive the foot of the elevator is attended with delay and trouble owing to the running in of wash and water. This has been overcome by designing a special form known as the "sinker," and shown in Fig. 16, where it will be seen that the intake-pipe has a swivel-joint which enables it to be readily accommodated to an uneven bed-rock. With this machine a place may be rapidly prepared for the erection of a larger elevator.

At the Waipori Deep Lead, at Waipori, an old bed of the Lammerlaw Creek is worked. The gravel deposit is 54 feet deep and from 25 to 30 feet wide. It was first worked by means of a "California pump" and later by means of a steam-winch. Ten thousand tons of gravel yielded 79 ounces of gold, worth £295, and this test was considered sufficiently satisfactory to warrant the erection of an elevator-plant. The water comes under a head of 390 feet and the proprietors have the right to 27 sluice-heads. The ditch, 5 feet by 2 feet, is constructed to carry 18 heads and has a gradient of 8 feet per mile, being constructed at a cost of 25 shillings per chain or £100 per mile. The ground is soft, being composed of a dense, peaty sod which is readily cut.

Where fluming is necessary, it has been built of 1-inch Ribo or red pine, and 3 by 2 feet in size. The fluming, not including trestles, costs 1*s.* 3*d.* per foot. From the penstock there are 6100 feet of piping, of which 2000 is 18-inch pipe (No. 14 gauge), and the

* This lead has been expended upon the curse of the country—the rabbit.

remaining 4000 feet, 15-inch pipe (No. 12 gauge). The pipes are double-riveted and provided with the usual air-valves at top of bends, mud-cocks in hollows and expansion-valves along straight courses.

The supply-pipe leading from the main is 11 inches in diameter, reduced to 9 before delivering the water at the jet of the elevator. This last differs from the ordinary type in being fitted with Robertson's patent atmospheric jet (see Fig. 5). The dimensions are: hopper, 16 inches; throat, 6 inches, with a 2¾-inch nozzle and a 12-

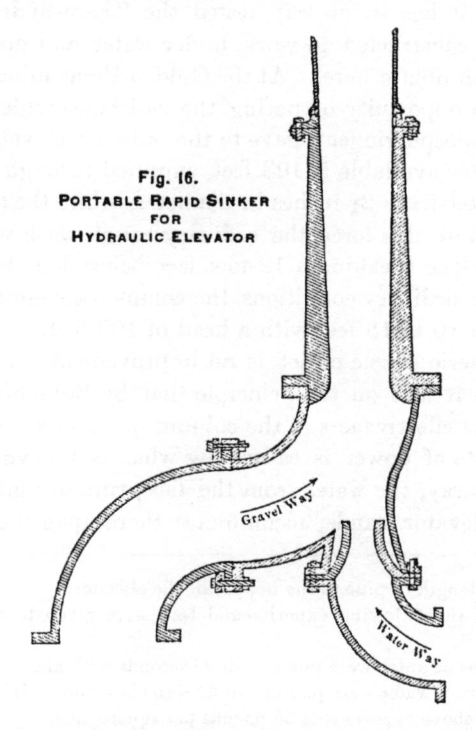

Fig. 16.
PORTABLE RAPID SINKER FOR HYDRAULIC ELEVATOR

inch uptake-pipe. The director, which breaks down the face of the gravel, is supplied by a 7-inch pipe. The brasses are 2 and 1½ inches respectively.

The material is delivered into sluice-boxes of the ordinary type—12 feet long, 3 feet wide, and 1 foot deep. The sides of the head or delivery-box are raised to 2 feet 6 inches. The first box receives 1 inch; the second, 3 inches; the third, 5 inches; the fourth, 6½ inches of fall per 12 feet. The rush of water keeps the first box clear. The average grade of the remaining boxes is 6½ inches.

Ordinary Venetian or "angle-iron on angle-iron" riffles are in use, and under them is placed common sacking. The end-boxes, and the first also, are supplied with perforated iron plate. The last two are lined with blanketing. Of the total gold obtained, 2 per cent. only is found below the first box.

At present, the uptake-pipe being 87 feet long, the vertical height lifted is 56 feet from the hopper, or 58 feet from the bed-rock. By utilizing the suction* another 10 feet will be available, making 66 feet in all. This lift is, in proportion to the power available, unusually high, but it has in no way tested the "aero-hydraulic" elevator, which is constructed to work under water and not under the conditions which obtain here. At the Golden Point mine, five miles away, I had an opportuity of noting the assistance which the addition of the "atmospheric jet" gave to the ordinary elevator. There the head of water available is 103 feet, supplied through an 11-inch pipe. The water-jet is 2¾ inches in diameter, while the uptake-pipe is 9 inches. With this force the ordinary rough wash was lifted to 32 feet, the work of breaking a 12-foot face being done by a 2½-inch nozzle. Under ordinary conditions, the common elevator would not lift higher than 10 to 15 feet with a head of 103 feet.

The atmospheric nozzle or jet is an improvement on the simple elevator; since it acts on the principle that by lightening the load you increase the effectiveness of the column of water.† In working elevators a waste of power is caused by what is known as "flooding;" that is to say, the water from the face, running into the basin in which the elevator stands, accumulates there upon the least stop-

---

* By adding a length of pipe to the hopper of the elevator.

† The results of the following experimental tests were given to me by the inventor:

No. 1. 2½ gallons of water were pumped in 20 seconds with air.

No. 2. 2½ gallons of water were pumped in 32 seconds without air.

Pressure in the above experiments 35 pounds per square inch.

No. 3. 2½ gallons were pumped in 2 minutes with air. A trial was made without air with the same quantity of water and no work was done. Pressure, 8 pounds to the square inch.

No. 4. 10 gallons were pumped with air in 1 minute, 30 seconds.

No. 5. 10 gallons were pumped without air in 2 minutes, 10 seconds.

In this test 5½ gallons were required to pump the 10 gallons with the air, and in pumping without air it took 10 gallons to do the same work. Pressure, 38 pounds per square inch. Sand, shot, spelter, etc., were lifted in the proportion of half solid matter to half water.

It was found that the machine emptied a given quantity of water with air faster than without air in the ratio of 1 to 1.33.

page,* and the hopper being soon under water the air is entirely shut off. Similarly, when the hopper is intermittently under water, the unequal irregular suction produces a fusilading of the piping by the pebbles, which is productive of great wear and tear. It is found that the admission of a little air, by keeping the hopper partly open, cannot be controlled, though it is desirable in that, while increasing the wear on the liner, it adds to the effectiveness of the machine. To avoid those difficulties and to obtain these advantages, there is introduced a constant air-supply by means of the addition shown in Fig. 5. An annular air-current passes from the outer air through a jet supplied by a pipe which extends above the water-line. This forms an air cushion round the water-jet and, by preventing it from feathering or breaking, tends to maintain a uniform velocity. The diminution of the suction is not sufficient to affect injuriously the feeding, while the lightening of the column aids the effectiveness of the machine. There is also a decrease in wear and tear, since the admission of a constant air-supply secures a constant speed to the ascending column of gravel and water.

Before leaving this part of the subject, a few general conclusions may be in place. The alluvial deposits of Otago are of remarkable extent, more particularly the lake- and river-beds along the course of the Clutha, Shotover and Kawarau rivers. These latter were not to any large extent accessible for mining previous to the introduction of the elevator, because the frequent rapid rise of the rivers prevents the satisfactory use of the dredge, and ordinary sluicing is rarely possible owing to the want of "dump." The most extensive deposits are in the flats formed from old lake-basins; but their level is such that the seepage from the river rapidly floods the miner's claim. To overcome the difficulty by ordinary pumping was not economically practicable. The hydraulic elevator, however, renders these deposits available for exploitation, and its introduction is rapidly spreading; the expense of the first installation of the costly plant required being met by the *consolidation* of small claims under companies provided with the necessary working-capital.

The elevator will have a great future in Otago; but, in the haste to use a new machine, there is a tendency to discard the assistance of methods which have previously rendered good service. Ground-sluicing has been put aside, and there is a tendency to make the elevator do work which can sometimes be less expensively done by

---

* As when a large stone sticks in the throat of the elevator.

simpler means.* In the blocking out of ground there is room for improvement. It is not very difficult to erect a good plant, and it is easy to commence the attack on a face of gravel, but the systematic blocking out of the claim tries the resources of those whose experience has been mainly confined to other forms of mining.

Dredging.

The first gold-seekers who prospected the sands of the Shotover and Molyneux rivers restricted their search to the easily accessible deposits which had accumulated under the shelter of rocky bars; and when the auriferous sands were found to extend under the waters of the river, they turned the stream by means of wing-dams. The rich alluvium was found, however, in places where this mode of operation was impracticable or too costly, and in such places, standing shovel in hand, they snatched with difficulty the golden sand, which increased in richness the less accessible it became. The simple shovel was useless in a fast current; hence the next step was to contrive a ladle or spoon with which to scoop up the river-bed. A piece of hide fastened to an iron rim was arranged behind the modified blade of a shovel, and this at the end of a long pole helped to increase still further the area available for work. Soon, however, the distance from the shore, and with it the increasing depth of water, prevented further advance. A barge or punt was then built, the pole was lengthened to 20 feet or more, the scoop was enlarged so as to hold a barrowful, and, the increased weight requiring other than mere hand-labor, a winch and tackle were rigged up. This now became the "spoon-dredge," the forerunner of the numerous types of bucket-dredges which have started a new branch of the mining industry of New Zealand.

The spoon-dredge served its purpose, but the numerous mines along the banks of the river had begun to send down tailings, which soon covered the bottom with a rapidly increasing thickness of valueless material. The spoon-dredge was not capable of coping with this fresh difficulty, but the rapidly flowing river suggested the greater power now needed to replace human muscle; a water-wheel took the place of the winch, and the "current-wheel-dredge" was invented. This consisted of a simple form of bucket-dredge, worked by an under-shot wheel placed at the side of a punt. The power needed to propel the machine severely restricted its usefulness, for it

* By this I refer particularly to the removal of the barren overburden, which can be done by ordinary ground-sluicing.

could not be employed in the back-waters, or indeed anywhere but in the full force of the river-current, and the richest parts of the channel had therefore to remain untouched. Steam was substituted for water-power, and the bucket-and-ladder type of dredge was advanced a stage further by the addition of revolving sizing-screens, winches for mooring and pumps for raising the water required to separate the fine gold-bearing silt from the coarse gravel. This brings us down to to-day, when the river-sands and the sea-beaches of the South Island are worked by a force of over 50 dredges, propelled by water, steam and electricity.

When I was at Dunedin, in 1890, it was difficult to find a man who had not located some acres of land on river-bank or sea-shore. Dredging properties were plentiful as "leaves in Vallombrosa" and "the potentialities of acquiring wealth beyond the dreams of avarice" were offered on every hand. Returning in March, 1891, I found that the number of dredges at work had largely increased, with a corresponding diminution in the dredging fever. The results had been disappointing.

The success of the Dunedin dredge was the main cause of the dredging fever. Directed by a man of great experience in this branch of engineering, and put to work in a part of the river Clutha which soon proved very rich, this dredge returned in ten months two-thirds of the paid-up capital of the company which owned it. One of the Shotover dredges had also been fairly successful, while on the sea-beach the Waipapa dredge, exploiting the titaniferous gold-bearing sand, had made, under many difficulties, very encouraging returns. It was at once found that the extent of ground available for this mode of working was almost unlimited; the river-channels and the sea-coast were soon covered with locations; claims were taken up and floated into companies, the vendor receiving, for ground which had cost him but a few shillings, a large sleeping interest in a concern to the working capital of which he contributed nothing. Large areas were assumed to be valuable on account of reported rich yields obtained by the diggers from spots representing a very small proportion only of the total field to be worked. No tests, or only unsystematic tests, were made. It was overlooked that the rapid rise of the river would prevent work during many months in the year. Moreover, in this form of mining the experience so far accumulated had been but small, and the saving of the gold was often attempted under conditions which were hopeless. Gradually the enterprising investor awoke to the stern fact

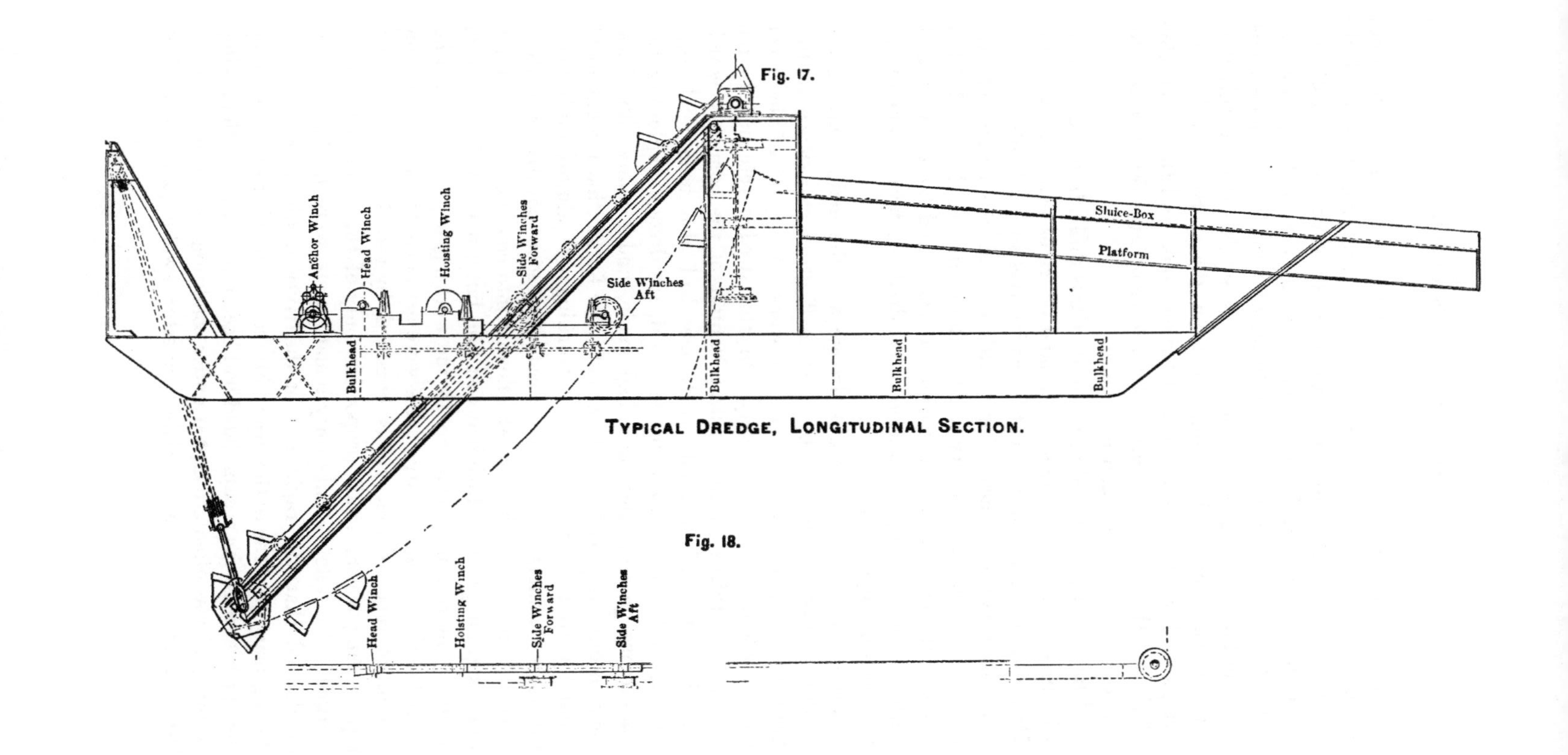

TYPICAL DREDGE, LONGITUDINAL SECTION.

that, like all other branches of mining, dredging required judgment, care and experience.

I will here describe two representative bucket-dredges. Figs. 17 and 18* show the general mode of construction. The pontoon in this case has a length of 80 feet, a width of 18¼ feet, and a depth of 4½ feet. The ladder is constructed to work to a depth of 20 feet below the surface of the water. The power to propel the machinery is derived from the river, and is transmitted as electricity.

The Dunedin dredge is working the bottom of the Clutha at a point about 3 miles above the town of Roxburgh. The deposit covering the rocky bed of the river consists here of 20 to 25 feet of barren drift, overlying a thickness of 2 to 2½ feet of gold-bearing wash. Sometimes the "pay" thins to 6 inches. The overburden of drift consists of small-sized gravel, but the pay-wash is composed of large boulders, among the interstices of which occurs the gold. Black sand is found in both drift and pay-wash, that of the drift originating, together with the bulk of the material in which it occurs, from the mines along the river. The gold is similar to that obtained elsewhere along the Clutha, consisting of fine, flat flakes in a black ironsand. Gold is not seen in the quartz of the pebbles.

This dredge has a double ladder of buckets holding 2 cubic feet each, 12 emptying per minute on each side, giving the machine a capacity of 106 cubic yards per hour. There are 31 buckets on each side, the material from which passes through the two perforated iron cylinders, the larger pebbles being ejected along an iron shoot which returns them to the river, while the fine gravel passes over a table 8 feet long by 7½ feet wide. The grade is 2 inches per foot, and cocoanut matting is used to arrest the gold. This table and the similar one on the other side discharge upon a middle table, 8 feet long by 4 feet wide, which is provided with iron-bar riffles. From here the gravel falls into the river. The above comprises the whole apparatus employed in the saving of the gold. The cocoanut matting is washed at regular intervals in a tank, the gold being subsequently separated from the black sand by panning.

The motive power for the machinery is derived from steam, the fuel used being the lignite which occurs at Coal Creek, 3 miles up the river. The price is 8 shillings per ton, delivered at the bank,

* The drawing I owe to the courtesy of Mr. Robert Hay, C.E., of Dunedin, who has designed many of the best dredges in use in Otago.

whence it is ferried by boat. The daily consumption is 3 tons. Electric light enables work to be uninterrupted.

The dredge had just re-started at the time of my visit, after a stoppage due to the rise in the river, which was even then flowing 9 knots per hour. During flood-time the rate of flow reaches 12 knots, while the average is 7 knots per hour. The working expenses are £250 per month, This includes the wages (£160) of 9 men and a dredge-master. The coal bill is £30. The balance of the £250 is taken up by repairs and supplies. The average of steady work is only 4 days per week of 6 working days. This does not allow for the interruption, often for weeks, due to floods. From December, 1889, to October, 1890, dividends amounting to £4080 were paid on a nominal capital of £7200, and a paid-up capital of £6240. After October, work ceased on account of the spring floods until the first week in Décember. The directors' report for the 6 months ending July 2, 1890, states that the gold obtained amounted to 884 ounces, 5 pennyweights, 8 grains,—value £3316—out of which £1680 was returned in dividends.

The Dunedin dredge is working the auriferous gravel lying upon the rock-bed of the swift-flowing Clutha; but the Waipori dredge, which we will next consider, is placed under much more favorable conditions. The Waipori river is a small stream, running through a flat valley surrounded by rolling foot-hills. A wide deposit lies in the hollow of the valley, and above it flows the present stream. The gold obtained does not lie upon the bed-rock of the original channel, which is much deeper, but at a horizon marked by coarse sand—the "false bottom" of the miners.

The dredge is of the bucket-and-ladder type. The stream is of insufficient depth to float it; but the dredge makes its own waterway by the removal of the gravel. The capacity is 7600 cubic yards per week of 125 to 130 working hours. The movement of the pontoon is directed by 4 winches operating wire ropes fastened to the shore. The engine is of 40 horse-power, supplying the power necessary to work the buckets themselves, as well as the pump which gives the water necessary for the tables. The contents of each bucket as it comes up are emptied into a revolving sizing-cylinder, constructed of horizontal iron bars. The fine silt goes straight to the tables, while the coarse passes to another sizing-cylinder (this time of perforated boiler-plate), which separates the coarse gravel and boulders. These fall into the stream from an iron shoot, while the fine goes to the tables, which are arranged on either side and are covered with coarse cocoanut matting, with linen underneath.

The following figures will indicate the cost of operations during the period of a year.

	£.	s.	d.
Repairs and Alterations, ...	222	9	1

This was chiefly in pins and bushes. The wear and tear is mostly about the buckets. The lips last for two years; and the pins on which the buckets are hinged, have an average life of 3 months.

	£.	s.	d.
Wages, ...	1192	16	4

This includes the pay of 8 men and 3 boys per day. One engineer, one winch-man and one boy to attend to the tables, on each shift, making 6 men and 3 boys per 24 hours. Then there are to be added the dredge-master and a blacksmith.

	£.	s.	d.
Material, ...	272	9	5

This includes the shed on the shore which serves as an office, forge and tool-house.

	£.	s.	d.
Coal, ...	507	12	0
Firewood, ...	377	6	0

The wood (manuka) costs £1 15s. per cord. Coal (from Westport) costs 55s. per ton, delivered.

	£.	s.	d.
Rent, ...	293	0	0

This is the amount (at 10s. per acre) paid to the government as rent for the claim.

During the year the first two months were taken up by the erection of the plant and in the remaining ten the gold obtained amounted in value to £3095, 8s. 6d.

The cost of the dredge, including fees to the engineer, was £3380, 14s. 3d.

The weekly cost upon the ground is £30 for labor, £20 for fuel, or (including repairs) about £65 in all.

Three typical weekly records are as follows:

First Week. Yield 21 ounces, 16 pennyweights. Time of actual work 140 hours.

Ground lifted 8400 yards. Depth of lift 10 feet. Firewood, 10 cords.

Second Week. Yield 33 ounces, 13 pennyweights, 18 grains. Time 126 hours.

Ground lifted, 8400 yards. Depth about 13 feet. Firewood, 10 cords.

Third Week. Yield 29 ounces, 12 pennyweights. Time 126 hours.

Ground lifted 7560 yards. Depth about 13 feet. Firewood, 10½ cords.

Up to that time the dredge had handled very little virgin ground, the material consisting chiefly of the tailings carried down from old sluicing-claims. Since then it has reached solid ground; and the results, with the same expense, have reached 50 ounces per week.

This dredge is working under conditions admirably suited to its capabilities. The working-cost is slight, the amount of capital tied up is very small, the area of the claim is large and the danger of floods *nil.* The depth at which it is working varies from 10 to 15 feet. The "false bottom" of compact coarse sand enables the buckets to get well under the pay-wash, and avoids a great deal of the wear and tear incident to working upon a rocky bed.

The material treated, while it is very similar to that of the mines on the Clutha, yet contains gold more shotty than that of the big river-claims. The gold of the Waipori flats came in large measure from the erosion of the neighboring quartz-lodes—the O. P. Q. and other reefs. While on the dredge I was shown a round, white quartz pebble, taken that morning from one of the buckets, which was of the size of a hand, and showed several splashes of gold as large as a small-finger nail. Such a find, very unusual in the alluvial mines of Otago, marks the somewhat exceptional conditions under which this deposit was probably formed.

Gold-Saving.

The methods used are of the crudest kind. There is no doubt that the river receives back more than half of the gold contained in the material raised by the dredge. The tables at Waipori are somewhat larger than those of the Dunedin dredge; but in both casesit appears the height of absurdity to think that they can save a large percentage of the fine, flaky gold carried along in such a large flow of water and amid so great a volume of sand and gravel. The report of the Directors of the Dunedin dredge company says that "the dredge-master reports that the gold-saving appliances are all that could be desired." This is the severest satire upon the childish efforts made to save the gold, fully sixty per cent. of which must be a mere passenger through the apparatus designed to arrest it. Over a hundred cubic yards of gravel, together with the water employed to transport it, are in this case handled by the dredge every hour, and of this a very large proportion passes over the surface of two tables covered with cocoanut matting whose dimensions are 8 feet by 7½ feet. There is no opportunity whatever given for a separation of the gold from the mass of heavy black iron-sand and the sediment in which it is enveloped. The surface over which it passes is far too small, and the distance over which it travels is altogether too short to enable it to be arrested by the simple means adopted. What is collected is in spite, rather than by reason, of the efforts made to catch it, and represents a small proportion only of the gold in the material, the larger part being lifted from the river only to be returned.

The Use of Mercury.

The use of mercury is not familiar to the alluvial miner of Otago, as it is to his brethren elsewhere. There exists a curious idea that it will not act in cold weather, due no doubt to the fact that the cleaning up in the mills and mines is invariably done with hot water. Of course there is a substratum of truth in this idea, since amalgamation is as a rule* retarded by cold and assisted by heat, but within narrow limits only and not to such an extent as to make the fact of any great practical importance. It certainly will not explain why the Otago digger has "left in the cold" one of the best friends of mining all the world over. From an examination of the conditions under which the gold of the alluvium of Otago occurs, particularly in the deposits of the Clutha and its tributaries, I am strongly of the opinion that the use of mercury will have to be resorted to if any large percentage of the precious metal is to be extracted.

In the case of the elevator-plant the example of California can be followed and mercury placed in the riffles of the sluice-boxes themselves. Then it requires but an extension of the idea of side-runs,

---

* In the mountains of Colorado, at an altitude of 8500 feet, the amalgamation at the stamp-mills meets with no obstacle in winter. On the contrary, it is curious to note that the mill-men of Gilpin county unite in asserting that the cold weather is beneficial to amalgamation, for the reason that warmth thins the mercury and causes it (with the vibration due to the falling stamps) to run off the plates.

preceded by under-currents, to permit the further use of the quicksilver in wells or traps. With the dredges the first alteration of existing methods which is demanded is the enlargement of the area over which the material passes. Dredges are now constructed at Dunedin to handle over 150 tons per hour. This requires a very large surface to effect even the roughest separation of the fine gold-bearing silt from the large mass of non-auriferous wash. The Welman dredge at Waipapa* which is supplied with tables 24 feet by 30 feet, having an incline of $\frac{3}{4}$-inch per foot, is designed on common-sense principles and supplies an example to the other dredges of Otago. But even the dispersion of the material over such an enlarged surface will not suffice to collect the finer particles of gold. Mechanical means must be assisted by chemical; gravitation must be followed by amalgamation.

The fineness of the gold in the New Zealand alluvium may be imagined when it is stated that by actual count it requires six to seven thousand particles of the gold as found on the west coast† to form a grain in weight. It will be said that, as a matter of practice, it has been noted that very little gold finds its way beyond the first strips of matting and that on washing, the bottom strips are seen to collect scarce any.‡ This is after the fashion of the millman who carefully assays his tailings but fails to note how much gold he is losing in the slimes which are carried down stream to cheer the the hearts of a tribe of Mongolians. The evidence obtained by such tests is an *ignis fatuus* to the miner, deluding him into a blissful ignorance of his losses. The fact is that in such cases the gold lost is in an entirely different condition from that saved, and methods

† This dredge was idle during the time of my stay in Otago, owing to a change which was being made in its construction. An interesting description appeared in the records of the Mining Conference held at Dunedin in 1890, written by Mr. Jas. Allen, a member of this Institute. This dredge is working the sea-beach, those portions where the titaniferous gold-bearing sand has been concentrated by the action of the prevailing winds.

‡ Best known by the Hokitika discoveries.

§ The following experiments were made at Waipapa:

First Experiment.—The bottom cloths—those 16 to 18 feet from the head—were washed separately from the remainder, with the result: Amalgam for eight days—top cloths, 89 ounces 16 pennyweights; bottom cloths, 4 pennyweights 5 grains.

Second Experiment.—The second row of cloths from the bottom, that is from 14 to 16 feet were separately washed. The amalgam from the top cloths was 90 ounces; the cloths from 14 to 16 feet yielded 8 pennyweights.

It is necessary to add here that the produce becomes amalgam only in cleaning up, mercury being used simply in collecting the residues from the washing of the plush—which last does the direct gold-saving.

which serve to arrest the latter are entirely insufficient to hold the former.

The character of the gold is in no way prejudicial to amalgamation. In both river-banks and sea-beaches, it is bright, of high caratage, and not "rusty," or mixed with minerals inimical to mercury. On the other hand its fine flakey character makes it particularly hard to arrest by purely mechanical methods. The thin plates of gold, especially when their edges are turned, as must often happen during the treatment the gravel undergoes, are especially adapted to be transported by water.

The material used for the gold-saving is cocoanut matting, which within certain limits answers admirably. Owing to its porous character, it is usually supplemented by linen placed along the woodwork of the tables or sluice-boxes. Some of the dredges employ plush. The choice is largely a matter of expense. At the Island Block it was found that the plush used in the side-runs caught too much of the black iron-sand and got quickly choked. At Waipapa, with a very large proportion of fine black sand, it answered well. The difference of experience is due probably to the quantity of water used, and the gradient over which the material passed.

The dredge has added largely to the area available for mining operations. In this branch of mining, Otago is opening up a new and important field. The practical result of the experience so far obtained proves that the bucket-dredge, though admirably simple and inexpensive, is best suited to the raising of auriferous alluvium ying upon a "false bottom." Upon a true bed-rock the wear is much increased, and the effectiveness much diminished. Everyone knows how difficult it is to scoop fine gold mixed with gravel *under water* by the aid of a shovel. It runs off. The bucket of the dredge is a modified shovel. For irregular bottoms, the suction-pump dredge, of which the Welman is a good example, will be found best adapted. In this case a powerful centrifugal pump draws up the water, gravel and gold, delivering them to the level of the tables. At Waipapa, stones 35 to 40 pounds in weight have been sent up by the pump; and it only requires an improvement in construction, giving durability and strength, to render it a most effective machine for this class of work. A dredge thus provided is able to sweep the bottom clean. After that, the extraction of the gold becomes the great question; and in this direction, as we have seen, there is a wide margin for improvement.

Time, however, will remedy these defects, and the Otago miner may meanwhile point with pride to the fact that he has shown the possibility of working the sands of the sea-shore at a profit, when they contain but two grains of gold per ton.

This concludes my notes upon a mining-field but little known on this side of the equator. The chief lesson it conveys is, that we should seek to profit by the experience of others. Otago has much to learn from California in lode-mining and quartz-milling; but California would do well to study the steps of Otago in hydraulic elevating and dredging. The miner should be the least conservative of men; his motto should be "pass it on;" the same difficulties should never require to be overcome twice; and thus should be avoided that worst of all wastes, the waste of experience.

NOTE BY THE SECRETARY.—Comments or criticisms upon all papers, whether private corrections of typographical or other errors, or communications for publication as "Discussion," or independent papers on the same or a related subiect, are earnestly invited.

SUBJECT TO REVISION.

[TRANSACTIONS OF THE AMERICAN INSTITUTE OF MINING ENGINEERS.]

The Development of Colorado's Mining Industry.*

BY T. A. RICKARD, STATE GEOLOGIST, DENVER, COLO.

(Colorado Meeting, September, 1896.)

THE history of this State is that of one generation. Thirty-six years only have elapsed since the birth of that beneficent industry whose footsteps were the first to traverse the wilderness of the prairies and penetrate the solitude of the mountains. So soon is history made. The men who halted on the rolling plain where Denver now stands and gazed westward at the snow-clad ranges in eager questioning of their possibilities of golden wealth, have lived to see a noble city built where once their camp-fires burned, and have participated in the discovery of magnificent series of productive mines amid the mountains which seemed at that time the Ultima Thule of their pilgrimage.

To those who investigate the workings of our complicated mining and metallurgical industries, the story of their evolution from humble beginnings will appear an instructive romance. The record that tells it presents features common to the growth of modern mining regions, but also bears aspects peculiar to those local conditions upon which the development of these industries is so essentially dependent.

In the summer of 1849 a party of seven Georgians were taking a herd of thoroughbred horses across the continent to California. They reached Camp Lyon, on the Arkansas, in October, and, meeting James Dempsey, a government guide, they were persuaded by him that it was too late to cross the mountains that season. His advice was followed, and, moving northward, they established a winter camp at the junction of Cherry creek and the Platte. Upon a sand-bar, lining the south side of the river, they built two cabins. During the closing months of that year they prospected the alluvial banks of Cherry creek,

* This paper was subsequently presented before the Colorado Scientific Society. The present version, however, is later in publication, and contains some additions.

but did not penetrate into the mountain canons for fear of the Indians. Gold was found in several places, and particularly at a point 16 miles up the stream. From the feathers of the wild geese which they shot, they made quills, serving as receptacles for small quantities of gold-dust.

This party consisted of Dr. Russell and his brother, Green Russell, A. T. Lloyd, G. W. Kiker, Charles Kiker and P. H. Clark. Early in 1850 they crossed the range by the Bridger pass and went on to California. They mined near Downieville and were successful. Mention was occasionally made of the gold found in western Kansas when on their way across the plains. The goose-quills were evidenced in proof of the story. In the spring of 1857 they, with others, sold out their interests in California and returned to Georgia. Before separating it was agreed among several of them that in the near future they would form a prospecting party to go to western Kansas and search for gold.

In May, 1858, eleven of them met at the Planters' House in St. Louis. In addition to the original seven there were present J. A. O'Farrell, three men of the name of Chastine and another called Fields. All save two were old Californians. Having organized, they went to Leavenworth by water and thence to Camp Harney along the military road. Late in July they left this frontier post, accompanied by an escort of twenty men under the command of Captain Lyon. In August the party reached the log cabins on the Platte. The banks were at that time covered with the wild cherries which gave the tributary stream its name. As soon as camp had been established, they went to the places where Russell and his friends had found gold in 1849. Sufficient was found to encourage them. Prospecting parties were then organized. One of these went northward until they came to a mountain stream full of large boulders. They went up to the forks of Boulder creek. In a small basin on the left hand branch they found gold, and called the spot Gold run. Another party went across the intervening ridges to Fall river, and over into Spring gulch. They did not descend into the valley of North Clear creek at that time, but crossed Quartz hill and found rich gravel at Russell gulch, named after the discoverer, Green Russell. It was too near winter to begin serious mining. They returned to camp. Six of the party went east to procure provisions, returning in the spring of the following year, 1859.

This was the year of golden discovery, By the close of 1858 rumors of rich diggings had crossed the plains, the rush had set in and crowds began to arrive. Among these came John Hamilton Gregory, who, with J. M. Cotton and his brother, William Cotton, went up Clear creek and discovered the outcrop of the Gregory lode on the 6th of May, 1859. This date is the birthday of Colorado's mining industry.

The discovery of the Gregory lode was immediately followed by that of other veins, whose production in the succeeding years made Gilpin county the leading gold mining district of the Rocky mountains.

In the meantime, bands of prospectors had scattered all over the neighboring hills, and were finding the gold depositories whose later developments made the counties of Clear Creek, Boulder, Gilpin, Summit, Park and Lake one great mining region.

Boulder was contemporaneous with Gilpin. In 1860 and 1861 the Columbia, Ni Wot, Horsfal and Hoosier veins were discovered and brought the Ward district into prominence. A dozen years later the tellurides of Gold hill were first recognized in the ores of the Red Cloud mine. The recognition of their true character led to the successful exploitation of the rich ore-bodies of the Magnolia, Melvina, Slide and John Jay mines, and to the growth of the hamlets of Sunshine, Salina, Providence and Magnolia. The Caribou district was born when Samuel Conger found the outcrop of the Poorman lode in the last days of 1869. The development of the Caribou vein began in the succeeding year.

The pioneers who followed up North Clear creek (or North Vasquez, as it was then called) and founded Gilpin's industry also wandered up the north fork of the stream and discovered the veins whose development gave wealth to Clear Creek county. In August, 1858, George Jackson did some prospecting about Vasquez forks, and in the winter of that year he penetrated to Idaho Springs and went up Chicago creek. On the 7th of January, 1859, he found rich gravel. This led to active search and successful work amid all the other tributaries of Clear creek. The diggers followed the stream to its head-waters amid the snows of the main range, and discovered the veins above Silver Plume and Georgetown. The sluicing of the soft outcrops of certain veins served as a link leading from placer to lode mining. The Whale above Idaho Springs was one of the finest so worked in 1861. In the upper country, near Empire, the gossan of the Griffith vein was successfully sluiced in 1859. The silver-ores were also recognized about that time, the Running lode, in Gilpin county, having attracted attention to the white metal because of the yield of a peculiar bullion rendered intelligible only when the presence of silver had been determined. The silver-mining industry of Upper Clear creek grew to important dimensions in the decade succeeding 1870. The Pelican and Dives on Republican mountain were discovered in 1868, but did not commence active production until 1871. The Pay Rock was found in 1872. The mines of Sherman mountain—the Terrible, Dunderberg, Cory City, etc.—began to be energetically worked early in the seventies; in Lower Clear creek, John Dumont began operating the Hukill in 1871. In 1878 the Hukill and Freeland were purchased by J. B. Mackay and associates.

The mines at the head of Virginia cañon woke to life by the opening up of the Specie Payment in 1876.

In the summer of 1859 a party of gold-seekers followed the Platte from the foot-hills through its gateway into the South Park, and, camping on the future site of Fairplay, they crossed over to the western slope and descended the head-waters of the Blue river. Near the place now covered by the town of Breckenridge, Reuben Spalding sunk the first hole that disclosed the riches of the placers of the Blue. In the following year the gravels of its tributary, the Swan, were prospected. Alma and Fairplay, on the eastern slope, sprang into life as the result of the alluvial mining which then began a productive existence. It was not until 1880, however, after the exhaustion of the first-found shallow gold-bearing gravel, that the veins of Summit county were exploited. The placers of the Blue river and its tributaries have yielded about $35,000,000.

So were born the mines of Clear Creek, Boulder, Summit and Park counties. But more wonderful discoveries were in store. Leadville was yet to be uncovered.

Among the scattering bands of placer diggers who spread

over the ranges in 1859, one party followed the Arkansas and camped on Georgia bar. In the following spring they continued up the river and divided at the junction of California gulch and the valley of the Arkansas. In a little valley leading from Iowa gulch they stopped at noon. In breaking through the snow to get water for their coffee, the creek had been reached, and in the sand John O'Farrell found some gold. The pieces of porphyry amid the gravel reminded the discoverer of similar conditions observed on the Feather river in California. Little did he guess the significance of those porphyry fragments, or the enormous wealth which that rock covered on the neighboring hills. This was April 6, 1860. George Stevens and party came soon after. Their discovery claims were just above the site of the A. Y. and Minnie mines. Then sprang up a placer-mining industry which lasted for fifteen years, and was only obscured by the greater discoveries which ushered in an era of prolific silver mining.

In the early sixties good gold veins were found on Printer Boy hill overlooking California gulch. These mines, of which the Printer Boy, Five-Twenty and Pilot were the chief, were productive for several years and foreshadowed the development —thirty years later—of the gold region rendered distinguished by the yield of the now celebrated Little Johnny property.

In 1874 W. H. Stevens and A. B. Wood came over the range from Fairplay, where they were mining, to build the ore-ditch. When examining California gulch, Wood found float consisting of carbonate of lead ore, and began digging on the south side, now known as Dome hill, on what was afterward the Rock claim. He sank a little shaft through the drift, which covered the outcrop, at a point subsequently worked by an open cut. He found ore in place, but low grade. This was in the fall of 1875. He made arrangements to have some work done that winter, and this led to the uncovering of the outcrop across California gulch up Iron hill. The next year the whole line of outcrop-claims was located on the supposed veins, and ore was taken in 1877 from the Rock claim to the smelter at Malta, which had been erected three years before to treat the ores of the Homestake mine.* Stevens got Harrison to erect reduction works in 1877, and in the following year Mr. James B. Grant

* Situated on the Saquache range opposite Leadville.

put up the establishment from which in later years grew the magnificent metallurgical industry of the Omaha & Grant Smelting and Refining Company.

In 1878, also, George Fryer sunk a hole on a hill north of Stray Horse gulch, and found carbonate ore, uncovering in this act the great ore-measure which, as "the first contact" proved, was one of the most remarkable bodies of ore known in mining geology. A month later Rische and Hook happened to sink a hole where the contact approached the surface, and found the ore-body which subsequently became the Little Pittsburgh mine, and the foundation of Tabor's fortune. Other discoveries followed fast, That year—1878—Leadville's output exceeded in value 3,000,000. In 1887 it culminated in an output estimated to be worth $13,500,000. Up to date its yield has been $215,000,000.

In the meantime other mining camps were being born. In the fall of 1872 veins were found near Rosita, in Custer county, and in 1877 the Bassick began to produce, creating the excitement of Silver cliff.

Among the progeny of the Leadville boom were Kokomo and Robinson, in the Ten-Mile district, whose checkered career belongs to later days.

In the meantime the mining industry of Colorado was going through the troubles of its early youth. The history of Gilpin county is so typical of this phase of its development, that I have reserved it for special description.

Following upon Gregory's discovery, other veins, subsequently famous, were found. On the 15th of May, the Bates lode was uncovered, on the 25th the Gunnell, Kansas and Boroughs. The Bobtail was discovered in June. The early mining operations consisted of the removal of the soft decomposed croppings which were washed in the sluices after the fashion of ordinary placer mining. The harder outcrops were crushed under trip-hammers until arrastras were introduced, to give way in their turn to primitive stamp-mills. By the 1st of July, 1860, there were 60 mills in operation. Everything proceeded serenely. But the gossan gave way to pyritic ore as the discovery shafts penetrated deeper. The saving of the gold became more difficult. In spite of these drawbacks, the richness of the upper portions of the first-found lodes was such as to leave a handsome margin and maintain a steadily growing population. In the winter of 1863, and the spring of 1864, several mines were sold in New York and Boston. A stock mania supervened, only to collapse suddenly in April. At this time also came the period of incoherent processes, with the promise of 100 per cent. extraction. Some of their progeny still survived, uttering claims nearly as magnificent. The inexperienced chemist, with his revelation for cheap ore-reduction, continued what the stock-jobber had begun. The mining industry of Gilpin was crippled unto death.

At this time the easily amalgamated surface-ores had in many instances become exhausted, giving place to hard pyritic material, which refused to yield up its contained gold. Extraction in the stamp-mills continued to become worse. Many mines were compelled to close down, others were operated at a ruinous loss of the gold in their ores. A depression fell upon the district, which was not removed until 1868, when a general revival began. The leasing of claims by working miners led to new discoveries, and the consolidation of adjoining territory diminished expenses. At this time the smelter came to the rescue of the baffled mill-man. In 1867 the Boston and Colorado Smelting Company was organized by Professor N. P. Hill. In June the first experimental plant was erected at Black Hawk. In January, 1868, the establishment opened for business. In 1873 the company ceased the shipment of matter to Swansea, and erected a refinery under the direction of Mr. Richard Pearce. In 1870 the smelter moved to Denver.

This represents a stage of progress common to all our mining regions. Crude milling methods gave way to fine reduction processes, and the latter, by their heavy charges, invite the mill-man to improve on his cheaper methods, so that competition is restored. The Black Hawk smelter saved the mining industry of Gilpin county when it was on the verge of utter collapse. And on the restoration of prosperity, the owners of the stamp-mills were enabled to carry on experiments whose expense was met by the sale of ore to the smelter, so enabling them to evolve a method of stamp-milling which was well adapted to the treatment of the heavy pyritic ores produced by their mines. In 1871 the problem had been solved, and to-day 500 stamps do excellent work with the low-grade ore of the district.

Gilpin county has produced about $68,000,000 to date.

While these problems were undergoing solution, mining was winning fresh territory southward, amid that great complex of mountains whose waters drain into the San Juan river. A party of pioneers, guided by Jim Baker, crossed the Sangre de Cristo range and reached the head-waters of the Animas in 1861. In spite of snow-slides and Indians the search for gold and silver was extended over the neighboring ranges. The ratification of the Brunot treaty, in 1873, marked the cession of this part of the territory by the Indians, and removed one of the most serious obstacles to the development of the region. In the meanwhile, mines were being opened up on every side. The Baker party tested the river gravels, and the evidences of their placer-workings still remain in many a secluded valley to tell of their first beginnings. In the spring of 1871 lode-mining may be said to have commenced by the discovery and location of the Little Giant vein, just above the present town of Silverton, by Miles T. Johnson. In the following year an arrastra was built, and the gold thus extracted out of the ore was taken for sale to Conejos, the nearest trading station. In 1874 Judge Green, of Cedar Rapids, Ia., commenced the erection of a smelter, the machinery of which came on burro-back from Colorado Springs, where the Denver and Rio Grande railroad had just reached. The ore supply of this smelter came principally from the Aspen, which at that time was the chief mine in the locality. In 1875 Mr. J. A. Porter, the metallurgist of Green & Co.'s smelter, introduced the syphon-tap, and in the following year he erected the first water-jacketed furnace built in Colorado.

The prospectors who made Silverton their headquarters scattered up the valley of Mineral creek, and on the watershed separating this tributary of the Animas from the Uncompaghre found the veins which gave fame to Red mountain. In 1879, Charles Newman and Harry Irving located the Carbon Lake claim, and did their annual assessment until 1882, when a discovery of copper-ore was made on the adjoining Congress claim and shipments began.

In August of the same year, Andrew Meldrum, while out hunting, stumbled upon the outcrop of galena which marked the now famous Yankee Girl vein. The American Belle and Guston were discovered shortly afterwards. Five years later this district became a magnificent producer of very rich copper-ores. In 1894 a smelter was erected at Silverton to treat the product of the region.

In the meanwhile, Gus Begole had crossed the western range from Silverton, and descending into the valley of the Dolores, located claims which later became the Yellow Jacket and Aztec mines. But the ore was too low in grade and the work soon ceased.

In 1878 John Glasgow and Sandy Campbell came northward from La Plata City, and began the successful development of the Grand View and Atlantic Cable, causing the growth of the town of Rico. The news was spread abroad that another Leadville had been found, and crowds trooped in across the hills during the summer of 1879. In the fall of 1880 the smelter began operations. Nevertheless the district would have gained but slight distinction, had not the ore-deposits of Newman hill been found. These began to be productive when, on the 6th of October, 1887, David Swickheimer struck the big ore-body of the Enterprise mine.

In June, 1870, gold was found in Wightman's gulch. This led to the location of the Little Annie in September, 1873, and the opening up of the Summitville district, which is tributary to Del Norte, whose position made it a natural gateway to the watershed of the Rio Grande.

Among the old districts recently revived is the La Plata mountain region, north of Durango and south of Rico. In 1873, Captain John Moss, representing Tiburcio Parrott and other San Francisco capitalists, came up the San Juan river from Arizona and penetrated the La Plata mountains, being attracted thither by the gold-bearing gravel of the streams. He followed the latter to their source and discovered a large number of veins. The Comstock, Morovitz, Euclid and Ashland claims were located at that time. La Plata City was founded, and great activity characterized the camp for a brief period. But the complex telluride ores proved refractory; and the arrastra was found to be powerless to extract the values. Great expectations had a sequel of small accomplishment. The district became depopulated. In 1894 new discoveries were made, and a revival took place, leading to more serious work, which now promises better things.

And so we come to recent times, no less stirring than the old. The history of the last decade centers round the discovery of Aspen, the stories of Creede and of Cripple Creek, the collapse of silver mining and the development of new gold-fields.

The first discovery in the Roaring Fork district, of which Aspen is the center, was made July 3, 1879, when Phillip W. Pratt and Smith Steel, coming from Gothic by the Maroon pass, found the Galena mine on West Aspen mountain. On the following day they located the Spar claim on Aspen mountain, and on the 5th Messrs. Allbright and Fuller located, at the foot of Smuggler mountain, the Little Rock claim, covering a part of the property of the present Smuggler mine. The Smuggler claim itself was located August 30th, by Charles Bennett. On account of the theft of the first four pages of the district recorder's book, it became necessary to make relocations, which now appear on the records at Gunnison, the district being then a part of Gunnison county.

The first mineral survey was made on the Monarch mine, October 12, 1879, by John Christian, of Leadville. The site of the present city of Aspen was devastated, in September of that year, by a forest-fire, in which many horses and pack-mules were lost. An Indian scare, following the Meeker massacre, caused most of the prospectors to leave the camp; but a sufficient number remained to build log-cabins on the site now occupied by Aspen, which was located as a placer-claim September 20, 1879, by Walter Clark.

It was not until ten years later that the new district won a commanding position. At that time (1889) the Aspen, Aspen Compromise and Compromise mines maintained a large output, and it was in 1891 that the big bonanza of the Mollie Gibson was uncovered. Aspen is credited with a production of 8,275,000 ounces of silver in the year 1892.

N. C. Creede found the float of the Holy Moses vein, on West Willow creek, a tributary of the Rio Grande, in 1889. As a consequence, the King Solomon district, as it was then called, began to attract the prospectors scattered in the mountains above Del Norte. No important results ensued until in June, 1891, D. H. Moffat and Capt. L. E. Campbell came to Wagon Wheel gap to visit the Holy Moses, on which they had secured an option. Creede was engaged to prospect for them. Shortly afterwards, Theodore Renniger found rich float on Bachelor mountain. He was unsuccessful in finding the vein in place until Creede came along, and, at a point 200 feet higher up on the hill-slope, discovered the outcrop of a large lode. He located the Amethyst claim and then informed Renniger, who took up on the same vein another claim, which he called the Last Chance. This was on the 8th of August, 1891. It should be added that, three years previously, J. C. McKenzie had located several claims on a heavy quartz outcrop, about 150 feet above the Amethyst. These were afterwards abandoned until the fall of 1891, when the Del Monte location covered them. *Débris* from the upper portions of the mountain had so obscured the outcrop of the Amethyst that its earlier recognition had been prevented. After Creede's discovery but little work was required to prove the existence of a magnificent lode, and, as a consequence, the camp sprang into tremendous activity, culminating in the boom of 1892. The immediate extension of the railway from Wagon Wheel gap stimulated a production which reached its maximum in an output of $3,100,000. Then in the summer of 1893 came the sudden fall in silver and a collapse from which Creede has not yet recovered.

Creede and Cripple Creek were rivals in attracting attention in 1891. Both have had strange vicissitudes. Our new gold-field lies on the southern slope of Pike's Peak, whose snow-clad crest was the beckoning guide of the pioneers of 1858. Although no gold discoveries of any moment were made in the early days among the streams, or on the hills lying at the foot of the old beacon mountain, it nevertheless gave its name to the mining excitement of that period. Time has, however, of late, justified the expectations of the tenderfeet of forty years ago.

The first recorded locations were made in February, 1891, but the clustering hills which are now dotted with productive mines, had been disturbed by the miner's pick as early as 1874. Silver-ores were found in a shaft located close to the present Elkton mine. Ores, rich in the white metal, had been found in late years, and the Moose made one shipment of 30 tons, carrying an average of 70 ounces of silver, in addition to the gold.

Ten years afterward, in April, 1894, the district was the scene of the queer fiasco which has gone into local history as the Mt. Pisgah excitement. A crowd of 4000 men were brought together on the rumored discovery of rich placer ground. Nothing was found save in the holes of the original locators. Man had endeavored to remedy nature's niggardliness and the poor rock had been artificially enriched. Lynching was threatened, but in the failure to catch the perpetrator a big picnic took place, after which the hill slopes again became the quiet cattle ranges for which they seemed best adapted. Those deluded prospectors let slip a great opportunity. Mt. Pisgah's dark front now overshadows the busy streets of the town of Cripple Creek, and, on the ridges opposite, line after line of smoking chimneys bespeak a long succession of productive mines.

During the spring of 1891 Cripple Creek began to receive respectful mention in mining circles, but the discoveries made at Creede during that summer diverted attention from a district whose previous experience had given it an unsavory reputation. When, however, the silver market collapsed in June, 1893, and mining seemed prostrated, the men of Aspen and Leadville turned with the energy of despair to the new gold field which previously had been pooh-poohed, and the concentrated activities of the State were directed to the development of Cripple Creek. With a rare good fortune, the new camp answered to the call, and, as explorations extended, there came a swift succession of rich discoveries which caused the output to spring from $583,000 in 1892 to $2,100,000 in 1893. This growth continued so that in 1894 the yield was $3,900,000, and in 1895 it reached $7,800,000. These figures are sufficiently eloquent of the development of a district which to-day is the largest producer of all the gold-mining camps of the United States.

During the past three years discoveries in other parts of the State have received mention, and large claims have been made in behalf of several new districts. West Creek, Cottonwood, Hahn's Peak and the Gunnison region may be cited. Of these the last named is much the most important. At the head-waters of certain streams, tributary to the Gunnison river, there have been found veins which are now supplying several stamp mills with pay-ore, and have given birth to the settlements of Vulcan, Spencer, Iris and Dubois.

While the new territory which has been won during later

ry of rich placer
ion of the original
nature's niggardli-
punished. Lynch-
ish the perpetrator
Aspen again became
and best adapted.
opportunity. Mt.
busy streets of the
opposite, line after
succession of pro-

k began to receive

attention from a dis-
an unsavory repu-
collapsed in June,
men Aspen and
ir to the new gold
d, and the concen-
to the development
ime, the new camp
extended, there came
caused the output
000 in 1893. This
was $3,900,000, and
are sufficiently
which to-day is the
camps of the United

other parts of the
ns have been made
Creek, Cottonwood,
be cited. Of these
At the head-waters
on river, there have
several stamp mills
ishments of Vulcan,

e seen during later

days adds its tribute of gold and silver to the yield of the old established mining centers, it must be noted that the latter have also accommodated themselves to new economic conditions. The Silverton district, in the southwest, for instance, which in 1892 yielded gold worth $155,624, and silver valued at $354,-125, gave in 1894 gold amounting to $360,320, and silver worth $235,000. This is typical of the changed conditions regulating the industry. It is a striking evidence of that resourcefulness which has enabled the State to overcome difficulties. The collapse of the silver-mining led directly to an impetus in the search for gold, and the variety of ores so puzzling in the infancy of our smelting industry, is to-day its chief aid, because it enables the attainment of that admixture of material which is the essential of successful reduction.

Colorado has yielded to date gold valued at $137,475,000, and silver having a coinage-value of nearly $400,000,000.

Thus, from humble beginnings, a great and complicated industry has been created. Its development may be summarized in four periods: The discoveries in Gilpin county and the adjoining camps in the granite rocks of the front range, the era of silver-mining in the carboniferous limestones of Leadville, Aspen and Rico, the development of the fissure veins in the andesites of the San Juan, and lastly, the revival of gold-mining consequent upon the uncovering of a great series of ore-deposits in the volcanic complex of the Pike's Peak region.

SUBJECT TO REVISION.

[TRANSACTIONS OF THE AMERICAN INSTITUTE OF MINING ENGINEERS.]

Gold-Milling in the Black Hills, South Dakota, and at Grass Valley, California.

BY T. A. RICKARD, DENVER, COLORADO.

(Atlanta Meeting, October, 1895.)

OUR *Transactions* contain two notable papers descriptive of the stamp-milling practice of the Black Hills and of Grass Valley, namely, the elaborate and complete treatise of Prof. H. O. Hofman, on "Gold-Milling in the Black Hills,"* and the paper describing the North Star mill, by E. R. Abadie, its superintendent.† Having visited and examined the stamp-mills of both localities, I venture to offer here such comment as more recent information or a different point of view has suggested.‡

I.—THE BLACK HILLS.

Prof. Hofman's paper was prepared in 1888, seven years ago, yet the methods in vogue to-day at Lead City and Terraville do not differ materially from those so carefully described by him.

The mining industry of the Black Hills is not so actively prosperous in 1895 as it was in 1888. Not so many stamps are dropping; some of the mines have been compelled to suspend operations owing to a falling off in the yield of ore, while others have passed into the comprehensive control of the Homestake management. Nevertheless, the mines and mills of the Belt produce more gold than any other single mining camp in the United States, with the exception of Cripple Creek, Colorado.

---

* *Trans.*, xvii., 498. † *Id.*, xxiv, 208.

‡ For more detailed descriptions and discussions, I would refer to a series of papers on "Variations in the Milling of Gold-Ores," which I have contributed to the *Engineering and Mining Journal* during the present year, and from which quotations are made, and statistical material is reproduced in the present paper. The illustrations herewith given will be found, with others, in the *Journal* of September 7 and 14, 1895, and are here used through the courtesy of the Scientific Publishing Company.

The Belt is that part of the region contiguous to the Homestake lode and its extensions, and reaching from Whitewood creek to Deadwood gulch. It is the center of the mining activity of the Black Hills, the only important mining region in the State of South Dakota.

The geological relations of the ores of the district have been referred to in our *Transactions* by W. B. Devereux* and F. B. Carpenter, respectively.† It is to be regretted that so important and interesting an ore-deposit as the Homestake vein should not have undergone long ago more detailed description at the hands of some one of our members. The commercial environment of mining enterprises often militates against scientific investigation.

While the writer has not had an opportunity of making a careful examination of the mines, a visit underground was sufficient to show the justification for the great ore-reduction establishments of the Homestake Mining Co. The ore occurs in large bodies of quartzified chloritic schists, conforming to the structure of the country and being a portion of it. The width of milling-ore varies from 50 to 400 feet. Most of the supply comes at present from above the 500-foot level; but the developments in lower levels, down to the 800-foot, indicate the continuity of the enormous ore-shoots of the mine. The huge excavations made along the outcrop of the lode (See Fig. 1), are a distinctive feature of the present topography of the Belt, and bear impressive witness to the immense quantities of millstuff that have gone under the stamps. Extensive bodies of ore are still held in reserve in these surface open-cuts.

Behind the Highland shaft-house there is a large cut, which gives a very excellent section of the geological formation, showing the uptilted edges of the gold-bearing schists, overlain by Cambrian sandstone. The latter, only a few feet thick at this point, is split by an intruding sheet of the porphyry which also overlies the whole formation and caps the hilltop. The lowermost member of the Cambrian series is a conglomerate, identified by its fossils as belonging to the Potsdam period, and said‡ to have been derived from the degradation of the gold-

* *Trans.*, x. 465. † *Id.*, xvii., 570.

‡ By Devereux, *Trans.*, x., 966, *et seq.* The development of the conglomerates of the Witwatersrand adds much interest to these Dakota deposits.

bearing lode under the outcrop of which it was formed by the seas of a very early geological time. These facts indicate for the Homestake vein an origin of remarkable geological antiquity.

The porphyry above mentioned is a felsite. At one time it was broken with the ore and sent to the mills in spite of its valueless character. Now it is used for filling up stopes. Underground it can be seen in dikes conformable to the vein-walls, and splitting the ore-bodies by its intrusion.

The first mill in the Black Hills was that erected by the Racine company, at the lower end of Lead City, in April, 1877. The beginning of the mining industry of the Black Hills dates back to June, 1876, when the Wheeler brothers found rich gravel in Deadwood gulch. The outcrops of the large quartz-lodes were early seen, but quickly discarded by the California and Montana miners, who ridiculed the idea of the profitable handling of ores which yielded on panning only from three to ten dollars per ton.

After the placer-mines had commenced to yield handsomely, and while the quartz-lodes were unappreciated, the early mining activity was diverted to the development of the lowermost beds of the Potsdam formation, the gold of which lay in a conglomerate. It was to reduce this conglomerate that the first stamp-mills were introduced. They were of the primitive Colorado pattern, and proved entirely unsuited to the extraction of gold from such material. They made a poor record, and were followed by the fast-dropping stamps, modeled on Californian practice, first introduced by Mr. Augustus J. Bowie, Jr., in his design of the Father de Smet mill,* the erection of which, at Terraville, was commenced in June, 1878. This plant was succeeded by the first of the Homestake Co.'s mills, which also had a narrow mortar and a rapid discharge. The Caledonia, erected the following year, used more roomy mortars and two inside amalgamating-places, more after the Colorado fashion. This mill is now idle, chiefly for lack of suitable ore-supply, and therefore a comparison between the two styles of working is not possible. Nevertheless, I do not hesitate to say that the Homestake mortar, deep and narrow, gives a com-

* See *Trans.*, x., 87.

bination for rapid pulverization and high percentage of extraction inside the battery which render it, for the ores of this district, far superior to any other mortar I know.

All the mills now at work on the Belt are under the direction of the Homestake management, with the exception of the little 10-stamp Columbus mill at Central City. When Prof. Hofman wrote his paper, there were 660 stamps dropping on the free-milling ores of the Belt; to-day the number is 550. The De Smet and Caledonia mines have been unable to survive a diminution in the tenor of the ores they produced, and the mills belonging to them have ceased operations, the former in 1892 and the latter in 1893. The Highland Co.'s mill has been lately increased by the addition of 20 stamps. The Golden Terra and Deadwood mills were consolidated six years ago, the 80 stamps of the Terra being placed behind the 80 stamps of the old Deadwood mill. The two large mills of the Homestake Co. have undergone steady enlargement, and in addition to the number of stamps given in the annexed table, there are 40 about to be added to the Golden Star mill, whose total will then be 200 stamps.

TABLE I.—*The Stamp Mills of the Belt, South Dakota.*

Name.	Date of Erection.	Location.	Number of Stamps.		Owners.
			1888.	1895.	
Homestake........	1878	Lead City.	80	100	The Homestake Mining Co.
Golden Star......	1879	Lead City.	120	160	
Highland..........	1880	Lead City.	120	140	The Highland Mining Co.
Deadwood.........	1879	Terraville.	80	160	The Deadwood-Terra Mining Co.
Golden Terra.....	1880	Terraville.	80		
Father de Smet..	1878	Central City.	100	100	The F. de S. Mining Co.
Caledonia..........	1879	Terraville.	80	80	The Caledonia Min. Co.
Columbus..........	1894	Central City.	......	10	....

The ore is dumped at the shaft's mouth into the rock-breakers. At the time of Prof. Hofman's investigations, all the Homestake mills were using the Blake, and the Caledonia had just introduced a Gates crusher. Since that time the Gates has replaced the Blake rock-breaker in every mill on the Belt. Furthermore, the rock-breaker is now placed in the shaft house instead of at the mill. This follows the tendencies of modern practice in California, where the crusher at the mine delivers the broken ore to the tramway, which carries it to the mill, or sometimes to a second rock-breaker. The latter arrangement relieves the stamps of the hard work of stone breaking, facilitates pulverization in the mortar and gives uniform conditions more favorable to successful amalgamation.

The transference of the rock-breaker from the mill to the mine is in itself a praiseworthy change. It enlarges the capacity of the ore-bins at the mill, and renders unnecessary the use of separate bins for coarse and fine. In small plants where the breaker, if at the mill, would not be driven by a separate engine, it does away with that irregularity in the working of the mill arising from the unequal consumption of power on the part of a rock-breaker. It renders easy the loading of the cars which bring the ore from the mine, a factor important in the case of aerial rope-ways carrying buckets of small capacity. But more important than these minor advantages is the almost entire cessation of the production of the dust so injurious to all the mechanism of the mill and always such a nuisance to those who work in it.

Most of the mills use the No. 6 Gates breaker. The Deadwood-Terra and Highland have two each. The Homestake mine has three, one of which is held in reserve to avoid delay due to repairs on either of the others. Experience has shown that the larger receiving capacity of the Gates and its greater crushing power render it more suitable for large milling-establishments than the Blake.

Prof. Hofman has described the various methods for transmitting the power to the different parts of the mill. Of these, the arrangement in the Golden Star mill resembles that in vogue in California and in the best Colorado mills. The driving-shaft is approximately level with the cam-shaft, and the connecting-belts are nearly horizontal. The latter are in a place easily accessible, well lighted and away from the dirt and water inseparable from the close neighborhood of the battery itself. Such an arrangement requires that the sills under the cam-floor shall be made stronger than if they simply supported the flooring, but the additional expense is trifling compared to the convenience of the plan. On the other hand, the placing of the counter-shaft immediately underneath the feeder-floor, in addition to the inaccessibility and inconvenience, the environment of dirt and water, the absence of light, etc., requires the use of tighteners injurious to the belting. The Star mill uses one belt while the Highland wears out three.

All the batteries are placed upon flat sites in two rows back to back, save at the Father de Smet, where the two rows of batteries face each other and discharge toward the center of the mill. The latter arrangement gives a larger storage-capacity to the bins overhead, but this advantage is obtained at the greater cost of darkening the amalgamating-tables.

The mortar is the most interesting feature of the Homestake mills. In Prof. Hofman's paper there are drawings of it, one of which is reproduced here in Fig. 2. Fig. 3, from a photograph furnished by Fraser and Chalmers, illustrates the latest design. The changes in the dimensions made since 1888, the date of Prof. Hofman's paper, are as follows:

	1888.	1895.
Weight, ...	5400 pounds.	7300 pounds.
Length of base, ...	54¾ inches.	56¾ inches.
Width " "	27¼ "	28¼ "
Height, ...	54½ "	58¼ "
Inside width at the level of the lip,	13½ "	12½

The most important change is the narrowing of the interior width at the level of the lip of the mortar, where a slight change is more important than in any other dimension. The measurement of a new mortar lying outside the Golden Star mill gave an inside width of 13 inches. At the Columbus mill it is 12 inches. The mortar, as now made, is provided with cast-steel false bottoms 2½ inches thick, with a cast-iron lining ⅞ inch thick along the sides and ½ inch thick upon the feed-hopper. The inside copper plates, placed along the front, are ¼ inch thick and are attached to chuck-blocks. The latter are wooden blocks, designed so as to serve as a false lip to the mortar, thereby raising the depth of the issue. A piece of 2-inch plank is bolted to a 1¾-inch board, the latter being made to project about 2 inches beyond the former, to which the copper plate is attached. The 2-inch plank had been replaced, at the time of Prof. Hofman's inspection, with ¼-inch iron, but has since been reverted to, because the slight increase in the distance between the chuck-block and the shoe obtained by the arrangement he describes was undesirable in a mortar characteristically narrow and designed for rapid crushing.

Two of these chuck-blocks are in use, one 9 and the other 7 inches high. When new dies have been introduced, the former is inserted, then making the distance from the bottom of the screen to the top of the die about 9 inches. As the dies wear down, the depth of discharge increases until, after about a fortnight, it becomes necessary to replace the high chuck-block with the lower one. The difference of 2 inches between the two is approximately equal to the diminution in the thickness of the die. After a further service of two or three weeks the dies are worn out, the depth of discharge has increased to 11 inches, new dies are inserted, and the high chuck-block is reintroduced. In this way some sort of an effort is made to maintain a rough uniformity in the depth of the issue, a factor the importance of which is generally overlooked or underestimated in stamp-milling.

It may be added that the copper plate on the high chuck-block is flat, while on the other it has a curved surface, and is mounted on slightly thicker wood, so as to bring it nearer the die. It is the intention of the mill-man to keep the bottom of the chuck-block about on a level with the bottom of the shoe, and to avoid so close a neighborhood to the ore on the die as would lead to the scouring of the amalgamated surface of the copper plate.

There is only one inside plate, made of plain copper and 5 inches wide. This is the one attached to the chuck-block. There is no back-plate, as was the case in the Caledonia mill (now idle), where the mortar was more roomy and of a different design. Free mercury is fed at intervals into the battery with the ore.

The Homestake mortar combines, to a notable degree, the two excellent features of a rapid discharge and a high percentage of amalgamation. Its width at the issue used to be 13½, was then diminished to 13, and in the newest design is 12 inches. The depth, by the introduction of chuck-blocks, is raised to from 9 to 11 inches. The mortar becomes thereby both narrower and deeper than the Californian pattern, its narrowness compelling a rapid expulsion of the pulp and giving the mill a capacity nearly twice that of the average Californian battery when working ore of similar character. At the same time the depth of the mortar prevents the scouring of the inside plate, and permits the arrest of the gold by this plate and by the free mercury added with the ore, so that the percentage of extraction follows closely in the wake of the roomy mortar of the Colorado mill, the crushing capacity of which is only one-quarter that of the Homestake. The following comparison will be of interest:

TABLE II.—*Comparison of Typical Mills.*

	Width at Issue.	Depth of Discharge.	Weight of Stamp.	Number of Drops per Minute.	Height of Drop.	Crushing-Capacity per Stamp per 24 hours.
	Inches.	Inches.	Lbs.		Inches.	Tons.
Golden Star, Deadwood, S. Dak......	12½	9 to 11	850	85	9½	4
Hidden Treasure, Black Hawk, Col..	24	13to15	550	30	17	1
North Star, Grass Valley, Cal........	17½	4	850	84	7	1½
Pearl, Bendigo, Australia..	15	3¼	840	74	7¼	2¼

It will be seen how closely the crushing-capacity is related to the interior width of the mortar at the level of the issue. Notwithstanding its rapid crushing, the Homestake mortar retains a percentage of the total gold extracted which compares well with any of the other districts, and is superior to some of them, though this factor will be affected by the variety of screen in use.

In 1888 the Homestake mills were using No. 7 diagonal-slot Russia iron punched screens. At the present time the mills uniformly employ the No. 8 size of the same variety. This means slightly finer pulverization. Prof. Hofman gives the crushing-capacity in 1888 at 4½ tons per stamp per day, with a 9-inch drop 85 times per minute. It is now about 4 tons per day, with a 9½-inch drop and the same speed. This shows the effect of the substitution of the No. 8 for the No. 7 screen.

The No. 8 screens are considered equivalent to 30-mesh wire. They break before they become worn out because of lines of weakness developed by the press used in their manufacture. They are, however, never retained in service until the

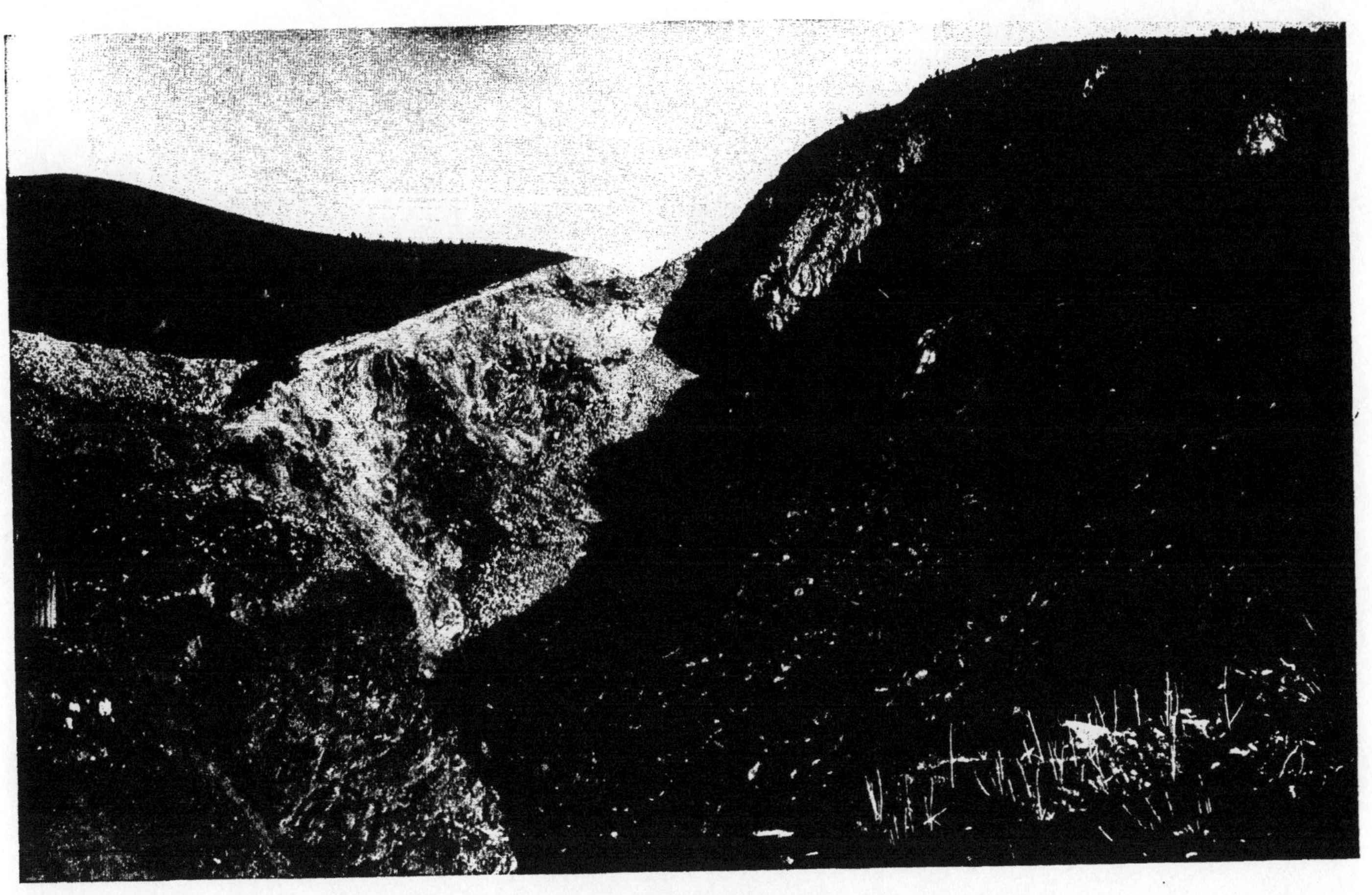

Deadwood-Terra Open Cut.

Fig. 2.

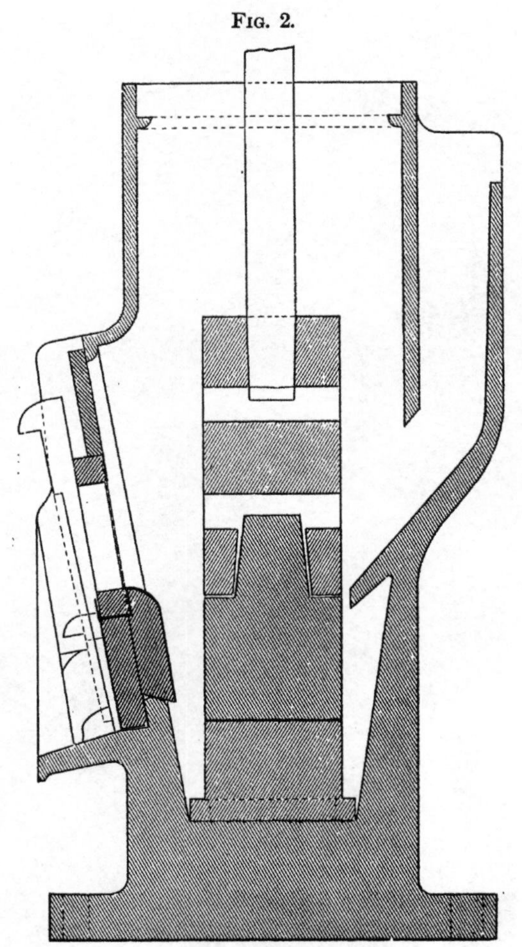

Scale, 1 inch = 1 foot.

The Homestake Mortar.

Fig. 3.

The Homestake Mortar.

FIG. 4.

apertures are much worn, because this would produce a coarse crushing detrimental to a uniform product. Their maximum service is about two weeks, but this is rarely attained, because breakage occurs after six to eight days. The cause of this is the accumulation, inside the battery, of wood chips, which have found their way into the ore from the timbers in the mine. They tend to dam up the pulp within the mortar, and so subject the screen to a pressure greater than it can bear. This results in a break or a burst. Where surface-ore is being milled, as at the Deadwood-Terra, the life of the screen is prolonged to an average of nine or ten days, because such ore comes from workings where there is but little timbering.

At the neighboring Columbus mill, 30-mesh brass wire is used, and the choking of the screens is minimized by having three sets for each battery, so that while one is in place the second is being dried out and the third cleaned with a wire brush. It would be better if the Homestake mills could find it practicable to use wire-cloth instead of punched Russia iron, since, apart from the advantage of a more uniform discharge, the pulverization is more regular, because the wires do not wear so easily as the punched holes, and the apertures therefore retain their size, and the screen does good work until simple breakage requires that it be patched or discarded.

The process of gold-extraction consists of amalgamation within the mortar upon outside plates and in traps, supplemented in a rudimentary way by an inadequate effort to concentrate the sulphides of the tailings.

The mortar becomes an amalgamating apparatus by the use of the inside copper plate and the addition of free mercury. About 50 per cent. of the total amalgam is obtained from these inside plates. At the Deadwood-Terra the proportion reaches 70 per cent.

To quote from my previous paper* on the subject:

"Mercury is fed into the battery in quantities proportioned to the richness of the ore and regulated by the condition of the amalgam on the apron-plates. At the four principal mills the rate at which the mercury is fed can be judged by the accompanying record, covering the two weeks previous to my visit.

"It will be observed that though the Golden Star and the Deadwood-Terra mills crush approximately the same amount of ore, the former uses more than

---

* *Eng. and Mining Journal*, p. 222, September 7, 1895.

twice as much mercury. This fact is explained by the wide difference in the richness of the ore, for while that crushed in the batteries of the Golden Star averages from $4 to $5 per ton, that which goes through the Deadwood-Terra ranges from $1.50 to $2 per ton.

TABLE III.—*Consumption of Mercury at the Homestake Stamp-Mills.*

Date.	Deadwood-Terra. 160 Stamps.		Golden Star. 160 Stamps.		Highland. 120 Stamps.		Homestake. 100 Stamps.	
	Lbs.	Ozs.	Lbs.	Ozs.	Lbs.	Ozs.	Lbs.	Ozs.
May 1, 1895	11	5	25	13	15	4	* 9	10
2,	10	2	* 17	14	* 12	9	12	2
3,	10	0	25	16	15	1	13	0
4,	12	2	26	13	17	10	12	12
5,	* 9	12	25	3	10	6	11	7
6,	10	6	24	8	16	8	12	2
7,	11	11	23	2	17	0	12	8
8,	11	6	24	12	19	3	12	15
9,	11	8	23	13	19	8	12	13
10,	10	14	25	3	18	12	12	13
11,	11	0	23	7	18	8	12	7
12,	10	12	24	6	17	4	11	1
Average per day	11	4	24	3	17	2	12	2

"For the year ending June 1, 1894, the Homestake mill used 2084 pounds and the Golden Star mill 3440 pounds, making a total of 5524 pounds avoirdupois, which, at the price obtaining that year, 42 cents, makes the value of the mercury used $2320.13. During that time 309,210 tons of ore were crushed, so that the consumption was at the rate of about 5 dwts. Troy or ¾ cent per ton. At the Deadwood-Terra 205 pounds were used in February, 1895, in treating 18,483 tons of ore. It is estimated that 22 per cent. of the amount of mercury used is lost."

On issuing from the battery the pulp falls from 6 to 10 inches before striking the aprons or first amalgamating-tables. This serves no particular purpose, while the actual damage possible to the plates by the scouring of their surface due to the impact of the pulp is obviated by the interposition of a splash-board, which breaks the fall of the sand and water. This splash-board might be placed at such an angle as would permit of its use as an amalgamating device by lining it with a copper plate.

In the Homestake Co.'s mills the aprons are 10 feet in length and 4½ feet wide. Those in the Highland mill are only 8 feet long. In all these mills two apron-plates deliver the pulp to one tail-plate having a size equal to one apron. At the Dead-

* Indicates clean-up days.

wood-Terra the aprons are somewhat larger, namely, 11 feet by 4 feet 8 inches, but the tail-plate is 8 feet long and only 16 inches wide. The latter is called, very appropriately, a sluice-plate, and is a truly absurd device for arresting the gold.

In the Homestake mills, both apron- and tail-plate have a slope of 1½ inches per foot, the minimum gradient at which the tables can clear themselves of the pulp. Both tail-plate and apron are dressed each morning, the aprons are cleaned up partially each day, and more completely deprived of their amalgam at the bi-monthly general clean-up, when both the tail-plates and the inside mortar plates are gone over.

The traps are intended to arrest escaping amalgam. The Golden Star mill (see Fig. 4) has two at the head of the tail-plate. They are 18 inches deep. They are preceded by a shallow trap or riffle 2 inches deep, which is stated to do better work because of the more regular passage of the pulp. These traps catch about 1 per cent. of the total amalgam. They are, to quote again,

"Cleaned up every two weeks. the accumulated pyrites are shovelled into buckets and then passed into a pan, which extracts all the free amalgam. The residues from the pan are then fed into a particular 5-stamp battery, provided with a No. :0 slot screen. They are passed through this battery twice and are then sent to the smelter, their final assay value being about $38 per ton.

"The above suggests the Australian practice of mercury wells, particularly employed at Clunes, with the obvious difference that the Homestake traps are not supplied with free mercury. It is claimed that if sufficient mercury is fed into the battery no free gold should escape, and the mercury in the traps would merely serve to thin the amalgam and make it easier of escape. The traps catch concentrates and amalgam only."

Then comes the concentration of the sulphides. Of this, not much can be said. The Deadwood-Terra mill makes no attempt to save the sulphides, since the ore comes from near the surface and contains nothing worthy of supplementary treatment. The Homestake mills and that of the Highland company are using the Gilpin county bumping-table. Seven years ago Prof. Hofman, noting the absence of any effort to save the valuable sulphides, suggested the employment of *Spitzlutten*, supplemented by *Spitzkasten*, preparatory to the re-crushing of the coarse sands and the concentration, on buddles, of the fine. This very sensible advice has not been followed. Instead, however, two blanket houses were erected, and, without any sizing or classification such as should precede all concentration, the blanketings were worth from $20 to $30 per ton. These have been idle for several years, and in their place eight bumping-tables were placed in both the Highland and Golden Star mills, while six were added to the Homestake mill. It is only necessary to add that the two larger mills have crushing capacities of 560 and 640 tons per day. and the Homestake about 400 tons, to indicate the absurdly disproportionate nature of the equipment, which can only be considered a badly-designed experiment. The results obtained are not by any means a proof of what could be done under proper conditions, as already stated in my previous criticism of this feature:

"During the year ending June 1, 1894, the two Homestake mills produced 915,010 pounds of concentrates whose assay value varied from $5 to $8 per ton. They consist of iron pyrite, arsenical pyrite and pyrrhotite. The ore contains from 3 to 5 per cent of sulphides, but only about 2 per cent. are saved. They are sent by rail to the Deadwood and Delaware Smelter, just below the town of Deadwood, where they are treated at a charge of half their assay value, and converted into an iron matte very low in copper and rich in gold, which goes to the Omaha and Grant Smelting and Refining Company, at Omaha, for further treatment."

At the present time experiments are being made with jigs in order to improve this part of the milling. It is to be regretted that a representative company, such as that operating the Homestake mines, should be so slow to adopt the best metallurgical practice, remaining satisfied with a manifestly inadequate equipment and a thoroughly unscientific treatment until the successful work of a Cornishman, treating its mill tailings a few miles down the creek, emphasized the desirability of doing something better.

Prof. Hofman has described in detail the method of sampling the mill-stuff by simple panning. At the present time the mill-work is checked by sampling the pulp as it leaves the apron-plates and fills a dipper at intervals of an hour for a period of five hours. This is done each afternoon. The gold is determined by fire-assay. No accurate knowledge of the completeness of the extraction can be obtained by so unsystematic a procedure.

The labor-costs are given in the accompanying table:

By comparing this with Table V. in Prof. Hofman's paper, it will be noticed that the mills have been enlarged without a proportionate increase of workmen. In 1888 the Homestake mill employed 20½ men to attend to 80 stamps and the Golden Star 23½ men for its 120 stamps. The chief change is in crusher-

TABLE IV.—*Labor Employed in the Principal Stamp-Mills.*

Name of mill / Number of stamps	DEADWOOD-TERRA. 160.			HOMESTAKE. 100.			GOLDEN STAR. 160.			HIGHLAND. 120.		
	Number.	Shift hours.	Wages, dollars.	Number.	Sift hrs.	Wages, dollars.	Number.	Shift hours.	Wages, dollars.	Number.	Shift hours.	Wages, dollars.
Foreman	$\frac{1}{2}$	A	7.00	$\frac{1}{3}$	B	8.00	$\frac{1}{3}$	B	8.00	$\frac{1}{3}$	B	8.00
Millwright	1	12	3.50	1	12	4.25	1	12	4.25	1	12	4.25
Pipefitter	......	......	......	$\frac{1}{2}$	C	3.50	$\frac{1}{2}$	C	3.50	......	D	......
Enginemen	2	12	3.00	2	12	3.50	2	12	3.50	2	12	3.50
Firemen	2	12	2.50	2	12	3.00	2	12E	3.00	2	12	3.00
Night foreman	......	......	......	$\frac{1}{3}$	F	5.00	$\frac{1}{3}$	F	5.00	$\frac{1}{3}$	F	5.00
Head amalgamators	2	10	3.50	1	10	4.00	1	10	4.00	1	10	4.00
Amalgamators	4	12	3.00	4	12	3.50	4	12	3.50	4	12	3.50
Crushermen	3	10	2.50	2	10	3.00	2	10	3.00	2	10	3.00
Oilers	2	12	2.50	2	12	3.00	2	12	3.00	2	12	3.00
Feeders	3	12	2.50	2	12	3.00	4	12	3.00	4	12	3.00
Laborers	1	10	2.00	2	10	2.50	4	10	2.50	3	10	2.50
Total	20$\frac{1}{2}$	......	59.00	19$\frac{1}{16}$	......	64.32	23$\frac{1}{6}$	......	75.32	21$\frac{2}{3}$	......	70.07

men, the number of whom, by the substitution of the large Gates for the small Blake rock-breakers, has been diminished from 5 and 6 men respectively to 2 for each mill. The Golden Star now employs 4 general laborers in place of 2, and the Homestake 2 in place of 1, and this is the only increase following the enlargement of the mill. The engine-men, firemen, amalgamators, etc., remain the same in number, while the general superintendence (foreman) has been diminished for any single mill by giving one man charge of both the Homestake mills as well as that of the Highland Co.'s.

The milling costs, per ton of ore, are as follows:

TABLE V.—*Cost of Stamp-Milling in the Black Hills, South Dakota.*

A.—THE HOMESTAKE MILL.			
Year	1887–1888.	1888–1889.	1893–1894.
Number of stamps	80.	80.	80.
Tons treated	96,790	106,780	104,995
Labor	$0.2561	$0.2395	$0.2543
Supplies	0.0130	0.0045	0.0105
Water	0.1729	0.1562	0.1986
Wood	0.2766	0.2230	0.0597
Coal	....	....	0.1784
Machinery	0.0922	0.0893	0.1097
Oil	0.0109	0.0084	0.0034
Candles	0.0016	0.0014	....
Quicksilver	0.0103	0.0053	0.0083
Lumber	0.0070	0.0139	0.0167
Timber	....	....	0.0155
Total cost per ton of ore	$0.8406	$0.7415	$0.8551

B.—THE GOLDEN STAR MILL.			
Year	1887–1888.	1888–1889.	1893–1894.
Number of stamps	120.	120.	160.
Tons treated	146,565	161,755	204,215
Labor	$0.2138	$0.1755	$0.1556
Supplies	0.0079	0.0044	0.0121
Water	0.1712	0.1622	0.2043
Wood	0.2739	0.1959	0.0346
Coal	....	....	0.1637
Machinery	0.1220	0.1088	0.1057
Oil	0.0084	0.0066	0.0021
Candles	0.0014	0.0014	....
Quicksilver	0.0252	0.0082	0.0071
Lumber	0.0054	0.0088	0.0160
Total cost per ton of ore	$0.8292	$0.6718	$0.7012

Prof. Hofman gave the costs for the year ending June, 1888.

It is remarkable how little the increased crushing-capacity of the mills, due to additional stamps, has diminished the costs per ton, and this in spite of the general cheapening of material inseparable from the much improved communication between Deadwood and the centers of industry. This applies especially to the Homestake mill. In the case of the larger plant the diminution of 12 cents per ton, as compared with 1887–88, did not follow the addition of 40 stamps, since previous to that the cost had been 3 cents less. Much of this discrepancy is traceable to the fact that the wood, fuel, water, castings, foundry-work, etc., are supplied to the mill by subsidiary companies.

As to the efficacy and completeness of the mills as ore-reduction plants, I find myself (always excepting the feeble attempt at concentration) to be very favorably impressed. It has been urged by unfriendly critics* that "haste and waste" is the chief characteristic of the milling-methods; but this has been said, I venture to believe, without proper regard to the conditions of the case.

Something has already been said, in the early part of this communication, regarding the mode of occurrence of the ore. The immense size of the ore-bodies and the extensive nature of the mine developments justify the scale of the Homestake Co.'s operations. The ore is mixed with a large proportion of country-rock—in fact, it is for the most part only gold-bearing schist, without any marked boundary or any very noticeable difference between what is worth exploitation (and consequently ore) and what is unprofitable to mine (and therefore waste or "country"). Of the sulphides, the most favorable association is that of arsenical pyrites. But the amount of sulphides present is small, and the gold is not very closely attached to them, so that the ore is notably "free-milling." In fact, in my opinion, it is more docile than even the quartz of the main Californian gold-belt. The results obtained in the mills confirm this view.

However, it is hard. The wear and tear of shoes and dies indicates this. The shoes wear at the rate of from 36 to 37

* Recently by Mr. C. G. Wamford Lock in "Gold Mining and Milling in the Black Hills," read before the Institution of Mining and Metallurgy, London, January 16, 1895.

pounds per 100 tons of ore, and the dies at the rate of 44 to 48 pounds. This represents a cost of only about 2 cents per ton, at the prices which the Homestake Company pays, as compared to 4 cents at Angels Camp* (California), 5½ cents in Gilpin county (Colorado), 4½ cents at Bendigo and 5½ at Clunes (in Australia), 9 cents at Grass Valley† (California), and 13 cents at Mammoth,‡ Pinal county, Arizona. These figures are instructive, but they depend largely on prices and freights. The following statement of wear in pounds of iron per 100 tons of ore is a better guide: At Angels Camp, 45 pounds; Grass Valley, 90; Gilpin county, 58; Bendigo, 144; Clunes, 157; Mammoth, 76; as against an average of 80½ pounds at the Homestake. The ore of Grass Valley is very hard indeed, while the lesser hardness of the material treated by the Australian mills is much more than offset by the absence of rock-breakers. The ore of Angels Camp and Gilpin county is comparatively soft.

At present the Homestake mills save, as far as I could learn, somewhere about 75 per cent. The ore is low-grade. During the year ending June, 1894, the Homestake and Golden Star mills treated 309,210 tons, yielding $1,390,610, equivalent to $4.49 per ton. Having in regard the low tenor of the ore, the immense reserves of it and the evident intention not to work for the good of posterity, it seems, from a commercial standpoint, a very proper thing to treat it with the utmost dispatch and rush it through the mills. Of course, slower treatment would give a higher extraction, but this would not compensate for the greater cost per ton due to diminished capacity. Moreover, the results are not bad; 75 per cent. is an extraction above the average even in mills treating a fraction of the quantity crushed per stamp in this district. The after-treatment, the saving of the sulphides, is a serious error, and the arrangement of the plates§ is a blemish; but the excellence of the design of the mortar, the ample rock-breaker capacity and the general arrangement of the mills is such that, taken as a whole, the milling practice is one of the best examples of the cheap treatment of a large mass of low-grade ore.

---

* In 1891. † In 1892. ‡ In 1893.

§ Placing one tail-plate below two aprons having a combined amalgamating-surface twice as great as the tail-plate.

II.—Grass Valley.

Mr. Abadie, in his excellent description of the work at the North Star mill, did not concern himself with the methods of his neighbors at Grass Valley. This was doubtless largely due to the fact that the mill of which he had charge was generally acknowledged as representative of the practice of the district.

The following additional matter, based on visits to this district made in December, 1886, May, 1891, and July, 1893, may, however, render the description of the milling-practice more complete.

The first mill erected in California was not built at Grass Valley in 1850, as stated* by Mr. Abadie. Mariposa county claims that distinction, and accords it to a mill of 8 stamps, each in its own separate mortar, erected on the Mariposa estate late in the summer of 1850.† It was not till the following January that a mill was erected on the west bank of Wolf Creek, nearly opposite the site of the present Empire mill at Grass Valley.

Mr. Abadie's drawings and photographs very completely illustrate the position and interior arrangement of the North Star mill. The drawings are particularly valuable. The most notable change made in the North Star mill since my first visit in 1886 has been the new arrangement of the amalgamating-tables. When going through the mill for the first time, I noticed the inadequacy of the amalgamating-surface and the uselessness of the narrow sluice-plates. Since then the shaking-tables have been replaced, in the case of the last two batteries added to the original 30 stamps, by new wide plates which are nearly 16 feet long. The old short apron and narrow sluice-plate have been thrown out. The other six batteries were, at the time of my last visit, and, it would appear,‡ up to the time of Mr. Abadie's description of the mill, still provided with the primitive apparatus which was first criticized by me nine years ago.

This is the great blemish of the North Star plant, which has followed in this respect the design of the typical Californian

* *Trans.*, xxiv., 208.

† Information derived from Mr. Melville Atwood and others of the pioneers.

‡ Page 215, vol. xxiv.

mill, whose narrow sluice-plates are a remnant of the apparatus originally borrowed by the quartz-miner from the placer-digger.

In order to discuss this question, a detailed description* of the passage of the pulp on its discharge from the mortar will be necessary. The screen-frame, which is 4 feet 4 inches long and 18 inches wide, has four partitions dividing the issue into five portions. Each division of the screen surface is 9 inches wide and 12¾ inches high. While this construction strengthens the screen, it robs it of 1 square foot of discharging area, and is not therefore to be commended.

The pulp then drops six inches and strikes the battery-plate, which is 4 feet 2 inches wide and 18 inches deep. It covers an iron apron which is bolted to the mortar.

Then there succeeds a trough from which the pulp passes through a distributor, consisting of a vertical iron partition pierced by 20 ¾-inch holes. Then follows a drop of 3½ inches to the apron-plate. The latter is 4 feet 5 inches wide † for 2½ feet, and then becomes narrowed for the remaining 2 feet, finishing with a width of 22 inches. Then come the sluice-plates, 22 inches wide and 12 feet long. They had a slope of 1 inch per foot at the time of my visit, but according to Mr. Abadie, the gradient is usually greater, viz., 1¼ inches. The aprons are given 1½ inches per foot.

Then come the copper-lined shaking-tables. The new plates are 16 feet long, of which the upper portion, of 2½ feet, is 53 inches wide and represents the former apron, and the remainder is 46 inches wide, representing an enlargement of the sluices.

Mr. Abadie says that these new plates "cannot be too highly recommended," which is quite true when we contrast them with the old arrangement; but one may be permitted to ask, Why this narrowing of the plate from 53 to 46 inches? I take the liberty of emphasizing this matter because it has for many years seemed to me that the California stamp-mill, otherwise the best

---

* Much of which appeared in my article on "Grass Valley," published in the issues of May 19 and 26, and June 2, 1894, of the *Engineering and Mining Journal*, New York.

† Some of my measurements, obtained by actual measurement in the mill, vary slightly from those since given by Mr. Abadie.

machine of its kind yet evolved by the ingenuity of man, suffers seriously from an unsuitable arrangement of the amalgamating surface. In tracing the evolution of the apparatus of the stamp-mill it will be found that the first gold-saving methods were modeled after placer-mining practice, and that in the term "sluice-plate" lingers the evidence of the transference of the sluice-box from the gulch into the mill. The arrangement of 4 feet of plate as wide as the mortar, followed by 10 or 12 feet of plate somewhat under 1 foot wide, was almost universal in California a few years ago, one mill copying another apparently without inquiring into the object of the arrangement. At the North Star an examination of the sluices showed the edges of copper plate to be abraded or scoured by the swift passage of the pulp.

The philosophy of the sluice-plate is not evident. The battery and apron have caught the coarser gold—that which it is easiest to arrest—and the object is to prevent the escape of particles which, because they were fine and difficult to catch, have not been stopped. Hence, the sluice-plate is put in, but it is more a launder for the convenient conveyance of the pulp to the vanners than a gold-saving device. How can we expect to catch gold which could not be arrested on a wide plate, by passing it over a very narrow one? The quantity of water and crushed ore is still the same, but it is crowded into a much lessened space, the speed of its passage is increased, and the depth of its flow augmented. In some mills the grade of the sluice-plate is actually greater than that of the apron.

Therefore the ordinary practice should be reversed, the apron should be succeeded by a wider rather than a narrower plate, additional facilities for the catching of the gold should be given by spreading the pulp so that every opportunity is afforded for its contact with the amalgamated surface of the plate.

The necessity for a uniform depth of battery discharge is hardly appreciated at Grass Valley. At the North Star no serious attempt was made to regulate it, so that it used to vary from 2 to 6 inches. Since my last visit it has become the practice to introduce cast-iron plates 2 inches thick underneath the dies. They have lessened the difference between the minimum and maximum depths of discharge, according to Mr. Abadie,* to 2 inches. Although his account does not state the fact, yet the context would indicate that these plates are introduced after the dies have been worn down, thereby restoring the height of the issue. At the Empire old dies used to be placed underneath as the dies in use were worn down, and this was found preferable to employing iron plates for false bottoms (as at the North Star), because the latter broke so often. If they are made to fit snugly, this breakage should be no great detriment, as the pieces will remain in place. In addition to these methods, the Empire mill uses the device of fixing wooden cleats to the bottom of the screen outside. The sand banks up inside the mortar, and protects the unused strip of screen until such time as, the dies having worn down, the cleats are removed and the issue lowered. At another mill, the W. Y. O. D. ("Work Your Own Diggings"), the dies are discarded before they have worn down deeply. The remnant is sold to the local foundry, and helps to pay for new dies. This is wise. The consumption of a few pounds of iron is a very small matter compared to the importance of maintaining the conditions best adapted to good work. When the depth of discharge varies between wide limits, the operation of the mill must be irregular. The minimum and the maximum depths cannot be equally favorable to the particular conditions required, and an effort should be made to find the exact depth best adapted to the treatment of the ores of the special mine, and that figure should then be maintained as far as is practicable.

In the matter of screens, the Grass Valley mills have, as it seems to me, taken a retrograde step. The general adoption of punched tin-plate in place of wire-cloth is defended on the ground of economy, Thus, Mr. Abadie says that "the life of a tin screen is about 30 days; the cost, one-fourth that of wire screens." As a rule, at Grass Valley, the brass-wire screens cost $1.55 apiece, while the tin-plate costs 55 cents per screen, the former giving a service of 25 days, the latter of 14 to 15 days. Steel-wire cloth was discarded because that which was used in this district had the defect of a shifting of the horizontal wires. The introduction of tin-plate into the Californian

* *Trans.*, vol. xxiv., p. 212.

mills dates several years back. I first saw it in use at the Utica mill, Angel's Camp, in 1886. In 1893 the Idaho was the only mill in Grass Valley which was not employing tin-plate in preference to wire-cloth. It is the usual custom to burn off the tin upon the blacksmith's forge, with the idea that the iron plate becomes annealed and toughened. Since tin amalgamates, its removal prevents the adherence of mercury to the screens.

The cost of screens per ton of ore varies from ¾ to 1 cent per ton, an item of expense which is trifling when compared to the importance of getting a screen which will properly size the pulp and aid in maintaining the conditions most favorable to good amalgamation. As between punched tin-plate and wire-cloth, the advantage in cheapness possessed by the former need not be considered unless it accompanies other more serious advantages; the difference of half a cent this way or that is as nothing when measured against good milling.

No one at Grass Valley had, as far as I could learn, made any careful experiments to determine the action upon the pulp of the use of the two varieties of screen. One would naturally expect* a more free issue and a more uniform crushing when using wire-cloth in place of punched plate,† because the former has a discharge-area approximately one-half of its surface,‡ while the latter, having apertures of the same size, has less of them per square inch.§

In making such tests it is necessary to be particularly careful, not only that the same kind of ore is fed and that the mortars are the same, but also that the shoes and dies are in the same state of wear, so that the depth of discharge is equal in both batteries. Moreover, attention should be paid to the possibility

---

* The only results of experiment published are those contributed by Mr. Thos. H. Leggett to the *Eng. and Mining Journal* of June 30, 1894, where it is shown that as between a round punched tin, No. 0 screen and a 40-mesh steel cloth, the former made nearly 11 per cent. less fines (passing through a 100-mesh screen) than the latter.

† A recent test at the Mammoth mill, Pinal county, Arizona, showed that as between a No. 6 slot-screen and an equivalent wire cloth, 24-mesh and 26 B. W. G. wire, the latter crushed 20 per cent. more than the former.

‡ With a width of mesh .027, a thickness of wire .01, and a gauge (B. W. G.) of 33, the discharge-area is just one-half the total surface.

§ See also *Trans.*, xxiii., 563.

of more fine stuff finding its way into one mortar than into the other. The outer batteries of a stamp-mill, where the rock-breaker is in the center, receive more than their share of the fines.

The milling-practice of the Grass Valley district has undergone important changes during the past ten years. The introduction of an inside amalgamating-plate—first done at the North Star in 1888—marked an important departure from the extreme type of fast-crushing California battery. The tendency to increase the percentage of inside amalgamation by deepening the discharge and inserting a chuck-block (which also carries an amalgamating-plate) is a notable feature. Thus from a rapid pulverizer the California mill has been made an improved amalgamator. This older, very shallow-drop mill, unable to use a plate inside the mortar, and relying solely on the outside tables, has become rare on the Pacific slope. It has become a recognized fact that rapid crushing will not compensate for poor amalgamation; that the sooner we catch our gold the better; and that the best feature of the stamp-mill, as compared to other pulverizers, is its capability to combine the crushing and amalgamating apparatus in one machine.

At the North Star and W. Y. O. D. mills nearly twice as much amalgam is obtained from the inside as from the outside of the mortar; while at the Empire the inside extraction is from 50 to 85 per cent. of the total saving.

Heavy silver-plating is recommended nowadays. The North Star plates carry 1 ounce of silver per square foot, those of the Empire have 2½ ounces, while at the W. Y. O. D., the newest plant, the amount has been increased to 5 ounces. One cannot remember too often that it is amalgam that catches gold, and that a good coating of gold-amalgam is better than all the nostrums in creation. In starting a new mill or introducing new plates a good coating of silver helps to get the tables into working order in a short time, the silver being gradually replaced by gold. There is no economy in poorly-plated tables or in short, narrow ones. No mill I have ever seen had too much amalgamating-surface; many of them have had too little. The value absorbed by the plates is an asset of the best kind; and there is no plate placed in a mill but will absorb some gold and be worth more when it is worn out and discarded than when it was first put in place. The following fact* will be of interest in this connection. The old plates of the 60 stamp-mill of the Montana Co., Ltd., at Marysville, Montana, after 4½ years' steady work, 1887–1891, were scraped and melted down. (This was, of course, after they had undergone the usual periodical clean-up.) The 12 plates, each 54 by 96 inches, yielded $90,-000. One plate gave as much as $8000. Even the small vanner plates,† 16 by 48 inches, yielded $500 each.

The importance of the careful sampling of the tailings as a check upon the mill-work is better appreciated at Grass Valley than in any other locality I have visited. At the Empire mill there is an automatic sampler, invented by Mr. Starr, the former superintendent, which does excellent service. The results indicate a saving of from 85 to 87 per cent., which is fairly representative of the district.

---

* Which I owe to Mr. R. T. Bayliss, the general manager of the Montana Co., Ltd.

† In the neighboring 50-stamp mill.

SUBJECT TO REVISION.

[TRANSACTIONS OF THE AMERICAN INSTITUTE OF MINING ENGINEERS.]

The Veins of Boulder and Kalgoorlie.

BY T. A. RICKARD, DENVER.

(New Haven Meeting, October, 1902.)

COMPARISONS frequently do good service by affording a means of distinguishing the features which are essential from those which are merely accidental. When one reads of the lode-structure of a new, and perhaps distant, gold-field, there arises the desire to compare the characteristics mentioned with those of the regions with which one is already familiar. It will therefore be interesting to refer to the other districts with which Cripple Creek has been likened.

A. THE VEINS OF BOULDER COUNTY, COLORADO.

Telluride-ores have been mined in Colorado since 1872. Previous to 1891, when Cripple Creek was discovered, the principal localities for such ores were the La Plata mountains and Boulder county. The former district is situated in the extreme southwestern part of the State. Boulder is about 90 miles due north of Cripple Creek, in that part of Colorado which was first opened up by the pioneers of 1859. The prevailing rock is the granite-gneiss of the Front range. The veins are notably associated with bands of chloritic gneiss, to which the French geological term of "protogine" was formerly applied. Dikes of quartz-porphyry and quartz-andesite occur at frequent intervals. In this respect Boulder resembles the adjacent well-known district of Gilpin county, but the ore-occurrence does not appear to be as intimately connected with the dikes as it undoubtedly is in Gilpin.

The veins of Boulder appear to be bands of crushed rock accompanying lines of fracture which have been healed by the impregnation of ore. Fig. 1 will illustrate this characteristic type. The Enterprise is the principal lode of the El Dora district, on the western border of the county. The enclosing formation is granitoid-gneiss, traversed by belts of protogine and mica schist. In this instance a band, 8 inches thick, of clay, called "gouge" by the miners, and consisting of rock crushed to the condition of mud, follows the hanging-wall, which appears as a distinct parting, separating the lode, A C, B D, from the enclosing country-rock. Next to the "gouge" comes a well-defined strip, 3 to 4 inches wide, of casing, which is made up of brecciated gneiss so laminated as to resemble a shale. The foot-wall is not very distinct, because the lode-matter merges insensibly into the country-rock. The ore, G H, appears in the form of dark threads of flinty quartz arranged in parallel lines amid the lode-stuff, which is essentially a granular matrix of partially-crushed granite-gneiss. The dark quartz carries finely-disseminated tellurides, chiefly petzite, which renders a width of 2 to 2½ feet sufficiently rich to yield an average of 2 ounces of gold per ton.

The Monongahela vein, shown in Fig. 2, occurs at Sunshine, in the central part of Boulder county, and resembles a score of similar lodes to be seen in the small mines which yield the bulk of gold from this interesting region, and are intermittently worked under lease by parties of experienced miners. Fig. 2 represents the eastern face of the 100-ft. level in the Monongahela workings as seen in April, 1897. The most prominent feature of the section is a dark band (A E, C H) of "hornstone," or flinty quartz, which, on closer examination, is found to be composed of two separate streaks, A F–C G and F E–G H, in both of which specks of petzite are clearly distinguishable. The upper portion, 3 inches wide, has numerous cavities, "vughs," lined with crystalline quartz and marcasite (white iron pyrites). The lower one, 2½ inches wide, is more compact. Between the pay-streak (A E, C H) and the foot-wall (B D) there is a thickness of 7 inches of dark-gray quartz, through which patches and streaks of granular protogine are distributed. This band carries only traces of gold. It marks the transition between ore and country-rock. There is no distinct hanging-wall. A narrow width of crushed country, A C, separates the pay-streak from the enclosing rock. The latter is traversed by several dark threads of flinty quartz, which may be regarded as feeders to the vein proper.

A little further west the pay-streak splits into two dark filaments of hornstone, separated by mottled quartz, and the lode

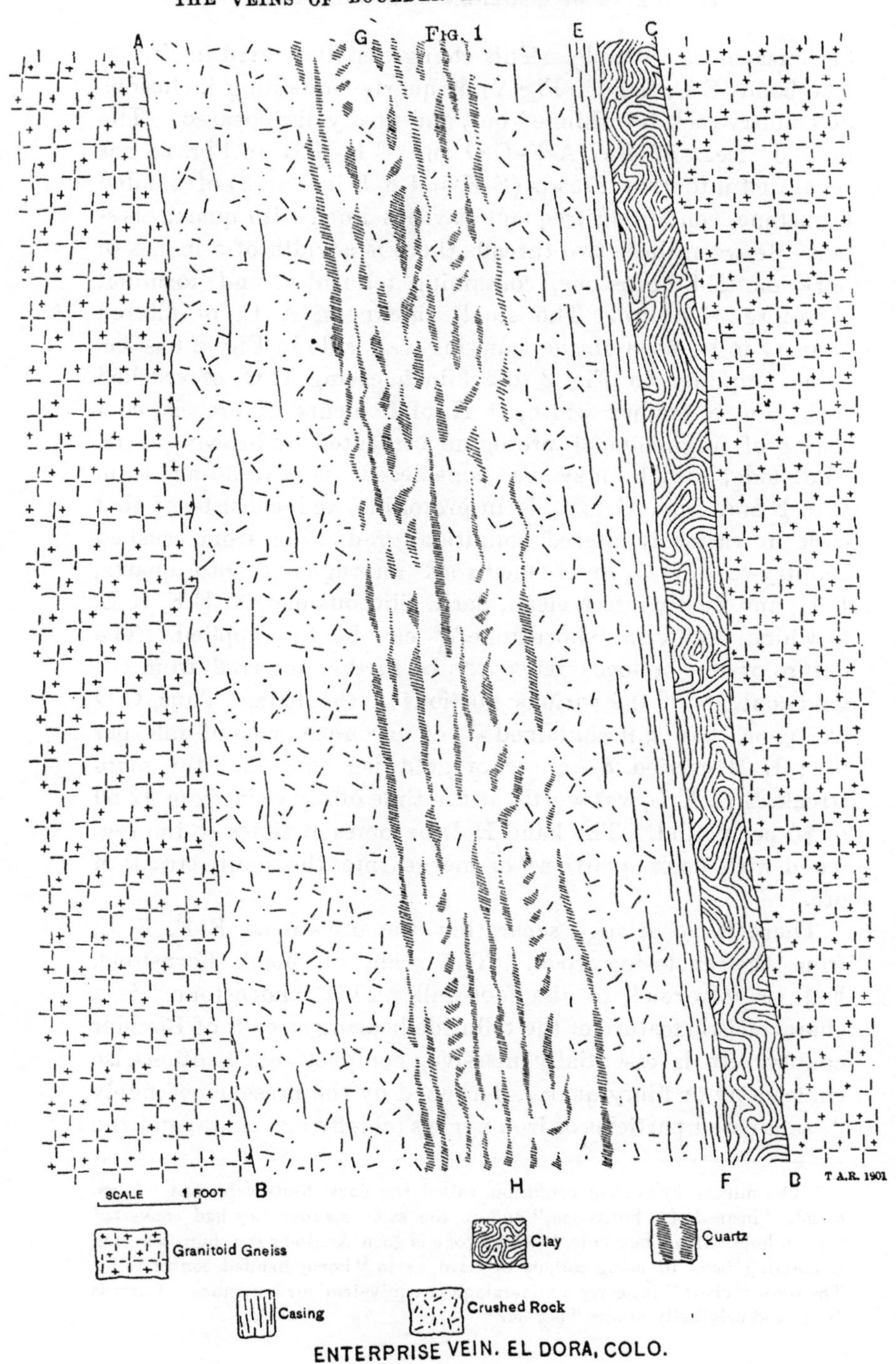

ENTERPRISE VEIN. EL DORA, COLO.

becomes more valuable. This change is illustrated in Fig. 3. The band (E B, H D, in Fig. 2) of quartz containing inclusions of country-rock has thinned out, and finally disappeared. The double streak of ore (A F–C G and F E–G H in Fig. 2) has separated into two threads (E K and F L in Fig. 3) of banded hornstone, containing frequent cavities lined with quartz-crystals. Between these two threads there is a width of 5 inches of dark mottled hornstone, containing tellurides, and forming, therefore, rich ore. The small thickness, A C, of altered country to be seen on the hanging-wall side in Fig. 2 has become amplified in Fig. 3 to a 3-inch casing, C G, of crushed rock, and a further width, D H, of 6 inches of breccia, consisting of fragments of protogine cemented by brown quartz. The lode, which may be considered as extending from C to F and from G to L, is in protogine, and consists of that rock in variously altered condition, graduating from crushed country-rock, C G, to crushed rock impregnated with quartz, D H, and finally into a clean, dark, siliceous matrix, E F, K L, to which the term "hornstone"* can be fitly applied. The nature of the ore-occurrence can be further inferred from the gold-contents of the various portions of the lode. Thus, C G was poor, that is, it contained $\frac{1}{10}$ of an ounce or so of gold per ton; H D carried 8 ounces of gold per ton; and the main streak, E F, K L, was worth, at the time of my visit, from $2.50 to $5 per pound. The band H D becomes at times so impregnated with dark quartz as to merge into the main streak of ore.

There is no selvage separating the ore-streak, E F, K L, from the enclosing rock. The seam of black hornstone, F L, is "frozen" to the foot-wall. This "hornstone" is a characteristic feature of the telluride-bearing streaks of Boulder county. It is essentially massive, compact silica, otherwise chalcedony or flinty quartz, darkened by the presence of finely disseminated particles of iron pyrites (chiefly marcasite) and the

* The miners, by evident confusion, called the dark, flinty veinstone "hornblende," instead of "hornstone," and in the same manner they had converted "vugh-hole" into "bug-hole." Hornstone is good Anglo-Saxon, horn meaning resembling horn in being callous or hard, as in "horny-handed son of toil." The word "chert" is nearly a mineralogical equivalent for hornstone. Chert is Irish, and originally meant "pebble."

FIG. 2

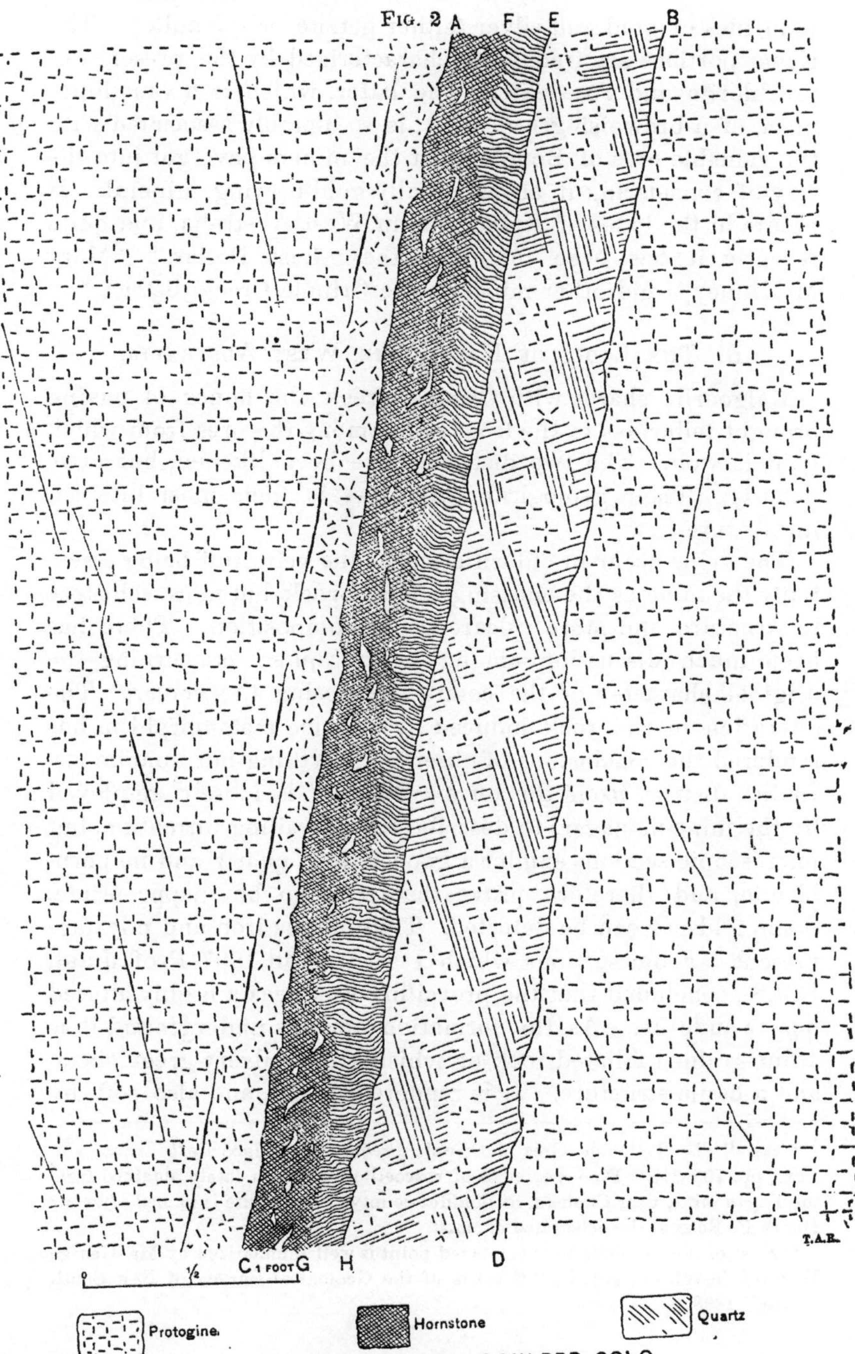

MONONGAHELA VEIN, BOULDER COLO.

tellurides of gold and silver (either petzite or sylvanite). The richer portions are frequently characterized by the presence of prochlorite and roscoelite. The latter, which is a vanadium-mica, of a brownish-green color, is so notably associated with the valuable ores of Boulder that the miners have got into the way of considering it an essentially gold-bearing mineral. It occurs in the lodes of Kalgoorlie, in West Australia, and quite recently it has been found in the "Last Dollar," "Mary McKinney," and other mines in the Cripple Creek district.*

B. The Lodes of Kalgoorlie, West Australia.

Kalgoorlie shares with Cripple Creek the honor of having brought tellurides to the front rank among the ores from which gold is won. The two districts have been likened, but erroneously. Their geological unlikeness is their most interesting feature.

The veins occur in chloritic schist, the vein-stuff being essentially the same as the encasing country-rock, but more schistose in structure and more calcareous in composition. There has been much discussion among petrographers concerning the original character of the country enclosing the veins.† The alteration in the rock induced by extreme metamorphism has rendered the conclusions of observers anything but unanimous. In the district itself the term "diorite" is loosely employed by the mine-managers to describe the prevailing formation, but microscopic sections exhibit a good deal of quartz and no hornblende, and therefore prove that term to be inappropriate. Some feldspar can be detected. Titaniferous iron and mica are present. From sections which I secured in 1897 Prof. Judd, F.R.S., concluded that the prevailing rock was a highly altered quartz-andesite. As the formation appears underground it is a fine-grained foliated rock or schist with a silvery-green sheen and a fissile structure. It is rendered tough, and also soft, by

* See "The Telluride Ores of Cripple Creek and Kalgoorlie," *Trans.*, vol. xxx., pp. 708–718. Fine specimens of roscoelite have been lately obtained from the Logan mine, near Crisman, in Boulder county. Roscoelite is named after Sir Henry E. Roscoe, the celebrated chemist.

† Most of the evidence on this mooted point is well summarized by Mr. George W. Card, in vol. vi., part i., "Records of the Geological Survey of New South Wales," 1898.

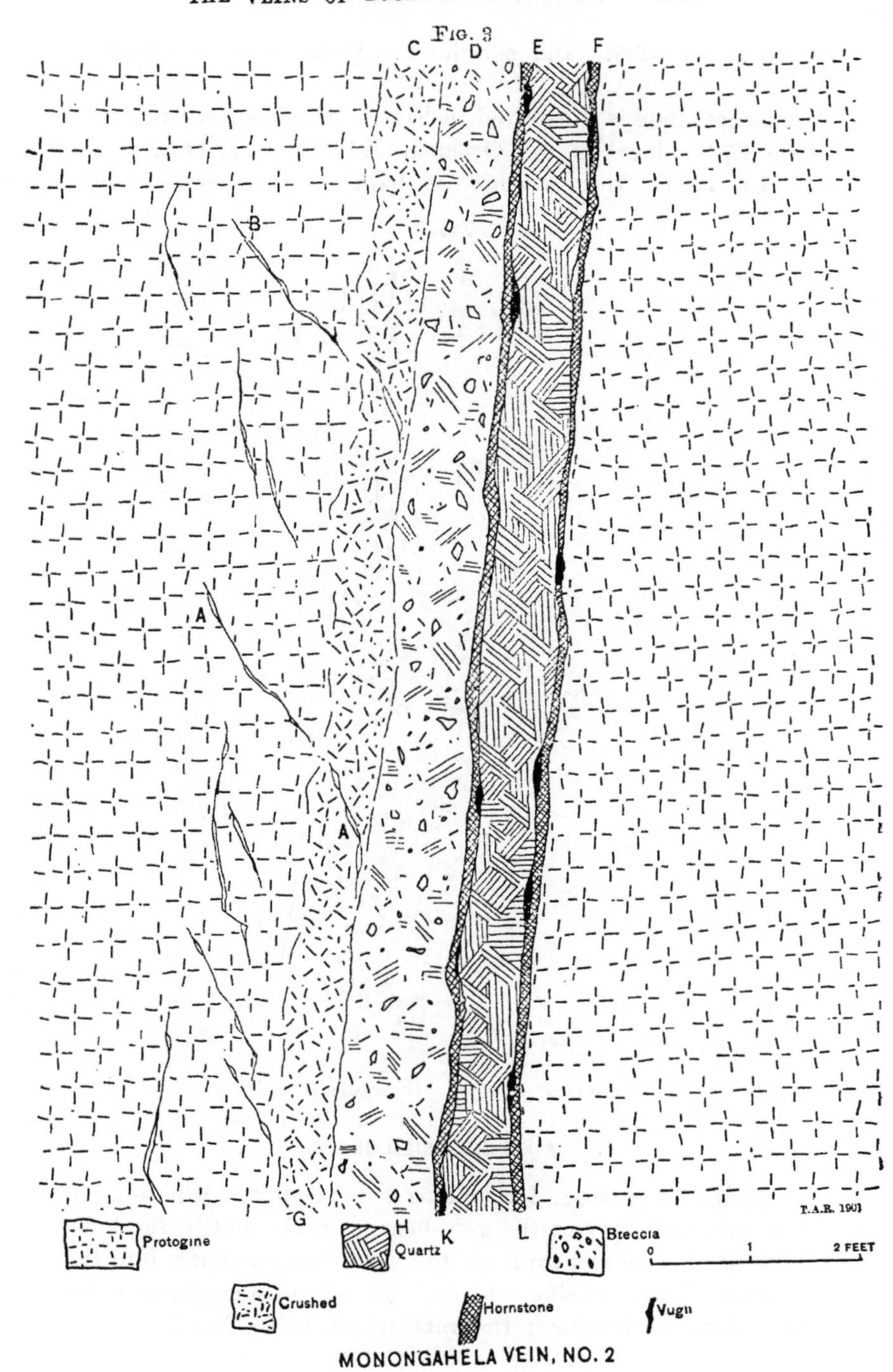

MONONGAHELA VEIN, NO. 2

reason of the calcite and sericite (or hydrous mica) which it carries.

A typical lode is illustrated in Fig. 4, which shows the face of the 270-ft. level in the Great Boulder Main Reef mine, on Oct. 18, 1897. It is a schistose band of rock with a well-

FIG. 4

A KALGOORLIE LODE

defined line of parting, A F. This "east wall," as it is termed, forms one boundary of the gold-bearing rock, that is, the lode, which has no defined limit on the west, but graduates into the country-rock on that side. There is an evident struggle between two systems of fracture; the rock which is ore-bearing, A C,

D F, is sheeted along lines which are nearly upright and parallel to the parting, A F, while the surrounding country has a contrary cleavage. The ore differs further from the surrounding rock in being traversed by veinlets of calcite and granules of quartz arranged along the lines of fracture. Iron pyrites is also liberally scattered through this calcareous vein-stuff. Petzite and calaverite gave the lode here illustrated an average value of 3½ ounces of gold per ton.

The telluride-ores of Kalgoorlie appear sometimes to occur in long, overlapping lenses, as is illustrated in Fig. 5, which represents a series of ore-bodies cut at the third level of the Lake View Consols mine. It is said that the original owners of this now famous property followed the line of one of these long, torpedo-like lenses, and, not knowing the nature of the occurrence, their workings ran out into barren country so as to necessitate a cross-cut, which, unfortunately, was put out on the wrong side, with the final result that they concluded, as the Cornishman would say, "that the lode had just naturally petered out." The ore tends to spread from one plane of foliation in the schist to the next one on the right, with the result that the longer axis of each lens makes a small angle with the strike of the country, and the successive lenses follow one another *en echelon.** An occasional "wall" may be prominent and the ore may follow it for a short distance, but it will leave it for another, equally well-marked. The "walls" or planes of parting do not limit the ore-channel; some of them are within it, some extend beyond it, to become faint as they are followed into the surrounding country. In the space separating the lenses there are stringers and small seams of ore which serve as connecting threads of discovery. They act as "leaders,"† guiding the observant miner from one ore-body to another.

The description of veins in two districts so far apart on the map as Boulder in Colorado and Kalgoorlie in West Australia emphasizes the diversity of structure characteristic of the lodes which carry gold. Both types of veins occupy fractures which

* A military term descriptive of the movement of troops advancing in diagonal step-like succession.

† This is an Australian term which is worthy of adoption. Leader is from the A. S. lædan, to guide, just as lode is from lād, a path. A lode is the occurrence of ore which guides or leads a miner.

Fig. 5

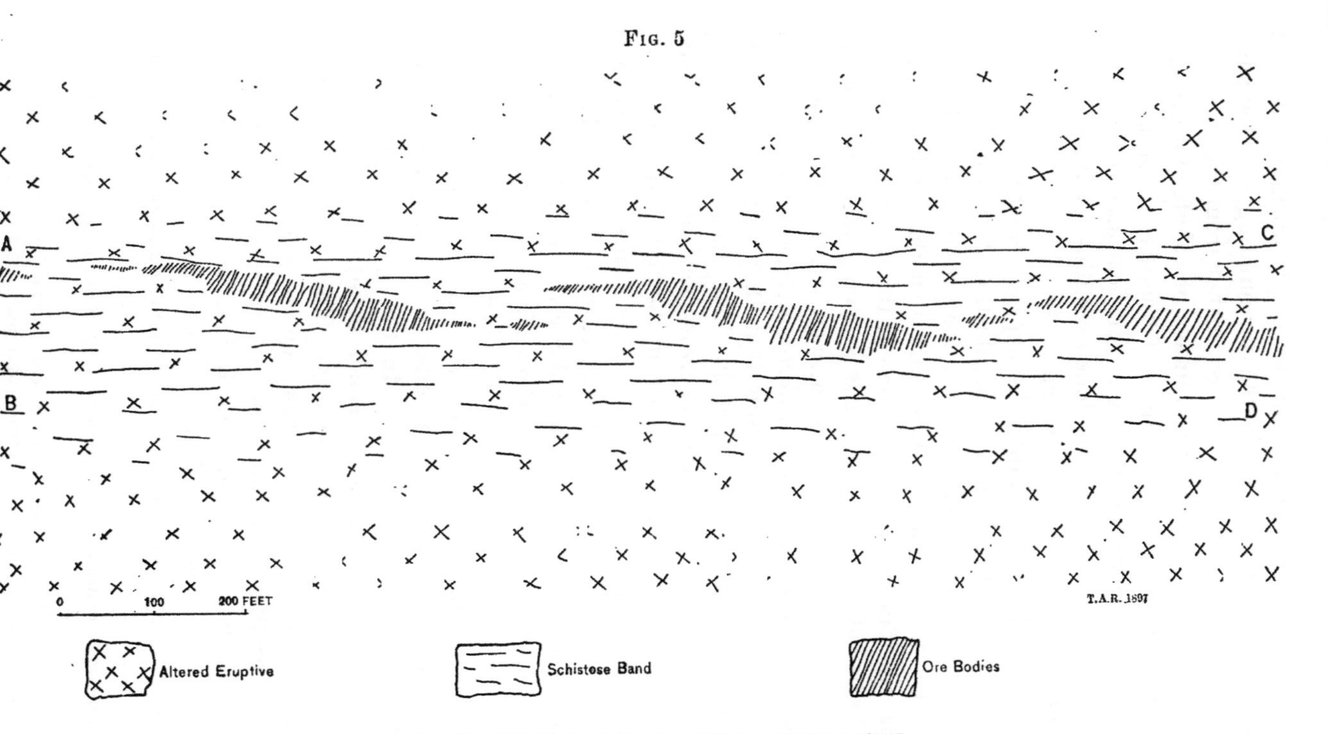

ORE OCCURRENCE IN THE LAKE VIEW CONSOLS MINE

have been healed by mineral solutions; the Boulder type is distinctly a segregation of amorphous quartz in the form of flint, with a cementing of the adjacent granite, which had been brecciated at the time of the formation of the vein fracture: the Kalgoorlie type represents a sheeting of the schistose country without any clean-cut fissuring and without a brecciation of the country, which, being more tough than a granular rock, such as granite, and less fissile, exhibits the effects of strain in a system of parallel sympathetic partings, along which calcite and quartz have been deposited, and, with them, the tellurides of gold. In the Boulder type the tendency is to produce tabular ore-bodies known as "shoots"; in the Kalgoorlie type the struggle between schistosity and sheeting, along a sheer zone, produces "lenses."

NOTE BY THE SECRETARY.—Comments or criticisms upon all papers, whether private corrections of typographical or other errors or communications for publication as "Discussions," or independent papers on the same or a related subject, are earnestly invited.

SUBJECT TO REVISION.

[TRANSACTIONS OF THE AMERICAN INSTITUTE OF MINING ENGINEERS.]

The Indicator Vein, Ballarat, Australia.

BY T. A. RICKARD, STATE GEOLOGIST, DENVER, COLORADO.

(Canadian Meeting, August, 1900.)

IN "The Genesis of Certain Auriferous Lodes"* Dr. Don makes a reference to a curious vein-formation known as the "Indicator," which characterizes a portion of the Ballarat mining district, in Victoria, Australia. During 1890 and again in 1898 I had an opportunity of making a few notes, which, although somewhat belated,† may be worth adding to the observations recorded in the *Transactions* of the Institute.

The country of the Ballarat gold-field consists of the Lower Silurian slates and sandstones, which also enclose the reefs of Bendigo; but it is noteworthy that the successive anticlinal folds, and accompanying saddles of quartz, which distinguish Bendigo are not characteristic of Ballarat, the geological structure of which presents greater complexity.

That part of the district known as Ballarat East became famous early in its history on account of the large nuggets which were found in the alluvium, and subsequently it became further known because of a peculiar persistent black seam, traversing the bed-rock underlying the alluvial deposits, which was found to be associated with rich bunches of gold-ore, not only in the gravel that capped it at the surface but also in the quartz-veins which crossed it underground. The name of "Indicator" was early given. In 1871 Mr. Morgan Llewellyn directed attention to it as influencing the distribution of the gold. In the government quarterly report for December, 1888,

* *Trans.*, xxvii., 564 *et seq.*

† The treatise contributed by Dr. Don reached me while making a second visit to the very localities among which he has gathered so much valuable evidence. Most of the lodes to which he refers came under my own observation in 1889–91, and though I have omitted until now to join other members in acknowledging the great usefulness of Dr. Don's contribution to the science of ore-deposits, it is solely due to the fact that I had hoped long ago to do so in a manner which he would most appreciate, that is, by contributing further data to the investigations in which he has been engaged.

Mr. E. J. Dunn described* a similar occurrence observed by him in the neighboring district of Wedderburn. In September, 1893, the writer contributed a note on the subject.† Mr. Ernest Lidgey, of the Geological Survey of Victoria, made a special report‡ on the Ballarat East gold-field, which was published in 1894. In 1895 and again in 1897 Mr. William Bradford added further data.§

Mr. Bradford, who lives at Ballarat and has made a careful study of the district, describes the Indicator as "a pyritic sheet, varying in width from one-eighth of an inch to about an inch, which has been formed in an almost parallel line with the

Fig. 1

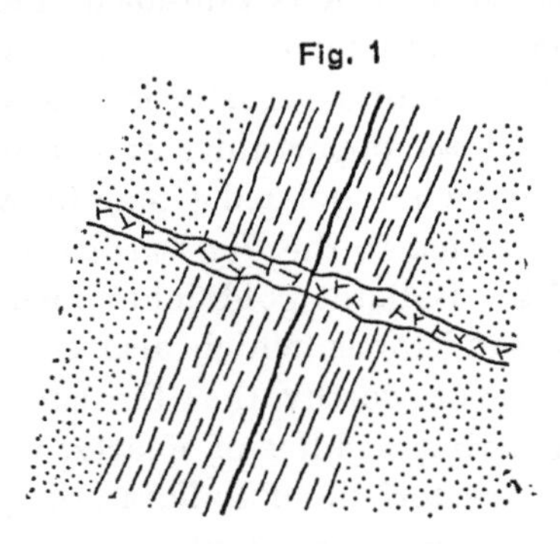

SANDSTONE SLATE

The Indicator. (After William Bradford.)

line of the strata. See sketch." The sketch referred to is reproduced in Fig. 1. Another good illustration from his later paper on the subject is reproduced, with slight modifications, in Fig. 2. The indicator is the dark thread G B, C H, which is dislocated from B to C by the fault-line A D. F B, C E is a flat seam of quartz which is also faulted, with enrichments at B and C, the points nearest to the Indicator and to the later line of fracture, A D. In his description of this occurrence he says:

* "Report on the Country in the Neighborhood of Wedderburn and Rheola," by E. J. Dunn. *Quarterly Rep. of the Min. Dept. of Victoria*, 1888.

† "Certain Dissimilar Occurrences of Gold-Bearing Quartz." *Proc. Col. Sci. Soc.*, vol. iv., pp. 329–331.

‡ "Report on the Ballarat East Gold-field," by Ernest Lidgey, assistant geological surveyor.

§ "The Indicator Feature in Some Gold-Occurrences," by William Bradford. *Trans. Austral. Inst. of Min. Engrs.*, vol. iv.

"Flat make of quartz displaced by a slide break. Nuggety gold was met with at the point where the Indicator intersected the quartz on the higher side of the break. Down the line of the break a layer of gold in the form of fine particles was found, and traced to the point where the lower part of the quartz occurrence abutted on the break."

Mr. Dunn, the distinguished geologist, whose work on the neighboring gold-field of Bendigo is well known to the members of this Institute, describes the Indicator at Wedderburn as being a "bed of rock of dark-grey to black color, and from

Fig. 2

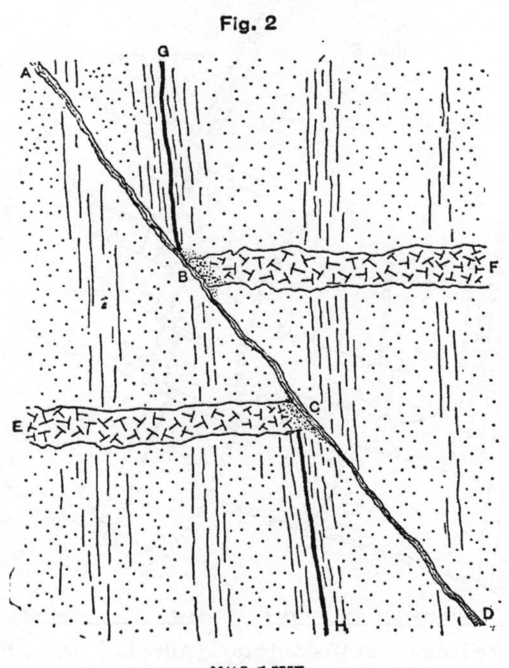

The Indicator Faulted. (After William Bradford.)

5 to 7 inches wide, made up of thinly laminated unctuous clay." His illustrations are reproduced in Fig. 3.

Mr. Lidgey, in his special report, says:

"These indicators are usually thin beds of dark-colored shales and slates, formed of a carbonaceous mud, containing a large percentage of iron sulphide. They are parallel, and, so far as worked, nearly vertical, any change from the vertical being usually due to the presence of faults."

The most persistent of the series, known as "The Indicator," he describes as "a narrow bed of dark slate, usually showing cleavage planes, and containing a large percentage of pyrite, distributed irregularly through it." In Fig. 4 a series of Mr. Lidgey's illustrations is reproduced.

Finally, in Fig, 5, I have added the drawing* exhibited by me in 1893 to the Colorado Scientific Society, from a sketch made at the 700-ft. level of the New Normanby mine on February 25, 1891.

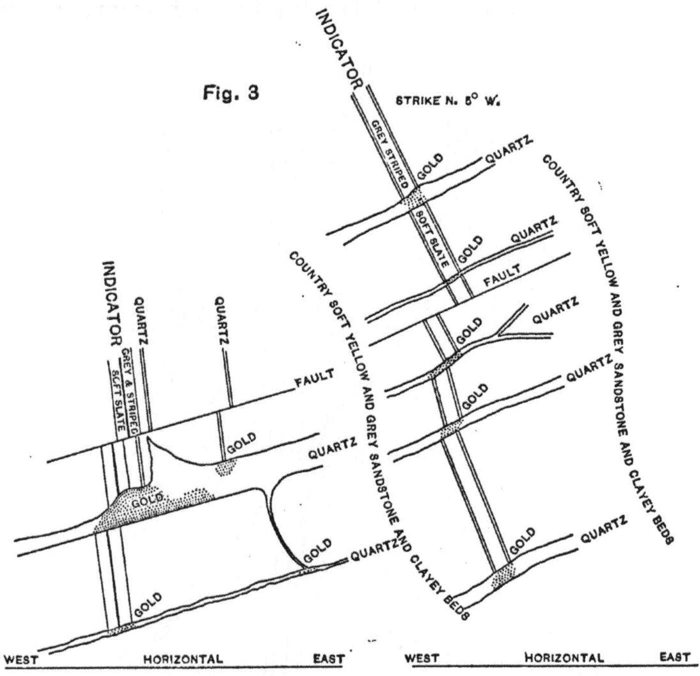

Plans of the Indicator at Wedderburn. (After E. J. Dunn, F. G. S.)

Whatever differences may be remarked in the foregoing descriptions are explained by the changes which this remarkable vein undergoes in its passage for seven or eight miles through a long series of mines, extending from Black Hill to Bunninyong.

Mr. Lidgey's report is accompanied by maps of the Ballarat East district, exhibiting the course of the Indicator through a series of claims belonging to various companies. A part of

* This drawing was reproduced by Dr. Don, on p. 7 of his treatise, and in his paper, *Trans.*, xxvii., 570.

these maps is reproduced in Figs. 6, 7 and 8*, which represent the plan, cross-section and longitudinal section of the ground near the Prince Regent mine. Further evidence is given in Figs. 9 and 10. Fig. 9 is a photograph† taken at the 500-ft. level of the No. 1 Llanberris mine. Fig. 10 is a diagrammatic interpretation which will, it is believed, aid the decipherment of the details in Fig. 9. The photograph shows the Indicator (at A B), and the foot-rule‡ affords a scale of measurement; yet, as in most underground photographs, the dark seams are confused by lines of shadow, and it is not possible to distinguish the slates from the sandstones, both being dark-grey and

Fig. 4

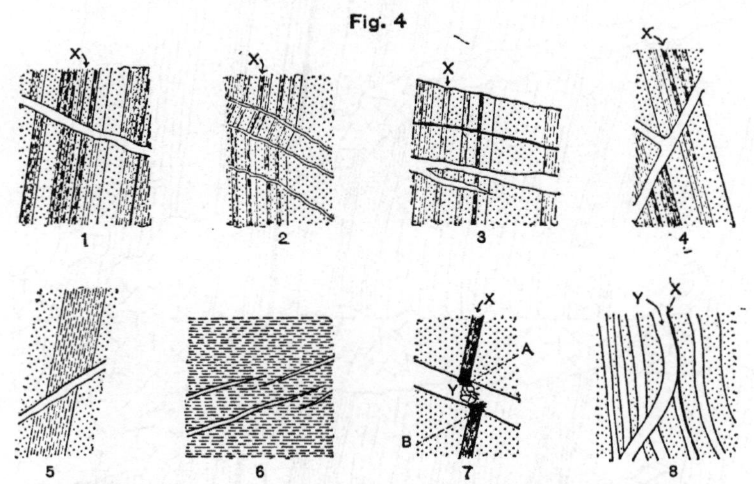

Diagrams Accompanying Report on Ballarat East. (After E. Lidgey.)

both showing cleavages. Nevertheless, it is a valuable piece of evidence, comparatively untainted with the subjective element which vitiates a sketch.

By way of supplementary testimony I add Figs. 11 and 12 from sketches made during my last visit to Ballarat, in January, 1898. In Fig. 11 the Indicator is shown as seen at the

---

* These three drawings must not, I think, be taken too literally; for I note that the stratification of the slates and sandstones is repeatedly shown as unbroken where traversed by cross-courses which, as is well known, fault the country. Such errors are doubtless due to the departmental draughtsman who prepared the drawings, and do not represent any error of observation on the part of Mr. Lidgey himself. It was the dislocation of the Indicator series that the drawings were intended to represent; and this they do satisfactorily.

† Thi excellent photograph was kindly given to me by Mr. Lidgey in 1898.

‡ Which, it should be noticed, is extended so as to cover a length of 2 feet.

770-ft. level of the Prince Regent mine, where it is dislocated by a flat seam of quartz which carries coarse gold at the cross-

Fig. 5

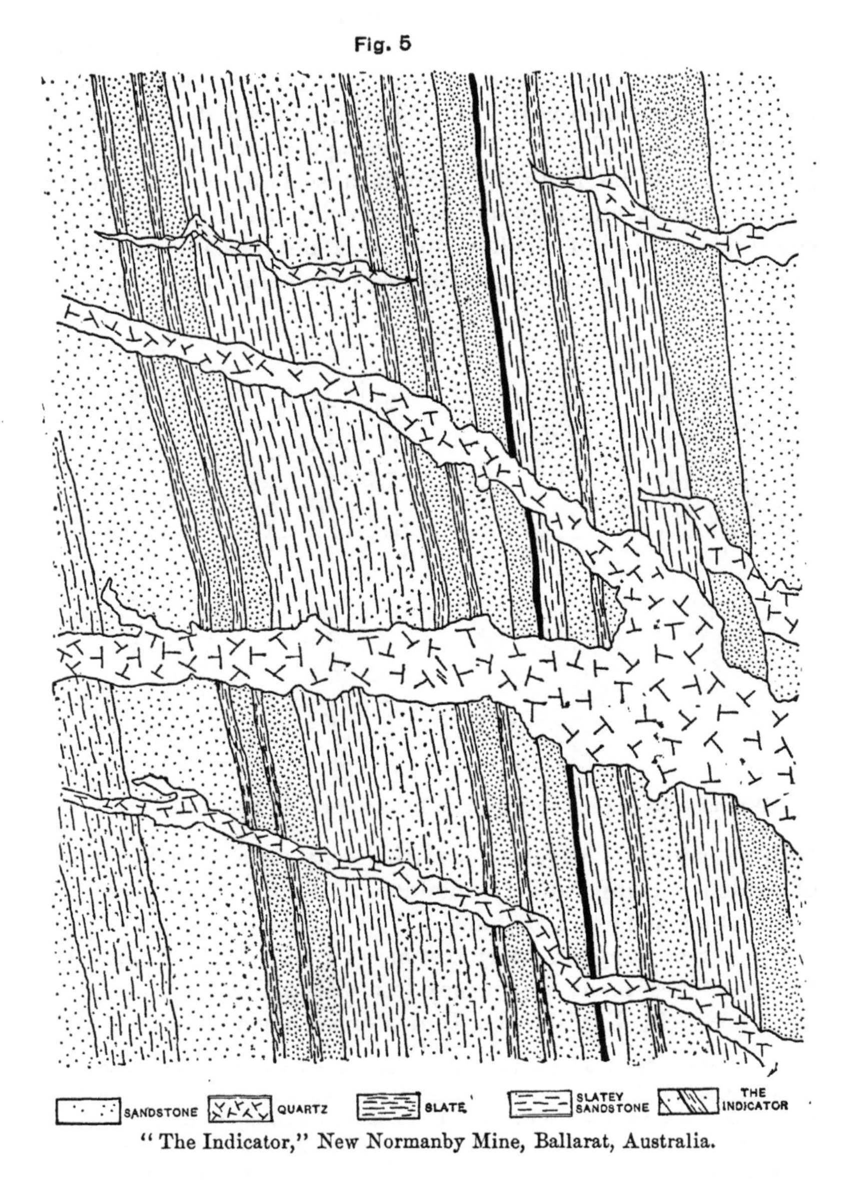

"The Indicator," New Normanby Mine, Ballarat, Australia.

ing. The Indicator evidently conforms to the bedding of the slate band, of which, indeed, it forms a part.

The Metropolitan lode, illustrated in Fig. 12, occurs in an

adjacent portion of the Ballarat district. It has, on a large scale,* certain of the features which characterize the geological structure of the Indicator series, and for this reason it is placed in evidence here. The Metropolitan lode consists of a series of spreading quartz-seams, to be considered as branches thrown out by nearly vertical veins, traversing a band (A B, C D) of thinly-bedded slates and sandstones, and dips slightly westward from the vertical. This band, which is enclosed by a series of massive beds of sandstone, forms a lode, or lode-channel, 80 ft. wide, in which the quartz-seams first referred to rarely exceed 7 or 8 ft. in thickness, and, having an easterly dip, cross the bedding almost at right angles. The cross-veins recognize the structure of the rocks they traverse by a succession of drops as they meet the successive seams of slate. The gold-occurrence is also modified by this structure, the richest quartz being found at the crossing of the slates, and especially upon the lower or eastern side. The quartz composing these seams is white "as a hound's tooth;" but occasionally, in the slate, it is besprinkled with a little arsenical pyrite.

The Indicator is essentially a very thin thread of black slate, which is remarkable on account of its extraordinary persistence, and also because the quartz seams which cross it are notably enriched at the place of intersection. In certain parts of Ballarat East the Indicator has coincided with lines of movement, and having, on this account, undergone attrition, it now appears as a seam of clay. At other places it is so impregnated with iron pyrites as to have the characteristics of a sulphide streak. When it is met by quartz-veins carrying galena and zinc-blende it partakes of their mineralization. At the end of the 700-ft. level in the New Normanby mine it was slightly over one-sixteenth of an inch thick, and consisted of crushed black slate, in which small crystals of pyrites were embedded. In a cross-cut nearby it appeared as a thin vein of quartz. This is an infrequent modification, and one which is found by experience to be unfavorable to the occurrence of rich ore. At the 770-ft. level of the Prince Regent mine the Indicator had the thickness of an English penny, and was freely studded with crystals of arsenical pyrites, tarnished by oxidation.

The belt of slates and sandstones constituting the prevailing

---

* The scale of Fig. 12 represents feet, and not inches, as in Fig. 11.

geological formation of the district is marked by a peculiarly persistent series of thin seams of black slate. The Indicator is the chief member of this series, by reason both of its persistence and of its economic importance as marking a line of particular enrichment. The others are variously known as the "Western Indicator," the "Pencil Mark," the "Nuggety Slate," "Dunk's Slate," "The Streaky Slate," etc.,—an odd nomenclature, originating with the miners, who have been quick to recognize these various "leaders," as they have been encountered in the workings underground. This series is crossed by a later system of quartz-veins and stringers, intersecting the bedding of the sedimentary rocks almost at right angles, and lying, therefore, approximately flat, with a slight dip northeast. Occasionally, large quartz-veins are seen following the planes of bedding; but such as do so, appear invariably to be poor. Those which cross the Indicator (and its companion-seams) are gold-bearing, but only at the places where they intersect the Indicator, and, as will have been noted in the illustrations, fault it. Therefore the workings follow the Indicator in such a way that the rock removed includes 2 or 3 ft.* on either side of it. The intersecting quartz-seams are so frequent that the entire band along the line of the Indicator is treated as a gold-bearing lode. Poor zones occur. Thus the Prince Regent mine has found no pay-ore on the Indicator between the 550-ft. level and the 770-ft. level. The Speedwell mine traversed a barren interval between 400 and 500 ft., but is said to have become profitable lower down. There is, however, at the present time, a general tendency toward impoverishment in the deeper workings of the group of mines occupying the Indicator belt.

The series of quartz-seams† which intersect the Indicator are not to be considered as isolated lenticles, but must be regarded as part of an extensive system of flat veins, which are lateral embranchments from the nearly vertical ones to be seen following the bedding of the slates and sandstones, as, for example, along B D, in Fig. 12. The workings of the mines exhibit frequent lines of faulting, or "slides," which dislocate the entire series referred to, *i.e.*, the Indicator, the quartz-seams en-

* As a rule a few inches on either side of the Indicator represents the entire width of the pay-ore, but 4 to 5 ft. is necessarily removed for convenience of exploitation.

† The Australian miner calls these "makes of spurs."

riched by it, and the enclosing slates and sandstones. These "slides" carry clay and exhibit the other characteristics of planes along which a movement of the rocks has taken place.

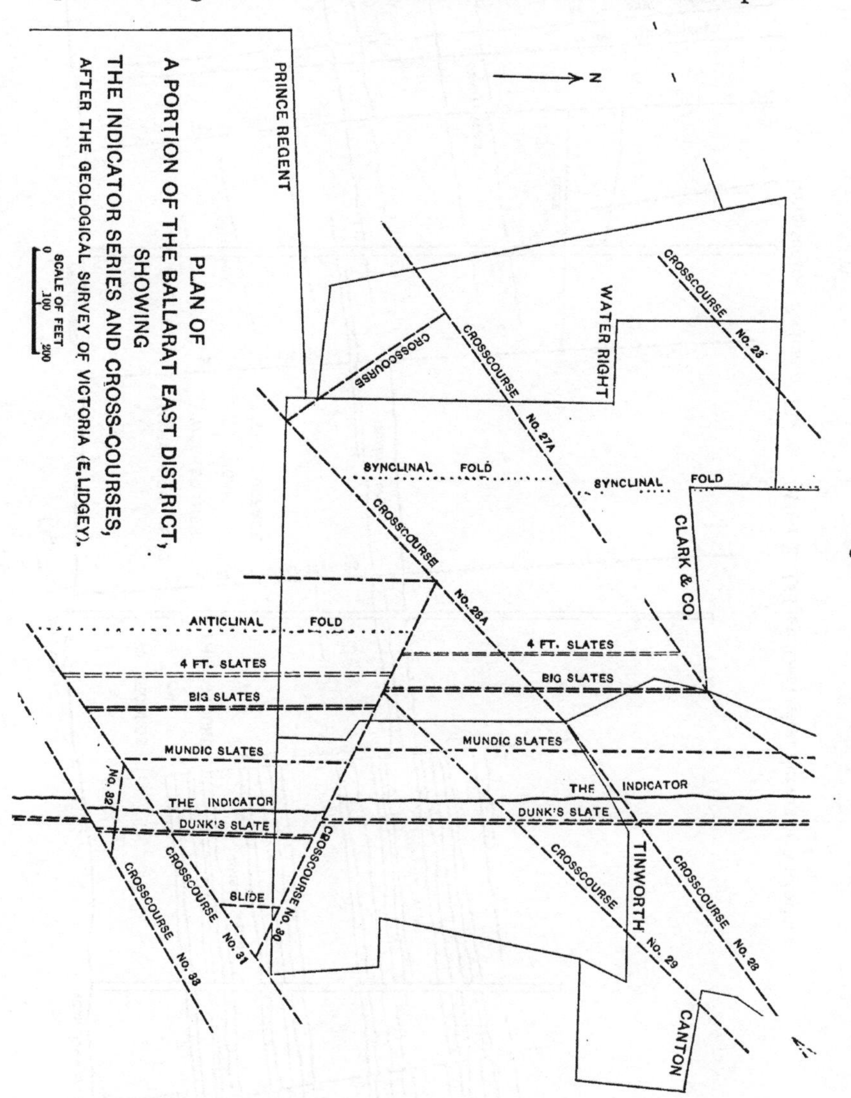

Fig. 6

Mr. William Chisholm, the manager of the Prince Regent mine, who is thoroughly familiar with the geological structure of this part of Ballarat, regards the impoverishment below the

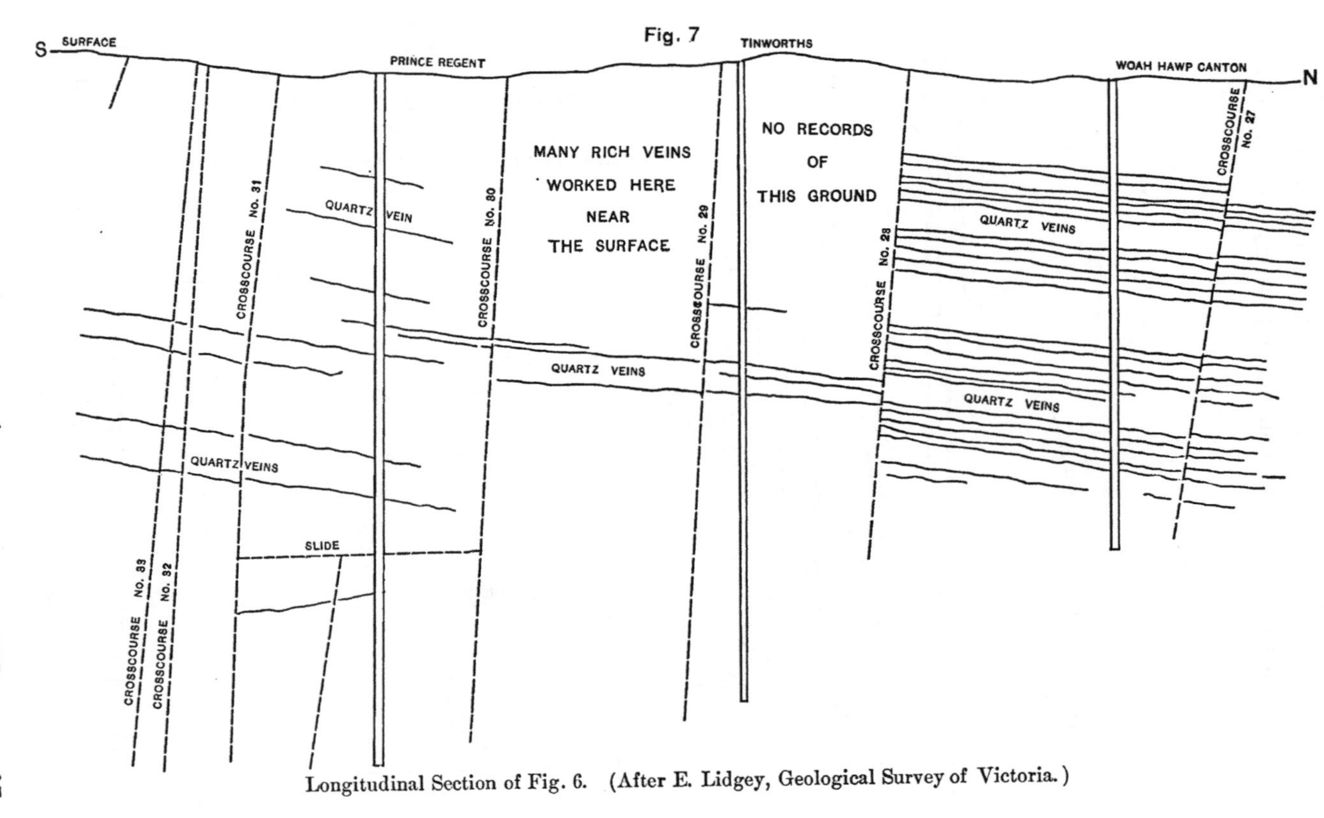

Longitudinal Section of Fig. 6. (After E. Lidgey, Geological Survey of Victoria.)

550-ft. level as connected with the "big slide" which cuts through the shaft 540 ft. below the surface and shifts the Indicator series 55 ft. westward. The managers of the neighboring properties also connect similar disturbances encountered in the workings of their mines with the diminution in the richness of the series of flat quartz-veins. Dr. Don takes this view, and confirms it with analyses of samples taken in the Prince Regent mine, which prove that the solid country enclosing the veins is not itself gold-bearing, while the clay accompanying

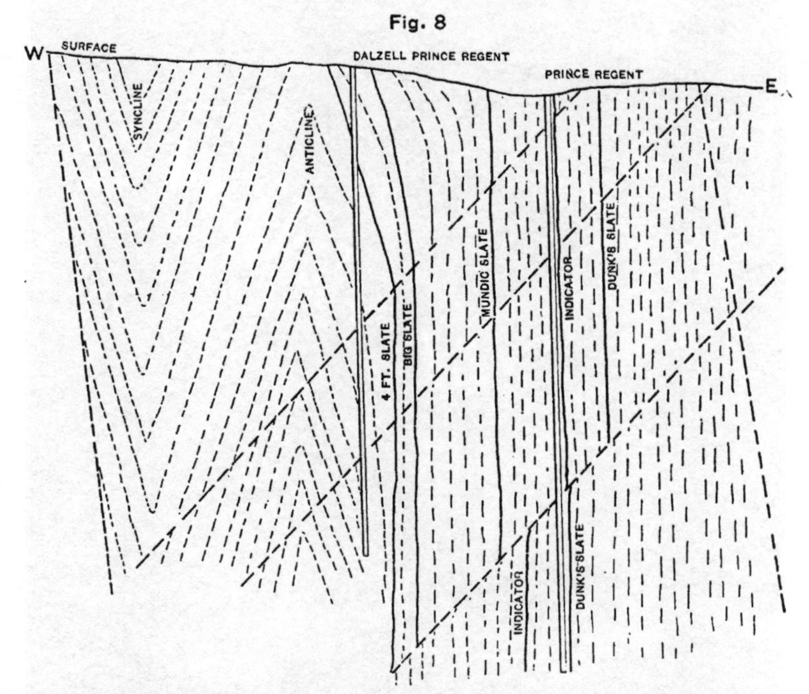

Cross-Section of Fig. 6. (After E. Lidgey, Geological Survey of Victoria.)

the fault-fracture is rich in gold. In both cases, the presence of gold bore no relation to the richness or poverty of the neighboring Indicator.

These views seem to me to be in accord with the evidence; but until the present horizon of impoverishment which characterizes the lower workings of the mines of Ballarat East is succeeded, deeper down, by a horizon of enrichment, one is scarcely warranted in speaking of "zones." My experience is, I believe, in accord with that of most of us, in recognizing the fact that gold-veins often get poorer in depth; and this occurrence is so frequent that a particular explanation, like the one

FIG. 9.

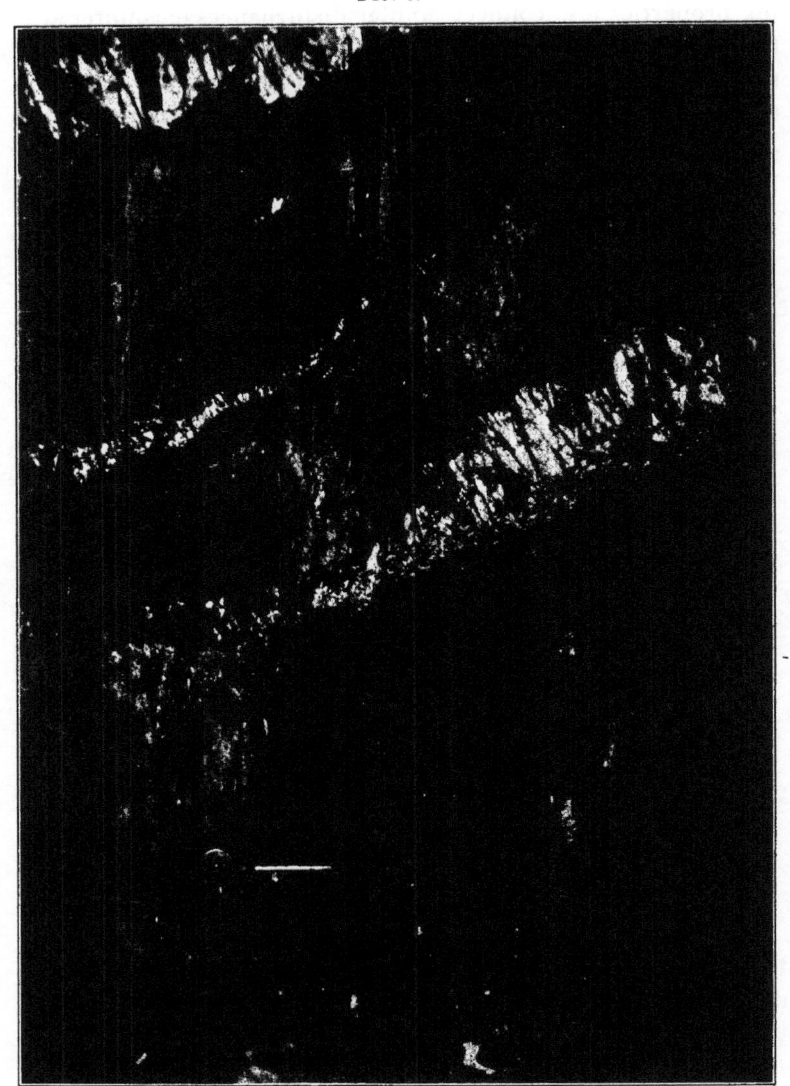

The Indicator, 500-ft. Level No. 1, Llanberris Mine.

above mentioned, gains no force by being connected with such a general fact. A zone of impoverishment is usually only a

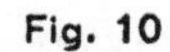

The Indicator. Diagram, Explaining Fig. 9.

euphuism calculated to obscure the frank recognition* of things as they are.

---

* I was glad to read Dr. Don's frank statement (*op. cit.*, p. 596) concerning this matter of impoverishment below the water-level. Science has to do with facts;

The enrichment of the selvage accompanying the fault-planes, as proved by Dr. Don, is very suggestive. If these faults served as passages for the upward circulation of the gold-bearing solutions which precipitated a part of their precious burden, *en route*, and the remainder when the Indicator seams were intersected, then we have a very pretty example of ore-

Fig. 11

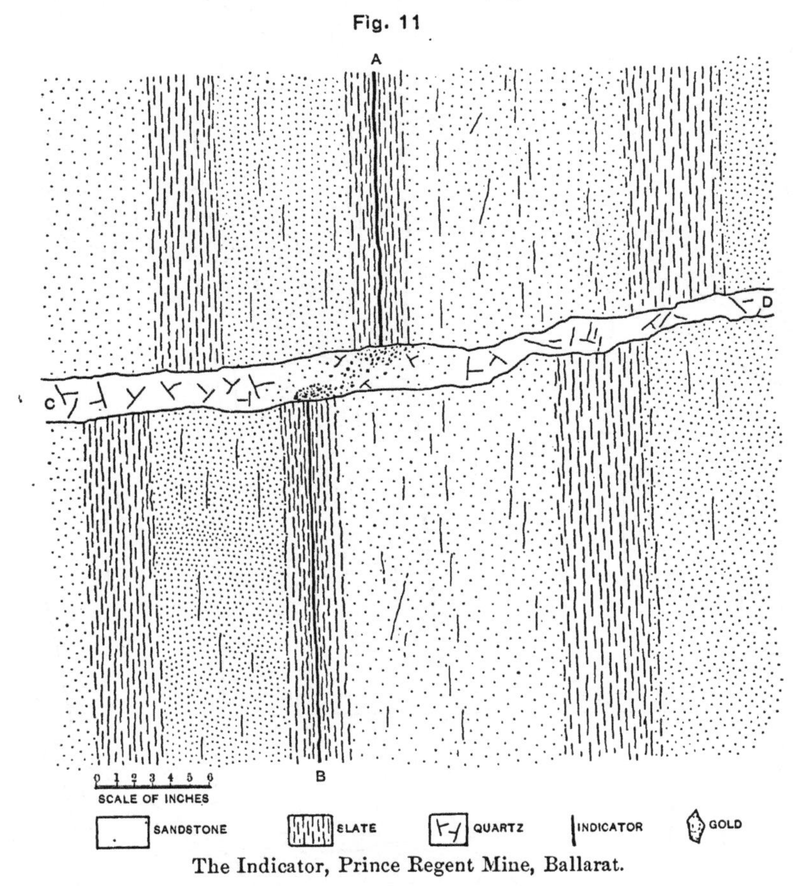

The Indicator, Prince Regent Mine, Ballarat.

deposition. As yet, however, the evidence does not go much further than to accentuate the coincidence between the occurrence of rich gold-ore and a casing of carbonaceous rock. The organic matter observed in the Indicator seam, and also in the clay following the faults, is of particular interest, because it agrees with similar observations made in other mining regions.

---

nor can the industry of mining which is based upon the application of science be aided by fancies, however alluring.

The graphitic slates of the neighboring district of Bendigo appear to have exerted a favorable influence on the deposition of gold. These, like those of Ballarat, are of Lower Silurian age, but the same coincidence marks the Devonian shales of Gympie, in Queensland, and the Jurassic slates of California. Small patches of peculiarly beautiful crystalline gold are found in the black Cretaceous shales of Farncomb hill, at Breckenridge,

Fig. 12

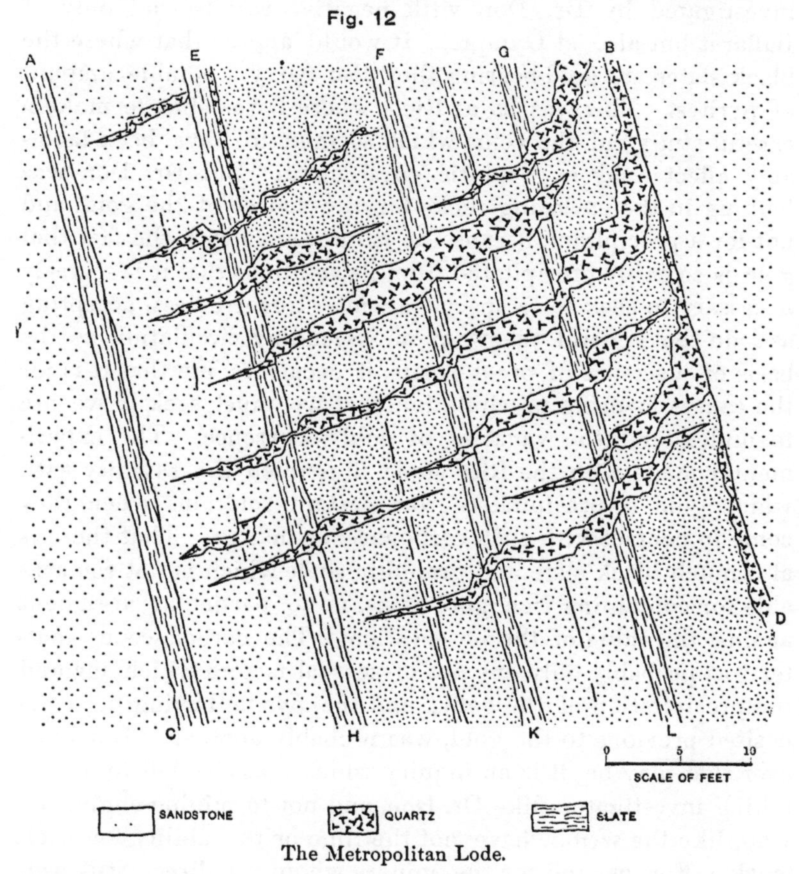

The Metropolitan Lode.

Colorado; and at Rico, in the same State, a series of black shales and limestones of Lower Carboniferous age enclose the richest portion of a very extensive series of veins. In these cases the blackness represents the carbonaceous remnant of vegetable matter which was mingled with the silt and slime out of which the slates and shales were formed. In the case of so persistent and thin a seam as the Indicator we have to suppose an extremely slow sedimentation, occurring in a large sheet of water undisturbed by cross-currents. The clay of the "slides," examined by Dr. Don, must have derived its carbonaceous constituent through mechanical action upon the edges of the slate-beds crossed by the line of faulting; the softer members of the series being, as I remember, the blackest.

This association of gold and carbonaceous matter has been investigated by Dr. Don with negative results, not only at Ballarat but also at Gympie. It would appear that where the black slates are gold-bearing they also carry a notable amount of pyrites. This was so at Rico, where I was able to make a careful study* of the veins which seemed to have been favorably affected by a graphitic encasement. But Dr. Don was looking in the enclosing rocks for the source of the gold, and not for a precipitant, because he is impressed with the fact that gold is so easily precipitated from its solutions that it is scarcely necessary to seek for a special agent. Nevertheless, accepting as I do the general theory of an ascending circulation as the basis of a science of ore-deposits, it seems to me that it is for the causes favoring precipitation that we must seek if we are to understand the vagaries of gold-occurrence. The experiments† made with the Rico shale illustrated the rapidity with which gold is precipitated from certain solutions when they come in contact with such carbonaceous matter. That there is also much black slate and graphitic shale which is not remarkable for containing rich gold-veins, is no particular argument against the value of the observation. The carbonaceous matter was probably only one out of several factors which favored the precipitation of the gold; and the pyrite, having been deposited previous to the gold, was probably another. But however that may be, it is an inquiry which must be left to pains-taking investigators like Dr. Don, and not to mining engineers, who, like the writer, have not the time or the ability for such work. For us, and for the miners whom we direct, it is well to emphasize the fact that certain observed conditions, structural or physical, are favorable to the finding of gold; and if, in our daily work, we can collect further observations of this kind we shall have done something to aid the endeavors of the specialists who are patiently trying to unravel the mysteries which beset the genesis of ore-deposits.

---

* "The Enterprise Mine, Rico, Colorado," *Trans.*, xxvi., 906.

† *Op. cit.*, p. 978.

[TRANSACTIONS OF THE AMERICAN INSTITUTE OF MINING ENGINEERS.]

The Alluvial Deposits of Western Australia.

BY T. A. RICKARD, STATE GEOLOGIST, DENVER, COLORADO.

(Buffalo Meeting, October, 1898.)

I.—General Geological Considerations.

The interior of West Australia is an arid table-land, elevated 1400 feet above the sea. This plateau is flanked to the south by the Tertiary limestones which fringe the Great Australian Bight. It is bordered northward by the Carboniferous beds of the Fitzroy river and westward by the granite of the Darling hills, while to the east this wide area, about 900 miles square, slopes downward imperceptibly into an undulating plain of sand, which stretches with dismal persistence across the boundary of South Australia. The waters of the ocean receded from this tract of land long ago; it is probably the oldest land-surface on the globe, and represents the basal wreck of a much larger continent. Fig. 1 is a map of this region.

The Coolgardie and Kalgoorlie gold-fields are situated in the southwestern part of the region. The rock-formation consists of granite penetrated by diorites and andesites. The latter are occasionally associated with tuffs, which have been readily mistaken for sedimentaries. There are no fossil-bearing rocks, such as would afford a datum-line from which to measure the relative geological age of the prevailing formation. On the extreme edges of the mining territory there are, it is true, remnants of sand-rock which are considered identical with the "Desert Sandstone" of Queensland, determined by Daintree to be of Mesozoic age. But even this formation has evidently been laid down so long subsequent to the underlying rocks that it serves merely to emphasize their much greater antiquity.

In their characteristics and in their relations to each other, the granite and the diorite of the Coolgardie region appear to me much to resemble the Laurentian granite and the Huronian schists of Ontario, in Canada.* Their age can only be vaguely described as Archæan.

On many parts of the earth's surface a long-continued, slow movement of continental uplift, interrupted by intermittent periods of rest or subsidence, has permitted the transfer of land to the sea by the erosion of the exposed parts and the deposition of their detritus as ocean sediment, thus causing the upbuilding of a mountain-system composed of a series of rocks belonging to successive epochs. In this particular region, on the contrary, all diversity is wanting, and a sameness of aspect wearies the observer. In the absence of an elevatory movement, more than sufficient to balance the slow degradation of the higher parts of the region, there has been no compensation for the effects of atmospheric erosion, so that this tract has become a dreary flat, strewn with the sandy wreck of weathered rocks.

The United States offers both a contrast and a parallel. In the Rocky Mountain region the movement of uplift which commenced in pre-Cambrian times has only been interrupted so as to permit of the laying-down of younger sediments; and the degradation of the high places has been compensated, and sometimes exceeded, by an elevation which has resulted in the formation of a mountain mass flanked by a long succession of strata now enclosing a great variety of mineral wealth. The interior of Australia can be likened to the Great Basin, occupied by Nevada and parts of Utah, Idaho and Arizona, between the Rocky Mountains themselves and the Sierra Nevada. There is only one large river in Australia which reaches the sea, namely, the Murray, which rises near the boundary-line of Victoria and New South Wales, and then flows toward the interior, to be saved by a backward sweeping curve, which permits the river at length to empty itself into the sea at the border of South Australia. There are many "lost rivers," like the Carson

* The mass of the granite is penetrated by the mass of the diorite, the latter being therefore the younger; but puzzling evidence is afforded by the fact that the diorite at the contact is sometimes traversed by embranchments of gran te, which are explainable on the supposition that subsequent metamorphism gave the granite a renewed mobility, permitting it to penetrate fractures in the diorite. At Rat Portage, Ontario, the overlying, younger Huronian schists are intercala:ed and penetiated by the older Laurentian granite at certain places along the line of contact.

FIG. 1.

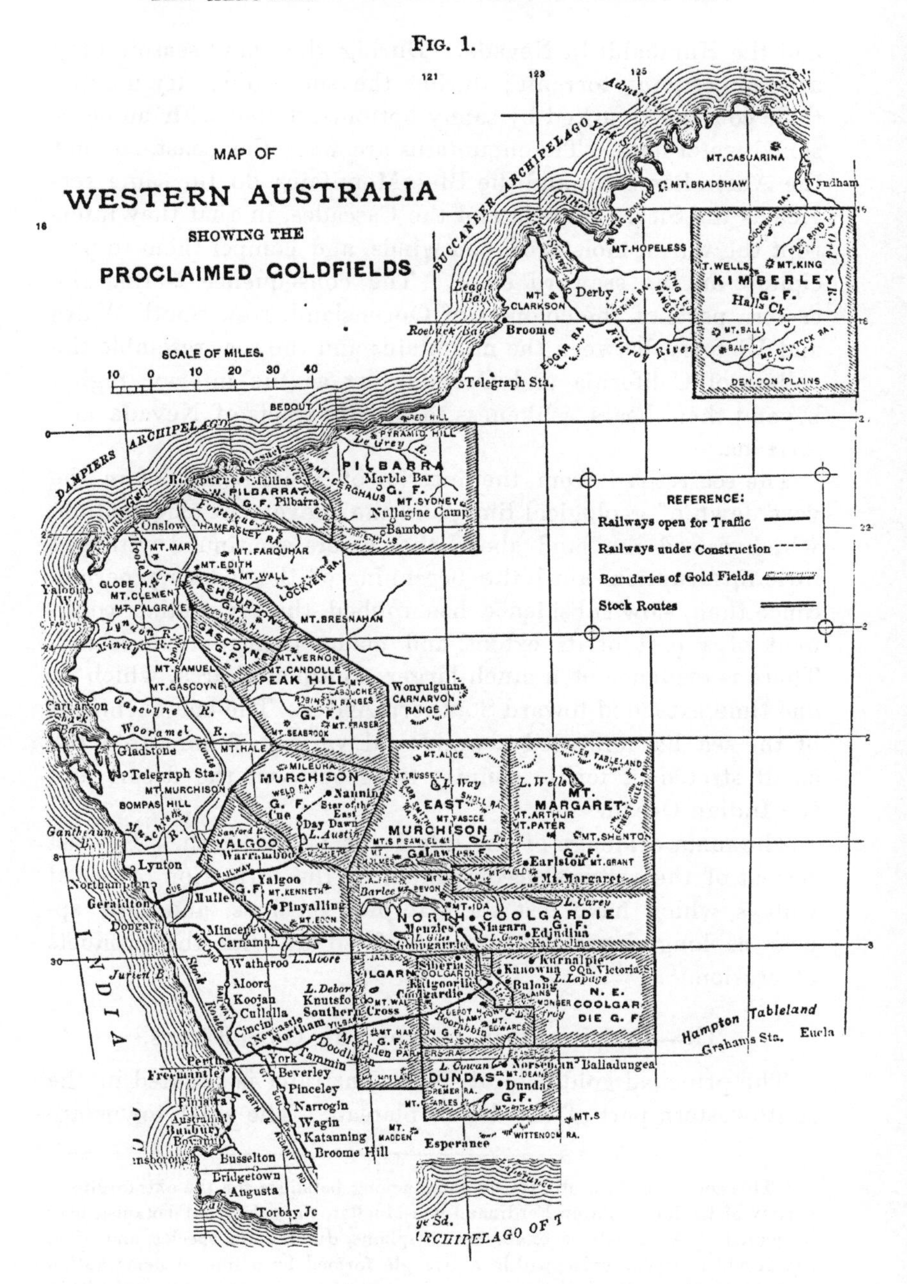
MAP OF
WESTERN AUSTRALIA
SHOWING THE
PROCLAIMED GOLDFIELDS
SCALE OF MILES.
10 0 10 20 30 40
REFERENCE:
Railways open for Traffic
Railways under Construction
Boundaries of Gold Fields
Stock Routes
KIMBERLEY G. F.
PILBARRA
W. PILBARRA G. F.
ASHBURTON G. F.
GASCOYNE G. F.
PEAK HILL G. F.
MURCHISON G. F.
EAST MURCHISON
YALGOO G. F.
MT. MARGARET G. F.
NORTH COOLGARDIE G. F.
N. E. COOLGARDIE G. F.
DUNDAS G. F.
Wyndham
Derby
Broome
Telegraph Sta.
Onslow
Carnarvon
Gladstone
Lynton
Geraldton
Northampton
Dongara
Mullewa
Yalgoo
Cue
Nannine
Day Dawn
Menzies
Niagara
Kalgoorlie
Coolgardie
Southern Cross
Northam
York
Beverley
Pinceley
Narrogin
Wagin
Katanning
Broome Hill
Perth
Freemantle
Pinjarra
Bunbury
Busselton
Bridgetown
Augusta
Torbay Je
Norseman
Dundas
Esperance
Hampton Tableland
Grahams Sta.
Eucla
BUCCANEER ARCHIPELAGO
DAMPIERS ARCHIPELAGO
INDIA

and the Humboldt in Nevada. During the rainy season they are tempestuous torrents; during the succeeding dry months their course is marked by sandy bottoms, dotted with an occasional water-hole. The mountains are near the coast, so that the Australian Alps and the Blue Mountains do the same service as the Sierra Nevada and the Cascades, in that they interrupt the warm, moisture-laden winds, and compel them to precipitate on the seaward slope. The consequence is that the eastern parts of the colonies of Queensland, New South Wales and Victoria, between the mountains and the sea, resemble the valleys of California and Oregon, just as the interior region beyond them bears a likeness to the dry tracts of Nevada and Arizona.

The sea retired from the interior of West Australia in the very dawn of geological time, and the movement of elevation, which raised the land above the waters, continued with but little interruption until the beginning of the Tertiary period. Since then, slow subsidence has robbed the Australian continent of a part of its extent, and made Tasmania an island. There is evidence of a much larger continental area, which at one time extended toward Southern Africa. The encroachment of the sea has crowded a wonderful variety of flora into the small stretch of fertile country lying between the desert and the Indian Ocean.*

The main drainage of the interior is to the south. The last retreat of the sea was accompanied by the formation of broad valleys, which have lost their former outlines, and now appear as long depressions, largely filled up with the products of erosion.

II.—The Physiography of the Gold-Fields.

The principal gold-field of West Australia is situated in the southwestern part of the desert plateau. The chief towns are

* This corner of Australia is celebrated among botanists for the extraordinary variety of its flora. Baron Ferdinand von Mueller, the celebrated botanist, may be quoted: "A marvellous exuberance of plants, different in species, and often gay or odd in aspect, exists within a triangle formed by a line of demarkation drawn from the south of Sharks Bay to the west of the Great Bight; and within this space are chiefly located those species which are exclusively restricted to West Australian territory."

Coolgardie and Kalgoorlie.* They are connected by 350 miles of railway with the coast. During the year 1897 the total rainfall amounted to $5\frac{2}{5}$ inches at the one place and $4\frac{3}{4}$ inches at the other.† In contrast to these figures, it may be added that the rate of evaporation in this region is estimated to be equivalent to 7 feet per annum.

The country consists of a sandy plain, the monotony of which is intensified by a series of alternating low rocky ridges and equally slight depressions, having a northwesterly direction. Most maps indicate the occurrence of lakes and the occasional course of a stream, but these are the mirages of the cartographer. The "lakes" are shallow basins with clay bottoms, in which, during the rainy season, a little water lingers, and the "streams" are sandy channels, where sinking will sometimes tap a trickling flow of brine.

The surface is devoid of vegetation, except in spring, when flowers‡ of a brilliant hue, but with the texture of hay, leap into brief existence. Animal life is infrequent. An occasional bustard may be provoked into leisurely flight, a troop of paroquets throws a momentary gleam athwart the dull gray of the bush, or a solitary kangaroo hops across the trail. These, however, are but infrequent interruptions to the sullen silence of the wilderness.

The real nakedness of the region is hidden by the "bush," consisting of scrub from 20 to 60 feet high, chiefly mulga and ti-tree.§ This covers all things as with a garment (see Fig. 2). The roads are cut through it with the monotonous regularity of a canal. One portion of the journey is but the counterpart of another. The sameness is wearisome beyond words. And

* These localities are not found save on recent maps. Kalgoorlie is in latitude 30° 45′ south and longitude 121° 30′ east, while Coolgardie lies in latitude 30° 57′ south and longitude 121° 10′ east. They are 25 miles apart.

† The rainfall at Denver is 14½; Alexandria, 10; Paris, 22; London, 35; Canton, 39; Calcutta, 76; Vera Cruz, 180; and at Cherrapongee, in Assam, 610 inches per annum.

‡ The "Everlastings," as they are usually called, belong chiefly to the genera Helichrysum, Helipterum, Waitzia, Podolepis and Angianthus. For about three months they appear as magnificent splashes of color, carpeting the desert with splendor. They are wholly devoid of perfume, and have the brittle texture of artificial flowers.

§ Both acacias. The characteristic tree-shrubs of the country belong to the genus Acacia. Many of them have a fragrant bloom in the spring.

when an elevated spot is attained the eye commands, from north to south, from east to west, one dark unbroken sea of trackless bush.

Gold-mining caused this desert to be invaded. The first discovery was made by Anstey, in 1887, at Yilgarn, which is 210 miles east from Perth, the capital of the colony. This started the Southern Cross mining district. Prospectors began to scatter further inland. In 1892 Bayley made the discovery which marked the birth of Coolgardie, and the commencement of an activity which culminated in the mining excitement of 1895. A series of rich finds, scattered over the surrounding desert, gave rise to the settlements of Menzies, Goongarrie, Kanowna, Kurnalpi, Kunanalling, Wagiemoola, and a score of other patches of corrugated-iron hideousness labeled with euphonious aboriginal names. In June, 1893, Patrick Hannan pegged out a discovery-claim at Kalgoorlie,* 25 miles east of Coolgardie. The find which he made was of no particular importance, and the neighboring area, like many others, became the scene of the purposeless digging which was at that time sufficient to give an impetus to a great deal of reckless company-promoting. However, just as, in Colorado, the Mt. Pisgah fiasco of 1884 preceded the real development, eight years later, of Cripple Creek, so the vagaries of irresponsible schemers led to the accidental opening up and the eventual recognition of the magnificent series of rich lodes that have now placed Kalgoorlie among the few great mining camps of the globe.

In 1897 West Australia produced 674,993 ounces of gold, to which Kalgoorlie alone contributed 306,000 ounces. During the same period the mines of the colony paid $2,400,420 in dividends, and out of this total Kalgoorlie is credited with $1,775,000. The growth of the industry is exhibited by the accompanying statistics:

Year.	Kalgoorlie. Ounces.	West Australia. Ounces.
1895,	36,000	231,513
1896,	103,000	281,265
1897,	306,000	674,993

It is estimated that the yield of the colony for the current

* See the interesting paper of my friend, Mr. George J. Bancroft, "Kalgoorlie, Western Australia, and Its Surroundings," read at the Atlantic City Meeting, February, 1898.

year will reach a million ounces, and that of this one-half will come from the Kalgoorlie district.

The development of a group of very rich telluride lodes amid this immense desert country, dotted over with the unhappy failures which were based on small pockets of specimen gold-quartz, did not happen without a sad expenditure of money and human life. With the whisper of every new discovery, crowds of reckless gold-seekers plunged madly into the outer desolation. Eager horsemen jostled the awkward camels, whose swinging gait carried them in turn past the mobs of diggers who trudged wearily to the scene of each successive excitement. One knows not whether to admire the pluck or to deride the foolishness of men who died of thirst and perished of fever in the mad search for gold. The incident known as "the Siberia rush" will be typical of early days. A man came into Coolgardie one night with a story that gold had been found at a locality thirty-odd miles to the north. Hundreds started off on horses or on camels; many went on foot, carrying their billies* and blankets upon their shoulders or trundling their packs in wheelbarrows. Some took the wrong direction, and of these many never reached their destination, but died miserably in the bush. Four hundred eventually reached Siberia.† The only water near the discovery was a soak,‡ seven miles distant. It was soon drained dry by the crowd of diggers. News came to Coolgardie that a water-famine was imminent. The superintendent of water-works, a government officer, instantly despatched a dozen camels§ to the succor of the adventurers. In the meantime, they, realizing the impending danger, had left the gold, and were making for the nearest condenser.‖ Many died on that return journey, and many more would have been lost save for the water brought by the camel-train. But in a few days there was another stampede in another direction. Thus the gold-fields were opened up. *Sic Etruria crevit.*

---

* The "billy" is the tinned-iron vessel, of from 2- to 4-quart capacity, in which the miner makes his tea and does his cooking.

† What a satire is the name! The locality has a mean annual temperature of 78°, and the summer heat is 112° to 120° in the shade.

‡ A "soak" is the morass of the desert, where water has accumulated in a depression, and is got by digging through the sand which covers it.

§ A camel carries two tins, each holding 20 gallons.

‖ All the drinking-water is the product of the "condenser," of which a description is given below. See Figs. 3, 4, 24, and 25.

The peculiar character of these "rushes" is directly traceable to the nature of the gold-occurrence. Gold is found lying on the very top of the ground, and the first surface-mining yields extraordinary profits. The search for the particles of gold scattered over the surface is called "specking." In the early days hundreds of ounces were thus picked up in a few hours by the men first to reach a rich spot. When the cream has thus been skimmed off, the sandy soil underneath is treated, the dirt being winnowed by pouring it from one pan into another. After that, actual digging begins, the shallow deposits being trenched and pitted in the search for those patches of rich ground through which the gold is found sporadically distributed. "Specking" is still a recognized occupation on Sundays,* even at the established mining centers. I have seen as many as a hundred men walking about with their hands in their pockets and their eyes intent on the ground, for all the world as if they were in disgrace. A five-ounce nugget may be found; and everyone hastens to the spot. Perhaps nothing more is picked up; or it may be that sufficient gold is discovered to attract troops of "dry-blowers" to the place. The "dry-blower" is brother to the "gulch-miner" of America and the "alluvial digger" of the eastern colonies of Australia. In the investigation of the methods of the dry-blower and the occurrence of the deposits out of which he wins the gold, I observed many facts of such interest, it seemed to me, as to warrant the preparation of this contribution to the *Transactions* of our Institute.

In mountainous regions the disintegration of the surface is mainly caused by the frost. Water penetrates into crevices and cracks, and, because its maximum density is at 4° C. and not at zero, it undergoes such expansion in the immediate approach to the freezing-point as to become a powerful lever for tearing the rocks apart. Thus is loosened that material out of which eventually the alluvial deposits of the valley are formed. In warmer climates the contraction and expansion of water is likewise a ceaselessly destructive agent. Even in a dry, hot, region, such as the interior of West Australia, the difference

---

* Sunday labor is generally forbidden throughout the Australian colonies; hence the opportunity to go "specking," as above described.

between the heat of day and the cold of night causes the dew to play an important part in moulding the physiography of the country. For it must be remembered that the changes of volume caused, in water as in other substances, by changes of temperature, are well-nigh irresistible in energy, however minute in amount. The cool nights, which alone make life bearable on the Coolgardie gold-fields, are thus beneficent in two ways. To them is ultimately owing the formation of those accumulations of gold-bearing dirt out of which many a prospector has dug himself a competence for life. The tables on page 10 from the government meteorological reports exhibit this variation in temperature.

In mountainous lands the melting of the winter snows yields the water employed in that process of concentration which begins as soon as the rock is shattered, and continues until each ultimate particle has been classified by the untiring machinery of nature. The " tailings " are the mud and sand which go to build new continents upon the ocean-floor, the " middlings " are the great masses of alluvium covering the plain, and the " heads " are the gold-bearing gravels of the mountain-valley. The soft is separated from the hard, the heavy from the light, until at length the metal once incased in quartz and enclosed within the rock is set free, to be collected wherever the eager stream has so abated its force as to permit the particles of gold to find a quiet resting-place.

It is a great sifting-process. The motion of the water is governed by the slope of the surface. In a flat country the conditions resemble those surrounding a mill so situated as to be incapable of getting rid of its accumulating tailings. Should the water-supply of the mill also prove insufficient to carry out its operations, then the analogy with a desert plateau is complete. The process of concentration must in both cases remain unfinished.

The gold-bearing gravels of California and Victoria, for example, usually rest on a hard bed-rock, whose water-worn surface speaks of the agency which made it so. The particles of gold and heavy iron-sand have been washed clean; the overlying pebbles and quartz gravel are comparatively free from material less resisting than themselves; and, if there be any clay, it is found in distinct layers, in positions testifying to the varia-

TABLE I.—*Meteorological Conditions at Coolgardie*, 1897.

Month.	Temperature.					Temperature of Dew-point.		Rainfall.	
	Mean Max.	Mean Min.	Highest Max.	Lowest Min.	Greatest Variation in One Day.	Mean. 9 A.M.	Mean. 3 P.M.	Total Inches.	Days
	°	°	°	°	°	°	°		
January	94.1	63.2	104.3	53.0	42.2	53.5	54.0	.56	4
February ..	89.4	58.5	104.6	47.4	47.3	51.8	53.6	.54	5
March........	85.6	57.7	98.4	50.0	41.0	49.2	55.0	.10	2
April	81.2	53.9	96.1	39.1	36.5	51.3	57.4	.01	1
May...........	71.4	46.6	88.4	38.2	41.5	46.9	53.5	.09	3
June..........	62.9	43.9	71.2	31.5	29.8	44.2	49.7	1.04	9
July...........	65.1	42.4	74.0	36.5	33.5	43.5	47.5	.34	6
August	63.8	41.5	81.0	33.0	34.6	40.8	43.7	1.08	10
September..	75.0	47.5	92.0	35.0	39.7	43.7	49.5	.29	6
October......	81.0	49.8	91.0	41.0	39.7	......	......	.06	2
November..	90.6	58.4	105.0	47.3	44.6	......	......	.09	1
December..	91.7	59.5	109.2	51.0	44.2	......	......	1.31	4
					1898.				
January	97.6	65.1	111.2	54.0	43.2	60.1	68.2	Nil.	...
February...	89.3	62.5	107.2	48.0	37.7	57.0	60.2	.27	1

TABLE II.—*Meteorological Conditions at Kalgoorlie*, 1897.

Month.	Temperature.					Temperature of Dew-point.		Rainfall.	
	Mean Max.	Mean Min.	Highest Max.	Lowest Min.	Greatest Variation in One Day.	Mean. 9 A.M.	Mean. 3 P.M.	Total Inches.	Days.
	°	°	°	°	°	°	°		
January.....	92.9	66.0	105.0	55.0	34.2	51.1	52.0	.38	2
February...	88.2	61.6	103.0	49.0	34.7	51.2	51.5	.02	1
March........	83.8	58.9	98.4	51.0	39.0	49.0	49.4	.52	6
April	80.0	55.9	95.4	38.8	82.8	48.0	46.8	.20	1
May	69.8	48.0	88.1	37.0	40.4	46.5	45.0	.10	1
June..........	62.9	47.4	73.2	36.2	30.2	46.4	49.1	1.26	14
July...........	64.4	43.3	74.0	33.2	29.8	42.8	43.7	.22	3
August	63.8	43.3	82.0	34.0	29.1	40.0	41.7	.65	9
September.	74.2	48.8	90.8	37.2	40.2	41.6	43.8	.41	5
October......	78.5	51.8	90.2	41.0	38.0	42.0	40.4	.11	2
November..	90.2	59.0	103.0	48.0	49.0	46.8	48.7	.06	1
December..	90.7	61.6	109.2	49.4	39.0	48.3	49.5	.82	4
					1898.				
January.....	98.0	66.8	113.2	55.0	41.0	51.3	50.3	.02	1
February...	90.5	63.4	109.8	48.2	38.6	51.2	49.9	.36	3

tions in the velocity and volume of the stream which laid down the deposit as a whole. It is an orderly arrangement of assorted material.

Compare with this the alluvium of the desert. A low ridge is crested with the outcrop of a gold-bearing quartz-vein which, amid that surrounding sea of dark-blue scrub, justifies its colonial designation, a "reef." On its flanks there is a thin cover of sandy soil which gradually thickens, a little lower down the slope, to a deposit of several feet. Sink a hole and you will find, first an inch or two of loose sand and dust, then a more solid layer of gravel and dirt, which in turn passes imperceptibly into a compact material consisting of fragments of rock and quartz, held firmly together by clay. It is called "cement," and it might better be termed an "agglomerate." It is an unclassified product of erosion, and lies close to the place of its origin, as a mere collection of unsorted *débris*.

The rock on which the deposit rests is so softened by decomposition that it is frequently taken for a part of the overlying detritus. If the hole be continued so as to become a well or shaft, it will penetrate through additional oxidized ground until this suddenly gives place to diorite or granite (the two prevailing formations) at a depth varying from 75 to 200 feet, which is the drainage-level of the region.

This deposit owes its existence to the wind and rain, assisted by gravity acting on a slightly inclined surface. Wind is ordinarily an insignificant geological agent, but in the constant and violent draughts of a high plateau there is a force which, working during long periods of time, is capable of producing notable results. In the vicinity of Coolgardie and Kalgoorlie, especially the latter, which has the less broken topography, the dust-storms are almost ceaseless, and bear forceful testimony to the amount of material which can be conveyed in the air. The wind careers over the country in gyrating whirls, to which the aborigines give the name of "willy-willy," nor have the white invaders ventured to call them otherwise; and as these whirlwinds go waltzing across the wretched town, they gather up in their skirts all the scattered refuse of a border civilization. I formed the impression that the wind had a prevailing direction from the southeast to the northwest, that is, from the nearest sea toward the heated interior; but the meteorological reports of the government do not confirm this supposition. In the accompanying table, Beaufort's scale of wind-force is employed. It will be observed that the meteorological reports confirm the impression that Kalgoorlie is more windy than Coolgardie, and that a quiet condition of the atmosphere is unusual in both districts. Nor is there any consistency of direction, as is proved by the observations made, for example, at

Kalgoorlie during the month of September, 1897. It is evident that the wind blew where it listed, and no man could tell whence it came. The following is Beaufort's scale of wind-force:

Number.	Description.	Speed of wind in miles per hour.
0	Calm.	3
1	Light air.	8
2	Light breeze.	13
3	Gentle breeze.	18
4	Moderate breeze.	23
5	Fre·h breeze.	28
6	Strong breeze.	34
7	Moderate gale.	40
8	Fresh gale.	48
9	Strong gale.	56
10	Whole gale.	65
11	Storm.	75
12	Hurricane.	90

TABLE III.—*Wind-Force*, 1897.

Month.	Coolgardie.				Kalgoorlie.			
	9 A.M.		3 P.M.		9 A.M.		3 P.M.	
	Max.	Min.	Max.	Min.	Max.	Min.	Max.	Min.
January	3	1	3	0	2	1	3	3
February	4	2	4	1	7	2	7	2
March	3	2	3	1	5	2	2	0
April	4	0	3	2	2	1	9	1
May	8	2	3	0	5	2	2	2
June	2	1	3	2	9	2	5	2
July	5	1	6	1	3	2	4	2
August	5	1	6	1	6	2	8	2
September	9	1	9	1	6	2	9	2
October	5	1	8	1	7	2	6	2
November	6	1	9	1	5	2	8	2
December	9	1	4	1	2	2	3	2

TABLE IV.—*Variation in Direction of the Wind at Kalgoorlie during September*, 1897.

9 A.M. Direction of wind.	Days.	3 P.M. Direction of wind.	Days.
N.	4	N.	4
N. to E.	5	N. to E.	4
E.	3	E.	1
S. to E.	5	S. to E.	3
S.	1	S.	4
S. to W.	6	S. to W.	7
W.	1	W.	5
N. to W.	2	N. to W.	2

Observations such as have been quoted in Tables III. and IV. necessarily fail to record the really characteristic play of the wind in this region. The whirl-storms, referred to already, spring up suddenly, rush madly across the plain, suck up everything lying loose on the surface, and as suddenly subside. These apparently erratic air-disturbances are responsible for the transport of the greater part of the material which weathering and erosion have disintegrated. So far as I know, measurements of the transporting-power of the wind have not been made in this particular region; but elsewhere in Australia scattered observations have been made; for example, that fences 4 feet high are buried by drifting sand in a period only slightly exceeding two years. In the Libyan desert, bordering the Nile valley, the same results can be seen. Thus, for instance, the sand-storms bury the temple of the Sphinx every summer, and the road built by Ismail Pasha, from the Mena House to the Pyramids, is filled with sand up to the level of the 6-foot parapet in less than ten days.*

In West Australia there is much evidence of that which geologists euphuistically term the Æolian agency. The wind has been stirring and sifting the material lying loose on the surface until it has become classified to a remarkable degree. In traveling over the country, one is soon called upon to notice the broken white quartz scattered over the ground, in big patches many acres wide. These alternate with stretches, steel-gray in the morning and blue-black toward the close of day, of ironstone fragments. Leave the trail; go a short distance into the bush; and you will find the surface covered with dust in which each step leaves an evident footprint. It is the veritable dust of ages, not the earth-smoke blown from man's restless to-and-fro. The wind has sorted the quartz, the ironstone and the dust. The latter has been scattered in contemptuous carelessness all over the face of the weary desolation, but the heavy ironstone remains not far from where it was broken off the decomposing diorite until, shattered and comminuted to powder, it also is winnowed by the dust-storm. The numerous veins, large and small, traversing the country have contributed the quartz which the wind has collected, so that it sometimes covers the ground with the glittering whiteness of a snow-drift.

---

* So my dragoman informed me when I was there last February.

From these stretches of ironstone and quartz one would naturally infer the occurrence, somewhere underneath the surface, of large masses of both. Owing to the extreme slowness with which denudation progresses in this arid region, and the consequent very gradual lowering of the zone of oxidation, the rocks exhibit, above the drainage-level, a marked intensity of chemical action. The granite is kaolinized to an almost incoherent clay, and the diorite is rendered abnormally heavy in iron by the surface-concentration of decomposition-products. And, as the rock is eroded, the quartz, on account of its hardness, persists, so that a series of small stringers eventually yields an accumulation suggestive of its derivation from a large mass. It is a process of concentration which, there is reason to believe, has also affected the gold-occurrence, the upper portions of veins being enriched by the deposition of the gold left behind from lode-matter which was long ago disintegrated and removed by erosion.

If we now turn to the "dry-blower" and watch him at his work we shall see the same processes utilized in the winnowing of the gold.

III.—Dry-Blowing.

In West Australia the absence of running water renders unavailable the cradle and the sluice-box of ordinary placer-mining, with the result that the prospector has learnt, intuitively, to utilize the agency which he sees incessantly at work in the nature around him. Wind replaces water. The method is simple. Taking two pans,* he places one of them on the ground, empty, while into the other he puts a shovelful of the "dirt," that is, the sandy detritus containing the gold. The material is shaken up so as to bring the big lumps on top, and then, resting the pan on one knee and holding it with his left hand he uses the right hand to skim off the coarse particles (as shown in Fig. 5). Then standing erect and facing at right angles to the direction of the wind, he slowly empties the full pan into the empty one at his feet (see Fig. 6). As the stream of dry dirt falls, the wind selects the fine and blows it in a cloud of dust to leeward. The operation is then reversed, the pan which has just been emptied being placed on the ground so as

* The Australian calls them "dishes."

to receive the contents of the other. This is repeated three or four times, according to the degree of concentration effected. In a strong breeze one operation may prove sufficient. To prevent the loss of the fine gold which is sometimes carried away with the dust it is customary to spread a piece of canvas on the ground, one end being placed under the pan and the other extending to leeward.* The next stage is to further winnow the material by tossing it up and down in the pan (see Fig. 7); the latter is held slanting forward, and is jerked so as to throw the dirt from the front to the back of the pan. The light particles are separated, as chaff is driven from grain. Then, giving the dish a vanning movement, the prospector again removes the coarser particles that come to the surface by skimming them off with his hand. There now remains about half a pint of material, and this is diminished by panning, just as in water, the dry particles having a mobility permitting this method of treatment. Finally he drops on his knee, and, holding the pan (see Fig. 8) so that it is tilted forward, he raises it up to his mouth and uses the breath of his lungs to complete the process. The particles of gold are seen fringing the edge of the iron sand. If the yield consist of only a few minute particles,† he puts his moist thumb on them, and so transfers them to his pocket; but if there be any coarse pieces—nuggets—they are put into the leather wallet attached to his belt.

In watching a dry-blower at work, it becomes evident that the operation, like every process of concentration when properly conducted, consists of sizing and classification. The wind removes the fine sand and the dust, the operator's hand skims off the larger lumps of dirt, so that there finally remains a collection of those heavy particles of ironstone which, as in ordinary placer-mining, accompany the gold.

Owing to the perfect dryness of the dirt and the heat imparted to the surface of the iron pan under a tropical sun, the material behaves with much of the mobility which it would have if water and not air were the vehicle employed. It reminds one of the behavior of a charge in the roasting-furnace, in which the hot air cushions each particle so as to give to the mass an apparent fluidity.

---

* As illustrated in the Figs. 5, 6, 7, and 8. These were reproduced by Mr. Henry Reed, of Denver, from photographs taken by me at Kalgoorlie.

† "A few small colors," as we would put it in the United States.

The rapidity and completeness of the operation depend on the strength and uniformity of the wind. There is a constant light breeze on the gold-fields, even during those happy intervals when the dust-whirls have temporarily subsided; but the cloudy mornings of the wet season and the sultry days of the hottest summer months are alike unfavorable to dry-blowing, because at such times the air is dead. Many of the mining-camps are situated on a slight rise of ground, overlooking those desolate sinks of salt and sand which are called "lakes;" and the difference of level is marked by a constant breeze which is a good friend to the dry-blower.

In the history of ordinary alluvial mining, washing with the pan was succeeded by the use of the sluice-box and the cradle. Similarly the "dishes" of the dry-blower are replaced by machines of several types, all of which, however, are based on the idea of a shaking movement in the presence of a current of air. The simplest contrivance is represented in Fig. 9. This machine is 2 feet wide and 4 feet long. A is a hopper with a sheet-iron bottom punched with 1-inch holes, B is a 12-mesh screen, C is an 18-mesh screen, and D is the final product of the operation. The dry-blower empties a shovelful of dirt into the hopper, places his hands on the two sides of the machine and shakes it from side to side. There is sufficient play in the frame itself to permit a movement which causes the material to pass through the series of screens and accumulate underneath. It is then treated by hand, as previously described and illustrated. One man will put through about 5 tons of loose dirt in a working-day of seven hours.

Another and more elaborate contrivance is exhibited in Fig. 10. It consists of a series of four trays, hung on a triangular frame, B C D. The trays are 22 inches in diameter. They are comparatively flat and have screen-bottoms, through the center of which an iron rod passes to the eccentric, G, which receives the required movement through the lever A E F, of which A is the handle. The trays are 5 inches apart and are held in place by wires, H H. The material to be treated is placed in the uppermost tray, which is a hopper pierced with inch-holes. No. 2 has $\frac{3}{8}$ holes, No. 3 has a 10-mesh screen and No. 4 has an 18-mesh screen. The lowest, No. 5, is flat and serves as a receptacle for the final product, which is dry-blown by hand, as heretofore described.

A Typical Scene in West Australia.

Fig. 3.

Buying Water at a Condenser.

Condenser at the Lake View and Junction Mine, Kalgoorlie.

Fig. 5.

Dry-blower at Work.

FIG. 6.

Dry-blower at Work.

Fig. 7.

Dry blower at Work.

FIG. 8.

Dry-blower at Work.

Fig. 9.

DRY BLOWING MACHINE

Fig. 10.

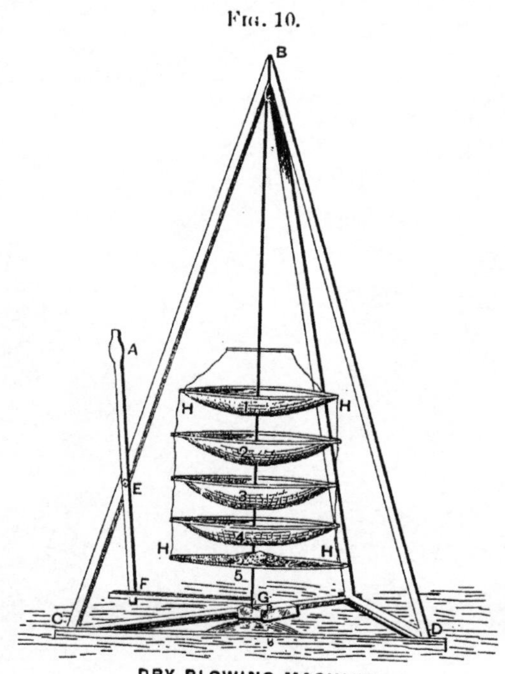

DRY BLOWING MACHINE

PLAN OF TRAY

Fig. 11.

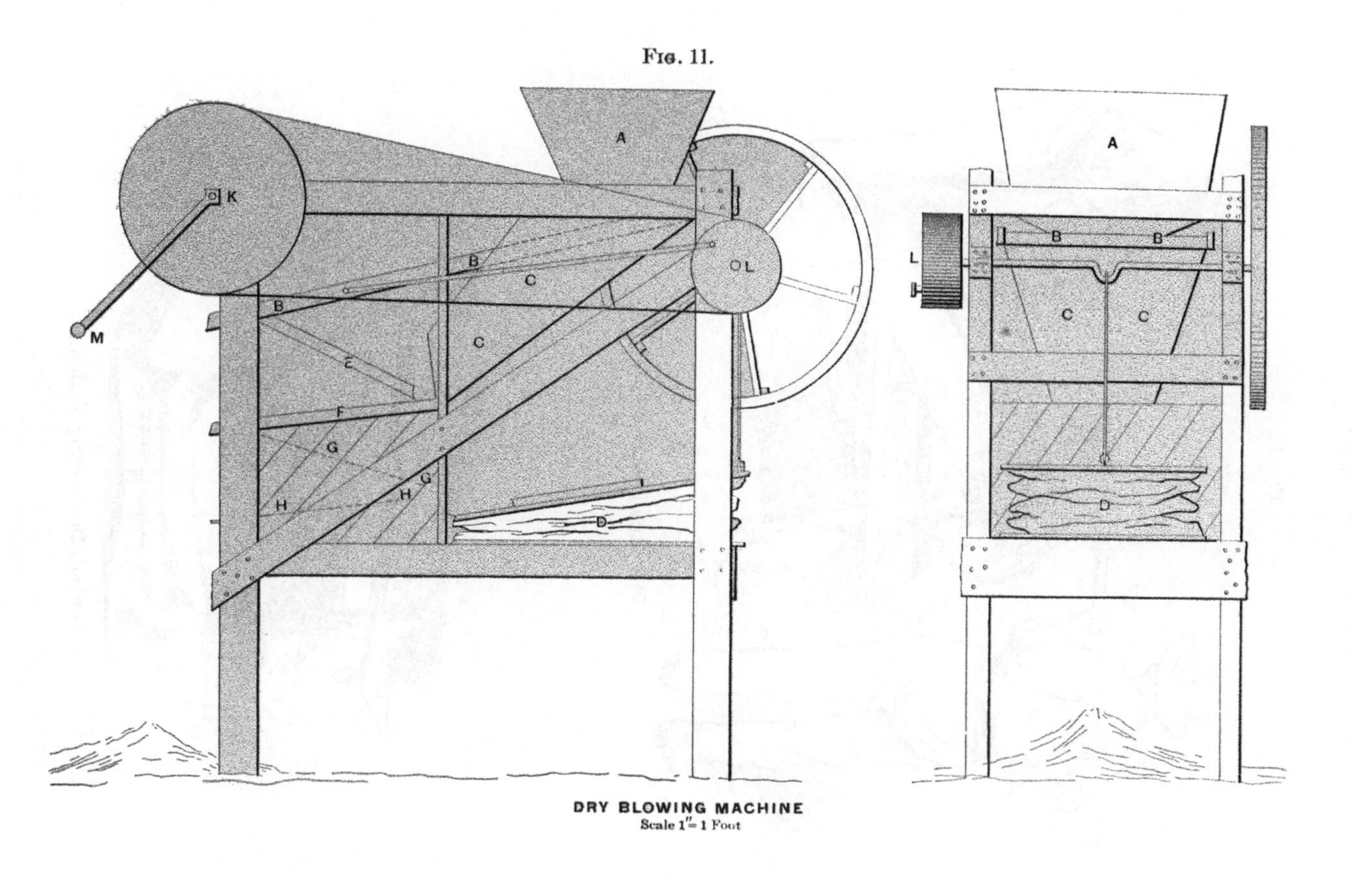

DRY BLOWING MACHINE
Scale 1" = 1 Foot

Fig. 12.

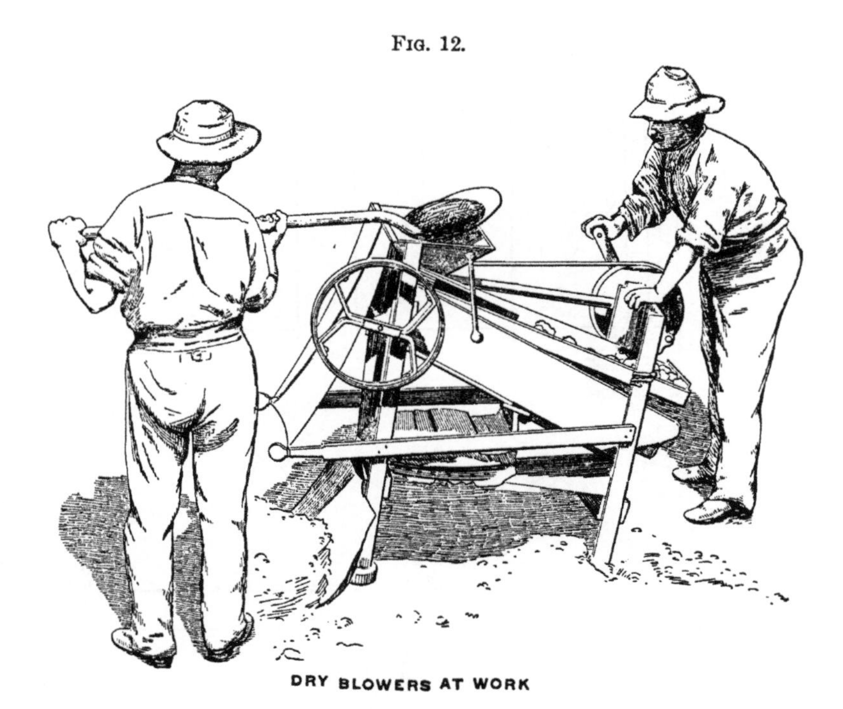

DRY BLOWERS AT WORK

Fig. 13.

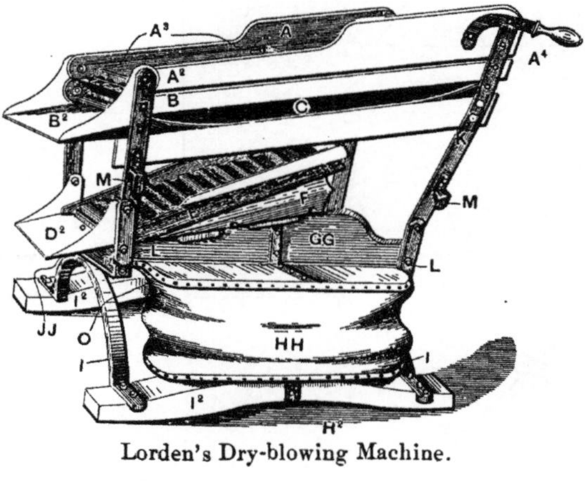

Lorden's Dry-blowing Machine.

Fig. 14.

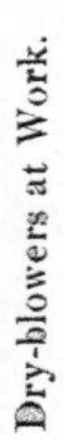

Dry-blowers at Work.

Dry-blowers at Kalgoorlie, West Australia.

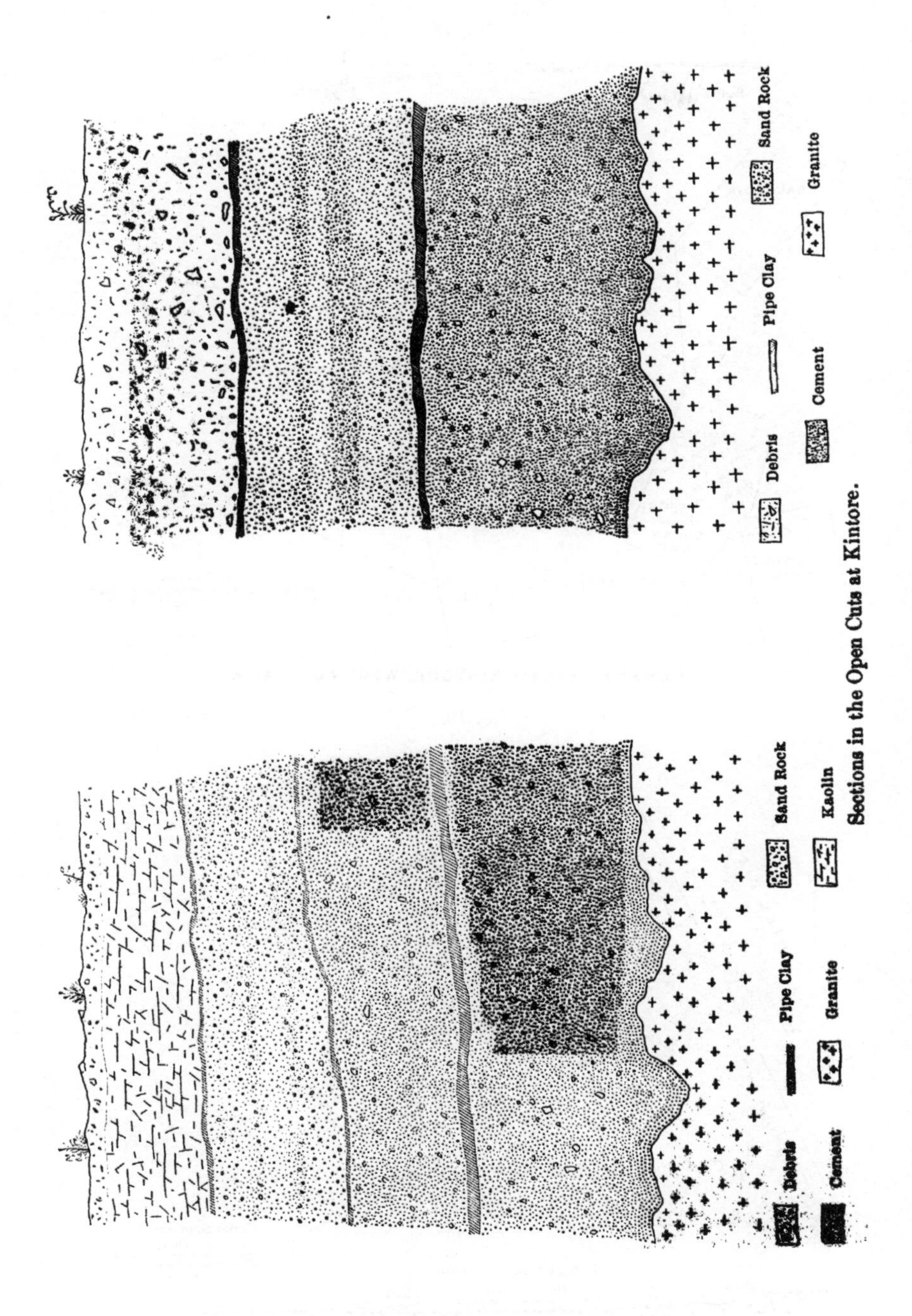

Sections in the Open Cuts at Kintore.

FIG. 18.

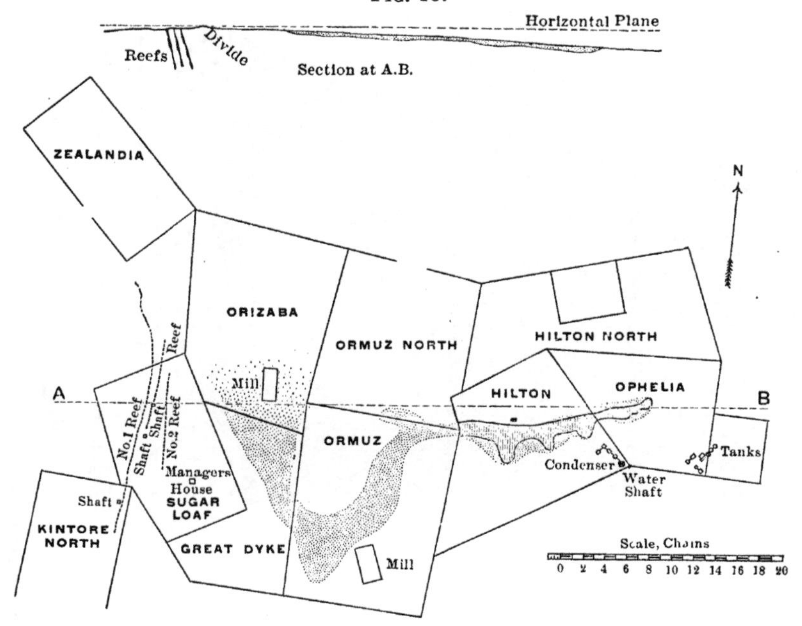

CEMENT DEPOSIT, KINTORE, WEST AUSTRALIA

FIG. 19.

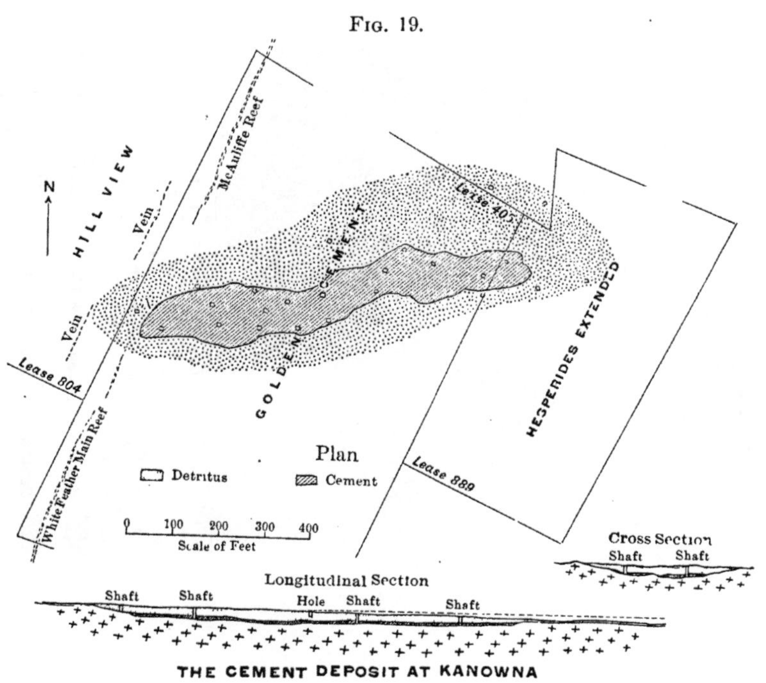

THE CEMENT DEPOSIT AT KANOWNA

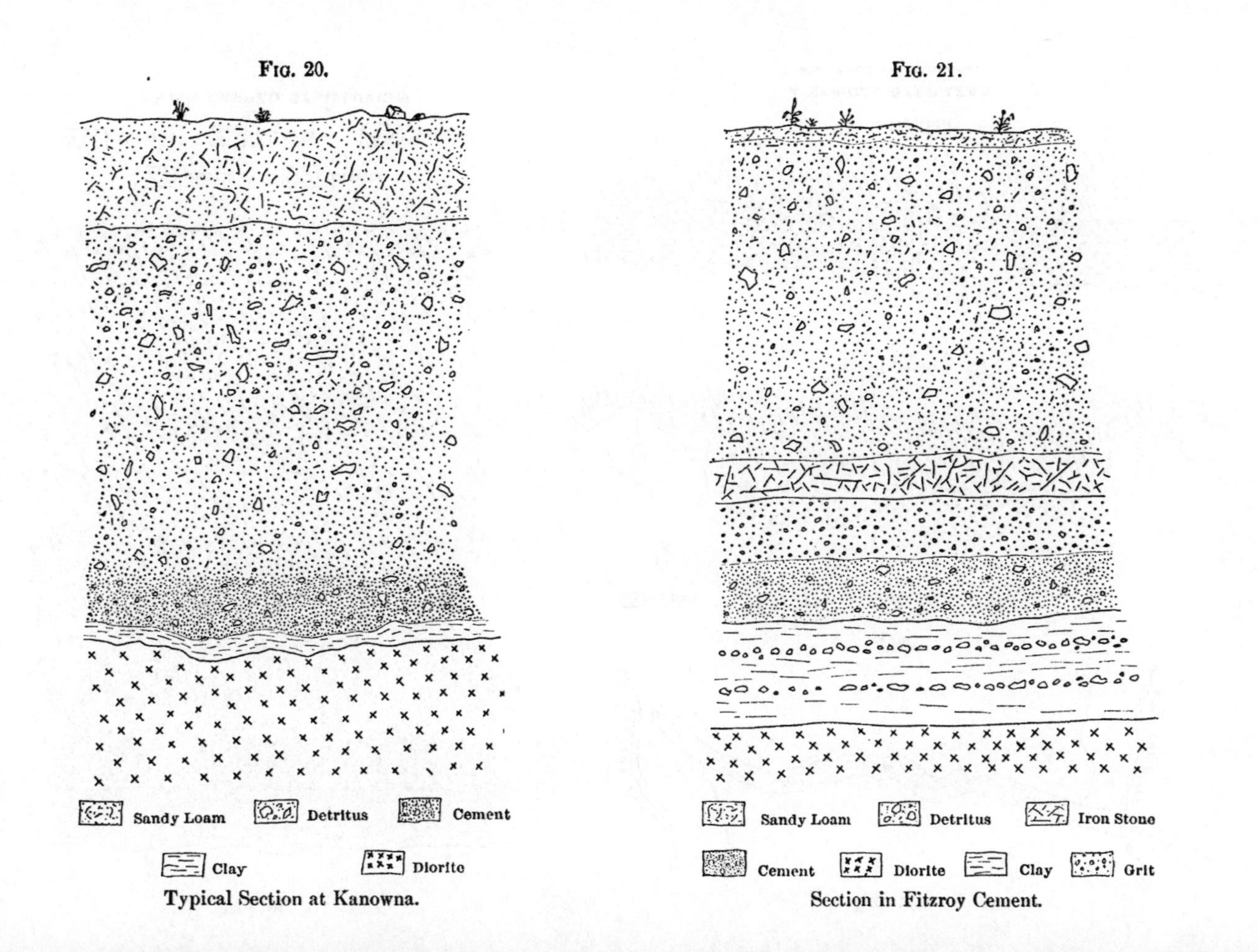

Typical Section at Kanowna.

Section in Fitzroy Cement.

FIG. 22.

FIG. 23.

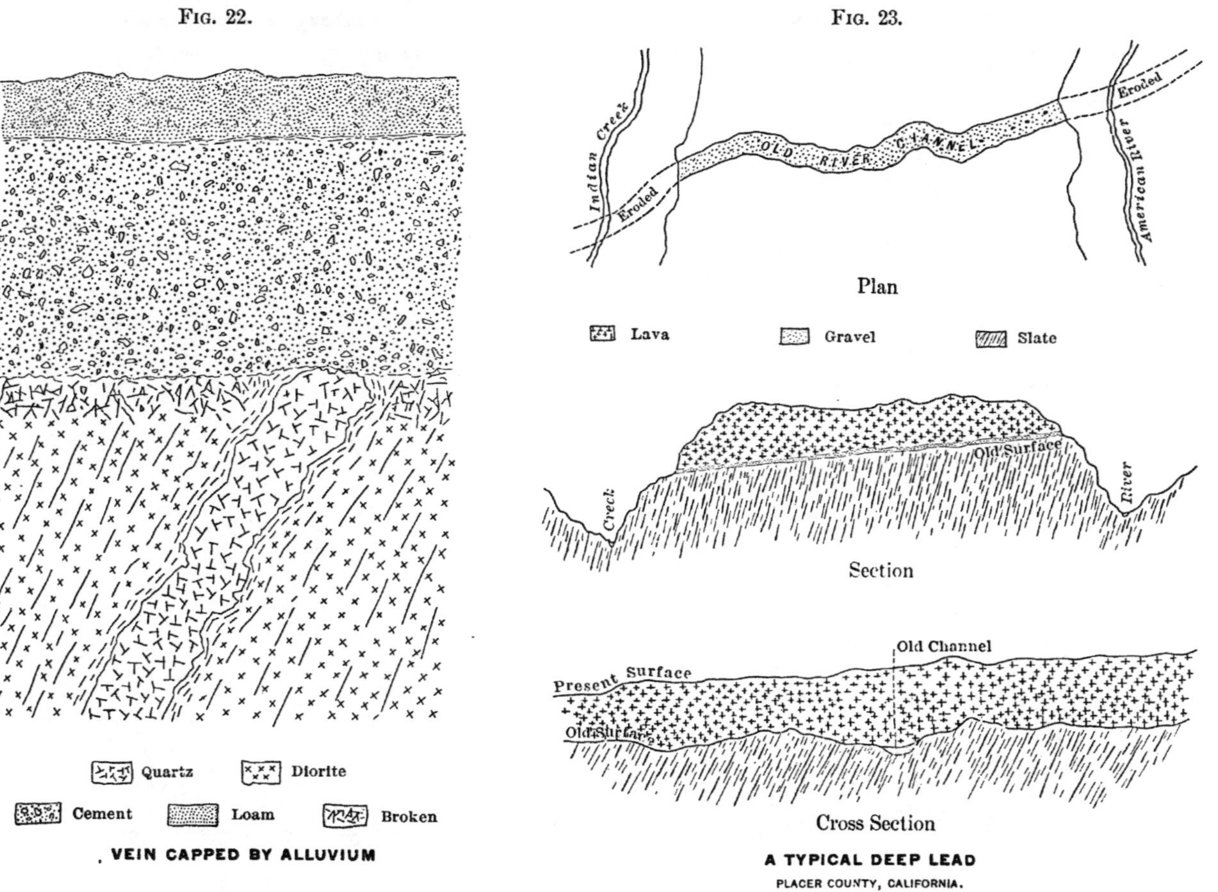

VEIN CAPPED BY ALLUVIUM

A TYPICAL DEEP LEAD
PLACER COUNTY, CALIFORNIA.

In these two contrivances no attempt is made to supplement the wind by an artificial air-current. The next step is to use a bellows. Fig. 11 shows such an arrangement. It consists of a hopper A and a series of screens, B, E, F, G and H. By turning the handle M the disk K is revolved; and this, by means of a belt, transmits its movement to the pulley, which shakes the screen B B through the eccentric-rod C, and at the same time operates the bellows D through the disk K. Fig. 12 illustrates the machine when in operation.

When the material is placed in the hopper A and the machine is set in motion, the large lumps run off over the grizzly or sizing-screen B B, the upper part of which is made of parallel wires $\frac{3}{16}$-inch apart, and the lower portion of 8-mesh. The finer stuff falls down into the shoots C C and E E, respectively, and reaches F, which is another (12-mesh) screen, supplied also with riffles. As it descends through the screens at G and H, both 18-mesh, the blast from the bellows keeps the material in agitation, and aids the requisite separation between the particles of gold and the dust. The final product is panned by hand.

One of the most popular dry-blowing machines is that made by Steve Lorden, at Freemantle. It is illustrated in Fig. 13. The essential parts are:

A. Feed-hopper, sheet-iron bottom, punched with inch-holes, hinged at A_2 and provided with riffles A_3, which arrest the heavier particles of gold, while the coarse lumps of dirt pass out of the machine over the shoot B_2 and the fine stuff falls through into

B. Second hopper, which has riffles and smaller perforations, repeating the process. To examine this hopper, the upper one, A, is sprung out of position at A_4.

C, C. Return-shoots which lead the reduced material for further sizing upon the sloping screen E, which also has a series of riffles, and is placed directly over the air-chamber F.

D_2 is a discharge-shoot for the screen E. G, G are air-channels from the bellows H, H.

H, H. Double blast-bellows, one on each side, which ride on carriers H_2, so arranged as to give the requisite play and to relieve the bellows from undue strain when in operation.

I, I. Rockers, bolted firmly to two foundation-blocks, I_2, which form the stand, the only part of the machine that is not in motion. The curve of the rockers allows all dirt to fall away from the pivot-pins J, J, by which the apparatus swings.

K, K. Standards, hinged at L, so that the machine folds up, as shown in Fig. 14.

M, M. Brackets for the insertion of two poles, by means of which a couple of men can carry the machine conveniently.

The operator holds one handle, at A_4, in each hand and rocks the machine, this serving simultaneously to put the bellows, hopper and screens all into movement. The machine weighs 124 pounds and has a capacity of from 10 to 14 loads (a load is roughly one ton) per day. The price at Freemantle is £16 or about $80.

The deposits exploited by the aid of these machines (see Fig. 15) are of a generally patchy character and lie at the upper ends of the depressions formed where the surface slopes away from ridges traversed by the veins of gold-bearing quartz. The prospector has an eye for the contour of the ground, and looks for the point where the rock-surface disappears under the fragmentary overburden which he calls "made ground" as distinguished from the underlying "bed-rock." In his search he is usually guided by seeing the outcrop of quartz, marking a possible source of detrital gold, and by the actual finding of specimens on the surface.

The distribution of the gold in these deposits reminds one of its position on a vanning-shovel. It may be traced up to the outcrop which yielded it, or it may be scattered in the sand for half a mile; but the rich and only workable part of the deposit will ordinarily be found at a distance of 30 or 40 feet from the reef.

Underneath these patches of superficial gold-bearing detritus there are found partially consolidated accumulations, which are more extensive and, quite apart from their greater economic value, are also of superior interest, because of their better-defined geological features.

IV.—The Cement-Deposits.

In the ordinary course of professional work I examined the two most important of these deposits, at Kintore and Kanowna, respectively. Since then a third formation of a similar character has been opened up, also at Kanowna. This I happened to see when in the stages of early development, and before it had been extensively exploited.

The deposit of cement at Kintore was one of the earliest worked. It is situated 23 miles northwest of Coolgardie, on the road to Menzies. The West Australian Proprietary Cement Company mined the ground with a success which was short-lived, because of the restricted quantity of material rich enough to pay the high costs of treatment. In four months, 7335 ounces of gold were obtained by treating 4115 tons in a stamp-mill, supplemented by cyanide-vats.

Enough work has been done to disclose the character of the deposit. Figs. 16 and 17 are representative sections obtained in the open cuts. Under a thin covering of sand and dust there occurs a bed of kaolin, ranging from a couple of inches to a foot in thickness; and this overlies from 15 inches to 2 feet of "sand-rock," which in turn gives place to the gold-bearing cement, which has an average thickness of 2½ feet. The last lies directly upon an irregular surface of decomposed granite.

The several layers composing the deposit are separated by seams of pipe-clay, which, like the kaolin, are simply the product of the decomposition of the constituents of the granite, particularly the feldspar. The sand-rock may be described as a coarse, incompletely consolidated sandstone or grit, consisting mainly of iron-stained particles of quartz, loosely cemented. The cement has a bluish-gray tinge, owing to the play of light on the quartz fragments. This, too, is not quite compacted, since fractures through the material do not break across the pebbles, which are harder than the clay binding them together. In this respect the cement differs, for example, from the South African "banket," to which it has been compared. From a distance, the cement looks like a coarse sandstone and exhibits a rough-joint structure. The materials of which it has been made up have undergone incipient sizing, so that different layers of varying coarseness are distinguishable.

The bed-rock is granite, so softened by decomposition as to be mistaken by the miners for a part of the alluvial deposit. It is kaolinized to a depth which the neighboring mine-shafts prove to reach a maximum of 130 feet. The surface on which the cement lies is marked by pot-holes having a maximum depth of 2 feet and a diameter reaching to 3 feet. These holes are filled with cement, which is usually poor save at the rims, where some of the richest mill-stuff has been obtained.

All the members of the deposit, from surface to bed-rock,

contain some gold, the kaolin being the poorest. The cement itself attains a maximum thickness of 5 feet. The richest parts occur in lateral embranchments from the main body of the deposit. The kaolin has become hardened and dried. It breaks like shale; and its true character is obscured by the down-filtering of red sand through cracks reaching to the soil overhead.

The deposit extends through a number of mining claims, as shown on the accompanying map (Fig. 18). It has been traced for a length of three-quarters of a mile. At the east end it begins as a narrow neck about 15 feet wide, and then enlarges steadily to 100 feet. Occasional bulges increase this width to a maximum of 250 feet. At the edge it gives place, as it thins out, to 2 or 3 feet of ironstone gravel, carrying 3 or 4 dwts. of gold per ton. The best part, economically, of the deposit lies in the Ophelia and Hilton claims, which, it will be noted, are situated at the lower end of the basin.

The bed-rock rises westward at the rate of 15 feet per thousand. This fact suggested that the origin of the gold-bearing cement was to be found in the reefs which were being profitably mined by the Sugar Loaf Company. The workings were 136 feet deep at the time of my visit in September, 1897. The veins traverse granite which has been kaolinized to 130 feet from the surface. They consist of white quartz and are narrow (4 to 12 inches), but they carry short shoots of very high-grade (3 to 10 ounces of gold per ton) ore. The gold occurs native, in flakes penetrating the cleavage-planes of the quartz like a golden mosaic, and also in coarser particles, which, under closer examination, prove to be octahedral crystals with curiously rounded edges. A comparison of these veins and their enclosing rock with the material composing the cement-deposit unquestionably indicates the derivation of the latter from the former. The cement carries gold which is exactly similar to that seen in the reefs; the quartz fragments in the alluvial are identical with the stone broken in the Sugar Loaf mine. Samples of both lie before me as I write, emphasizing the conclusion just stated. In the cement occur particles of quartz showing gold. The loose gold in the cement has been but slightly worn, and the quartz pieces are rather subangular than rounded, so that they can hardly be termed "pebbles;" and the deposit itself is better defined as an agglomerate than as a conglomerate.

The binding-material, the overlying layer of kaolin and the sand-rock capping the gold-bearing stratum of cement, all exhibit very clearly their derivation from a decomposed granite, similar to that which encloses the reefs and forms the bed-rock of the alluvium itself.

The topography confirms this supposed relationship. The highest point along the major axis of the cement-deposit is a very low ridge separating the workings of the Sugar Loaf from the alluvial ground. The house of the manager of the Sugar Loaf is on this divide. The reefs are 462 feet westward, and only 15 feet lower where they appear at the surface. The cement deposit begins on the Great Dyke lease, at a point 530 feet eastward, and only 8 feet lower. The cement then extends on a gentle down-slope of 15 feet per thousand for a distance of 3500 feet.*

It occupies a very shallow depression, and in its structure bears internal evidence of considerable geological age, suggesting that it was formed at a time when the Sugar Loaf reefs reached the surface at a level superior to the slight ridge now separating them. Reference to the map and longitudinal section (Fig. 18) will aid the above description.

Another deposit of similar character has been explored at Kanowna, 25 miles northeast of Kalgoorlie, and about 60 miles east of the locality just described.

The discovery was made in 1893. Each digger secured a claim 50 feet square, and sunk a shaft to the gold-bearing cement which the dry-blowers had uncovered in the course of their prospecting. The deposit became in due time "gophered" with holes and shafts, so that the boundaries of the cement were accurately determined. In 1895 an English company secured the property and consolidated the claims into larger leases. The expectations held out freely by responsible engineers that an extremely profitable enterprise could be based on the remnants of gold-bearing ground were wholly dissipated in the succeeding two years.†

The deposit lies in a shallow trough, the longer axis lying

* I am indebted to the courtesy of Mr. Alexander Brand and Mr. T. G. Paisley, the managers of the two companies, for the measurements quoted.

† A gross blunder was made, simply through insufficient and unsystematic sampling.

east and west. The body of gold-bearing cement has a length of about 700 feet and a maximum width of 105 feet (see Fig. 19). Vertical sections exhibit an overburden of sandy loam, from a few inches to 2½ feet thick. This was the material worked by the dry-blowers. Then comes a layer of detritus, called "wash" by the miners, composed of fragments of ironstone and quartz imbedded in clay, and reaching to a maximum of 25 feet from the surface. This overlies the cement itself, 6 inches to 5 feet thick, and easily distinguished from its roof of detritus and its floor of clay. The cement consists of particles of quartz in a greenish clay. Near the rim of the trough the quartz occurs in larger and more angular pieces. A typical section, obtained from a pillar in the old workings, is given in Fig. 20.

The gold-contents are irregular. The whole body of cement probably averaged one ounce per ton; but only the richest parts were worked, and these carried many ounces to the ton; so that the remnants now accessible average, from the surface down, about 3½ dwts.* The clay carries 2 dwts. per ton. The material was treated at neighboring stamp-mills.

When the neighboring mines, the White Feather Reward and the White Feather Main Reef, were visited, it seemed as natural to deduce this cement-deposit from the erosion of gold-bearing quartz-veins as it had been to relate the Sugar Loaf reefs at Kintore with the deposit worked by the West Australian Proprietary Cement Company. Further investigation confirmed this inference.

The McAuliffe vein (worked by the W. F. Reward mine) and the Main Reef (worked by the W. F. Main Reef mine) traverse diorite at or near the line where large dikes of granite porphyry penetrate. The two reefs are probably identical, and have a strike which takes them right across the longer axis of the cement-deposit at a point near the head of the trough in which it lies (see map, Fig. 19). A shaft recently sunk to a depth of 200 feet by the Golden Cement Company, at a point marked A, reached this reef by means of a crosscut, and found a comparatively barren quartz-vein, carrying small spots of rich ore. The enclosing rock was diorite, and the quartz itself was encased on both sides by bands of clay fully 2 feet thick.

---

* Information which I owe to the courtesy of Mr. S. H. Williams, the manager.

On comparing the veins and their encasing rock, as seen in the workings of the two mines on opposite sides of the alluvial deposit, it is not found necessary to go further for the origin of the latter. The cement is underlain by a clay which is essentially steatite, and is as readily traceable to the neighboring diorite as the kaolin at Kintore was deducible from the granite. The green color of the cement is imparted by chlorite, derived from the decomposition of the epidote in the diorite. The "ironstone" of the detritus overlying the gold-bearing part of the deposit consists of fragments of altered diorite. The quartz in the cement and the gold accompanying it are both identical with those of the reefs close by.

Here also the topography confirms the suggested explanation. The cement lies in a shallow depression, at the upper rim of which the quartz reefs cross the country. Furthermore, these reefs traverse a low divide, which in a rough way separates the deposit from another, which slopes in the opposite direction. The latter is known as the Fitzroy cement. In this deposit rich discoveries were made during October, 1897. A typical section is exhibited in Fig. 21.

Apart from their importance as depositories of gold, the cements have played an interesting part in the development of the gold-fields, because they often cover the tops of reefs. In Fig. 22 such an occurrence* is illustrated. A similar feature proved a serious hindrance to the recognition of the lodes at Kalgoorlie, where worthless quartz veins were worked for some time before a trench traversing the cap of cement accidentally uncovered the top of one of the rich deposits of telluride ore, which did not outcrop, on account of the comparative softness of the lode.

V.—Theories of Origin.

Of course, several theories have been mooted, the most fanciful of which have naturally been those of the working miner himself. The fact that the gold-bearing cement is in places overlain by a considerable thickness of partially-consolidated rock has led to the supposition that the deposit was a "deep lead;" while the resemblance to a conglomerate has caused more than one Africander to liken it to the "banket" of the Transvaal.

---

* It is the top of the Lady Shenton reef, at Menzies.

The latter is an immense shore-deposit of gold-bearing conglomerates, now covered by a series of later sediments. Its dimensions, comparative homogeneity and persistence are in striking contrast to the narrow, restricted, irregular patches of detritus which have been described as occurring in West Australia. This total unlikeness renders unnecessary any discussion of a fancied similarity of origin.

But because this alluvium disappears under an overburden of rock,* the Australian digger easily fancies he is working a deposit similar to the "deep leads" with which he became familiar at Ballarat, for example. A distinguished government geologist from a neighboring colony visited Kanowna in October, 1897, and gave authority to the term "deep lead" by using it himself. "Deep" it may be, for that is a comparative adjective, but a "deep lead" in the technical sense it most assuredly is not. On the Forest Hill divide,† in Placer county, California, and at Creswick, in the Ballarat district, Victoria, the typical "deep leads" occur. They are, as is well known, old (Miocene) gold-bearing river channels, which have been saved from erosion by a cap of lava. The lava probably overflowed the original surface as a steaming mud, and is now found consolidated into a volcanic rock sufficiently hard to need little timbering when penetrated by underground workings. The cement deposits of West Australia occur under an overburden of "made ground;" that is to say, both the deposit itself and all the material under which it dips are of distinctly detrital origin, the products of weathering and erosion accumulated in shallow depressions of the much-decomposed surface of granite or diorite.

It is the placer of a country destitute of running water. The climatic conditions and the physiography of the Coolgardie gold-fields have been carefully described, in order to make it evident why these deposits differ from those of more favored countries, like California or New Zealand. Surely it is not in keeping with the scientific method to seek for fantastic or far-fetched explanations, when processes in operation to-day

---

* Using the word in its geological sense. Mud is "rock" as much as granite.

† I append a drawing (Fig. 23) of a typical deep-lead recently examined by me in this particular locality.

are able to supply an adequate understanding of the observed facts.

The quartz of the cement is subangular; it has evidently undergone very little attrition, and suggests, therefore, that it has not traveled far. On comparing it with the matrix of neighboring veins, an identity appears obvious. The examination of the topography renders highly probable the derivation of the one from the other. The cementing material is similarly found to be the clay resulting from the decomposition of the rock encasing the quartz-veins, and varying according to the composition of that rock, whether it be granite or diorite. Finally, the gold particles which have rendered the cement worth mining are found to be identical in fineness and physical appearance with the gold of the neighboring veins, and their scarcely-rounded edges invite the conclusion that the gold also has not been borne far from the place of its origin. The comparatively unclassified condition of the deposits is in keeping with the evidence afforded by the material of which they are composed. The absence of running water on this desert plateau has prevented any such sifting-process as in other regions leads to the deposition of well-defined layers of clay, gravel and gold upon a clean bed-rock. It is an exceptional illustration of the working of those agencies to whose unceasing play is due the configuration of the earth's surface; it is geological action in its most instructive form.

VI.—Water-Supply.

The early history of the gold-fields of West Australia is the record of a struggle to exist amid conditions which were inimical to human life on account of the scarcity of water. That great necessity has been, in some sort, satisfied by the energetic action of the government, supplemented by private enterprise. The gold-fields are now dotted over with condensing-plants, which turn the brine of the wells into water fitted for the use of man and beast. Existence is thus rendered endurable; but the mining industry is still handicapped by an item of cost unknown in more favored regions.

The water of the country is salt, sometimes almost to the point of saturation. Sea-water contains 3½ per cent. of salts, three-quarters of this percentage being common salt, the chloride of sodium. At Menzies, in September, 1897, I found one* of the two important mines of that district using water which contained 17 per cent. of salts, and the manager informed me that in December evaporation increased the amount to 30 per cent.† For this liquid he paid 25 shillings per thousand gallons. It came from a neighboring "soak." The condensed (distilled) water, bought for use in the boilers, cost £1 per hundred gallons. Milling in a ten-stamp-mill was carried on at an average cost of 30 shillings or $6 per ton, the item of water alone amounting to 13 shillings or $3.25 per ton.

Under these conditions a wet mine becomes a source of revenue. Many properties at Kalgoorlie, unable to find pay-ore, lessened the expenses of development by selling their water to those that had mills. The price varied according to the season. At the Great Boulder Main Reef mine, for example, the lowest price paid for water during 1897 was £3.10s. per thousand gallons, and the highest £11.5s. This was piped from neighboring shafts, and had not passed through the condenser; it therefore had the character of sea-water, but it was four times as saline.

An analysis of the water of the Great Boulder Proprietary mine gave the following results. The sample was turbid, and it was found that the matter in suspension amounted to 5.25 grains per gallon, or .075 gramme per liter. The clear water on analysis yielded:

	Grammes per liter.
Silica (SiO_2),	0.038
Alumina and ferric oxide (Al_2O_3 and Fe_2O_3),	0.024
Lime (CaO),	1.878
Magnesia (MgO),	8.106
Soda (Na_2O),	48.470
Carbonic anhydride (CO_2),	0.064
Sulphuric anhydride (SO_3),	6.026
Chlorine (Cl),	67.230
	131.836
Deduct oxygen equivalent to chlorine,	15.150
	116.686
Combined water, organic matter and loss,	8.534
Total solids,	125.220

* The Queensland Menzies mine.

† The water of the Dead Sea varies from 20 to 26 per cent. salts, and of this, 10 per cent is common salt.

The chief salts probably present were, therefore:

	Grammes per liter.
Calcium carbonate, $CaCO_3$,	0.145
Calcium sulphate, $CaSO_4$,	4.365
Magnesium sulphate, $MgSO_4$,	5.189
Magnesium chloride, $MgCl_2$,	15.144
Sodium chloride, NaCl,	91.467

Expressed in grains per gallon, the results appear more striking. The proportion of common salt amounts to no less than 6402.7 grains per gallon. Ordinary drinking-water contains about 3 grains of common salt per gallon.

The water-supply of the region comes from two sources, namely, that which has collected amid the purely superficial deposits of *débris* and drift covering the actual rock-surface, and, secondly, that which has penetrated through the decomposed rock down to the zone where oxidation ceases.

Wherever a depression occurs, the prevailing rocks, granite and diorite, are overlaid with a variable thickness of their own detritus, which allows the collection of rain-water and affords protection from too rapid evaporation. These are known as "soaks." The government geologist defines them as "valleys silted up with a thin covering of recent superficial deposits more or less saturated with water."* At Hampton Plains, 8 miles from Coolgardie, a supply of condensing-water has been obtained from beds of this nature. The section† was as follows:

	Thickness. Feet.	Depth. Feet.
Clay, with ironstone gravel,	27	27
Fine sand,	30	57
Coarse yellow sand,	43	100
Clay,	4	104
Land wash,	11	115
Kaolin,	8	123
Bed-rock,	39	162

Water was struck in the third stratum, described as coarse yellow sand. The bed-rock was granite.

During the wet season, some of the depressions filled with such accumulations of detritus will receive more water than

* Report in connection with the water-supply of the gold-fields. 1897. A. Gibb Maitland.

† From the report just referred to.

they can hold, and then the eye becomes gladdened for a few days by the sight of water running over the surface. As it becomes diminished by evaporation it disappears from view, but will be found to linger in the rock-holes, where the supply is maintained by the slow drainage of the surrounding porous area. These are called by the aborigines "guamma" holes. The life of such natural wells depends, of course, on the dimensions of the water-bearing depression and upon the relative porosity of the deposit which it drains.

These supplies are in their nature uncertain. The next and more important source of supply is found at the ordinary drainage-level of the country, namely, the horizon where the oxidation of the surface ceases and the relatively hard unoxidized rock offers a partial barrier to the free descent of the waters which have percolated through the overlying formation. The depth of this zone will depend upon the permeability of the superficial rock-formation; it varies from 40 feet at Earlston to 202 feet at Kalgoorlie. A characteristic section is that given by the well put down on Reserve 3096, Coolgardie, where the Gold-fields Water Supply Department sunk 165 feet and found 7 feet of sand, 47 feet of conglomerate and 111 feet of decomposed granite. The water was found at the base of the last, just above the unaltered granite.

Condensers.—Frequent mention has been made of the condenser. This is a characteristic feature of every mining settlement in the interior of Western Australia, and occupies the position accorded to the public well of a European village. Without this process of distillation, which renders the brackish water of the wilderness fit for human consumption, the mining industry could never have progressed beyond the merest prospecting.

The introduction of condensers is traceable to the sailors who took a prominent part in the early exploring expeditions. The name itself suggests this; for a landsman would be likely to call the condenser a "still." When the rush to the new gold-fields occurred, the government, by erecting condensers at intervals along the main lines of travel, did much to diminish the loss of life otherwise inevitable in times of wild excitement by reason of the recklessness of those who joined the stampede without due care for the great necessity of life in a tropical desert.

The process of converting brine into drinkable water is simple. The salt water is put into a boiler and converted into steam, which is then condensed in vessels presenting a maximum of cooling-surface. Ship-tanks, having a capacity of 400 gallons, are commonly employed as boilers, and the condensing apparatus is constructed out of the corrugated iron which is everywhere employed for roofing-purposes. The tanks used as boilers are usually set on edge in pairs, as shown in Figs. 24 and 25. The average product of each 400-gallon tank is about 300 gallons of distilled water daily. Two vertical short iron pipes draw off the steam, which passes into condensing chambers or "coolers." The latter were originally plain circular tanks, which were increased in capacity by the addition of sections, increasing the height. Sheets of corrugated iron were bent round until the ends met, and these were united by riveting and soldering. Several such sections were united, and a tubular tower, about 30 feet high and 3 feet in diameter, was the result. The top and bottom were closed by a flat sheet of iron. A 6-inch pipe connected the towers, and steam traveled up the one tower and down the next.

This type has been superseded of late by annular chambers. An outer corrugated iron cylinder, 2.5 feet in diameter, surrounds an inner 1.5 ft. cylinder, so as to leave an annular space, 6 inches wide, which becomes the condensing-chamber. No attempt is made to supplement the cooling effect of the surrounding air, though a brush shelter is sometimes erected so as to ward off the direct rays of the sun.*

The daily expenditure includes the labor of two men, one on each shift, and the cost of the fuel consumed. A typical condenser is that erected on the Lake View and Boulder Junction mine, shown in Fig. 4. This plant has a capacity of 1,500 gallons of condensed water per day and cost £100. The water treated comes from the mine and shaft and has a specific gravity of 1.03385, the total solids amounting, according to the analyses of Mr. E. S. Simpson, to 4.9308 per cent. and the chlorine to 2.3933 per cent. The cost averages, during the cool season, 5 shillings, and during the summer 6.5 shillings

* For many of these details I am indebted to Mr. Frank G. Grace, Kalgoorlie, and to Mr. Edward S. Simpson, Government Assayer, Perth.

Fig. 24.

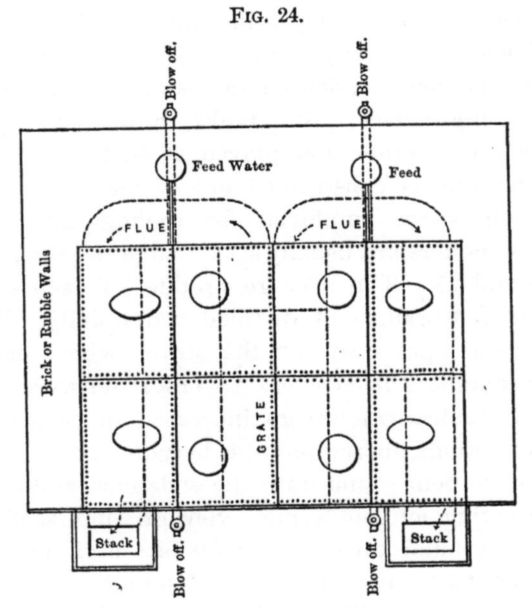

Plan of Ordinary Condenser.

Fig. 25.

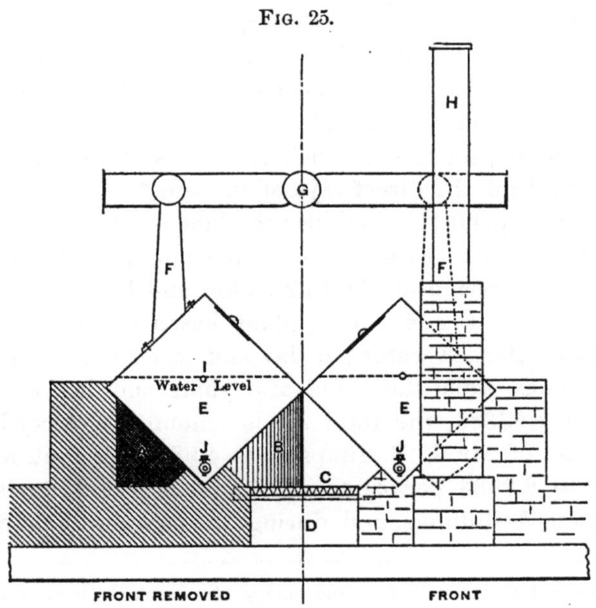

Section and Elevation of Ordinary Condenser.

per hundred gallons of condensed water. The condensed water is sold at from 10 to 12 shillings per hundred gallons. A first-class condenser would consist of at least 8 boilers, each having 400 gallons capacity and a daily output of 300 gallons, of condensed water, giving the plant a daily yield of 2,400 gallons. The salt water is usually bought for from 2 to 4 shillings per thousand gallons. Fuel costs 20 shillings per cord, and is consumed at the rate of 1 cord per thousand gallons of condensed water. Thus the cost would be:

2 men at £4 per week of 6 days,	27 shillings
2½ cords of wood,	50 "
3,200 gallons of salt water,	10 "
Total,	87 shillings

This would be at the rate of 36 shillings per thousand gallons. During November, 1897, condensed water sold for 100 to 150 shillings per thousand gallons.

Concerning the use of salt water in the stamp-mills and leaching works of Western Australia, I would say that its density is an obstacle to amalgamation because of the facility with which slimes are created. The finer particles of gold are thus prevented from settling on the copper plates of the tables, and are carried away in the tailings. At Kalgoorlie the cyanide works use the natural brine successfully; its magnesia being precipitated by lime, so as to prevent the decomposition of the stock-solution.

In order to aid the development of the mining industry, which is still severely handicapped by the want of a sufficient supply of water, the government of West Australia has decided to carry out a hydraulic enterprise of great magnitude. It is proposed to supply 5,000,000 gallons of fresh water per day to the Coolgardie gold-field by building a pipe-line from the Darling range, where the Helena river will be impounded by a concrete dam 100 feet in height and 650 feet long. The source of supply is 320 feet above sea-level, while the service-reservoir at Coolgardie will be on Mt. Burgess, at a height of 1670 feet, or 1350 feet higher. These two reservoirs will be connected by 330 miles of 30-inch steel-pipe. Nine pumping-stations will be required. The appropriation for this work is $12,500,000. The annual operating expenses will probably approximate $1,600,000, and water will be delivered through a hundred miles of distributing pipes at a cost of 3 shillings and 6 pence, or 84 cents, per thousand gallons.

It will occur to the reader to ask whether boring for artesian wells has been attempted. Yes; in obedience to the public demand, the government put down a bore at Coolgardie which reached a depth exceeding 2000 feet and found—granite. The geological conditions render an artesian flow of water highly improbable. Nevertheless, in this colony, as elsewhere among the arid tracts of Australia, there is a whispered hope of finding a subterranean river. As the Carsón and the Humboldt are swallowed up by the alkali wastes of Nevada, so in the desert plains of this southern continent there are many streams which flow into the interior and lose themselves in the sand or find for themselves an undergound channel.* This fact has given rise to conceptions, more poetic than scientific, of a great subterranean river yet to be discovered, and destined in days to come to make the desert break forth into fertility. It is a dream. No irrigation can turn the wastes of quartz-sand into waving fields of wheat. Time, geological time, covering a period to measure which the duration of a man's life is an inadequate unit, can alone render the wilderness fit for human habitation.

* This occurs in Queensland, where the geological conditions are quite different. Several very successful bore-holes have been put down, an abundant artesian flow being obtained at depths of 2000 feet and over.

SUBJECT TO REVISION.

[TRANSACTIONS OF THE AMERICAN INSTITUTE OF MINING ENGINEERS.]

The Formation of Bonanzas in the Upper Portions of Gold-Veins.

BY T. A. RICKARD, DENVER, COLORADO.

(Richmond Meeting, February, 1901.)

INTRODUCTORY.

THE presentation to the Institute, eight years ago, of the paper of Pošepny on "The Genesis of Ore-Deposits" has borne fruit in much fresh investigation, as is evidenced, for example, by the group of very valuable papers, by distinguished members of the United States Geological Survey, read at the Washington meeting—discussions of general principles particularly suggestive to those who are engaged in mining.

Pošepny, in the discussion of his famous treatise, said that the present writer seemed to look at every new conception in ore-deposition "from the sole standpoint of its immediate usefulness in mining."* Protesting mildly against "sole" and "immediate," I accept the impeachment. It calls for no defence.

THE DEVELOPMENT OF RECENT THEORIES.

Given the idea of an underground water-circulation as the chief factor in the deposition of ore, the next step in the inquiry as to the genesis of such deposits is the endeavor to determine which particular part of the general water-circulation is responsible for the results. Around this question have centered the controversies of a generation, and to these controversies we owe the gradual clarification of our ideas upon the processes of ore-formation. It is unnecessary to sketch here their progress from Werner to Le Conte, who combated in 1883 the extreme views of the lateral-secretionists, and in 1893 opposed the narrow interpretation of the ascensionist-theory. The generally accepted opinions of to-day are a well-deserved tribute to his philosophic discrimination.

---

* *Trans.*, vol. xxiv., 996.

Thanks to Prof. Van Hise and Mr. Slichter, whose work he utilizes, we have now arrived at a comprehensive conception of the underground circulation, which emphasizes the conclusion that sulphide-ores are generally deposited by ascending waters. In estimating the importance of this conclusion, it is to be remembered that, apart from placers and iron-mines, the largest portion, by far, of the ores exploited by the miner are sulphides. Morever, it has been shown that the other, equally essential, parts of the circulation, namely, its lateral and descending portions, particularly the latter, also play their part, to which many "secondary enrichments" are due.

This approach toward an understanding of the processes of secondary enrichment in ore-deposits is an extremely important advance in the application of geology to the exploitation of mines. For such enrichments pre-eminently constitute the ore-masses valuable to man. Chemistry and physics may unite in determining the conditions favorable to the precipitation of gold; geology may unravel the intricacies of rock-structure, but it does not come within the province of these sciences to decide whether a gold-vein will prove rich enough for profitable mining. Nature knows no ratio of sixteen to one, or any other standard of monetary value. Therefore, the determination of the particular conditions favorable to the mere occurrence of gold-ores remains but a barren discovery until it includes some suggestion as to the search for the richest portions. To the geologist, material carrying 2 dwts. of gold per ton is as truly an auriferous deposit as if it contained 12 dwts. per ton; but, under existing economic conditions, the miner may regard the former as only fit for macadam, and the latter as potential of fortune.

When the science of ore-deposits, therefore, has predicted with certainty the places where gold can be found, it has fulfilled a conclusive test of a true theory. But this means to the miner no more than the restriction of his search for profitable gold-deposits to those places where there is any gold at all—a restriction which, after all, amounts to little, for the progress of scientific inquiry and practical exploration has rather enlarged than diminished the field of the distribution of this metal. A greater service will be the determination of the conditions which control the formation and distribution of those particular portions of the multitudinous deposits of gold which constitute the secondary enrichments of the geologist and the bonanzas of the miner.

Such a desired consummation seems now to be nearer of attainment. The practical result of the papers of Messrs. Van Hise, Emmons and Weed will be to direct attention to the one line of inquiry most useful to the miner. Unquestionably the theories of secondary enrichment have been largely suggested by the experience of the men whom the geologists have met at the mines; and the invaluable assistance thus given to mining engineers is a pleasant outcome of such an exchange of views.

The Enrichment of Gold-Veins Near the Surface.

A quartz lode carrying gold in association with pyrite is here taken as the type of deposit under discussion. In lodes of this kind, it is a common experience to find bodies of rich oxidized ores extending to a variable depth from the surface. In this general phenomenon of enrichment two processes must be separately recognized, namely, relative enrichment by a method of natural concentration and positive enrichment by the deposition of additional gold through secondary reactions.

Enrichment by Concentration.

The iron sulphide accompanying the gold is removed by weathering. Weathering is a process of chemical decomposition and mechanical disintegration in which oxidation is aided by the shattering of the rock due to the alternate expansion and contraction of the water present in its pores, seams and cavities. The depth to which these effects extend will depend upon the facilities afforded for the penetration of surface-waters carrying free oxygen; and it will be regulated by the local groundwater-level. The results observed usually cease at the groundwater-level because at that horizon the descending surface-waters become mingled with the larger body of neutralized water, and so lose their free oxygen. When, however, they can find channels permitting a relatively rapid passage, they may not become at once diffused, and may thus continue their oxidizing action even below that level. But the actual lowering of the groundwater-level, by a change of surface altitude or hydrostatic conditions, affords the chief factor in enlarging the scope of such oxidizing action on the part of the surface-waters.

The chemistry of the process is pretty well understood, and need not be discussed here.

In the case of enrichment by concentration, the evidence indicates that the leaching and removal of the pyrite has been affected without shifting the gold, which remains behind in its native state. I have specimens from Idaho and West Australia exhibiting crumbly native sulphur, within the cubic cavities vacated by the pyrite, and in those from West Australia there is also gold in fine crystals which are readily shaken loose. The removal of pyrite; the occurrence of fine particles of gold in the vacant casts produced by this removal, and the formation of a sintery honeycombed mass of iron-stained quartz are familiar aspects of the process of natural concentration.

Weathering, then, by removing the baser and more soluble constituents of the vein, decreases the weight without diminishing the volume of the ore, which thus becomes so much the richer *per ton*. Iron-stained gossan, rich in gold, is a familiar occurrence in mining, and the frequent development of such material has had a far-reaching effect in determining the character of the industry. Apart from the richness of such oxidized ore, its metallurgical docility greatly enhances its value. In comparison with the unaltered and relatively refractory pyritic ores, the oxidized material is not only easier to crush, but also easier to treat by amalgamation, chlorination, etc. Hence the contrast which is occasionally offered between the early successes of the discoverers of a gold-vein and the subsequent troubles of the mining company which buys their property. The gossan of the gold-vein has been the source of a large part of the world's store of the precious metal; and to it we owe the successful beginnings of many districts, which, if they had been compelled to commence operations upon refractory pyritic ore, would have waited long for their active development.

Secondary Enrichments Due to Descending Surface-Waters.

The diagnosis of the general process by which these are formed by descending waters has been stated in clear terms in the contributions of Messrs. Van Hise, Emmons and Weed.

The occurrence of restricted bodies of extraordinarily rich gold-bearing quartz has been a startling feature of gold-mining in all countries. From them fortunes have been made with picturesque suddenness; and by means of them the inexperienced have been led into sanguine expectations, the failure of which has brought disasters not less romantic, though much less welcome to their victims. Such instances have furnished matter for proverbs concerning the uncertainty of mining; but they are soon forgotten. Nevertheless, the uncertain occurrences of rich ore on which they are based present an important feature of the ore-deposits in all gold-mining districts, though they are more particularly characteristic of desert regions, such as the area of the Great Basin, stretching between the Rocky Mountains and the Sierra Nevada, and also those arid parts of Australia and West Australia which have yielded so much of the wealth of the colonies.

The outcrop of a gold-vein is not always the richest portion. The sintery gossan formed at the immediate surface may be poor in gold, and yet may be succeeded near, or even below, the water-level, by extremely rich masses of half-decomposed pyritic ore. In such cases it would appear that the gold had been leached out of the oxidized portion of the lode, and had migrated in the wake of the iron until precipitated, so as to form the secondary enrichment now under discussion.

In considering the formation of these bonanzas, one of the first problems presented is the question of the mode of occurrence of the gold in the pyritic quartz of the lode. The evidence as yet available indicates that the gold does not exist in chemical combination with the iron sulphide of the pyrite, but usually occurs in minute filaments or crystal aggregates distributed through the substance, and especially along the structural planes, of the pyrite. In my collection I have a handful of fragments of pyrite obtained from the Orphan Boy mine, in Boulder county, Colo. This mine was the beginning and end of a mining excitement which happened, in the spring of 1892, in connection with a locality named Copper Rock. Under a magnifying-glass the specimens exhibit little crystals of gold, which, by the rounding of their edges, appear in places as globules distributed over the facets and in the crevices of the pyrite.

The behavior of such gold-ore under metallurgical treatment

also suggests strongly that its usual mode of occurrence is analogous to the above example. When gold-bearing pyrite is treated by cyanidation, the gold may be leached out without deformation of the pyrite or any other change in its appearance except the acquisition by its facets of a pitted surface suggesting cavities left by the removal of a soluble constituent. Moreover, there are many mining districts yielding gold from pyritic veins in which the native metal is rarely seen. The ores of Gilpin county, in Colorado, for example, contain an average of from 10 to 15 per cent. of iron and copper pyrites; and I know from frequent trial that when crushed and washed in a pan, such material, even though very rich, will not yield a "color," that is, a speck of visible metallic gold. Nevertheless, in the stamp-mill these ores yield their gold to amalgamation, indicating by their behavior in this respect that the gold is in a condition of such freedom as to permit its separation by a crude mechanical process, and its subsequent ready combination with mercury so as to form an amalgam.

The gold which occurs thus in the pyrite of the quartz-vein is soluble in many natural reagents, some of which are formed in the very process of weathering which leaches the pyrite, while others are known to be present in the surface-waters which circulate through the lode-fractures under observation at the present day. By whatever means it is dissolved, the gold is then supposed to be carried by the surface-waters in their descent toward the groundwater-level, where it is precipitated under conditions to be discussed in due course.

Solvents.

In the process of weathering, the pyrite yields many subordinate compounds, such as sulphuretted hydrogen, sulphurous and sulphuric acid, and proto- and sesqui-sulphates of iron. Of the latter, the sesqui-sulphate, $Fe_2(SO_4)_3$, is a solvent for gold, and has been cited by Wurtz and Le Conte in early discussions concerning the origin of masses of native gold in oxidized ores. Dr. Richard Pearce, in later years, has frequently drawn attention to the probability that this sesqui-sulphate is a factor in the process of gold-deposition.*

* Presidential Address, *Proc. Colo. Sci. Soc.*, vol. iii., part ii. (1889), p. 244.

The gold-deposits in the cavernous quartzite of Battle Mt., Colo.,* have characteristics which appear to confirm this view. In these ores large pieces of native gold, of a nuggety appearance, but really crystalline in structure, have been found associated with horn-silver and the sesqui-sulphate of iron. The latter occurs in lumps, mixed with clay; and although these are very rich in gold, the gold occurs in a form not to be detected by careful panning. Analyses of several large lots of ore showed the presence of 12 per cent. of the hydrated sesqui-sulphate of iron.†

But other solvents, capable of doing this work, also occur in nature, and, although the amount of any one of them to be detected in existing surface-waters may be minute, we have to remember that the processes of nature are permitted so much more time than those of the laboratory that the dilution of the solution is compensated by the quantity of it.

Most writers refer to chlorine as a possible reagent. Such a reference is suggested not only because it is a prominent reagent in the metallurgical practice of to-day, but also by the fact that it has a wide distribution throughout nature in the form of common salt. This is most apparent in arid regions where evaporation causes concentrated solutions to be formed. Thus, in the deserts of West Australia the water encountered in the mines is always brackish, and frequently contains more salt than the sea.‡ The water of the Great Boulder Proprietary mine, at Kalgoorlie, in 1897, contained 6402 grains of common salt per gallon.§ A considerable amount of magnesium chloride was also present. In some of the water used in the stamp-mills, and obtained from temporary "lakes,"|| the salts were present up to the point of saturation and the liquid carried further salts in suspension, so that the amount reached as high as 30 per cent., rendering the term "brine" more suitable than "water." This liquid contained 17 per cent. of salts

---

* F. Guiterman, "Gold Deposits in the Quartzite Formation of Battle Mountain, Colorado," *Proc. Colo. Sci. Soc.*, vol. iii., part iii. (1890), pp. 264–268.

† *Ibid.*, p. 266.

‡ Sea-water contains 3½ per cent. of salts, three-quarters of which is common salt, the chloride of sodium.

§ This is equivalent to 9 per cent.

|| "Sinks" or salt-marshes. They form an important feature of the physiography of West Australia.

in solution even when most diluted by recent rains, and it therefore afforded a parallel to the Dead Sea, the waters of which contain from 20 to 26 per cent. of salts, of which 10 per cent. is common salt. These excessive percentages are not due to the presence of deposits of salt in the rocks of the district, but simply to the concentration brought about by the excessive evaporation* which takes place in a hot, arid climate.

Mine-waters frequently contain a noteworthy quantity of chlorine, as chloride of sodium. At the Mammoth mine, in Pinal county, Arizona, the water carries five grains of salt per gallon, while the well-water, used in the stamp-mill, situated in the valley below the mine, contains twice as much.† This would be equivalent to six grains of free chlorine per gallon. The larger amount contained in the water from the well, as compared with that in the drainage of the mine, suggests the results of surface-leaching. Even in mountainous districts, such as Cripple Creek, Colo., the mine-waters carry chloride of sodium to a noteworthy extent. The water of the Independence mine contains three grains per gallon.

Another suggestive feature is offered by the abundance of horn-silver or cerargyrite, the chloride of silver, throughout the dry tracts of Arizona, New Mexico and Nevada.‡ Prof. Penrose emphasizes this interesting fact, and connects it with the bodies of salt water which still survive in places as "sinks" and "lakes."§ Furthermore, the oxi-chloride of copper, atacamite (which derives its name from the Atacama desert, between Chili and Peru), is frequent in these regions. Another and more uncommon mineral may also be mentioned in this connection. In the Mammoth mine, already cited, and in the well-known Vulture mine, both in Arizona, the precious metals are associated with vanadinite, which contains chlorine as a chloro-vanadate of lead, $3Pb_3(VO_4)PbCl_4$.|| Thus the chlorides

---

* The rate of evaporation, in the region mentioned, has been estimated to be as much as 7 ft. per annum.

† As I am informed by Mr. T. G. Davey.

‡ The general occurrence of horn-silver in the outcrops of lodes throughout the southern parts of Arizona and New Mexico has originated the term "chloriding" which the miners employ as a synonym for "prospecting," which, by the way, the Australian calls "fossicking."

§ R. A. F. Penrose, Jr., "The Superficial Alteration of Ore-Deposits," *The Journal of Geology*, vol. ii., No. 3, 1894.

|| Dana.

of copper, lead and silver are found in the oxidized ores of these regions, while the corresponding combination of gold is absent. The explanation is obvious. The chloride of gold is an unstable and readily soluble compound, while the minerals formed by the corresponding combination with the baser metals are comparatively insoluble in water, especially the chloride of silver, for the abundance of which there is therefore a good reason. It remains but to add that, in several Arizona mines which I have sampled, the ores above the water-level carried a notable proportion of silver with very little gold, while in depth the silver contents have diminished and the gold has increased, especially in the vicinity of the water-level.*

Of the many reagents which would liberate the chlorine from salt, it is only necessary to mention ferric sulphate and sulphuric acid, both derived from the ordinary oxidation of pyrite. The hydrochloric acid thus formed would yield free chlorine in the presence of manganese oxides,† which are very prevalent in the upper portion of gold-lodes, in the form of the black earthy mineral, psilomelane.

There are other possible solvents which need not be discussed here.

Precipitants.

Whatever the solvents which leach out the gold from the superficial portions of the vein, there is assuredly no lack of precipitants. It is probable that the gold does not migrate far before encountering conditions which compel deposition. Even when it is eventually carried to a considerable distance it is most likely that such removal is effected by alternating stages of precipitation and solution.

Organic matter is a probable precipitant for the gold in such surface-waters. It exists deeper than hasty observation would suggest. At the Great Boulder Main Reef mine, at Kalgoorlie, I saw the roots of trees which, in their energetic search for moisture, had attained a depth of 85 ft. below the surface; and at the Sugar Loaf mine, near Kunanalling (also

* I may instance two well-known mines, the Mammoth and the Commonwealth.

† See the experiments made by Dr. Don, to test this matter, *Trans.*, xxvii., p. 599.

in West Australia), I saw a similar occurrence at a depth of 74 ft.*

Another agency which, under certain chemical conditions, is a probable factor in reducing the gold from surface-waters, is pyrite itself. Thus, the gold dissolved from the decomposed pyrite at the surface may be precipitated upon the unoxidized pyrite deeper down. Among the exhibits belonging to the Colorado Scientific Society is a bottle containing cubes of pyrite, on the faces of which crystals of gold are to be seen. They are the result of one of Mr. Pearce's experiments. The gold of a Cripple Creek ore was dissolved by using common salt, sulphuric acid and psilomelane as reagents, the chlorine being thus obtained in a manner analogous to conditions which probably occur in nature. This solution was placed in a small bottle, and to it were added a few large pure crystals of pyrite from the St. Louis mine, at Leadville. After several months the gold became precipitated in the manner described. In this connection the story of Daintree's experiment, which I have quoted before,† is worth repeating. In 1871, Daintree commenced a series of experiments at Dr. Percy's laboratory at the Royal School of Mines, London. In a number of small bottles he placed a solution of chloride of gold, and to each he added a crystal of one of the common metallic sulphides, such as pyrite, blende, galena, etc. At the time when Daintree died, a few years later, no results could be discerned; but one of the bottles, containing the gold solution and a crystal of common pyrite, was removed to Dr. Percy's private laboratory, in Gloucester Crescent, and there, in 1886, the experiment was completed by the discovery of a cluster of minute crystals of gold upon the smooth surface of the pyrite. The experiment had occupied fifteen years; and on account of its very length it may be said to have more nearly approached the actual conditions occurring in nature.

In a case like that of the "Indicator," at Ballarat, which I

---

* Since writing the above I have read Professor Vogt's very valuable contribution, and I note that he mentions having seen, among the mineral exhibits at Paris, specimens of such roots, from the Great Boulder Main Reef mine, on which gold had actually been precipitated. "Problems in the Geology of Ore-Deposits," pamphlet edition, page 43.

† *Trans.*, xxii., 313.

have lately described again,* it may be questioned whether it is the pyrite in the thin seam of graphitic slate or the carbonaceous matter of the latter which causes the precipitation of the gold. Even if the pyrite was the decisive factor, it must be remembered that it, in turn, probably owed its previous deposition to the action of the carbonaceous precipitant in the Indicator seam. This would apply also to the beds of black slate which have had so marked an influence on the occurrence of gold in the Gympie district,† Queensland, but it would not, I think, be applicable to the Rico deposits,‡ where pyrite is not an especial constituent of the black shales, as compared with the sandstone beds of the same stratified series.

Solution and Precipitation.

It is to be noted that in the two examples of ore-forming processes which have been considered, the gold in the superficial part of the vein is supposed, in one case, to remain in the gossan after the pyrite has been removed, while in the other instance the gold also is dissolved and carried elsewhere. This may appear contradictory. It is a good illustration of the perplexities arising from the application of chemical hypotheses to the theory of ore-deposition.

Nature knows no interval of inaction; solution is going on at one time, precipitation at another. The gold is constantly the object of one or the other activity. After the pyrite is removed, or while it is still undergoing leaching, the gold is being dissolved, but more slowly than the baser metals. That which remains to enrich the gossan may well be supposed to be the survival from a larger quantity of gold which has been undergoing slow solution. The gold which was deposited deeper down, from the surface-waters, may, as erosion takes away the upper part of the vein, eventually find itself close to the surface and undergo re-solution. It is a question whether the mining of to-day breaks in upon the gold-deposits at one stage or another of a continuous process. The miner finds the balance of gold left on deposit from a current account in Nature's

* "The Indicator Vein, Ballarat, Australia." *Trans.*, xxx., 1004.

† J. R. Don. "The Genesis of Certain Auriferous Lodes." *Trans.*, xxvii., 577-580.

‡ The Enterprise Mine, Rico, Colorado." *Trans.*, xxvi., 906.

bank. Solution and precipitation are everywhere in action; it is the excess of one or the other which determines the formation of ores.

The Distribution of Ore-Bonanzas.

The shifting of the zone of oxidation is a principal factor in determining the distribution of rich ores. By the erosion of the superficial portions of the vein, in common with the enclosing rock, the further downward penetration of the oxidizing agencies is facilitated. The depression of the groundwater-level lowers the zone at which precipitation of gold, from descending surface-waters, takes place, while, on the other hand, when a change in the hydrostatic level causes the groundwater to rise, the zone of deposition moves up. In both cases the tendency is to give vertical extension to the rich mass of secondary gold-ore, and thus to produce the occurrence which the miners term a "shoot."

Erosion is followed by another result, in itself of great importance to gold-mining. The steady removal of the superficial part of the vein causes the lower portion, which has been enriched at or below the groundwater-level, to undergo a relative elevation by being brought nearer to the surface. In this way the bonanza-zone, in process of time, may become the outcrop. This appears to me to explain the formation of the extraordinarily rich bunches of specimen-quartz, such as made West Australia famous in 1894 and 1895, and started the mining stampedes of other days elsewhere. In many instances fortunes have been gathered almost at the grass-roots from veins which, on systematic development, have proved unprofitable. The gold-quartz veins of West Australia traversed rocks of great geological antiquity which have not, during late geological periods, undergone any notable disturbance. We do not know at what period the veins were formed; but, even though their formation dates no further back than the beginning of the Tertiary, they have since been continuously exposed to the same quiet forces of erosion which have leveled the region until it appears as an arid table-land strewn with the wreckage of geological time.

Whatever the alternations of slow depression and elevation which have affected this region, as part of a continental area, it is certain that erosion has been long at work with patient constancy. Throughout this period chemical agencies have been active in the zone of weathering, near the surface, removing the gold to the zone of precipitation, near the groundwater. Whatever the slight changes which have marked the level of the groundwater from time to time, erosion has continued uninterruptedly, and therefore it has steadily gained, with the result that the enriched portion of the vein has been brought nearer and nearer to the actual surface, until it finally appears as the outcrop which rewards the search of the prospector.

The Localization of Ore-Shoots.

To the miner the localization of these richer portions of the vein is of more immediate practical interest than the theory of their origin. A gold-vein is not a homogeneous mass of auriferous quartz, of tabular form, penetrating the rocks like a sheet of paper, but rather as an irregular occurrence of ore, the composition and shape of which are very variable, because they are the result of chemical agencies and structural conditions of great complexity. While the traces of the agencies which precipitated the ore are obscure, because they have been largely obliterated by subsequent chemical action, the relation between the vein and its encasing rock can often be traced by observation. In this direction the miner obtains great aid from the geologist. The transactions of this Institute and the publications of the U. S. Geological Survey contain numerous clear expositions of such structural relations. The monographs on the Leadville and Eureka mining districts may be especially instanced as affording striking examples of the direct application of geology to underground work.

Australia.—One of the best examples of the localization of rich ore came under my notice in 1890 in the Bright mining district. Bright is geographically in the Australian Alps, and geologically in the Upper Silurian slates and sandstones. Though these rocks have undergone metamorphism, and exhibit a well-developed cleavage, yet their bedding has not been obliterated. The veins cross the bedding-planes of the enclosing country both in strike and dip. When investigating the distribution of the ore in the mines of this district, I found that the ore-shoots had a pitch corresponding with the line of intersection between vein and country. This was well illustrated at the Shouldn't Wonder mine, 7 miles from the town of Bright. The lode was a simple quartz vein from 15 to 24 in. wide, carrying a small percentage of pyrite. It had a strike of N. 28° W. of N. and a dip to the NE. of about 75°, while the country dipped SW. 79° and had a strike of N. 55° W. The plane of the vein cut across the beds of the country and the intersections thus produced were to be seen along the foot-wall of the lode as lines, pitching 42° to 46° southward. While the foot-wall was more regular than the hanging, and therefore exhibited this feature best, yet the hanging also carried lines corresponding with those observed on the opposite wall.

The boundaries of the ore-shoots in the mine followed these lines; and the longitudinal section of the workings, as seen on the mine-maps, proved also that these lines of intersection had an inclination which coincided with the trend of the ore-bodies, as stoped out between the four successive upper levels of the property.

At the Myrtle mine, in the same district, there was the same correlation between the pitch of the ore-bodies and the line of intersection of the wall of the lode with the bedding-planes of the enclosing country. The stratification was distinct, the rocks consisting of altered, silicified slates of a gray to gray-blue tint. In the stopes above the 700-ft. level the pay-ore was separated from the normal valueless quartz of the lode by a small step, due to the irregular fracture of the vein in crossing two beds of unequal hardness. It marked the line of intersection between lode-plane and country bedding, and also proved to be the boundary of the pay-shoot. In the different portions of the mine the variation in the dip of the country produced variations in the angle of the lines of intersection, and also in the pitch of the ore-shoots.

It is not often that the formation traversed by a vein has such a simple structure as was presented by these Silurian sedimentary rocks; but it is probable that in other districts also the pitch of the ore-bodies may have been determined by structural conditions of a similar kind, which have been obscured, however, by metamorphism.

Colorado.—Experience has shown that the intersection of fractures favors the occurrence of rich ore-bodies. An interesting example was afforded by the Moon-Anchor mine, at

Cripple Creek, in 1899. This is illustrated in Fig. 1. The ore in the mine occurs in a lode-channel marked by a band of fractured andesite breccia. At the 400-ft. level a small dike (EF) of granite, 2 to 6 in. thick, intersects the lode-channel at a place where a counter-fracture (CD) also traverses it. A triangle is produced by these intersections, and the ore is

FIG. 1.

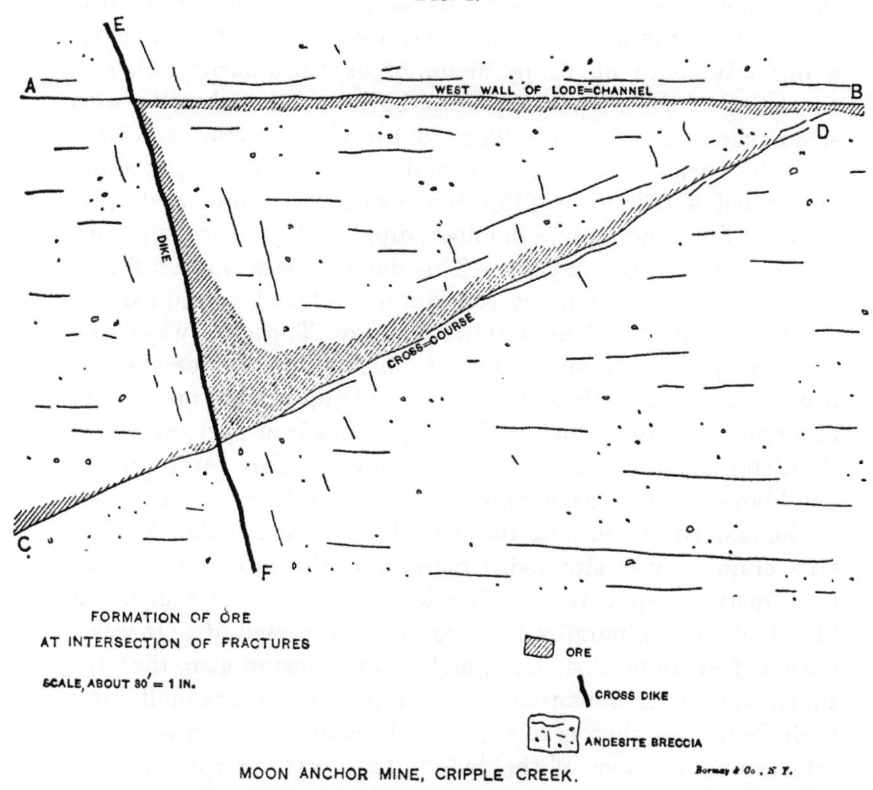

proved to surround a block of ground which is also mineralized, but not sufficiently so to be regarded in its entirety as pay-ore. At the crossing of the dike and cross-fractures a very rich body of telluride-ore was encountered.

This reminds me of the Yankee Girl ore-body, mentioned by Emmons.* This body of ore was of phenomenal richness,

* "The Secondary Enrichments of Ore-Deposits." Page 19 of pamphlet edition.

many ten-ton lots being shipped which carried 7 or 8 ounces of gold and 3000 to 4000 ounces of silver per ton. The ore was also rendered remarkable by carrying the rare mineral stromeyerite, a sulphide of silver and copper. Mr. Emmons speaks of the bonanza turning into low-grade pyritic ore as depth was attained. I may add* that this change was not gradual, but sudden, and coincident with certain structural relations. At the surface, the vein consisted of comparatively low-grade ore, which led to the finding of a nearly vertical "chimney," averaging only 25 to 30 ft. in diam., of extraordinarily rich ore, consisting of the copper sulphides, bornite and erubescite, with stromeyerite and barite. The gold in the ore was associated with the barite. From the second to the sixth level, at about 500 ft. below the surface, this bonanza proved immensely productive; then, suddenly, a flat floor, dipping W. and accompanied by clay, crossed the deposit. This flat vein was worked for 90 ft., from the south drift at the No. 6 level, and contained ore similar to that of the Yankee Girl chimney. The latter was found again deeper down, and out of its former line of descent, but it was much diminished in richness, and appeared to merge into the general body of low-grade copper and iron pyrites† which characterized the lode at the tenth level. This mine and its neighbors, the Robinson and Guston, are idle now. They are in the andesite breccia of the San Juan region. The Yankee Girl chimney was situated, I believe, at the crossing of three lode-fractures, appearing as breaks in the andesite, which was bleached and mineralized where they traversed it. It was a curious feature of this mine, and of the Guston also, that the short, very rich bonanzas of the upper levels gradually lost their definition, that is to say, they became no richer than the intervening portions of the lode. This was interpreted as a "lengthening" of the ore-shoots, which may be true, viewed in one way; but I think that it should be more properly regarded as an impoverishment of the lode, marked by a disappearance of the bonanzas. The surface-waters of these mines are very acid, as Mr. Emmons remarks. At the Yankee Girl mine it became necessary to encase the pipes in redwood, brought

* From notes made during an examination of the mine in January, 1892.

† Assaying 20 to 60 ozs. silver, 1 to 4 dwts. gold, 5 to 15 per cent. copper.

from California. I found that the water issuing from a shallow adit (73 ft. below the collar of the shaft) readily precipitated copper on scrap iron. Ore-forming agencies were evidently still at work.

California.—In California, especially in that mining region which follows the foothills of the Sierra Nevada and traverses the counties of Amador, Calaveras and Tuolumne, the occurrence of pockets of rich ore, full of native gold, is a notable feature of the superficial parts of the quartz-veins. These pockets appear to be confined to the zone between the surface and the water-level, and to be dependent upon the results produced by the small cross-veins which encounter the main lodes. In 1887 I had the pleasure of extracting, in two hours, a little over 170 ounces of gold, worth about about $3000, from one of these pockets. It was at the Rathgeb mine, near San Andreas, in Calaveras county. The main lode consisted of 5 to 8 ft. of massive "hungry-looking" quartz, the foot-wall of which was a beautiful augite-schist and the hanging a hard diabase. The water-level was 160 ft. below the surface. Down to this point, the country was oxidized, the hanging-wall exhibiting only slight alteration, while the schist of the foot-wall was softened and decomposed almost to a clay. This was traversed by numerous small veins, which appeared to act as "feeders," forming bunches of rich ore where they encountered the main lode. At the 120-ft. level, south from the shaft, there were some old workings; and the examination of these led to the discovery of a small seam, about one-sixteenth of an inch thick, filled with red clay which carried a good deal of native gold, as was proved by washing it in a pan. An experienced miner was put to work, with instructions to follow this small streak. It varied in thickness, and occasionally opened out into small lenticular cavities, containing a clay in which the gold was distributed like the raisins in a pudding. Each of these "pockets" yielded several hundred dollars' worth of gold. At length the streak widened to 6 or 8 inches of quartz, lined with clay. The amount of red clay commenced to increase; coarse gold became more frequent; and a big discovery was hourly expected. It was finally made. The vein suddenly became faulted, and at the place of faulting there was a soft, spong, wiry mass of gold and clay—more gold than clay. The first handful I broke, while yet the stope was thick with powder-smoke, contained three ounces of gold. Within the next two hours this pocket gave us $3000, and during the following week it yielded over

FIG. 2.

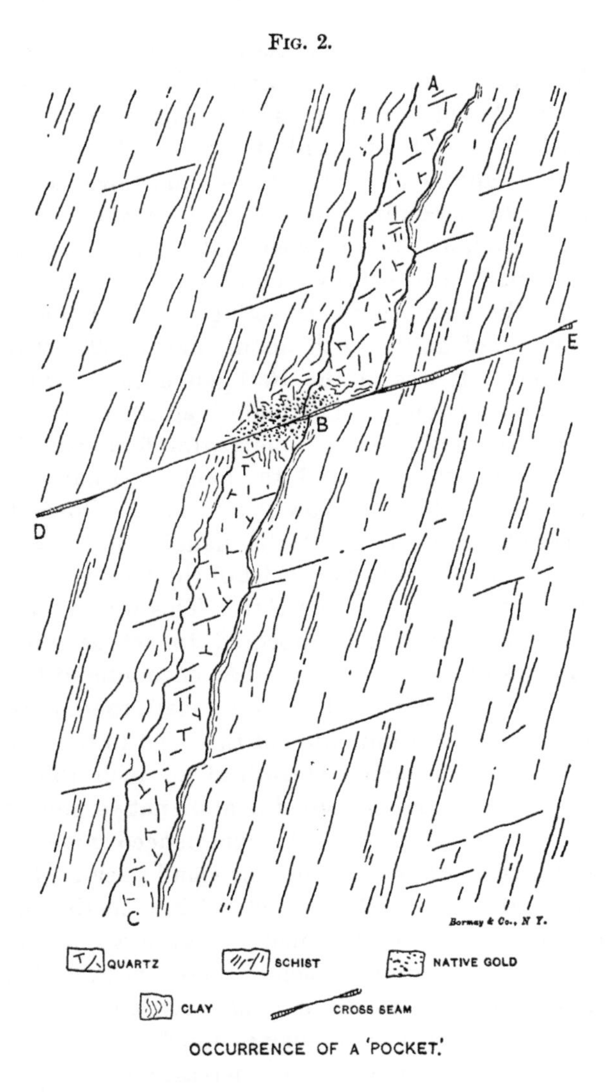

OCCURRENCE OF A 'POCKET.'

$20,000, an amount which was obtained at a total cost of less than $200. When it had been worked out, it was easy to observe the conditions which determined its occurrence at this place, as Fig. 2 will explain. The vein, AC, had been faulted about its own width, namely, 10 inches, by a small cross-seam, DE, and at this intersection, B, the pocket lay. The gold was spongy and was intermixed with quartz. The clay which penetrated the whole mass was partly red and ochreous, and partly a gray gelatinous material. In the quartz, and associated with the gold, there were acicular black crystals of pitch-blende (uraninite), together with uranium ochre. This association of gold with uranium is uncommon.

New Zealand.—Intersections which coincide with enrichments form a notable characteristic of the Hauraki gold-field* in the north island of New Zealand. In this district the occurrence of patches of native gold is an important feature of the regular mining operations. When I was there, in 1891, each stamp-mill had its "specimen-stamp," a single stamp working in a separate mortar, and employed solely for the treatment of specimen-ore. These rich patches occur at the places where the "reefs" or lodes cross bands of flinty quartz. The latter are known among the miners as "flinties." They vary in thickness from a few inches to mere threads of chalcedonic quartz. They are barren in themselves, but have a favorable effect on the gold-veins. The latter are also intersected by cross-veins, producing an enrichment similar to that caused by the "flinties." Fig. 3 is a sketch of one of these intersections, as seen by me in the Moanataeri mine. The lode, AB, consists of a series of small seams of quartz, conforming to the structural lines of the enclosing country, which is hornblende-andesite. The cross-vein, CD, is a band of soft gray decomposed rock, which also carries a number of small quartz-seams, but only near its crossing with the main lode, AB. The line of CD is parallel to a large fault, to be seen elsewhere in the mine-workings. The "leaders," or quartz-seams, in AB are gold-bearing, and exhibit marked enrichment at the intersection with CD.

The prevailing formation of this mining district is an andesite, which is traversed by soft bands of decomposition, called "sandstone" by the miners. The latter, when penetrated by quartz-seams, are favorable to the finding of ore. The gold-occurrence is essentially sporadic and dependent upon local en-

* It is also known as the Thames district.

richments, such as have been described. The district is surrounded by thermal springs, and is near the well-known volcanic region of Tarawera, which was active in 1884. The mine-waters are heavily mineralized and very acid, so that the metal screens used in the mills are quickly corroded. Tellurides and selenides of gold have been detected in the ores; but the pre-

FIG. 3.

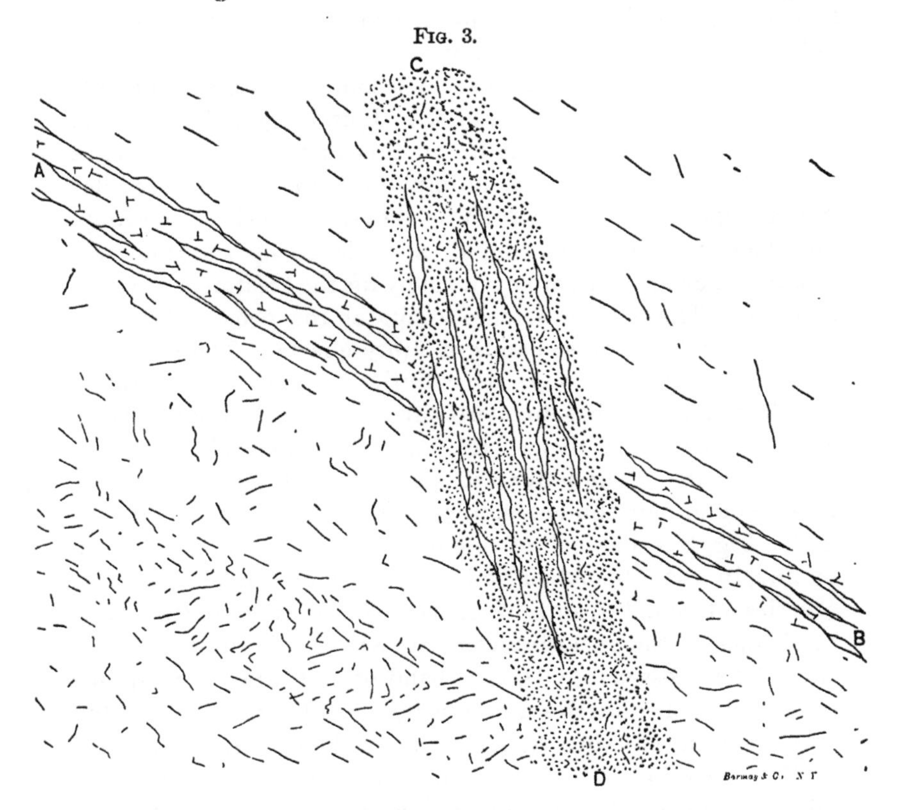

ENRICHMENT AT INTERSECTION MOANATAERI MINE, NEW ZEALAND.

cious metal is usually found native and in coarse particles, which are frequently coated with native arsenic. The district is one which, I think, if thoroughly examined, would afford many suggestions regarding ore-deposition.*

* The best description which has come under my notice is "The Geology of the Thames Goldfield," by James Park, read before the Auckland Institute, 1894. See also "On the Rocks of the Hauraki Goldfields," by F. W. Hutton, *Proc. Austral. Assn. Adv. Sci.*, 1888; and J. R. Don, "The Genesis of Certain Auriferous Lodes," *Trans.*, xxvii., 584–589.

Concluding Remarks.

It is to be hoped that the recent recognition of the agencies which bring about the formation of enrichments by surface-waters will not cause too violent a swing in the direction of a sweeping advocacy of the general efficiency of descending solutions to form ore-bodies. The study of the problems of ore-occurrence has been hindered in the past by such reactions from one extreme view to its opposite. Therefore, in concluding this contribution to the discussion of the results produced by descending surface-waters, I would emphasize the wider agency of ascending solutions in forming the ore-masses amid which such secondary enrichments are occasionally found. It is agreed that the sulphide-ores are primarily deposited from ascending waters; it is also likely that such a result is repeated. A region once subjected to fracturing, which has permitted the subsequent passage of mineral-bearing solutions, is likely, at a later period, to be subjected to a repetition of these activities. The geological history of many mining regions gives clear evidence of a repeated disturbance of structure. This is indicated by the existence of several systems of fractures crossing each other, the later ones dislocating the earlier. It is probable that each period was marked by mineralization, the character of which may have varied. The banded arrangement of the lodes of certain districts, such as Freiberg, Rico and Butte, suggests this. Enrichment may have been caused by mere addition; the introduction of other metals may have changed the average composition of the ore in the lode so that it is now extremely valuable, whereas before it may have had no economic importance; a silver-ingredient may have been added to the gold-contents, or the addition of copper may have made a deposit doubly valuable by improving its metallurgical character. I hope the present discussion on ore-deposition will prove as inspiring to further investigation as did Pošepny's paper of 1893, and that data concerning the possible secondary enrichment of sulphide-ores by the repetition of ascending solutions will be sought for. There is nothing like a working theory to sharpen the observation. Theories do not alter facts, but they often lead us to find new ones.

In cordially welcoming the splendid treatise of Professor Van Hise I need make no reservation. When Pošepny made

2867

clear the essential character of the upper or "vadose" water-circulation, he did us a great service; and when he combated "lateral secretion" he overthrew a very narrow interpretation of ore-formation, which was calculated to hinder seriously our progress toward the understanding of these difficult problems. But Pošepny was carried so far by his controversy with Sandberger as to over-emphasize the sole agency of ascending currents. At that time, in 1893, I demurred to this extreme view and said, "the word circulation is the key to the whole matter."* By this I meant that the entire underground water-circulation played a part in the formation of ore, and that to swing from one portion of that circulation to another, restricting oneself to the agency of either, would not (so it seemed to me from experience in the mines) solve the problem.

It does not appear to me that Professor Van Hise has erred by exaggerating any particular view of the subject. His elucidation of the water-circulation as a complete system is based on a broad conception of the whole matter. Of course, in indicating the work done by an agency hitherto largely overlooked, he was compelled to place some emphasis on certain neglected features of the descending portion of the water-circulation, and thus to give it some prominence in his masterly analysis. This makes the consideration of the question of secondary enrichments by surface-waters one of the most valuable parts of his treatise.

Regarding this question of secondary enrichment, it is to be pointed out that all ore-deposits are "secondary," the ore as found by the miner being merely the last term of a series of solutions and precipitations through which its substance has passed in a constant shifting due to the underground water-circulation. However, the last stage of the journey is the only one of immediate importance to the miner; and the determination of the causes which brought it there is, to him, far the most interesting aspect of the general inquiry. That Mr. Emmons should also have investigated and illuminated the problem is matter of much pleasure to a great many, engaged in mining throughout the West, to whom his geological contributions have always seemed to possess a practical bearing and value unfortunately not always found in scientific descriptions of geological phenomena.

---

* *Trans.*, xxiv., 950.

Lightning Source UK Ltd.
Milton Keynes UK
UKHW021555030119
334851UK00014B/1609/P

9 781334 477256